The chemistry of the

carbon–nitrogen double bond

THE CHEMISTRY OF FUNCTIONAL GROUPS

A series of advanced treatises under the general editorship of Professor Saul Patai

The chemistry of alkenes (2 volumes)
The chemistry of the carbonyl group (2 volumes)
The chemistry of the ether linkage
The chemistry of the amino group
The chemistry of the nitro and nitroso groups (2 parts)
The chemistry of carboxylic acids and esters
The chemistry of the carbon–nitrogen double bond
The chemistry of amides
The chemistry of the cyano group
The chemistry of the hydroxyl group (2 parts)
The chemistry of the azido group
The chemistry of acyl halides
The chemistry of the carbon–halogen bond (2 parts)
The chemistry of the quinonoid compounds (2 parts)
The chemistry of the thiol group (2 parts)
The chemistry of the hydrazo, azo and azoxy groups (2 parts)
The chemistry of amidines and imidates
The chemistry of cyanates and their thio derivatives (2 parts)
The chemistry of diazonium and diazo groups (2 parts)
The chemistry of the carbon–carbon triple bond (2 parts)
Supplement A: The chemistry of double-bonded functional groups (2 parts)
Supplement B: The chemistry of acid derivatives (2 parts)

$$>C{=}N{-}$$

The chemistry of the **carbon–nitrogen double bond**

Edited by

SAUL PATAI
The Hebrew University
Jerusalem, Israel

1970

JOHN WILEY & SONS

CHICHESTER — NEW YORK — BRISBANE — TORONTO

An Interscience ® Publication

Library of Congress Catalog Card No. 70-104166

ISBN 0471 66942 3

Reprinted September 1979

Printed in Great Britain by
Page Bros (Norwich) Ltd,
Mile Cross Lane, Norwich

Contributing authors

Jean-Pierre Anselme	University of Massachusetts, Boston, U.S.A,
R. Bonnett	Queen Mary College, London, England.
Albert Bruylants	Université de Louvain, Belgium.
David J. Curran	University of Massachusetts, Amherst, Massachusetts, U.S.A.
Shlomo Dayagi	Tel-Aviv Univeristy, Israel.
Yair Degani	Weizmann Institute of Science, Rehovoth, Israel.
K. Thomas Finley	Eastman Kodak Company, Rochester, New York, U.S.A.
Kaoru Harada	University of Miami, Florida, U.S.A.
Henning Lund	University of Aarhus, Denmark.
C. G. McCarty	West Virginia University, Morgantown. U.S.A.
Mrs. E. Feytmants-de Medicis	Université de Louvain, Belgium.
R. J. Morath	College of St. Thomas, St. Paul, Minnesota, U.S.A.
C. Sandorfy	Université de Montréal, Canada.
Sidney Siggia	University of Massachusetts, Amherst, Massachusetts, U.S.A.
J. W. Smith	Bedford College, London, England.
Gardner W. Stacy	Washington State University, Pullman, Washington, U.S.A.
L. K. J. Tong	Eastman Kodak Company, Rochester, New York, U.S.A.
Gunnar Wettermark	Institute of Physical Chemistry, University of Uppsala, Sweden.

Foreword

The general plan of the present volume is again the same as described in the Preface to the series, printed on the following pages.

This volume was originally planned to consist of seventeen chapters, out of which three did not materialize. These should have been chapters on the "Directing and Activating Effects" and on the "Syntheses and Uses of Isotopically Labelled Azomethine Groups", both in which cases the scarcity of the information available on the subjects was the decisive factor, and a chapter on the "Biological Formation and Reaction of the Azomethine Groups".

SAUL PATAI

Jerusalem, September 1969

The Chemistry of the Functional Groups
Preface to the series

The series 'The Chemistry of the Functional Groups' is planned to cover in each volume all aspects of the chemistry of one of the important functional groups in organic chemistry. The emphasis is laid on the functional group treated and on the effects which it exerts on the chemical and physical properties, primarily in the immediate vicinity of the group in question, and secondarily on the behaviour of the whole of the group in question, and secondarily on the behaviour of the whole molecule. For instance, the volume *The Chemistry of the Ether Linkage* deals with reactions in which the C—O—C group is involved, as well as with the effects of the C—O—C group on the reactions of alkyl or aryl groups connected to the ether oxygen. It is the purpose of the volume to give a complete coverage of all properties and reactions of ethers in as far as these depend on the presence of the ether group, but the primary subject matter is not the whole molecule, but the C—O—C functional group.

A further restriction in the treatment of the various functional groups in these volumes is that material included in easily and generally available secondary or tertiary sources, such as Chemical Reviews, Quarterly Reviews, Organic Reactions, various 'Advances' and 'Progress' series as well as textbooks (i.e. in books which are usually found in the chemical libraries of universities and research institutes) should not, as a rule, be repeated in detail, unless it is necessary for the balanced treatment of the subject. Therefore each of the authors is asked *not* to give an encyclopaedic coverage of his subject, but to concentrate on the most important recent developments and mainly on material that has not been adequately covered by reviews or other secondary sources by the time of writing of the chapter, and to address himself to a reader who is assumed to be at a fairly advanced post-graduate level.

With these restrictions, it is realized that no plan can be devised for a volume that would give a *complete* coverage of the subject with *no* overlap between chapters, while at the same time preserving the readability of the text. The Editor set himself the goal of attaining *reasonable* coverage with *moderate* overlap, with a minimum of cross-references between the chapters of each volume. In this manner, sufficient freedom is given to each author to produce readable quasi-monographic chapters.

The general plan of each volume includes the following main sections:

(a) An introductory chapter dealing with the general and theoretical aspects of the group.

(b) One or more chapters dealing with the formation of the functional group in question, either from groups present in the molecule, or by introducing the new group directly or indirectly.

(c) Chapters describing the characterization and characteristics of the functional groups, i.e. a chapter dealing with qualitative and quantitative methods of determination including chemical and physical methods, ultraviolet, infrared, nuclear magnetic resonance, and mass spectra; a chapter dealing with activating and directive effects exerted by the group and/or a chapter on the basicity, acidity or complex-forming ability of the group (if applicable).

(d) Chapters on the reactions, transformations and rearrangements which the functional group can undergo, either alone or in conjunction with other reagents.

(e) Special topics which do not fit any of the above sections, such as photochemistry, radiation chemistry, biochemical formations and reactions. Depending on the nature of each functional group treated, these special topics may include short monographs on related functional groups on which no separate volume is planned (e.g. a chapter on 'Thioketones' is included in the volume *The Chemistry of the Carbonyl Group*, and a chapter on 'Ketenes' is included in the volume *The Chemistry of Alkenes*). In other cases, certain compounds, though containing only the functional group of the title, may have special features so as to be best treated in a separate chapter as e.g. 'Polyethers' in *The Chemistry of the Ether Linkage*, or 'Tetraaminoethylenes' in *The Chemistry of the Amino Group*.

This plan entails that the breadth, depth and thought-provoking nature of each chapter will differ with the views and inclinations of the author and the presentation will necessarily be somewhat uneven.

Moreover, a serious problem is caused by authors who deliver their manuscript late or not at all. In order to overcome this problem at least to some extent, it was decided to publish certain volumes in several parts, without giving consideration to the originally planned logical order of the chapters. If after the appearance of the originally planned parts of a volume, it is found that either owing to non-delivery of chapters, or to new developments in the subject, sufficient material has accumulated for publication of an additional part, this will be done as soon as possible.

It is hoped that future volumes in the series 'The Chemistry of the Functional Groups' will include the topics listed below:

The Chemistry of the Alkenes (published)
The Chemistry of the Carbonyl Group (published)
The Chemistry of the Ether Linkage (published)
The Chemistry of the Amino Group (published)
The Chemistry of the Nitro and Nitroso Group (Part 1, published, Part 2, in preparation)
The Chemistry of Carboxylic Acids and Esters (published)
The Chemistry of the Carbon–Nitrogen Double Bond (published)
The Chemistry of the Cyano Group (in press)
The Chemistry of the Amides (in press)
The Chemistry of the Carbon–Halogen Bond
The Chemistry of the Hydroxyl Group (in press)
The Chemistry of the Carbon–Carbon Triple Bond
The Chemistry of the Azido Group (in preparation)
The Chemistry of Imidoates and Amidines
The Chemistry of the Thiol Group
The Chemistry of the Hydrazo, Azo and Azoxy Groups
The Chemistry of Carbonyl Halides
The Chemistry of the SO, SO_2, $—SO_2H$ *and* $—SO_3H$ *Groups*
The Chemistry of the $—OCN$, $—NCO$ *and* $—SCN$ *Groups*
The Chemistry of the $—PO_3H_2$ *and Related Groups*

Advice or criticism regarding the plan and execution of this series will be welcomed by the Editor.

The publication of this series would never have started, let alone continued, without the support of many persons. First and foremost among these is Dr. Arnold Weissberger, whose reassurance and trust encouraged me to tackle this task, and who continues to help and advise me. The efficient and patient cooperation of several staffmembers of the Publisher also rendered me invaluable aid (but unfortunately their code of ethics does not allow me to thank them by name). Many of my friends and colleagues in Jerusalem helped me in the solution of various major and minor matters and my thanks are due especially to Prof. Y. Liwschitz, Dr. Z. Rappoport and Dr. J. Zabicky. Carrying out such a long-range project would be quite impossible without the non-professional but none the less essential participation and partnership of my wife.

The Hebrew University, SAUL PATAI
Jerusalem, ISRAEL

Contents

CHAPTER **1**

General and theoretical aspects

C. SANDORFY

Université de Montréal, Canada

I. INTRODUCTION

In previous volumes of this series the carbon–carbon double bond and the carbonyl group were dealt with extensively. The azomethine group is, in many respects, intermediate between the two, and we shall find it useful to compare the properties of these three groups.

Certain analogies and differences are obvious. All three groups have two electrons in π orbitals and these account for some of their most

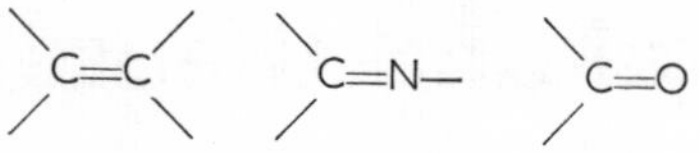

$$\diagdown\!\!\!\!\diagup C{=}C \diagup\!\!\!\!\diagdown \qquad \diagdown\!\!\!\!\diagup C{=}N{-} \qquad \diagdown\!\!\!\!\diagup C{=}O$$

characteristic properties. While carbonyl groups are necessarily end groups, both atoms of the C=C and C=N groups can be located at internal positions in chains or rings. The nitrogen and oxygen atoms in these groups possess lone pairs of electrons which account for other distinctive properties of the azomethine and carbonyl groups.

The C=N group occurs in many organic molecules of fundamental importance which were extensively studied by quantum chemical as well as by ultraviolet and infrared spectroscopic methods. Surprisingly, the C=N unit itself received less attention and we shall make an attempt to remedy at least partly this situation. The biological importance of the C=N group is also very great and this is one more reason to give it careful and detailed attention.

II. SOME PHYSICO-CHEMICAL DATA

The author envies his distinguished colleagues who wrote the corresponding chapters in the first two volumes of this series relating to the C=C and C=O groups[1,2]. In fact, data on such quantities as dipole moments, bond energies and interatomic distances are plentiful in the literature for these groups. Unfortunately this is not the case for the C=N group. The unstable character of the simplest azomethine compounds is probably the main reason for this. An attempt has been made to gather together all available data, however.

A. Internuclear Distances

The distance usually quoted for the C=N double bond is 1·29–1·31 Å for the non-conjugated group and 1·35 or 1·36 Å for aza-aromatic compounds (see, for example, pages M 131, M 141, M 150, M 162, M 179, M 217 in reference 3).

Layton, Kross and Fassel[4] found a linear relationship between the infrared CN stretching frequency and the internuclear distance. From their very carefully prepared diagram, which is based on a great number of data, one can see that typical distances are 1·47 Å for the

C—N bond, 1·29–1·31 for the C=N bond and 1·35 or 1·36 for the aza-aromatic C=N bond. The C—N bond length in linear conjugated molecules of the C=C—C=N type is not significantly longer than the length of the isolated C=N bond. This is similar to what was found for 1,3-butadiene.

The following table contains some representative data.

TABLE 1. Some typical bond lengths.

C—C (Ethane)	1·537	C—N	1·47	N—N	1·47	C—O	1·40
C=C (Benzene)	1·397	C=N (ring conj.)	1·36			C=O (strongly conj.)	1·29
C=C (Ethylene)	1·338	C=N	1·30	N=N	1·24	C=O	1·21
C≡C (Acetylene)	1·205	C≡N	1·16	N≡N	1·09		

The values for ethane, ethylene, benzene and acetylene are from Stoicheff[5], Dowling and Stoicheff[6], Langseth and Stoicheff[7] and Christensen, Eaton, Green and Thompson[8] respectively. (See also Costain and Stoicheff[9].) These were obtained from high resolution rotational Raman spectra and are better than ±0·003. The other data quoted in the table are less accurate and represent averages of data obtained by different methods on different molecules.

The length of the C=N bond is difficult to determine. On the assumption that there is complete resonance in melamine:

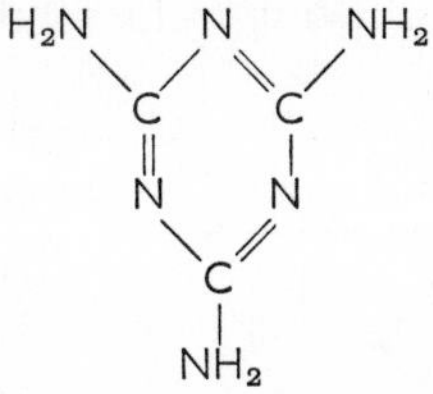

```
H2N       N       NH2
   \    /   \\   /
     C         C
     ||        |
     N         N
       \     //
          C
          |
         NH2
```

and taking all CN distances equal (1·35 Å), Palmer[10] computed it by supposing that C—C in benzene bears the same relation to C—C (1·54 Å) and C=C (1·33 Å) as C—N in melamine does to C—N (1·47 Å) and C=N(X).

Then

$$\frac{1{\cdot}47 - 1{\cdot}35}{1{\cdot}47 - X} = \frac{1{\cdot}54 - 1{\cdot}39}{1{\cdot}54 - 1{\cdot}33}$$

whence $X = 1{\cdot}30$. With (C═C) = 1·339 we obtain 1·304 for (C═N). Bond length–bond order curves are available for CN, CO and OO bonds (Orville-Thomas[11], Morris and Orville-Thomas[12]) and naturally, CC bonds. (See, for example, Daudel[13], Coulson[14].)

B. Overlap Integrals

Overlap integrals provide some insight into bonding conditions. In general, they are larger the shorter the bond. An increase in polar character decreases their value. The following (π–π) overlap integrals are taken from the famous paper of Mulliken, Rieke, Orloff and Orloff[15] where they are given as a function of the internuclear distance and the nuclear charges. Slater atomic orbitals were used to compute them.

We used the bond distances given in Table 1 and $Z_{eff} = 3{\cdot}25$ for carbon, 3·90 for nitrogen and 4·55 for oxygen.

TABLE 2. Some typical overlap integral values.

C═C (Benzene)	0·246	C═N (ring conj.)	0·208	C═O (strongly conj.)	0·186
C═C (Ethylene)	0·271	C═N	0·230	C═O	0·218
C≡C (Acetylene)	0·335	C≡N	0·297		

C. Dipole Moments

The reader would no doubt be interested in knowing the value of the bond moment for a hypothetical C═N bond. Smyth[16,17] estimates it to 0·9 D. The following values taken from his book are of interest (Table 3).

TABLE 3. Some typical bond dipole moments.

C═C	0·0 D
C═N	0·9 D
C═O	2·3 D
C═S	2·6 D
N═O	2·0 D
C≡N	3·5 D

Smith[18] computed 0·45 D for the C—N bond (if the bond moment of H—C is 0·4 D). Because of this and our considerations of the previous section, we believe that the contribution of the π and σ bonds to the 0·9 D of C═N is roughly the same.

The following data measured by Everard and Sutton[20] are given by McClellan[19]:

N-n-butylethylidenamine	1·61 ± 0·03 D
N-ethylbutylidenamine	1·67 ± 0·05 D
N-n-butylpropylidenamine	1·61 ± 0·02 D
N-n-propylbutylidenamine	1·55 ± 0·02 D
N-isopropylbutylidenamine	1·61 ± 0·02 D
N-n-butylbutylidenamine	1·61 ± 0·03 D
N-isobutylbutylidenamine	1·55 ± 0·03 D

Sastry and Curl[21,22] measured the dipole moment of *N*-methylmethylenimine $CH_2{=}NCH_3$ in the gas phase by microwave methods and found it to be 1·53 D.

From these data, taking the C—H, C—N and $—CH_3$ bond and group moments equal to 0·4, 0·45 and 0·4 D, it is easily estimated that the C=N bond moment must be close to 1 D*.

D. Bond Energies

As Cottrell puts it, 'The bond energy term as thermochemical bond energy is a quantity assigned to each of the bonds in a molecule, so that the sum over all bonds is equal to the heat of atomization of the molecule (a positive quantity, on the thermodynamic sign convention).'

This is a delicate concept. A good discussion of the implications can be found in Cottrell's book[23].

The heat of atomization is the enthalpy change, ΔH, for the reaction

$$A_lB_mC_n\cdots \longrightarrow lA + mB + nC + \cdots$$

but the states of the molecules on the left and the atoms on the right should be exactly specified. Furthermore, a generalized use of bond energies requires the supposition that they are constant from molecule to molecule. Clearly this transferability can only be approximate. Cottrell's data imply the definition of the bond energy sum as 'the heat of atomization in the ideal gas state at 298·16°K to the atomic elements in their ground states at this temperature'. The numerical values of the bond energy terms involving carbon depend on the knowledge of L_C, the heat of sublimation of carbon. Cottrell[23] used 170·9 kcal/mole. The values given by Cottrell and quoted here were calculated from the original data of Coates and Sutton[24].

* The dipole moment of $Me_2CHCH{=}NCH_2CHMe_2$ has been recently measured by M. Asselin in D. W. Davidson's laboratory on a sample prepared by D. Vocelle. The result was 1·56 ± 0·05 D.

TABLE 4. Some typical bond energies.

$E_{(C-C)}$ =	82·7 kcal/mole	$E_{(C-O)}$ =	85·5 kcal/mole
$E_{(C=C)}$ =	145·8 kcal/mole	$E_{(C=O)}$ =	179·0 kcal/mole (in ketones)
$E_{(C\equiv C)}$ =	199·6 kcal/mole		
$E_{(C-N)}$ =	72·8 kcal/mole	$E_{(N-N)}$ =	39·0 kcal/mole
$E_{(C=N)}$ =	147·0 kcal/mole	$E_{(N=N)}$ =	101·0 kcal/mole
$E_{(C\equiv N)}$ =	212·4 kcal/mole	$E_{(N\equiv N)}$ =	225·8 kcal/mole

$E_{(C=N)}$ seems to vary by about 10 kcal/mole from molecule to molecule. As can be seen it is not very different from $E_{(C=C)}$.

We believe that the best we can do under the circumstances is to give data which were obtained in the same laboratory under similar experimental conditions.

Another collection is found in Palmer's book[10] (Table 5).

TABLE 5. Some bond energy data.

$E_{(C-C)}$	83 kcal/mole (ethane)
$E_{(C=C)}$	146 kcal/mole (ethylene)
$E_{(C\equiv C)}$	200 kcal/mole (acetylene)
$E_{(C-N)}$	69 kcal/mole (amines)
$E_{(C=N)}$	142 kcal/mole (isobutilidene-n-butylamine)
$E_{(C\equiv N)}$	214 kcal/mole (acetonitrile)
$E_{(C-O)}$	82·6 kcal/mole (dimethylether)
$E_{(C=O)}$	175 kcal/mole (acetone)

Palmer also gives some detailed examples of the calculation of bond energies from thermochemical data. $E_{(C=N)}$ is considered as one of the less well-known bond energies, but it is satisfactory from the point of view of this article that it has been determined for an aliphatic Schiff base.

III. QUANTUM CHEMICAL ASPECTS

A. The Nitrogen Atom; Hybridization

The nitrogen atom has one more electron than carbon and this simple fact, with the underlying arrangement of atomic orbitals, is perhaps at the origin of the great structural versatility and adaptability of nitrogen compounds.

The configuration of the nitrogen atom in its ground state is

$$\text{N} \quad 1s^2 2s^2 2p^3$$

The three 'equivalent' p electrons yield three terms*, which are compatible with the Pauli principle: ^4S, ^2D and ^2P. Of these, according to Hund's rule and in agreement with observation, ^4S has the lowest energy, so this is the ground term of the nitrogen atom. In this the total spin projection S is $\frac{3}{2}$, since $2S + 1 = 4$, that is, all these p electrons have parallel spins. This means that the configuration is actually

$$\mathrm{N} \quad 1s^2 2s^2 2p_x 2p_y 2p_z$$

and the atom is prepared for forming three covalent links at right angles to each other.

The valence angles are, however, significantly different from 90° in most nitrogen compounds, indicating that the N atom enters into molecules with a modified orbital arrangement. There is nothing surprising in this. The 'pure' s, p, . . . wave functions were solutions of the Schrödinger equation for the hydrogen atom, and although the successful interpretation of the periodic system and the whole volume of atomic spectroscopy show that the orbital picture obtained in the hydrogen atom problem is essentially applicable to higher atoms, there is no reason to expect that atoms can be built into molecules with their unchanged atomic orbital arrangement. This is, indeed, not the problem. The problem is rather to know if we can build good approximate molecular wave functions from the pure s,p, . . . functions or if it is better to use combinations of them for that purpose. The two approaches are actually equivalent, but for the understanding of the principles of stereochemistry and for the adapting of the wave functions to molecular symmetry, the hybridized atomic orbitals have certain advantages or at least a higher descriptive value.

Just as in the case of carbon, the 'pure' $2s$, $2p_x$, $2p_y$, $2p_z$ orbitals of nitrogen can be combined to yield tetrahedral, trigonal and digonal hybrides. There are, however, certain differences.

(a) In the case of carbon, the ^3P ground state is issued from configuration $1s^2\ 2s^2\ 2p_x\ 2p_y$; one of the p orbitals is empty and the $2s$ orbital is doubly occupied. Hybridization increases the number of covalencies from two to four.

In the case of the ^4S ground state of nitrogen, all three $2p$ orbitals are singly occupied and $2s$ is doubly occupied. Since in this case we have five electrons to place instead of four, we shall have three orbitals ready to enter into covalent bonds before as well as after hybridization and the number of valencies does not change. (This applies in the case of oxygen also.)

* The reader may like to consult G. Herzberg, *Atomic Spectra and Atomic Structure*, Dover Publications, New York, 1944, p. 130.

(b) The possibility of having a lone pair of electrons on the nitrogen atom is a highly important fact from the point of view of purely chemical as well as spectral and other physical properties of nitrogen-containing molecules. Hybridization offers very good qualitative information on this. Let us compare the configuration of nitrogen under the different possible hybridizations:

			Example
Unhybridized	N	$1s^2\, 2s^2\, 2p_x\, 2p_y\, 2p_z$	N
Tetrahedral	N	$1s^2\, \mathrm{te}_1\, \mathrm{te}_2\, \mathrm{te}_3\, \mathrm{te}_4{}^2$	NH_3
Trigonal	N	$1s^2\, \mathrm{tri}_1\, \mathrm{tri}_2\, \mathrm{tri}_3\, 2p_z{}^2$	aniline
or	N	$1s^2\, \mathrm{tri}_1\, \mathrm{tri}_2\, \mathrm{tri}_3{}^2\, 2p_z$	pyridine
Digonal	N	$1s^2\, \mathrm{di}_1\, \mathrm{di}_2\, 2p_y\, 2p_z{}^2$	
or	N	$1s^2\, \mathrm{di}_1\, \mathrm{di}_2{}^2\, 2p_y\, 2p_z$	C≡N

We notice, and this will be essential in what follows, that in the case of trigonal hybridization (which prescribes a coplanar arrangement) the lone pairs occupy either a symmetrical unhybridized $2p$ orbital whose axis is perpendicular to the plane defined by the axes of the trigonal hybrids, or an asymmetrical hybrid orbital whose axis is in the plane, leaving only one π electron in the unhybridized $2p$ orbital. Molecules like pyridine and other aza-aromatic compounds have this latter type of hybridized nitrogen and this also applies to the azomethine group, the main subject of this book.

Thus we think of the C=N group as having both C and N atoms trigonally hybridized, with each of these atoms contributing one π electron and the nitrogen having an asymmetrical lone pair orbital in the plane of the group, the valence angles being close to 120°.

C=N

Somewhat more mathematically we may describe the conditions as follows. Suppose that we could solve exactly the Schrödinger equation for the nitrogen atom and the $2s$, $2p$, . . . wave functions would be exact wave functions. Since the Schrödinger equation is linear, any linear combination of these solutions is itself a solution. We are interested in trigonal hybridization with the molecular plane being the xy plane.

Then

$$\begin{aligned} \mathrm{tri}_1 &= a_1 s + b_1 x + c_1 y \\ \mathrm{tri}_2 &= a_2 s + b_2 x + c_2 y \\ \mathrm{tri}_3 &= a_3 s + b_3 x + c_3 y \end{aligned} \tag{1}$$

with $\quad s \equiv \psi_{2s} \quad x \equiv \psi_{2p_x} \quad$ and $\quad y \equiv \psi_{2p_y}$

We have nine coefficients to determine. The hybrid wave functions must be normalized and mutually orthogonal, as were the original 'pure' functions. Thus we can write

$$
\begin{aligned}
a_1^2 + b_1^2 + c_1^2 &= 1 \\
a_2^2 + b_2^2 + c_2^2 &= 1 \\
a_3^2 + b_3^2 + c_3^2 &= 1 \\
a_1a_2 + b_1b_2 + c_1c_2 &= 0 \\
a_1a_3 + b_1b_3 + c_1c_3 &= 0 \\
a_2a_3 + b_2b_3 + c_2c_3 &= 0
\end{aligned} \tag{2}
$$

There are six relationships for nine unknowns. If we want three identical hybrids (but for their orientation in space) they must clearly have the same admixture of the 2*s* orbital which has spherical symmetry.

Then

$$a_1 = a_2 = a_3$$

We shall choose the axes as shown:

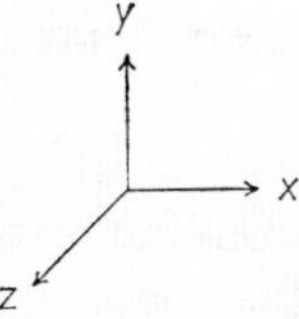

and require that the axis of tri_1 lie in the *x* axis. Then ψ_{2p_y} cannot contribute to tri_1, since the *xz* plane is its nodal plane and since its positive and negative lobes are symmetrically placed with respect to this plane. It follows that

$$c_1 = 0$$

With $a = 1/\sqrt{3}$ and using the relationships of equation (2), we finally obtain:

$$
\begin{aligned}
\text{tri}_1 &= \frac{1}{\sqrt{3}}\psi_{2s} + \frac{\sqrt{2}}{\sqrt{3}}\psi_{2p_x} \\
\text{tri}_2 &= \frac{1}{\sqrt{3}}\psi_{2s} - \frac{1}{\sqrt{6}}\psi_{2p_x} + \frac{1}{\sqrt{2}}\psi_{2p_y} \\
\text{tri}_3 &= \frac{1}{\sqrt{3}}\psi_{2s} - \frac{1}{\sqrt{6}}\psi_{2p_x} - \frac{1}{\sqrt{2}}\psi_{2p_y}
\end{aligned} \tag{3}
$$

Once equations (1) and (2) were given it would have been possible to obtain equation (3) by simple group theoretical considerations[26].

However, the three hybrids are identical only if there is threefold symmetry around the atom and all three angles are equal (D_3 because of the requirement of coplanarity), as in CH_3^+ or BF_3. If this condition is not fulfilled, the perturbation by the other atoms will cause the hybrids to be different. Then the coefficients a, b, c, can be determined from the experimentally known angles[25,29].

The reader who wishes to know more about hybridization is referred to the relevant works by Coulson[14], Cotton[26], Pitzer[27], Eyring, Walter and Kimball[28] and especially Julg[25,29].

We conclude this section in drawing attention again to the possibility of having a lone pair in either a π or an sp^2 hybrid orbital on trigonally hybridized nitrogen. We believe that this is of fundamental chemical and biological importance, and together with the variability of the angles of hybridization it makes possible the formation of nitrogen containing molecules with all the delicate differences in physico-chemical properties necessary to produce the phenomena of life.

B. The π Electrons of the Azomethine Group; Empirical Methods

I. The Hückel method

There are only two π electrons in an azomethine group and they are responsible for many characteristic properties of this group. Our first task will be to examine their behaviour in the light of the various approximate methods used in quantum chemistry. We start with the simple Hückel molecular orbital method.

Most organic chemists to-day are familiar with the LCAO–MO method in its simplest form*. In the case of ethylene, with only the two π electrons considered, the molecular orbitals are of the form:

$$\phi = c_A\chi_A + c_B\chi_B \tag{4}$$

where $c_A = \pm c_B$ for reasons of symmetry. χ_A and χ_B are atomic orbitals, that is, wave functions which an electron would have in the free atoms A or B; ϕ depends only on *one* electron's coordinates. The energy of an electron in molecular orbital ϕ is obtained by use o

* For simple worked examples the reader may refer among others to Streitwieser[30], Sandorfy[31], Salem[32] and Julg[29].

the variational method, which minimizes the energy with respect to the coefficients c_A and c_B. Supposing the wave functions are real

$$E = \frac{\int (c_A\chi_A + c_B\chi_B)H(c_A\chi_A + c_B\chi_B)\,d\tau}{\int (c_A\chi_A + c_B\chi_B)^2\,d\tau} \tag{5}$$

$$E = \frac{c_A^2 \int \chi_A H \chi_A\,d\tau + 2c_Ac_B \int \chi_A H \chi_B\,d\tau + c_B^2 \int \chi_B H \chi_B\,d\tau}{c_A^2 \int \chi_A^2\,d\tau + 2c_Ac_B \int \chi_A\chi_B\,d\tau + c_B^2 \int \chi_B^2\,d\tau} \tag{6}$$

In equation (6) $\int \chi_A H \chi_A\,d\tau$ is called a Coulomb integral, because it represents the energy due to the electrostatic forces between one electron and the core formed by the nuclei and the other electrons in their spheres that we do not want to consider explicitly. The integral $\int \chi_A H \chi_B\,d\tau$ is called the resonance integral. It contains two atomic orbitals and has no classical analogue. It would vanish if the two atoms were at infinite distance and it may be taken as a rough measure of the strength of the bond. $\int \chi_A\chi_B\,d\tau$ is an overlap integral. The contribution to this integral of a given volume element is different from zero only if both χ_A and χ_B are different from zero in that volume element, that is, where the two orbitals overlap. It is usually roughly proportional to the resonance integral and to the energy of the bond. The following notations are widely used:

Coulomb integral $\alpha_r \equiv H_{rr} \equiv \int \chi_r H \chi_r\,d\tau$

Resonance integral $\beta_{rs} \equiv H_{rs} \equiv \int \chi_r H \chi_s\,d\tau$

Overlap integral $S_{rs} \equiv \int \chi_r\chi_s\,d\tau$

Indices r and s represent atomic orbitals, and because of the requirement of normalization

$$S_{rr} = \int \chi_r\chi_r\,d\tau = 1$$

We then have in the case of ethylene

$$E = \frac{c_A^2H_{AA} + 2c_Ac_BH_{AB} + c_B^2H_{BB}}{c_A^2S_{AA} + 2c_Ac_BS_{AB} + c_B^2H_{BB}} \tag{7}$$

In order to find the minimum value of the energy with the given wave function, we consider it as a function of the coefficients c_A and c_B, compute the derivatives of equation (7) with respect to these coefficients separately and equate the derivatives to zero. Then we obtain the set of equations

$$c_A(H_{AA} - ES_{AA}) + c_B(H_{AB} - ES_{AB}) = 0$$
$$c_A(H_{AB} - ES_{AB}) + c_B(H_{BB} - ES_{BB}) = 0$$

In the simple Hückel approximation the overlap integrals are neglected[33]. (Also all Coulomb integrals are taken equal for carbon atoms as are all the resonance integrals for C—C bonds.) Resonance integrals between non-bonded atoms are neglected.

Then we have, with the above notations

$$\begin{aligned} (\alpha - E)c_A + \beta c_B &= 0 \\ c_A\beta + (\alpha - E)c_B &= 0 \end{aligned} \tag{8}$$

or with $y = (\alpha - E)/\beta$

$$\begin{aligned} c_A y + c_B &= 0 \\ c_A + c_B y &= 0 \end{aligned} \tag{9}$$

and $y = \pm 1$
Therefore

$$\begin{aligned} E_1 &= \alpha + \beta \qquad (y = -1) \\ E_2 &= \alpha - \beta \qquad (y = +1) \end{aligned}$$

where E_1 is the energy of the orbital of lower energy.
From equation (15), after normalization

$$\begin{aligned} \phi_1 &= 0{\cdot}707\chi_A + 0{\cdot}707\chi_B \\ \phi_2 &= 0{\cdot}707\chi_A - 0{\cdot}707\chi_B \end{aligned} \tag{10}$$

Since in the ground state both electrons are in ϕ_1:

E_2 ———

E_1 —×—×— (11)

$$E_{(\text{total})} = 2\alpha + 2\beta$$

The electronic charge densities[34] are simply

$$q_A = 2c_A^2 = 1 \quad \text{and} \quad q_B = 2c_B^2 = 1$$

and the bond order[35,36]

$$P_{AB} = 2c_A c_B$$

Then, with the normalized coefficients

$$E = \int (c_A\chi_A + c_B\chi_B)H(c_A\chi_A + c_B\chi_B)\,d\tau = (c_A^2 + c_B^2)\alpha + 2c_A c_B\beta$$

and, for the two electrons

$$E_{(\text{total})} = (q_A + q_B)\alpha + 2P_{AB}\beta = 2\alpha + 2\beta \tag{12}$$

where P_{AB} is the Coulson bond order. As to the actual values of α and β, it is preferable to choose suitable empirical values. This can remedy at least a part of the errors inherent in the method.

Thorough criticisms of the LCAO–MO method in its various approximations have been given by many authors and it would be out of place to do it here. The method proved highly useful in interpreting and correlating many chemical and spectral facts, however, and it certainly gives more information about the physico-chemical properties of molecules than the traditional method of dashes and + and − signs. Yet the reader knows that the latter was sufficient to produce that prodigious world called organic chemistry. So let us use the Hückel LCAO–MO method as an empirical device of wave mechanical inspiration with some confidence*. The method is expected to give better results when it is trying to explain 'trends', that is the gradual change in properties from molecule to molecule, than in explaining the behaviour of one given molecule†.

a. Estimation of parameters. For many purposes it is not actually necessary to know the numerical value of α and β. As we have seen, the electronic charge densities and bond orders, for example, can be obtained without them. In comparing different molecules, however, we must know how α and β *vary*.

It is reasonable to suppose that if we replace a carbon atom by a nitrogen atom, the greatest part of the variation of α_r would be due to the difference in interaction of the π electron with the carbon and nitrogen cores. It is therefore logical to link α_r to quantities like the ionization potential or electronegativity of the respective atom.

Electronegativity was first introduced by Pauling[37] as a semi-quantitative concept. Pauling's electronegativities were linked to bond energies and the partial ionic character of bonds. The values of his original electronegativities (X) for some elements are:

H	2·1	(2·0)
C	2·5	
N	3·0	(3·15)
O	3·5	(3·60)
F	4·0	(4·15)
Cl	3·0	(3·10)

where the numbers in brackets are the somewhat modified values computed by Bellugue and Daudel[38]. If $D_{(A-A)}$ and $D_{(B-B)}$ are the energies of covalent

* We recommend that those who are not engaged in research in quantum chemistry read the review of the approximations involved in MO theory by Streitwieser (reference 30, pp. 99–101).

† This chapter has been written in the belief that most of those who will read it are organic chemists. It is thought that they may like to have some more information on hand relating to quantum chemical methods. This is the purpose of the paragraphs given in small print. The theoretical chemist will of course find these elementary.

bonds A—A and B—B and $D_{(A-B)}$ that of A—B (which has a partial ionic character), then the difference between $D_{(A-B)}$ and the arithmetic mean of $D_{(A-A)}$ and $D_{(B-B)}$ is

$$\Delta = D_{(A-B)} - \tfrac{1}{2}\{D_{(A-A)} + D_{(B-B)}\} \tag{13}$$

and

$$X_A - X_B = 0{\cdot}208\sqrt{\Delta}$$

where Δ is in electron volts. Then X_A, for example, is obtained by comparing such differences relating to two or more A—Y bonds. Thus the units for the above given values for Pauling's electronegativities are $(\text{ev})^{\frac{1}{2}}$ multiplied by an empirical factor.

Mulliken[39,40] found a more rigorous way of defining electronegativities. He proposed the electronegativity to be taken for the arithmetic mean of the first ionization potential and the electron affinity of the atom. Approximate theoretical justification for this was given by him and by Moffitt[41].

In LCAO language we can say that atoms A and B are equally electronegative if in equation (4) c_A^2 and c_B^2 are equal. For this it is necessary, however, that the Coulomb integrals α_A and α_B be the same. Thus Mulliken took the negative of the Coulomb integral (α and β are inherently negative) as the absolute theoretical electronegativity. The Coulomb integral is not a purely atomic integral, and therefore electronegativities can be assigned to individual atoms in an approximate manner only. In fact it should and does vary, in general, from one molecule to another. Now Mulliken has shown that the interatomic terms in α are either small or cancel each other in a fairly good approximation, and he obtained for the electronegativity equation (14):

$$X_A = \tfrac{1}{2}(I_A + E_A) \tag{14}$$

The electronegativity scale based on this formula parallels the one of Pauling so that

$$(X_A - X_B)_{\text{Pauling}} \approx 0{\cdot}36(X_A - X_B)_{\text{Mulliken}} \tag{15}$$

where Mulliken's values are expressed in ev and Pauling's in $(\text{ev})^{\frac{1}{2}}$ (see above).

It would be advantageous to change the units so that the Coulomb integral α is taken equal to Mulliken's electronegativity, but in units of the resonance integral β (or a suitably chosen β_0 if more than one different β is used for a molecule). Since according to equation (15):

$$X_{\text{Mulliken}} \approx 2{\cdot}8X_{\text{Pauling}}$$

it would be easy to say that

$$X_{\text{Mulliken}} \approx \beta X_{\text{Pauling}}$$

if β was equal to 2·8 ev and X_{Pauling} dimensionless. This may be accidentally true for a given C—C π bond and may serve for some qualitative reasoning. At any rate, since for many purposes α and β may remain indetermined, it is customary to keep α_C of a carbon atom as a reference without giving it a numerical value and then take for nitrogen, for example,

$$\alpha_N - \alpha_C = (X_N - X_C)_{\text{Pauling}} h\beta_0$$

or

$$\alpha_N = \alpha_C + \Delta X h \beta_0 \qquad \text{where } h \text{ is a parameter} \tag{16}$$

Using this expression, the secular equation is easier to solve. In the case of nitrogen ΔX would be about 0·5, and in the case of oxygen, about 1·0, using Pauling's original electronegativities. To start with one can put $h = 1$. (For the parameters recommended by Streitwieser, see reference 30, p. 135.)

Mulliken pointed out that the ionization potentials and electron affinities entering the definition of electronegativity (equation 14) are not those of the free atom but should be taken in the appropriate valence state[39,40]. A valence state is not a state of the free atom. When we say that an atom is in a valence 'state', we mean that its electronic configuration and energy are what they would be if the atom could be taken out of the molecule with all bonds severed but with no change in the arrangement of its orbitals. In general, these configurations contain equivalent electrons (same n and l quantum numbers), and therefore more than one state is issued from them since the total angular momentum and spin quantum numbers (L and S) may take several values.

In forming a valence 'state' the centre of gravity of all spectroscopic states issued from the given configuration is taken. (To configuration s^2p^2 of carbon three spectroscopic states belong: ^3P, ^1D and ^1S; to sp^3: ^5S, ^3D, ^3P, ^1D, ^3S and ^1P.) In the case of nitrogen, to configuration s^2p^3 belong ^4S, ^2D and ^2P, and to sp^4: ^4P, ^2D; ^2S, ^2P. In most cases these states are not all precisely known.

They may be taken from experiment or computed by the theory of atomic spectra developed by Slater[42], Condon and Shortley[43] and Racah[44] and fitted to available spectroscopic data. Many of these states are not 'pure' and their observed energies belong to a mixture involving identical terms issued from other configurations (configuration interaction). Furthermore, if the atomic orbitals are hybridized in the valence state then a promotion energy has to be added to the energy computed as the centre of gravity of the spectroscopic states, and the calculation of the promotion energy involves certain approximations. Naturally, ionization potentials and electron affinities (and thus electronegativities) are in general different for different oxidation states and vary also with hybridization. So, for instance, the electronegativity of carbon is supposed to increase in the order tetrahedral < trigonal < digonal[14].

The best sources for finding *valence state* ionization potentials and electron affinities at present are the publications of Pilcher and Skinner[45] (and previous ones by Pritchard and Skinner[46]) and by Hinze and Jaffé[47].

The data in Table 6 are taken from their papers. They are all in ev, s ≡ $2s$, x ≡ $2p_x$, etc., te = tetrahedral, tri = trigonal.

It appears clearly from these data that I_P and E_A cannot be considered as a property of an atom as a whole, but as a characteristic of an orbital of a given type in the respective valence state of that atom. It is by no means indifferent, for example, if in C tri_1 tri_2, tri_3 z the electron is taken away from (or added to) a trigonal hybrid orbital or a $2p_z$ orbital. Since this is so it is necessary to define orbital electronegativities rather than atomic electronegativities. Hinze and Jaffé[47] defined as orbital electronegativities half the sum (in ev) of the ionization potential and electron affinity for the relating valence state and orbital. In order to preserve the linear relationship between Mulliken's and Pauling's electronegativity scales they worked out a formula (equation 17) converting their orbital electronegatives X_M, which are basically of the Mulliken type, into Pauling-type electronegativities (X_P).

$$0{\cdot}168(2X_M - 1{\cdot}23) = X_P \tag{17}$$

The ratio X_M/X_P is approximately equal to 3·2 ev, [instead of 2·8 from equation (15)]. The values they obtained for X_M and X_P for the cases of interest to us are given in Table 6. If, following Mulliken, we take $X_M = \alpha$ for the Coulomb integral to be used in the Hückel LCAO–MO method, then for the π orbital in trigonal hybridization we have for carbon $\alpha_C = 5{\cdot}60$ ev and for nitrogen, with the lone pair in a trigonal hybrid, $\alpha_N = 7{\cdot}95$ ev. The respective 'Pauling' values of Hinze and Jaffé are $X_P = 1{\cdot}68$ and 2·47.

The concept of orbital electronegativity has been considerably developed through works by Sanderson[48], Iczkowski and Margrave[49], Klixbüll-Jörgensen[50], Hinze, Whitehead and Jaffé[51] and Klopman[52]. We shall use this concept in the course of this discussion. Concerning

TABLE 6. Valence state ionization potentials and electron affinities.

	Valence state and process	I_P (*P–S*)	I_P (*H–J*)	Valence state and process	E_A (*P–S*)	E_A (*H–J*)	$2X_M$	X_P (*H–J*)
C	$\mathrm{sxyz} \rightarrow \mathrm{xyz}$	20·78	21·01	$\mathrm{sxyz} \rightarrow \mathrm{s^2\,xyz}$	8·89	8·91	29·92	4·84
	$\mathrm{sxyz} \rightarrow \mathrm{sxy}$	11·31	11·27	$\mathrm{sxyz} \rightarrow \mathrm{sx^2\,yz}$	0·87	0·34	11·61	1·75
	$\mathrm{tri_1\,tri_2\,tri_3\,z} \rightarrow \mathrm{tri_1\,tri_2\,tri_3}$	11·22	11·16	$\mathrm{tri_1\,tri_2\,tri_3\,z} \rightarrow \mathrm{tri_1\,tri_2\,tri_3\,z^2}$	0·62	0·03	11·19	1·68
	$\mathrm{tri_1\,tri_2\,tri_3\,z} \rightarrow \mathrm{tri_1\,tri_2\,z}$	15·56	15·62	$\mathrm{tri_1\,tri_2\,tri_3\,z} \rightarrow \mathrm{tri_1^2\,tri_2\,tri_3\,z}$	2·34	1·95	17·58	2·75
	$\mathrm{te_1\,te_2\,te_3\,te_4} \rightarrow \mathrm{te_1\,te_2\,te_3}$	14·57	14·61	$\mathrm{te_1\,te_2\,te_3\,te_4} \rightarrow \mathrm{te_1^2\,te_2\,te_3\,te_4}$	1·79	1·34	15·95	2·48
N	$\mathrm{s^2\,xyz} \rightarrow \mathrm{s^2\,xy}$		13·94	$\mathrm{s^2\,xyz} \rightarrow \mathrm{s^2\,x^2\,yz}$		0·84	14·78	2·28
	$\mathrm{sx^2\,yz} \rightarrow \mathrm{x^2\,yz}$	27·70	26·92	$\mathrm{sx^2\,yz} \rightarrow \mathrm{s^2\,x^2\,yz}$	13·80	14·05	40·98	6·70
	$\mathrm{sx^2\,yz} \rightarrow \mathrm{sx^2\,y}$	14·78	14·42	$\mathrm{sx^2\,yz} \rightarrow \mathrm{sx^2\,yz^2}$	1·34	2·54	16·96	2·65
	$\mathrm{tri_1\,tri_2\,tri_3\,z^2} \rightarrow \mathrm{tri_1\,tri_2\,z^2}$	20·32	19·72	$\mathrm{tri_1\,tri_2\,tri_3\,z^2} \rightarrow \mathrm{tri_1^2\,tri_2\,tri_3\,z^2}$	4·14	4·92	24·63	3·94
	$\mathrm{tri_1^2\,tri_2\,tri_3\,z} \rightarrow \mathrm{tri_1\,tri_2\,tri_3\,z}$	15·19						
	$\mathrm{tri_1^2\,tri_2\,tri_3\,z} \rightarrow \mathrm{tri_1^2\,tri_2\,tri_3}$	14·51	14·12	$\mathrm{tri_1^2\,tri_2\,tri_3\,z} \rightarrow \mathrm{tri_1^2\,tri_2\,tri_3\,z^2}$	1·20	1·78	15·90	2·47
	$\mathrm{tri_1^2\,tri_2\,tri_3\,z} \rightarrow \mathrm{tri_1^2\,tri_2\,z}$		20·60	$\mathrm{tri_1^2\,tri_2\,tri_3\,z} \rightarrow \mathrm{tri_1^2\,tri_2\,tri_3^2\,z}$		5·14	25·74	4·13
	$\mathrm{tri_1\,tri_2\,tri_3\,z^2} \rightarrow \mathrm{tri_1\,tri_2\,tri_3\,z}$	12·25						
	$\mathrm{te_2^2\,te_2\,te_3\,te_4} \rightarrow \mathrm{te_1^2\,te_2\,te_3}$	19·35	18·93	$\mathrm{te_1^2\,te_2\,te_3\,te_4} \rightarrow \mathrm{te_1^2\,te_2^2\,te_3\,te_4}$	3·63	4·15	23·08	3·68
	$\mathrm{te_1^2\,te_2\,te_3\,te_4} \rightarrow \mathrm{te_1\,te_2\,te_3\,te_4}$	14·31						

the numerical value of the resonance integral β the situation is confused. Estimates can be based on comparing calculated and experimental values of heats of hydrogenation, heats of formation or electronic excitation energies. These quantities themselves usually depend on certain assumptions and are affected to a variable extent by the approximations introduced into the Hückel MO method. Estimates vary from perhaps 1·5 to 3·5 ev for carbon π electrons (cf. reference 30, pp. 103–110).

The situation is much better in the case of the *ratios* of the β values in going from one molecule to the other. As Mulliken first observed, the β_{rs} vary roughly linearly with the overlap integrals S_{rs} (computed with Slater orbitals) which themselves vary with the internuclear distance for the same pair of atoms and are easy to compute. If we take for reference (β_0) the resonance integral for a π–π bond in ethylene, the following data illustrate the relationship between the internuclear distance (Å), S_{rs} and β_{rs} for two π electrons:

TABLE 7. π Internuclear distances, overlap integrals and relative values of resonance integrals.

(Å)	S_{rs}	β/β_0
1·205	0·335	1·24
1·339	0·271	1·00
1·397	0·246	0·91
1·537	0·193	0·71

With some caution the β–S relationship can be extended to heteroatomic molecules. Also, if the axes of the two π orbitals are twisted relative to one another by an angle of ϑ the relationship:

$$\beta = \beta_0 \cos \vartheta$$

is approximately valid[53,54]. With the overlap integrals taken from the tables of Mulliken, Rieke, Orloff and Orloff[15] we have for the respective non-conjugated π–π bonds:

C=C	0·271	1·00
C=N	0·230	0·85
C=O	0·218	0·80

2. Calculations on the azomethine group

Let us now make a simple calculation for the two π electrons of the azomethine group. We shall use simply:

$$\alpha_N = \alpha_C + 0{\cdot}5\beta$$

and

$$\beta = \beta_{CN} = \beta_0 \tag{18}$$

The secular equation is:

$$\begin{aligned} c_C(\alpha_C - E) + c_N\beta &= 0 \\ c_C\beta + c_N(\alpha_C + 0{\cdot}5\beta - E) &= 0 \end{aligned} \tag{19}$$

or if we put:

$$y = \frac{\alpha_C - E}{\beta}$$

$$\begin{aligned} c_C y + c_N &= 0 \\ c_C + c_N(y + 0{\cdot}5) &= 0 \end{aligned} \tag{20}$$

Equation (19) yields $y_1 = -1{\cdot}281$ and $y_2 = +0{\cdot}781$. Since β is a negative quantity (as is α) y_1 corresponds to the ground state (E_0) and

$$E_0 = \alpha_C + 1{\cdot}281\beta \tag{21}$$

for one π electron. In the Hückel approximation the *double* of this represents the contribution of the two π electrons to the total energy. The charge distribution is

$$\overset{0{\cdot}76}{C} \text{——} \overset{1{\cdot}24}{N} \quad \text{or} \quad \overset{+0{\cdot}24}{C} \text{——} \overset{-0{\cdot}24}{N}$$

This predicts a fairly large dipole moment of about 1·50 D if the C═N distance is taken as 1·30 Å. (One effective charge and an internuclear distance of 1 Å would mean 4·8 D.) Clearly the amount of charge transferred from the carbon to the nitrogen is a function of $\alpha_N - \alpha_C$ and it is almost certainly less than 0·24.

It is instructive to rewrite the ground state wave function in terms of valence-bond structures. For ethylene we obtain

$$\psi_{tot} = (0{\cdot}707\chi_A + 0{\cdot}707\chi_B)\ (1)\ (0{\cdot}707\chi_A + 0{\cdot}707\chi_B)\ (2)$$

where 1 and 2 represent the coordinates of electron 1 and 2 respectively.

$$\begin{aligned} \psi_{tot} = {} & 0{\cdot}5\chi_A(1)\chi_A(2) + 0{\cdot}5\chi_A(1)\chi_B(2) \\ & + 0{\cdot}5\chi_A(2)\chi_B(1) + 0{\cdot}5\chi_B(1)\chi_B(2) \end{aligned} \tag{22}$$

This means that the three possible structures have the following weights (their share in $\psi^2_{tot} = 1$, neglecting overlap):

$C_A^-\ C_B^+$	$\dot{C}_A\ \dot{C}_B$	$C_A^+\ C_B^-$
25%	50%	25%

The share of the ionic structures is highly exaggerated, a well-known defect of the simple Hückel MO method. For the azomethine group we obtain, with the above normalized coefficients,

$$\begin{aligned}\psi_{\text{tot}} = {} & (0{\cdot}616\chi_C + 0{\cdot}788\chi_N)(1)(0{\cdot}616\chi_C + 0{\cdot}788\chi_N)(2) \\ & 0{\cdot}380\chi_C(1)\chi_C(2) + 0{\cdot}485\chi_C(1)\chi_N(2) + 0{\cdot}485\chi_C(1)\chi_N(2) \\ & + 0{\cdot}620\chi_N(1)\chi_N(2)\end{aligned} \tag{23}$$

$C^- \; N^+$	$\dot{C} \; \dot{N}$	$C^+ \; N^-$
14·5%	47%	38·5%

It is seen that the difference between the weight of C^+ N^- and C^- N^+ is equal to the effective negative charge on the nitrogen (24%). So this can be considered as a measure of the ionic character of the bond.

Since the Hückel MO method attaches too great an importance to ionic structures, an often-used method to improve the situation is the so called charge-effect method or ω technique. It is based on the idea that the electronegativity (and Coulomb integral) of an atomic orbital depends on the electronic charge in that orbital. Thus if, as in our example, an atom in the molecule attracted to itself a charge equal to 0·24 electrons its electronegativity is diminished since it now attracts the other electrons less. This was pointed out by P. and R. Daudel[55], Laforgue[56] and Wheland and Mann[57]. If q_r is the electron charge density in a given orbital (1·24 in our example, not $-0{\cdot}24$) then the 'original' Coulomb integral can be modified according to

$$\alpha_r = \alpha_0 + (1 - q_r)\omega\beta_0 \tag{24}$$

where ω is a dimensionless parameter chosen to give agreement with experimental data. With the modified α_r we repeat the Hückel calculation which will yield a second q_r, which can again be used[30] to obtain a twice improved value for α_r, and so forth. This is an iterative procedure which can be repeated until α_r and q_r are in conformity with each other. This amounts to a certain self-consistency and somewhat better treatment of electronic repulsion in the simple Hückel MO method[58].

Most of the earlier authors used $\omega = 1$, but Streitwieser[58] recommends $\omega = 1{\cdot}4$ for carbon as well as for heteroatoms. This number is based on agreement with observed ionization potentials and with results obtained by the semi-empirical Pariser–Parr–Pople method[59,60].

With $\omega = 1{\cdot}4$ and $\alpha_N = \alpha_C + 0{\cdot}5\beta$ we have:

0·895	1·105
C ——	N

It is interesting in this respect to invoke an idea originally due to Sanderson[48] which then became a crucial point in orbital electronegativity theories. According to it, since the electronegativity of an atom in a molecule depends on its charge density q, the electronegativities of both atoms bonded together must become equal. Subsequently this idea has been applied to the bonding atomic orbitals rather than the whole electron distribution around the atoms. Iczkowski and Margrave[49] observed that it is possible to obtain electron affinities from the extrapolation of successive ionization potentials using an equation of the form

$$E(q) = aq + bq^2 + cq^3 + dq^4 + \cdots \quad (25)$$

where q is the number of electrons in the AO[31] and a, b, c, are constants. Then the assumption is made that q is actually the electron density of the AO so that it may have either integer or fractional values. The derivation of equation (25) with respect to q gives:

$$\frac{\mathrm{d}E}{\mathrm{d}q} = a + 2bq + 3cq^2 + 4\,\mathrm{d}q^3 + \cdots \quad (26)$$

The constants can be determined from observed ionization potentials. If we plot $\mathrm{d}E/\mathrm{d}q$ against q then the area below the function for two values of q, say q_1 and q_2, gives the energy for the addition of $q_2 - q_1$ electrons.

If, as usual, only the first two terms in equation (25) need be considered

$$E = aq + bq^2 \quad (27)$$

and

$$\frac{\mathrm{d}E}{\mathrm{d}q} = a + 2bq \quad (28)$$

From these we obtain with $q = 1$

$$E = a + b = I_v \quad (29)$$

the ionization energy (I_v) of the simply filled orbital in the given valence state.

With $q = 2$ we have:

$$E = 2a + 4b = I_v + E_v \quad (30)$$

the sum of the ionization energy (I_v) and the electron affinity (E_v). From equations (29) and (30) we obtain that:

$$a = \frac{3I_v - E_v}{2} \qquad \text{and} \qquad b = \frac{E_v - I_v}{2}$$

In this approximation the electronegativity becomes:

$$X = \left(\frac{\mathrm{d}E}{\mathrm{d}q}\right)_{q=1} = a + 2b = \frac{I_v + E_v}{2} \tag{31}$$

which is the same as Mulliken's definition.

It is then tempting to use equation (28) and find the charge density making the orbital electronegativities of the C and N orbitals equal. For carbon with Hinze and Jaffé's values[47]

$$\begin{aligned} X_C &= \left(\frac{\mathrm{d}E}{\mathrm{d}q}\right) = a + 2bq = \frac{3I_v - E_v}{2} + (E_v - I_v)q \\ &= \tfrac{1}{2}(3 \times 11{\cdot}16 - 0{\cdot}03) + (0{\cdot}03 - 11{\cdot}16)q \end{aligned}$$

and for nitrogen

$$\begin{aligned} X_N &= \left(\frac{\mathrm{d}E}{\mathrm{d}q}\right) = a + 2bq = \frac{3I_v - E_v}{2} + (E_v - I_v)q \\ &= \tfrac{1}{2}(3 \times 14{\cdot}12 - 1{\cdot}18) + (1{\cdot}18 - 14{\cdot}12)q \end{aligned}$$

With $\bar{q} = 1 - q$ being the effective charge

$$\begin{aligned} &(1 - \bar{q})(0{\cdot}03 - 11{\cdot}16) + \tfrac{1}{2}(3 \times 11{\cdot}16 - 0{\cdot}03) \\ &\qquad = (1 + \bar{q})(1{\cdot}78 - 14{\cdot}12) + \tfrac{1}{2}(3 \times 14{\cdot}12 - 1{\cdot}78) \end{aligned}$$

whence $\bar{q} = +0{\cdot}100$.

Then our charge distribution is:

$$\overset{0{\cdot}90}{\mathrm{C}} \text{———} \overset{1{\cdot}10}{\mathrm{N}}$$

Mulliken's electronegativities are then with $q_C = 0{\cdot}9$ and $q_N = 1{\cdot}1$ 6·72 ev for both C and N, which is almost the same as the mean value of the respective electronegativities computed from Mulliken's original $(I_v + E_v)/2$ (5·60 ev for C and 7·95 ev for N).

The last charge distribution is practically the same as the one obtained with the ω technique with $\omega = 1{\cdot}4$ and $\alpha_N = \alpha_C + 0{\cdot}5\beta$. It yields a π-dipole moment equal to 0·62 D with an internuclear distance of 1·30 Å. The author believes that this is reasonable.

3. Inclusion of overlap

The situation changes when overlap integrals are included in the LCAO–MO method.

In the case of the two π-electron system of ethylene the secular equation becomes:

$$\begin{aligned} (\alpha - E)c_A + (\beta - ES)c_B &= 0 \\ (\beta - ES)c_A + (\alpha - E)c_B &= 0 \end{aligned} \tag{32}$$

We can divide throughout by $(\beta - ES)$ and put, according to a suggestion by Wheland[61]

$$y = \frac{\alpha - E}{\beta - ES} \tag{33}$$

Then

$$\begin{aligned} c_A y + c_B &= 0 \\ c_A + c_B y &= 0 \end{aligned} \tag{34}$$

This is the same as we had in the treatment without overlap integrals, so we do not need to make a new calculation. The y and the coefficient are the same as before except for normalization. The condition for the latter is now

$$(c_A \chi_A + c_B \chi_B)^2 = c_A{}^2 + c_B{}^2 + 2c_A c_B S = 1 \tag{35}$$

whence with $S = 0{\cdot}271$

$$c_A{}^2 = 0{\cdot}393 \qquad c_B{}^2 = 0{\cdot}393 \qquad 2c_A c_B S = 0{\cdot}213$$

and we obtain the following diagram (for the two π electrons):

```
0·786   0·426   0·786
  C ─────────── C
```

Thus we now have an overlap charge of 0·426 electrons whose centre can be located at the midpoint of the internuclear line. For the C═N group we can write:

$$\begin{aligned} (\alpha_C - E)c_C + (\beta - ES)c_N &= 0 \\ (\beta - ES)c_C + (\alpha_N - E)c_N &= 0 \end{aligned} \tag{36}$$

If we want to express α_N as α_C + some supplement we encounter difficulties in dividing by $(\beta - ES)$. Therefore, according to Mulliken[40] and Wheland[61] we introduce a new resonance integral

$$\gamma = \beta - S\alpha \qquad (\beta = \gamma + S\alpha) \tag{37}$$

Then

$$(\beta - ES) = \gamma + S(\alpha - E) \tag{38}$$

Substituting equation (38) into (36) and dividing throughout by γ we obtain

$$\begin{aligned} \left(\frac{\alpha_C - E}{\gamma}\right)c_1 + \left(1 + S\frac{\alpha_C - E}{\gamma}\right)c_2 &= 0 \\ \left(1 + S\frac{\alpha_C - E}{\gamma}\right)c_1 + \left(\frac{\alpha_C - E}{\gamma} + \delta\right)c_2 &= 0 \end{aligned} \tag{39}$$

where $\alpha_N = \alpha_C + \delta$ and δ is expressed in γ units.

Introducing, according to Wheland,

$$\frac{\alpha - E}{\beta - S\alpha} = \frac{\alpha - E}{\gamma} = \frac{y}{1 - Sy}$$

we obtain, multiplying through by $(1 - Sy)$

$$c_{\mathrm{C}}y + c_{\mathrm{N}} = 0$$
$$c_{\mathrm{C}} + c_{\mathrm{N}}(y + \delta(1 - Sy)) = 0$$

If for the sake of simplicity we take $S = 0{\cdot}23$ and $\delta = 0{\cdot}5\gamma$, we obtain

0·564 0·359 1·077
C ——— N

These charges seem to be too great. Intuitively, since γ has a larger value than β we should use a lower value for δ. With $\delta = 0{\cdot}2\gamma$ we obtain:

0·712 0·371 0·916
C ——— N

If we divided the overlap charge equally between C and N we should have again about 0·1 electron for the effective charges, as was found above. Actually the centroid of the overlap charge is expected to be somewhat closer to the nitrogen atom, giving a little more negative charge to the latter. It is interesting that the overlap charge changed only slightly in going from C=C to C=N. In fact the main difference is that about one tenth of an electronic charge was transferred from C to N.

C. The π Electrons of the Azomethine Group; Semi-empirical Methods

The simple Hückel molecular orbital method implies many approximations. In spite of this it had a remarkable success in interpreting the physico-chemical properties of π-electron systems; nevertheless, it is necessary to remedy some of its fundamental weaknesses.

(a) The Pauli exclusion principle was not adequately taken into account since no spin wave functions were introduced. Therefore, spin was not allowed to exert any influence on the energy levels. An obvious consequence is that excited states where two electrons are in singly occupied orbitals will have the same energy whether they are singlet or triplet; the method cannot separate them.

(b) The method does not take explicit account of the mutual repulsion energy between the electrons, although one might say that this energy is 'averaged out' and included in the core potential, which in turn is expressed in terms of parameters determined by comparison with experiment.

In order to obey the Pauli principle we must multiply the wave functions depending on the space coordinates of the electrons by adequate spin functions, take into account all the possible permutations of electrons between the molecular orbitals and form a properly antisymmetrized combination of them with respect to the exchange of the coordinates of any two electrons. This leads us to so-called Slater determinants. One of them is sufficient to describe a configuration where all the electrons are placed in closed shells. Then we have to extend the Hamiltonian operator to include all the interactions between the charged particles forming the system.

Even so, while the mutual repulsion between the electrons is allowed to exert an influence on the energies, it will not directly affect the wave functions. This will allow electrons of different spin to find themselves in the same volume element too often. (The Pauli principle does not keep *these* separate.) The so-called correlation problem arising from this fact is one of the major topics of modern quantum chemistry[62]. It will not be dealt with in this chapter. It is partially allowed for through configuration interaction. Four configurations are possible for our two-electron system. In the one of lowest energy both electrons will be in ϕ_1. This is necessarily a singlet. Then we can excite an electron to ϕ_2, thereby freeing the spins. The resulting configuration may be either singlet or triplet. Finally both electrons can be put into ϕ_2.

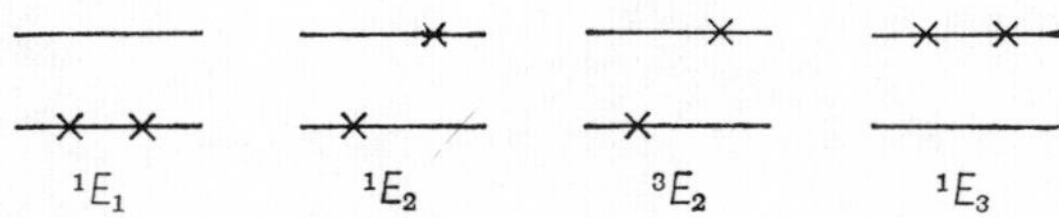

The antisymmetrized total wave functions corresponding to these configurations are, with the bar meaning β spin:

$$^1\Psi_1 = \frac{1}{\sqrt{2}}\,[\phi_1(1)\overline{\phi_1(2)} - \overline{\phi_1(1)}\phi_1(2)]$$

$$^1\Psi_2 = \frac{1}{2}\,[\phi_1(1)\overline{\phi_2(2)} - \overline{\phi_1(1)}\phi_2(2) + \overline{\phi_1(2)}\phi_2(1) - \phi_1(2)\overline{\phi_2(1)}]$$

$$^3\Psi_2 = \frac{1}{\sqrt{2}}\,[\phi_1(1)\phi_2(2) - \phi_1(2)\phi_2(1)]$$

$$^1\Psi_3 = \frac{1}{\sqrt{2}}\,[\phi_2(1)\overline{\phi_2(2)} - \overline{\phi_2(1)}\phi_2(2)] \tag{40}$$

The Hamiltonian of the system can be written as follows:

$$H = T(1) + H_0(1) + T(2) + H_0(2) + \frac{e^2}{r_{12}} \tag{41}$$

where $T(1)$ and $T(2)$ represent the kinetic energies of electrons (1) and (2), $H_0(1)$ and $H_0(2)$ the potential energies arising from the interactions of electrons (1) and (2) with the core that remains of the system if we formally remove the two π electrons. e^2/r_{12} is the repulsion potential between the two electrons (r_{12} is the distance between the electrons).

The energies of the configurations are then computed from the variational formula.

$${}^1E_1 = \int {}^1\psi_1 H {}^1\psi_1 \, d\tau \tag{42}$$

and so on.

We shall not go into details. This treatment is essentially the same as the one given by Parr and Crawford[63] for ethylene and it was given in full elsewhere[31]. Substituting the wave functions given under equation (40) and the Hamiltonian under equation (41) into (42) we obtain the energies in terms of molecular integrals. Among these there are molecular core integrals of the type

$$H_i^{\text{core}} = \int \phi_i(1)[T(1) + H_0(1)]\phi_i \, d\tau \tag{43}$$

and molecular integrals of the Coulomb type

$$J_{ij} = \int \phi_i(1)\phi_i(1) \frac{e^2}{r_{12}} \phi_j(2)\phi_j(2) \, d\tau \tag{44}$$

and of the exchange type

$$K_{ij} = \int \phi_i(1)\phi_j(1) \frac{e^2}{r_{12}} \phi_i(2)\phi_j(2) \, d\tau \tag{45}$$

For a closed-shell configuration the well-known expression for the total electronic energy is

$$E = 2 \sum_i H_i^{\text{core}} + \sum_i \sum_j (2J_{ij} - K_{ij}) \tag{46}$$

where the summation is extended to all occupied molecular orbitals i, j and $J_{ii} = K_{ii}$.

The next step is to introduce the expression of the molecular orbitals ϕ_i in terms of atomic orbitals. Then the molecular integrals are transformed into integrals over atomic orbitals. The core Hamiltonian introduces so-called penetration integrals of types

$$(P^+ : qq) = \int \chi_q(1)[T(1) + H_0(1)]\chi_q(1) \, d\tau$$

and

$$(P^+:ql) = \int \chi_q(1)[T(1) + H_0(1)]\chi_l(1)\, d\tau \tag{47}$$

where q and l represent atomic orbitals centred on the respective atoms. It is customary to separate in $H_0(1)$ the term relating to the electron's 'own' atom, assuming that $H_0 = \sum_q H_q^+$.

Then $[T(1) + H_p^+(1)]$ can be considered as an operator whose atomic orbital χ_P is an eigenfunction,

$$[T(1) + H_P^+(1)]\chi_p = W_{2p}\chi_p \tag{48}$$

with the eigenvalue W_{2p} representing the valence-state ionization potential of an electron in χ_p. This is one of the Goeppert–Mayer and Sklar[64] approximations. (It is hard to justify.) Thus we eliminate the kinetic term and the one-centre penetration integral $(P^+:pp)$ from the expression of the molecular core integral.

The e^2/r_{12} term of the Hamiltonian introduces a number of integrals of type

$$(pp/pp),\ (pp/qq),\ (pp/pq),\ (pq/pq)$$

where

$$(pp/qq) = \int \chi_p(1)\chi_p(1)\, \frac{e^2}{r_{12}}\, \chi_q(2)\chi_q(2)\, d\tau \tag{49}$$

and so on. These atomic integrals are of the one-centre Coulomb, two-centre Coulomb, Coulomb-exchange and exchange type respectively. In a two-electron problem their number is moderate but in larger molecules it becomes prohibitively great and there will be many three- and four-centre integrals among them.

This is one of the reasons why the above-described antisymmetrized molecular orbital (ASMO) method as such is seldom used. Another reason is that this method does not include any other minimization of the energy than the one applied in obtaining the simple Hückel orbitals. Still another reason is that serious objections can be made against the theoretically computed values of certain integrals, especially those of the (pp/pp) and (pp/qq) type when they are used in the framework of this method due to electron correlation and core polarization effects.

Better wave functions are obtained by applying the Hartree–Fock self-consistent-field (SCF) method or the configuration interaction method or a combination of the two. Roothaan has shown[65] how the SCF procedure can be applied to Slater determinants formed from LCAO molecular orbitals while the energy and charge distribution in each molecular orbital depends on the electron distributions represented by all the molecular orbitals. For a closed-shell state the expression to minimize is equation (46). Since the molecular integrals themselves depend on the coefficients of the atomic orbitals in the molecular integrals the procedure is of an iterative nature. In the configuration interaction method, improved wave functions are obtained by taking linear combinations of the wave functions of configurations of the same multiplicity and symmetry:

$$\Psi = s_1\psi_1 + s_2\psi_2 + \cdots \tag{50}$$

and the energy is minimized with respect to the s_k.

Pariser and Parr[66] and Pople[68] introduced some simplifying features into the ASMO–CI and SCF methods respectively. They proposed the neglect of differential overlap, that is, the neglect of all atomic integrals containing a product over two different atomic orbitals [like $\chi_p(1)\chi_q(1)$] either once or twice. This is a sweeping simplification whose meaning and justification was and still is discussed by many authors. Furthermore, in the Pariser–Parr–Pople method the Coulomb integrals (and these are the only two-electron integrals which remain) are adjusted empirically and the core integrals are treated as parameters.

One-centre Coulomb integrals which have a certain physical meaning[66,67] are equated to the difference of the related valence state ionization potentials and electron affinities.

$$(pp/pp) = I - A \tag{51}$$

They can also be adjusted by diminishing the Z_{eff} number of the orbitals using spectral data[69,70].

The two-centre Coulomb integrals are determined according to Pariser and Parr[66,71] in the following way. For distances $r \geq 2{\cdot}8$ Å they are computed from the uniformly charged sphere formula:

$$(pp/qq) = (7{\cdot}1975/r)\left\{\left[1 + \left(\frac{1}{2r}\right)^2(R_p - R_q)^2\right]^{-\frac{1}{2}} + \left[1 + \left(\frac{1}{2r}\right)^2(R_p + R_q)^2\right]^{-\frac{1}{2}}\right\}\text{ ev} \tag{52}$$

in which

$$R_p = \frac{4{\cdot}579}{Z_p} \times 10^{-8}\text{ cm}$$

where Z_p is the effective nuclear charge according to Slater.

For distances smaller than 2·8 Å they are determined from the equation

$$ar + br^2 = \tfrac{1}{2}[(pp/pp) + (qq/qq)] - (pp/qq) \tag{53}$$

in which the constants a and b are obtained by fitting values obtained from equation (52) for $r = 2{\cdot}80$ Å and $r = 3{\cdot}70$ Å.

Another way of adjusting the two-centre Coulomb integrals was proposed by Mataga and Nishimoto[72]. These authors assumed the values given by the expression:

$$(pp/qq) = \frac{e^2}{a + r_{pq}} \tag{54}$$

The parameter a is determined, in the homonuclear case, from the equation

$$\frac{e^2}{a} = (pp/pp) = I_p - A_p \tag{55}$$

and for the case of two different atoms by taking as e^2/a the arithmetic mean of the e^2/a of carbon and nitrogen for example. They found $a_{CC} = 1{\cdot}328$ Å, $a_{CN} = 1{\cdot}212$ Å and $a_{NN} = 1{\cdot}115$ Å. The values obtained in this manner are somewhat lower than the ones determined by Pariser and Parr's formulas.

In the case of two π electrons, the core integrals become

$$\begin{aligned}
H_1^{\text{core}} &= \tfrac{1}{2}\int(\chi_C + \chi_N)H_0(\chi_C + \chi_N)\,d\tau \\
&= \tfrac{1}{2}[\int\chi_C H_0\chi_C\,d\tau + \int\chi_N H_0\chi_N\,d\tau + \int\chi_C H_0\chi_N\,d\tau + \int\chi_N H_0\chi_C\,d\tau \\
&= \tfrac{1}{2}[\alpha_C + \alpha_N + \beta_{CN} + \beta_{NC}] \qquad (56)\\
H_2^{\text{core}} &= \tfrac{1}{2}\int(\chi_C - \chi_N)H_0(\chi_C - \chi_N)\,d\tau \\
&= \tfrac{1}{2}(\alpha_C + \alpha_N - \beta_{CN} - \beta_{NC})
\end{aligned}$$

where from equation (49)

$$\begin{aligned}
\alpha_C &= \int\chi_C(1)[T(1) + H_0(1)]\chi_C\,d\tau \\
&= \int\chi_C[T + H_C{}^+ + H_N{}^+]\chi_C\,d\tau \qquad (57)\\
&= W_{2p} + (H_N{}^+:\text{cc})
\end{aligned}$$

In the polyelectronic case there will be a sum of integrals of the type $(H_a{}^+:bb)$. According to Goeppert–Mayer and Sklar's second approximation the neutral (spherical) Hamiltonian is often introduced instead of the ionic Hamiltonian.

β is treated as an adjustable parameter to fit experimental data.

The similarity between the Hückel and Pariser and Parr methods may be seen clearly. The parameters do not have necessarily the same numerical values, however. The α and β of the Hückel method are imagined to contain a term representing the energy due to the repulsion of the other electrons whereas in the Pariser and Parr method they do not, since these interactions are included explicitly.

Readers who are interested in details of the methods we have mentioned will find derivations and worked examples in the books of Parr[73], Daudel, Lefebvre and Moser[74], Daudel[75] and Salem[76].

We now return to the two π-electron problem of the >C=N— group. We tried several choices of parameters. The one we show gives the first π–π, singlet–singlet transition energy correctly. It is approximately 7 ev (see p. 47). This can be obtained by the following selection of parameters:

W_{2p} carbon	11·22 ev
W_{2p} nitrogen	14·51 ev

both equal in absolute value to the valence state ionization potentials given by Pilcher and Skinner[45].

The Z_{eff} values were for carbon 3·25, for nitrogen 3·90, and $\beta_{CN} - 2{\cdot}7$ ev. (Pariser and Parr[66] originally used $-2{\cdot}58$ ev[77].)

The initial Hückel coefficients were:

$$\phi_1 = 0{\cdot}667\chi_C + 0{\cdot}745\chi_N$$
$$\phi_2 = 0{\cdot}745\chi_C - 0{\cdot}667\chi_N$$

These values, however, are not important. Only the number of iterations needed to achieve self-consistency depends on the original coefficients.

The Coulomb integrals were taken to be:[66]

$$(CC/CC) = 10{\cdot}53 \text{ ev}$$
$$(CC/NN) = 7{\cdot}82 \text{ ev}$$
$$(NN/NN) = 12{\cdot}27 \text{ ev}$$

Self-consistency was obtained after six iterations. The total electronic energy turned out to be $-29{\cdot}24$ ev. Taking account of the repulsion of the nuclei it became $-36{\cdot}85$ ev. The transition energies calculated with the orbitals obtained from the iteration of the ground state were:

first singlet–triplet	3·97 ev
first singlet–singlet	7·05 ev

Next, configuration interaction was applied. The diagonal elements in the secular determinant represent the energies of the three singlet configurations

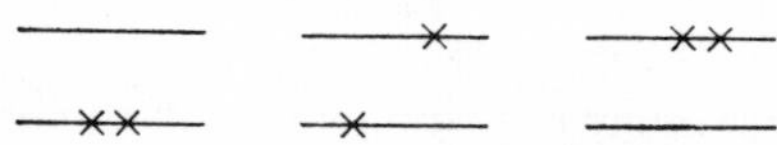

in the order given and the off-diagonal ones the interaction terms between them:

$$\begin{bmatrix} -36{\cdot}85 - E & 0{\cdot}00 & 1{\cdot}54 \\ 0{\cdot}00 & -29{\cdot}80 - E & 0{\cdot}87 \\ 1{\cdot}54 & 0{\cdot}87 & -25{\cdot}59 - E \end{bmatrix} = 0 \qquad (58)$$

The numbers are given to two decimal places only; the exact value of the 1–2 off-diagonal term was actually 0·000267. This again depends on the original choice of the Hückel coefficients but the final results do not. In ethylene the singly excited configuration has different symmetry from the two others, so that this term and the 2–3 term (here 0·87) are exactly zero. The ground state interacts more strongly with the doubly excited state and this causes a lowering of its dipole moment.

The charges are, for the resulting three singlet states:

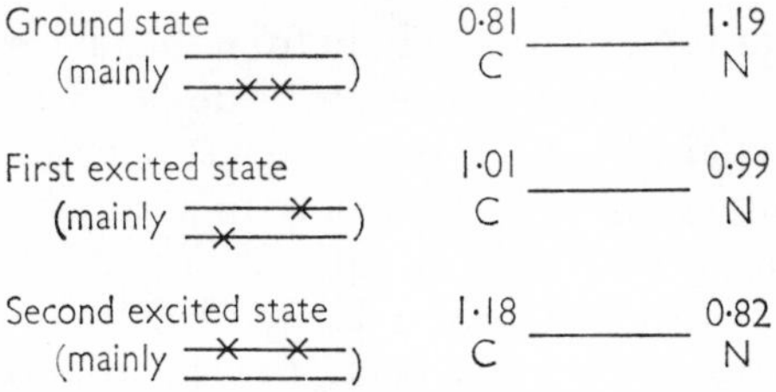

	C	N
Ground state (mainly)	0·81	1·19
First excited state (mainly)	1·01	0·99
Second excited state (mainly)	1·18	0·82

In the ground state this would lead to a dipole moment about double what we arrived at previously, i.e. 1·12 D. The excitation energies are 7·09 ev (or 57,219 cm^{-1}) and 11·83 ev (or 95,446 cm^{-1}). The respective oscillator strengths are $f = 0{\cdot}37$ and 0·009. We have no experimental guide relating to the latter transition which would be symmetry forbidden in ethylene.

The triplet state is unique and its energy is not affected by configuration interaction. The singlet–triplet separation given above is certainly too great.

It is possible to obtain the lower dipole moment by lowering the W_{2p} of the nitrogen atom to about 13·5 ev but then the excitation energy becomes too high. We did not try to introduce further refinements.

D. The σ Electrons; a Hückel-type Calculation

The σ framework has not until now been explicitly considered. In the present section a Hückel-type calculation concerning the σ electrons will be presented. Early works on σ-electron systems were based on bond or group orbitals[78]. The first 'individual electron' calculations on saturated hydrocarbons and their derivatives based on modified Hückel methods were made by Sandorfy and Daudel[80], Sandorfy[81], Fukui, Kato and Yonezawa[82,83], Klopman[84,85] and were followed by the 'extended Hückel method' of Hoffmann[86], who used the Wolfsberg

Helmholtz parametrization[87], and by the more elaborate treatment of Pople and Santry[79], who made a thorough study of the causes of delocalization in σ-electron systems.

We are going to consider the following model:

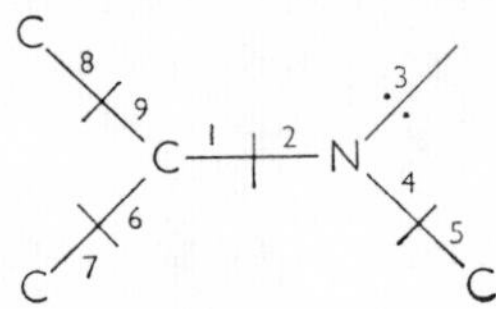

The molecular orbitals will be built from the nine sp^2 hybrid atomic orbitals shown on the above diagram. Number 3 is the orbital containing the lone pair. Since our main interest is in the C═N group, the other carbon atoms are of interest only in as much as they make up the environment of the C═N group and their orbitals not shown in the diagram are disregarded. Also disregarded are the two π electrons and the 1*s* electrons, the latter being thought to form the core with the nuclei.

The following parametrization will be adopted:

(a) Coulomb integrals will be taken equal to the arithmetic mean of the valence state ionization potentials and electron affinities with values given by Hinze and Jaffé[47]. Thus, for the carbon sp^2 orbitals:

$$\alpha_1 = \alpha_5 = \alpha_6 = \alpha_7 = \alpha_8 = \alpha_9 = \frac{15{\cdot}62 + 1{\cdot}95}{2} = 8{\cdot}79 \text{ ev}$$

For the nitrogen sp^2 orbitals the corresponding value would be

$$\alpha_2 = \alpha_4 = \frac{20{\cdot}60 + 5{\cdot}14}{2} = 12{\cdot}87 \text{ ev}$$

It will turn out, however, that with this value an unrealistically great amount of negative charge would accumulate on the nitrogen atom. For this reason, this integral will be varied in order to explore its influence on the charge distribution.

There is an uncertainty concerning the choice of α_3, the Coulomb integral relating to the orbital containing the lone pair. The corresponding ionization potential found in the tables of Pilcher and Skinner[45] is 15·19 ev and we may use this for α_3, taking the electron affinity as zero. However, from the lone pair orbital electronegativities given by Hinze, Whitehead and Jaffé[51] it can be inferred that this parameter should have a much lower value, perhaps 5 ev. Anyway, we varied α_3 too, our calculation being of an exploratory nature.

The resonance integrals, β, were estimated in the following way. The π–π β_{CN} was assumed to be 2·70 ev since this gave us reasonable results in our π-electron calculation. Then we supposed that all β values are proportional to the overlap integrals so that

$$2{\cdot}70:0{\cdot}23 = x:0{\cdot}718$$

and

$$\beta_{12} = 8{\cdot}43 \text{ ev.}$$

We computed the overlap integrals from the tables of Mulliken, Rieke, Orloff and Orloff[15]. The presumed internuclear distances were:

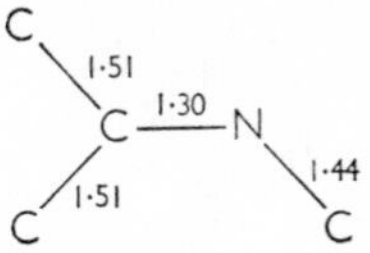

C–C 1·51, C–N 1·30, N–C 1·44, C–C 1·51

and all angles were taken for 120°. $Z_C = 3{\cdot}25$, $Z_N = 3{\cdot}90$.

Then the overlap integrals were:

$S_{11} = 1{\cdot}000$	$S_{16} = 0$	$S_{22} = 1{\cdot}000$
$S_{12} = 0{\cdot}718$	$S_{17} = 0{\cdot}096$	$S_{23} = 0$
$S_{13} = 0{\cdot}149$	$S_{18} = 0{\cdot}096$	$S_{24} = 0$
$S_{14} = 0{\cdot}149$	$S_{19} = 0$	$S_{25} = 0{\cdot}059$
$S_{15} = 0{\cdot}216$		$S_{26} = 0{\cdot}083$
$S_{27} = 0{\cdot}184$	$S_{33} = 1{\cdot}000$	$S_{44} = 1{\cdot}000$
$S_{28} = 0{\cdot}184$	$S_{34} = 0$	$S_{45} = 0{\cdot}647$
$S_{29} = 0{\cdot}083$	$S_{35} = 0{\cdot}149$	$S_{46} = 0{\cdot}114$
	$S_{36} = -0{\cdot}116$	$S_{47} = 0{\cdot}040$
	$S_{37} = -0{\cdot}045$	$S_{48} = -0{\cdot}045$
	$S_{38} = -0{\cdot}040$	$S_{49} = -0{\cdot}116$
	$S_{39} = 0{\cdot}114$	
$S_{55} = 1{\cdot}000$	$S_{66} = 1{\cdot}000$	$S_{77} = 1{\cdot}000$
$S_{56} = -0{\cdot}078$	$S_{67} = 0{\cdot}690$	$S_{78} = 0{\cdot}146$
$S_{57} = 0{\cdot}118$	$S_{68} = 0{\cdot}096$	$S_{79} = 0{\cdot}096$
$S_{58} = 0{\cdot}012$	$S_{69} = 0{\cdot}000$	
$S_{59} = -0{\cdot}005$		
$S_{88} = 1{\cdot}000$	$S_{99} = 1{\cdot}000$	
$S_{89} = 0{\cdot}690$		

The reader who is accustomed to π-electron problems will be interested to see that many of the long range overlap integrals have quite appreciable values.

This is a fact which may help in the understanding of some distant atom effects in saturated organic molecules. Overlap integrals between hybrids occurring in σ-electron problems can be considerably greater than those entering π-electron problems.

Here are some of the charge distributions we obtained.

I. $\alpha_3 = 15{\cdot}19$ ev $\qquad \alpha_2 = \alpha_4 = 9{\cdot}00$ ev

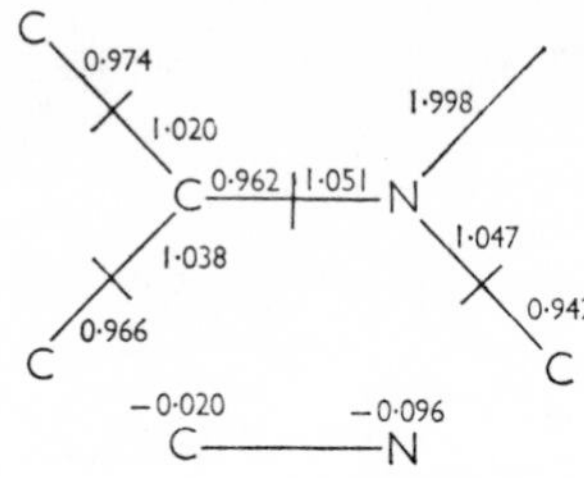

C 0·974 1·020 C 0·962 1·051 N 1·998 1·047 0·942 C 1·038 0·966 C

−0·020 C—N −0·096

II. $\alpha_3 = 10{\cdot}00$ ev $\qquad \alpha_2 = \alpha_4 = 9{\cdot}00$ ev

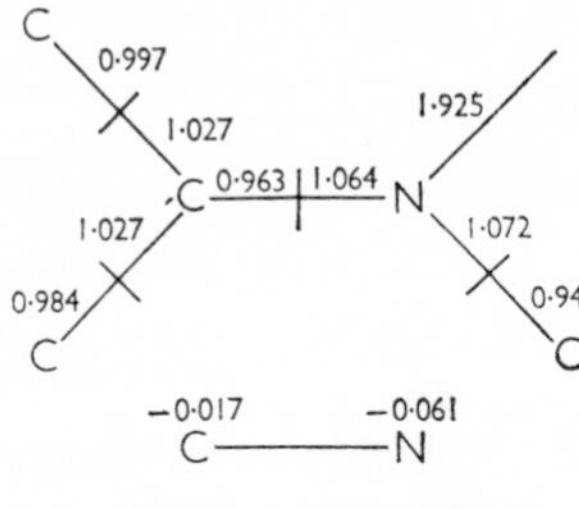

C 0·997 1·027 C 0·963 1·064 N 1·925 1·072 0·940 C 1·027 0·984 C

−0·017 C—N −0·061

III. $\alpha_3 = 8{\cdot}00$ ev $\qquad \alpha_2 = \alpha_4 = 9{\cdot}00$ ev

C 1·031 1·041 C 0·969 1·078 N 1·738 1·127 0·966 C 1·023 1·025 C

−0·033 C—N +0·057

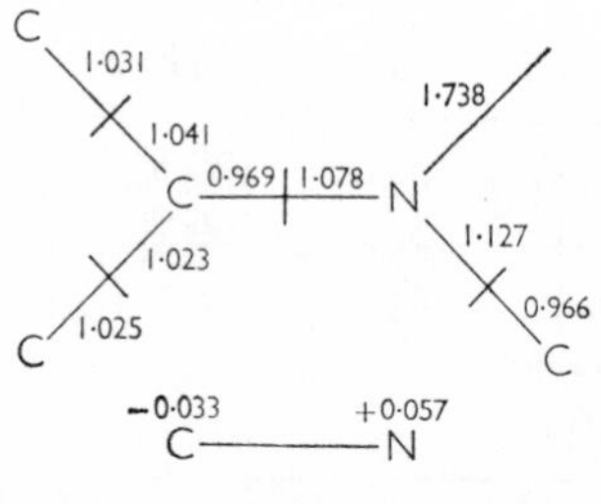

From some other calculations we only give the charge distribution around the nitrogen atom.

TABLE 8. Charge densities in σ orbitals on the N atom.

α_3	$\alpha_2 = \alpha_4$	q_2	q_3	q_4
15·19	10·00	1·237	1·999	1·275
10·00	10·00	1·251	1·908	1·298
8·00	10·00	1·272	1·630	1·362
15·19	11·00	1·480	1·999	1·536
10·00	11·00	1·491	1·881	1·550
8·00	11·00	1·516	1·442	1·609
15·19	12·87	1·901	1·999	1·894
10·00	12·87	1·900	1·808	1·894
8·00	12·87	1·906	0·971	1·904
7·00	12·87	1·909	0·568	1·911
7·00[a]	12·87	1·851	0·513	1·740

[a] Overlap integrals between non-neighbours neglected.

The first observation that emerges from these data is the great sensitivity of the charge distribution to the choice of parameters α_3 and $\alpha_2 = \alpha_4$. Only I and II seem to give reasonable results with an effective negative charge between 0·05 and 0·10 on the nitrogen atom. (See the small diagrams in I, II and III in which the numbers represent $q_1 + q_6 + q_9$ for carbon and $q_2 + q_3 + q_4$ for nitrogen, respectively.) Higher values of α_2 yield unreasonably high charges on the nitrogen atom. We have to conclude that in these Hückel-type calculations the Coulomb integral of N should not be much higher than the one of C.

$$(\alpha_C = 8{\cdot}37 \qquad \text{and} \qquad \alpha_N = 9{\cdot}00 \text{ ev})$$

The effect of α_3 is of course very great on the charge distribution in the lone pair orbital. Since this orbital would serve as a proton acceptor for hydrogen bonding and other intermolecular associations, it would seem to be contrary to experiment to have an appreciable positive charge on it, even though the centre of the charge distribution is far from the nucleus. For this reason, according to our results α_3 should not be lower than 10 ev.

We should now put together our σ and π calculations. We could make a calculation choosing β values to take σ–π interaction into account. We could also consider the effect of the q_π and q_σ on each other's Coulomb integrals in an iterative matter, as in the ω technique. However, in view of the approximate character of these calculations we shall refrain from making further refinements. Let us consider, for the time being, our π and σ charges as additive. (Cf. the diagram on p. 20 and II on p. 34.)

Pariser–Parr–Pople type calculations have been made on saturated molecules by Pohl, Rein and Appel[88], Pople, Santry and Segal[89], Kaufman[90], Klopman[91], Skancke[92], Katagiri and Sandorfy[93] and others. A deeper look at the problem was recently made by Cook, Hollis and McWeeny[94]. These methods have not been applied to our problem.

E. $>C{=}N^{+}<$ and $>C{=}N^{-}$

The ions that are produced from the $>C{=}N$ group by protonation and deprotonation of the nitrogen atom play an important role in chemical reactions. The knowledge of the charge distribution in these ions and the theoretical parameters which are necessary to obtain them are therefore of considerable importance. A successful attempt in this direction has been made by Brown and Penfold[95].

They treated the problem by both the Pariser and Parr method completed by configurational interaction, and by the SCF method with Pople's approximations.

Their choice of parameters was somewhat different from the one we have used above, but most of the differences are not essential. The Coulomb integral for N^- constitutes a certain problem. Brown and Penfold used the 'virtual hydrogen atom approximation'. α_N for a negatively charged nitrogen is different from that for neutral NH because the value of the penetration integral (H_N:NN) for the penetration of the $2p\pi$ nitrogen orbital into a neutral hydrogen atom changes greatly when the proton is removed. A similar problem arises for N^+ if a proton is added to the nitrogen atom. For N^- the situation is intermediate between that of a negatively charged nitrogen atom, in which both electrons which originally formed the N—H bond are accommodated in an sp^2 nitrogen orbital, and that of a neutral nitrogen atom with 'a nearby electron occupying the 1*s* orbital left behind by the removed proton'.

Similarly 'N^+ could be treated as a positively-charged nitrogen atom or as a neutral nitrogen atom with an adjacent proton'. The numerical values of the integrals were given by Brown and Penfold[95]. We reproduce here the charge distributions they obtained. They refer to the π-electron distribution.

It is seen that extra charges on the nitrogen cause a reversal of the electronegativities for N^- and rather large effective π-electron charges for both N^- and N^+.

TABLE 9. π-Electron charge densities[95].

	SCFMO		ASMOCI	
	0·938	1·062	0·953	1·047
$CH_2{=}NH$	C ——	N	C ——	N
	0·385	1·615	0·425	1·575
$CH_2{=}NH_2^+$	C ——	N	C ——	N
	1·529	0·471	1·480	0·520
$CH_2{=}N^-$	C ——	N	C ——	N

Based on these semi-empirical calculations Brown and Penfold gave the approximate supplements to the Hückel Coulomb integrals as

$+1{\cdot}4$ for N^+, $\quad +0{\cdot}1$ for N, $\quad -1{\cdot}1$ for N^-

With these values the charge distributions are the same as the ones obtained by the Pariser–Parr–Pople calculations.

We have to mention that the internuclear distances in these calculations were

C—H 1·071 Å N—H 1·014 Å C=N 1·36 Å

The latter value corresponds to the C=N distance in heterocycles rather than to a C=N group in an aliphatic environment. The charges we arrived at earlier in this chapter were somewhat higher than the ones of Brown and Penfold. However, even ours, which are of the order of 0·1 electron, are still much lower than traditional values.

IV. INFRARED AND RAMAN SPECTRA

By far the greatest number of data to be found in the literature relate to the C=N stretching frequency. In most cases it is a strong and fairly sharp band (ε about 100–300), located at somewhat lower frequencies than the bands of carbonyl groups in similar environments and close to C=C stretching frequencies. The following typical values summarize these facts. They relate to groups in a purely aliphatic environment in the absence of strain, steric hindrance or other complicating factors and to dilute solutions in so-called neutral solvents.

	ν (cm^{-1})	ε (litre $mole^{-1}$ cm^{-1})
C=O	1715	400–1000
C=N	1670	100– 300

The corresponding force constants are in the harmonic oscillator approximation

C=O	11·9 dynes cm^{-1}
C=N	10·6 dynes cm^{-1}

Infrared spectra of purely aliphatic imines are scarce in the literature. The values shown in Table 10 were measured in 1968 by J. Cobo in the author's laboratory.

TABLE 10. C=N infrared stretching frequencies (non-conjugated).

$$\begin{matrix} R' \\ \quad \diagdown \\ \qquad C{=}N{-}R''' \\ \quad \diagup \\ R'' \end{matrix}$$

R′	R″	R‴	$\nu_{\text{liq. or solid}}$ (cm^{-1})	ν_{CCl_4} (cm^{-1})
CH_3	H	C_2H_5	1671	—
CH_3	H	C_3H_7	1673	1673
C_2H_5	H	C_3H_7	1670	—
C_3H_7	H	C_3H_7	1670	1670
$(CH_3)_2CH$	H	$CH_2C_6H_5$	1661	1668
CH_3	CH_3	C_6H_{11}	1660	—
C_2H_5	CH_3	C_3H_7	1661	1659

Fabian and Legrand[96] found 1673 cm^{-1} for C_2H_5—CH=N—C_3H_7 in CCl_4. Kahovec[97,98] and Kirmann and Laurent[99] measured the Raman spectra of 18 aldimines (R″=H) and found in all of them a band between 1674 and 1665 cm^{-1}.

Generally speaking, there is very little difference between infrared and Raman frequencies and between the spectra of pure liquids and solids and their solutions in CCl_4, or other not very associative solvents. The C=N frequency in trialkyl ketimines (no H on the two double bonded atoms) is somewhat lower, about 1660 cm^{-1}.

If there are one or more groups conjugated with the C=N group the frequency is usually lower. The data in Table 11 were also measured by J. Cobo.

It may be seen that phenyl groups lower the C=N frequency more than the vinyl groups and that a phenyl group on the nitrogen causes further lowering. Similarly two phenyl groups on the carbon give a low frequency, showing that the axes of the π orbitals in the C=N and phenyl groups are far from being perpendicular to each other.

In 1956 Fabian, Legrand and Poirier[100] published an extensive review on the infrared and Raman spectra of imines. The following two tables are reproduced from their paper. Table 12 gives a general view of C=N band frequencies and Table 13 gives C=N stretching frequencies in amidines. Reference 100 should be consulted for details.

TABLE 11. C=N infrared stretching frequencies (conjugated).

$$\begin{matrix} R' \\ & \diagdown \\ & & C{=}N{-}R''' \\ & \diagup \\ R'' \end{matrix}$$

R′	R″	R‴	$\nu_{\text{liq. or solid}}$ (cm^{-1})	ν_{CCl_4} (cm^{-1})
$CH_2{=}CH$	H	C_6H_{11}	1655	1655
$CH_3CH{=}CH$	H	C_6H_{11}	1657	—
$CH_3CH{=}CH$	H	C_3H_7	1656	1658
C_6H_5	H	$CH_2CH(CH_3)_2$	1640	1646
C_6H_5	H	C_6H_5	1625	1628
C_6H_5	CH_3	C_6H_5	1627	1639
C_6H_5	C_6H_5	C_6H_5	1611	1619

Oximes have somewhat wider frequency-ranges but do not differ greatly from the imines in their C=N frequency. In hydrazones the band does not usually appear[101].

$$\begin{matrix} R' \\ & \diagdown \\ & & C{=}N{-}NH_2 \\ & \diagup \\ R'' \end{matrix} \qquad \begin{matrix} R' \\ & \diagdown \\ & & C{=}N{-}NH{-}C_6H_5 \\ & \diagup \\ R'' \end{matrix}$$

This is probably due to the compensating action of the amino group making the change of the dipole moment during the vibration near-zero. Where two C=N groups are conjugated, e.g.

$$(CH_3)_2C{=}N{-}N{=}C(CH_3)_2$$

the frequencies are about 1660 in the infrared and about 1625 in the Raman spectrum. Since these molecules have at least an approximate centre of symmetry, the two bands correspond in all likelihood to the antisymmetrical stretching mode appearing in the infrared and to the symmetrical mode appearing in the Raman spectrum.

In general, C=N vibrations exhibit a lesser degree of localization than C=O vibrations. This is expected, since whereas the oxygen atom is always an end atom in C=O, the nitrogen atom in imines is not and its stretching is therefore more likely to involve motions by neighbour atoms. However, the proper understanding of this observation would require extensive theoretical work perhaps in the style of Bratoz and Besnainou[102,103]. We deliberately refrain from attempting to interpret small differences in frequencies. These are due to a combination of various electronic, mechanical and steric changes and all possible explanations would be purely speculative.

TABLE 12. Ranges for C=N stretching frequencies[100].

	Non-conjugated	Monoconjugated	Diconjugated
—CH=N—	R—CH=N—R′ 1665–1674 cm^{-1} (19 compounds)	C_6H_5—CH=N—R′ 1629–1656 cm^{-1} (20–30 compounds)	C_6H_5—CH=N—C_6H_5 1626–1637 cm^{-1} (7 compounds)
—C(—)=N—	R—C(R″)=N—R′ 1649–1662 cm^{-1} (6 compounds)	C_6H_5—C(R″)=N—R′ 1640–1650 cm^{-1} (2 compounds)	C_6H_5—C(CH_3)=N—C_6H_5 1630 solid 1640 solution
>C=NH	(R)(R′)C=NH 1640–1646 cm^{-1} (associated) (5 compounds)	(C_6H_5)(R)C=NH 1620–1633 cm^{-1} (associated) (7 compounds)	
—CH=NOH	R—CH=NOH 1652–1673 cm^{-1} (associated) (8 compounds)	C_6H_5—CH=NOH 1614–1645 cm^{-1} (associated) (8 compounds)	
—C(—)=NOH	(R)(R)C=NOH 1652–1684 cm^{-1} (10 compounds)	(C_6H_5)(R′)C=NOH 1620–1640 cm^{-1} (3 compounds)	
—C(S—)=N—	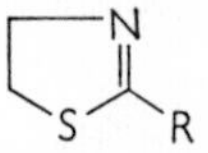 N, S, R (ring) 1627–1640 cm^{-1} (5 compounds)	C_6H_5—C(SCH_3)=NCH_3 1622 cm^{-1} 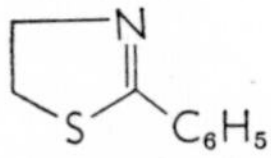 N, S, C_6H_5 (ring) 1607–1613 cm^{-1} (2 compounds)	C_6H_5—C(SCH_3)=N—C_6H_5 1611 cm^{-1}
—C(NH—R)=N—	CH_3—C(NHC_2H_5)=N—C_2H_5 1675 Raman 1685 I.r.	C_6H_5—C($NHCH_3$)=N—CH_3 1635–1651 cm^{-1} according to solvent	C_6H_5—C(NHR)=N—C_6H_5 1620–1630 cm^{-1} (11 compounds)
—C(N<)=N—		C_6H_5—C(NR′R″)=N—R 1614–1621 cm^{-1} (2 compounds)	C_6H_5—C(NRR′)=N—C_6H_5 1582–1597 cm^{-1} (4 compounds)
—C(O—)=N—	1664–1690 cm^{-1} (oxazolones)	1645–1667 cm^{-1} (oxazolones)	C_6H_5—C(OCH_3)=N—C_6H_5 1666 cm^{-1}

TABLE 13. C=N stretching frequencies in amidines[100].

N-monosubstituted amidines		N-disubstituted amidines	
$R-C(=N-R')-NH-R''$	Raman: 1675 cm^{-1} (dioxan)	$R-C(=N-R')-NR^1R^2$	
$C_6H_5-C(=N-R')-NH-R''$	I.r.: 1635 ($CHCl_3$), sh. 1655; 1651 (CCl_4), sh. 1640	$C_6H_5-C(=N-R')-NR^1R^2$	I.r.: 1614–1618 ($CHCl_3$) 1621–1624 (CCl_4)
$R-C(=N-C_6H_5)-NH-R''$	Raman: 1636–1652 ($CHCl_3$) [2 bands]	$R-C(=N-C_6H_5)-NR^1R^2$	Raman: 1629–1623 ($CHCl_3$)
$C_6H_5-C(=N-C_6H_5)-NH-R''$	I.r.: 1620–1627 ($CHCl_3$) 1627–1630 (CCl_4)	$C_6H_5-C(=N-C_6H_5)-NR^1R^2$	I.r.: 1586–1595 ($CHCl_3$ or CCl_4)

Fabian and Legrand[104] carried out a very useful study on the molecular extinction coefficients of the C=N stretching vibration. This is how they summarize their results: 'The intensity of absorption due to —C(Z)=N— is highest when Z represents an oxygen containing group or one containing a second nitrogen (imino ethers or amidines). It decreases when Z is a sulphur-containing group, an alkyl or a hydrogen (imino thioethers, ketimines or aldimines). The presence of an oxygen or nitrogen containing substituent on the nitrogen of the imino group (oximes and hydrazones) further weakens the absorption.'

Numerically, the following ranges can be given:

(a) $\epsilon > 350$ when Z is NRR′ (amidines or OR (imino ethers)
(b) $\epsilon \approx 250$ when Z is R (ketimines) or SR (imino thioethers[105])
(c) ϵ is between 180 and 140 when Z is a hydrogen (aldimines). The conjugated aldimines tend to have extinction coefficients closer to the higher limit of this range
(d) $\epsilon < 30$ for oximes[106]
(e) ϵ is practically zero for hydrazones.

Fabian, Delaroff and Legrand[107] found unusually high extinction coefficients (~900) for *N*-alkyl-*N′*-phenylbenzamidines. Leonard and Paukstelis[108] gave infrared frequencies for the positive ion

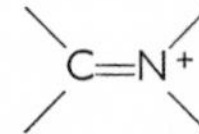

mostly for chlorate salts. They all have strong infrared C=N$^+$ stretching bands. We reproduce here some of their data (Table 14).

Some of the frequency differences are hard to explain. It is clear, however, that C=N$^+$ groups have in general higher frequencies than C=N groups. This is in accordance with the theoretical results of Brown and Penfold[95], supposing proportionality between bond energies and force constants. Conjugation clearly lowers the frequency as expected but the effect of the various hydrocarbon radicals linked to the azomethine carbon atom is not predictable.

Many C=N$^+$ containing molecules were studied by Edwards and Marion and coworkers[109-111] and by Witkop and his collaborators[112-114]. Witkop indicated a range of 1639–1626 cm^{-1} for aromatic Schiff bases and 1672–1646 cm^{-1} for their salts.

The C=N bonds in oximes were listed in Table 12. The reader may be interested to consult the book by Colthup, Daly and Wiberley[115] for a good review of the C=N vibrations.

TABLE 14. C=N$^+$ infrared stretching frequencies[108].

Compound	Structure	ν (cm^{-1})
N-isopropylidenedimethylaminium	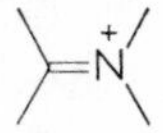	1687
N-isopropylidenepyrrolidinium	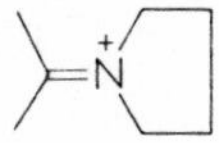	1690
N-2-butylidenepyrrolidinium	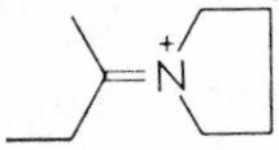	1680
N-3-pentylidenepyrrolidinium	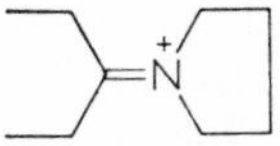	1665
N-cyclohexylidenepyrrolidinium	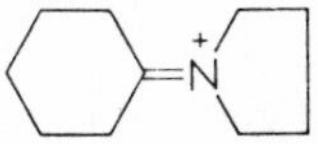	1665
N-cyclohexylidenemorpholinium	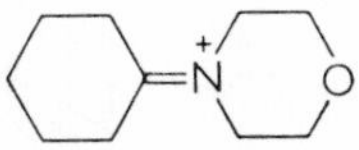	1640
N-1-methylcinnamylidenepyrrolidinium	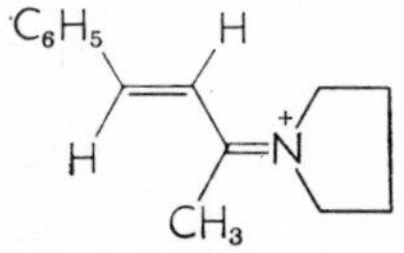 C_6H_5, H, H, CH_3	1622

Many molecules containing the C=N double bond exhibit cases of chelation or metal complex formation.

A typical example for the former is *N*-salicylidene-2-aminopropane (SA)

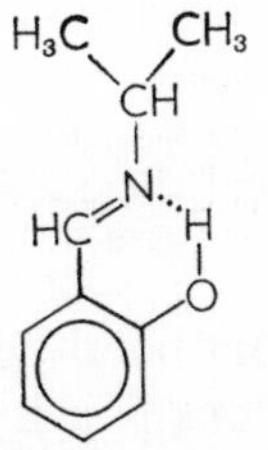

whose structure has been confirmed by Teyssie and Charette[116]. A broad, concentration-independent band centred at about 2700 cm^{-1} gives evidence of a strong intramolecular hydrogen bond. The C=N stretching band is at 1634 cm^{-1} in this case, showing the joint effect of conjugation and hydrogen bonding. In the liquid state only small fractions of these molecules form dimers with O—H···O or O—H···N type hydrogen bonds.

Teyssie and Charette also measured the infrared spectra of a number of metal complexes of SA and similar compounds. Here are a few of the C═N stretching frequencies they obtained:

SA	$(SA)_2$ Co	$(SA)_2$ Ni	$(SA)_2$ Cu	$(SA)_2$ Zn	$(SA)_2$ Pd	
1634	1607	1612	1617	1619	1624	(cm^{-1})

In a more recent paper Kovacic[117] reported the infrared spectra of copper complexes of salicylideneanilines:

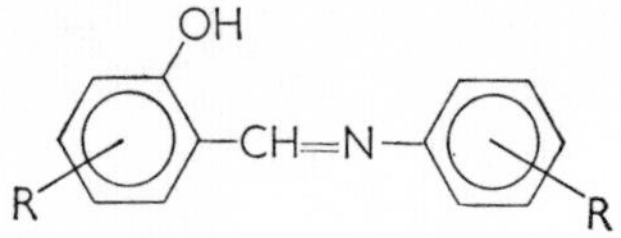

These have the general structure

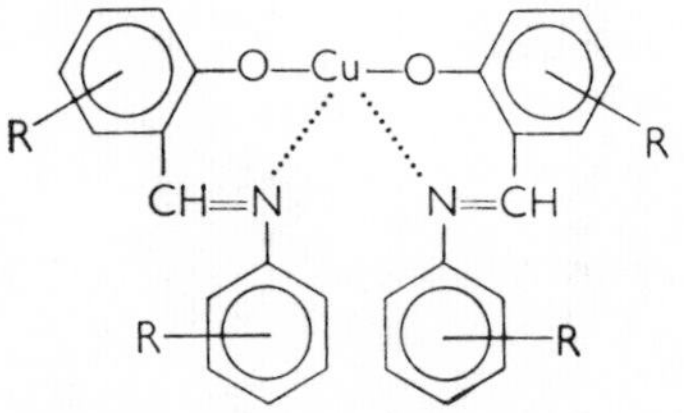

In Nujol mulls the C═N stretching frequency is in the range 1616–1603 cm^{-1} while in methylene chloride solution it is between 1612 and 1602. In these complexes the phenolic C—O vibration appears at 1330–1310 whereas in the free anils it is found between 1288 and 1265. In the salicylidene anilines themselves an internal hydrogen bond is formed,

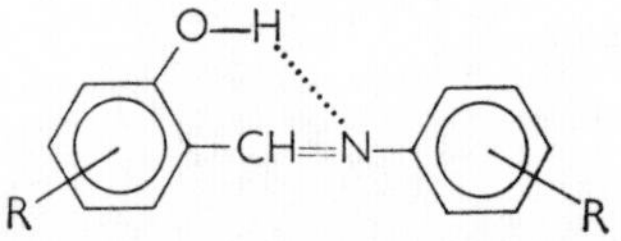

causing a weak and broad OH band between 3100 and 2700 cm^{-1}. Clearly the nitrogen with its lone pair is an excellent proton acceptor, as it is in aza-aromatic compounds in which the nitrogen atom has essentially the same hybridization.

In polymeric salicylidene anilines Marvel, Aspey and Dudley[118] found the C═N absorption at 1637–1616 cm^{-1} in the free bases and at 1656–1606 in the metal chelates.

As pointed out at the beginning of this section, the near-totality of infrared and Raman studies relating to imines concerned themselves with the C=N stretching frequency. There is little that can be done about it at this stage. A brief group theoretical treatment might, however, be useful at this point.

The

```
C
 \
  C=N
 /    \
C      C
```

model has only C_s symmetry since the carbon atom attached to the nitrogen is not collinear with the C=N bond.

The character table is rather simple:

C_s	E	σ_h		
A'	1	1	x, y, R_Z	x^2, y^2, z^2, xy
A''	1	-1	z, R_X, R_Y	yz, xz
Γ	15	5		
Γ_s	4	4		
Γ_b	4	4		
Γ_{op}	4	-4		

Γ is the reducible representation based on the $3N$ rectangular coordinates, where $N = 5$ is the number of atoms. It reduces to

$$\Gamma = 10A' + 5A''$$

Subtracting the components of translation and rotation which are shown in the character table we obtain:

$$\Gamma_{vib} = 7A' + 2A''$$

Thus we expect seven totally symmetrical and two non-totally symmetrical vibrations.

Some futher insight is gained if we use internal coordinates. We expect $N - 1 = 4$ stretching vibrations, $2N - 3 = 7$ in-plane vibrations and $N - 3 = 2$ out-of-plane vibrations[119].

Γ_s is the reducible representation based on the stretching coordinates, or more exactly the increments of the four internuclear distances between neighbours, Γ_b the one based on the four in-plane bond angles (their increments) and Γ_{op} on the four out-of-plane angles, which are of course zero at the vibrational equilibrium. Three of these latter correspond to the out-of-plane motions of the three peripheral carbon atoms and one to the twisting of the C=N bond in which one of these two atoms moves upwards and one downwards.

We obtain

$$\Gamma_s = 4A'; \Gamma_b = 4A'; \Gamma_{op} = 4A''$$

This follows since stretching and in-plane bending coordinates cannot change sign upon reflection in the molecular plane. The out-of-plane bending coordinates must, however.

This, however, would give 12 normal vibrations, although we can have only 9. Thus there are three redundancies. Since the total number of in-plane vibrations is 7 and since there cannot be redundancy among the stretching coordinates, one of the in-plane bending modes must be eliminated. This is naturally linked to the interdependence of the angle increments around the central carbon atom. There must be then two more redundancies among the four out-of-plane modes we obtained, as shown by Γ_{vib}.

Finally we have the following normal vibrations:

(a) Four stretching modes, one of them will be essentially a C=N motion. The three others will contain various amounts of C—C and N—C motions. These are likely to be located in the 1300–1000 cm^{-1} range.

(b) There will be three in-plane bending modes, two of them of mainly $\widehat{CCC}$ type and one of mainly $\widehat{CNC}$ type.

(c) There will be two out-of-plane vibrations which are expected to be strong because they involve a considerable change of dipole-moment perpendicular to the molecular plane. Because of the low symmetry of the molecule, we cannot tell without a complete normal coordinate analysis how much out-of-plane bending and how much twisting they will contain. Naturally, if there are one or more hydrogens replacing the peripheral carbons the corresponding frequencies will be much higher.

V. ELECTRONIC SPECTRA

A. Non-conjugated Azomethine Group

Little is known about the electronic spectrum of the >C=N-group itself in a purely aliphatic environment.

Since there are two π electrons, an sp^2 lone pair and, of course, σ electrons, there should be $\pi^* \leftarrow n$, $\sigma^* \leftarrow n$, $\pi^* \leftarrow \pi$ and $\sigma^* \leftarrow \sigma$ transitions, both singlet–singlet and singlet–triplet†. The whole spectrum is expected, however, to lie in the far ultraviolet and, probably because of the unstable character of these compounds, it has not been reported.

Recently G. Bélanger, D. R. Salahub and P. Sauvageau in the author's laboratory measured the spectrum of $CH_3CH{=}NC_3H_7$ prepared by D. Vocelle between 2000 and 1200 Å (Figure 1) in the gas phase, using the procedure by Lombos, Sauvageau and Sandorfy[120].

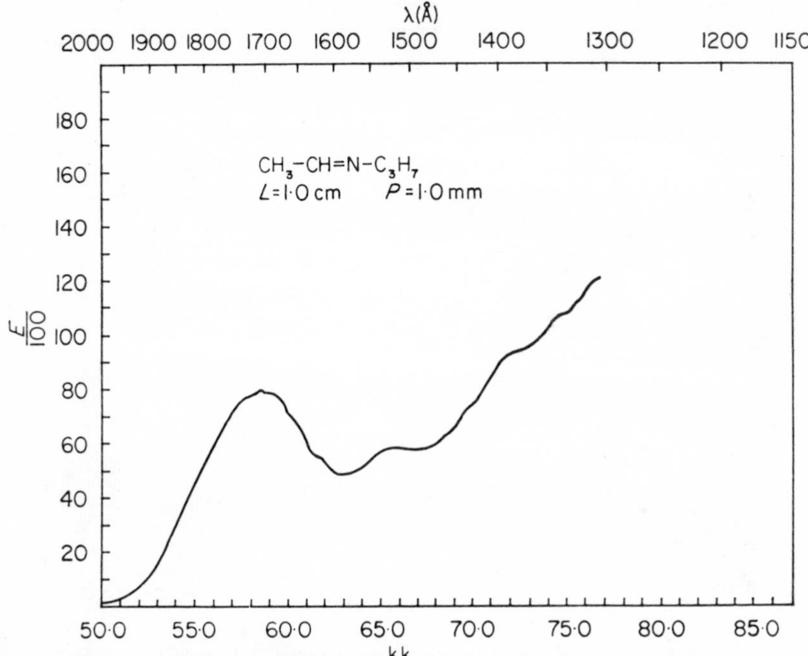

FIGURE 1. The far ultraviolet absorption spectrum of $CH_3CH{=}NC_3H_7$.

Unfortunately the spectrum consists of broad bands, making the interpretation difficult. There is a strong, diffuse band centred at about 1700 Å, having a molecular extinction coefficient of nearly 8000. Since the alkyl groups absorb only from 1600 Å down this band must be due to the C═N group. The main contributor to its intensity is

† In this article, Herzberg's convention of always writing the upper state first in transition symbolism is used.

in all likelihood the first $\pi^* \leftarrow \pi$ transition. Ethylene has a broad band centred at 1620 Å in the gas phase but extending to 2000 Å[121,122]. The $\pi^* \leftarrow \pi$ absorption in simple carbonyl compounds like formaldehyde is at higher frequencies, about 1550 Å[123].

Since we could not detect any absorption bands at frequencies lower than 2000 Å we shall tentatively admit that the $\pi^* \leftarrow n$, $\sigma^* \leftarrow n$ or σ–π transitions must be under the shoulder of the strong band, the $\pi^* \leftarrow n$ band perhaps between 2000 and 1900 Å and the $\sigma^* \leftarrow n$ band

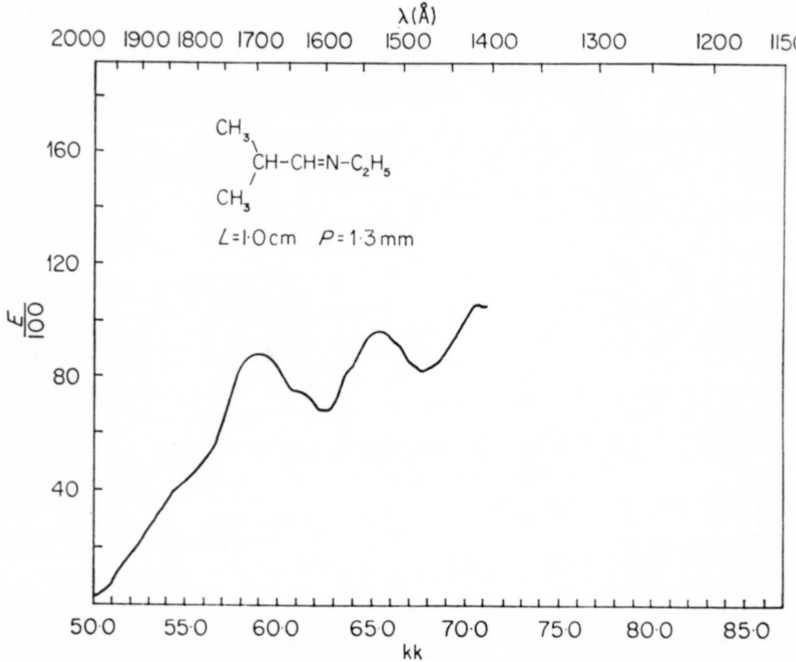

FIGURE 2. The far ultraviolet absorption spectrum of $(CH_3)_2CHCH{=}NC_2H_5$.

at about 1800 Å. The asymmetrical shape of the strong band would appear to confirm its composite character. At frequencies higher than 1700 Å there should be Rydberg transitions and transitions due to the σ electrons. We cannot go further. The energy of the lone pair compared with the energy of the π electrons or its ionization potential is not known.

$(CH_3)_2CHCH{=}NC_2H_5$ has a similar spectrum (Figure 2) but the shoulders between 50,000 and 55,000 cm^{-1} are more pronounced.

Platt[124] and Sidman[125] estimated that the $\pi^* \leftarrow n$ transition lies at 2100 Å if the $\rangle$C=N-group carries only aliphatic substituents, at 2500 if conjugated with a vinyl group and at 2900 if on a benzene ring.

It is interesting to compare the position of the $\pi^* \leftarrow n$ band for various unconjugated lone pair containing groups:

$\rangle$C=N—	2000–1900 Å (?)
—NO_2	2700
$\rangle$C=O	2800
—N=N—	3700
$\rangle$C=S	5500
—N=O	6800

We cannot expect that C=N and C=O should behave in a very similar manner. Among other things, hybridization is different in the two groups[130] and the π electrons are more strongly bonded in C=O (cf. Berthier and Serre[2]). No singlet–triplet bands were found up to the present time.

B. Conjugated Azomethine Group

Much more is known about the spectra of compounds in which the C=N group is substituted by aromatic rings.

Many of these spectra were measured in the early forties by Hertel and Schinzel[126], Kiss and Auer[127] and Kiss, Bacskai and Varga[128]. In the much more recent work of Jaffé, Yeh and Gardner[129], which deals mainly with azobenzene derivatives, there is a new interpretation of the spectrum of benzalaniline, $C_6H_5CH{=}NC_6H_5$, which is compared with the spectra of stilbene and azobenzene. The spectra of these three compounds are given in Figure 3. Stilbene has, of course, no $\pi^* \leftarrow n$ transition. In *trans*-azobenzene there are two of them, and both seem to contribute to the broad weak band at about 4400 Å. In benzalaniline, the energy of the lone pair is likely to be close to the average of the two n levels formed by the two lone pairs of azobenzene but the relating excited π orbital must be considerably higher. Therefore the $\pi^* \leftarrow n$ transition of benzalaniline is expected to lie at higher frequencies. Unfortunately it is hidden under the envelope of the next $\pi^* \leftarrow \pi$ band. Jaffé, Yeh and Gardner[129] located it at 3600 Å with $\epsilon \approx 100$.

The other bands in these spectra are $\pi^* \leftarrow \pi$. The apparent peaks are given in Table 15.

TABLE 15. Apparent peaks in the spectra of stilbene, benzalaniline and azobenzene[125].

	$\pi^*\leftarrow n$ (Å)	$\pi^*\leftarrow\pi$ (Å)		
Stilbene		3000	2280	2000
Benzalaniline	3600	3000	2600	
Azobenzene	4400	3200	2300	
Conjugate acid of azobenzene	3100	4200	2900	2350

Jaffé and coworkers carried out Hückel calculations and applied Platt's scheme to interpret these spectra. The lower frequency $\pi^* \leftarrow \pi$ peak is due to transitions between π orbitals largely localized in the central double bond. The frequency peak is due to transitions between orbitals largely localized in the phenyl rings. Other bands due to

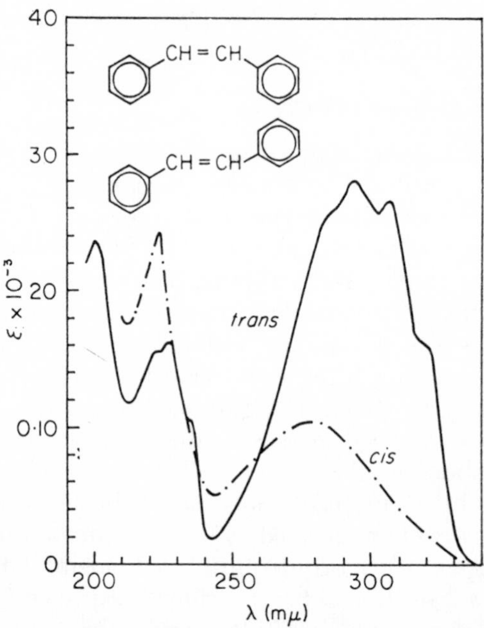

FIGURE 3a. Spectra of *trans* and *cis*-stilbene in heptane and ethanol respectively. Reprinted by permission from H. Suzuki, *Bull. Chem. Soc. Japan*, **33**, 381 (1960).

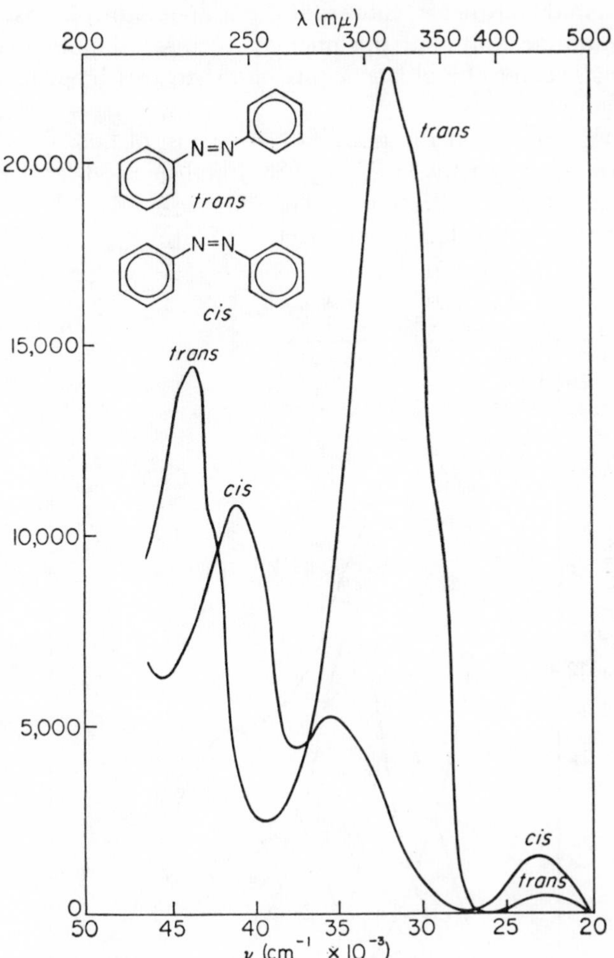

FIGURE 3b. The spectra of *cis*- and *trans*-azobenzene in 95% ethanol. Reprinted by permission from H. H. Jaffé and M. Orchin, *Theory and Applications of Ultraviolet Spectroscopy*, John Wiley and Sons, New York, 1962[134].

weaker transitions of mixed type probably contribute to the intensity of the lower frequency $\pi^* \leftarrow \pi$ bands.

The data given in the table are approximate and represent band centres. Further details are given in references 130 and 131. The *cis* isomer of benzalaniline does not seem to have been studied.

The 2600 Å bond in the spectrum of benzalaniline has a much lower intensity than the corresponding bands for stilbene and azobenzene.

Jaffé, Yeh and Gardner[129] show that this is due to the lower symmetry of benzalaniline and that the phenyl → central double bond type transitions become the greater contributors to this band in the case of benzalaniline.

Kanda[132] measured the gas phase spectrum of benzalaniline. He found two band systems at 2940–2833 Å and at 2488–2351 Å with a vibrational interval of 273 cm^{-1}. The $\pi^* \leftarrow \pi$ bands of benzalaniline are not very sensitive to solvent effects.

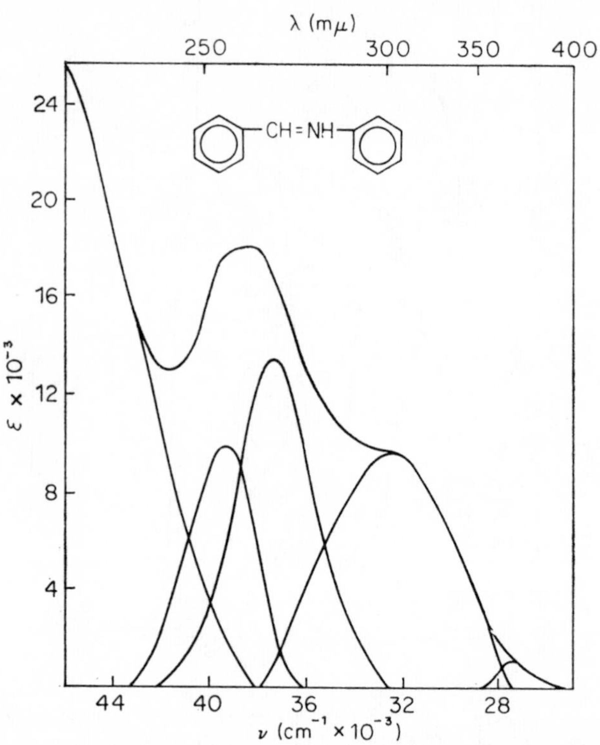

FIGURE 3c. The spectrum of benzalaniline in 95% ethanol. Reprinted by permission from H. H. Jaffé and R. W. Gardner, *J. Mol. Spectr.*, **2**, 120 (1958).

Charette, Faltlhansl and Teyssie[133] studied the ultraviolet spectra of a series of *N*-salicylidenealkylamines (SA) and their aryl-substituted derivatives in different solvents. Spectacular changes occur when the inert solvents are replaced by hydrogen bonding solvents. We reproduce here the spectra of *N*-salicylidene-2-aminopropane in cyclohexane, methanol and formamide (Figure 4)[134].

In cyclohexane there are three bands in the ultraviolet at 2170, 2540 and 3180 Å in order of decreasing intensity. In methanol the band at 2170 Å remains practically unchanged while the intensity of the two other bands decreases. At the same time two new bands appear at 2760 and 4000 Å. This by far exceeds the usual solvent shifts. The phenomenon seems to be the same in all hydrogen bonding solvents, only the relative intensity of the old and new bands changes. All the above-mentioned bands are clearly $\pi^* \leftarrow \pi$ bands as shown by

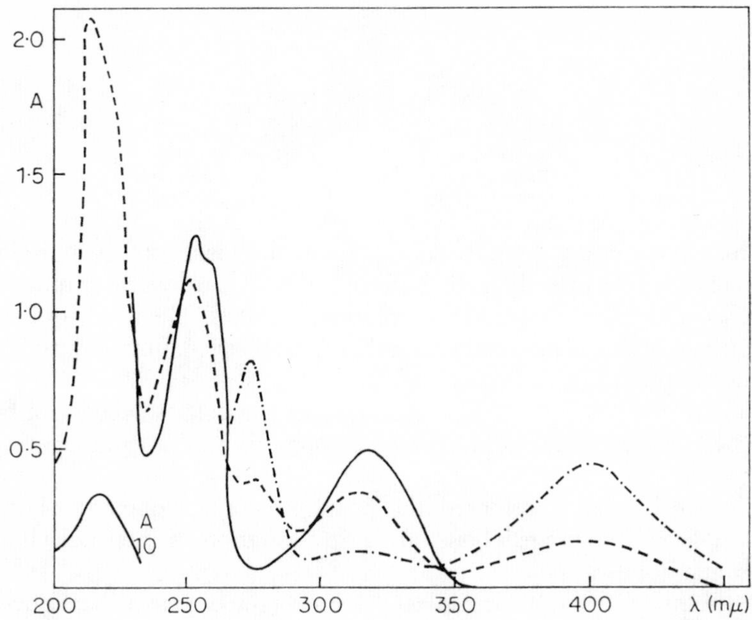

FIGURE 4. Absorption spectra of SA in cyclohexane (———); methanol (– – – –); formamide (–·–·–·–). Concentration 10^{-4}M; b = 1 cm[134].

their high intensities and red shift character in order of increasing solvent polarity. The new bands do not correspond to the positive or negative ions of the molecule. The anion is obtained by dissolving SA in 0·1N sodium methylate. There is a strong band centred at 3500 Å in its spectrum instead of the two new bands described above. The cation SA—H^+ is obtained in glacial acetic acid or by addition of gaseous HCl to a methanol solution. It has two strong bands at 2750 Å and 3500 Å.

Spectra of many other *ortho*-hydroxy compounds, mainly those of naphthalene and phenanthrene derivatives, were examined by Voss[135] and by Muszik[136]. Examples are:

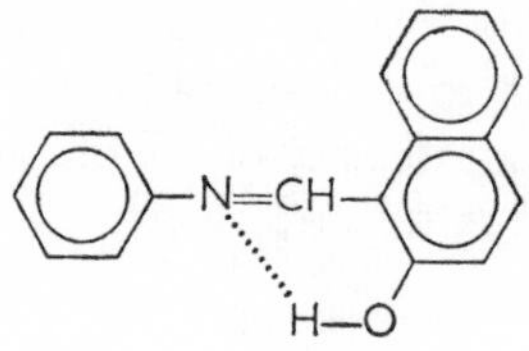

and

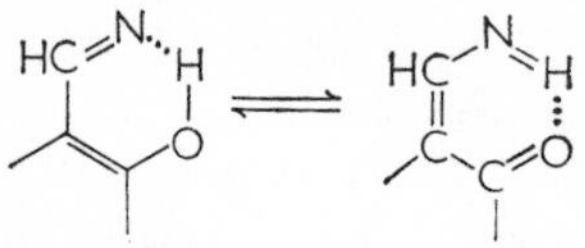

The solvent and temperature changes in the spectra of these molecules can be interpreted on the basis of NH/OH tautomeric equilibria which are highly sensitive to environmental effects. The spectra taken in different solvents intersect each other in isosbestic points confirming this assumption.

The long wavelength bands appearing in the polar solvents can be assigned to the NH form. *Cis–trans* isomerism is a complicating factor[136].

Charette[133] also considered the possibility of the opening of the intramolecular hydrogen bond and the formation of intermolecular hydrogen bonds in polar solvents.

If the hydroxy group is replaced by a methoxy group the above-mentioned spectacular changes upon changing the solvent do not occur. On the other hand, the *p*-hydroxy derivatives exhibit such changes while the *m*-hydroxy derivatives do not, showing again the involvement of tautomeric forms.

Bruyneel, Charette and de Hoffmann[134] explained the kinetics of the hydrolysis of *o*-, *m*-, and *p*-hydroxy-*n*-benzylidene-2-aminopropanes, invoking tautomeric equilibria.

Mason[137], in a study of the electronic spectra of hydroxy derivatives of *N*-heteroatomic systems, observed that these systems can exist, in general, as any one of four different species: the neutral enol (**1**), the cation (**2**), the anion (**3**) and the zwitterion (**4a**) or amide (**4b**). For example:

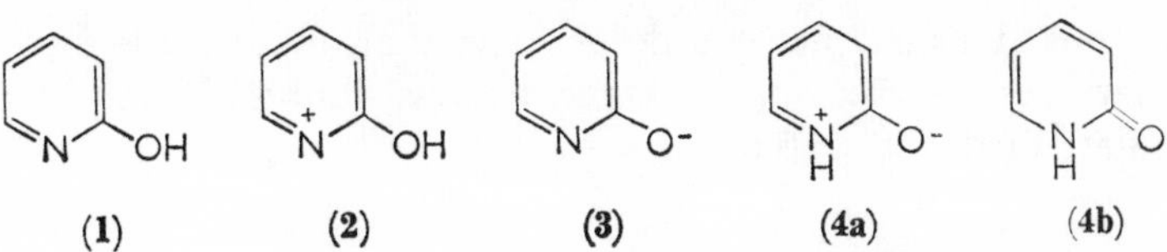

He found that the lowest energy transition of a given compound lies at wavelengths which show the order zwitterion > anion > cation > enol (the enol having the highest frequency) and that the spectrum of the zwitterion shifts considerably towards the red on changing from aqueous to non-polar solvents. Chatterjee and Douglas[138] carried out similar studies on *o*-hydroxy aromatic Schiff bases and found that the order for these compounds is zwitterion > cation ≈ anion > enol. They measured the spectra of *N*-methylsalicylaldimine, *N*-salicylideneaniline and *N*-methoxybenzylidene-*o*-hydroxyaniline and followed them by Hückel molecular orbital calculations.

According to Voss[135] the compounds of the type

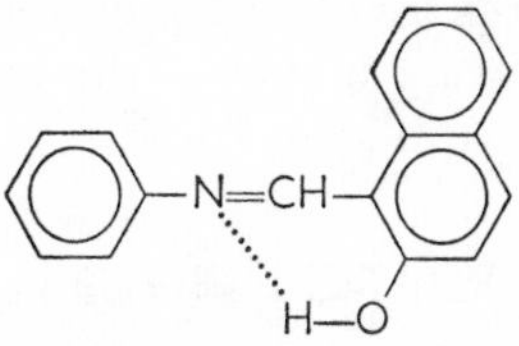

are not fluorescent at room temperature. At low temperature in solution or in rigid glasses, however, the NH-form is fluorescent. Muszik[136] found that the NH form of the molecules of type

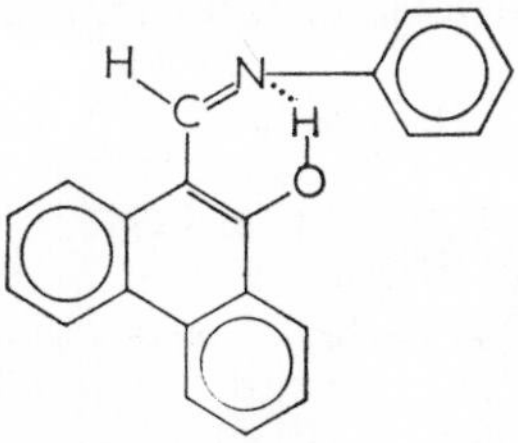

is fluorescent and that the fluorescence intensity increases in solvents favouring the NH form.

In conclusion, it is obvious that still very much remains to be done. We have no definite understanding of either the electron distribution in azomethine compounds, or of their electronic spectra, or of their vibrations. This chapter can do no more than to assemble some preliminary results, ask questions and express doubts. If only it could stimulate further research!

VI. ACKNOWLEDGEMENTS

The author wishes to express his very sincere thanks to his numerous colleagues who helped him in obtaining the results described in this chapter. They were Dr. D. Vocelle, who kindly prepared the aliphatic and other C=N compounds mentioned; Mr. J. J. Cobo, G. Bélanger, P. Sauvageau D. R. Salahub and M. Asselin, who measured their infrared and ultraviolet spectra; D. R. Salahub also made critical comments on the theoretical part (all from Département de Chimie, Université de Montréal).

Dr. D. W. Davidson, from the National Research Council of Canada, Ottawa, gave the benefit of his experience in dipole moment measurements.

Dr. G. Bessis gave indispensable computational help in the quantum chemical calculations during a stay of the author at the Centre de Mécanique Ondulatoire Appliqée, Paris.

VII. REFERENCES

1. C. A. Coulson and E. T. Stewart in *The Chemistry of Alkenes* (Ed. S. Patai), Interscience, London, 1964.
2. G. Berthier and J. Serre in *The Chemistry of the Carbonyl Group* (Ed. S. Patai), Interscience, London, 1966.
3. *Tables of Interatomic Distances and Configuration in Molecules and Ions* (Ed. L. E. Sutton.) Special publication No. 11, London, The Chemical Society, 1958.
4. E. M. Layton, Jr., R. D. Kross and V. A. Fassel, *J. Chem. Phys.*, **25**, 135 (1956).
5. B. P. Stoicheff, *Can. J. Phys.*, **40**, 358 (1962).
6. J. M. Dowling and B. P. Stoicheff, *Can. J. Phys.*, **37**, 703 (1959).
7. A. Langseth and B. P. Stoicheff, *Can. J. Phys.*, **34**, 350 (1956).
8. M. T. Christensen, D. R. Eaton, B. A. Green and H. W. Thompson, *Proc. Roy. Soc.* (*London*), **A 238**, 15 (1956).
9. C. C. Costain and B. P. Stoicheff, *J. Chem. Phys.*, **30**, 777 (1959).
10. W. G. Palmer, *Valency*, Second edition, Cambridge University Press, 1959, p. 125.
11. W. J. Orville-Thomas, *Chem. Rev.*, **57**, 1179 (1957).
12. F. E. Morris and W. J. Orville-Thomas, *J. Mol. Spectr.*, **6**, 572 (1961).
13. R. Daudel, R. Lefebvre and C. M. Moser, *Quantum Chemistry*, Interscience, New York, 1959.

14. C. A. Coulson, *Valence*, 2nd Ed., Oxford University Press, 1961.
15. R. S. Mulliken, C. A. Rieke, D. Orloff and H. Orloff, *J. Chem. Phys.*, **17**, 1248 (1949).
16. C. P. Smyth, *J. Am. Chem. Soc.*, **60**, 183 (1938).
17. C. P. Smyth, *Dielectric Behaviour and Structure*, McGraw-Hill, New York, 1955.
18. J. W. Smith, *Electric Dipole Moments*, Butterworths, London, 1955.
19. A. L. McClellan, *Tables of Experimental Dipole Moments*, W. H. Freeman, San Francisco, 1963.
20. K. B. Everard and E. L. Sutton, *J. Chem. Soc.*, 2318 (1949).
21. K. V. L. N. Sastry and R. F. Curl, *J. Chem. Phys.*, **41**, 77 (1964).
22. *Digest of Literature on Dielectrics*, **30**, National Academy of Sciences, Washington, 1966.
23. T. L. Cottrell, *The Strength of Chemical Bonds*, 2nd Ed., Butterworths, London, 1958.
24. G. E. Coates and L. E. Sutton, *J. Chem. Soc.*, 1187 (1948).
25. A. Julg, *Chimie Théorique*, Dunod, Paris, 1964, p. 30.
26. See, for example, F. A. Cotton, *Chemical Applications of Group Theory*, Interscience, New York, 1963.
27. K. S. Pitzer, *Quantum Chemistry*, Prentice-Hall, New York, 1953.
28. H. Eyring, J. Walter and G. E. Kimball, *Quantum Chemistry*, John Wiley and Sons, New York, 1944.
29. A. Julg and O. Julg, *Exercices de chimie quantique*, Dunod, Paris, 1967.
30. For a review see A. Streitwieser, Jr., *Molecular Orbital Theory for Organic Chemists*, John Wiley and Sons, New York, 1961.
31. C. Sandorfy, *Electronic Spectra and Quantum Chemistry*, Prentice-Hall, Englewood Cliffs, New Jersey, 1964.
32. L. Salem, *The Molecular Orbital Theory of Conjugated Systems*, W. A. Benjamin, New York, 1966.
33. E. Hückel, *Z. Physik*, **70**, 204 (1931).
34. G. W. Wheland and L. Pauling, *J. Am. Chem. Soc.*, **57**, 2086 (1935).
35. C. A. Coulson, *Proc. Roy. Soc.* (*London*), **A 161**, 413 (1939).
36. C. A. Coulson and H. C. Longuet-Higgins, *Proc. Roy. Soc.* (*London*), **A 191**, 39 (1947).
37. L. Pauling, *J. Am. Chem. Soc.*, **54**, 3570 (1932).
38. J. Bellugue and R. Daudel, *Rev. Sci.*, **84**, 541 (1946).
39. R. S. Mulliken, *La Liaison Chimique*, Publications du Centre National de la Recherche Scientifique, Paris, 1948, p. 197.
40. R. S. Mulliken, *J. Chim. Phys.*, **46**, 497; 675 (1949).
41. W. Moffitt, *Proc. Roy. Soc.* (*London*), **A 202**, 548 (1950).
42. J. C. Slater, *Phys. Rev.*, **34**, 1293 (1929).
43. E. U. Condon and G. H. Shortley, *Theory of Atomic Spectra*, Cambridge University Press, 1935.
44. G. Racah, *Phys. Rev.*, **62**, 438 (1942).
45. G. Pilcher and H. A. Skinner, *J. Inorg. Nucl. Chem.*, **24**, 937 (1962).
46. H. O. Pritchard and H. A. Skinner, *Chem. Rev.*, **55**, 745 (1955).
47. J. Hinze and H. H. Jaffé, *J. Am. Chem. Soc.*, **84**, 540 (1962).
48. R. T. Sanderson, *Science*, **114**, 670 (1951).
49. R. P. Iczkowski and J. L. Margrave, *J. Am. Chem. Soc.*, **83**, 3547 (1961).

50. C. Klixbüll-Jörgensen, *Orbitals in Atoms and Molecules*, Academic Press, New York, 1962, p. 85.
51. J. Hinze, M. A. Whitehead and H. H. Jaffé, *J. Am. Chem. Soc.*, **85**, 148 (1963).
52. G. Klopman, *J. Am. Chem. Soc.*, **86**, 1463 (1964).
53. R. G. Parr and B. L. Crawford, *J. Chem. Phys.*, **16**, 526 (1948).
54. M. J. S. Dewar, *J. Am. Chem. Soc.*, **74**, 3345 (1952).
55. P. and R. Daudel, *J. Phys. Radium, Ser.* 8, **7**, 12 (1946).
56. A. Laforgue, *J. Chim. Phys.*, **46**, 568 (1949).
57. G. W. Wheland and D. E. Mann, *J. Chem. Phys.*, **17**, 264 (1949).
58. A. Streitwieser, *J. Am. Chem. Soc.*, **82**, 4123 (1960).
59. R. Pariser and R. G. Parr, *J. Chem. Phys.*, **21**, 466; 767 (1953).
60. J. A. Pople, *Trans. Faraday. Soc.*, **49**, 1375 (1953).
61. G. W. Wheland, *J. Am. Chem. Soc.*, **63**, 2025 (1941).
62. O. Sinanoglu, *Adv. Chem. Phys.*, **6**, 315 (1964).
63. R. G. Parr and B. L. Crawford, *J. Chem. Phys.*, **16**, 526 (1948).
64. M. Goeppert-Mayer and A. L. Sklar, *J. Chem. Phys.*, **6**, 645 (1938).
65. C. C. J. Roothaan, *Rev. Mod. Phys.*, **23**, 69 (1951).
66. R. Pariser and R. G. Parr, *J. Chem. Phys.*, **21**, 466; 767 (1953).
67. R. Pariser, *J. Chem. Phys.*, **21**, 568 (1953).
68. J. A. Pople, *Trans. Faraday. Soc.*, **49**, 1375 (1953).
69. I. Fischer-Hjalmars in *Molecular Orbitals in Chemistry, Physics and Biology* (Eds. P.-O. Löwdin and B. Pullman), Academic Press, New York, 1964, p. 361.
70. W. A. Bingel, *Z. Naturforsch.*, **9a**, 675 (1954).
71. R. G. Parr, *J. Chem. Phys.*, **20**, 1499 (1952).
72. N. Mataga and K. Nishimoto, *Z. Phys. Chem.*, **13**, 140 (1957).
73. R. G. Parr, *Quantum Theory of Molecular Electronic Structure*, W. A. Benjamin, New York, 1964.
74. R. Daudel, R. Lefebvre and C. Moser, *Quantum Chemistry, Methods and Applications*, Interscience, New York, 1959.
75. R. Daudel, *Structure Electronique des Molécules*, Gauthier-Villars, Paris, 1962.
76. L. Salem, *The Molecular Orbital Theory of Conjugated Systems*, W. A. Benjamin, New York, 1966.
77. F. Halverson and R. C. Hirt, *J. Chem. Phys.*, **19**, 711 (1951).
78. R. Daudel, *Structure Electronique des Molécules*, Gauthier-Villars, Paris, 1962, pp. 115–140.
79. J. A. Pople and D. P. Santry, *Mol. Phys.*, **7**, 269 (1963–1964).
80. C. Sandorfy and R. Daudel, *Compt. Rend.* **238**, 93 (1954).
81. C. Sandorfy, *Can. J. Chem.*, **33**, 1337 (1955).
82. K. Fukui and H. Kato and T. Yonezawa, *Bull. Chem. Soc. Japan*, **33**, 1197 (1960).
83. K. Fukui, H. Kato and T. Yonezawa, *Bull. Chem. Soc. Japan.*, **34**, 442 (1961).
84. G. Klopman, *Helv. Chim. Acta.*, **45**, 711 (1962).
85. G. Klopman, *Helv. Chim. Acta.*, **46**, 1967 (1963).
86. R. Hoffmann, *J. Chem. Phys.*, **39**, 1397 (1963).
87. M. Wolfsberg and L. Hemholtz, *J. Chem. Phys.*, **20**, 837 (1952).
88. H. A. Pohl, R. Rein and K. Appel, *J. Chem. Phys.*, **41**, 3385 (1964).

89. J. A. Pople, D. P. Santry and G. A. Segal, *J. Chem. Phys.*, **43**, S129; 136 (1965).
90. J. J. Kaufman, *J. Chem. Phys.*, **43**, S152 (1965).
91. G. Klopman, *J. Am. Chem. Soc.*, **86**, 1463, 4450 (1964).
92. P. N. Skancke, *Arkiv Fysik*, **29**, 573 (1965); **30**, 449 (1965).
93. S. Katagiri and C. Sandorfy, *Theoret. Chim. Acta*, **4**, 203 (1966).
94. D. B. Cook, P. C. Hollis and R. McWeeny, *Mol. Phys.*, **13**, 553 (1967).
95. R. D. Brown and A. Penfold, *Trans. Faraday. Soc.*, **53**, 397 (1957).
96. J. Fabian and M. Legrand, *Bull. Soc. Chim. France*, 1461 (1956).
97. L. Kahovec, *Z. Physik. Chem.*, **B 43**, 364 (1939).
98. L. Kahovec, *Acta Phys. Austriaca*, **1**, 307 (1948).
99. A. Kirmann and P. Laurent, *Bull. Soc. Chim. France*, 1657 (1939).
100. J. Fabian, M. Legrand and P. Poirier, *Bull. Soc. Chim. France*, 1499 (1956).
101. L. A. Jones, J. C. Holmes and R. B. Seligman, *Anal. Chem.*, **28**, 191 (1956).
102. S. Bratoz and S. Besnainou, *J. Chim. Phys.*, **56**, 555 (1959).
103. S. Bratoz and S. Besnainou, *J. Chem. Phys.*, **34**, 1142 (1961).
104. J. Fabian and M. Legrand, *Bull. Soc. Chim. France*, 1461 (1956).
105. J. D. S. Goulden, *J. Chem. Soc.*, 997 (1953).
106. L. H. Cross and A. C. Rolfe, *Trans. Faraday. Soc.*, **47**, 354 (1951).
107. J. Fabian and V. Delaroff and M. Legrand, *Bull. Soc. Chim. France*, 287 (1956).
108. N. J. Leonard and J. V. Paukstelis, *J. Org. Chem.*, **28**, 3021 (1963).
109. O. E. Edwards and L. Marion, *Can. J. Chem.*, **30**, 627 (1952).
110. O. E. Edwards, F. H. Clarke and B. Douglas, *Can. J. Chem.*, **32**, 235 (1954).
111. O. E. Edwards and T. Singh, *Can. J. Chem.*, **32**, 465 (1954).
112. B. Witkop and T. W. Beiler, *J. Am. Chem. Soc.*, **76**, 5589; 5597 (1954).
113. B. Witkop and J. B. Patrick, *J. Am. Chem. Soc.*, **75**, 4474 (1953).
114. B. Witkop, *Experientia*, **10**, 420 (1954).
115. N. B. Colthup, L. H. Daly and S. E. Wiberley, *Introduction to Infrared and Raman Spectroscopy*, Academic Press, New York, 1964, pp. 282–284.
116. P. Teyssie and J. J. Charette, *Spectrochim. Acta*, **19**, 1407 (1963).
117. J. E. Kovacic, *Spectrochim. Acta.*, **23A**, 183 (1967).
118. C. S. Marvel, S. A. Aspey and E. A. Dudley, *J. Am. Chem. Soc.*, **78**, 4905 (1956).
119. E. B. Wilson, J. C. Decius and P. C. Cross, *Molecular Vibrations*, McGraw-Hill, New York, 1955, p. 241.
120. B. A. Lombos, P. Sauvageau and C. Sandorfy, *J. Mol. Spectr.*, **24**, 253 (1967).
121. J. R. Platt, H. B. Klevens and W. C. Price, *J. Chem. Phys.*, **17**, 466 (1949).
122. P. G. Wilkinson and R. S. Mulliken, *J. Chem. Phys.*, **23**, 1895 (1955).
123. G. Fleming, M. M. Anderson, A. J. Harrison and L. W. Pickett, *J. Chem. Phys.*, **30**, 351 (1959).
124. J. R. Platt in *Radiation Biology* (Ed. A. Hollaender), Vol. 3, Chap. 2, McGraw-Hill, New York, 1956.
125. J. W. Sidman, *Chem. Rev.*, **58**, 689 (1958).
126. E. Hertel and M. Schinzel, *Z. Physik. Chem.*, **B 48**, 289 (1941).
127. A. Kiss and G. Auer, *Z. Physik. Chem.*, **A189**, 344 (1941).
128. A. Kiss, G. Bacskai and E. Varga, *Acta Phys. Chem.*, Szeged, **1**, 155 (1943).

129. H. H. Jaffé, S. J. Yeh and R. W. Gardner, *J. Mol. Spectr.*, **2**, 120 (1958).
130. J. N. Murrell, *The Theory of the Electronic Spectra of Organic Molecules*, Methuen, London, 1963.
131. H. H. Jaffé and M. Orchin, *Theory and Applications of Ultraviolet Spectroscopy*, John Wiley and Sons, New York, 1962, Chap. 12.8.
132. Y. Kanda, *Chem. Abstr.*, **46**, 9982 (1951).
133. J. J. Charette, G. Faltlhansl and P. Teyssie, *Spectrochim. Acta*, **20**, 597 (1964).
134. W. Bruyneel, J. J. Charette and E. de Hoffmann, *J. Am. Chem. Soc.*, **88**, 3808 (1966).
135. W. Voss, *Dissertation*, Stuttgart, Technische Hochschule, 1960.
136. J. A. Muszik, *Dissertation*, Stuttgart, Technische Hochschule, 1965.
137. S. F. Mason, *J. Chem. Soc.*, 1253 (1959).
138. K. K. Chatterjee and B. E. Douglas, *Spectrochim. Acta*, **21**, 1625 (1965).

CHAPTER 2

Methods of formation of the carbon–nitrogen double bond

SHLOMO DAYAGI

Tel-Aviv University, Israel

and

YAIR DEGANI

Weizmann Institute of Science, Rehovoth, Israel

I. INTRODUCTION

In this article we have tried to give a descriptive review of the various routes leading to the formation of the carbon–nitrogen double bond. In view of the vast number of reported syntheses of C═N bonds appearing in the literature, several limitations had to be observed.

Only reactions resulting in the formation of stable and well-defined products were considered. Cases in which azomethines are only intermediary or transient were omitted, unless these can be isolated under appropriate conditions.

Compounds in which the C═N bond describes only one form of a mesomeric system, such as in hetero aromatic rings, diazo compounds, etc., were not considered.

The material was primarily arranged according to mechanistic types of formation of the C═N bond, and not by the types of the compounds containing this bond. First to be discussed are methods in which the C═N bond is formed by the binding together (through condensation, coupling, etc.) of separate carbon and nitrogen components. Next, methods which involve an existing carbon–nitrogen bond, are considered and finally, rearrangement reactions leading to C═N bonds are discussed.

The references to the literature were largely limited to open-chain and simple alicyclic azomethine compounds. Alkaloids or steroids are only occasionally included. Excluded from this review are C═N containing compounds which are discussed elsewhere in this or other volumes of this series, such as quinonimines, isocyanates and *aci*-nitro compounds. Also omitted for this reason are the numerous heterocyclic compounds containing C═N bonds which are prepared by 1,3-dipolar cycloadditions.

II. REACTIONS OF CARBONYL GROUPS WITH AMINO GROUPS AND RELATED REACTIONS

The condensation of amines with aldehydes and ketones has numerous applications, for preparative uses (e.g. heterocyclic compounds); for the identification, detection and determination of aldehydes or ketones; for the purification of carbonyl or amino compounds (e.g. amino acids in protein hydrolysates[1]); or for the protection of these groups during complex or sensitive reactions (e.g. amino acids during peptide synthesis[2]).

In this section will also be discussed condensations of derivatives of carbonyl compounds which give the parent compound during the reaction and interchange reactions of azomethines with other amines or carbonyl compounds, which proceed by essentially the same mechanistic pattern as the carbonyl condensations themselves.

A. *Condensations of Aldehydes and Ketones with Amines*

I. Primary amines

The condensation of primary amines with carbonyl compounds was first reported by Schiff[3], and the condensation products are often referred to as Schiff bases. The reaction was reviewed[4,5].

The experimental conditions depend on the nature of the amine and especially of the carbonyl compound which determine the position of the equilibrium

$$RR'CO + R''NH_2 \rightleftharpoons RR'C{=}NR'' + H_2O \qquad (1)$$

Usually, it is advisable to remove the water as it is formed by distillation or by using an azeotrope-forming solvent[6,7]. This is necessary with diaryl or arylalkyl ketones, but aldehydes and dialkyl ketones can usually be condensed with amines without removing the water. Aromatic aldehydes react smoothly under mild conditions and at relatively low temperatures in a suitable solvent or without it. In condensations of aromatic amines with aromatic aldehydes, electron-attracting substituents in the *para* position of the amine decrease the rate of the reaction, while increasing it when on the aldehyde[8]. In both cases, a linear sigma-rho relationship was observed.

With ketones, especially with aromatic ones, higher temperatures, longer reaction times, and a catalyst are usually required, in addition to the removal of the water as it is formed.

The reaction is acid catalysed. However, only aldehydes and ketones which do not aldolize easily in acidic media can be condensed with amines in the presence of strong acid catalysts (e.g. concentrated protonic acids[9], BF_3 etherate[10], $ZnCl_2$[7,9,11–13] or $POCl_3$[14]). For methyl ketones, only weak acids should be used, while for methylene ketones, which are less sensitive to acid-catalysed aldolizations, stronger acids may be used as catalysts[4]. Ultraviolet irradiation is reported[12] to promote the formation of azomethines from aldehydes. This is explained[13] as a light-promoted autoxidation of part of the aldehyde to the corresponding acid, which in turn acts as a catalyst.

Aromatic aldehydes and aliphatic or aromatic ketones give with amines quite stable azomethines. However, those from primary aldehydes which contain a —CH_2CH=N— group undergo very easily aldol-type condensations, so that in the reactions of such aldehydes with amines, polymers are usually formed[15]. The condensation can be stopped at the dimer or the trimer stage. For example, acetaldehyde gives with aniline a mixture of two isomeric dimers, 'Eckstein bases', which probably have the following structures[16]:

$$CH_3\underset{\substack{|\\ NHPh}}{CH}CH_2CH{=}NPh \quad \text{and} \quad CH_3\underset{\substack{|\\ NHPh}}{CH}CH{=}CHNHPh$$

The same products are also obtained by the reaction of aniline with aldol. Other aldehydes give similar dimers[16–18]. Secondary aldehydes, whose azomethines are incapable of forming α,β-unsaturated imines which would result in polymerization, give the monomeric imines[17]. Primary aliphatic aldehydes can give azomethines with various amines if the reaction is carried out at 0°c, and the product distilled from KOH[19,20].

Acetone and 2-butanone react with aromatic amines to give substituted dihydroquinolines[21,22]:

$$PhNH_2 + (CH_3)_2CO \longrightarrow \text{2,2,4-trimethyl-1,2-dihydroquinoline}$$

Isopropylidene amines, however, can be easily prepared by the method of Kuhn[23], which uses a complex of the amine hydriodide with silver iodide, which is soluble in DMF. This forms with acetone an insoluble complex of the azomethine with AgI. The free base can be isolated from the complex by the addition of KCN or triethylamine:

$$(CH_3)_2CO + RNH_2{\cdot}HI{\cdot}2AgI \xrightarrow[(DMF)]{} (CH_3)_2C{=}NR{\cdot}HI{\cdot}2AgI \xrightarrow{KCN} (CH_3)_2C{=}NR$$

R can be an aliphatic or an aromatic group, and even a hindered one (e.g. 2,6-dichlorophenol).

Isopropylideneaniline is formed also by the reaction of aniline with acetoacetic ester[24], through a route similar to the alkaline hydrolysis of acetoacetic ester:

$$\begin{array}{ccccc}
CH_3COCH_2COOEt & \xrightarrow{PhNH_2} & CH_3COCH_2NHPh & \xrightarrow{PhNH_2} & (CH_3)_2CO + (PhNH)_2CO \\
\downarrow{\scriptstyle PhNH_2} & & \downarrow{\scriptstyle PhNH_2} & & \downarrow{\scriptstyle PhNH_2} \\
CH_3C(=NPh)CH_2COOEt & \xrightarrow{PhNH_2} & CH_3C(=NPh)CH_2CONHPh & \xrightarrow{PhNH_2} & (CH_3)_2C{=}NPh + (PhNH)_2CO
\end{array}$$

α,β-Unsaturated ketones do not condense with amines or ammonia to azomethines, but rather add them to the double bond to form β-amino ketones[25].

α-Bromo ketones react with alkyl amines to give α-hydroxyimines; epoxides are formed as intermediates[26].

$$PhC(=O)C(CH_3)_2Br \xrightarrow{EtNH_2} PhC(NHEt)\overset{O}{—}C(CH_3)_2 \longrightarrow PhC(=NEt)C(CH_3)_2OH$$

The reactions of formaldehyde with amines were reviewed[4,15]. The $N{=}CH_2$ function is very sensitive to polymerizations, and except for isolated cases[19,27], e.g. $R_3CN{=}CH_2$, only trimers which have the *s*-triazene skeleton, or linear polymers, could be obtained; methylenimines are proposed[4,15] as intermediates in their formation. Schiff bases of α-amino acids are usually not stable enough for isolation. However, those derived from *o*-hydroxybenzaldehyde and related aldehydes are stabilized by chelate formation, and can be isolated[28].

In the formation of azomethines, both *anti* and *syn* isomers may be formed. However, as the energy barrier between them is low, the isolation of a pure isomer is impossible as a rule; there are only few proven exceptions to this rule, e.g. the case of

$$Cl{-}C_6H_3(NH_2){-}C(Ph){=}NCH_2CH_2N(CH_2CH_2)_2O$$

where two isomers whose configurations were proved by u.v., i.r. and n.m.r. spectra were isolated[29].

It must be borne in mind that although usually an azomethine is formed by the condensation of an amine and an aldehyde or ketone, in a few cases the tautomeric enamine is more stable and is the one which is preferably obtained; for example, enamines which are stabilized by intramolecular chelate formation[30]:

$$CH_3COCH_2CO_2Et + PhCH_2NH_2 \longrightarrow$$ (chelated enamine: H_3C, OEt, $PhCH_2$N–H···O)

2. Ammonia

Ammonia does not give azomethines with formaldehyde or with primary aliphatic aldehydes[4], but either 'aldehyde ammonia' addition compounds or polymers, of which hexamethylenetetramine is one of the best known.

Aromatic and secondary or tertiary aliphatic aldehydes, however, give condensation products of three moles of aldehyde with two moles of ammonia[31-33]:

$$3\ RCHO + 2\ NH_3 \longrightarrow RCH(N{=}CHR)_2$$

These compounds, especially those derived from aliphatic aldehydes, are unstable, and decompose on heating or distillation:

(a) $$ArCH(N{=}CHAr)_2 \xrightarrow{\Delta}$$ (imidazole ring bearing Ar, Ar, Ar; N, NH)

(b) $$(CH_3)_2CH[N{=}CHCH(CH_3)_2]_2 \rightleftharpoons (CH_3)_2C\text{–}CH\text{–}N{=}CHCH(CH_3)_2 \longrightarrow$$ (H, NH, $(CH_3)_2C{=}CH$)

$$(CH_3)_2C{=}CHN{=}CHCH(CH_3)_2 + (CH_3)_2CHCH{=}NH$$

$-NH_3$

$$\left[\begin{matrix} (CH_3)_2CHCH{=}NH \\ \Updownarrow \quad + \\ (CH_3)_2C{=}CHNH_2 \end{matrix}\right] \rightarrow (CH_3)_2CHCH_2NH_2\text{–}NH\text{–}(CH_3)_2C{=}CH$$

(c) $$R_3CCH(N{=}CHCR_3)_2 \longrightarrow R_3CCH{=}NCH_2CR_3 + R_3CCN$$

Aromatic ketones give imines with ammonia under drastic conditions, e.g. by bubbling dry NH_3 through the molten ketone[34], by heating alcoholic solutions of the reactants in sealed tubes to high temperature[35], by using $AlCl_3$ as catalyst at high temperatures[36] or by passing NH_3 with the ketone over ThO_2 at 300–400°[37]. Ammonium salts (e.g. thiocyanate) may serve as an ammonia source in reactions with reactive ketones[38].

When the imine is capable of cyclization a cyclic imine is isolated:

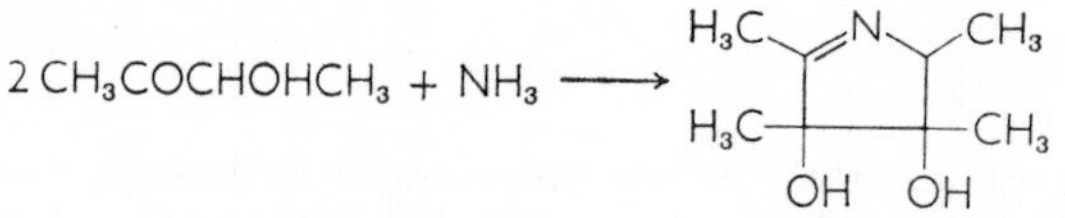

$$2\,CH_3COCHOHCH_3 + NH_3 \longrightarrow \text{(cyclic imine: ring N, } H_3C, CH_3, H_3C, CH_3, OH, OH) \quad \text{(ref. 39)}$$

$$2\,(CH_3)_2C(OH)COCH_3 + NH_3 \longrightarrow \text{(oxazoline: } H_3C, CH_3, O, H_3C, C(OH)(CH_3)_2, H_3C, N) \quad \text{(ref. 39)}$$

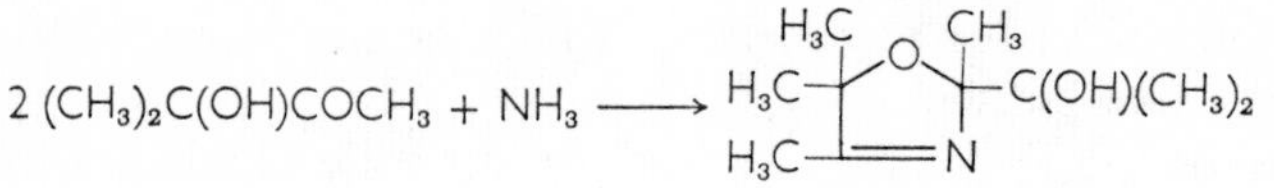

$$(CH_3)_2C(OH)COCH_3 + NH_3 + PhCHO \longrightarrow \text{(oxazoline: } H_3C, Ph, O, H_3C, H_3C, N) \quad \text{(ref. 40)}$$

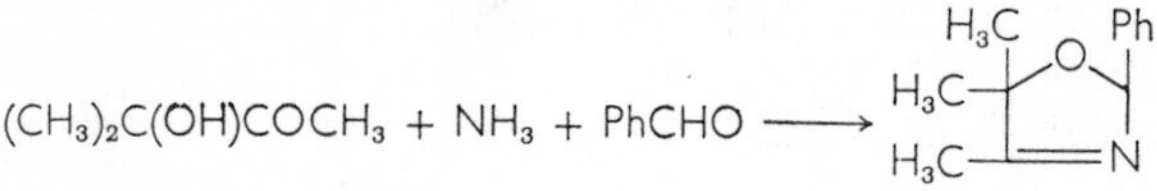

3. Secondary amines

The reaction of secondary amines with carbonyl compounds cannot lead to azomethines without a rearrangement. However, when a salt of the amine is treated with an aldehyde or a ketone, an immonium salt is obtained:

$$R_2NH{\cdot}HX + R'_2CO \longrightarrow R_2\overset{+}{N}{=}CR'_2\,X^-$$

Since azomethinium perchlorates are less hygroscopic than other salts and are easier to isolate and recrystallize, the amine perchlorates are often preferred for this reaction[41,42]. Fluoroborates, while less convenient, are less explosive and safer to use[41].

In some cases, a neutral amine may be used when zwitterion formation is possible, e.g.[43]

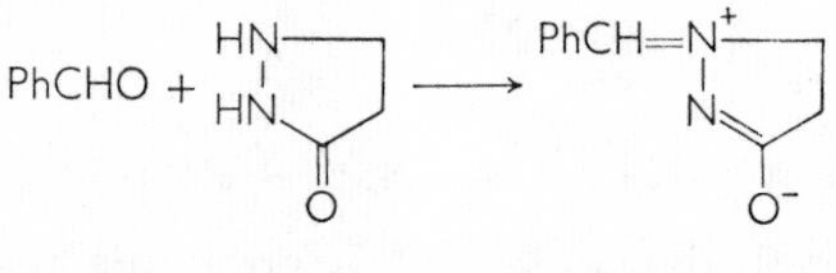

$$PhCHO + \text{(pyrazolidin-3-one: HN–HN–C=O ring)} \longrightarrow \text{(PhCH}{=}\overset{+}{N}\text{ ring with N}{=}C{-}O^-)$$

Some secondary cyclic amines may give a neutral azomethine either by ring opening[44] or by rearrangement[45]:

$$ArCHO + HN\triangleleft \longrightarrow ArCH{=}NCH_2CH_2N\triangleleft$$

$$ArCHO + \text{(hexahydro-1,2,4-triazine-3,6-dione)} \longrightarrow ArCH{=}N\text{–}N\text{(imidazolidine-2,4-dione)}$$

B. Condensations of Aldehydes and Ketones with Other Amino Groups

1. Amino group attached to oxygen

In these condensations, oximes, *O*-alkyl oximes, *O*-acyl oximes, and compounds in which the C=N—O group is part of a cyclic system are formed. The reaction is usually easier than that with amines and milder conditions are required. The equilibrium here is in favour of the oxime formation. Hydroxylamine and its derivatives, which are sensitive and decompose in their free form, are supplied as their salts, which are then completely or partially neutralized by the addition of a base or by basic ion exchangers[47], or by carrying out the reaction in pyridine[48]. The latter method is used especially for the oximation of difficultly soluble ketones[49] and of steroidal ketones[46,50].

The basicity of the reaction medium is of high importance. The dependence of the rate of the reaction versus the pH of the solution[13,51] shows a rate maximum at a pH close to neutral. The addition of an aqueous base to hydroxylamine hydrochloride or sulphate produces by a buffering effect a pH close to the optimum. Sometimes the reaction is carried out in a buffer solution (e.g. in aqueous NaH_2PO_4[52]). Although the reaction is acid catalysed, it is only rarely carried out in the presence of strong acid (e.g. conc. aqueous HCl[53]).

Sodium hydroxylamine disulphonate, which is easily prepared by bubbling SO_2 through a solution of sodium nitrite and sodium bisulphite in water, is frequently used without isolation[54]:

$$NaNO_2 + NaHSO_3 + SO_2 \longrightarrow HON(SO_3Na)_2$$

$$HON(SO_3Na)_2 + RR'CO \longrightarrow RR'C{=}NOH + 2\ NaHSO_4$$

Only a few reactions of hydroxylamine with carbonyl compounds in strong basic media are reported. They are useful for the preparation of sterically hindered oximes[55]. The hindered ketone (e.g. acetyl- or benzoyl mesitylene) in a solution of potassium *t*-amylate in *t*-amyl alcohol is allowed to stand for a long time (32 to 450 days; a 'lethargic reaction') at room temperature with hydroxylamine hydrochloride.

The oximes, which could not be obtained otherwise[56] because the ordinary reaction is very slow[57], are thus obtained in good yields. When the reaction was tried under reflux, much lower yields were obtained, probably due to the decomposition of free hydroxylamine at these conditions. The reaction of hydroxylamine with ketones in very strongly basic media proceeds through a mechanism which is different from normal amino–carbonyl condensations. Here, the attacking agent is probably either the anion **1** or **2**[55].

$$NH_2OH \underset{}{\overset{OH^-}{\rightleftharpoons}} \overset{-}{N}HOH \rightleftharpoons NH_2O^-$$
$$(1) \qquad (2)$$

Alternative methods to prepare oximes of highly hindered ketones are by the action of hydroxylamine on a ketimine[58] (see Section II.G) and by the reaction of hydroxylamine with the hindered ketone under very high pressures (about 9500 atmospheres)[59], which gives good yields (about 70%) of oximes like that of hexamethylacetone. The positive effect of the pressure on the rate of reaction is explained by the assumption that the rate-determining step of the reaction leads to a highly polar transition complex from reactants that are neutral or much less polar[59a].

Whereas amines or ammonia usually add to the carbon–carbon double bond of α,β-unsaturated ketones, hydroxylamine condenses normally in acidic media with the carbonyl group to give oximes; in basic media, both condensation and addition occur, together with secondary reactions[25]. On the other hand, *O*-methylhydroxylamine only adds to the double bond of α,β-unsaturated ketones, while with β-diketones it gives mono-*O*-methyloxime[25,60].

Ketones which are stable in strong acids (e.g. fluorenone) give oximes by reaction with nitromethane in hot (190–200°) polyphosphoric acid[61]. Nitromethane is hydrolysed to formic acid and hydroxylamine, and the latter in turn reacts with the ketone to give the oxime.

Oximes can be isolated in two configurations, *anti* and *syn*. The isolation of one or the other may be achieved by changing the experimental conditions of the reaction. As the *anti* isomer is usually more stable[62] thermodynamically, the *syn* isomer can be isomerized easily to the *anti* form, as by the action of acids[63]. One of the methods for the elucidation of the configuration of a given oxime is to isolate and identify the product of its Beckmann rearrangement. Here the reader must bear in mind that all the configurations which were assigned by that method to ketoximes up to 1921 are opposite to the presently accepted ones. This was due to the erroneous mechanism which was suggested for the Beckmann rearrangement until it was clarified by Meisenheimer[64].

N-Alkylhydroxylamines give nitrones with aldehydes and ketones[65–67].

$$R_2CO + R'NHOH \longrightarrow \underset{\displaystyle\overset{\downarrow}{O}}{R_2C{=}NR'} + H_2O$$

When the aldehyde or ketone have α-hydrogens, an aldol-type condensation of the nitrone follows its formation[67,68], e.g.

$$(CH_3)_2CO + PhNHOH \longrightarrow [(CH_3)_2C{=}N(O)Ph] \longrightarrow (CH_3)_2C(N(OH)Ph)CH_2C(CH_3){=}N(O)Ph$$

Cyclic nitrones are more stable against aldolization and can be dimerized with basic catalysts, e.g. sodamide[69]. When the aldolization product is not desired, a condensation of the hydroxylamine with an acetal instead of a ketone is favourable[65].

N-triphenylmethylhydroxylamine gives with benzaldehyde an *O*-trityl derivative[70]. Probably a nitrone is formed initially, and then a rearrangement leads to the less hindered and more stable *O*-trityl oxime:

$$PhCHO + Ph_3CNHOH \longrightarrow [\underset{\displaystyle\overset{\downarrow}{O}}{PhCH{=}NCPh_3}] \longrightarrow PhCH{=}NOCPh_3$$

2. Amino group attached to nitrogen

Reactions with hydrazine. In the reactions of carbonyl groups with hydrazine, either one or two of the available amino groups may condense to form hydrazones or azines respectively.

Aldehydes and dialkyl ketones react readily, usually just by shaking the reactants in water or in alcohol, and in most cases only azines are obtained[71]. In order to obtain the hydrazone, the reaction should be carried out in a large excess of hydrazine[72] and in the total absence of acids. Alternatively, the hydrazones may be prepared indirectly by the action of hydrazine on the azine[72,73]. Hydrazones from benzaldehyde or substituted benzaldehydes with electron-donating substituents are very unstable, and azines are slowly precipitated from their alcoholic solution. *p*-Dimethylaminobenzalhydrazone cannot be obtained at all.

On the other hand, *p*-nitrobenzalhydrazone is stable and is unchanged by refluxing its alcoholic solution. It forms the azine, however, by the addition of a little acid[72,74].

Aryl alkyl ketones on refluxing in alcohol or acetic acid give without an acidic catalyst a hydrazone; in the presence of mineral acids, an azine is formed[75,76].

Diaryl ketones need more drastic conditions[77,78]; in some cases a water-removing agent[60,79] has to be added, or the reaction has to be carried out at high temperatures[78,80]. Usually, only hydrazones are obtained, but at higher temperatures in an autoclave an azine may be formed[75]. The azine, however, can be prepared from the hydrazone, either by acidifying its alcoholic solution and allowing it to stand at room temperature[79], or by refluxing the hydrazone with an excess of ketone[74,81,82]. By the last method, mixed azines may also be prepared[74,82].

β-Diketones and hydrazine give usually monohydrazones[83], while γ-diketones give an internal cyclic azine, a 4,5-dihydropyridazine derivative[84]:

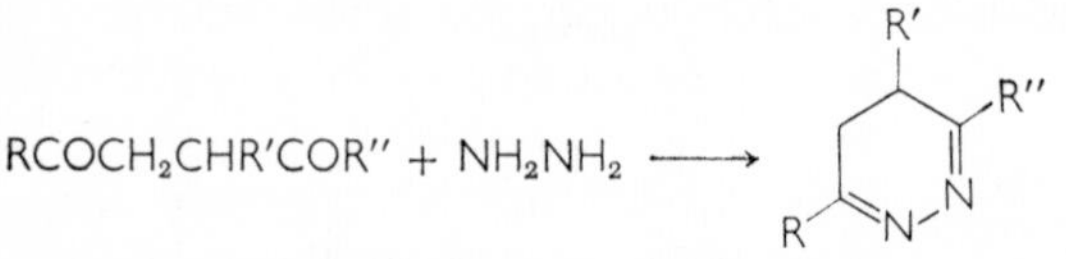

$$RCOCH_2CHR'COR'' + NH_2NH_2 \longrightarrow$$

cis-1,2-Cyclopropane dicarboxaldehyde gives with hydrazine a trimer of the corresponding internal azine[85]:

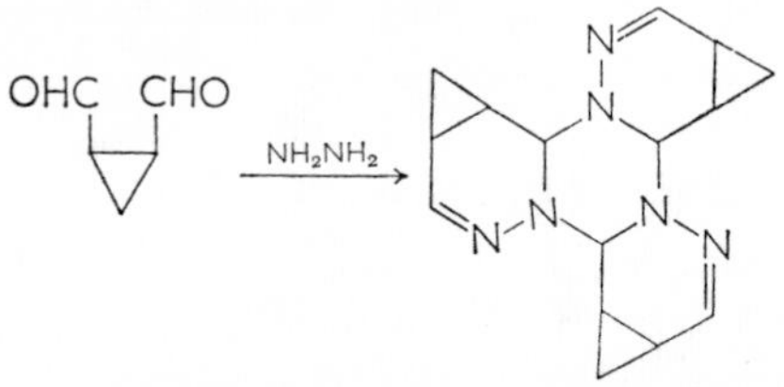

$$\xrightarrow{NH_2NH_2}$$

β-Halo ketones give with hydrazine a five-membered cyclic hydrazone[86]:

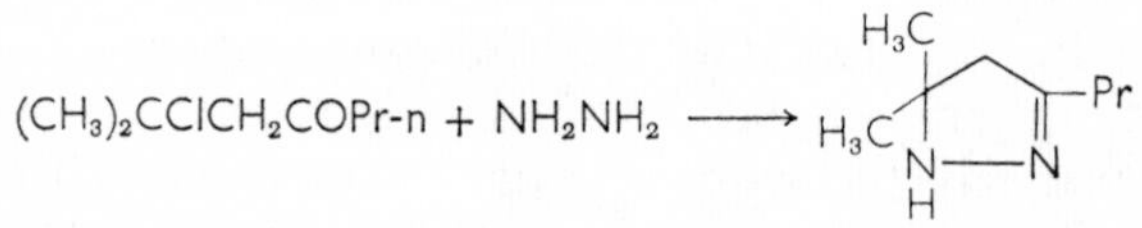

$$(CH_3)_2CClCH_2COPr\text{-}n + NH_2NH_2 \longrightarrow$$

Cyclic keto nitrones react with hydrazine to give a rearranged hydrazone[87]:

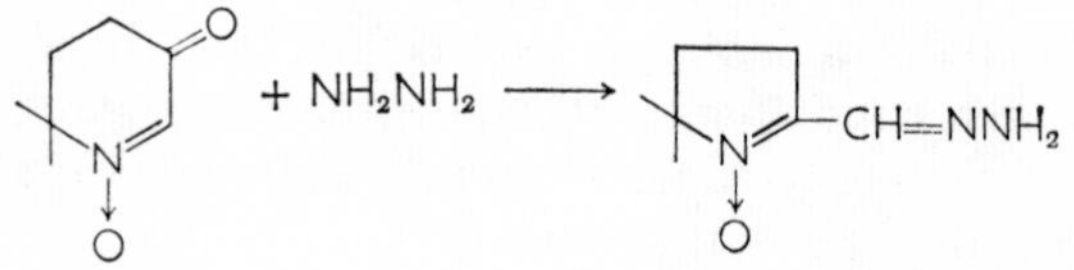

$$+ NH_2NH_2 \longrightarrow \quad -CH{=}NNH_2$$

Reactions with substituted hydrazines. The reactions of alkyl and *N,N*-dialkylhydrazines with aldehydes and ketones give usually normal substituted hydrazones[72,88]. The *N*-hydrogen of the alkyl hydrazones formed from aromatic aldehydes is labile, and might add to the carbonyl group of another molecule of the aldehyde when the latter is in excess; the product depends on the experimental conditions:

$$ArCHO + CH_3NHNH_2 \xrightarrow[HCl]{MeOH} CH_3NN{=}CHAr \text{ (with } ArCHOCH_3 \text{ on N)} \qquad \text{(ref. 43)}$$

$$\xrightarrow{EtOH} \begin{array}{c} CH_3NN{=}CHAr \\ | \\ CHAr \\ | \\ CH_3NN{=}CHAr \end{array} \qquad \text{(ref. 89)}$$

An enormous number of aryl hydrazones was prepared for the identification of carbonyl compounds. Phenylhydrazones, especially the nitro-substituted ones, are very stable towards hydrolysis, even by strong acids, and are usually prepared in strongly acidic media, usually alcoholic HCl or H_2SO_4.

As with other amines, the reactions of aryl hydrazines with formaldehyde does not give the simple methylene hydrazines. *p*-Tolylhydrazine is reported to give with formaldehyde in acetic acid *N*-methylene-*p*-tolylhydrazine[90] together with its polymer. The same reactants give in water

$$\begin{array}{c} p\text{-}ToNN{=}CH_2 \\ | \\ CH_2 \\ | \\ p\text{-}ToNN{=}CH_2 \end{array}$$

analogous to the product from methylhydrazine and aromatic aldehydes[89], and in ethanol $[(CH_2{=}N)(To)NCH_2]_2O$[90].

β-Dimethylamino ketones give with phenylhydrazine, through an elimination of dimethylamine, a five-membered cyclic hydrazone[91]. The intermediate open-chain hydrazone was also isolated[92].

$$PhCOCH_2CH_2N(CH_3)_2 \xrightarrow{PhNHNH_2} \underset{\underset{NHNPh}{\|}}{PhC}CH_2CH_2N(CH_3)_2 \xrightarrow{-Me_2NH}$$

$$\left[\underset{\underset{NNHPh}{\|}}{PhC}CH{=}CH_2\right] \longrightarrow \text{1-phenyl-3-phenyl-2-pyrazoline (Ph–C=N–N(Ph) ring)}$$

A number of monotosylhydrazones of α-diketones, which serve as starting materials for the preparation of α-diazo ketones, were prepared from α-diketones and tosylhydrazine[83a,93].

Semicarbazide and thiosemicarbazide are also widely used as 'carbonyl reagents'. The semicarbazones are generally easier to hydrolyse than the corresponding oximes or hydrazones. As with the formation of oximes, there is an optimal pH for the formation of semicarbazones[13,51,94,95].

The formation of semicarbazones from ketones and semicarbazide is catalysed by aniline[96]. Here, the mechanism is different from the normal general acid-catalysed formation of semicarbazones. First, an anil is formed which is then interchanged with the semicarbazide.

$$(1)\quad R_2CO + PhNH_2 \xrightleftharpoons{\text{slow}} R_2C{=}NPh + H_2O$$

$$(2)\quad R_2C{=}NPh + NH_2NHCONH_2 \xrightleftharpoons{\text{fast}} R_2C{=}NNHCONH_2 + PhNH_2$$

The rate of the attack of semicarbazide on Schiff bases is faster by several orders of magnitude than on the free carbonyl; the reason is probably the much higher basicity of azomethines in comparison with their parent carbonyl compounds.

Osazones and related compounds. Since the first isolation of sugar osazones by Fischer[97], their structure and the mechanism of their formation has intrigued many chemists.

Osazones may be prepared by the action of an excess of a substituted hydrazine on an α-diketone, but in most cases they are prepared from α-hydroxy, α-halo, α-methoxy, α-acetoxy or α-dimethylamino ketones.

The reaction of aryl hydrazine with α-hydroxy ketones or aldehydes (including sugars) may lead to either an aryl hydrazone or to an osazone, or to a mixture of both. The results depend upon the experimental conditions, on the relative amounts of the reactants, and on the structure of both the hydrazine and the carbonyl compound. Usually, in the presence of strong acids, hydrazones are obtained[98–101]. In mild acidic media, such as acetic acid, both osazones and hydrazones might be obtained. In acetic acid–water mixtures, the higher the concentration of HOAc, the higher is the proportion of osazone to hydrazone[101]. Hydrazones are also formed in neutral aqueous or alcoholic solutions[99,100]. In alcohols or in water, substituted phenylhydrazines give hydrazones with both aldoses and ketoses when the substituent is an electron-attracting group (nitro, carboxy, carboethoxy, bromo), whereas when the substituent is an electron-repelling group (methyl, methoxy), hydrazones are obtained only from aldoses and not from ketoses. Phenylhydrazine reacts with both types[100].

α-Halo ketones, too, may give with hydrazines either hydrazones or osazones, or both, depending on the structure of the halo ketone and the hydrazine and on the experimental conditions. Brady's reagent (an aqueous methanolic solution of 2,4-dinitrophenylhydrazine containing H_2SO_4) gives with some halo ketones[102] osazones, whereas with others, α-halo hydrazones are obtained[79,80]. The α-halogen in these hydrazones is very labile, and gives in methanol α-methoxy hydrazones[102,103], and in acetic acid α-acetoxy hydrazones[104]. α,β-Unsaturated hydrazones may be formed by elimination[103]. An excess of DNP transforms them all to osazones. The best procedure to obtain hydrazones is therefore to carry out the reaction in a concentrated mineral acid (e.g. 12N hydrochloric acid)[105].

Chloral[106] and dichloroacetaldehyde[107] also gives osazones with aryl hydrazines, as do α-methoxy ketones[108] and α-dimethylamino ketones.

α-Halo ketones give glyoximes with hydroxylamine in a similar reaction[109]:

$$RCOCHXR' + NH_2OH \longrightarrow \begin{array}{l} RC{=}NOH \\ \quad | \\ R'C{=}NOH \end{array}$$

Alkylamines give with α-bromo ketones α-hydroxy imines[26], while fluorinated ketones give only additions to the C=O bond to form amino alcohols without the formation of a C=N bond[110].

The problem of the structure of the osazones was reviewed[111,112]. The currently accepted structure of the osazones was first proposed by Fieser and Fieser[113] and was later proved by chemical, spectral (u.v. and n.m.r.), polarographic methods[112,114] and by x-ray analysis[115] to be a 'quasi-aromatic' system:

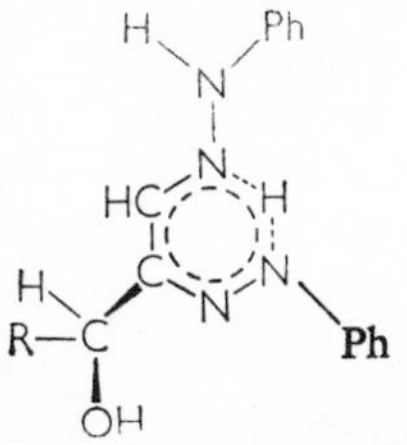

H Ph
N
N
HC H
C N
H N Ph
R—C
OH

This structure explains why, in the osazonization of sugars by phenylhydrazine, the oxidation of the hydroxy groups of the sugar is confined to only one hydroxy group adjacent to the carbonyl. Indeed, in the reaction of sugars with *N*-methyl-*N*-phenylhydrazine, where no chelation and formation of the 'quasi-aromatic' system is possible, all the primary and secondary hydroxyl groups are oxidized and converted to hydrazones, and alkazones are formed[116]:

$$\begin{array}{c} CHO \\ | \\ (CHOH)_n \\ | \\ CH_2OH \end{array} \quad \text{or} \quad \begin{array}{c} CH_2OH \\ | \\ CO \\ | \\ (CHOH)_{n-1} \\ | \\ CH_2OH \end{array} \quad + NH_2N(CH_3)Ph \longrightarrow \begin{array}{c} CH{=}NN(CH_3)Ph \\ | \\ [C{=}NN(CH_3)Ph]_n \\ | \\ CH{=}NN(CH_3)Ph \end{array}$$

The 'quasi-aromatic' structure explains also the differences in spectra and in chemical properties between sugar and non-sugar osazones[112]: The $C_{(3)}$ hydroxy group which is absent in the non-sugar osazones stabilizes the 'quasi-aromatic' ring formation.

An intriguing problem in the osazone formation is the mechanism of the oxidation of the hydroxy groups by the aryl hydrazine. Aryl hydrazines are known to oxidize primary or secondary alcohols containing at least one aromatic or ethylenic substituent to give aldehydes or ketones, isolable as their hydrazones[117]. A mechanism similar to the one proposed for the oxidation, i.e.

$$\begin{array}{l} X \\ \;\;\diagdown \\ \;\;\;CHOH \\ \;\;\diagup \\ Y \end{array} + ArNHNH_2 \xrightarrow{H^+} \left[\begin{array}{c} X\diagdown \;\; C\cdots H\cdots \overset{+}{N}H_3 \\ Y\diagup \;\; | \qquad\quad | \\ O\cdots H\cdots NHAr \end{array} \right] \longrightarrow \begin{array}{l} X \\ \;\;\diagdown \\ \;\;\;C{=}O \\ \;\;\diagup \\ Y \end{array} + NH_4^+ + NH_2Ar$$

was also proposed for the formation of osazones[118], and the only difference is the substitution of the hydroxyl by an aryl hydrazine group prior to the oxidation step:

$$\begin{array}{l} X \\ \;\;\diagdown \\ \;\;\;CHOH \\ \;\;\diagup \\ Y \end{array} + ArNHNH_2 \longrightarrow \begin{array}{l} X \\ \;\;\diagdown \\ \;\;\;CHNHNHAr \\ \;\;\diagup \\ Y \end{array}$$

$$\begin{array}{l} X \\ \;\;\diagdown \\ \;\;\;CHNHNHAr \\ \;\;\diagup \\ Y \end{array} + ArNH\overset{+}{N}H_3 \longrightarrow \left[\begin{array}{c} X\diagdown \;\; C\cdots H\cdots \overset{+}{N}H_3 \\ Y\diagup \;\; | \qquad\quad | \\ ArNHN\cdots H\cdots NHAr \end{array} \right] \longrightarrow \begin{array}{l} X \\ \;\;\diagdown \\ \;\;\;C{=}NNHAr \\ \;\;\diagup \\ Y \end{array} + NH_4^+ + NH_2Ar$$

This mechanism explains why the reaction is acid catalysed, and why the formation of osazones is favoured with phenylhydrazines with electron-attracting substituents[119], an effect which is opposite to the substituent effect in the formation of hydrazones. However, α-hydroxy ketones and related compounds may be converted to osazones through other mechanisms, one involving unsaturated azo intermediates of the type C=C—N=NAr[103,120], which were isolated in some cases[120].

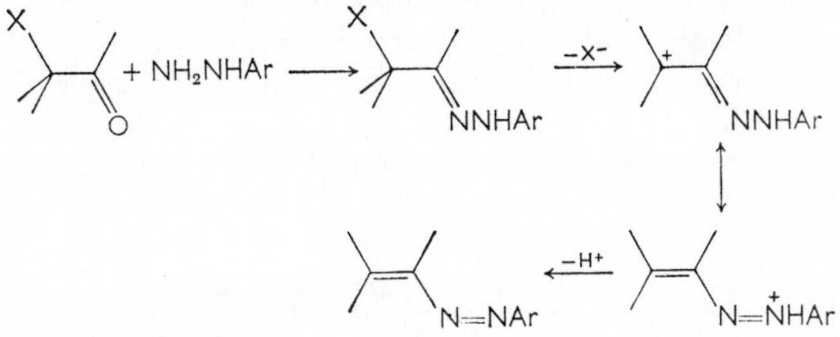

X, + NH_2NHAr ⟶ X, NNHAr $\xrightarrow{-X^-}$ (+) NNHAr ⟷ N=$\overset{+}{N}$HAr $\xrightarrow{-H^+}$ N=NAr

3. Other amino groups

Aldehydes and ketones react normally with the following types of amino compounds:

Carbamates, $NH_2NHCOOR$[121], aminoguanidine[122], sulphenamides[123], nitramine[124], chloramine[125], phosphinic acid hydrazides[126] and triazane derivatives[127].

Very few cases of condensations of carbonyl groups with amidic amino groups are reported. Urea reacts with benzoin to give a glyoxalone derivative[128].

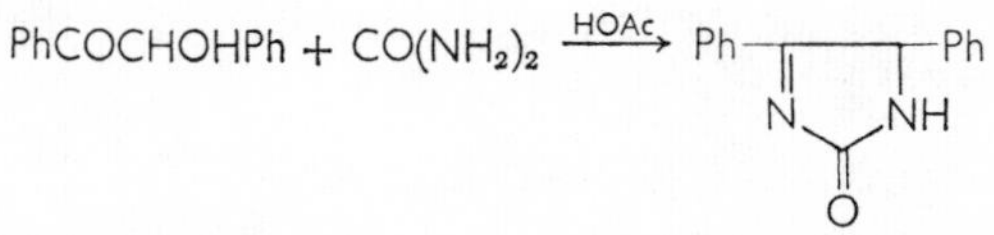

$$PhCOCHOHPh + CO(NH_2)_2 \xrightarrow{HOAc}$$ Ph—C=C—Ph (N, NH, C=O ring)

Sulphonamides react with aldehydes under the influence of Lewis acids ($ZnCl_2$ or $AlCl_3$)[129].

$$ArSO_2NH_2 + RCHO \longrightarrow ArSO_2N{=}CHR$$

An interesting reaction between an amidic NH_2 group and an amidic carbonyl group was reported. Ureas condense with dimethylformamide in the presence of various acyl chlorides[130].

$$RNHCONH_2 + HCON(CH_3)_2 \xrightarrow[0^\circ-10^\circ]{RCOCl} RNHCON{=}CHN(CH_3)_2 \cdot HCl$$

C. *Reactions of Amino Groups with Amidic, Esteric and Related Carbonyl Groups*

The amidic carbonyl group does not react normally with amines. Very few cases are reported of the reaction of hydroxylamine with amides to give amidoximes, e.g.

$$\text{(3-phenyl-1,2,4-oxadiazol-5-yl)}-C(=NOH)-NH_2 \quad \text{(ref. 131)} \qquad \text{or} \qquad NH_2C(=NOH)COOH \quad \text{(ref. 132)}$$

However, thioamides are much more reactive, and aromatic thioamides[133], thioureides[134], thiolactams[135] or thiourethanes[136] may be condensed with amines[137], ammonia[136], hydroxylamine[133,134] or hydrazine[195].

One case is reported of a reaction of benzoyl chloride with an amine[138]:

$$\text{2,4-diamino-1,5-benzenedithiol (HS, SH, } H_2N, NH_2) + PhCOCl \xrightarrow{NaOH} \text{2,6-diphenylbenzobisthiazole}$$

Substituted amides are condensed in three steps with amines, with $POCl_3$ as a condensing agent[139]:

$$RC(=O)NR'R'' + POCl_3 \longrightarrow RC(OPOCl_2)=\overset{+}{N}R'R''\ Cl^- \xrightarrow{R'''NH_2} RC(NHR''')=\overset{+}{N}R'R''\ Cl^- \xrightarrow{OH^-} RC(=NR''')-NR'R''$$

Similarly, amines react with the product formed from *N,N*-disubstituted amides and aryl sulphonyl chlorides, with the formation of amidinium salts[140]:

$$HCONMe_2 + ArSO_2Cl \longrightarrow Me_2\overset{+}{N}=CHOSO_2ArCl^- \xrightarrow{RNH_2} RNHCH=\overset{+}{N}Me_2\ ArSO_3^-$$

Hydrazones react with similar amide–sulphonyl products to give amide hydrazones[141]:

$$RR'C=NNH_2 + ArSO_2OC(R'')=NR'''_2\ Cl^- \longrightarrow RR'C=NNHC(R'')(OSO_2R)-NR'''_2 \longrightarrow$$

$$RR'C=NNHCR''=\overset{+}{N}R'''_2\ ArSO_3^- \xrightarrow{OH^-} RR'C=NN=CR''NR'''_2$$

D. Other Carbonyl–Amino Condensations

N,N-Bis(trimethylsilyl)amines react with ketones, and new C=N bonds are formed by the elimination of trimethylsilyl groups:

$$RR'CO + (Me_3Si)_2NR'' \xrightarrow[\text{or } Zn/Cd]{ZnCl_2} RR'C=NR'' \qquad \text{(ref. 142)}$$

$$R_2CO + (Me_3Si)_2NNa \longrightarrow R_2C=NSiMe_3 \qquad \text{(ref. 143)}$$

E. The Mechanism of the Carbonyl–Amino Condensation

The mechanism was discussed in detail in another volume of this series[13]. It is a two-step mechanism, consisting of an initial addition of the amine to the carbonyl to form a carbinolamine, followed by dehydration to give the C=N bond. Both steps are specific and general acid catalysed. The differences between the various types of amino compounds in regard to their condensations with carbonyl groups are quantitative rather than qualitative, and stem from differences in the relative magnitude of the rate and equilibrium constants of both steps. For additional information and for references regarding the mechanism the reader is advised to turn to the review mentioned[13].

F. Reactions of Amino Groups with Potential Carbonyls

Carbonyl derivatives, which are easily transformed to aldehydes and ketones, condense with amino compounds sometimes even more readily and in better yield than the parent carbonyl compounds themselves.

1. Hydrates, acetates, ketals and other esters

gem-Dihydroxy compounds[144] and their acetates[145] react with amino groups as readily as their parent carbonyls.

Acetals and ketals react with amino groups[146] either on refluxing in a solvent or on removal of the alcohol formed in the reaction by distillation. The reaction has a marked advantage when the parent carbonyl compounds are unstable. *N*-Benzalsulphonamides can be prepared from benzaldehyde acetals by distillation of the alcohol as it is formed[147].

N-Alkylhydroxylamines give with ketals the corresponding nitrones smoothly[65]. Benzohydroxamic acid reacts with benzaldehyde acetal on heating *in vacuo* and benzaldoxime is formed[148]. It is assumed that first the nitrone *N*-benzoylbenzaldoxime is formed, which then acts as a benzoylating agent towards another molecule of benzohydroxamic acid:

$$\text{PhCONHOH} + \text{PhCH(OEt)}_2 \xrightarrow[\text{vac.}]{80°} [\text{PhCON(O)=CHPh}] \xrightarrow{\text{PhCONHOH}} \text{PhCH=NOH} + \text{PhCONHOCOPh}$$

By using the hydrochloride of the hydroxamic acid, a rearrangement of the nitrone takes place:

$$\text{PhCONHOH·HCl} + \text{PhCH(OEt)}_2 \xrightarrow[\text{vac.}]{80°} \text{PhCON(O)=CHPh} \longrightarrow \text{PhCOON=CHPh}$$

Under the same conditions, benzophenone ketal gives a cyclic product with benzohydroxamic acid:

$$PhCONHOH + Ph_2C(OEt)_2 \longrightarrow Ph_2C\begin{matrix} O-N \\ \| \\ O-C-Ph \end{matrix}$$

Thioacetals and disulphides behave like acetals[149], e.g.

$$RCH_2SSCH_2R + NH_2NH_2 \longrightarrow 2\,RCH{=}NNH_2$$

Ortho esters react like acetals with ammonia[150], amino compounds[151], and sulphonamides[152], and imidoates are formed. An excess of the amino component causes the substitution of the OR group, and amidines are formed[151]:

$$PhNH_2 + CH(OEt)_3 \longrightarrow PhN{=}CHOEt \xrightarrow{PhNH_2} PhN{=}CHNHPh$$

Urea and *N*-alkylureas react with ortho formates[153]:

$$2\,RNHCONH_2 + HC(OEt)_3 \longrightarrow RNHCON{=}CHNHCONHR$$

In the presence of acetic anhydride, only imidoates of the type RNHCON=CHOEt are formed[154].

Ethyl orthocarbonate reacts similarly with sulphonamides[155]:

$$TsNH_2 + C(OEt)_4 \xrightarrow[EtOH]{\text{dist. of}} TsN{=}C(OEt)_2$$

2. Enols, enol ethers and phenols

Enolic forms of aldehydes or ketones (e.g. α,β-diketones) react with amino compounds like normal carbonyls[156]. Enol ethers (vinyl ethers) require acid catalysts[157], probably to hydrolyse the ether prior to condensation:

$$RCH{=}CHOCH_3 + NH_2OH\cdot HCl \xrightarrow{BF_3-Et_2O} RCH_2CH{=}NOH$$

Certain phenols, especially polycyclic ones, behave like their keto tautomers and give with aryl hydrazines the corresponding aryl hydrazones, which can be transformed by the route of the Fischer indole synthesis into substituted carbazoles[158], e.g.

OH + PhNHNH₂ ⟶ NNHPh ⟶ NH

3. gem-Dihalides

Azomethines are formed from *gem*-dichloro or dibromo compounds and amino compounds, usually in an excess of the amine. The reaction has found special uses in cases when the reaction with the parent carbonyl compound is sluggish, e.g. with diaryl ketones[159,160]. Ketones give usually only the *anti* isomer, whereas from *gem*-dihalides, both *syn* and *anti* isomers may be obtained[14,160].

Isocyanide dichlorides (iminophosgenes) behave like other dichlorides towards amines, and give carbodiimides by their reaction under nitrogen[161]:

$$RNH_2{\cdot}HCl + ArN{=}CCl_2 \xrightarrow[N_2]{180°} RN{=}C{=}NAr$$

N-Arylidenesulphonamides are obtained from benzal chloride[162], benzotrichloride[163] or benzophenone dichloride[164] and aryl sulphonamides.

Diiodomethane reacts with secondary amines to give methyleneimmonium iodides[15]:

$$CH_2I_2 + HN\langle\text{piperidine}\rangle \longrightarrow \langle\text{piperidine}\rangle\overset{+}{N}{=}CH_2 \quad I^-$$

G. The Interchange of Existing C=N Bonds

1. With amino compounds

In these reactions, the formation of a new C=N bond depends on the equilibrium

$$R_2C{=}NR' + R''NH_2 \rightleftharpoons R_2C{=}NR'' + R'NH_2 \quad (2)$$

Equilibrium constants for the displacement of substituted benzalanilines by aryl amines were measured and compared. 'Relative displacement abilities' of a number of amines were obtained by comparison with that of sulphanilamide, which was taken as one. The following values are given: *p*-anisidine, 30–38; aniline, 14–15; *m*-nitroaniline, 1·15; sulphanilamide, 1·0; *p*-nitroaniline, 0·36. The equilibrium was found to be governed by the empirical equation

$$a \log K_{OH} + b \log w + c = 0$$

where w is the displacement ability of the amine and K_{OH} its ionization constant.

Thus hydroxylamine, hydrazines or semicarbazide give good yields of oximes, hydrazones or semicarbazones, respectively, by reacting with anils[96,165]. Hydrazones are similarly formed from oximes and hydrazines[166]. Removal of one of the components of the equilibrium will shift the reaction into completion in favour of one product. Hence, most C=N compounds can be easily obtained from ketimines by their reaction with amino compounds, when NH_3 is eliminated by heating[137,167,168]. The last method is especially useful for the preparation of azomethines of sterically hindered ketones which cannot be prepared directly from the parent ketone[58].

The conversion of hydrazones to azines by the effect of acid catalysts can be regarded as a displacement reaction which follows the mechanism[79]

$$R_2C{=}NNH_2 \xrightleftharpoons{H^+} R_2\overset{+}{C}{-}NHNH_2 \xrightleftharpoons{R_2C=NNH_2} R_2C(-NHNH_2)(-\overset{+}{N}H_2N{=}CR_2) \rightleftharpoons$$

$$R_2C(-\overset{+}{N}H_2NH_2)(-NHN{=}CR_2) \xrightleftharpoons{-NH_2NH_2} R_2C^+(-NHN{=}CR_2) \xrightleftharpoons{-H^+} R_2C{=}NN{=}CR_2$$

The reverse path describes the also-known conversion of an azine to a hydrazone by an excess of hydrazine[72,73].

Nitrones also yield azomethines with a variety of amino compounds[169]. On the other hand, nitrones are formed by the interchange of an imine with *N*-substituted hydroxylamine[70b].

Phenylhydrazones with difluoramine yield *N*-fluoroimines through a mechanism which involves a cyclic intermediate[170]:

$$(CH_3)_2C{=}NNHPh + HNF_2 \longrightarrow (H_3C)_2C(NHNHPh)(N{-}F{-}F) \xrightarrow{-HF} (H_3C)_2C\overset{N-NHPh}{\underset{N-F}{\triangle}}$$

$$\longrightarrow (CH_3)_2C{=}NF + [Ph\overset{+}{N}H{=}\overset{-}{N} \longrightarrow PhH + N_2]$$

2. With carbonyl compounds

The formation of a new C=N bond by this interchange reaction is reported with acetone or 2-butanone imines[171], hydrazones[172] and selenocarbazones[173], which react with aldehydes or ketones:

$$(CH_3)_2C{=}NR + R'_2CO \longrightarrow (CH_3)_2CO + R'_2C{=}NR$$

The acetone or 2-butanone is removed by distillation.

The main application of the reaction is the generation of ketones from their oximes, which are usually quite stable to hydrolysis, by the reaction of the oximes with formaldehyde[51,53,174]. The initially formed methylenehydroxylamine is unstable and is immediately decomposed to ammonia and formic acid.

$$R_2C{=}NOH + HCHO \longrightarrow R_2C{=}O + [CH_2{=}NOH \xrightarrow{H_2O} HCOOH + NH_3]$$

H. Azomethines by the Cleavage of C=C Bonds with Amines

Compounds containing carbon–carbon double bonds activated by strong electron-attracting groups react with amino compounds to give azomethines[175,176]:

$$RCH{=}CX_2 + R'NH_2 \longrightarrow RCH{=}NR' + CH_2X_2$$

An intermediate addition product of the type $RCH(NHR')CHX_2$ was isolated in the reaction of benzalacetylacetone with aniline.

Secondary amines might also displace the active methylene component to form an immonium salt[41]:

$$(CH_3)_2C{=}CHCOCH_3 + HN\langle\text{ring}\rangle\cdot HClO_4 \longrightarrow (CH_3)_2C{=}\overset{+}{N}\langle\text{ring}\rangle\ ClO_4^-$$

Styrene derivatives undergo hydrazinolysis with hydrazine in the presence of sodium hydrazide, and hydrazones are formed[177]:

$$PhCH{=}CHR + NH_2NH_2 \xrightarrow{NaNHNH_2} PhCH_3 + RCH{=}NNH_2$$

Under similar experimental conditions, hydrazones are formed from β-halo-α-phenylalkanes, probably by elimination to yield an alkene with subsequent hydrazinolysis of the latter[178].

III. NITROSO–METHYLENE CONDENSATIONS

Aromatic nitroso compounds react with active methylene groups to form anils. The condensation, known as the 'Ehrlich–Sachs reaction'[179], is catalysed by bases such as sodium hydroxide, potassium carbonate and piperidine. The reaction is often accompanied by the formation of nitrones as by-products. Both products probably emerge from a common aldol-type intermediate, which either loses water to form the anil or undergoes dehydrogenation yielding, the nitrone[180].

$$ArNO + CH_2XY \longrightarrow ArN(OH)CHXY \begin{cases} \xrightarrow{-H_2O} ArN{=}CXY \\ \xrightarrow[{[O]}]{} Ar\overset{\overset{O}{\uparrow}}{N}{=}CXY \end{cases}$$

The oxidative side reactions are mainly ascribed to the unreacted nitroso compound, which is thereby reduced to the corresponding amine or azoxy compound[181].

$$\mathrm{ArN(OH)CHXY + 2\,ArNO \longrightarrow Ar\overset{\overset{\displaystyle O}{\uparrow}}{N}{=}CXY + Ar\overset{\overset{\displaystyle O}{\uparrow}}{N}{=}NAr}$$

Examples of methylene compounds which condense with aromatic nitroso compounds to give anil or nitrone or a mixture of both are benzyl cyanides[182], malonic esters[183], α-methylene ketones[184], fluorenes[185] and cyclopentadienes[186]. Bis-methylene compounds such as tetralones may form dianils[187]:

$$\text{5-methoxy-7-methyl-1-tetralone} + p\text{-}ONC_6H_4N(CH_3)_2 \longrightarrow \text{(2,4-bis}[NC_6H_4N(CH_3)_2]\text{ derivative with } CH_3O,\ H_3C,\ =O)$$

Compounds possessing reactive methyl groups, such as 2,4-dinitrotoluene and 9-methylacridine (the latter also under acid catalysis[188]), may react similarly with aromatic nitroso compounds, yielding anils and/or nitrones. The aldonitrones formed in such cases may be transformed to the isomeric anilides[189,190].

The anil–nitrone product ratio in these condensations seems to be affected by a variety of factors, including reaction medium, catalyst, time and temperature, and no satisfactory generalization can be reached. Thus, in the reaction of *p*-dimethylaminonitrosobenzene with aryl acetonitriles, weak bases such as piperidine and low reaction temperatures favoured nitrone formation[12], whereas sodium hydroxide or anhydrous potassium carbonate and high temperatures yielded mainly an anil[191,192]. On the other hand, in the condensation of 2,4-dinitrotoluene with *o*-nitrosotoluene, piperidine favoured anil formation, while sodium carbonate lead to the formation of a nitrone. In some condensations, the absence of water was found to increase anil formation up to 99%[192]. Prolonged reaction time was shown to increase nitrone–anil ratio in the reaction of 2,4-dinitrotoluene and *p*-dimethylaminonitrosobenzene. With 9-methylacridine, the anil–

nitrone product ratio varied with the nature of the substituents on the aromatic nitroso compound[193]. Using an excess of the nitroso compound increases nitrone formation[188,191]. Air oxidation is probably also a contributing factor in the formation of nitrone. It has been suggested[194] (admittedly without strong basis) that aliphatic active methylene compounds tend to produce predominantly anils, whereas activated toluenes are more likely to give nitrones.

Methylene compounds activated by a quaternary pyridinium group react with aromatic nitroso compounds, to yield solely nitrones. The pyridinium group is expelled in the process. The reaction, known as the Kröhnke reaction, was reviewed by Kröhnke in 1953 and 1963[195]. Formation of nitrones in this reaction is not due to oxidation, but to an intramolecular step in which the lone electron pair on the nitrogen of the aldol-type intermediate displaces the pyridine from the neighbouring carbon:

$$RCH_2-\overset{+}{N}C_5H_5\ X^- \xrightarrow{\text{base}} R\bar{C}H-\overset{+}{N}C_5H_5 \xrightarrow{ArNO}$$

$$R-CH(-\overset{+}{N}C_5H_5)(^-O-NAr) \longrightarrow \begin{matrix} RCH \\ \| \\ O{\leftarrow}NAr \end{matrix} + NC_5H_5$$

Pyridine can also be displaced from the intermediate hydroxylamine derivative by nucleophiles. Thus the treatment of benzylpyridinium halide with *p*-nitrosodimethylaniline[196] or with nitrosophenolates[197] followed by aqueous potassium cyanide gives high yields of α-cyanoanils.

$$ArCH_2-\overset{+}{N}C_5H_5\ Br^- + ON-C_6H_4-N(CH_3)_2 \xrightarrow{KCN}$$

$$ArC(CN){=}N-C_6H_4-N(CH_3)_2 + NC_5H_5 + KBr$$

However, these α-cyanoanils are prepared more simply and conveniently from the appropriate aryl acetonitriles[198,199].

IV. FORMATION OF OXIMES VIA C-NITROSATIONS

Compounds containing an active methylene or methyl group are readily nitrosated by a variety of nitrosating agents. The initially formed *C*-nitroso intermediates undergo usually a rapid prototropic isomerization to oximes:

$$RCH_2R' \longrightarrow RCH(NO)R' \longrightarrow RCR'{=}NOH$$

The nitrosation requires the presence of an electron-attracting group on the atom adjacent to the one being nitrosated. Nitrosation is usually carried out with nitrous acid or one of its esters. Excess nitrosating agent has to be avoided since it reacts with the oximino compound, e.g.

$$RR'C{=}NOH + EtONO \longrightarrow RR'C{=}O + N_2O + EtOH$$

C-Nitrosation has been reviewed, covering the literature until 1953[200].

A. Nitrosation of Primary and Secondary Carbons

The ketonic carbonyl strongly activates an adjacent carbon for nitrosation. In methyl alkyl ketones the methylene group is nitrosated in preference to the methyl group[201]:

$$RCH_2COCH_3 + MeONO \xrightarrow{HCl} \underset{\underset{NOH}{\|}}{RC}COCH_3$$

With identical methylene groups on both sides of the ketone, such as in symmetrical alicyclic ketones, α,α'-dioximino compounds may be formed[202,203]. Chloro ketones such as chloroacetone[204] and phenacyl chloride[205] are readily nitrosated to the corresponding hydroxamic acid chlorides, whereas oximation of α-bromo ketones is reported to occur preferentially at methylene groups not containing Br[206].

$$CH_3COCH_2Cl \longrightarrow CH_3COCCl{=}NOH$$
$$PhCH_2COCH_2Br \longrightarrow Ph\underset{\underset{NOH}{\|}}{C}COCH_2Br$$

If a carboxyl group is situated on an atom adjacent to the active methylene, very ready decarboxylation follows the nitrosation. A cyclic mechanism, anologous to that attributed to decarboxylation of β,γ-unsaturated carboxylic acids, has been postulated for the nitrosative decarboxylation[207].

$$\text{O=N}\cdots\text{CR}_2\text{–C(=O)–O–H (cyclic)} \longrightarrow HO{-}N{=}CR_2 + CO_2$$

A single alkoxycarbonyl group does not activate an adjacent methylene group for nitrosation. When an additional activating group is present, as in arylacetic esters, β-keto esters, malonic esters and cyanoacetic esters, oximation occurs readily[200]. Oximation of aliphatic mononitriles has not been reported, but glutamic and adipic nitriles, in which a methylene is situated in α-position to a single cyano group, have been oximated in quantitative yields[208].

Primary nitroalkanes are nitrosated to nitrolic acids. The reaction is usually carried out by treating the nitronate salt with acidified alkali nitrite[209] or with N_2O_4[210].

$$RCHNO_2Na + NaNO_2 \xrightarrow{H_2SO_4} RC(NO_2){=}NOH$$

Hydrocarbons possessing a methylene group of pronounced reactivity such as cyclopentadiene[211] and fluorene[212] can undergo base-catalysed nitrosation. Relatively inactive compounds such as alkyl benzenes and alkyl pyridines can be nitrosated if a sufficiently strong base, e.g. alkali amide in liquid ammonia, is used as catalyst[213]. Under such conditions various alkyl heteroaromatics, e.g. methyl pyrimidine[214] and pyridazine-*N*-oxides[215], have been oximated.

B. Nitrosation at Tertiary Carbons

While compounds containing a nitroso group on a tertiary carbon are relatively stable (usually as dimers), nitrosation of active methine groups carring two electron-attracting groups may lead to C—C bond cleavage. Thus nitrosation of α-substituted β-keto esters results in the formation of α-oximino esters[200].

$$RCOCH(R')COOR'' \longrightarrow R'C({=}NOH)COOR''$$

α-Substituted β-keto lactones behave similarly[216]. The cleavage is catalysed by bases or by acids.

$$RC(=O)\text{—}C(N{=}O)(R')\text{—}COOR'' \xrightarrow{OEt^-} R\text{—}C(OEt)(O^-)\text{—}C(N{=}O)(R')\text{—}COOR'' \longrightarrow RC(=O)OEt + {}^-O\text{—}N{=}C(R')\text{—}COOR''$$

$$RC(=O)\text{—}C(N{=}O)(R')\text{—}COOR'' \xrightarrow{H^+} R\text{—}C(=O)\text{—}C(N{=}\overset{+}{O}H)(R')\text{—}COOR'' \xrightarrow{H_2O} RC(=O)OH + HO\text{—}N{=}C(R')\text{—}COOR''$$

α-Substituted β-keto acids, like unsubstituted ones, undergo decarboxylation rather than deacylation. Decarboxylation also takes place upon nitrosation of alkylmalonic and cyanoacetic esters[200]. α-Substituted β-diketones are converted to α-oximino ketones[217].

Nitrosations of aliphatic carbons carrying diazo groups are accompanied by the loss of nitrogen. Thus diazo ketones react with nitrous acid or with nitrosyl chloride to give the corresponding hydroxamic acid or its chloride, respectively[218]:

$$RCOCHN_2 \xrightarrow{HNO_2} RCOC(=NOH)-OH$$

$$RCOCHN_2 \xrightarrow{NOCl} RCOCCl=NOH$$

C. Nitrosation at Aliphatic Unsaturated Carbon

Few cases have been recorded in which compounds were nitrosated at unsaturated carbons to give oximes. These include nitrosation of hydrazones to give azo oximes (nitrosazones)[219] and the nitrosation of an enamine to an α-imino oxime[220]:

$$ArNHN{=}CHR \xrightarrow[NaOEt]{RONO} ArN{=}NCR{=}NOH$$

$$\text{(benzoxazinone)}{=}CHCOOEt \xrightarrow[CCl_3COOH]{i\text{-}AmONO} \text{(benzoxazinone)}{-}C(=NOH)COOEt$$

D. The Mechanism of the C-Nitrosation

Nitrosation agents may be represented as X—NO (X=OH, OR, NO_2, halogen, etc.) where X functions as a carrier of a potential nitrosonium ion, NO^+ [207,221]. Under basic catalysis, the carbanion formed from the active methylene compound attacks the nitrogen of the nitroso group, and a *C*-nitroso compound is formed:

$$R_2CH^- + X{-}N{=}O \longrightarrow R_2CHNO + X^-$$

The acid-catalysed nitrosation has been envisioned[222] as an electrophilic attack of a nitrosonium ion or of its carrier on the enolic form of the active methylene compound:

$$RCH_2{-}\overset{\overset{\displaystyle O}{\|}}{C}R' \underset{}{\overset{NO^+}{\rightleftharpoons}} RCH{-}\overset{\overset{\displaystyle OH}{|}}{\underset{+}{C}}R' \xrightarrow{-H^+} RCH{-}\overset{\overset{\displaystyle O}{\|}}{C}R'$$
(with NO attached to the CH carbon in the second and third structures: $\underset{\displaystyle NO}{|}$)

This mechanism would require the rate of the nitrosation and the rate of the enolization under equal conditions to be the same. Since this has not been experimentally observed, it was suggested that the nitrosonium electrophile initially attacks the carbonyl oxygen of the keto form[223]:

$$NO^+ + RCOCH_2R' \xrightarrow{slow} R{-}\overset{\overset{\displaystyle O{-}NO}{|}}{\underset{+}{C}}{-}CH_2R' \xrightarrow{fast} R{-}\overset{\overset{\displaystyle O}{\|}}{C}{-}\overset{\overset{\displaystyle NO}{|}}{C}HR' \longrightarrow etc.$$

E. Free-Radical Nitrosations

Non-activated aliphatic saturated hydrocarbons undergo photochemical nitrosation to give oximes in high yields[224]. The reaction is carried out by the irradiation (at 325–600 mμ) of the hydrocarbon in the presence of nitrosyl chloride and HCl or of nitrogen monoxide, chlorine and HCl. NOCl or Cl_2 generate free chlorine radicals, which in turn initiate the formation of hydrocarbon radicals. The latter subsequently react with NO and *C*-nitroso derivatives are formed:

$$NOCl \xrightarrow{h\nu} NO + Cl^{\cdot}$$
$$Cl_2 \xrightarrow{h\nu} 2\,Cl^{\cdot}$$
$$Cl^{\cdot} + RH \longrightarrow R^{\cdot} + HCl$$
$$R^{\cdot} + NO \longrightarrow R{-}NO$$

The HCl formed catalyses the isomerization of the nitrosoalkane to the corresponding oxime. Owing to the lack of specificity, the reaction has been largely limited to cycloalkanes, such as cyclohexane. The photo-nitrosation of the latter to cyclohexanone oxime has found important industrial application in the production of caprolactam, monomer of Nylon 6[225].

Nitrosation of hydrocarbons in the presence of nitrogen monoxide, using high energy γ-rays, has also been achieved[226]. In this case the hydrocarbon free radical is formed directly by the irradiation.

F. Transnitrosations

Active methylene compounds have been oximated without catalysis by aromatic *N*-nitrosamines[186]:

$$\rangle CH_2 + Ph_2NNO \longrightarrow \rangle C{=}NOH + Ph_2NH$$

The effect of various *p*-substituents in the phenyl rings of Ph_2NNO on the yield of the reaction[227] support the assumption that the N—N bond cleavage depends on the electron attraction of the phenyl groups. Aliphatic *N*-nitrosamines, in which the N—N bond is strengthened by electron donation from the alkyl groups, require acid catalysis for the transnitrosation reaction[186].

Under the simultaneous influence of light and an acid, *N*-nitrosamines may undergo intramolecular transnitrosation to yield amidoximes[228,229]:

$$RCH_2N(NO)R' \xrightarrow[H^+]{h\nu} \underset{\underset{NOH}{\|}}{RC}NHR'$$

For example, *N,N*-dibenzylnitrosamine photo-isomerizes to the corresponding amidoxime in 90% yield. A suggested mechanism involves photo-elimination of the species $\overline{\underline{N}}$—OH from the hydrogen-bonded, 1:1 complex of the *N*-nitrosamine and the acid, followed by its addition to the resulting imine. When a methine group is present adjacent to the amino nitrogen, the imine is the final product of the elimination[228]:

$$RR'CHN(NO)R'' \xrightarrow[H^+]{h\nu} RR'C{=}NR''$$

In the presence of an olefin, photo-addition of the *N*-nitrosamine to the carbon–carbon double bond takes place in high yield, the dialkylamino group being attached to the less substituted carbon atom[230]:

$$RCH_2{=}CH_2 + R'R''NNO \xrightarrow[H^+]{h\nu} R\underset{\underset{NOH}{\|}}{C}CHNR'R''$$

By a similar reaction, 10-substituted anthrone oximes were prepared[231].

$$\text{anthracene} + R_2NNO \xrightarrow[H^+]{h\nu} \text{10-(H)(NR}_2\text{)-anthrone oxime (C=NOH)}$$

In unsymmetrically substituted olefins, in which the intermediary tertiary nitroso compound cannot tautomerize to an oxime, the reaction results in a carbon–carbon bond fission[231]:

$$R_2C{=}CR'_2 + R''NNO \xrightarrow[H^+]{h\nu} [R_2C{-}CR'_2 \text{ with } N{=}O \text{ and } \overset{+}{N}HR''] \longrightarrow R_2C{=}NOH + CR'_2{=}\overset{+}{N}R'_2$$

V. DIAZONIUM SALT–METHYLENE CONDENSATIONS

Aliphatic compounds containing an activated methylene group couple with diazonium salts to form aryl hydrazones[232,233]. The reaction is regarded as an electrophilic attack of the diazonium ion on the carbanion derived from the methylene compound, forming an unstable azo compound which spontaneously tautomerizes to a hydrazone:

$$Ar{-}N{\equiv}N^+ + \bar{C}H(X)(Y) \longrightarrow [Ar{-}N{=}N{-}CH(X)(Y)] \longrightarrow Ar{-}NH{-}N{=}C(X)(Y)$$

The reaction is usually carried out in cold aqueous solutions buffered with sodium acetate, but the pH of the medium can be lowered for strongly activated methylene compounds. The activating effect of the substituents (X and Y) for the coupling can be arranged according to Hünig and Boes[234] in the decreasing order:

$$—NO_2 > —CHO > —COCH_3 > —CN > —CO_2Et$$
$$> —CONH_2 > —COOH > —SO_2CH_3 > —SOCH_3 > —C_6H_5$$

Thus, for example, the reactivity of various β-dicarbonyl compounds towards ferrocenyl diazonium salts was found to be[235]:

$$CH_3COCH_2COCH_3 > CH_3COCH_2CO_2Et$$
$$\approx CH_3COCH_2CONHPh > CH_2(CO_2Et)_2$$

Other activating groups are the tertiary sulphonium and quaternary pyridinium ions, and heterocyclic residues such as 2-pyridyl and 2-quinolyl.

When one of the activating groups is a carboxyl, coupling is accompanied by decarboxylation, resembling the 'nitrosative decarboxylation':

$$CH_3COCH_2COOH + ArN_2^+ \longrightarrow \left[CH_3CO{-}CH(N{=}N{-}Ar)\text{—}C(=O)O{-}H \right] \longrightarrow CH_3COCH{=}NNHAr + CO_2$$

Active methyl groups situated at the 2 or 4 position to the ring nitrogen of heterocyclic compounds such as α-picoline[236] and 9-methylacridine[237] can also couple with a diazonium salt to yield a hydrazone. The activity of the methyl group is increased if the heteroatom is first quaternarized to the onium salt.

A methinyl carbon carrying at least two activating groups reacts with diazonium salts to form an unstable azo compound, which is subsequently transformed to an aryl hydrazone by hydrolytic cleavage of one of the electron-attracting groups.

$$RCH(X)(Y) + ArN_2^+ \longrightarrow \left[R{-}C(X)(Y){-}N{=}N{-}Ar \right] \xrightarrow{H_2O} R{-}C(X){=}N{-}NHAr + YOH$$

The reaction is known as the Japp–Klingemann reaction, and has been reviewed[233,238]. Groups that usually undergo splitting are carboxyl, acyl, and alkoxycarbonyl. The scission of the latter probably follows a prior saponification.

Examples:

$$RCOCH(R')COOH + ArN_2^+ \longrightarrow ArNHN{=}C(R')COR + CO_2$$

$$RCOCH(R')CO_2Et + ArN_2^+ \longrightarrow ArNHN{=}C(R')CO_2Et + RCOOH$$

$$EtOCOCH(R)COOEt + ArN_2^+ \longrightarrow ArNHN{=}C(R)COOEt$$

In diacylacetic esters, the acyl group corresponding to the weaker acid is more readily cleaved. For example[239]:

$$CH_3COCH(COPh)COOEt + ArN_2^+ \longrightarrow ArNHN{=}C(COPh)COOEt + CH_3COOH$$

$$EtCOCH(COCH_3)COOEt + ArN_2^+ \longrightarrow ArNHN{=}C(COCH_3)COOEt + EtCOOH$$

The behaviour of various 2-substituted cyclohexanones illustrates the tendencies of various activating groups to undergo splitting relative to that of the ring carbonyl. The reaction can follow two courses, either side-chain splitting or a ring opening. With 2-carboxyl and 2-formyl groups, the side-chain is preferentially cleaved, whereas the corresponding ethoxycarbonyl and acetyl residues are more strongly bound and consequently the cyclohexanone ring is opened[238,240].

$$\text{2-(COOH)cyclohexanone} \xrightarrow{ArN_2^+} \text{2-(=NNHAr)cyclohexanone} + CO_2$$

$$\text{2-(COOEt)cyclohexanone} \xrightarrow{ArN_2^+} EtOCOC({=}NNHAr)CH_2CH_2CH_2CH_2COOH$$

$$\text{2-(CHO)cyclohexanone} \xrightarrow{ArN_2^+} \text{2-(=NNHAr)cyclohexanone} + HCOOH$$

$$\text{2-(COCH}_3\text{)cyclohexanone} \xrightarrow{ArN_2^+} CH_3COC({=}NNHAr)CH_2CH_2CH_2CH_2COOH$$

The reaction of benzenediazonium ions with tribenzoylmethane was formerly thought to proceed via an *O*-azo derivative of the enolic form (**3**)[241]:

$$(PhCO)_3CH + PhN_2^+ \longrightarrow \underset{(\mathbf{3})}{(PhCO)_2C{=}C(Ph)ON{=}NPh} \xrightarrow{\text{heat}} \underset{(\mathbf{4})}{(PhCO)_2C(Ph)N{=}NPh}$$

This suggested a general alternative route to diazo coupling of β-diketo esters and triketones[232,238]. Renewed structural assignments of the reaction products have, however, shown[242] that the *C*-azo compound **4** is formed directly as the primary coupling product. It can subsequently undergo competing rearrangements to the enol-benzoate **5** and the triketone **6**.

$$\mathbf{4} \longrightarrow PhCO{-}C{=}C(OCOPh){-}Ph\ \text{with}\ N{=}NPh\ \text{on}\ C \quad (\mathbf{5})$$

$$\mathbf{4} \longrightarrow PhCO{-}C({=}N{-}N(COPh)Ph){-}COPh \quad (\mathbf{6})$$

The *C*-azo intermediates of the normal Japp–Klingemann reaction have also been isolated in several cases[234–246], by performing the reaction at low temperatures and in weakly acid media. The subsequent hydrolytic cleavage of the *C*-azo derivative to an aryl hydrazone is catalysed by either base or acid. The mechanism is probably analogous to that of the cleavage of *C*-nitroso esters.

$$RC({=}O){-}C(N{=}NAr)(R'){-}COOEt \xrightarrow{OH^-} RC(OH)(O^-){-}C(N{=}NAr)(R'){-}COOEt \longrightarrow RC(=O)OH + ArNHN{=}C(R'){-}COOEt$$

$$RC({=}O){-}C(N{=}\overset{+}{N}HAr)(R')COOEt \xleftarrow{H^+} RC(=O){-}C(N{=}NAr)(R'){-}COOEt; \quad RC(=O){-}C(N{=}\overset{+}{N}HAr)(R')COOEt \longrightarrow RC(=O)OH + ArNHN{=}C(R'){-}COOEt$$

Other nucleophilic reagents (like ethanol, phenol and aniline) also catalyse the cleavage[244].

Primary nitroalkanes couple in their *aci*-nitro form with diazonium salts to give aryl hydrazones of α-nitro aldehydes[232]:

$$RCH_2NO_2 + ArN_2^+ \longrightarrow R{-}C({=}NNHAr){-}NO_2$$

The reaction is usually carried out by adding the nitronate salt to the weakly acid diazonium solution. With nitromethane the main product is a nitro formazane[247].

$$CH_3NO_2 + 2\,ArN_2^+ \longrightarrow ArNHN{=}C(NO_2){-}N{=}NAr$$

Secondary nitroalkanes yield stable azo compounds. In the reaction of aryl dinitromethanes and diaryl nitromethanes with benzenediazonium ion, migration of a nitro group occurs to give a *p*-nitrophenylhydrazone[232]:

$$PhCH(NO_2)_2 + PhN_2^+ \longrightarrow PhC(NO_2){=}NNH{-}C_6H_4{-}NO_2$$

Diazoalkanes couple with diazonium salts to form an unstable azo diazonium ion which rapidly loses nitrogen. In the presence of excess chloride anion, the apparently formed azo carbonium ion yields an azo chloride which tautomerizes to the hydrazone derivative[248]:

$$ArN_2^+ + CH_2N_2 \longrightarrow [Ar{-}N{=}N{-}CH_2{-}\overset{+}{N}{\equiv}N \xrightarrow{-N_2} Ar{-}N{=}N{-}\overset{+}{C}H_2] \xrightarrow{Cl^-} ArN{=}NCH_2Cl \longrightarrow ArNH{-}N{=}CHCl$$

Methanol can also act as a nucleophile to give the methoxy analogue ArNHN=CHOCH$_3$. In the absence of such nucleophiles, the intermediary carbonium ion rearranges to give a cyanamide which may be alkylated by another mole of the diazoalkane.

$$[Ar{-}N{=}N{-}\overset{+}{C}H_2 \rightleftharpoons Ar{-}NH{-}N{=}\overset{+}{C}H \longrightarrow Ar{-}\overset{+}{N}H{-}N{=}CH\ (\text{ring}) \longrightarrow Ar{-}NH{-}CH{=}\overset{+}{N}] \xrightarrow{-H^+} Ar{-}NH{-}C{\equiv}N \xrightarrow{CH_2N_2} Ar{-}N(CH_3){-}C{\equiv}N$$

Aryl hydrazones may also be formed by diazonium coupling with unsaturated carbon atoms in tertiary enamines and various vinyl compounds. These reactions were reviewed[232,233].

VI. ADDITIONS TO CARBON–CARBON DOUBLE OR TRIPLE BONDS

Simple acetylenes require high temperatures and pressures for their reactions with amines. The enamine formed undergoes a tautomeric shift to form an azomethine, e.g.[249]

$$CH{\equiv}CH + EtNH_2 \longrightarrow [CH_2{=}CHNHEt] \longrightarrow CH_3CH{=}NEt$$

In the cases where aldol-type condensations are possible, the yields are only moderate due to the formation of higher amino compounds[249]. In the presence of mercuric chloride, an aldol-type product can be isolated[250]:

$$CH{\equiv}CH + ArNH_2 \xrightarrow{HgCl_2} ArNHCH(CH_3)CH_2CH{=}NAr$$

Acetylenes activated with strong electronegative groups add amines much more readily[251], and the enamine might undergo a prototropic shift to give a C=N bond, e.g.[60]

$$PhC{\equiv}CCOPh + CH_3ONH_2 \longrightarrow PhC(NHOCH_3){=}CHCOPh \rightleftharpoons PhC({=}NOCH_3)CH_2COPh$$

β-Chloroacetylenes add alkali amides with eliminations of an alkali chloride, and unsaturated imines are formed[252].

$$(CH_3)_2CClC{\equiv}CH + (CH_3)_3CNHLi \longrightarrow (CH_3)_2C{=}CHCH{=}NC(CH_3)_3$$

Activated ethylenes add amines in a reaction accompanied by dehydrogenation[253]:

$$(CF_3)_2C{=}CHBu + EtNH_2 \xrightarrow{Et_2O} (CF_3)_2CHCBu{=}NEt$$

In the presence of sodium hydroxide, hydrazine adds similarly to conjugated dienes, and the hydrazo compound formed undergoes dehydrogenation to an azine[254]:

$$2\,CH_2{=}C(CH_3)C(CH_3){=}CH_2 + NH_2NH_2 \xrightarrow{NaNHNH_2} (CH_3)_2C{=}C(CH_3)CH{=}NN{=}CHC(CH_3){=}C(CH_3)_2$$

The reaction of aryl ethylenes with nitroso compounds leads to the cleavage of the double bond and to the formation of nitrones[255,256]. The proposed mechanism involves an addition to the double bond, followed by cleavage[256]:

$$2\,PhCH{=}CH_2 + 2\,PhNO \longrightarrow \underset{\text{(Ph–N(→O)–, N–Ph linked through O)}}{Ph{-}CH{-}CH_2} \longrightarrow Ph{-}CH{=}N(O){-}Ph + \left[CH_2{=}N(O)Ph \longrightarrow PhNHCHO\right]$$

Acetylenes add two molecules of nitroso compounds to form dinitrones[257]:

$$PhC{\equiv}CPh + 2\,PhNO \longrightarrow PhN(O){=}C(Ph)C(Ph){=}N(O)Ph$$

Aromatic nitro compounds undergo a photochemical reaction with acetylenes to form anils, which then might react further[258].

$$PhNO_2 + PhC{\equiv}CPh \xrightarrow{h\nu} [PhNO + Ph_2C{=}C{=}O] \longrightarrow Ph_2C{=}NPh + CO_2 \xrightarrow{Ph_2C{=}C{=}O} \begin{matrix} Ph_2C\text{—}N\text{—}Ph \\ | \quad\ \ | \\ Ph_2C\text{—}C{=}O \end{matrix}$$

The additions of nitrosyl chloride and related nitrosyl compounds to olefins or to acetylenes at low temperatures give nitroso compounds, which, depending on their structure, may dimerize or undergo a prototropic shift to give an oxime[259,260]. At higher temperatures, oxidation and chlorination may take place.

Azides add to double bonds activated by aromatic systems[261], by ether groups (vinyl ethers)[262] or by amino groups (enamines)[263], forming a triazole ring, which loses nitrogen spontaneously or by heating to form an azomethine:

$$\text{indene} + ArN_3 \xrightarrow{\text{room temp.}} \text{triazoline (N-Ar)} \xrightarrow{\Delta} \text{indan}{=}NAr$$

$$\text{dihydropyran} + PhSO_2N_3 \longrightarrow \text{tetrahydropyran}{=}NSO_2Ph$$

$$\text{1-}NR_2\text{-cyclohexene} + PhN_3 \longrightarrow \text{triazoline (}NR_2\text{, N—Ph, N=N)} \xrightarrow{\Delta} \text{cyclopentyl—}C({=}NPh)\text{—}NR_2$$

α-Azidoethylenes undergo on heating or by irradiation an internal cycloaddition with elimination of nitrogen, and azirenes are formed, sometimes together with ketenimines, their rearrangement products, e.g.

$$PhC(N_3){=}CH_2 \xrightarrow{\Delta} \text{Ph-azirine (}H_2\text{, N)} \qquad \text{(ref. 264)}$$

$$CF_3CF{=}CF_2 \xrightarrow{N_3^-} CF_3CF{=}CFN_3 \xrightarrow{\Delta} F_3C\text{—}\overset{F}{C}\text{(azirine, N)—}F \qquad \text{(ref. 265)}$$

$$CH_3C(N_3){=}CRCOOEt \xrightarrow[\text{in PhH}]{h\nu} H_3C\text{-azirine(}R\text{, N)-}CO_2Et + CH_3N{=}C{=}CRCOOEt \qquad \text{(ref. 266)}$$

Nitrogen fluoride adds to double bonds, and in a basic solvent the addition product loses HF to form either an imino nitrile or a diimine:

$$\mathrm{RCH{=}CH_2} \xrightarrow{N_2F_4} \mathrm{RCH(NF_2){-}CHNF_2} \xrightarrow[\text{or pyridine}]{\text{NaF}} \mathrm{RC({=}NF){-}CN} \quad \text{(ref. 267)}$$

$$\mathrm{PhCH{=}CHPh} \xrightarrow{N_2F_4} \mathrm{PhCH(NF_2){-}CH(NF_2)Ph} \xrightarrow{Et_3N} \mathrm{PhC({=}NF){-}C({=}NF)Ph} \quad \text{(ref. 268)}$$

VII. FORMATION OF C=N BONDS THROUGH YLIDS

A. *With Carbonyl Compounds*

Aldehydes and ketones react with iminophosphoranes to give Schiff bases and imines[110,269-271].

$$\mathrm{Ph_3P{=}NR + R'R''CO \longrightarrow R'R''C{=}NR + Ph_3PO}$$
$$\mathrm{Ph_3P{=}NH + RR'CO \longrightarrow RR'C{=}NH + Ph_3PO}$$

Analogously, ketenes give ketenimines[269,271].

The mechanism of the reaction is of the Wittig type, involving the formation of a betaine intermediate[272]:

$$\mathrm{R_3P{=}NR' + R''_2CO \longrightarrow R_3\overset{+}{P}{-}N(R'){-}CR''_2{-}O^- \longrightarrow R_3PO + R''_2C{=}NR'}$$

Instead of iminophosphoranes, phosphoramide anions may be used[271,273], e.g.

$$\mathrm{(EtO)_2P(O)NHR \xrightarrow{NaH} (EtO)_2P(O)\bar{N}R \xrightarrow{R'CHO} RN{=}CHR' + (EtO)_2PO_2^-}$$

Phosphazines similarly form azines[274,275]; the method is especially valuable for the formation of unsymmetrical azines.

Sulphur ylids, such as thionylaniline[147,276,277] or sulphur diimides[278] also react with aldehydes to form azomethines:

$$\mathrm{R\bar{N}{-}\overset{+}{S}{=}O + R'CHO \longrightarrow RN(R'CH{-}O^-){-}\overset{+}{S}{=}O \longrightarrow RN{=}CHR' + SO_2}$$

$$\mathrm{RN{=}S{=}NR + PhCHO \longrightarrow PhCH{=}NR + RNSO \xrightarrow{PhCHO} 2PhCH{=}NR + SO_2}$$

B. *With Diazo Compounds and Azides*

Phosphoranes give azines with diazo compounds[279,280]:

$$\mathrm{PhCOCHN_2 + Ph_3\overset{+}{P}{-}\overset{-}{C}HR \longrightarrow \left[Ph_3\overset{+}{P}{-}CH(R){-}N{=}N{-}\overset{-}{C}HCOPh\right] \longrightarrow RCH{=}NN{=}CHCOPh + Ph_3P}$$

Azides give Schiff bases with phosphonium ylids[281]; the triphenylphosphine which is formed in the reaction might give an iminophosphorane with an excess of the azide.

$$Ph_3\overset{+}{P}—\overset{-}{C}HPh + PhN_3 \xrightarrow{-N_2} Ph_3\overset{+}{P}—CHPh(—\overset{-}{N}Ph) \longrightarrow Ph_3P + PhCH{=}NPh$$

$$(Ph_3P + PhN_3 \xrightarrow{-N_2} Ph_3P{=}NPh)$$

C. With Isocyanates

Isocyanates in the presence of catalytic amounts of triphenylphosphine, -arsine or -stibine oxides react with elimination of CO_2, and yield carbodiimides[271,282,283]. The reaction proceeds via the formation of an iminophosphorane:

$$PhNCO + Ph_3PO \longrightarrow \left[\begin{matrix} Ph\overset{-}{N}—C{=}O \\ Ph_3\overset{+}{P}—O \end{matrix}\right] \longrightarrow PhN{=}PPh_3 + CO_2$$

$$Ph_3P{=}NPh + PhNCO \longrightarrow \left[\begin{matrix} Ph_3\overset{+}{P}—N—Ph \\ {}^{-}O—C{=}NPh \end{matrix}\right] \longrightarrow Ph_3PO + PhN{=}C{=}NPh$$

The relative reactivity of these oxides as catalysts was found to be $Ph_3As > Ph_3P > Ph_3Sb$[283]. Other oxides that catalyse the reaction are 1-ethyl-3-methyl-3-phospholine-1-oxide[282,284], 1-ethoxy-2-phospholine oxide[285], 1-ethyl- or 1-phenyl-3-phospholine oxide, triethylphosphine oxide, and even pyridine oxide and dimethylsulphoxide[283]. No carbodiimides were obtained with catalytic amounts of diphenylsulphone, diphenylsulphoxide, trimethylamine oxide or 4-nitropyridine oxide[283].

According to the proposed mechanism, iminophosphoranes also give carbodiimides with isocyanates[269,271]. The reaction was used to prove the presence of an unstable iminophosphorane by the carbodiimide it forms with an isocyanate[286]:

$$CH_3NH_2{\cdot}HCl + PCl_5 \longrightarrow \begin{matrix} Cl_3P—NCH_3 \\ | \quad\quad | \\ CH_3N—PCl_3 \end{matrix} \longrightarrow [Cl_3P{=}NCH_3] \xrightarrow{CH_3NCO} CH_3N{=}C{=}NCH_3$$

In another variation of the reaction, CO_2 reacts with iminophosphoranes to give isocyanates, which then react with excess of the iminophosphorane to form carbodiimides[269,287].

Phosphoramide anions can also be used with isocyanates to give asymmetrical carbodiimides[271,273]:

$$(EtO)_2P(O)NHR \xrightarrow[2.\ R'NCO]{1.\ NaH} RN{=}C{=}NR'$$

The formation of carbodiimides by these reactions is discussed in a recent review[288].

Isocyanates react similarly with phosphoranes to form ketenimines[269], e.g.

$$Ph_3\overset{+}{P}{-}\overset{-}{C}Ph_2 + PhNCO \longrightarrow Ph_2C{=}C{=}NPh + Ph_3PO$$

However, when using phosphonium monoaryl methylide, a prototropic shift causes the formation of an amide[289]:

$$Ph_3\overset{+}{P}{-}\overset{-}{C}HAr + PhNCO \longrightarrow \left[\begin{matrix} Ph_3\overset{+}{P}{-}CHAr \\ | \\ {}^{-}O{-}C{=}NPh \end{matrix} \longrightarrow \begin{matrix} Ph_3\overset{+}{P}{-}\overset{-}{C}Ar \\ | \\ HO{-}C{=}NPh \end{matrix}\right] \longrightarrow Ph_3\overset{+}{P}{-}\overset{-}{C}ArCONHPh$$

D. With Nitroso Compounds

The products of the reactions of ylids with nitroso compounds depend on the nature of the ylids. Phosphoranes give with nitroso compounds azomethines and a phosphine oxide[290-292]:

$$RR'C{=}PPh_3 + PhNO \longrightarrow RR'C{=}NPh + Ph_3PO$$

Similar results are obtained with phosphazines[292]. On the other hand, sulphoranes[291] and arsonanes[293] both give nitrones with nitrosobenzene:

$$RR'\overset{-}{C}{-}\overset{+}{S}Me_2 + PhNO \longrightarrow RR'C{=}\overset{\overset{O}{\uparrow}}{N}Ph + S(CH_3)_2$$

$$RR'\overset{-}{C}{-}\overset{+}{A}sPh_3 + PhNO \longrightarrow RR'C{=}\overset{\overset{O}{\uparrow}}{N}Ph + Ph_3As$$

Both types of products are obtained through the same class of betaine intermediate; the anil is formed by a Wittig-type oxygen transfer:

$$\begin{matrix} \overset{+}{Z}{-}\overset{-}{C}\lessdot \\ O{=}N{-}Ph \end{matrix} \longrightarrow \left[\begin{matrix} \overset{+}{Z}{-}C\lessdot \\ | \quad | \\ \overset{-}{O}{-}N{-}Ph \end{matrix}\right] \begin{matrix} \longrightarrow \overset{+}{Z}{-}\overset{-}{O} + \rangle{=}N{-}Ph \\ \longrightarrow Z + \rangle{=}\overset{+}{N}(O^-){-}Ph \end{matrix}$$

A mechanistic possibility in the formation of the nitrone is an intermediate containing an oxazirane ring[272,291]:

$$Ph_2\overset{-}{C}{-}\overset{+}{S}Me_2 + PhNO \longrightarrow \left[\begin{matrix} O^- \\ | \\ Ph_2C{-}N{-}Ph \\ | \\ {}^{+}SMe_2 \end{matrix}\right] \longrightarrow \left[Ph_2C\overset{O}{\frown}N{-}Ph\right] \longrightarrow Ph_2C{=}\overset{\overset{O}{\uparrow}}{N}Ph$$

The rearrangement of the oxazirane ring to give a nitrone is well known[193,291]; however, no oxazirane derivative has yet been isolated from reactions of this type.

α-Sulphonyl carbanions behave in a very similar manner towards nitrosobenzene, giving a nitrone[294]:

$$\text{9-(PhSO}_2\text{)fluorene (9-H)} \xrightarrow{\text{NaH}} \text{9-(PhSO}_2\text{)fluorenyl anion} \xrightarrow{\text{PhNO}} \left[\text{spiro-oxazirane, Ph—N}\right] \longrightarrow \text{fluorenylidene}{=}\text{N(Ph)}\rightarrow\text{O}$$

Diazo compounds also behave like ylids towards nitroso compounds, giving a nitrone[166,291]:

$$Ph_2\bar{C}—\overset{+}{N}{\equiv}N + PhNO \longrightarrow \left[\begin{matrix} Ph_2C—\overset{+}{N}{\equiv}N \\ | \\ Ph—N—O^- \end{matrix}\right] \xrightarrow{-N_2} \left[\begin{matrix} Ph_2C \\ | \quad \diagdown O \\ Ph—N \diagup \end{matrix}\right] \longrightarrow \begin{matrix} Ph_2C \\ \| \\ Ph—N\rightarrow O \end{matrix}$$

The reaction of pyridinium salts with nitroso compounds in the presence of a base, the Kröhnke reaction[195] (see Section III), can be regarded as proceeding through the intermediate formation of a pyridinium ylid:

$$C_5H_5\overset{+}{N}—CH_2R\ X^- \xrightarrow{OH^-} \left[C_5H_5\overset{+}{N}—\bar{C}HR\right] \xrightarrow{R'NO} \left[C_5H_5\overset{+}{N}—CHR—N(R')—O^-\right] \longrightarrow R'—\overset{+}{N}(O^-){=}CHR + C_5H_5N$$

In some cases, a pyridinium ylid could be isolated, e.g. 9-fluorenylpyridinium[295]:

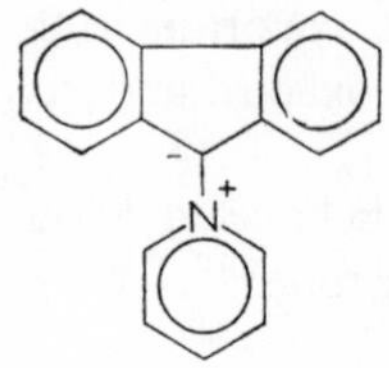

$$\text{(fluorenylide)}^{-}\text{—}\overset{+}{N}C_5H_5$$

E. Other Reactions of Ylids

Iminophosphoranes add to the triple bond of dimethyl acetylene dicarboxylate[296,297]:

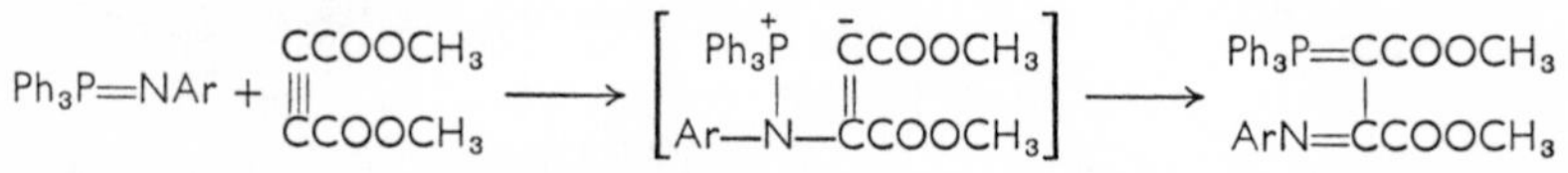

$$Ph_3P{=}NAr + \begin{matrix} CCOOCH_3 \\ ||| \\ CCOOCH_3 \end{matrix} \longrightarrow \left[\begin{matrix} Ph_3\overset{+}{P} & \overset{-}{C}COOCH_3 \\ | & || \\ Ar{-}N{-} & CCOOCH_3 \end{matrix} \right] \longrightarrow \begin{matrix} Ph_3P{=}CCOOCH_3 \\ | \\ ArN{=}CCOOCH_3 \end{matrix}$$

The reaction proceeds readily when Ar is phenyl or *p*-bromophenyl; when Ar contains a strong electron-attracting substituent, the reaction does not take place, probably owing to the reduced nucleophilicity of the phosphoranes.

A similar reaction is the addition of phosphazines to dimethyl acetylenedicarboxylate[297].

Dimethyloxosulphonium methylide reacts with 1,3-dipoles to give a mixture of products containing the C=N bond. The main products are formed by the following scheme, illustrated for benzonitrile oxide[298].

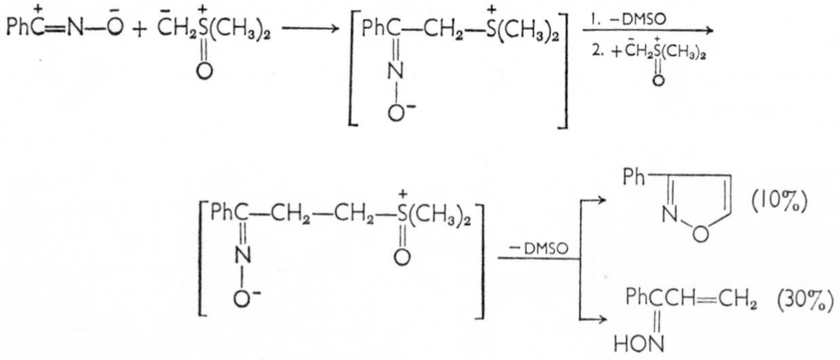

$$Ph\overset{+}{C}{=}N{-}\overset{-}{O} + \overset{-}{C}H_2\overset{+}{\underset{\underset{O}{||}}{S}}(CH_3)_2 \longrightarrow \left[Ph\underset{\underset{\underset{O^-}{|}}{N}}{\underset{||}{C}}{-}CH_2{-}\overset{+}{S}(CH_3)_2 \right] \xrightarrow[2.\ +\overset{-}{C}H_2\overset{+}{\underset{\underset{O}{||}}{S}}(CH_3)_2]{1.\ -DMSO}$$

$$\left[Ph\underset{\underset{\underset{O^-}{|}}{N}}{\underset{||}{C}}{-}CH_2{-}CH_2{-}\overset{+}{\underset{\underset{O}{||}}{S}}(CH_3)_2 \right] \xrightarrow{-DMSO} \begin{cases} \text{3-phenylisoxazoline (Ph, N, O ring)} & (10\%) \\ Ph\underset{\underset{HON}{||}}{C}CH{=}CH_2 & (30\%) \end{cases}$$

Nitrile oxides react with phosphoranes and iminophosphoranes[299] to give the 1,3-cycloaddition products, which undergo elimination or rearrangement on heating to give ketenimines or carbodiimides, respectively.

$$Ar\overset{+}{C}{=}N{-}\overset{-}{O} + \overset{+}{C}H_2{-}\overset{-}{P}Ph_3 \longrightarrow \begin{matrix} Ar{-}C{=}N{-}O \\ | \quad\quad | \\ CH_2{-}PPh_3 \end{matrix} \xrightarrow{\Delta} CH_2{=}C{=}NAr + Ph_3PO$$

$$Ar\overset{+}{C}{=}N{-}\overset{-}{O} + Ph\overset{+}{N}{-}\overset{-}{P}Ph_3 \longrightarrow \begin{matrix} Ar{-}C{=}N{-}O \\ | \quad\quad | \\ PhN{-}PPh_3 \end{matrix} \xrightarrow{\Delta} ArN{=}C{=}NPh + Ph_3PO$$

VIII. TAUTOMERIZATION OF AMIDES AND THIOAMIDES AND RELATED REACTIONS

The formation of amidates and thioamidates from amides and thioamides was reviewed[300].

A. Alkylation and Acylation

Alkylation of amides by alkyl halides in the presence of a base gives usually *N*-alkyl amides. However, when dimethyl sulphate is used as a methylation agent, *O*-methyl derivatives may be obtained from amides[301], anilides[302], ureas[303] or lactams[304]:

$$RCONHR' + (CH_3)_2SO_4 \longrightarrow RC(OCH_3){=}NR'{\cdot}HSO_4CH_3$$

N,N-Dialkylamides are similarly alkylated to immonium salts[305] with dimethyl sulphate or with trimethyloxonium fluoroborate. Alkyl halides, however, may *O*-alkylate hydroxamic esters[306,307]. Amides react with ethyl chloroformate to form, after decarboxylation, an *O*-alkylation product[308]:

$$RCONH_2 + ClCOOEt \longrightarrow \left[\begin{matrix} NH{\cdot}HCl \\ \| \\ RCOCOOEt \end{matrix}\right] \xrightarrow{-CO_2} \begin{matrix} NH{\cdot}HCl \\ \| \\ RCOEt \end{matrix}$$

N-Substituted amides of γ-halo acids undergo intramolecular alkylations[309]. In acidic, neutral or mildly basic media (pH 2–10), *O*-alkylation to a tetrahydrofuran derivative takes place, whereas in strongly basic media, pyrrolidone derivatives are formed by *N*-alkylation, together with the product of the *O*-alkylation, in a ratio of nine to one[310]:

$$\text{(N-phenylpyrrolidone)} \xleftarrow{pH > 10} BrCH_2CH_2CONHPh \xrightarrow{pH\ 2\text{–}10} \text{(2-phenyliminotetrahydrofuran)}$$

Kinetic data for the above reaction[310] in water suggest an S_Ni mechanism:

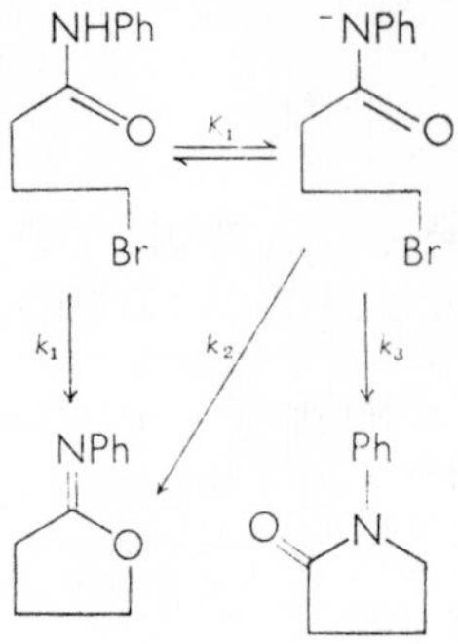

The observed rate constant for the disappearance of the amide, k_0, was found to obey the following relationship:

$$k_0 = k_1 + [(k_2 + k_3)K_1/K_w][OH^-]$$

where K_w was the ion product of water, taken as 10^{14}.

A related reaction is the intramolecular O-alkylation of N-(bromoethyl)amides, catalysed by methoxide ions[311]:

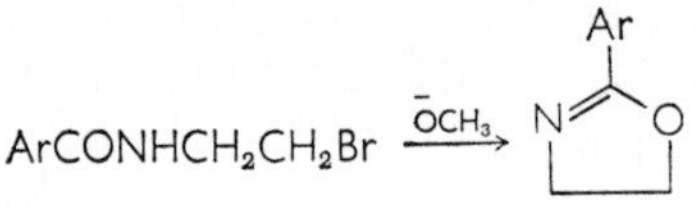

$$ArCONHCH_2CH_2Br \xrightarrow{\bar{O}CH_3}$$

Electron-withdrawing groups on the aryl group facilitate the reaction. Again an S_Ni mechanism was proposed:

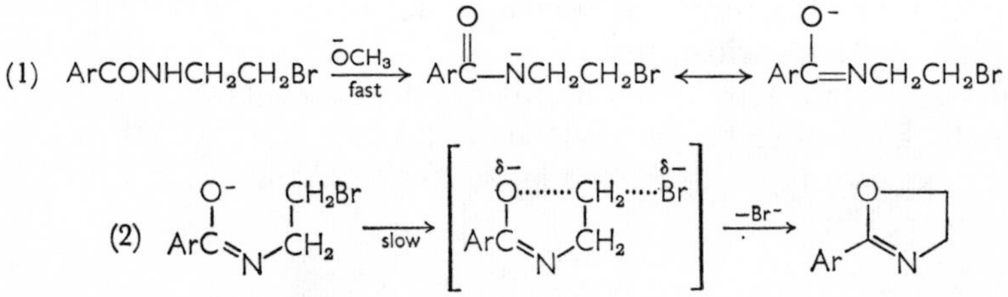

(1) $ArCONHCH_2CH_2Br \xrightarrow[\text{fast}]{\bar{O}CH_3} ArC(=O)\bar{N}CH_2CH_2Br \longleftrightarrow ArC(O^-){=}NCH_2CH_2Br$

(2) $ArC(O^-){=}N{-}CH_2{-}CH_2Br \xrightarrow{\text{slow}} [\,ArC(\overset{\delta-}{O}\cdots CH_2\cdots \overset{\delta-}{Br}){=}N{-}CH_2\,] \xrightarrow{-Br^-}$

Phosphonate groups may also serve as leaving groups in the reaction, e.g. in the cyclization of $ArCONHCH_2CH_2OPO(OR)_2$[312]. N-(β-hydroxyethyl)amides give similar products in the presence of aluminium oxide at 300–500°, while *N*-(γ-hydroxypropyl)amides give under the same conditions six-membered heterocyclic compounds[313]:

$$RCONHCH_2CH_2CH_2OH \xrightarrow[300-500^\circ]{Al_2O_3} \text{(6-membered ring: O–C(R)=N–CH}_2\text{CH}_2\text{CH}_2\text{–)}$$

Thioamides are much more susceptible to *S*-alkylations than amides to *O*-alkylations. Alkyl chlorides and bromides give *S*-alkyl products with thioamides[314,315], *N*-alkyl, *N*-aryl and *N,N*-dialkylthioamides[315,316], thioureas[317], thiohydroxamates[318], and thiourethanes[319], without a catalyst. Thioureas with ethyl chloroformate, in the presence of triethylamine, yield *S*-carbethoxyisothioureas[320], which are less susceptible towards decarboxylation than their *O*-analogues.

S-Arylation of thioureas takes place through their reaction with diazonium salts in mildly basic solution[321].

$$PhNHCSNHPh + ArN_2^+ \xrightarrow{NaOAc} PhN{=}C(SAr)NHPh$$

Acylation of amides with acyl chlorides in the presence of a strong base gives both *N*- and *O*-acyl derivatives[322].

Another method of *O*-alkylation of amides is the reaction of their silver salts with alkyl halides. Amides and anilides give imidoates by *O*-alkylation, whereas *N*-methylamides show *N*-alkylation[323]. γ-Halopropionamides give an internal *O*-alkylation with silver fluoroborate[324]. *O*-Acylation might also be carried out by the silver salt method[325].

Diazomethane gives *O*-methylation with certain amido groups, e.g. with hydroxamic acids[326] or other amides[327]:

$$RCONHOH \xrightarrow{CH_2N_2} RC(OCH_3){=}NOCH_3$$

B. Cyclization of Amides

N-Acylamino acids, under the influence of acetic anhydride or acetyl chloride, undergo cyclization with the elimination of water to form azolactones (oxazolones)[328]:

$$\begin{array}{l} RCH{-}COOH \\ \;| \\ NHCOR' \end{array} \xrightarrow{Ac_2O} \text{(oxazolone ring: R–CH–C(=O)–O–C(R')=N–)}$$

Oxazole derivatives are formed also from keto amides when the product is stabilized by aromatization[329].

$$\text{(2-benzamidotropone)} + Me_2SO_4 \longrightarrow \text{(2-phenylcyclohepta[d]oxazolium)} \quad CH_3OSO_3^-$$

Monoamides of those dicarboxylic acids which are capable of forming a cyclic anhydride give with carbodiimides isoimides of the corresponding anhydride[330,331]:

$$HOOC\text{–}CH{=}CH\text{–}CONHPh \xrightarrow{N,N'\text{-dicyclohexylcarbodiimide}} \text{(5-phenylimino-2(5H)-furanone, =NPh)}$$

N-(Hydroxymethyl)amides add to ethylenes to form an oxazine derivative[332]:

$$RCONHCH_2OH + R'R''C{=}CH_2 \longrightarrow \text{(6-R',6-R''-2-R-5,6-dihydro-4H-1,3-oxazine)}$$

N-(β-Arylethyl)amides, under the influence of acidic catalysts (P_2O_5, $POCl_3$, PCl_5, $ZnCl_2$, $AlCl_3$, $SOCl_2$), undergo cyclization to form a dihydroisoquinoline (Bischler–Napieralsky reaction)[333]:

$$PhCH_2CH_2NHCOR \xrightarrow{-H_2O} \text{(1-R-3,4-dihydroisoquinoline)}$$

Oximes which are capable of forming by Beckmann rearrangement, a suitable amide, also undergo a Bischler–Napieralsky cyclization (i.e. $PhCH_2CH_2CR{=}NOH$. *N*-arylsulphonyloximes of the same type (e.g. $PhCH_2CH_2CR{=}NOSO_2Ph$) give the same product on heating only, without a catalyst.

C. Fixation of Enolic Forms in Thioamides

Heavy-metal ions form complexes with the enolic forms of thioamides[334] or hydroxamic acids[335]. The chelating complexation stabilizes the enolic form:

$$2\ \text{(2-pyridyl)}\text{—CSNHR} + Cu^{2+} \longrightarrow \text{Cu chelate (S—Cu—S, N}\rightarrow\text{Cu, C=NR)} + 2\,H^{+}$$

Oxidation of thioamides which affect the sulphur atom may give either a disulphide[336]:

$$\text{PhCSNHR} \xrightarrow[\text{or } I_2,\ H^+]{K_3Fe(CN)_6,\ OH^-} \text{PhC(=NR)—S—S—C(=NR)Ph}$$

or a hydroperoxide of the enolic form[337]:

$$\text{ArNHCSOAr}' \xrightarrow{H_2O_2} \text{ArN=C(SOH)OAr}'$$

D. Substitution of the Enolic Hydroxyl Group

Anilides react with PCl_5 to form imidoyl chlorides in a reaction that is the first step in the Sonn and Müller aldehyde synthesis[338,339].

$$\text{RCONHPh} \xrightarrow{PCl_5} \text{RCCl=NPh}$$

When R is aliphatic, the chloride is unstable, and decomposes spontaneously[340]. Benzohydroxamates[341] and hydrazides[342] give analogously imino chlorides:

$$\text{PhCONHOEt} \xrightarrow{PCl_5} \text{PhCCl=NOEt}$$

$$\text{PhCONHNHPh} \xrightarrow{PCl_5} \text{PhCCl=NNHPh}$$

N,*N*-Disubstituted amides give with $POCl_3$[302,343] or with phosgene[344] an immonium salt in the first step of the Vilsmeyer–Haack reaction:

$$HCON(CH_3)_2 \xrightarrow{POCl_3} ClCH{=}\overset{+}{N}(CH_3)_2Cl^-$$

$$\text{(piperidino)NCOEt} \xrightarrow{COCl_2} \text{(piperidinium)N}^+\text{=CClEt}\quad Cl^-$$

Grignard reagents react with anilides to give anils[345].

$$\mathrm{RMgBr + R'CONHPh \longrightarrow RR'C{=}NPh}$$

IX. ADDITION REACTIONS TO NITRILES, ISONITRILES, NITRILE OXIDES AND RELATED COMPOUNDS

A. Addition to Nitriles

I. Reduction

Controlled catalytic hydrogenation of nitriles to imines is usually very difficult; only low yields of imines are obtained when it is attempted to stop the reaction at the imine stage, due to secondary reactions[346]. Formamidines are obtained by the catalytic or electrolytic reaction of cyanamides[347].

Reduction by lithium aluminium hydride is much more controllable, and in some cases, imines can be obtained in fair yields[348], although usually a mixture of a primary amine and an azomethine is obtained[349].

$$\mathrm{PhCN \xrightarrow{LAH} PhCH_2NH_2 + \underset{59\%}{PhCH{=}NCH_2Ph} + NH_3}$$

Nitriles are reduced by $SnCl_2$ and HCl (the first step of the Stephen aldehyde synthesis) to give the tin chloride complex of the imine[339,350]. The reaction probably proceeds through addition of HCl to the carbon–nitrogen triple bond, followed by reduction of the iminoyl chloride:

$$\mathrm{RC{\equiv}N \xrightarrow{HCl} RCCl{=}NH \xrightarrow{SnCl_2} (RCH{=}\overset{+}{N}H_2)\,SnCl_6^{2-}}$$

This complex is either hydrolysed to an aldehyde, or is converted to the free imine by the addition of an excess of anhydrous ammonia or triethylamine[351].

Hydrogenation of nitriles with Raney nickel in hydrazine gives the azine of the corresponding aldehyde[352]. The reaction probably proceeds through an addition of the hydrazine to the nitrile to give an imino hydrazide, which is then reduced to the hydrazone from which the azine is formed:

$$\mathrm{RCN + NH_2NH_2 \longrightarrow R\overset{\overset{NH}{\|}}{C}NHNH_2 \xrightarrow[-NH_3]{[H]} RCH{=}NNH_2 \longrightarrow RCH{=}NN{=}CHR}$$

Similarly, nitriles are reduced in the presence of semicarbazide to the semicarbazones of the corresponding aldehydes[169].

2. Addition of alcohols and thiols

The preparation of imidoates from alcohols or phenols and nitriles (Pinner reaction) was reviewed[300]. The reaction is usually carried out in dry ether or dioxan in the presence of dry HCl, and the imine hydrochloride is separated[300,353]. Instead of HCl as a catalyst for the addition, an oxonium salt may be used[354].

$$RCN \xrightarrow{R'_3O^+BF_4^-} RC{\equiv}\overset{+}{N}R'\ BF_4^- \xrightarrow{R''OH} R\overset{OR''}{C}{=}\overset{+}{N}HR'\ BF_4^-$$

Vicinal diols add to cyanogen chloride to form cyclic iminocarbonates[355]:

$$\begin{matrix} >C{-}OH \\ | \\ >C{-}OH \end{matrix} + ClCN \longrightarrow \begin{matrix} -C{-}O \\ | \quad\quad \backslash \\ | \quad\quad C{=}NH{\cdot}HCl \\ | \quad\quad / \\ -C{-}O \end{matrix}$$

Phenols give iminocarbonates $(ArO)_2C{=}NH{\cdot}HCl$ with BrCN or ClCN[356].

γ-Hydroxy nitriles undergo under the influence of HCl an internal addition to form iminolactones[357]:

CN CHOH(CH₃) steroid, HO — HCl → HCl·HN=C–O–CH(CH₃) iminolactone

Nitriles react with enolizable ketones to give the addition product of the aldol[358]:

$$RCN + \text{cyclohexanone} \xrightarrow{AlCl_3} \text{2-(1-(OCR{=}NH)cyclohexyl)cyclohexanone}$$

The addition of alcohols to nitriles can also take place under basic (ethoxide) catalysis[359]. Imidoate formation is promoted by the influence of electron-attracting groups on the nitrile. Cyanogen can add either one or two molecules of an alcohol in basic media[360,361]:

$$ROH + (CN)_2 \longrightarrow NCC(OR){=}NH \xrightarrow{ROH} HN{=}C(OR)C(OR){=}NH$$

Thiols also add to nitriles or to cyanogen, under acidic or basic conditions[360,362,363], forming thioimidoates.

3. Addition of amino compounds

The addition of amines to nitriles to give amidines requires acidic conditions, e.g. $AlCl_3$[364]. With cyanogen, diamidines are formed[365],

$$RR'NH + (CN)_2 \longrightarrow RR'N\overset{\overset{NH}{\|}}{C}-\underset{\underset{NH}{\|}}{C}NRR'$$

whereas with cyanamides, guanidines are formed[366]. Dicyanamides give biguanides with nitriles[367]:

$$ArNH_2 + NaN(CN)_2 \xrightarrow{HCl} ArNH\underset{\underset{NH}{\|}}{C}NH\underset{\underset{NH}{\|}}{C}NHAr$$

The addition of α-amino acids[368] or of *o*-aminobenzoic acids[369] to BrCN is accompanied by HBr elimination, and oxazoles or oxazines respectively are formed. The reaction with α-amino acids was used in peptide synthesis[368]:

$$RCH(NH_2)COOH + BrCN \longrightarrow$$

$$\begin{array}{l}RCH\text{—}CO \\ \;|\quad\quad\;\; | \\ HN\diagdown\!\!\diagup O \\ \quad\;\; \| \\ \quad\; NH\end{array} \xrightarrow{R'CH(NH_2)COOH} NH_2CONHCHRCONHCHR'COOH$$

The amino groups of thioamides[370] or sulphonamides[371] also add to cyano groups under strongly acidic conditions:

$$PhCSNH_2 + CH_3CN \xrightarrow{HCl} CH_3C(NH_2){=}NCSPh\cdot HCl$$

$$NH_2SO_2NH_2 + CH_3CN \xrightarrow{HCl} NH_2SO_2NHC(CH_3){=}NH\cdot HCl$$

The reaction of *o*-dicyanobenzene with alcoholic HCl to give a cyclic derivative[372] proceeds probably through hydrolysis of one cyano group to an amido group which then in turn adds to the other cyano group:

$$C_6H_4(CN)_2 \xrightarrow{EtOH,\ HCl} C_6H_4(C{=}NH)(NH)(C{=}O)$$

Hydrazines[169,373] and hydroxylamine[169,306,362,374,375] add to nitriles to give iminohydrazides and iminohydroxamic acids, respectively:

$$RNHNH_2 + R'CN \longrightarrow RNHNHCR'{=}NH \text{ or } RNHN{=}CR'NH_2$$

$$NH_2OH + RCN \longrightarrow HN{=}CRNHOH \text{ or } RC(NH_2){=}NOH$$

Excess hydrazine may displace the imino group[376]. Hydrazides of cyanoacetic acid undergo an internal addition[377], e.g.

$$PhN(COCH_2CN){-}NHPh \xrightarrow{Na_2CO_3} \text{PhN—NPh (ring with C=O and C=NH)}$$

The same product may be obtained by heating the hydrazide of chloroacetic acid with KCN.

β-Dialkylamino cyanides undergo an internal addition reaction on heating with HCl, with the subsequent elimination of an alkyl chloride[378]:

$$Ph_2C(CN)CH_2CH_2NEt_2 \xrightarrow{HCl} \left[Ph_2C(CH_2CH_2NEt_2){-}C(Cl){=}NH\cdot HCl\right] \longrightarrow \left[\text{Ph, Ph-pyrrolidine, C=NH, }\overset{+}{N}Et_2\right]Cl^- \longrightarrow \text{Ph, Ph-pyrrolidine (C=NH, NEt)} + EtCl$$

4. Aldol-type additions of aliphatic nitriles

Under the influence of strong bases (e.g. sodamide), acetonitrile forms an aldol-type dimer[379,380].

$$CH_3CN \xrightarrow{NaNH_2} \bar{C}H_2CN \xrightarrow{CH_3CN} CH_3C({=}NH)CH_2CN$$

Suitable dinitriles undergo a similar internal condensation with ethoxides[380] or sodamide derivatives[381] as catalysts:

$$(CH_2)_2(CH_2CN)_2 \xrightarrow{\bar{O}Et} \text{2-iminocyclopentanecarbonitrile (ring with C=NH and CN)}$$

Certain trinitriles form analogously bicyclic compounds[382]:

$$\text{CN} \quad \text{CN} \quad \text{CN} \xrightarrow{NaNH_2} \text{HN} \quad \text{N(H)} \quad \text{N} \quad \text{NH}$$

5. Addition of Grignard reagents

One of the best methods to prepare imines is by the addition of a Grignard reagent to a nitrile[5,383]. In the regular work-up by aqueous reagents, a ketone might be formed. Hence, in order to obtain the imine, the decomposition is carried out with dry HCl or with anhydrous ammonia[384], or preferably with absolute methanol[385]. Sterically hindered ketimines such as phenyl 2,2,6-trimethylcyclohexyl ketimine[386], phenyl *t*-butyl ketimine[387] or dimesityl ketimine[136] are more stable to hydrolysis, so they may be obtained in aqueous media.

The kinetics of the addition reaction were studied[388] and a mechanism was proposed in which diarylmagnesium is the reactive species:

$$Ar_2Mg + PhCN \longrightarrow \left[\begin{array}{c} X\text{—}C_6H_4^{\delta-} \cdots\cdots \overset{\delta+}{C}\text{—Ph} \\ \qquad\qquad \vdots\!\!\parallel \\ Ar\text{—}Mg^{\delta+} \cdots\cdots N^{\delta-} \end{array} \right] \longrightarrow Ar_2C{=}NMgPh$$

Since the rate is strongly dependent on the $MgBr_2/Ar_2Mg$ ratio, a transient complex was proposed, i.e.

$$Ar_2Mg{\cdot}MgBr_2 + PhCN \longrightarrow \left[\begin{array}{c} X\text{—}C_6H_4^{\delta-} \cdots\cdots \overset{\delta+}{C}\text{—Ph} \\ \qquad\qquad \parallel \\ Ar\text{—}Mg \cdots\cdots N \\ \vdots \qquad\qquad \vdots \\ Br\text{———}Mg\text{—}Br \end{array} \right] \longrightarrow Ar_2C{=}NMgPh{\cdot}MgBr_2$$

A ρ value of $-2{\cdot}85$ for the $\sigma\rho$ relationship suggests the carbanionic character of the aryl group in the aryl magnesium bromide.

6. Addition of aromatic compounds

Aromatic compounds which are reactive towards electrophilic reagents (e.g. phenols or phenol ethers) add to nitriles, usually in the presence of a Lewis-acid catalyst (Hoesch synthesis). The reaction was reviewed[389]. When phenols are used, imino ethers are formed as by-products as a result of a Pinner type reaction. α,β-Unsaturated nitriles add the aromatic compounds at their carbon–carbon double bond rather than at the nitrile group.

Certain heterocyclic compounds, e.g. pyrroles, can also be condensed with aliphatic[390,391] and aromatic[391] nitriles. Usually, the imines are not isolated but are hydrolysed directly to give the corresponding ketones.

The Gattermann aldehyde synthesis[392] also involves an addition to a nitrile; however, the unstable aldimine thus formed is not isolated.

B. *Addition to Isonitriles*

Due to their divalent character, isonitriles undergo insertion reactions rather than addition. Isonitriles insert into the O—H bond of unsaturated alcohols under the catalytic influence of cuprous chloride[393] or of saturated alcohols with copper(I or II) oxide[394] to give an imino ether in almost quantitative yields:

$$\text{RNC} + \text{R'OH} \xrightarrow{\text{CuO}} \text{RN{=}CHOR'}$$

Analogously, isonitriles give amidines with amines[395]:

$$\text{RNC} + \text{R'R''NH} \xrightarrow[\text{or CuCN}]{\text{CuCl}} \text{RN{=}CHNR'R''}$$

Other reported insertions of isonitriles are the following: with hydroxylamine to give amidoximes[396]; with halogens to give iminophosgenes[397]; with dry hydrohalogenic acid or acyl halides to give imidoyl chloride derivatives[398] and with sulphenyl chlorides to give imidoyl chloride thio ethers[399]:

$$\text{PhNC} + \text{NH}_2\text{OH} \longrightarrow [\text{PhN{=}CHNHOH}] \longrightarrow \text{PhNHCH{=}NOH}$$

$$\text{RNC} + \text{Br}_2 \longrightarrow \text{RN{=}CBr}_2$$

$$\text{RNC} + \text{HX} \longrightarrow \text{RN{=}CHX}$$

$$\text{PhNC} + \text{CH}_3\text{COCl} \longrightarrow \text{PhN{=}CClCOCH}_3$$

$$\text{RNC} + \text{R'SCl} \longrightarrow \text{RN{=}C(Cl)SR'}$$

The reaction of isonitriles with Grignard reagents constitutes a special case of the insertion reaction[400]. Cyclohexyl isocyanide and phenylmagnesium bromide give an iminomagnesium compound, which yields on hydrolysis mostly a variety of dimerization products in proportions which depend on the relative amounts of the reactants and on the temperature:

$$\mathrm{RNC} + \mathrm{PhMgBr} \longrightarrow \left[\mathrm{RN{=}C(Ph)(MgX)} \longrightarrow \mathrm{RN(MgBr)C(Ph){=}C(Ph)N(MgBr)R} \right] \xrightarrow{H_2O}$$

$$\begin{matrix}\mathrm{PhC{=}NR} \\ | \\ \mathrm{PhC{=}NR}\end{matrix} + \begin{matrix}\mathrm{PhCHNHR} \\ | \\ \mathrm{PhC{=}NR}\end{matrix} + \begin{matrix}\mathrm{PhCHNHR} \\ | \\ \mathrm{PhCHNHR}\end{matrix} + \begin{matrix}\mathrm{PhCH_2} \\ | \\ \mathrm{PhCH_2}\end{matrix}$$

Isocyanides also insert into immonium salts[401]:

$$\mathrm{R_2\overset{+}{N}{=}CR'_2Y^-} + \mathrm{R''NC} \longrightarrow \mathrm{R_2NCR'_2C(Y){=}NR''}$$

Some nitroso compounds yield with isonitriles a cyclic, 1,2-oxazetidine derivative, which on heating *in vacuo* decomposes into a isocyanate and a carbodiimide[402].

$$\mathrm{CF_3NO} + 2\,\mathrm{CH_3NC} \longrightarrow \begin{matrix}\mathrm{CF_3N{-}C{=}NCH_3} \\ |\quad\;\; | \\ \mathrm{O{-}C{=}NCH_3}\end{matrix} \xrightarrow[\text{vacuum}]{400^\circ} \begin{matrix}\mathrm{CF_3N{=}C{=}NCF_3} \\ +\,\mathrm{CH_3NCO}\end{matrix}$$

C. Additions to Nitrile Oxides, Fulminates, Isocyanates and Related Compounds

Nitrile oxides form in some of their 1,3-dipolar additions C=N bonds. In most cases the nitrile oxides are obtained *in situ*, and are not isolated. For example, with an olefinic compound cycloaddition to an oxazole derivative takes place[403]:

$$\mathrm{RCNO} + \mathrm{{>}C{=}C{<}} \longrightarrow \text{(3-R-isoxazoline ring: R–C=N–O)}$$

With HCl, a chloro oxime is formed[404]; with amines, amidoximes[405] and with a Grignard reagent a ketoxime[406].

Nitrile oxides give with phosphoranes a betaine intermediate, which, depending on the structure of the reactants, can yield on heating either an azirene or a ketenimine[407]:

$$\mathrm{PhNCO} + \mathrm{(CH_3)_2C{=}PPh_3} \longrightarrow \text{Ph–(azirine ring, N)–}(\mathrm{CH_3})_2$$

$$\mathrm{EtOCONCO} + \mathrm{CH_3C(COOEt){=}PPh_3} \longrightarrow \text{EtOCO–(azirine ring, N)–}(\mathrm{CH_3})(\mathrm{COOEt})$$

$$\mathrm{PhNCO} + \mathrm{CH_3C(COOEt){=}PPh_3} \longrightarrow \mathrm{CH_3C(COOEt){=}C{=}NPh}$$

Fulminates also add various reactants with the formation of C=N bonds. Thus, mercury fulminate undergoes Friedel–Crafts type reactions with aromatic compounds[408]:

$$ArH + Hg(ONC)_2 \xrightarrow{AlCl_3} ArCH{=}NOH$$

Fulminic acid itself trimerizes to the so-called 'metafulminic acid' which is a dioxime of isoxazolinedione[409].

$$3\ HCNO \longrightarrow \text{(isoxazoline ring: HON=C5, C4=NOH, C3=N, O1)}$$

The free acid also adds chlorine[410] and HCl[411] to give dichloroformaldoxime and formohydroxamoyl chloride, respectively:

$$HCNO + Cl_2 \longrightarrow Cl_2C{=}NOH$$

$$HCNO + HCl \longrightarrow ClCH{=}NOH$$

Nitrile ylids also undergo 1,3-cycloadditions[412]:

$$PhC{\equiv}\overset{+}{N}{-}\overset{-}{N}Ph \begin{cases} + RCH{=}CH_2 \longrightarrow \text{(pyrazoline: R, PhN, N=C–Ph)} \\ + PhCHO \longrightarrow \text{(oxadiazoline: Ph, O, Ph–N, N=C–Ph)} \\ + PhCH{=}NPh \longrightarrow \text{(triazoline: Ph, N–Ph, Ph–N, N=C–Ph)} \end{cases}$$

or non-cyclic additions to amines, alcohols, thiols, phenols or carboxylic acids[412]:

$$PhC{\equiv}\overset{+}{N}{-}\overset{-}{N}Ph + XH \longrightarrow PhC(X){=}NNHPh$$

when X is RO, RS, RNH, or RCOO.

Isocyanates react with carbonyl compounds[413], including amidic carbonyl groups[414], with elimination of CO_2 and the formation of azomethines and amidines respectively. The reaction probably proceeds through a 1,2-cycloaddition to form a four-membered ring as an intermediate:

$$PhNCO + ArCHO \longrightarrow \left[\begin{array}{l} PhN\text{—}C{=}O \\ \;\;|\qquad\;| \\ ArCH\text{—}O \end{array}\right] \longrightarrow PhN{=}CHAr + CO_2$$

$$CH_3C_6H_4SO_2NCO + HCON(CH_3)_2 \longrightarrow CH_3C_6H_4SO_2N{=}CHN(CH_3)_2 + CO_2$$

Alkyl chlorides with a carbonyl or an aryl group in the α- or β-position undergo cycloaddition with thiocyanates; the reaction proceeds through a nitrile salt intermediate[415].

$$PhCH_2CH_2Cl + RSCN \xrightarrow{SnCl_4} [PhCH_2CH_2\text{—}\overset{+}{N}{\equiv}C\text{—}SR] \longrightarrow \text{1-SR-3,4-dihydroisoquinoline}$$

$$PhCHClCOPh + RSCN \xrightarrow{SnCl_4} [PhCH(\overset{+}{N}{\equiv}C\text{—}SR)\text{—}C(Ph){=}O] \longrightarrow \text{4,5-diphenyl-2-SR-oxazole}$$

$$(CH_3)_2CClCH_2COCH_3 \xrightarrow{SnCl_4} \text{4,4,6-trimethyl-2-SR-4H-1,3-oxazine}$$

Organic cyanates add in basic media various compounds having a labile hydrogen, such as alcohols, phenols, thiols, amines, active methylene compounds, etc., e.g.[356,416]

$$ROCN + R'OH \longrightarrow ROC({=}NH)OR'$$

$$ROCN + R'SH \longrightarrow ROC({=}NH)SR'$$

$$ROCN + R'NH_2 \longrightarrow ROC({=}NH)\text{—}NHR' \xrightarrow{ROCN} ROC({=}NH)\text{—}N(R')C({=}NH)OR$$

$$ROCN + CH_2XY \longrightarrow ROC({=}NH)\text{—}CHXY \rightleftharpoons ROC(NH_2){=}CXY$$

X. OXIDATION OF AND ELIMINATION FROM NITROGEN COMPOUNDS

A. Dehydrogenation and Oxidation of Amines

Compounds containing C═N bonds are only seldom prepared by oxidation or dehydrogenation of amines. Even when C═N bonds are formed initially, and further oxidation is avoided, the imines formed are usually hydrolysed to carbonyl compounds or converted into secondary reaction products.

Primary amines of the type RCH_2NH_2 undergo dehydrogenations on a copper chromite–nickel catalyst in the presence of K_3PO_4, and a mixture of products is formed, among them azomethines[417]. The following reaction scheme accounts for some of the products formed:

$$RCH_2CHO \xrightarrow{-H_2} RCH_2CH_2OH$$

$$\uparrow H_2O$$

$$RCH_2CH_2NH_2 \xrightarrow{-H_2} RCH_2CH{=}NH \xrightarrow[-NH_3]{RCH_2CH_2NH_2} RCH_2CH_2NHCH_2CH_2R \xrightarrow{-H_2}$$

$$RCH_2CH{=}NCH_2CH_2R \xrightarrow{RCH_2CH{=}NH} \underset{\displaystyle \overset{\|}{CHCH_2R}}{RCCH{=}NCH_2CH_2R} \xrightarrow{H_2}$$

$$\underset{\displaystyle \overset{|}{CH_2CH_2R}}{RCHCH{=}NCH_2CH_2R}, \text{ etc.}$$

Catalytic dehydrogenation of secondary amines containing an α-hydrogen with Ni, Pt or Cr catalysts yields azomethines[418]:

$$R_2CHNHR' \xrightarrow{-H_2} R_2C{=}NR'$$

The dehydrogenation can also be carried out by sulphur[419], by amyl disulphide[420], by selenium[421] or by sodamide in liquid ammonia[422].

Dehydrogenation of a secondary amine can also be effected by an organic hydrogen acceptor in a reaction involving hydride transfer. For example, hexamethylenetetramine gives an azomethine by the reaction with a secondary amine[423]:

$$(ArCH_2)_2NH + C_6H_{12}N_4 \longrightarrow ArCH{=}NCH_2Ar$$

Among the compounds which may be used as hydrogen abstractors are formamide, formanilide, *N,N'*-diphenylformanilide, formic acid and derivatives of diaminomethane[424]. One of the steps in the Sommelet reaction is also regarded as a dehydrogenation through a hydride transfer mechanism[425]:

$$ArCH_2NH_2 + [CH_2{=}NH] \longrightarrow ArCH{=}NH + CH_3NH_2$$

An interesting case of a dehydrogenation is the reaction of a lithium derivative of a secondary amine with *o*-bromoanisole[426]:

$$R_2CHNR'Li + o\text{-}BrC_6H_4OCH_3 \longrightarrow \left[o\text{-}CH_3OC_6H_4\text{–}Br\cdots Li\text{–}NR'\text{–}CR_2\text{–}H \right] \longrightarrow C_6H_5OCH_3 + LiBr + R_2C{=}NR'$$

Aliphatic amines are oxidized by permanganate, usually in acetone solution. Primary carbinylamines as a rule give aldehydes, and the imines are not isolated[427]. Secondary carbinylamines R_2CHNH_2 give an imine $R_2C{=}NH$[428], while in neutral aqueous acetone or *t*-butanol either a Schiff base or an azine is obtained, depending on the amount of the oxidant[429]. Cyclohexylamine is oxidized by potassium permanganate in the presence of an excess of formaldehyde[430] or acetaldehyde[431] to nitrosocyclohexane or to cyclohexanone oxime. Substituted benzyl amines $ArCH_2NH_2$ give with a neutral permanganate solution a number of products, among them azomethines of the type ArCH=NCH(Ar)NHCOAr; the intermediate is suggested to be an unstable imine, ArCH=NH[432].

Secondary amines containing α-hydrogens (R_2CHNR') give by oxidation with permanganate[433,434] or with MnO_2[435] stable azomethines.

Oxidation of amines with hydrogen peroxide[436] in the presence of sodium tungstate[437], or with persulphate[438], does not stop at the imine stage, but proceeds further to give an oxime (from primary amines)[437,438] or a nitrone (from secondary amines)[436].

t-Butyl peroxide oxidizes primary amines to give, via an imine, the corresponding ketones[439]. *t*-Butyl hydroperoxide oxidizes primary or secondary amines to axomethines:

$$RR'CHNH_2 \xrightarrow{t\text{-BuOOH}} RR'C{=}NCHRR' \qquad \text{(ref. 440)}$$

$$(R_2CH)_2NH \xrightarrow{t\text{-BuOOH}} R_2C{=}NCHR_2 \qquad \text{(refs. 440, 441)}$$

Kinetic[440] and e.s.r.[441] data suggest a free-radical mechanism[441]:

$$(1) \qquad (RCH_2)_2\bar{N}H + t\text{-BuOOH} \longrightarrow (RCH_2)_2\bar{N}H\cdots HOOBu\text{-}t$$

$$(2) \qquad (RCH_2)_2\bar{N}H\cdots HOOBu\text{-}t \longrightarrow (RCH_2)_2\bar{N}OH + t\text{-BuOH}$$

$$(3) \qquad (RCH_2)_2\bar{N}\text{—}OH + t\text{-BuOOH} \longrightarrow (RCH_2)_2\bar{N}\text{—}O^{\bullet} + t\text{-BuO}^{\bullet} + H_2O$$

$$(4) \qquad (RCH_2)_2\bar{N}\text{—}O^{\bullet} + (RCH_2)_2\bar{N}H \longrightarrow (RCH_2)_2\bar{N}\text{—}OH + RCH_2\bar{N}H\text{—}\dot{C}HR$$

$$(5) \qquad RCH_2\bar{N}H\dot{C}HR + t\text{-BuO}^{\bullet} \longrightarrow RCH_2N{=}CHR + t\text{-BuOH}$$

Hypochlorites oxidize both primary[442] and secondary[443] amines to imines or azomethines; the reaction proceeds via the formation of an *N*-chloramine, which then loses HCl to form the imine, e.g.[442]

$$R_2CHNH_2 \xrightarrow{t\text{-BuOCl}} R_2CHNHCl \xrightarrow{-HCl} R_2CH{=}NH$$

α-Amino acids undergo a similar oxidation with hypohalites, accompanied by decarboxylation[444]:

$$\underset{\displaystyle NHR''}{RR'CCOOH} \xrightarrow{NaClO} \underset{\displaystyle NClR''}{RR'CCOONa} \longrightarrow RR'C{=}NR'' + CO_2 + NaCl$$

Mercuric acetate in dilute acetic acid oxidizes cyclic secondary[445] or tertiary[446,447] amines to imines or to immonium salt respectively. The mechanism is a concerted β-elimination by the attack of an acetate ion on the mercury–amine complex:

$$\text{(quinolizidine)} + Hg(OAc)_2 + AcO^- \longrightarrow \text{(complex: } {}^-OAc \text{ attacking H; N}\cdots{}^+HgOAc) \longrightarrow \text{(immonium, } \overset{+}{N}{=}C)$$

Other oxidants reported to convert amines to imines are chromic acid[434], ferric chloride[434], silver oxide[448], $S_2O_8^{2-}/Ag^+$[449] and lead tetracetate[450]. Oxygen difluoride oxidizes primary amines to oximes[451]. The oxidation of tertiary amines with chlorine dioxide gives an aldehyde through the formation of intermediary immonium salts[452]:

$$ArCH_2N(CH_3)_2 \xrightarrow{ClO_2} ArCH{=}\overset{+}{N}(CH_3)_2 \xrightarrow{H_2O} ArCHO + (CH_3)_2\overset{+}{N}H_2$$

Oxidation by electrophilic agents. Tertiary amines or amine oxides containing at least one α-hydrogen are attacked by electrophilic reagents to form ammonium salts, which in turn yield immonium salts by elimination. The following scheme[453] illustrates some of the reactions of this type:

$(RCH_2)_3N$ → (Br_2 or NBS) → (a) $(RCH_2)_3\overset{+}{N}Br$ → ($-HBr$) → $(RCH_2)_2\overset{+}{N}{=}CHR$

$(RCH_2)_3N$ → (HNO_2 or $NO^+BF_4^-$) → (b) $(RCH_2)_3\overset{+}{N}NO$ → ($-[HNO]$) → $(RCH_2)_2\overset{+}{N}{=}CHR$

$(RCH_2)_3N$ → ($C(NO_2)_4$) → (c) $(RCH_2)_3\overset{+}{N}NO_2$ → ($-HNO_2$) → $(RCH_2)_2\overset{+}{N}{=}CHR$

$(RCH_2)_3N{\rightarrow}O$ → (H^+) → (d) $(RCH_2)_3NOH$ → ($-H_2O$) → $(RCH_2)_2\overset{+}{N}{=}CHR$

$(RCH_2)_3N{\rightarrow}O$ → (SO_2) → (e) $(RCH_2)_3\overset{+}{N}OSO_2^-$ → ($-HSO_3^-$) → $(RCH_2)_2\overset{+}{N}{=}CHR$

(a) ref. 454; (b) ref. 453; (c) ref. 455; (d) ref. 456; (e) ref. 457.

To the same class of eliminations belongs the reaction of substituted diaminomethanes with chlorine or with acyl halides[458]:

$$(R_2N)_2CH_2 + Cl_2 \longrightarrow R_2\overset{+}{N}(Cl)CH_2NR_2\ Cl^- \longrightarrow R_2\overset{+}{N}{=}CH_2\ Cl^- + R_2NCl$$

$$(R_2N)_2CH_2 + R'COCl \longrightarrow R_2\overset{+}{N}(COR')\text{—}CH_2NR_2\ Cl^- \longrightarrow R_2\overset{+}{N}{=}CH_2\ Cl^- + R_2NCOR'$$

B. Oxidation of Hydroxylamine Derivatives and Other Compounds

Some monosubstituted hydroxylamines may be oxidized to oximes or to *C*-nitroso compounds (or their dimers[462]) by benzoquinone[459], by nitrates[460] or by oxygen catalysed by copper(II) salts[461].

N,*N*-Disubstituted hydroxylamines are oxidized by miscellanous oxidants, including atmospheric oxygen, to nitrones:

$$RCH_2N(R')OH \xrightarrow{[O]} RCH{=}N(O)R'$$

More information on the choice of conditions and on the products of the oxidations can be found in a recent review[193]. In the oxidation of *N*-acylhydroxylamines, the acyl group is easily eliminated during the reaction, and an oxime if formed[143,462,463]. In the case of *N*-aroylhydroxylamines, the oxidation may be accompanied by a rearrangement of the produced *N*-acyloxime (nitrone) to an *O*-acyloxime[148]:

$$PhCON(OH)CHPh_2 \xrightarrow{HgO} Ph_2C{=}NOCOPh$$

Careful oxidation of substituted hydrazines with $FeCl_3$ or HgO may give hydrazones or azines[464]. Bromine can also be used[465] and the mechanism of the oxidation is probably similar to that proposed for tertiary amines[454]:

$$(CH_3)_2NNHCH_3 \longrightarrow [(CH_3)_2\overset{+}{N}{=}NCH_3]Br^- \longrightarrow [(CH_3)_2\overset{+}{N}HN{=}CH_2]Br^-$$

The oxidation of *N*-benzyl-*N*-phenylhydrazine is accompanied by a rearrangement to a hydrazone[466]:

$$PhCH_2N(Ph)NH_2 \xrightarrow{HgO} [PhCH_2\overset{+}{N}Ph{=}\overset{-}{N}] \longrightarrow [PhCH{=}\overset{+}{N}(Ph)\overset{-}{N}H] \longrightarrow PhCH{=}NNHPh$$

Azo compounds containing α-hydrogens are oxidized to azines by free radicals[467]:

$$RR'CHN{=}NCHRR' \xrightarrow[\text{or } Cl_3C^\bullet]{RS^\bullet} RR'C{=}NN{=}CRR'$$

C. Oxidative and Reductive Eliminations

Tertiary amines containing *N*-methyl groups (e.g. in alkaloid systems) are reported to undergo *N*-demethylation by oxidation with lead tetracetate[468].

$$\text{(indoline N-}CH_3\text{)} \xrightarrow{Pb(OAc)_4} \text{(indolenine, C=N)}$$

The very stable perfluoroazalkanes undergo a reductive defluorination with ferrocene[469]:

$$CF_3CF_2CF_2CF_2NF_2 \xrightarrow{-2\,F} CF_3CF_2CF_2CF{=}NF$$

$$(C_2F_5)_2NF \xrightarrow{-2\,F} C_2F_5N{=}CFCF_3$$

The mechanism proposed for the reaction involves as first step a one-electron reduction of the N—F bond, followed by elimination of an α-fluorine as an anion:

$$R_2C(F){-}N(R')F \xrightarrow[-F^-]{e^-} R_2C(F){-}\dot{N}R' \xrightarrow{e^-} R_2C(F){-}\bar{N}{-}R' \xrightarrow{-F^-} R_2C{=}NR'$$

Manganese pentacarbonyl hydride can also be used for the defluorination[470].

D. Elimination from Substituted Amines

Amines substituted on the nitrogen by anionic leaving groups X eliminate HX easily, and a C=N bond is formed. *N*-haloamines[267,268,442–444,471] are converted to azomethines by alkalis or just by heating; *N,N*-dihaloamines give nitriles, e.g.[267]

$$RCH{=}CH_2 + N_2F_4 \longrightarrow RCH(NF_2){-}CH_2(NF_2) \xrightarrow{NaF} RC({=}NF){-}CN$$

N-nitroso[472] or nitroamines[473] give azomethines by losing nitroxyl or nitrous acid, respectively.

N-Aryl sulphonamides, especially *N*-tosyl derivatives, undergo elimination under the influence of a strong base[474].

$$RCH_2N(R')SO_2Ar \xrightarrow{base} RCH{=}NR'$$

This reaction requires an easily removable α-hydrogen; compounds of the type $RCH_2N(Ph)Ts$ undergo elimination by alkoxides in toluene at room temperature when R is a fairly strong electron-attracting group (PhCO, *p*-nitrophenyl), but not when R is H or a phenyl group. Sulphonamides of primary amines (RCH_2NHTs) do not undergo this elimination, since the *N*-hydrogen is abstracted by the base more easily than the *C*-hydrogen[475]. Tosylhydrazines undergo a similar elimination[476]. In some cases, the *N*-tosyl compound need not be isolated prior to the elimination, and a hydrazone is formed directly from a substituted hydrazine by the reaction with tosyl chloride[465,477]:

$$(CH_3)_2NNHCH_3 \xrightarrow{TsCl} (CH_3)_2NN{=}CH_2$$

In the reaction of *O*-acetyl arene sulphohydroxamic acids with a base, both the arene sulphonyl and the acetyl groups are eliminated, and an oxime is obtained[460]:

$$ArCH_2N(OAc)SO_2Ar' \xrightarrow{OH^-} ArCH_2\overset{O^-}{\overset{|}{N}}SO_2Ar' \longrightarrow ArCH_2NO \longrightarrow ArCH{=}NOH$$

The elimination of the benzene sulphonyl group from *N′*,*N′*-diethylbenzene sulphonhydrazide by a strong base is accompanied by a rearrangement to a hydrazone[478]:

$$PhSO_2NHNEt_2 \longrightarrow CH_3CH{=}NNHEt$$

Eliminations of HX from substituted amines in which the X group is on the α-carbon are also well known. The elimination of water or an alcohol from α-hydroxy or alkoxyamines is the final step in the condensation of amines with carbonyl compounds or their acetals, respectively. Acid-catalysed elimination of an alcohol from α-alkoxy tertiary amines gives an immonium salt[15,479].

$$R_2C(OR')NR''_2 \xrightarrow{H^+} R_2C{=}\overset{+}{N}R''_2 + R'OH$$

Immonium salts are also reported to be formed by elimination of cyanide ions from α-cyano tertiary amines[480]:

$$\text{1-(1-cyanocyclopentyl)piperidine} \xrightarrow{AgNO_3} \text{cyclopentylidene}{=}\overset{+}{N}\text{(piperidinium)} \quad NO_3^-$$

An amine having two strong electron-attracting groups in the β-position will lose an active methylene compound to give an azomethine[177]:

$$PhCH(NHPh)CH(COCH_3)_2 \longrightarrow PhCH{=}NPh + CH_2(COCH_3)_2$$

Very few eliminations from α-halo amines are reported; among them is the elimination of HF from hexafluoromethyl amine[481] in the gaseous phase:

$$(CF_3)_2NH \xrightarrow[140\text{–}150°]{KF} CF_3N{=}CF_2$$

The reduction of *gem*-chloronitroso compounds to oximes may proceed through an elimination of HCl from an intermediate α-chlorohydroxylamine[482]:

$$R_2CClNO \xrightarrow{H_2} [R_2CClNHOH] \xrightarrow{-HCl} R_2C{=}NOH$$

An interesting elimination reaction leading to the formation of hydrazones is that of hydrazo derivatives of the onium salts of certain heterocyclics[483]:

$$\underset{\underset{R}{|}}{\overset{+}{N}}\!\!\!=\!C\text{—}Y\ \ X^- \xrightarrow{NH_2NHR'} \underset{\underset{R}{|}}{\overset{+}{N}}\!\!\!=\!C\text{—}NHNHR'\ \ X^- \xrightarrow{-HX} \underset{\underset{R}{|}}{N}\text{—}C{=}NNHR'$$

X = Hal, $EtSO_4^-$, BF_4^-; Y = Hal, OR, SR; R' = H, COPh, $COOCH_3$, SO_2Ph, CHO

When R′ is hydrogen, azines are formed. In pyridinium systems, attack may take place either in the α or in the γ position, e.g.

$$Cl\text{—}C_5H_4\overset{+}{N}\text{—}CH_3 + PhCONHNH_2 \xrightarrow{EtOH} PhCONHNH\text{—}C_5H_4\overset{+}{N}\text{—}CH_3 \xrightarrow{KOH} PhCONHN{=}C_5H_4N\text{—}CH_3$$

Azomethines are formed by elimination of amines from derivatives of diaminomethane[15,484]. The mechanism of the acid-catalysed reaction is described[15] as

$$ArNHCH_2NHAr \underset{}{\overset{H^+}{\rightleftharpoons}} Ar\overset{+}{N}H_2CH_2NHAr \rightleftharpoons ArNH_2 + \overset{+}{C}H_2NHAr \overset{-H^+}{\rightleftharpoons} CH_2{=}NAr$$

Phenyl isocyanate may be used as an amine acceptor[485]:

$$(RNH)_2CHPh + PhNCO \longrightarrow RN{=}CHPh + RNHCONHPh$$

N-alkylamides are dehydrated to ketenimines[486]:

$$R_2CHCONHR' \xrightarrow[\text{pyridine}]{P_2O_5} R_2C{=}C{=}NR'$$

and thiosemicarbazones are catalytically desulphurized to formamidrazones[487]:

$$PhCH{=}NNHCSNH_2 \xrightarrow{H_2/Ni} PhCH{=}NN{=}CHNH_2$$

Reactions of ureas, thioureas, isothioureas and related compounds to give carbodiimides were reviewed in detail recently[288], and therefore will not be included in the present treatment.

XI. REDUCTION OF NITRO COMPOUNDS

The reduction of an aliphatic nitro compound containing an α-hydrogen may be stopped at the oxime stage.

$$RR'CHNO_2 \xrightarrow{[H]} RR'C{=}NOH$$

Direct reduction with Zn and acetic acid gives only poor yields due to complete reduction to amines[488], while stannous chloride gives better results[489]. Nitrous acid also transforms certain nitro compounds into oximes[490] by displacement of the nitro group by the nitroso group:

$$CH_3CH(NO_2)COOEt + HNO_2 \longrightarrow [CH_3CH(NO)COOEt] \longrightarrow CH_3C({=}NOH)COOEt$$

Catalytic hydrogenation of α-chloro nitro compounds gives good yields of oximes[491]:

$$\text{1-chloro-1-nitrocyclohexane} \xrightarrow{H_2,\ Pd/C} \text{cyclohexanone}{=}NOH$$

Nitro compounds are reduced to oximes by their reaction with substituted benzyl halides, and the aldehyde corresponding to the benzyl halide is obtained[491]. The reaction is thus related to the Sommelet reaction.

$$ArCH_2X + RR'CHNO_2 \xrightarrow{NaOEt} [RR'C{=}N(O)OBz] \longrightarrow RR'C{=}NOH + ArCHO$$

The reduction of nitro compounds with trialkyloxonium salts[492], gives a mixture of the oxime and the *O*-alkyl oxime:

$$C_6H_{11}{-}NO_2 + Et_3O^+ \cdot BF_4^- \longrightarrow [C_6H_{10}{=}N(O)OEt] \longrightarrow C_6H_{10}{=}NOH\ (74\%) + C_6H_{10}{=}NOEt\ (19\%) + CH_3CHO$$

Sodium salts of *aci*-nitro compounds react with diethyl ether in the presence of HCl, and a chloro oxime is formed[493]. The nature of the reducing agent was not reported, but it is probably again an oxonium salt formed from the ether with HCl.

$$PhCH{=}NO_2^- \xrightarrow{Et_2O,HCl} PhCCl{=}NOH$$

Nitro olefins yield different products, depending on the choice of the reducing agent. With Zn–acetic acid[494] or on catalytic reduction[495] (especially in acidic media)[496], a saturated oxime is formed. The catalytic hydrogenation of nitrostyrene with Pd in ethanolic HCl gives a mixture of phenylacetaldehyde oxime and phenylnitrosoethane dimer; the relative amount of the oxime increases with the concentration of the HCl[497]. Reduction of nitro olefins with $SnCl_2$ in HCl gives an oxamoyl chloride[498]:

$$R_2C{=}CRNO_2 \xrightarrow{SnCl_2/HCl} R_2CClCR{=}NOH$$

whereas with lithium aluminium hydride in the cold, an imine is formed[499]:

$$PhCH{=}CHNO_2 \xrightarrow{LAH} PhCH_2CH{=}NH$$

While the reaction of benzene with β,γ-unsaturated nitro compounds in the presence of $AlCl_3$ gives the normal addition product $R_2CHCRPhCR_2NO_2$, α,β-unsaturated nitro compounds give oxamoyl chlorides under similar conditions[500,501]. The proposed mechanism involves intramolecular oxidation–reduction[501]:

$$R_2C{=}CHNO_2 \xrightarrow[AlCl_3]{PhH} [R_2CPhCH{=}NO_2H] \longrightarrow R_2CPhC(OH){=}NOH \xrightarrow{AlCl_3} R_2CPhCCl{=}NOH$$

In favour of the mechanism is the isolation of a hydroxy oxime derivative in the reaction of *aci*-nitro compounds with benzoyl chloride[322].

$$CH_3CH{=}NO_2^- + PhCOCl \longrightarrow CH_3C(OH){=}NOCOPh$$

Reduction of γ-nitro ketones with Zn in aq. NH_4Cl causes cyclization to cyclic nitrones, Δ^1-pyrroline-1-oxide derivatives[445a,502].

$$H_2C(NO_2)CH_2CH_2COR \longrightarrow \text{2-R-}\Delta^1\text{-pyrroline-1-oxide } (N{\rightarrow}O)$$

XII. FORMATION OF AZOMETHINES BY REARRANGEMENTS AND PHOTOCHEMICAL REACTIONS

A. Prototropic Shifts

The simplest type of rearrangement leading to a C=N bond is the prototropic shift, which is spontaneous in many cases. Among the important classes of this type of shifts are: (a) The azo–hydrazo shift, CH—N=N ⇌ C=NNH[503] (see Section III); (b) The enamine–imine or immonium salt transformation C=CNH ⇌ CHC=N or C=CN= ⇌ CHC=$\overset{+}{N}$=[504]; (c). The *C*-nitroso–oxime shift (see Section IV); (d). The transformation between two isomeric azomethines, CHN=C ⇌ C=NCH, where the equilibrium is in favour of the more conjugated imine[504]. The base-catalysed shift of the last type is believed to proceed through an azaallylic anion as one intermediate[505]:

$$PhCH(CH_3)N{=}CAr_2 \underset{}{\overset{OH^-}{\rightleftharpoons}} PhC(CH_3)\cdots\bar{N}\cdots CAr_2 \overset{H^+}{\rightleftharpoons} PhC(CH_3){=}NCHAr_2$$

B. Rearrangements through Nitrene Intermediates

Compounds containing C=N bonds may be obtained also by the Beckmann, Hoffmann, Lossen, Curtius and Stieglitz rearrangements. Since these were reviewed recently[466b], the main types leading to C=N bonds will be outlined only:

I. Azide decomposition, thermal or photolytic

(a) $RCH_2N_3 \xrightarrow{-N_2} [RCH_2\bar{N}:] \longrightarrow RCH{=}NH$ (ref. 506)

(b) $RCOCR'R''N_3 \xrightarrow{-N_2} [RCOCR'R''\bar{N}:] \longrightarrow RCOCR'{=}NR''$ (ref. 507)

(c) $R_3CN_3 \xrightarrow{-N_2} [R_3C\bar{N}:] \longrightarrow R_2C{=}NR$ (ref. 508)

The photochemical reaction is catalysed by triplet sensitizers and is believed to involve an azide triplet[509];

(d) $RCH{=}CR'N_3 \xrightarrow{-N_2} [RCH{=}CR'\bar{N}:] \longrightarrow RC\overset{\displaystyle N}{—}CHR' + RCH{=}C{=}NR'$ (aziridine ring: RC—CHR' bridged by N)

(See Section VI; refs. 264–6)

(e) $CF_3CHFCF_2N_3 \xrightarrow{-N_2} [CF_3CHFCF_2\text{—}\bar{N}:] \longrightarrow CF_3CHFN{=}CF_2$ (ref. 510)

2. Decomposition of tetrazoles[288]

$$ArN\!-\!CR\ (\text{tetrazole}) \longrightarrow \left[Ar\!-\!N\!=\!C(R)N_3 \xrightarrow{-N_2} ArN\!=\!C(R)\ddot{N}^- \right] \longrightarrow \begin{cases} X\text{-benzimidazole (2-R, 1-H)} \\ ArN\!=\!C\!=\!NR \end{cases}$$

3. Rearrangement of triaryl methylamine derivatives (Stieglitz rearrangement)[70a,419,511]

$$\left.\begin{array}{l} Ph_3CNHOH \xrightarrow{PCl_5} \\ Ph_3CNHCl \xrightarrow{OH^-} \\ Ph_3CNCl_2 \xrightarrow{\Delta} \end{array}\right\} \longrightarrow [Ph_3C\bar{N}:] \longrightarrow Ph_2C\!=\!NPh$$

4. Beckmann-type rearrangements

(a) $ArC(CN)\!=\!NOTs \xrightarrow{NaOEt} [ArC(CN)(OEt)\bar{N}:] \longrightarrow$
$EtOC(CN)\!=\!NAr \xrightarrow{OEt^-} (EtO)_2C\!=\!NAr$ (ref. 512)*

(b) $CH_3C(\!=\!NOH)CH(CH_3)_2 \xrightarrow{PCl_5} CH_3CCl\!=\!NCH(CH_3)_2$ (ref. 513)

5. Rearrangements to hydrazones via diazenes

$$>N\!-\!\bar{N}: \longleftrightarrow >N^+\!=\!N^-$$

(a) $RCH_2NR'NH_2 \xrightarrow{HgO} [RCH_2NR'\bar{N}:] \longrightarrow$
$RCH_2N\!=\!NR' \longrightarrow RCH\!=\!NNHR'$ (ref. 466a)

(b) $RCH_2NR'NHTs \xrightarrow{OH^-} [RCH_2NR'\ddot{\bar{N}}Ts \xrightarrow{-Ts^-} RCH_2NR'\bar{N}:] \longrightarrow$
$RCH_2N\!=\!NR' \longrightarrow RCH\!=\!NNHR'$ (refs. 478, 514)

(c) pyrrolidine NH + Na_2ONNO_2 (Angeli's Salt) $\longrightarrow$ [pyrrolidine $N\!-\!\bar{N}:$] $\longrightarrow$ tetrahydropyridazine (ref. 515)

6. Reactions of diazo compounds

In some of their reactions, diazo compounds behave as nitrenes, e.g.[516]

$$R_2CN_2 \xrightarrow{h\nu} R_2C: + N_2$$

$$R_2C: + R_2C\!=\!N\!-\!\bar{N}: \longrightarrow R_2C\!=\!NN\!=\!CR_2$$

* Recently a mechanism involving a nucleophilic attack by an ethoxide ion was proposed for this reaction [267b].

7. Rearrangements via imidonium ion ($R_2\overset{+}{N}:$) intermediates: acid-catalysed decompositions of azides[517]

(a) cyclopentyl–N_3 $\xrightarrow{H^+}$ [cyclopentyl–$\overset{+}{N}H-N_2$ $\xrightarrow{-N_2}$ cyclopentyl–$\overset{+}{\ddot{N}}H$] ⟶ tetrahydropyridine + cyclopentyl=NH (ref. 518)

79·8% 2·2%

(while irradiation of cyclopentyl azide gives 55% imine[506b]).

(b) $Ar_2CHN_3 \xrightarrow{H_2SO_4} [Ar_2CH\overset{+}{\ddot{N}}H] \longrightarrow ArCH{=}NAr$ (ref. 519)

(c) $R_2CO + HN_3 + EtOH + HCl \longrightarrow RC(OEt){=}NR\cdot HCl$ (refs. 300, 520)

(d) cyclohexanone + HN_3 + EtOH + HCl ⟶ 2-ethoxy-azacycloheptene (N=C–OEt) (ref. 521)

(e) $CH_3(CH_2)_3N_3 + (CH_3)_3O^+\ BF_4^- \longrightarrow [CH_3(CH_2)_3N(CH_3)N_2^+ \longrightarrow$
$CH_3(CH_2)_3\overset{+}{\ddot{N}}CH_3] \longrightarrow \underset{80\%}{CH_3CH_2CH_2CH{=}NCH_3} + \underset{10\%}{CH_3(CH_2)_3N{=}CH_2}$
$+\ CH_3CH_2CH_2\overset{+}{N}(CH_3){=}CH_2\ BF_4^-$ (ref. 522)

C. *Rearrangements through Other Free Radicals*

Tertiary *N*-nitroso amines containing α-hydrogens are decomposed with rearrangement by irradiation, and azomethines are formed[228,229,523].

$$(RCH_2)_2NNO \xrightarrow{h\nu} NO + (RCH_2)_2N^{\bullet} \longrightarrow (RCH_2)_2NH + RCH_2N{=}CHR$$

The photolysis in acidic media gives amidoximes, which are obtained as follows[524]:

$$\underset{N=O\cdots H^+}{RN-CH_2R'} \longrightarrow RN{=}CHR' + [HNO] \longrightarrow \underset{^+N-OH}{RNHCHR'} \longrightarrow \underset{NOH}{RNHCR'}$$

N-Nitrosamides behave similarly[525].

Irradiation of nitrites causes migration of a nitroso group to a γ-methyl or σ-methylene group: an oxime is formed by tautomerization (Barton reaction)[357,526].

$$\text{ONO}\quad \text{CH}_2\text{R} \xrightarrow{h\nu} \text{OH}\quad \text{CHRNO} \longrightarrow \text{OH}\quad \text{CR{=}NOH}$$

The reaction proceeds through the initial formation of an alkoxy radical, followed by *stereospecific* intramolecular hydrogen abstraction and recombination of the carbon radical with NO, when a nitroso monomer or dimer or an oxime is formed. The reaction was mainly used in steroidic systems.

Azo compounds containing α-cyano groups are decomposed to ketenimines by refluxing in an inert solvent[527]:

$$RR'C(CN)N{=}NC(CN)RR' \xrightarrow{-N_2} [2\ RR'\dot{C}CN] \longrightarrow RR'C{=}C{=}NC(CN)RR' + RR'C(CN)C(CN)RR'$$

D. Other Rearrangements

N-alkyl or acyl oximes (nitrones) isomerize to their corresponding *O*-alkyl or acyl oximes, in some cases spontaneously, or under the influence of heat or acid catalysts:

$$PhCONHOH{\cdot}HCl + PhCH(OEt)_2 \longrightarrow [PhCON(O){=}CHPh] \longrightarrow PhCOON{=}CHPh \quad \text{(ref. 148)}$$

$$Ph_2CHN(OH)COPh \xrightarrow{HgO} [Ph_2C{=}N(O)COPh] \longrightarrow Ph_2C{=}NOCOPh \quad \text{(ref. 148)}$$

$$Ph_2C{=}N(O)CHPh_2 \xrightarrow[\text{or } H^+]{200^\circ} Ph_2C{=}NOCHPh_2 \quad \text{(refs. 70b, 528)}$$

N-Benzoylaziridines undergo a benzoyl group migration[529] to form hydrazones:

$$RR'C\langle N(COPh)N(COPh) \rangle \longrightarrow RR'C{=}NN(COPh)_2$$

β-Nitroso nitro compounds rearrange to oximes in basic media[530]:

$$R_2C(NO)CHR'NO_2 \xrightarrow{OH^-} R_2C(NO)CR'{=}NO_2^- \longrightarrow \left[\text{cyclic intermediate: } R_2C{-}CR' \text{ bridged by } {}^-O{-}N{-}O{-}N{\rightarrow}O \right] \longrightarrow$$

$$\underset{NO_2}{R_2C}{-}\underset{NO^-}{CR'} \longrightarrow R_2C(NO_2)CR{=}NOH$$

Nitroso dimers containing α-hydrogens are transformed to *N*-acylhydrazones with acid catalysts[531].

$$RCH_2N(O){=}N(O)CH_2R \xrightarrow{HCl} [RCH{=}N(O)N(OH)CH_2R] \longrightarrow RCONHN{=}CHR$$

XIII. REFERENCES

1. E. Brand and M. Sandberg, *Org. Synth.*, Coll. Vol. **2**, 49 (1943).
2. E. Dane, F. Drees, P. Konrad and T. Drockner, *Angew. Chem.*, **74**, 873 (1962); J. C. Sheehan and V. J. Grenda, *J. Am. Chem. Soc.*, **84**, 2417 (1962).
3. H. Schiff, *Ann. Chem.*, **131**, 118 (1864).
4. M. M. Sprung, *Chem. Rev.*, **26**, 297 (1940).
5. R. W. Layer, *Chem. Rev.*, **63**, 489 (1963).
6. R. B. Moffett and W. M. Hoehn, *J. Am. Chem. Soc.*, **69**, 1792 (1947); R. B. Moffett, *Org. Synth.*, Coll. Vol. **4**, 605 (1963); M. Freifelder, *J. Org. Chem.*, **31**, 3875 (1966).
7. R. Grewe, R. Hamann, G. Jacobsen, E. Nolte and K. Riecke, *Ann. Chem.*, **581**, 85 (1953).
8. E. F. Pratt and M. J. Kamlet, *J. Org. Chem.*, **26**, 4029 (1961).
9. J. H. Billman and K. M. Tai, *J. Org. Chem.*, **23**, 535 (1958).
10. M. E. Taylor and T. J. Fletcher, *J. Org. Chem.*, **26**, 940 (1961).
11. G. Reddelien, *Ber.*, **43**, 2476 (1910).
12. J. Kolsa, *Arch. Chem.*, **287**, 62 (1954).
13. R. L. Reeves in *Chemistry of the Carbonyl Group* (Ed. S. Patai), Interscience, 1966, p. 567.
14. D. Y. Curtin and J. W. Hausser, *J. Am. Chem. Soc.*, **83**, 3474 (1961).
15. E. C. Wagner, *J. Org. Chem.*, **19**, 1862 (1954).
16. A. Werner and H. Buss, *Ber.*, **27**, 1280 (1894); A. Eibner, *Ann. Chem.*, **318**, 58 (1901); **328**, 121 (1903).
17. W. v. Miller and J. Plöchl, *Ber.*, **25**, 2020 (1892).
18. M. S. Kharasch, I. Richlin and F. R. Mayo, *J. Am. Chem. Soc.*, **62**, 494 (1940).
19. M. D. Hurwitz, *U.S. Pat.* 2,582,128 (1952); *Chem. Abstr.*, **46**, 8146 (1952).
20. R. Tiollais, *Bull. Soc. Chim. France*, 708 (1947).
21. E. Knoevenagel, *Ber.*, **54**, 1722 (1921).
22. G. Reddelien and A. Thurm, *Ber.*, **65**, 1511 (1932).
23. R. Kuhn and H. Schretzmann, *Chem. Ber.*, **90**, 557 (1957).
24. R. M. Roberts and M. B. Edwards, *J. Am. Chem. Soc.*, **72**, 5537 (1950).
25. N. H. Cromwell, *Chem. Rev.*, **38**, 83 (1946).
26. C. L. Stevens, P. Blumbergs and M. Munk, *J. Org. Chem.*, **28**, 331 (1963).
27. S. J. Angyal, G. B. Barlin and P. C. Wailes, *J. Chem. Soc.*, 3512 (1951).
28. O. Gerngross and A. Olcay, *Chem. Ber.*, **96**, 2550 (1963).
29. S. C. Bell, G. L. Conklin and S. J. Childress, *J. Am. Chem. Soc.*, **85**, 2868 (1963).
30. H. P. Schad, *Helv. Chim. Acta*, **38**, 1117 (1955); G. O. Dudek and G. P. Volpp, *J. Am. Chem. Soc.*, **85**, 2697 (1963).
31. A. Lipp, *Ber.*, **14**, 1746 (1881); *Ann. Chem.*, **211**, 344 (1882).
32. R. H. Hasek, E. U. Elam and J. C. Martin, *J. Org. Chem.*, **26**, 1822 (1961).
33. J. W. Clark and A. L. Wilson, *U.S.Pat.* 2,319,848 (1943); *Chem. Abstr.*, **37**, 6275 (1943).
34. G. H. Harris, B. R. Harriman and K. W. Wheeler, *J. Am. Chem. Soc.*, **68**, 648 (1946).
35. C. Thomae, *Arch. Pharm.*, **243**, 395 (1905).

36. H. H. Strain, *J. Am. Chem. Soc.*, **52**, 820 (1930).
37. G. Mignonac, *Compt. Rend.*, **169**, 237 (1919).
38. G. Kabas, *J. Org. Chem.*, **32**, 218 (1967).
39. J. R. Gaines and D. D. Lidel, *J. Org. Chem.*, **28**, 1032 (1963).
40. J. R. Gaines and G. R. Hansen, *J. Heterocyclic Chem.*, **1**, 96 (1964).
41. N. J. Leonard and J. V. Paukstelis, *J. Org. Chem.*, **28**, 3021 (1963).
42. E. P. Blanchard, Jr., *J. Org. Chem.*, **28**, 1397 (1963).
43. J. C. Howard, G. Gever and P. H. L. Wei, *J. Org. Chem.*, **28**, 868 (1963).
44. Y. Oshiro, K. Yamamoto and S. Komori, *Yuki Gosei Kagaku Kyokai Shi*, **24**, 945 (1966); *Chem. Abstr.*, **66**, 37706 (1967).
45. J. Gante, *Chem. Ber.*, **97**, 1921 (1964).
46. Y. Sato and H. Kaneko, *Steroids*, **5**, 279 (1965).
47. S. Dokic and M. Cakara, *Kem. Ind. (Zagreb)*, **13**, 261 (1964); *Chem. Abstr.*, **61**, 10545 (1964).
48. I. J. Finar and G. H. Lord, *J. Chem. Soc.*, 3314 (1957); A. T. Troshchenko and T. P. Lobanova, *Zh. Org. Khim.*, **3**, 501 (1967); *Chem. Abstr.*, **67**, 2581 (1967).
49. W. E. Bachmann and C. H. Boatner, *J. Am. Chem. Soc.*, **58**, 2097 (1936).
50. G. Rosenkranz, O. Mancera, F. Sondheimer and C. Djerassi, *J. Org. Chem.*, **21**, 520 (1956).
51. W. P. Jencks, *J. Am. Chem. Soc.*, **81**, 475 (1959).
52. B. V. Joffe and K. N. Zelenin, *Izv. Vysshikh Uchebn. Zavedenii, Khim. i Khim. Tekhnol.*, **6**, 78 (1963); *Chem. Abstr.*, **59**, 6244 (1963).
53. A. Lapworth, *J. Chem. Soc.*, 91, 1133 (1907).
54. W. L. Semon and V. R. Damerell, *J. Am. Chem. Soc.*, **46**, 1290 (1924); W. L. Semon, *Org. Synth.*, Coll. Vol. **1**, 318 (1941); J. C. Eck and C. S. Marvel, *Org. Synth.*, Coll. Vol. **2**, 76 (1943); H. Koopman, *Rec. Trav. Chim.*, **80**, 1075 (1961).
55. D. E. Pearson and O. D. Keaton, *J. Org. Chem.*, **28**, 1557 (1963).
56. R. G. Kadesch, *J. Am. Chem. Soc.*, **66**, 1207 (1944).
57. F. Greer and D. E. Pearson, *J. Am. Chem. Soc.*, **77**, 6649 (1955).
58. C. R. Hauser and D. S. Hoffenberg, *J. Am. Chem. Soc.*, **77**, 4885 (1955).
59. J. H. Jones, E. W. Tristram and W. F. Benning, *J. Am. Chem. Soc.*, **81**, 2151 (1959).
59a. J. Buchanan and S. D. Hamann, *Trans. Faraday Soc.*, **49**, 1425 (1953).
60. A. H. Blatt, *J. Am. Chem. Soc.*, **61**, 3494 (1939).
61. F. A. L. Anet, P. M. G. Bavin and M. J. S. Dewar, *Can. J. Chem.*, **35**, 180 (1957).
62. W. D. Phillips, *Ann. N.Y. Acad. Sci.*, **70**, 817 (1958).
63. W. Theilacker and L. H. Chou, *Ann. Chem.*, **523**, 143 (1936).
64. J. Meisenheimer, *Ber.*, **54**, 3206 (1921).
65. J. Scheiber and P. Brandt, *Ann. Chem.*, **357**, 25 (1907); J. Scheiber and H. Wolf, *Ann. Chem.*, **362**, 64 (1908); V. Bellavita, *Gazz. Chim. Ital.*, **65**, 755, 889 (1935); E. Bayland and R. Nery, *J. Chem. Soc.*, 3141 (1963); E. W. Cummins, *French Pat.*, 1,437,188 (1966); *Chem. Abstr.*, **66**, 10726 (1967).
66. O. Exner, *Collection Czech. Chem. Commun.*, **16**, 258 (1951).
67. F. H. Banfield and J. Kenyon, *J. Chem. Soc.*, 1612 (1926).
68. J. Thesing and H. Mayer, *Chem. Ber.*, **89**, 2159 (1956).

69. R. F. C. Brown, V. M. Clark, I. O. Sutherland and A. Todd, *J. Chem. Soc.*, 2109 (1959).
70. (a) J. Stieglitz and B. N. Leech, *J. Am. Chem. Soc.*, **36**, 272 (1914).
(b) A. C. Cope and A. C. Haven, Jr., *J. Am. Chem. Soc.*, **72**, 4896 (1950).
71. H. H. Hatt, *Org. Synth.*, Coll. Vol. **2**, 395 (1943); C. G. Overberger and J. J. Monagle, *J. Am. Chem. Soc.*, **78**, 4470 (1956); F. J. Allan and G. G. Allan, *J. Org. Chem.*, **23**, 639 (1958).
72. R. L. Hinman, *J. Org. Chem.*, **25**, 1775 (1960).
73. T. Curtius and H. Franzen, *Ber.*, **35**, 3234 (1902); H. Staudinger and A. Gaule, *Ber.*, **49**, 1897 (1916).
74. T. Curtius and A. Lublin, *Ber.*, **33**, 2460 (1900).
75. E. R. Blout, V. W. Eager and R. M. Gofstein, *J. Am. Chem. Soc.*, **68**, 1983 (1946).
76. I. Fleming and J. Harley-Mason, *J. Chem. Soc.*, 5560 (1961); S. Goldschmidt and B. Acksteiner, *Chem. Ber.*, **91**, 502 (1958).
77. R. Hüttel, J. Riedl, M. Martin and K. Franke, *Chem. Ber.*, **93**, 1425 (1960).
78. N. Schapiro, *Ber.*, **62**, 2133 (1929); O. Grummitt and A. Jenkins, *J. Am. Chem. Soc.*, **68**, 914 (1946).
79. H. H. Szmant and C. McGinnis, *J. Am. Chem. Soc.*, **72**, 2890 (1950).
80. L. I. Smith and K. L. Howard, *Org. Synth.*, Coll. Vol. **3**, 351 (1955).
81. H. Goetz and H. Juds, *Ann. Chem.*, **698**, 1 (1966).
82. R. A. Braun and W. A. Mosher, *J. Am. Chem. Soc.*, **80**, 3048 (1958).
83. (a) M. P. Cava, R. L. Litle and D. R. Napier, *J. Am. Chem. Soc.*, **80**, 2257 (1958);
(b) P. Yates and B. L. Shapiro, *J. Org. Chem.*, **23**, 759 (1958).
84. R. G. Jones, *J. Org. Chem.*, **25**, 956 (1960).
85. G. Meier and T. Sayrac, *Chem. Ber.*, **101**, 1354 (1968).
86. A. P. Mashchetyakov, L. V. Petrova and V. G. Glukhovtsev, *Izv. Akad. Nauk SSSR*, 114 (1961); *Chem. Abstr.*, **55**, 17526 (1961).
87. R. F. C. Brown, L. Subrahmanyan and C. P. Whittle, *Australian J. Chem.*, **20**, 339 (1967).
88. R. H. Wiley, S. C. Slaymaker and H. Kraus, *J. Org. Chem.*, **22**, 204 (1957); R. H. Wiley and G. Irick, *J. Org. Chem.*, **24**, 1925 (1959); R. F. Smith and L. E. Walker, *J. Org. Chem.*, **27**, 4372 (1952); G. Adembri, P. Sarti-Fantoni and E. Belgodere, *Tetrahedron*, **22**, 3149 (1966); *French Pat.* 1,455,835 (1967); *Chem. Abstr.*, **67**, 11327 (1967).
89. O. L. Brady and G. P. McHugh, *J. Chem. Soc.*, **121**, 1648 (1922).
90. M. Leonte, *Anal. Stiint. Univ. "Al I. Cuza, iasi.,"* Sect. **I**, 4197 (1958); *Chem. Abstr.*, **56**, 4645 (1962).
91. A. Jacob and J. Madinaveitia, *J. Chem. Soc.*, 1929 (1937).
92. H. B. Nisbet, *J. Chem. Soc.*, 126 (1945).
93. M. P. Cava and R. L. Litle, *Chem. Ind. (London)*, 367 (1957).
94. H. P. Figeys and J. Nasielski, *Bull. Soc. Chim. Belge*, **75**, 601 (1966).
95. J. B. Conant and P. D. Bartlett, *J. Am. Chem. Soc.*, **54**, 2881 (1932).
96. E. H. Cordes and W. P. Jencks, *J. Am. Chem. Soc.*, **84**, 826 (1962).
97. E. Fischer, *Ber.*, **17**, 579 (1884); **20**, 821 (1887).
98. G. H. Stempel, Jr., *J. Am. Chem. Soc.*, **56**, 1351 (1934).
99. H. H. Stroh, *Chem. Ber.*, **90**, 352 (1957); H. H. Stroh and H. Lamprecht, *Chem. Ber.*, **96**, 651 (1963); H. H. Stroh and P. Golüke, *Z. Chem.*, **7**, 60 (1967).

100. H. H. Stroh, *Chem. Ber.*, **91**, 2645 (1958).
101. H. H. Stroh and B. Ihlo, *Chem. Ber.*, **96**, 658 (1963).
102. M. D. Soffer and M. A. Jevnik, *J. Am. Chem. Soc.*, **77**, 1003 (1955).
103. F. Ramirez and A. F. Kirby, *J. Am. Chem. Soc.*, **74**, 4331 (1952).
104. F. Ramirez and A. F. Kirby, *J. Am. Chem. Soc.*, **75**, 6026 (1953).
105. A. Ross and R. N. King, *J. Org. Chem.*, **26**, 579 (1961).
106. I. C. Torres and S. Brosa, *Annales Soc. Españ. Fis. Quim.*, **32**, 509 (1934); *Chem. Abstr.*, **28**, 6104 (1934).
107. F. D. Chattaway and L. H. Farinholt, *J. Chem. Soc.*, 96 (1930).
108. H. Adkins and A. G. Rossow, *J. Am. Chem. Soc.*, **71**, 3836 (1949).
109. R. Scholl and G. Matthiopoulos, *Ber.*, **29**, 1550 (1896); A. Hantzsch and W. Wild, *Ann. Chem.*, **289**, 285 (1896); R. Belcher, W. Hoyle and T. S. West, *J. Chem. Soc.*, 2743 (1958).
110. N. P. Gambaryan, E. M. Rokhlin, Yu. V. Zeifman, C. Ching-Yun and I. L. Knunyants, *Angew. Chem.* (*Intern. Ed. Engl.*), **5**, 947 (1966).
111. G. Henscke and H. J. Binte, *Chimia*, **12**, 103 (1958).
112. L. Mester, *Angew. Chem.*, (*Intern. Ed. Engl.*), **4**, 574 (1965).
113. L. Fieser and M. Fieser, *Organic Chemistry*, D. C. Heath, Boston, 1944, p. 353.
114. L. Mester, E. Moczar and J. Parello, *J. Am. Chem. Soc.*, **87**, 596 (1965).
115. K. Bjamer, S. Dahn, S. Furberg and S. Petersen, *Acta Chim. Scand.*, **17**, 559 (1963).
116. O. L. Chapman, W. J. Welstead, Jr., T. J. Murphy and R. W. King, *J. Am. Chem. Soc.*, **86**, 732 (1964).
117. E. A. Braude and W. F. Forbes, *J. Chem. Soc.*, 1762 (1951); S. Patai and S. Dayagi, *J. Org. Chem.*, **23**, 2014 (1958).
118. G. J. Bloink and K. H. Pausacker, *J. Chem. Soc.*, 661 (1952).
119. J. Kenner and E. C. Knight, *Ber.*, **69**, 341 (1936); G. J. Bloink and K. H. Pausacker, *J. Chem. Soc.*, 1328 (1950).
120. L. Caglioti, G. Rosini and F. Rossi, *J. Am. Chem. Soc.*, **88**, 3865 (1966).
121. N. Rabjohn and H. D. Barnstorff, *J. Am. Chem. Soc.*, **75**, 2259 (1953); L. A. Carpino, *J. Am. Chem. Soc.*, **79**, 98 (1957); A. Alemagna, T. Bacchetti and S. Rossi, *Gazz. Chim. Ital.*, **93**, 748 (1963).
122. T. Ueda and M. Furukawa, *Chem. Pharm. Bull.*, **12**, 100 (1964).
123. J. A. Barltrop and K. J. Morgan, *J. Chem. Soc.*, 3075 (1957); D. Kaminsky, J. Shavel, Jr. and R. I. Meltzer, *Tetrahedron Letters*, 859 (1967).
124. J. W. Suggitt, G. S. Myers and G. F. Wright, *J. Org. Chem.*, **12**, 373 (1947).
125. H. C. Ramsperger, *J. Am. Chem. Soc.*, **51**, 918 (1929).
126. G. Tomaschweski and G. Geissler, *Chem. Ber.*, **100**, 919 (1967).
127. A. Wohl, *Ber.*, **33**, 2759 (1900); A. Wohl and H. Schiff, *Ber.*, **35**, 1900 (1902).
128. B. B. Corson and E. Freeborn, *Org. Synth.*, Coll. Vol. **2**, 231 (1943).
129. J. Lichtenberger, J. P. Freury and B. Barelle, *Bull. Soc. Chim. France*, 669 (1955); J. E. Abrashanowa, *J. Gen. Chem.* (*Russ.*), **27**, 1993 (1957).
130. K. Thinius and W. Lahr, *Z. Chem.*, **6**, 315 (1966).
131. E. Dario and S. Dugone, *Gazz. Chim. Ital.*, **66**, 139 (1936).
132. H. Schiff, *Ann. Chem.*, **321**, 357 (1902).
133. F. Tiemann, *Ber.*, **19**, 1668 (1886); A. Spilker, *Ber.*, **22**, 2767 (1889); O. Goldbeck, *Ber.*, **24**, 3658 (1891).

134. V. Carneckij, S. Chladek, F. Sorm and J. Smrt, *Collection Czech. Chem. Commun.*, **27**, 87 (1962).
135. K. Baker and H. E. Fierz-David, *Helv. Chim. Acta*, **33**, 2011 (1950).
136. G. S. Skinner and H. C. Vogt, *J. Am. Chem. Soc.*, **77**, 5440 (1955).
137. M. E. Baguley and J. A. Elvidge, *J. Chem. Soc.*, 709 (1957).
138. C. Finzi and G. Grandolini, *Gazz. Chim. Ital.*, **89**, 2543 (1959).
139. H. Rapoport and R. M. Bonner, *J. Am. Chem. Soc.*, **72**, 2783 (1950), H. H. Bosshard and H. Zollinger, *Helv. Chim. Acta*, **42**, 1659 (1959); H. Bredereck and K. Bredereck, *Chem. Ber.*, **94**, 2278 (1961).
140. C. L. Dickinson, W. J. Middleton and V. A. Englhardt, *J. Org. Chem.*, **27**, 2470 (1962); J. D. Albright, E. Benz and A. E. Lanzilotti, *Chem. Commun.*, 413 (1965).
141. W. Hoyle, *J. Chem. Soc.*, 690 (1967).
142. N. Duffaut and J. P. Dupin, *Bull. Soc. Chim. France*, 3205 (1966).
143. C. Krüger, E. G. Rochow and U. Wannagat, *Chem. Ber.*, **96**, 2132 (1963).
144. E. Fahr, *Ann. Chem.*, **627**, 213 (1959).
145. M. Portelli, *Gazz. Chim. Ital.*, **93**, 1600 (1963).
146. J. Hoch, *Compt. Rend.*, **201**, 560 (1935); C. A. Grob and P. W. Schiess, *Helv. Chim. Acta*, **43**, 1546 (1960); J. P. Horwitz and A. J. Tomson, *J. Org. Chem.*, **26**, 3392 (1961); T. Mukaiyama and K. Sato, *Bull. Chem. Soc. Japan*, **36**, 99 (1963).
147. R. Albrecht, G. Kresze and B. Mlakar, *Chem. Ber.*, **97**, 483 (1964).
148. O. Exner, *Collection Czech. Chem. Commun.*, **21**, 1500 (1956).
149. T. Kametani, K. Fukumoto, Y. Satoh, T. Teshigawara and O. Umezava, *Bull. Chem. Soc. Japan*, **33**, 1678 (1960).
150. E. C. Taylor and W. A. Ehrhart, *J. Am. Chem. Soc.*, **82**, 3138 (1960); E. C. Taylor, W. A. Ehrhart and M. Kawanisi, *Org. Synth*, **46**, 39 (1966).
151. R. M. Roberts and R. H. DeWolfe, *J. Am. Chem. Soc.*, **76**, 2411 (1956); E. C. Taylor and W. A. Ehrhart, *J. Org. Chem.*, **28**, 1108 (1963).
152. H. L. Yale and J. T. Sheehan, *J. Org. Chem.*, **26**, 4315 (1961); B. Loev and M. F. Kormendy, *Can. J. Chem.*, **42**, 176 (1964).
153. C. W. Whitehead, *J. Am. Chem. Soc.*, **75**, 671 (1953).
154. C. W. Whitehead and J. J. Traverso, *J. Am. Chem. Soc.*, **77**, 5872 (1955).
155. R. F. Meyer, *J. Org. Chem.*, **28**, 2902 (1963).
156. E. Wenkert, B. S. Bernstein and J. H. Udelhofen, *J. Am. Chem. Soc.*, **80**, 4899 (1958); K. Brückner, K. Irmscher, F. v. Werder, K. H. Bork and H. Metz, *Chem. Ber.*, **94**, 2897 (1961).
157. G. Casnati, A. Quilico, A. Ricca and P. Vita Finzi, *Gazz. Chim. Ital.*, **96**, 1064 (1966).
158. B. Robinson, *Chem. Rev.*, **63**, 373 (1963).
159. F. J. Moore, *Ber.*, **43**, 563 (1910); W. Theilacker and O. R. Leichtle, *Ann. Chem.*, **572**, 121 (1951).
160. A. Hantzsch and F. Kraft, *Ber.*, **24**, 3511 (1891).
161. E. Kuehle, *Ger. Pat.* 1,149,712 (1963); *Chem. Abstr.*, **59**, 12704 (1963).
162. V. L. Dubina and S. I. Burmistrov, *Zh. Org. Khim.*, **3**, 424 (1967); *Chem. Abstr.*, **66**, 115403 (1967).
163. V. L. Dubina and S. L. Burmistrov, *Zh. Org. Khim.*, **2**, 1841 (1966); *Chem. Abstr.*, **66**, 55171 (1967).

164. B. A. Porai-Koshits and A. L. Remizov, *Svornik Statei Obshch. Khim.*, **2**, 1570, 1577, 1590 (1953); *Chem. Abstr.*, **49**, 5367 (1955).
165. S. Bodforss, *Ann. Chem.*, **455**, 41 (1927); F. Knotz, *Monatsh. Chem.*, **89**, 718 (1958); E. H. Cordes and W. P. Jencks, *J. Am. Chem. Soc.*, **84**, 826 (1962).
166. W. Platner, *Ann. Chem.*, **278**, 359 (1894); H. Staudinger and K. Mischer, *Helv. Chim. Acta*, **2**, 554 (1919); F. Kröhnke and E. Börner, *Ber.*, **69**, 2006 (1936); S. P. Findlay, *J. Org. Chem.*, **21**, 644 (1956).
167. C. G. Raison, *J. Chem. Soc.*, 2858 (1957); R. Cantarel and F. Souil, *Compt. Rend.*, **246**, 1436 (1958); R. Cantarel and J. Guenzet, *Bull. Soc. Chim. France*, 1549 (1960).
168. A. Pinner, *Ber.*, **17**, 184 (1884).
169. A. Giner-Sorolla, I. Zimmermann and A. Bendich, *J. Am. Chem. Soc.*, **81**, 2515 (1959).
170. C. L. Bumgardner and J. P. Freeman, *Tetrahedron Letters*, 5547 (1966).
171. W. E. Haury, *U.S. Pat.* 2,692,284 (1954); *Chem. Abstr.*, **49**, 15946 (1955).
172. A. E. Arbuzov and Y. P. Kitaev, *Tr. Kazan. Khim. Tekhnol. Inst. im. S. M. Kirova*, **23**, 60 (1957); *Chem. Abstr.*, **52**, 9980 (1958); F. Sparatore and F. Pagani, *Gazz. Chim. Ital.*, **91**, 1294 (1961).
173. R. Huls and M. Renson, *Bull. Soc. Chim. Belge*, **65**, 684 (1956).
174. W. Hückel and M. Sachs, *Ann. Chem.*, **498**, 166 (1932).
175. L. C. Rinzema, J. Stoffelsma and J. F. Arens, *Rec. Trav. Chim.*, **78**, 354 (1959); W. Gustowski and T. Urbanski, *Roczniki Chem.*, **57**, 437 (1963).
176. S. Buhemann and E. R. Watson, *J. Chem. Soc.*, **85**, 456 (1904).
177. T. Kauffmann, *Angew. Chem.*, **76**, 206 (1964); T. Kauffmann, H. Henkler, E. Rauch and K. Lötzsch, *Chem. Ber.*, **98**, 912 (1965).
178. T. Kauffmann, W. Burkhardt and E. Rauch, *Angew. Chem. (Intern. Ed. Engl.)*, **6**, 170 (1967).
179. P. Ehrlich and F. Sachs, *Ber.*, **32**, 2341 (1899).
180. A. Schönberg and R. Michaelis, *J. Chem. Soc.*, 627 (1937).
181. F. Kröhnke, *Ber.*, **71**, 2583 (1938).
182. F. Sachs and E. Bry, *Ber.*, **34**, 118 (1901); F. Barrow and F. J. Thorneycroft, *J. Chem. Soc.*, 769 (1939).
183. T. K. Walker, *J. Chem. Soc.*, **125**, 1622 (1924).
184. A. Schönberg and R. C. Azzam, *J. Chem. Soc.*, 1428 (1939).
185. E. Bergmann, *J. Chem. Soc.*, 1628 (1937).
186. C. H. Schmidt, *Angew. Chem.*, **75**, 169 (1963).
187. J. E. Davis and J. C. Roberts, *J. Chem. Soc.*, 2173 (1956).
188. O. Tsuge, M. Mishinohara and M. Toshiro, *Bull. Chem. Soc. Japan*, 36, 1477 (1963).
189. L. Chardonnens and P. Heinrich, *Helv. Chim. Acta*, 27, 321 (1944).
190. I. Tanasescu and I. Nanu, *Ber.*, **75**, 650 (1942).
191. R. C. Azzam, *Proc. Egypt. Acad. Sci.*, **9**, 89 (1953); *Chem. Abstr.*, **50**, 16685, (1956).
192. A. McGookin, *J. Appl. Chem.*, **5**, 65 (1955).
193. J. Hamer and A. Macaluso, *Chem. Rev.*, **64**, 473 (1964).
194. P. A. S. Smith, *Open-Chain Nitrogen Compounds*, Vol. 2, W. A. Benjamin, 1966, p. 368.
195. F. Kröhnke, *Angew. Chem.*, **65**, 612 (1953); **75**, 317 (1963).
196. F. Kröhnke, *Chem. Ber.*, **80**, 303 (1947).

197. J. W. Ledbetter, D. N. Kramer and F. M. Miller, *J. Org. Chem.*, **32**, 1165 (1967).
198. W. Schulze and H. Willitzer, *J. Prakt. Chem.*, **21**, 168 (1963).
199. D. M. W. Anderson and F. Bell, *J. Chem. Soc.*, 3708 (1959).
200. O. Touster, *Org. Reactions*, **7**, 327 (1953).
201. W. L. Semon and V. R. Damerel, *Org. Synth.*, Coll. Vol. **2**, 204 (1943); A. F. Ferris, *J. Org. Chem.*, **24**, 1726 (1959).
202. D. C. Batesky and N. S. Moon, *J. Org. Chem.*, **24**, 1694 (1959).
203. A. F. Ferris, F. E. Gould, G. S. Johnson and H. Strange, *J. Org. Chem.*, **26**, 2602 (1961).
204. G. Hesse and G. Krehbiel, *Chem. Ber.*, **88**, 130 (1955).
205. N. Levin and W. H. Hartung, *Org. Synth.*, Coll. Vol. **3**, 191 (1955).
206. C. C. Singhal, *Agra Univ. J. Res.*, **11**, 241 (1962); *Chem. Abstr.*, **58**, 5558 (1963).
207. A. T. Austin, *Sci. Prog.*, **49**, 619 (1961).
208. J. Hannart and A. Bruylants, *Bull. Soc. Chim. Belges*, **72**, 423 (1963).
209. E. Bamberger, *Ber.*, **33**, 1781 (1900).
210. L. J. Khmelnitskii, S. S. Novikov and O. V. Lebedev, *Izv. Akad. Nauk SSSR*, 2019 (1960); *Chem. Abstr.*, **55**, 19833 (1961).
211. J. Thiele, *Ber.*, **33**, 666 (1900).
212. C. F. Koelsch, *J. Org. Chem.*, **26**, 1291 (1961).
213. S. E. Forman, *J. Org. Chem.*, **29**, 3323 (1964).
214. Y. Ashani, H. Edery, J. Zehavy, Y. Künberg and S. Cohen, *Israel J. Chem.*, **3**, 133 (1965).
215. M. Ogata, *Chem. Pharm. Bull.*, **11**, 1517 (1963).
216. M. Fedorchuk and F. T. Semeniuk, *J. Pharm. Sci.*, **52**, 733 (1963); *Chem. Abstr.*, **63**, 11350 (1965).
217. E. J. Moriconi and F. J. Creegan, *J. Org. Chem.*, **31**, 2090 (1966).
218. U. S. Seth and S. S. Deshapande, *J. Indian Chem. Soc.*, **29**, 539 (1952).
219. E. Bamberger and W. Pemsel, *Ber.*, **36**, 57 (1903).
220. E. Bickert and H. Kössel, *Ann. Chem.*, **662**, 83 (1963).
221. T. A. Turney and G. A. Wright, *Chem. Rev.*, **59**, 497 (1959).
222. N. V. Sidgwick, *The Organic Chemistry of Nitrogen*, 3rd Ed., Revised and rewritten by I. T. Miller and H. D. Springall, Clarendon Press, Oxford, 1966, p. 312.
223. K. Singer and P. M. Vamplew, *J. Chem. Soc.*, 3052 (1957).
224. E. Müller, H. Metzger, D. Fries, U. Heuschkel, K. Witte, E. Waidelich and G. Schmid, *Angew. Chem.*, **71**, 229 (1959); E. Müller, H. G. Padeken, M. Salamon and G. Fiedler, *Chem. Ber.*, **98**, 1893 (1965).
225. E. Müller and G. Schmid, *Chem. Ber.*, **94**, 1364 (1961).
226. C. Matasa, *Chem. Ind.* (*Paris*), **91**, 57 (1964); *Chem. Abstr.*, **63**, 9831 (1965).
227. D. B. Parihar, *Chem. Ind.* (*London*), 1227 (1966).
228. E. M. Burges and E. M. Lavanish, *Tetrahedron Letters*, 1221 (1964).
229. Y. L. Chow, *Tetrahedron Letters*, 2333 (1964).
230. Y. L. Chow, *J. Am. Chem. Soc.*, **87**, 4642 (1965).
231. Y. L. Chow, *Chem. Commun.*, 330 (1967).
232. S. M. Parmeter, *Org. Reactions*, **10**, 1 (1959).
233. E. Enders in *Methoden der Organischen Chemie* (*Houben-Weyl*), 4th Ed., Vol. X/3, 1965, p. 467.

234. S. Hünig and O. Boes, *Ann. Chem.*, **579**, 28 (1953).
235. V. N. Drozd, *Bull. Acad. Sci. SSSR*, 1855 (1965) (Engl. transl.).
236. A. A. Kharkharov, *Zh. Obshch. Khim.*, **23**, 1175 (1958); *Chem. Abstr.*, **47**, 12390 (1953).
237. A. E. Porai-Koshits and A. A. Kharkharov, *Bull. Acad. Sci. URSS*, 79 (1944); *Chem. Abstr.*, **39**, 1631 (1945).
238. R. R. Phillips, *Org. Reactions*, **10**, 143 (1959).
239. C. Bülow and E. Hailer, *Ber.*, **35**, 915 (1902).
240. H. J. Teuber, *Chem. Ber.*, **98**, 2111 (1965).
241. C. Dimroth and C. Hartmann, *Ber.*, **40**, 2404, 4460 (1907); **41**, 4012 (1908).
242. D. Y. Curtin and M. J. Poutsma, *J. Am. Chem. Soc.*, **84**, 4887, 4892 (1962).
243. H. Hennecka, H. Timmler, R. Lorenz and W. Geiger, *Chem. Ber.*, **90**, 1060 (1957).
244. H. C. Yao and P. Resnick, *J. Am. Chem. Soc.*, **84**, 3514 (1962).
245. B. Eistert, *Ann. Chem.*, **666**, 97 (1963); *Chem. Ber.*, **96**, 2290, 2304, 3120 (1963).
246. B. Heath-Brown and P. G. Philpott, *J. Chem. Soc.*, 7185 (1965).
247. D. S. Tarbel, C. W. Todd, M. C. Paulson, E. G. Lindstrom and V. P. Wystrach, *J. Am. Chem. Soc.*, **70**, 1381 (1948).
248. K. Clusius, H. Hurzeler, R. Huisgen and H. J. Koch, *Naturwissenschaften*, **41**, 213 (1954); R. Huisgen, *Angew. Chem.*, **67**, 439 (1955); R. Huisgen and H. J. Koch, *Ann. Chem.*, **591**, 200 (1955).
249. C. W. Kruse and R. F. Kleinschmidt, *J. Am. Chem. Soc.*, **83**, 213 (1961).
250. N. S. Kozlov and L. U. Pinegina, *Tr. Perm. Sel.-Khoz. Inst.*, **29**, 83 (1965); *Chem. Abstr.*, **67**, 10922 (1967).
251. E. Winterfeldt, *Angew. Chem.* (*Intern. Ed. Engl.*), **6**, 423 (1967).
252. N. R. Easton, R. D. Dillard, W. J. Doran, M. Livezey and D. E. Morrison, *J. Org. Chem.*, **26**, 3772 (1961).
253. I. L. Knunyants, L. S. German and B. L. Dyatkin, *Izv. Akad. Nauk SSSR*, 221 (1960); *Chem. Abstr.*, **54**, 20870 (1960).
254. T. Kauffmann, H. Müller and C. Kosel, *Angew. Chem.*, **74**, 284 (1962).
255. A. Alessandri, *Gazz. Chim. Ital.*, **54**, 426 (1924); G. Bruni and E. Geiger, *Rubber Chem. Technol.*, **1**, 177 (1928).
256. N. F. Hepfinger and C. E. Griffin, *Tetrahedron Letters*, 1361, 1365 (1963).
257. A. Alessandri, *Gazz. Chim. Ital.*, **55**, 729 (1925); **57** ,195 (1927); R. Pummerer and W. Gündel, *Chem. Ber.*, **61**, 1591 (1928); W. Gündel and R. Pummerer, *Ann. Chem.*, **529**, 11 (1937).
258. M. L. Scheinbaum, *J. Org. Chem.*, **29**, 2200 (1964).
259. L. J. Beckham, W. A. Fessler and M. A. Kise, *Chem. Rev.*, **48**, 319 (1951); R. Biela, I. Hahnemann, H. Panovsky and W. Pritzkow, *J. Prakt. Chem.*, **33**, 282 (1966); A. D. Treboganov, R. S. Astakhova, A. A. Kraevskii and N. A. Preobrazhenskii, *Zh. Org. Khim.*, **2**, 2178 (1966); *Chem. Abstr.*, **66**, 75730 (1967).
260. H. K. Wiese and P. E. Burton, *U.S. Pat.* 3,270,043 (1966); *Chem. Abstr.*, **66**, 10587 (1967); A. Nenz and G. Ribaldone, *Chim. Ind.* (*Milan*), **49**, 43 (1967); *Chem. Abstr.*, **66**, 104647 (1967).
261. J. Jaz and A. Gerbaux, *Angew. Chem.* (*Intern. Ed. Engl.*), **6**, 248 (1967).
262. R. Huisgen, L. Möbius and G. Szeinües, *Chem. Ber.*, **98**, 1138 (1965); R. E.

Harmon and D. L. Rector, *Chem. Ind.* (*London*), 1264 (1965); J. E. Franz, M. W. Dietrich, A. Henshall and C. Osuch, *J. Org. Chem.*, **31**, 2847 (1966).
263. R. Fusco, G. Bianchetti and D. Pocar, *Gazz. Chim. Ital.*, **91**, 849, 933 (1962); G. Bianchetti, P. D. Croce and D. Pocar, *Tetrahedron Letters*, 2043 (1965).
264. G. Smolinsky, *J. Am. Chem. Soc.*, **83**, 4483 (1961); *J. Org. Chem.*, **27**, 3557 (1962).
265. R. E. Banks and G. J. Moore, *J. Chem. Soc.* (*C*), 2304 (1966).
266. R. R. Harvey and K. W. Ratts, *J. Org. Chem.*, **31**, 3907 (1966).
267. (a) A. L. Logothetis and G. N. Sausen, *J. Org. Chem.*, **31**, 3689 (1966); (b) T. E. Stevens, *J. Org. Chem.*, **32**, 670 (1967).
268. F. A. Johnson, C. Haney and T. E. Stevens, *J. Org. Chem.*, **32**, 466 (1967).
269. H. Staudinger and J. Meyer, *Helv. Chim. Acta*, **2**, 635 (1919); H. Staudinger and E. Hauser, *Helv. Chim. Acta*, **4**, 861 (1921).
270. R. Appel and A. Hauss, *Chem. Ber.*, **93**, 405 (1960); *Z. Anorg. Allgem. Chem.*, **311**, 290 (1961); R. R. Schmidt, *Angew. Chem.*, **76**, 991 (1964); L. Horner and A. Gross, *Ann. Chem.*, **591**, 117 (1955).
271. W. S. Wadsworth, Jr., and W. D. Emmons, *J. Am. Chem. Soc.*, **84**, 1316 (1962).
272. A. W. Johnson, *Ylid Chemistry*, Academic Press, 1966, p. 227; A. W. Johnson and S. C. K. Wong, *Can. J. Chem.*, **44**, 2793 (1966).
273. W. S. Wadsworth, Jr. and W. D. Emmons, *J. Org. Chem.*, **29**, 2816 (1964).
274. G. Wittig and W. Haag, *Chem. Ber.*, **88**, 1654 (1955).
275. H. J. Bestmann, *Angew. Chem.*, **72**, 326 (1960); H. J. Bestmann and H. Fritzsche, *Chem. Ber.*, **94**, 601 (1961).
276. A. Senning, *Acta Chem. Scand.*, **18**, 1958 (1964); G. Kresze, D. Sommerfeld and R. Albrecht, *Chem. Ber.*, **98**, 601 (1965).
277. R. Albrecht, and G. Kresze, *Ger. Pat.* 1,230,785 (1966); *Chem. Abstr.*, **66**, 46230 (1967).
278. D. H. Clemens, A. J. Bell and J. L. O'Brien, *Tetrahedron Letters*, 1481, 1491 (1965).
279. G. Märkl, *Tetrahedron Letters*, 811 (1961).
280. G. Wittig and M. Schlosser, *Tetrahedron*, **18**, 1023 (1962).
281. H. Hoffmann, *Chem. Ber.*, **95**, 2563 (1962).
282. J. J. Monagle, T. W. Campbell and H. F. McShane, Jr., *J. Am. Chem. Soc.*, **84**, 4288 (1962).
283. J. J. Monagle, *J. Org. Chem.*, **27**, 3851 (1962).
284. T. W. Campbell and J. J. Monagle, *J. Am. Chem. Soc.*, **84**, 1493 (1962); *Org. Synth.*, **43**, 31 (1963); T. W. Campbell, J. J. Monagle and V. S. Foldi, *J. Am. Chem. Soc.*, **84**, 3673 (1962).
285. K. Hunger, *Tetrahedron Letters*, 5929 (1966).
286. U. Ulrich and A. A. Sayigh, *Angew. Chem.*, **74**, 900 (1962).
287. A. Messmer, I. Pinter and F. Szago, *Angew. Chem.*, **76**, 227 (1964).
288. F. Kurzer and K. Douraghi-Zadeh, *Chem. Rev.*, **67**, 107 (1967).
289. S. Trippett and D. M. Walker, *J. Chem. Soc.*, 3874 (1959).
290. U. Schöllkopf, *Angew. Chem.*, **71**, 260 (1959).
291. A. W. Johnson, *J. Org. Chem.*, **28**, 252 (1963).
292. A. Schönberg and K. H. Borowski, *Chem. Ber.*, **92**, 2602 (1959).
293. A. W. Johnson and J. O. Martin, *Chem. Ind.* (*London*), 1726 (1965).

294. A. W. Johnson, *Chem. Ind. (London)*, 1119 (1963).
295. F. Kröhnke, *Chem. Ber.*, **83**, 253 (1950).
296. G. W. Brown, R. C. Cookson, I. D. R. Stevens, T. C. W. Mak and J. Trotter, *Proc. Chem. Soc.*, 87 (1964).
297. G. W. Brown, R. C. Cookson and I. D. R. Stevens, *Tetrahedron Letters*, 1263 (1964).
298. A. Umani-Ronchi, P. Bravo and G. Guadiano, *Tetrahedron Letters*, 3477 (1966); G. Guadiano, A. Umani-Ronchi, P. Bravo and M. Acompora, *Tetrahedron Letters*, 107 (1967).
299. R. Huisgen and J. Wulff, *Tetrahedron Letters*, 917, 921 (1967).
300. R. Roger and D. G. Neilson, *Chem. Rev.*, **61**, 179 (1961).
301. A. Bühner, *Ann. Chem.*, **333**, 289 (1904).
302. H. Bredereck, R. Gompper, K. Klemm and H. Rempfer, *Chem. Ber.*, **92**, 837 (1959).
303. J. W. Janus, *J. Chem. Soc.*, 355 (1955).
304. R. E. Benson and T. L. Cairns, *Org. Synth.*, Coll. Vol. **4**, 588 (1963).
305. S. Petersen and E. Tietze, *Ann. Chem.*, **623**, 166 (1959); H. Bredereck, F. Effenberger and G. Simchen, *Angew. Chem.*, **73**, 493 (1961).
306. A. T. Fuller and H. King, *J. Chem. Soc.*, 963 (1947).
307. W. Lossen, *Ann. Chem.*, **252**, 170 (1889).
308. W. Hechelkammer, *Ger. Pat.* 948,973 (1956); *Chem. Abstr.*, **53**, 6088 (1959).
309. C. J. M. Stirling, *J. Chem. Soc.*, 255 (1960); H. E. Zaugg, R. G. Michaels, A. D. Schaefer, A. M. Wenthe and W. H. Washburn, *Tetrahedron*, **22**, 1257 (1966).
310. B. A. Cunningham and G. L. Schmir, *J. Org. Chem.*, **31**, 3751 (1966).
311. H. W. Heine, *J. Am. Chem. Soc.*, **78**, 3708 (1956).
312. C. Zioudrou and G. L. Schmir, *J. Am. Chem. Soc.*, **85**, 3258 (1963).
313. W. Seeliger and W. Thier, *Ann. Chem.*, **698**, 158 (1966).
314. A. Bernthsen, *Ann. Chem.*, **197**, 341 (1879); F. E. Condo, E. T. Hinkel, A. Fassero and R. L. Shriner, *J. Am. Chem. Soc.*, **59**, 230 (1937); B. Böttcher and F. Bauer, *Ann. Chem.*, **568**, 218 (1950); R. Boudet, *Compt. Rend.*, **239**, 1803 (1954); H. Bredereck, R. Gompper and H. Seiz, *Chem. Ber.*, **90**, 1837 (1957); M. Bercot-Vatteroni, *Ann. Chim. (Paris)*, **7**, 312 (1962).
315. R. Boudet, *Bull. Soc. Chim. France*, 377 (1951).
316. D. A. Peak and F. Stansfield, *J. Chem. Soc.*, 4067 (1952); P. A. S. Smith and J. M. Sullivan, *J. Org. Chem.*, **26**, 1132 (1961).
317. J. J. Donleavy, *J. Am. Chem. Soc.*, **58**, 1004 (1936).
318. T. Bachetti and A. Alemagna, *Rend. Inst. Lombardo Sci.*, **91 I**, 30 (1957); *Chem. Abstr.*, **52**, 11749 (1958).
319. E. Fromm and M. Bloch, *Ber.*, **32**, 2212 (1899); M. Delepine, *Bull. Soc. Chim. France*, [3], **29**, 53 (1903).
320. R. F. Coles and H. A. Levine, *U.S. Pat.* 2,942,025 (1960); *Chem. Abstr.*, **54**, 24464 (1960).
321. E. P. Nesyanov, M. M. Besprozvannaya and P. S. Pelkis, *Zh. Org. Khim.*, **2**, 1891 (1966); *Chem. Abstr.*, **66**, 55138 (1967).
322. M. Busch and K. Schulz, *J. Prakt. Chem.*, **150**, 173 (1938).
323. G. D. Lander, *J. Chem. Soc.*, **83**, 320 (1903).
324. H. Peter, M. Brugger, J. Schreiber and A. Eschenmoser, *Helv. Chim. Acta*, **46**, 577 (1963).

325. W. Lossen, *Ann. Chem.*, **281**, 169 (1894).
326. F. Arndt and H. Scholz, *Ann. Chem.*, **510**, 62 (1934).
327. F. Arndt and C. Martius, *Ann. Chem.*, **499**, 228 (1932).
328. R. M. Herbst and D. Shemin, *Org. Synth.*, Coll. Vol. **2**, 1 (1943); J. S. Buck and W. S. Ide, *Org. Synth.*, Coll. Vol. **2**, 55 (1943); H. E. Carter, *Org. Reactions*, **3**, 198 (1946); Y. Iwakura, F. Toda and H. Suzoki, *J. Org. Chem.*, **32**, 440 (1967).
329. *Neth. Pat.* 6,600,777 (1966); *Chem. Abstr.*, **66**, 18707 (1967).
330. R. J. Cotter, C. K. Sauers and J. M. Whelan, *J. Org. Chem.*, **26**, 10 (1961).
331. W. R. Roderick and P. L. Bhatia, *J. Org. Chem.*, **28**, 2018 (1963).
332. W. Seeliger and W. Diepers, *Ann. Chem.*, **697**, 171 (1966).
333. W. M. Whaley and T. R. Govindachari, *Org. Reactions*, **6**, 74 (1951).
334. G. V. Glazneva, E. K. Mamaeva and A. P. Zeif, *Zh. Obshch. Khim.*, **36**, 1499 (1966); *Chem. Abstr.*, **66**, 18652 (1967).
335. K. Nagata and S. Mizukami, *Chem. Pharm. Bull.*, **14**, 1249, 1263 (1966); **15**, 61 (1967).
336. H. Rivier and J. Zeltner, *Helv. Chim. Acta*, **20**, 691 (1937); J. R. Shaeffer. C. T. Goodhue, H. A. Risley and R. E. Stevens, *J. Org. Chem.*, **32**, 392 (1967),
337. W. Walter and K. D. Bode, *Ann. Chem.*, **698**, 122 (1966).
338. A. Sonn and E. Müller, *Ber.*, **52**, 1927 (1919); C. L. Stevens and J. C. French, *J. Am. Chem. Soc.*, **75**, 657 (1953); P. A. S. Smith and N. W. Kalenda, *J. Org. Chem.*, **23**, 1599 (1958).
339. E. Mosettig, *Org. Reactions*, **8**, 240 (1954).
340. J. von Braun, F. Jostes and A. Heymons, *Ber.*, **60**, 92 (1927); J. von Braun, F. Jostes and W. Münch, *Ann. Chem.*, **453**, 113 (1927).
341. W. Lossen, *Ann. Chem.*, **252**, 216 (1889).
342. H. von Pachmann and L. Seeberger, *Ber.*, **27**, 2121 (1894); R. Stolle, *J. Prakt. Chem.*, [2], **75**, 416 (1907).
343. H. H. Bosshard and H. Zollinger, *Helv. Chim. Acta*, **42**, 1659 (1959); L. A. Paquette, B. A. Johnson and F. M. Hinga, *Org. Synth.*, **46**, 18 (1966); G. Hazebroucq, *Ann. Pharm. Franc.*, **24**, 793 (1966); C. Jutz and W. Müller, *Chem. Ber.*, **100**, 1536 (1967).
344. H. Eilingsfeld, M. Seefelder and H. Weidinger, *Chem. Ber.*, **96**, 2671 (1963).
345. M. Montagne and G. Rousseau, *Compt. Rend.*, **196**, 1165 (1933).
346. V. Grignard and R. Escourrou, *Compt. Rend.*, **180**, 1883 (1925).
347. E. Ichikawa and K. Odo, *Yuki Gosei Kagaku Kayokai Shi*, **24**, 1241 (1966); *Chem. Abstr.*, **67**, 54786 (1967).
348. W. Nagata, *Tetrahedron*, **13**, 287 (1961).
349. L. M. Soffer and M. Katz, *J. Am. Chem. Soc.*, **78**, 1705 (1956).
350. H. Stephen, *J. Chem. Soc.*, **127**, 1874 (1925); P. S. Pyrylalova and E. N. Zilberman, *Izv. Vyssh. Ucheb. Zaved., Khim. Khim. Tekhnol.*, **9**, 912 (1966); *Chem. Abstr.*, **67**, 27049 (1967).
351. T. L. Tolbert and B. Houston, *J. Org. Chem.*, **28**, 695 (1963).
352. S. Pietra and C. Trinchera, *Gazz. Chim. Ital.*, **85**, 1705 (1955); W. W. Zajak, Jr. and H. H. Denk, *J. Org. Chem.*, **27**, 3716 (1962).
353. N. I. Shirokova, T. V. Krasanova and I. V. Alexandrov, *Z. Priklad. Khim.*, **33**, 746 (1960); *Chem. Abstr.*, **54**, 20871 (1960); E. N. Zilberman and A. E. Kulikova, *Zh. Obshch. Khim.*, **29**, 3039 (1959); *Chem. Abstr.*, **54**, 11979 (1960); P. Reynaud and R. C. Moreau, *Bull. Soc. Chim. France*, 2002

(1960); F. C. Schaefer and G. A. Peters, *J. Org. Chem.*, **26**, 412 (1961); G. Sosnovsky and P. Schneider, *Tetrahedron*, **19**, 1313 (1963).

354. H. Meerwein, P. Laasch, R. Mersch and J. Spille, *Chem. Ber.*, **89**, 209 (1956).
355. R. W. Addor, *J. Org. Chem.*, **29**, 738 (1964).
356. M. Hadayatullah, *Bull. Soc. Chim. France*, 416, 422, 428 (1967).
357. D. H. R. Barton, J. M. Beaton, L. E. Geller and M. M. Pechet, *J. Am. Chem. Soc.*, **82**, 2640 (1960).
358. H. A. Bruson, E. Riener and T. Riener, *J. Am. Chem. Soc.*, **70**, 483 (1948).
359. F. Cramer, K. Pawelzik and H. J. Baldauf, *Chem. Ber.*, **91**, 1049 (1958).
360. H. M. Woodburn and C. E. Sroog, *J. Org. Chem.*, **17**, 371 (1952).
361. H. M. Woodburn, A. B. Whitehorse and B. G. Paulter, *J. Org. Chem.*, **24**, 210 (1959).
362. J. Houben and R. Zivadinovitsch, *Ber.*, **69**, 2352 (1936).
363. H. M. Woodburn and B. G. Paulter, *J. Org. Chem.*, **19**, 863 (1954).
364. P. Oxley, M. W. Partridge and W. F. Short, *J. Chem. Soc.*, 1110 (1947).
365. W. S. Zehrung III and H. M. Woodburn, *J. Org. Chem.*, **24**, 1333 (1959).
366. M. Frankel, Y. Knobler and G. Zvilichovsky, *J. Chem. Soc.*, 3127 (1963).
367. J. T. Shaw and F. J. Gross, *J. Org. Chem.*, **24**, 1809 (1959); A. B. Sen and P. R. Singh, *J. Indian Chem. Soc.*, **39**, 41 (1962).
368. J. Schreiber and B. Witkop, *J. Am. Chem. Soc.*, **86**, 2441 (1964).
369. K. Lempert and G. Doleschall, *Monatsh. Chem.*, **95**, 950 (1964).
370. J. Goerdeler and H. Porrmann, *Chem. Ber.*, **94**, 2859 (1961).
371. D. Zoellner and A. Mewsen, *Z. Anorg. Allgem. Chem.*, **349**, 19 (1966).
372. J. Kranz, *French Pat.* 1,446,964 (1966); *Chem. Abstr.*, **66**, 55509 (1967).
373. R. Engelhardt, *J. Prakt. Chem.*, [2], **54**, 143 (1896); R. L. Hinman and D. Fulton, *J. Am. Chem. Soc.*, **80**, 1895 (1958).
374. R. Lenaers, C. Moussebois and F. Eloy, *Helv. Chim. Acta*, **45**, 441 (1962).
375. F. Eloy and R. Lenaers, *Chem. Rev.*, **62**, 155 (1962).
376. P. Schmidt, K. Eichenberger and M. Wilhelm, *Helv. Chim. Acta*, **45**, 996 (1962).
377. M. A. McGee, H. D. Morduch, G. T. Newbold, J. Redpath and F. S. Spring, *J. Chem. Soc.*, 1989 (1960).
378. F. F. Blicke, A. J. Zambito and R. E. Stenseth, *J. Org. Chem.*, **26**, 1826 (1961).
379. H. Adkins and G. M. Whitman, *J. Am. Chem. Soc.*, **64**, 150 (1942); A. Dornow, I. Kühlke and F. Baxmann, *Chem. Ber.*, **82**, 254 (1949).
380. K. Ziegler, H. Ohlinger and H. Eberle, *Ann. Chem.*, **504**, 94 (1933).
381. K. Ziegler, H. Ohlinger and H. Eberle, *Ger. Pat.* 591,269 (1934); *Chem. Abstr.*, **28**, 2364 (1934).
382. T. Takata, *Bull. Chem. Soc. Japan*, **35**, 1438 (1962).
383. J. B. Culberston, D. Butterfield, O. Kolewe and R. Shaw, *J. Org. Chem.*, **27**, 729 (1962); S. J. Storfer and E. I. Becker, *J. Org. Chem.*, **27**, 1868 (1962); F. A. E. Schilling, *U.S. Pat.* 3,280,196 (1966); *Chem. Abstr.*, **66**, 10849 (1967).
384. C. Moureu and G. Mignonac, *Ann. Chim. (Paris)*, [9], **14**, 322 (1920); G. E. P. Smith, Jr., and F. W. Bergstrom, *J. Am. Chem. Soc.*, **56**, 2095 (1934); J. B. Cloke, *J. Am. Chem. Soc.*, **62**, 117 (1940); P. L. Pickard and D. J. Vaugham, *J. Am. Chem. Soc.*, **72**, 5017 (1950).

385. P. L. Pickard and T. L. Tolbert, *J. Org. Chem.*, **26**, 4886 (1961).
386. H. L. Lochte, J. Horeczy, P. L. Pickard and A. D. Barton, *J. Am. Chem. Soc.*, **70**, 2012 (1948).
387. P. L. Pickard and D. J. Vaughan, *J. Am. Chem. Soc.*, **72**, 876 (1950).
388. H. Edelstein and E. I. Becker, *J. Org. Chem.*, **31**, 3375 (1966).
389. P. E. Spoerri and A. S. DuBois, *Org. Reactions*, **5**, 387 (1949).
390. H. Fischer, K. Schneller and W. Zerweck, *Ber.*, **55**, 2390 (1922); H. Fischer, B. Weiss and M. Schubert, *Ber.*, **56**, 1194 (1923); J. Houben and W. Fischer, *Ber.*, **66**, 339 (1933).
391. R. Seka, *Ber.*, **56**, 2058 (1923).
392. W. E. Truce, *Org. Reactions*, **9**, 37 (1959).
393. T. Saegusa, I. Ito and S. Kobayashi, *Tetrahedron Letters*, 521 (1967).
394. T. Saegusa, Y. Ito, S. Kobayashi, N. Takeda and K. Hirota, *Tetrahedron Letters*, 1273 (1967).
395. T. Saegusa, Y. Ito, S. Kobayashi, K. Hirota and H. Yoshioka, *Tetrahedron Letters*, 6121 (1966).
396. M. Passerini, *Gazz. Chim. Ital.*, **57**, 452 (1927); M. Okano, Y. Ito, T. Shono and R. Oda, *Bull. Chem. Soc. Japan*, **36**, 1314 (1963).
397. M. Lipp, F. Dallacker and I. Meier-Köcker, *Monatsh. Chem.*, **90**, 41 (1959); H. W. Johnson, Jr., and P. H. Daughetee, Jr., *J. Org. Chem.*, **29**, 246 (1964).
398. J. U. Neff, *Ann. Chem.*, **280**, 291 (1894).
399. H. J. Havlik and M. M. Wald, *J. Am. Chem. Soc.*, **77**, 5171 (1955).
400. I. Ugi and U. Fetzer, *Chem. Ber.*, **94**, 2239 (1961).
401. I. Ugi, *Angew. Chem.*, **74**, 9 (1962).
402. S. P. Makarov, V. A. Shpanskii, V. A. Ginsburg, A. I. Shchekotichin, A. S. Filatov, L. L. Martinova, I. V. Pavlovskaya, A. F. Golovaneva and A. Y. Yakubovich, *Dokl. Akad. Nauk SSSR*, **142**, 596 (1962); *Chem. Abstr.*, **57**, 4528 (1962).
403. G. B. Bachman and L. E. Strom, *J. Org. Chem.*, **28**, 1150 (1963).
404. H. Wieland, *Ber.*, **40**, 418 (1907); C. Grundmann and J. M. Dean, *J. Org. Chem.*, **30**, 2809 (1965).
405. G. Zinner and H. Günther, *Angew. Chem.*, **76**, 440 (1964).
406. H. Wieland, *Ber.*, **40**, 1667 (1907); H. Wieland and B. Rosenfeld, *Ann. Chem.*, **484**, 236 (1930).
407. H. J. Bestmann and R. Kunstmann, *Angew. Chem.* (*Intern. Ed. Engl.*), **5**, 1039 (1966).
408. R. Scholl and F. Kacer, *Ber.*, **36**, 322 (1903).
409. H. Wieland, *Ann. Chem.*, **475**, 54 (1929).
410. L. Birckenbach and K. Sennewald, *Ann. Chem.*, **489**, 7 (1931); *Chem. Ber.*, **65**, 546 (1932).
411. R. Scholl, *Ber.*, **27**, 2816 (1894).
412. A. L. Nussbaum and C. H. Robinson, *Tetrahedron*, **17**, 3 (1962).
413. H. Staudinger and R. Endle, *Ber.*, **50**, 1042 (1917); E. Niwa, H. Aoki, H. Tanaka, K. Munakata and M. Namiki, *Chem. Ber.*, **99**, 3932 (1966).
414. C. King, *J. Org. Chem.*, **25**, 352 (1960).
415. M. Lora-Tamayo, R. Madranero, D. Gracian and V. Gomez-Parra, *Tetrahedron Suppl.*, **8**, 305 (1966).

416. K. A. Jensen, M. Due, H. Holm and C. Wentrup, *Acta Chem. Scand.*, **20**, 2091 (1966); E. Grigat and R. Pütter, *Angew. Chem.* (*Intern. Ed. Engl.*), **6**, 206 (1967).
417. R. E. Miller, *J. Org. Chem.*, **25**, 2126 (1960).
418. V. E. Haury, *U.S. Pat.* 2,421,937 (1947); *Chem. Abstr.*, **41**, 5892 (1947); A. A. Balandin and N. A. Vasyunina, *Dokl. Akad. Nauk SSSR*, **103**, 831 (1955); *Chem. Abstr.*, **50**, 9283 (1956).
419. J. J. Ritter, *J. Am. Chem. Soc.*, **55**, 3322 (1933).
420. C. M. Rosser and J. J. Ritter, *J. Am. Chem. Soc.*, **59**, 2179 (1937).
421. E. A. Calderon, *Anales Asoc. Quim. Arg.*, **35**, 149 (1947); *Chem. Abstr.*, **42**, 7744 (1948).
422. C. R. Hauser, W. R. Brasen, P. S. Skell, S. W. Kantor and A. E. Brodhag, *J. Am. Chem. Soc.*, **78**, 1653 (1956).
423. J. C. Duff and V. I. Furness, *J. Chem. Soc.*, 1512 (1951).
424. P. J. McLaughlin and E. C. Wagner, *J. Am. Chem. Soc.*, **66**, 251 (1944).
425. S. J. Angyal, *Org. Reactions*, **8**, 197 (1954).
426. R. A. Benkeser and C. E. DeBoer, *J. Org. Chem.*, **21**, 281 (1956); H. S. Mosher and E. J. Blanz, Jr., *J. Org. Chem.*, **22**, 445 (1957).
427. S. Goldschmidt and V. Voeth, *Ann. Chem.*, **435**, 265 (1923).
428. S. Goldschmidt and W. Beuchel, *Ann. Chem.*, **447**, 197 (1926).
429. H. Shechter and S. S. Rawalay, *J. Am. Chem. Soc.*, **86**, 1706 (1964).
430. I. Okamora, R. Sokurai and T. Tanabe, *Chem. High Polymers* (*Japan*), **9**, 279, 284 (1952); *Chem. Abstr.*, **48**, 4225 (1954).
431. I. Okamora and R. Sakurai, *Chem. High Polymers* (*Japan*), **9**, 434 (1952); *Chem. Abstr.*, **48**, 9079 (1954).
432. H. Shechter, S. S. Rawalay and M. Tubis, *J. Am. Chem. Soc.*, **86**, 1701 (1964).
433. W. Herzberg and H. Lange, *Ger. Pat.* 482,837 (1925); *Chem. Abstr.*, **24**, 626 (1925); A. Skita, *Chem. Ber.*, **48**, 1685 (1915).
434. O. Bayer in *Methoden der Organische Chemie* (*Houben-Weyl*), 4th. Ed., Vol. VII/1, 1954, p. 206.
435. E. F. Pratt and T. P. McGovern, *J. Org. Chem.*, **29**, 1540 (1964).
436. A. A. R. Sayigh and U. Ulrich, *J. Org. Chem.*, **27**, 4662 (1962).
437. K. Kahr and C. Berther, *Chem. Ber.*, **93**, 132 (1960).
438. E. Bamberger and R. Seligmann, *Ber.*, **36**, 701 (1903); I. Okamura and R. Sakurai, *Chem. High Polymers* (*Japan*), **9**, 10, 230 (1952); *Chem. Abstr.*, **48**, 9933, 11794 (1954).
439. E. S. Huyser, C. J. Bredeweg and R. M. VanScoy, *J. Am. Chem. Soc.*, **86**, 4148 (1964).
440. H. E. DeLaMare, *J. Org. Chem.*, **25**, 2114 (1960).
441. G. M. Coppinger and J. D. Swalen, *J. Am. Chem. Soc.*, **83**, 4900 (1961).
442. L. Hellerman and A. G. Sanders, *J. Am. Chem. Soc.*, **49**, 1742 (1927); W. E. Bachmann, M. P. Cava and A. S. Dreiding, *J. Am. Chem. Soc.*, **76**, 5554 (1954).
443. H. Lettre and L. Knof, *Chem. Ber.*, **93**, 2860 (1960); A. Brossi, F. Schenker and W. Leimburger, *Helv. Chim. Acta*, **47**, 2089 (1964).
444. K. Langheld, *Ber.*, **42**, 2360 (1909).
445. (a) A. R. Battersby and R. Binks, *J. Chem. Soc.*, 4333 (1958);
(b) R. Bonnett, V. M. Clark, A. Giddy and A. Todd, *J. Chem. Soc.*, 2087, 2094 (1959).

446. J. Knabe and G. Grund, *Arch. Pharm.*, **296**, 820 (1963).
447. F. Bohlmann and C. Arndt, *Chem. Ber.*, **91**, 2167 (1958); N. J. Leonard, L. A. Miller and P. D. Thomas, *J. Am. Chem. Soc.*, **78**, 3463 (1956).
448. R. I. Fryer, G. A. Archer, B. Brust, W. Zally and L. H. Sternbach, *J. Org. Chem.*, **30**, 1306 (1965).
449. R. G. R. Bacon, H. J. W. Hanna, D. J. Munro and D. Stewart, *Proc. Chem. Soc.*, 113 (1962).
450. A. Stojiljkovic, V. Andrejevic and M. L. Mihailovi, *Tetrahedron*, **23**, 721 (1967).
451. R. F. Merritt and J. K. Ruff, *J. Am. Chem. Soc.*, **86**, 1392 (1964).
452. D. H. Rosenblatt, L. A. Hull, D. C. De Luca, G. T. Davis, R. C. Weglein and H. K. R. Williams, *J. Am. Chem. Soc.*, **89**, 1158 (1967); L. A. Hull, G. T. Davis, D. H. Rosenblatt, H. K. R. Williams and R. C. Weglein, *J. Am. Chem. Soc.*, **89**, 1163 (1967).
453. P. A. S. Smith and H. G. Pars, *J. Org. Chem.*, **24**, 1325 (1959); P. A. S. Smith and R. N. Loeppky, *J. Am. Chem. Soc.*, **89**, 1147 (1967).
454. J. Meisenheimer, *Ber.*, **46**, 1148 (1913); H. Böhme and W. Krause, *Chem. Ber.*, **84**, 170 (1951); D. Beke and E. Eckhart, *Chem. Ber.*, **95**, 1059 (1962).
455. E. Schmidt and H. Fischer, *Ber.*, **53**, 1529 (1920).
456. M. Polonovski, *Bull. Soc. Chim. France*, **41**, 1186 (1927).
457. H. Z. Lecher and W. B. Hardy, *J. Am. Chem. Soc.*, **70**, 3789 (1948).
458. H. Böhme, E. Mundlos and O. E. Herboth, *Chem. Ber.*, **90**, 2003 (1957); H. Böhme and K. Hartke, *Chem. Ber.*, **93**, 1305 (1960).
459. I. W. Kissinger and H. E. Ungnade, *J. Org. Chem.*, **25**, 1471 (1960).
460. P. A. S. Smith and G. E. Hein, *J. Am. Chem. Soc.*, **82**, 5731 (1960).
461. N. B. Shitova, K. I. Matveev and M. M. Danilova, *Kinet. Katal.*, **7**, 995 (1966); *Chem. Abstr.*, **66**, 54790 (1967).
462. T. Emery and J. B. Nielands, *J. Am. Chem. Soc.*, **82**, 4903 (1960).
463. A. Angeli, L. Alessandri and M. Aiazzi-Mancini, *Atti Reale Accad. Linzei*, [5], **20 I**, 546 (1911); *Chem. Abstr.*, **5**, 3403 (1911); P. Grammaticakis, *Compt. Rend.*, **224**, 1066 (1947).
464. H. Franzen and F. Kraft, *J. Prakt. Chem.*, [2], **84**, 127 (1911).
465. S. Wawzonek and W. McKillip, *J. Org. Chem.*, **27**, 3946 (1962).
466. (a) M. Busch and K. Lang, *J. Prakt. Chem.*, [2], **144**, 291 (1936);
(b) R. A. Abramovitch and B. A. Davies, *Chem. Rev.*, **64**, 149 (1964).
467. E. C. Kooyman, *Rec. Trav. Chim.*, **74**, 117 (1955).
468. M. F. Bartlett, B. F. Lambert and W. I. Taylor, *J. Am. Chem. Soc.*, **86**, 729 (1964).
469. R. A. Mitsch, *J. Am. Chem. Soc.*, **87**, 328 (1965).
470. R. E. Banks, R. N. Haszeldine and R. Hatton, *J. Chem. Soc. (C)*, 427 (1967).
471. M. F. Grundon and B. E. Reynolds, *J. Chem. Soc.*, 2445 (1964).
472. O. Fischer, *Ann. Chem.*, **241**, 331 (1887); T. Curtius, *J. Prakt. Chem.* [2], **62**, 83 (1900); J. Thiele, *Ann. Chem.*, **376**, 239 (1910).
473. C. Holstead and A. H. Lamberton, *J. Chem. Soc.*, 1886 (1952).
474. W. Paterson and G. R. Proctor, *Proc. Chem. Soc.*, 248 (1961); W. M. Speckamp, H. de Koning, U. K. Pandit and H. O. Huisman, *Tetrahedron*, **21**, 2517 (1965); E. Negishi and A. R. Day, *J. Org. Chem.*, **30**, 43 (1965).

475. W. Paterson and G. R. Proctor, *J. Chem. Soc.*, 485 (1965).
476. M. S. Newman and I. Ungar, *J. Org. Chem.*, **27**, 1238 (1962).
477. G. Zinner and W. Ritter, *Arch. Pharm.*, **296**, 681 (1963).
478. D. M. Lemal, F. Menger and E. Coats, *J. Am. Chem. Soc.*, **86**, 2395 (1964).
479. T. D. Stewart and W. H. Bradley, *J. Am. Chem. Soc.*, **54**, 4172 (1932); T. D. Stewart and H. P. Kung, *J. Am. Chem. Soc.*, **55**, 4813 (1933); J. P. Mason and M. Zief, *J. Am. Chem. Soc.*, **62**, 1450 (1940).
480. H. G. Reiber and T. D. Stewart, *J. Am. Chem. Soc.*, **62**, 3026 (1940).
481. K. A. Petrov and A. A. Neimysheva, *Zh. Obshch. Khim.*, **29**, 2165, 2169, 2695 (1959); *Chem. Abstr.*, **54**, 10912 (1960).
482. E. Müller, H. Metzger and D. Fries, *Chem. Ber.*, **87**, 1449 (1954).
483. S. Hünig and H. Herrmann, *Ann. Chem.*, **636**, 21 (1960); S. Hünig and K. H. Oette, *Ann. Chem.*, **640**, 98 (1961); S. Hünig and F. Müller, *Ann. Chem.*, **651**, 73, 89 (1962); S. Hünig, H. Balli, H. Conrad and A. Schott, *Ann. Chem.*, **676**, 36 (1964); S. Hünig, S. Geiger, G. Kaupp and W. Kniese, *Ann. Chem.*, **697**, 116 (1966).
484. C. Eberhardt and A. Welter, *Ber.*, **27**, 1804 (1898); A. Eibner, *Ann. Chem.*, **302**, 349 (1898); C. A. Bischoff and F. Reinfeld, *Ber.*, **36**, 41 (1903).
485. J. Goerdeler and H. Ruppert, *Chem. Ber.*, **96**, 1630 (1963).
486. C. L. Stevens and G. H. Singhal, *J. Org. Chem.*, **29**, 34 (1964).
487. R. Stolle, *J. Prakt. Chem.* [2], **73**, 277 (1906); F. D. Chattaway and A. J. Walker, *J. Chem. Soc.*, **127**, 2407 (1925).
488. K. Johnson and E. F. Degering, *J. Am. Chem. Soc.*, **61**, 3194 (1939).
489. J. von Braun and E. Danziger, *Ber.*, **46**, 103 (1913).
490. N. Kornblum and J. H. Eicher, *J. Am. Chem. Soc.*, **78**, 1494 (1956).
491. H. B. Hass and M. L. Bender, *J. Am. Chem. Soc.*, **71**, 1767 (1949); *Org. Synth.*, Coll. Vol. **4**, 932 (1963).
492. L. G. Donarama, *J. Org. Chem.*, **22**, 1024 (1957).
493. C. D. Nenitzescu and D. A. Isacescu, *Bull. Soc. Chim. Romania*, **14**, 53 (1932); *Chem. Abstr.*, **27**, 964 (1933).
494. L. Bouveault and A. Wahl, *Compt. Rend.*, **134**, 1145 (1902); **135**, 41 (1902).
495. E. P. Kohler and N. L. Drake, *J. Am. Chem. Soc.*, **45**, 1281 (1923); B. Reichert and W. Koch, *Arch. Pharm.*, **273**, 265 (1935).
496. H. B. Hass, A. G. Susie and R. L. Heider, *J. Org. Chem.*, **15**, 8 (1950); W. K. Seifert and P. C. Condit, *J. Org. Chem.*, **28**, 265 (1963).
497. L. K. Freidlin, E. F. Litvin and V. M. Chursina, *Kinet. Katal.*, **7**, 1093 (1966); *Chem. Abstr.*, **66**, 54775 (1967).
498. A. Dornow, H. D. Jordan and A. Müller, *Chem. Ber.*, **94**, 67, 76 (1961).
499. R. T. Gilsdorf and F. F. Nord, *J. Am. Chem. Soc.*, **72**, 4327 (1950).
500. A. Lambert, J. D. Rose and B. C. L. Weedon, *J. Chem. Soc.*, 42 (1949).
501. C. D. Hurd, M. E. Nilson and D. M. Wikholm. *J. Am. Chem. Soc.*, **72**, 4697 (1950).
502. R. F. C. Brown, V. M. Clark and A. Todd, *Proc. Chem. Soc.*, 97 (1957); M. C. Kloetzel, F. L. Chubb, R. Gobran and J. L. Pinkus, *J. Am. Chem. Soc.*, **83**, 1128 (1961).
503. A. J. Bellamy and R. D. Guthrie, *J. Chem. Soc.*, 3528 (1965).

504. G. Stork, A. Brizzolara, H. Landesman, J. Szmuszkovicz and R. Terrell, *J. Am. Chem. Soc.*, **85**, 207 (1963); T, C. Bruice and R. M. Topping, *J. Am. Chem. Soc.*, **85**, 1480, 1488 (1963); G. H. Alt and A. J. Speziale, *J. Org. Chem.*, **29**, 794 (1964); D. J. Cram and R. D. Guthrie, *J. Am. Chem. Soc.*, **87**, 397 (1965).
505. D. J. Cram and R. D. Guthrie, *J. Am. Chem. Soc.*, **88**, 5760 (1966).
506. (a) C. L. Arcus and M. M. Coombs, *J. Chem. Soc.*, 4319 (1954); (b) D. H. R. Barton and L. R. Morgan, Jr., *J. Chem. Soc.*, 622 (1962); (c) E. Koch, *Tetrahedron*, **23**, 1747 (1967).
507. J. H. Boyer and D. Straw, *J. Am. Chem. Soc.*, **74**, 4506 (1952); **75**, 1642, 2683 (1953).
508. D. H. R. Barton and L. R. Morgan, Jr., *Proc. Chem. Soc.*, 206 (1961); P. A. S. Smith and J. H. Hall, *J. Am. Chem. Soc.*, **84**, 480 (1962); W. H. Saunders and E. A. Caress, *J. Am. Chem. Soc.*, **86**, 861 (1964); R. Kreher and D. Kühling, *Angew. Chem.*, **76**, 272 (1964).
509. F. D. Lewis and W. H. Saunders, *J. Am. Chem. Soc.*, **89**, 645 (1967).
510. I. L. Knunyants, E. G. Bykhovskaya and V. M. Frosin, *Proc. Acad. Sci. USSR*, **132**, 513 (1960).
511. B. A. Stagner and I. Vosburgh, *J. Am. Chem. Soc.*, **38**, 2069 (1916); B. A. Stagner and A. F. Morgan, *J. Am. Chem. Soc.*, **38**, 2095 (1916); J. K. Senior, *J. Am. Chem. Soc.*, **38**, 2718 (1916).
512. T. E. Stevens, *J. Org. Chem.*, **28**, 2436 (1963).
513. P. A. S. Smith, *Open-Chain Nitrogen Compounds*, Vol. 2, W. A. Benjamin, 1966, Chap. 8.
514. P. Carter and T. S. Stevens, *J. Chem. Soc.*, 1743 (1961).
515. D. M. Lemal and T. W. Rave, *J. Am. Chem. Soc.*, **87**, 393 (1965).
516. W. Kirmse, L. Horner and H. Hoffmann, *Ann. Chem.*, **614**, 19 (1958); H. Reimlinger, *Chem. Ber.*, **97**, 339 (1964); H. E. Zimmermann and D. H. Paskovich, *J. Am. Chem. Soc.*, **86**, 2149 (1964); C. G. Overberger and J.-P. Anselme, *J. Org. Chem.*, **29**, 1188 (1964).
517. P. A. S. Smith in *Molecular Rearrangements* (Ed. P. De Mayo), Interscience, 1963, Vol. 1, Chap. 8, p. 457.
518. J. H. Boyer, P. C. Canter, J. Hamer and R. K. Putney, *J. Am. Chem. Soc.*, **78**, 325 (1956).
519. C. H. Gudmundsen and W. E. McEwen, *J. Am. Chem. Soc.*, **79**, 329 (1957); A. N. Nesmeyanov and M. I. Nybinskaya, *Bull. Acad. Sci. USSR*, 761 (1962).
520. H. Wolff, *Org. Reactions*, **3**, 307 (1946).
521. A. G. Knoll, *Ger. Pat.* 521,870 (1929); *Chem. Abstr.*, **25**, 3364 (1931).
522. W. Pritzkow and G. Pohl, *J. Prakt. Chem.* [4], **20**, 132 (1963); N. Wiberg and K. H. Schmid, *Angew. Chem.*, **76**, 381 (1964).
523. C. H. Bamford, *J. Chem. Soc.*, 12 (1939).
524. Y. L. Chow, *Can. J. Chem.*, **45**, 53 (1967).
525. Y. L. Chow and A. C. H. Lee, *Can. J. Chem.*, **45**, 311 (1967).
526. D. H. R. Barton and J. M. Beaton, *J. Am. Chem. Soc.*, **83**, 750 (1961); D. H. R. Barton, J. M. Beaton, L. E. Geller and M. M. Pechet, *J. Am. Chem. Soc.*, **83**, 4076, 4083 (1961); A. Nickon, J. R. Mahajan and F. J. McGuire, *J. Org. Chem.*, **26**, 3617 (1961); A. L. Nussbaum and C. H. Robinson, *Tetrahedron*, **17**, 35 (1962).

527. C. H. S. Wu, G. S. Hammond and J. M. Wright, *J. Am. Chem. Soc.*, **82**, 5386, 5394 (1960); A. Nickon, J. R. Mahajan and F. J. McGuire, *J. Org. Chem.*, **27**, 4053 (1962); G. S. Hammond and J. R. Fox, *J. Am. Chem. Soc.*, **86**, 1918 (1964).
528. M. Martynoff, *Ann. Chim.* (*Paris*), [11], **7**, 424 (1937).
529. E. Schmitz, D. Habisch and C. Gründemann, *Chem. Ber.*, **100**, 142 (1967).
530. C. A. Burkhard and J. F. Brown, Jr., *J. Org. Chem.*, **29**, 2235 (1964).
531. R. Hohn and H. Schaefer, *Tetrahedron Letters*, 2581 (1965).

CHAPTER **3**

Analysis of azomethines

DAVID J. CURRAN AND SIDNEY SIGGIA

University of Massachusetts, Amherst, Massachusetts, U.S.A.

I. INTRODUCTION

The methods for detection and measurement of azomethines are scattered through the literature. First, the diverse nomenclature is responsible for the scatter; material can be found under *azomethine*, *Schiff base*, *anil*, *imine* and *ketimines*. Also, specific compounds such as benzylidene aniline can also be found as benzalaniline (or benzaniline), benzalanil (or benzanil) or as the Schiff base of benzaldehyde and aniline. Material is also found under measurement of aldehydes and amines where these compounds were first transformed to a measurable azomethine. In addition, there are diverse analytical tools available to detect and measure azomethines.

This chapter is broken down into the various approaches amenable to analysing compounds via the azomethine group. In their order of presentation, these are: chemical methods, electrochemical, infrared absorption spectroscopy, mass spectroscopy, nuclear magnetic resonance, and fluorescence and other photochemical approaches. Ultraviolet absorption is not included in this chapter since it is discussed in Chapters 1 and 4.

II. QUALITATIVE CHEMICAL METHODS

There has been very little done on qualitative chemical methods designed specifically for azomethine compounds. In most cases, hydrolysis of the Schiff base to the corresponding amine and carbonyl compounds with subsequent identification of these entities is the approach of choice.

Tarugi and Lenci[1] devised a test for Schiff bases which involved treatment of the sample with phenol and hypochlorite; an intense blue colour results. However, nitro compounds, amino acids, amino aldehydes and some primary amines also give the test.

Marcarovici and Marcarovici[2] did some microscopic identification of azomethines using microcrystallity parameters.

Feigl and Liebergott[3] devised a colour test for aromatic aldehydes using thiobarbituric acid which results in orange products. Azomethines which cleave in acid to yield aromatic aldehydes also give the test. Used to test the method were benzalazine*, benzylideneaniline, resorcylaldoxime*, salicylazine*, 4-hydroxybenzaldoxime*, and *m*-nitrobenzaldazine*.

III. QUANTITATIVE CHEMICAL METHODS

The quantitative chemical methods for measuring the azomethine group fall into two categories: (1) those based on the basic properties of the group and (2) those based on the hydrolysis of the azomethine to the parent carbonyl compound and measurement of this group.

A. Acidimetric Titration Methods

The basic nature of the azomethine group has long been known; thus the name Schiff base. However, the general low level of interest in this

* Oximes and azines are not normally considered azomethines; however, they do contain the C=N group and were included in the reference cited.

group did not encourage work in its measurement. Also, there were the problems of (a) the weakly basic nature of the group which made accurate analysis difficult; (b) differentiating the azomethine from its original amine; and (c) the ease of hydrolysis of the azomethine back to its original amine. These problems were difficult to overcome until the advent of non-aqueous titrations.

Wagner, Brown and Peters[4] first encountered a non-aqueous titration which detected the azomethine group. The investigators were concerned with differentiating primary amines from their secondary and tertiary counterparts. They accomplished this, forming the azomethine by reacting the primary amine with salicylaldehyde. The resulting Schiff base had a lower basicity than the secondary and tertiary amines and could be detected by potentiometric titration in isopropanol as shown in Figure 1.

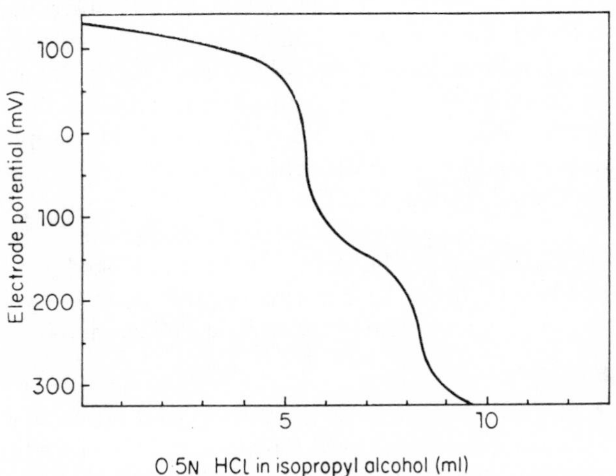

FIGURE 1. Titration of a mixture of secondary and tertiary butyl amines with the azomethine of n-butylamine (with salicylaldehyde). The break at 250 mv indicates the end-point of the azomethine titration. Reprinted with permission, reference 4.

Siggia, Hanna and Kervenski[5] extended the work of Wagner, Brown and Peters to aromatic amine systems still using salicylaldehyde but substituting a mixture of glycol–isopropanol solvent for solubility purposes.

Both of the above cases worked well for differentiating primary from secondary and tertiary amines because the azomethine had such a reduced basicity. One merely ran a titration before and after the addition of salicylaldehyde and the difference in the stronger base content of both systems was equivalent to the primary amine content.

Siggia and Segal[6] took advantage of this decreased basicity between the azomethine and its parent amine to determine carbonyl compounds in various situations. Lauryl amine was added to the sample; the carbonyl component formed the Schiff base which lowered the lauryl amine content. Glycol–isopropanol mixture was the solvent used. Potentiometric titration differentiated the unreacted lauryl amine from the azomethine. This work revealed the lability of the azomethine linkage to acid. The investigators were limited to the use of salicylic acid as titrant; mineral acids caused hydrolysis back to the parent compounds.

All the above workers, though measuring the azomethine grouping, were not concerned with this purpose and hence the methods were not optimized in this direction. Fritz[7], while investigating the use of different solvents for titrating amines, found that titration with perchloric acid using acetonitrile as solvent gave better differentiation between the azomethine and its parent or related amines than did the solvents used by Wagner et al. and Siggia et al. He proposed that primary amines could now be determined by a direct measure of the azomethine formed, rather than by the decrease in strong base (requiring two titrations) used by the earlier investigators.

Freeman[8] elaborated on the work of Fritz and looked at three procedures already established for titrating organic bases but which had not been thoroughly investigated for titration of azomethines.

I. Procedure A

Involves titration of the azomethine in glacial acetic acid and is essentially the titration Fritz[9] used for amines. This procedure yields good titration for azomethines, however it will not distinguish the azomethines from the parent and other amines. In acetic acid both types of compounds have the same order of basicity. Figure 2 shows the titration of such a mixture.

METHOD: *Dissolve a sample containing the order of 3 mequiv. of azomethine in 50 ml of glacial acetic acid. Titrate with 0·1N perchloric acid in acetic acid. The endpoint can be determined potentiometrically or visually, using methyl violet indicator to a blue-green endpoint. For the potentiometric titration an ordinary pH meter can be used except that millivolts are read. Glass versus calomel electrodes have been used, though Fritz recommends glass versus a silver wire with a thin coating of silver chloride*[9].

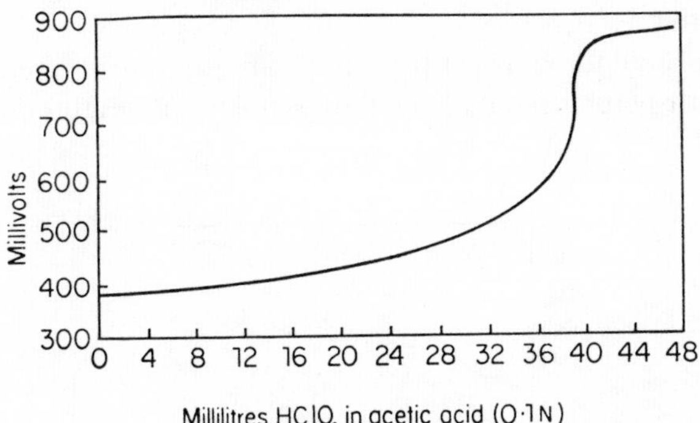

FIGURE 2. Titration of 1 to 1 mixture of aniline and *N*-benzylideneaniline in acetic acid.

2. Procedure B

Involves potentiometric titration of the Schiff base in chloroform with perchloric acid in dioxan. Figure 3 shows a typical curve with definite differentiation between azomethine and parent amine.

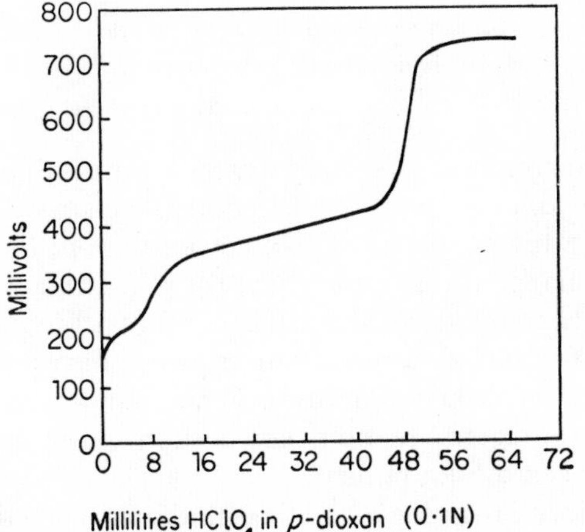

FIGURE 3. Titration of *N*-benzylidene-n-hexylamine containing 10% n-hexylamine in chloroform.

METHOD: *Dissolve a sample containing about 0·3 mequiv. of total base in 25 ml of chloroform and titrate potentiometrically (as in procedure A) with 0·01N perchloric acid in dioxan. The titrant is standardized against purified diphenyl guanidine in chloroform via potentiometric titration.*

3. Procedure C

Involves potentiometric titration of the sample in acetonitrile as solvent with perchloric acid (in dioxan solution). Figure 4 shows an example of such a titration.

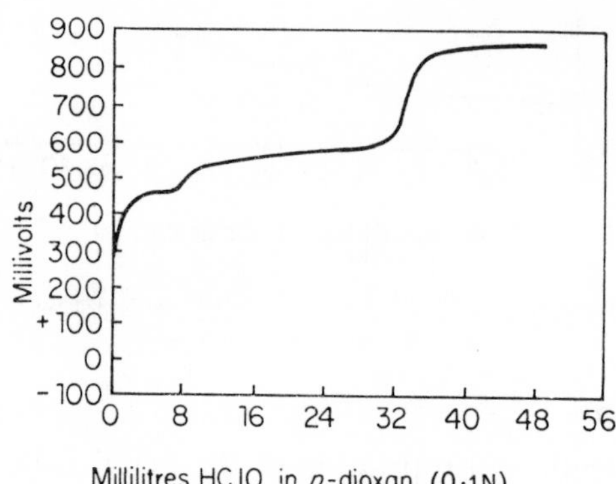

FIGURE 4. Titration of *N*-benzylideneaniline containing 10% aniline in acetonitrile. Reprinted with permission, Ref. 8.

METHOD*: *Dissolve the sample containing about 3 mequiv. of total Schiff base and amine in 50 ml of acetonitrile and titrate potentiometrically with 0·1N perchloric acid in dioxan. Run a 50 ml blank on each batch of acetonitrile.*

With the exception of N^4-*p*-methoxybenzylidinesulphathiazole, the potential change at the equivalence point using procedure A was so great (70 to 100 mv per 0·1 ml of volumetric solution) that no record of the potential change was necessary. For the sulphathiazole Schiff base, the potential change was of the order of 10 to 15 mv per 0·1 ml of titrant, and its exact determination required recording to note its position. However, when procedure C was employed for this substance, a much greater break in the titration curve occurred which obviated the need to plot points.

When faced with titrating a sample containing azomethine where no free amine or water is suspected, procedure A is recommended since it has a sharp endpoint, easily detected by indicator alone. Any free amine would be included in the analysis. Water in the sample could cause hydrolysis when the sample is dissolved in the acetic acid.

Procedure B can be used for azomethines in mixtures with amines

* Figures 2–4 and the method of procedure C are reprinted from reference 8, with permission.

(some water can be tolerated). However, from the work reported by Freeman it will not work well in cases where the amine portion of the azomethine is aromatic. These azomethines are very weak bases. Note that Figure 3, which illustrates procedure B, uses n-hexylamine as the amine portion of the azomethine, while Figure 4 for procedure C has aniline as the amine portion. Procedure C is thus recommended for the most general utility of the acidimetric methods.

Table 1 shows some azomethine compounds used by Freeman with one of the acidimetric methods and one of the hydrolytic methods.

TABLE 1. Comparison of volumetric and gravimetric results.*

Schiff Base	Procedure A (%)	Nitrogen (%)	2,4-Dinitrophenylhydrazone (%)
N-n-Butylidene-n-butylamine	99·6	99·6	—
N-Benzylidene-n-butylamine	99·5	99·6	100·0
N-Benzylidene-n-hexylamine	99·4	99·4	100·2
N-*p*-Chlorobenzylidene-n-hexylamine	100·0	99·9	100·9
N-*p*-Methoxybenzylidene-benzylamine	98·9	98·9	99·8
N-*p*-Methoxychlorobenzyl-idenebenzylamine	99·4	99·5	100·1
N-Benzylideneaniline	99·7	99·8	100·2
N-n-Butylideneaniline	99·0	99·1	—
1,2-Bis(benzylideneamino)-ethane	99·7	99·8	100·1
N^4-*p*-Methoxybenzylidene-sulphathiazole	97·8	97·6	98·3

* Reprinted in part with permission, reference 8.

Pichard and Iddings[10a] titrated ketimines potentiometrically, much in the same manner as procedure A above. They also used glacial acetic acid as solvent, perchloric acid as titrant, glass versus platinum as electrodes or crystal violet as indicator. They successfully titrated diethyl ketimine, ω-cyclohexylpentyl-2-butyl ketimine and 2-butyl-*o*-tolyl ketimine. However, amines and other basic substances are usually also titrated in the above system.

Pohloudek-Fabini et al.[10b] used non-aqueous titrations to determine halogenated and non-halogenated azomethines. Thiocyanate-substituted Schiff bases did not titrate and were determined gravimetrically with 2,4-dinitrophenylhydrazine.

Hara and West[10c] tried titrating azomethine compounds via high frequency non-aqueous titrations using pyridine as a solvent. The end-points were poor, however.

B. Hydrolytic Methods

Azomethines easily hydrolyse, especially under acidic conditions, back to the original amine and carbonyl compound:

$$\begin{matrix} R \\ \quad\diagdown \\ \quad C{=}NR'' \\ \quad\diagup \\ R' \end{matrix} \xrightarrow{H_3O^+} \begin{matrix} R \\ \quad\diagdown \\ \quad C{=}O + R''NH_2 \\ \quad\diagup \\ R' \end{matrix}$$

where R′ and/or R″ can be hydrogen.

The analytical approaches involving hydrolysis of the azomethine proceed through determination of the resultant carbonyl compound. Freeman[8] utilized the 2,4-dinitrophenylhydrazine precipitation, the reagent being dissolved in 2N hydrochloric acid, creating the acidity needed for hydrolysis. Hillenbrand and Pentz[11] used a bisulphite addition analysis system and a hydroxylamine system.

1. 2,4-Dinitrophenylhydrazine method[8]

To about 1 mequiv. of azomethine in 50 ml of ethyl alcohol is added a 50% excess of a 1% solution of 2,4-dinitrophenylhydrazine in 2N hydrochloric acid. A precipitate generally forms immediately and is allowed to stand overnight at room temperature to agglomerate. It is then filtered, washed with 2N hydrochloric acid and dried to constant weight. Drying above 100°C is not recommended due to possible decomposition of hydrazone. Vacuum drying at 50–70°C is preferred. Table 1 shows results obtained with this method.

2. Bisulphite method[11]

Twenty-five ml of distilled water are added to each of four 500 ml glass-stoppered Erlenmeyer flasks. Two flasks serve for blank determination. Into each of the other flasks are introduced weighed samples, each containing 1–2 mequiv. of azomethine. The flasks are placed in an ice bath at 0–5°C for 10 minutes. Two drops of methyl red indicator (0·6% aqueous solution) are added to all flasks and the contents are neutralized with 0·6N sulphuric acid from a burette. On addition of the acid the flasks should be swirled to assure complete neutralization of the azomethine without an excess of acid. The flasks are returned to the ice bath and allowed to stand for 10 minutes. Twenty-five ml of 0·2N sodium bisulphite is added into each flask, keeping the tip of the pipette just below the surface of the solution. The addition of the reagent should be spaced to allow equal reaction times for blanks and sample. The flasks are then removed from the ice bath and allowed to stand at room temperature for exactly 30 minutes, swirling the contents from time to time. The flasks are then returned to the ice bath for 5 minutes. Then they are again removed and to each is added approximately 25 grams of crushed ice. To each is added 2 ml of starch indicator (1·0% aqueous solution) and the contents are titrated immediately with 0·1N iodine to the first blue colour.

Compounds run successfully by this method were ethylimine polymer, butylimine, *N*-butylidenebutylamine, *N*-decylidenedecylamine, amylimine, *N*-amylideneamylamine, *N*-(2-butylactylidene)-3-butylactylamine, *N*-ethylideneaniline, *N*-butylideneaniline.

In cases where solubility difficulties occur, a 3:2 mixture of isopropanol and water can be used.

3. Hydroxylamine method[11]

Cases occur where the aldehyde released on hydrolysis cannot be determined with bisulphite. In these cases hydroxylamine can be used.

The sample is added to 15 ml of isopropanol and 10 ml of water. One ml of bromophenol blue indicator (0·04% in methanol) is added to each sample and blank, and the sample is nearly neutralized with 6N hydrochloric acid. The final neutralization is completed with 0·5N hydrochloric acid. To both blank and sample is added 35 ml of 0·5N hydroxylamine hydrochloride which has been neutralized with 0·5N sodium hydroxide to the bromophenol blue indicator endpoint. Exactly 3 ml of standard 0·5N hydrochloric acid is added to each blank and sample and they are allowed to stand at room temperature for 1 hour. The solutions are then titrated with standard 0·5N sodium hydroxide to a blue-green endpoint.

This method has been successfully applied to ethylimine, *N*-isopropylideneisopropylamine, butylimine, *N*-butylidenebutylamine, 2-ethylbutylimine, 2-ethylhexylimine, *N*-(2-ethylbutylidene)-2-ethylbutylamine and α-methylbenzylimine.

IV. POLAROGRAPHY

Studies of the polarographic reduction of the azomethine group, first reported by Zuman[12,13] in 1950, are complicated by hydrolysis when the polarogram is obtained in an aqueous or mixed aqueous alcohol solvent. Von Kastening, Holleck and Melkonian[14] utilized the reduction wave of the azomethine group to examine the hydrolysis of *N*-benzylideneaniline and found that the reaction was catalysed in strong acid solution by hydrogen ion and in strong base by hydroxyl ion. These results are in agreement with a similar study by Dezelic and Dursun[15] who used the Schiff bases of aniline and toluidine with pyrrole-2-aldehyde and their hydrochlorides. In each case an intermediate pH region was found where the reaction proceeded uncatalysed. In these studies, and in all studies reported in the literature, the azomethine group reduced at potentials more positive than those required for the corresponding carbonyl compounds from which the Schiff bases were derived. In many cases, the carbonyl reduction wave was also accessible and the polarographic method offered a very useful technique for studying the formation and hydrolysis of Schiff bases in electrolytic solutions. Dezelic and Durzun obtained a first-order rate constant for the hydrolysis of the Schiff base of aniline and pyrrole-2-aldehyde at 20°c and pH 9·06 of $6{\cdot}26 \times 10^{-3}$ min^{-1}. Bezuglyi et al.[16] have reported a polarography study of the reaction of aniline with benzaldehyde and some of its derivatives. Rate constants, equilibrium constants and activation energies are given.

Zuman[12,13,17,18] has obtained equilibrium constants for the reaction of a number of carbonyl compounds with various amines. Included were the Schiff bases of pyruvic acid with ammonia, glycine, alanine, ethanolamine, histidine and histamine; acetone with ammonia, and glycine; cyclopentanone, cyclohexanone and methylcyclohexanone with NH_3, $MeNH_2$, $EtNH_2$, $HOCH_2CH_2NH_2$, $NH_2CH_2CO_2H$ and $MeCH(NH_2)COOH$; and others.

Holleck and von Kastening[19] examined the polarographic reduction of benzylideneaniline in MeOH–H_2O buffers. In acid solution, pH 6, only the current–voltage curve for benzaldehyde could be observed. The anil showed two waves at intermediate pH but they merged to one wave in strongly basic solution. This behaviour is in agreement with the work on the pyrrole-2-aldehyde compounds, but Lund[20] found two waves at pH 13 for the reduction of benzophenone anil. The pH dependence of the half-wave potential of the first wave of *N*-benzylideneaniline was examined by Holleck and Kastening and found to fit the equation: $E_{\frac{1}{2}} = -0{\cdot}350{-}0{\cdot}075$ pH. The process was described as a one-electron reversible reduction involving the transfer of one proton. The second wave was irreversible and independent of pH. The overall reduction process was described as follows:

$$C_6H_5CH{=}NC_6H_5 + 1e^- + H^+ \rightleftharpoons \left[\begin{array}{c} C_6H_5CH{=}N\dot{H}C_6H_5 \\ \updownarrow \\ C_6H_5\dot{C}H_2{-}\dot{N}C_6H_5 \end{array}\right] \xrightarrow{1e^-} \left[\begin{array}{c} C_6H_5\bar{C}H{-}NHC_6H_5 \\ \updownarrow \\ C_6H_5CH_2{-}\bar{N}C_6H_5 \end{array}\right] \xrightarrow{H^+} C_6H_5CH_2NHC_6H_5$$

The reaction at high pH was suggested as being similar to the reduction of aldehydes, but this would not seem to apply to ketimines in view of Lund's observation on the reduction of benzophenone anil.

Further, Zuman[12] points out a half-wave dependence on pH at constant ionic strength in the pH range 8·5–10 of 90 to 100 mv/pH. This suggests that the conjugate acid of the Schiff base is the electroactive species and Zuman formulated the reduction process as follows:

$$R_2C{=}NR + H^+ \rightleftharpoons R_2C{=}\overset{+}{N}RH$$

$$R_2C{=}\overset{+}{N}RH + 2e^- + 2H^+ \rightleftharpoons R_2CH{-}NRH_2$$

Only one wave was observed in his work. The kinetics of the ion combination reaction might be expected to govern the nature of the polarographic limiting current, at least in some region of pH, but in all of the systems studied by Zuman this was observed only for the glycine case at a pH greater than 10·2. For the other cases, the limiting current was diffusion controlled. This was explained by assuming that the rate of ion combination was fast compared with the rate of diffusion.

The polarography of *N*-salicylideneaniline has been studied by Vainshtein and Davydovskaya[21,22] in EtOH–H_2O buffers and they obtained diffusion controlled currents for this compound. It was also observed that the reduction waves of salicylaldehyde in the same buffer solutions were 15–20% higher than those obtained from hydrolysis of the anil. The difference was attributed to stability of the intermediate amino alcohol formed in the hydrolysis reaction, the intermediate not being reducible. It would appear that further studies are necessary before any general statements can be made concerning the polarographic reduction of the azomethine group in aqueous solution.

Nevertheless, the group is reducible and use can be made of this fact. Although there are no reports in the literature on the use of polarography for the quantitative determination of azomethine compounds, Schiff bases have been used for the analysis of carbonyl compounds. Zuman[12] stated that polarographic determinations of acetone, cyclohexanone, and the oxidation product of ascorbic acid were possible utilizing the reaction of these compounds with ammonia in an ammonium sulphate electrolyte. Conditions had to be carefully controlled because of the volatility of the supporting electrolyte and the necessity for removal of oxygen. In the case of acetone and cyclohexanone, the equilibrium constants were small so that a low ratio of current to carbonyl concentration was obtained. Glycine was later recommended as the amine[17] and finally, methylamine[18]. Van Atta and Jamieson[23] investigated n-butylamine as the reagent for determination of acetone. A well defined wave at $E_{\frac{1}{2}} = -1{\cdot}58$ v versus s.c.e. was found in a solution 1·7M in n-butylamine and 1M in ammonium chloride. Equilibrium was established at least as quickly as the solution and cell could be prepared for polarization, and oxygen did not have to be removed. An equilibrium constant of $8{\cdot}4 \times 10^{-2}$ was estimated for the imine formation reaction. The limiting current was determined as the difference in current between that measured at $-1{\cdot}75$ v and that measured at $-1{\cdot}45$ v. Plots of limiting current versus concentration of acetone were linear. The minimum concentration measurable was 0·05 volume % and pure acetone was analysed by appropriate dilution. The average accuracy of the method was $\pm 1\%$ relative.

Hall[24] described a general method for aliphatic aldehydes and ketones using hexamethylenediamine (HMD) as the reagent. For aldehydes a 2% solution of the amine was used as the reagent and also as the supporting electrolyte. For ketones, optimum conditions were found using a 20% solution of the amine. A difference in half-wave potentials of nearly two-tenths of a volt was observed between the aldimines and the ketimines in 20% HMD. The former were the more easily reduced. Relative errors in the analysis of acetone and n-heptaldehyde were a few percent or less using samples sizes up to a few milligrams.

Polarographic reduction of an azomethine compound is the basis for a Soviet patent concerning purity control of 1,6-hexanediamine[25].

There have been several polarographic studies of substituent effects in aromatic azomethines. Dmitrieva and coworkers[26] obtained differential oscillopolarograms of 19 compounds in dimethylformamide using 0·05M tetramethylammonium iodide as the supporting electrolyte. Peak potentials were measured versus a mercury pool reference electrode over the range −0·4 to −1·5 v. The peak potential for *N*-benzylideneaniline was −1·40 v. Substitutions included *ortho* and *meta* chloro in the aniline ring and *ortho* hydroxy in the benzylidene ring. Other compounds included those obtained by replacement of the phenyl ring in the benzylidene portion by 1-naphthyl and 2-hydroxy-1-naphthyl and a series of bisazomethines with substitutions and ring replacements similar to those of the monoazomethines. The peak potential for α-naphthylmethylene aniline was −1·27 v. This positive shift was attributed to increased polarizability of the molecule due to lengthening of the conjugated chain. Substitution of a hydroxy group in the *ortho* position of the aldehyde moiety led to more positive potentials in all cases and was explained in terms of intramolecular hydrogen bonding. Substitution of chlorine in the *ortho* or *meta* position of the aniline residue facilitated reduction of the monoazomethines and corresponded to negative induction effects of the chloro substitution in these positions. Bisbenzalaniline was reported as having three peaks in the oscillopolarogram at −1·22, −1·28 and −1·36 v. The shift to positive potentials with respect to benzylideneaniline is clear but the source of three peaks is not well understood. Stepwise reduction of the azomethine group would account for two peaks, but three (or four) peaks would seem to require different peak potentials for each of the azomethine groups in the bis compound. Chloro substitution in the aniline residue of the bis compounds produced positive shifts in peak potential. The magnitude of the shift depended on the position of substitution. The shift increased in the order 2, 3′; 2, 2′; 3, 3′. Steric hindrance in the 2, 2′ molecule was used to explain the trend. Levchenko and coworkers[27] used classical polarography in DMF for a study of 30 Schiff bases of the type RCH═NR′ where R was phenyl, 2-hydroxyphenyl and 2-hydroxy-1-naphthyl and R′ was C_6H_5, *p*-$C_6H_5C_6H_4$, *p*-$C_6H_5C_6H_4C_6H_4$, *p*-$C_6H_5CH{=}CHC_6H_4$, *p*-$C_6H_5CH_2CH_2C_6H_4$, *p*-$C_6H_5CH_2C_6H_4$, *p*-$C_6H_5OC_6H_4$, *p*-$C_6H_5SC_6H_4$, *p*-$C_6H_5NHC_6H_4$ and *p*-$C_6H_5COC_6H_4$. The half-wave potential for benzylideneaniline was reported as −1·38 v versus the mercury pool reference electrode, which was in good agreement with the derivative oscillopolarographic value of −1·40 v. The results of this study were in agreement with the work of Dmitrieva et al. Lengthening of the conjugation chain shifted the half-wave potential in the positive direction, as did formation of intramolecular hydrogen bonds. Introduction of an ethylenic linkage between the phenyl rings in the *o*-hydroxy compounds shifted the half-wave potential in the positive direction, while introduction of the electron donating saturated side chain produced a negative shift with respect to *o*-hydroxybenzalaniline.

Two studies of substitution effects in anils in aqueous EtOH solvent are available. Both Uehara[28] and Dmitrieva, Kononenko and Bezuglyi[29] used benzalaniline as their reference compound. Uehara reported positive shifts in half-wave potential when the aldehyde phenyl ring was replaced with polynuclear arenes and a negative shift when alkyl aldehydes were used. *Para* substitution in the phenyl ring of the benzaldehyde portion of the molecule produced positive or negative shifts depending on whether the substitution group was an electron acceptor or an electron donor, respectively. This is in agreement with Dmitrieva's work in the same solvent and with the work in DMF. Dmitrieva reported linear plots of $E_{\frac{1}{2}}$ versus $\sigma_{\rho-\chi}$ of slope 0·23 and 0·13 at pH 10·5 and 6·8, respectively. Uehara claimed that any substitution of the phenyl ring in the aniline moiety produced no appreciable shift in $E_{\frac{1}{2}}$, but Dmitrieva showed data for large negative shifts in $E_{\frac{1}{2}}$ for *para* substitution by methyl, methoxy and hydroxy groups. Slight positive shifts were observed for *para* substitution by the acceptor groups Cl, Br and SO_2NH_2. *Ortho* or *meta* substitution by methyl or methoxy also produced negative shifts of $E_{\frac{1}{2}}$. The results of Dmitrieva's work were interpreted as confirming evidence for noncoplanarity of the two benzene rings and for the suggestion that the nature of the substituent on the aniline ring exerts an influence on the conformation of the molecule.

V. INFRARED SPECTROSCOPY

All frequencies reported for the C═N stretching vibration in compounds of the type X—C(—Y)═N—Z where X, Y and Z may be hydrogen, alkyl or aryl occur in the region 1680–1603 cm^{-1}. Factors affecting the position of the C═N stretching absorption within this band include the physical state of the compound, the nature of the substituent groups, conjugation with either carbon or nitrogen, or both, and hydrogen bonding. The region is considerably narrower than the

1689–1471 cm^{-1} region described for all compounds containing the C=N group[30]. An extensive review which covers the literature on the i.r. and Raman spectroscopy of the carbon–nitrogen double bond through early 1956 was presented by Fabian, Legrand and Poirer[31]. Their findings on azomethines are included here along with work published since the time of their review.

A. Aldimines

Fabian et al.[31] concluded that azomethines of the saturated aliphatic aldimine type absorb in the region 1674–1665 cm^{-1}, although most of the evidence was based on Raman studies. This was confirmed by Suydam[32] who examined the $\nu_{s_{C=N}}$ (i.r.) of twenty-five compounds and reported a range of 1672–1664 cm^{-1} and by Steele[33] who found a range of 1680–1666 cm^{-1} for compounds with alkyl groups on the nitrogen and the carbon. Suydam's work showed that neither chain length nor chain branching of the groups joined to either the carbon or the nitrogen had an appreciable effect on the frequency of absorption; a conclusion the reviewers had reached earlier. Further, the frequency was reported as the same for the pure liquids as for chloroform solutions. Fabian and Legrand[34] found that *N*-(n-propylidene)-n-propylamine absorbed at 1673 cm^{-1} in CCl_4 and 1671 cm^{-1} in $CHCl_3$. It would appear that the effect of these solvents on the absorption frequency is slight to negligible. The formylidene derivative of 2-amino-3-methyl-3-butanol is an exception to the above range of frequencies. The absorption band for this compound is at 1653 cm^{-1}. The shift to lower frequencies was attributed to the lack of an alkyl group on the carbon atom of the C=N group since an analogous shift to lower frequencies is observed for the corresponding carbon–carbon double bond situation[31].

If the aliphatic chain on either the carbon or nitrogen atom includes a phenyl group not in conjugation with the C=N group, a slight reduction of the frequency range to 1669–1653 cm^{-1} is found and branching in the remaining aliphatic chain produces a slight frequency increase[32].

When the aliphatic chain contains a single ethylenic double bond in conjugation with the azomethine linkage, a small reduction in the frequency range to 1664–1658 cm^{-1} occurs. If further conjugation of the C=C group is achieved with a phenyl ring, a substantial shift in frequency to 1639–1637 cm^{-1} is observed[32].

For compounds of the type Ar—CH=N—R, Fabian et al.[31] found a frequency range of 1650–1638 cm^{-1} when Ar is an unsubstituted phenyl group, but Suydam[32] reported a smaller range of 1650–1645 cm^{-1}. Steele[33] states that the frequency is lowered to 1650 cm^{-1} when the phenyl group is on either the nitrogen or the carbon atom. Nitro or halogen substitution in the phenyl ring widens the range somewhat to 1656–1631[31] cm^{-1}.

Compounds of the type Ar—CH=N—Ar have been the object of considerable recent interest. In a review article[31], the frequency region assigned to these compounds was 1637–1626 cm^{-1}. Clougherty, Sousa and Wyman[35] examined seventeen anils and found a frequency range of 1631–1613 cm^{-1} as shown in Table 2. Chemical evidence was obtained for the band assignment, since the absorption disappeared when selected compounds were reduced to the corresponding *N*-benzylanilines with sodium borohydride. The agreement among workers on the peak frequencies of the anils is excellent, as illustrated in Table 3.

TABLE 2. C=N Stretching frequencies in aromatic Schiff bases.* (Measurements made on Beckman IR-3 spectrophotometer, NaCl optics.)

Compound	Frequency[a] (cm^{-1})
N-benzylideneaniline[b]	1631
N-(2-hydroxy)benzylideneaniline[b]	1622
N-(4-hydroxy)benzylideneaniline	1629[c]
N-(4-methoxy)benzylideneaniline[b]	1630
N-(2-nitro)benzylideneaniline	1621[c]
N-(4-acetylamino)benzylideneaniline	1629[c]
N-(4-dimethylamino)benzylideneaniline[b]	1626
N-benzylidene-2-aminophenol	1629
N-benzylidene-2-anisidine	1631[d]
N-(4-methoxy)benzylidene-2-anisidine	1627[d]
N-benzylidene-4-anisidine[b]	1629
N-(4-methoxy)benzylidene-4-anisidine	1626[c]
N-benzylidene-4-toluidine	1628[d]
N-benzylidene-*N'*-dimethyl-4-phenylenediamine[b]	1627
N-(2-hydroxy)benzylidene-2-aminophenol	1624[c]
N-(4-dimethylamino)benzylidene-2-aminophenol	1613
N,N'-dibenzylidene-4-phenylenediamine	1628

[a] In CCl_4 solution.
[b] Compounds reduced using $NaBH_4$.
[c] In $CHCl_3$ solution.
[d] As KBr pellets.

* Reprinted with permission, reference 35.

TABLE 3. Reproducibility of the absorption frequency for C=N stretching.

Frequency (cm^{-1})	Solvent	Reference
	N-benzylideneaniline	
1631	CCl_4	27
1630	CCl_4	26
1630	$CHCl_3$	28
1628	$CHCl_3$	26
	N-(2-hydroxy)benzylideneaniline	
1622	$CHCl_3$	28
1622	CCl_4	27
1619	CCl_4	29
1622	mineral oil	29

Inspection of Table 3 reveals that a frequency shift of about −8 cm^{-1} occurs when hydroxy is substituted in the 2-position of the benzylidene phenyl ring. Freedman[36] seems to be the first to have discussed the influence of intramolecular hydrogen bonding on the frequency of the C=N stretching vibration. Using *N*-benzylideneaniline as a reference, the shift in frequency for a number of substituted compounds (primarily hydroxy substituted) are shown in Table 4. The bathochromic shift mentioned above is in accord with that expected for a chelated hydrogen bonded system but the analogous shift in the C=O stretch for salicylaldehyde is well known to be much

TABLE 4. C=N Stretching frequencies.[a]*

R^1	R^2	System	$\nu_{C=N}$[b] (cm^{-1})	Δ_ν
H	H	$R^1C_6H_4$–CH=N–$C_6H_4R^2$	1630(s)	
p-OH	H		1629(vw)	−1
o-OH	H		1622(s)	−8
H	*p*-OH		1630(s)	0
H	*o*-OH		1626(s)	−4
o-OH	*o*-OH		1621(s)	−9
H	*o*-HOC_6H_4	$R^1C_6H_4$–CH=NCH_2R^2	1657(s)	+27
o-OH	*o*-HOC_6H_4		1634(s)	−4
H	—CH_2N=CHC_6H_5		1646(s)	+16
o-OH	—CH_2N=CHC_6H_4— *o*-OH		1634(s)	+4

[a] Dilute solutions in chloroform measured in 1·0 mm cell.
[b] s = strong, vw = very weak.
* Reprinted with permission, reference 36.

larger. The weaker intramolecular hydrogen bond produced by the five-membered non-conjugated ring in *N*-benzylidene-2-aminophenol is indicated by a -4 cm^{-1} shift. Further evidence for intramolecular hydrogen bonding is obtained by the shift to higher frequencies by the *N*-alkyl compounds when the resonance system is eliminated by alkyl replacement of the phenyl ring. Most of this shift can, in turn, be eliminated by proper hydroxy group substitution as shown in Table 4. *N*-(2-hydroxy)benzylideneaniline could exist in either the benzenoid or quinoid form:

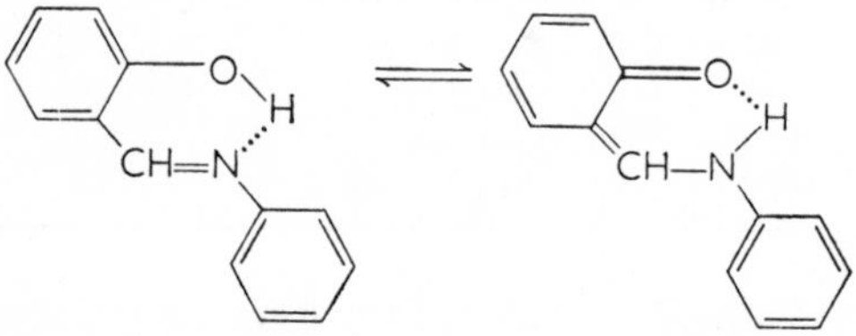

CH=N ⇌ CH—N

Minkin and coworkers argue strongly for the benzenoid structure on the basis of the u.v. and i.r. behaviour[37] and the dipole moments[38] of a number of anils of *o*-hydroxyaldehydes. They show that the frequency of the *N*-benzylideneaniline compound is only very slightly affected by further substitution in the benzylidene or aniline rings, and that therefore the six-membered chelate ring is strongly stabilized since it resists substituent effects which change the acidity of the hydroxy group and the basicity of the nitrogen atom. Further, they show that salicylideneaniline hydrochloride has a band displaced to 1652 cm^{-1}— a result which should not be observed if the tautomer has the quinoid structure. Bands due to free OH stretching were not observed even when the solution was very dilute. Instead, broad diffuse bands at 2800–2900 cm^{-1} were found which were assigned to an intramolecularly bound OH···N group. The phenolimine structure is also claimed by Shigorin and coworkers[39] for the *o*-hydroxy anils of the naphthyl and benzyl series.

Heinert and Martell[40] have examined the i.r. spectra of the Schiff bases formed from 3-hydroxypyridine-4-aldehyde and 3-hydroxypyridine-2-aldehyde and the amino acids glycine, alanine, valine, phenylalanine and glutamic acid. In addition, the *o*-methoxy and unsubstituted Schiff bases were studied and found to have a C=N stretching frequency at 1640–1630 cm^{-1}. The spectra of the *o*-hydroxy Schiff bases were noteworthy in that the band at 1650–1625 cm^{-1} was unusually intense, and a new band at 1510 cm^{-1} appeared. On the strength of these observations and the observation by Freedman[36] that the bathochromic shift for hydrogen-bonded anils is much smaller than would be expected, Heinert and Martell assigned the 1650–1625 cm^{-1} band to a carbonyl stretching vibration of an amide (amide I band) and the 1510 cm^{-1} band to a vinylogue C=C stretching vibration. Thus it is proposed that these Schiff bases exist in the enamine structure:

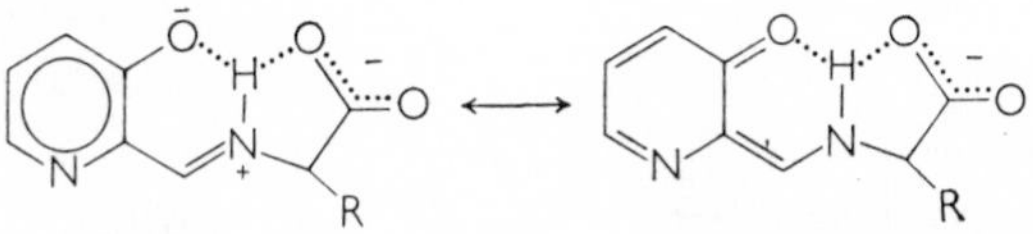

The possibility that the two phenyl rings of the anils may not be coplanar is well recognized and the object of some debate. One of the principal arguments for or against steric hindrance has been summarized by Ledbetter, Kramer and Miller[41], and rests on the interpretation of the reduced intensity of the long wavelength $\pi \rightarrow \pi^*$ transition in the ultraviolet. These workers prepared a series of α-cyano-*N*-benzylidene anils and examined the effect of the cyano group on the $\nu_{s_{C=N}}$ and the long wavelength u.v. transition. The steric effect of the cyano group would influence the coplanarity of the phenyl rings, but it is also in conjugation with the azomethine double bond and the aniline ring. These two factors have opposite effects on the frequency of the C=N stretch. Comparison of the frequency of *N*-benzylideneaniline and that of α-cyano-*N*-benzylideneaniline revealed a bathochromic shift of 20 cm^{-1}. This shift, together with their u.v. data, led these workers to conclude that steric hindrance to coplanarity in the α-cyano compounds is at best intermediate and that an extrapolation to *N*-benzylideneaniline suggests that steric hindrance there must be less than intermediate. Feytmans-de Medicis[42] has examined the *syn–anti* isomerism of α-cyano-*N*-(3-methyl-4-dimethylamino)-benzylideneaniline. In absolute alcohol, isomerization is complete and both forms yield identical ultraviolet spectra with an absorption maximum at 459 mμ (molar absorptivity = 17,100). As expected, the isomers are resolved in the i.r.; the ν_{CN} for the *syn* form was assigned at 2210 cm^{-1} while that for the *anti* form was 2250 cm^{-1} using Nujol mulls.

B. Ketimines

In contrast to the recent research on the i.r. spectroscopy of aldimines, very little appears to have been published on the i.r. absorption of ketimines since the review of Fabian, Legrand and Poirer[31].

TABLE 5. I.R. C═N stretching frequencies of ketimines.

Azomethine type or compound	$\nu_{sC=N}$ (cm^{-1})
$R(R)C{=}N{-}H$ (dialkyl)	1646–1640
$R(Ar)C{=}N{-}H$ (alkylaryl)	1633–1620
$Ph(Ph)C{=}N{-}H$ (diaryl)	1603
$R(R'')C{=}N{-}R'$ (trialkyl)	1622–1649
$R(Ph)C{=}N{-}R'$ (phenyldialkyl)	1650–1640
$CH_3(i\text{-}C_4H_9)C{=}N{-}Ph$ (dialkylphenyl)	1658
$Ph(CH_3)C{=}N{-}Ph$ (phenylalkylphenyl)	1628 (solid) 1640 (CCl_4, $CHCl_3$, shoulder at 1628)
$Ph(Ph)C{=}N{-}CH_2CH_2OH$	1616
$Ph(Ph)C{=}N{-}Ph$ (triphenyl)	1614

Their findings are summarized in Table 5. The shift of the absorption band to lower frequencies as conjugation is increased is evident from the table.

Staab and Voegtle[43] studied a series of double aldimines and ketimines. In some cases, the absorption bands occur in frequency regions close to those which would be predicted from information

TABLE 6. C=N Stretching frequencies of some double aldimines and ketimines.

(R)(R′)C=N—CH(Ph)—CH(Ph)—N=C(R)(R′)

Compound	Group (R, R′)	$\nu_{sC=N}$ (cm^{-1})
1	CH_3(Ph)C=	1630
2	$(CH_2)_6$ C= (ring)	1635
3	$(CH_2)_7$ C= (ring)	1630
4	(Ph—CH_2)(Ph—CH_2)C=	1640
5	(Ph)(Ph)C=	1610
6	2,4,6-(CH_3)$_3C_6H_2$—C(H)=	1630

available for the mono compounds. For example, compounds of the type

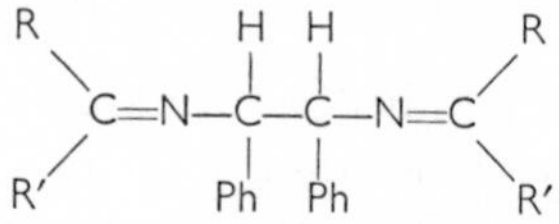

$$\begin{array}{c} R \quad\quad\quad H \quad H \quad\quad\quad R \\ \diagdown \quad\quad | \quad | \quad\quad \diagup \\ C{=}N{-}C{-}C{-}N{=}C \\ \diagup \quad\quad | \quad | \quad\quad \diagdown \\ R' \quad\quad Ph \quad Ph \quad\quad R' \end{array}$$

had an absorption band at 1660 cm^{-1} in most cases when R and R′ were aliphatic groups. This is in accord with the frequency region for trialkyl type monoketimines (Table 5). Frequencies for other groups on the double compound are shown in Table 6. The frequency for compound **5** is also in accord with Table 5 but those for compounds **1** and **4** occur at lower frequencies than Table 5 would predict.

VI. MASS SPECTROMETRY

The mass spectrometry of azomethine compounds has only been recently investigated. Elias and Gillis[44] studied a series of substituted *N*-benzylideneanilines using a 70 ev electron beam energy and an inlet temperature of 200°c. The molecular ion was the base peak in all cases except for *ortho* substituted compounds. *Meta* and *para* substituted compounds all underwent loss of the azomethine proton to yield an $(M - 1)^+$ peak of variable intensity, and peaks typical of aromatic structures were observed. The results suggested that fission was more easily accomplished at the ring–nitrogen bond rather than at the ring–carbon bond. The spectral data are shown in Table 7.

Ortho substituted compounds gave rise to spectra decidedly different from the *meta* and *para* compounds. Peaks were interpreted in terms of four-membered ring formation. Compounds formed from *o*-methoxybenzaldehyde were especially interesting since the base peak corresponded to the parent amine. Formation of the amine ion radical from the molecular ion was confirmed by a metastable peak, and a mechanism involving two hydrogen transfers was proposed.

Fischer and Djerassi have studied the mass spectrometry of a number of alkylalkyl azomethines[45]. Included are many of the compounds derived from methylamine, ethylamine, and n-butylamine with butanal, pentanal, heptanal, 4-heptanone, 5-nonanone, cyclopentanone and cyclohexanone, plus some deutero substitutions in some of the above compounds. As expected, the molecular ions were of weak intensity except for those derived from the cyclic ketones. In general the fragmentation patterns were analogous to those obtained for the corresponding carbonyl compounds, but some features were noteworthy.

TABLE 7. Principal peaks for mass spectra of Schiff bases.*

(Structure: R-substituted phenyl–N=CH–phenyl substituted with R′)

Compound	R	R′	*m/e* with percentage in parentheses
7	H	H	181(100), 180(98), 104(4), 90(5), 77(57)
8	*p*-CH_3	H	195(100), 194(78), 193(6), 118(11), 107(5), 106(7), 91(34), 77(7)
9	*m*-CH_3	H	195(100), 194(81), 193(8), 118(11) 91(43), 77(7)
10	*o*-CH_3	H	195(100), 194(80), 193(7), 180(6), 119(9), 118(92), 117(16)
11	H	*m*-CH_3	195(100), 194(98), 193(6), 178(6), 136(6), 104(10), 91(19), 77(45)
12	*p*-OCH_3	H	211(100), 210(14), 197(14), 196(95), 167(13), 77(9)
13	H	*p*-OCH_3	211(100), 210(92), 167(8), 77(34)
14	*p*-NO_2	H	226(100), 225(35), 196(14), 180(11), 179(20), 153(9), 152(12), 151(5), 149(5), 138(38) 108(10), 106(46), 105(45), 103(5), 77(53)
15	H	*p*-NO_2	226(100), 225(10), 180(8), 179(40), 178(5), 152(8), 104(19), 77(45)
16	*o*-OH	H	197(88), 196(74), 195(14), 167(6), 121(8), 120(100), 104(14)
17	H	*o*-OH	197(100), 196(77), 168(7), 167(5), 120(12), 104(6)
18	*o*-OH	*o*-OH	213(100), 212(82), 211(10), 196(9), 184(7), 121(8), 120(84).
19	H	*o*-OCH_3	211(11), 180(5), 119(38), 104(5), 94(7), 93(100), 91(35)
20	*o*-OH	*o*-OCH_3	227(63), 226(14), 225(5), 212(7), 211(14), 196(7), 183(7), 120(32), 119(6), 110(7), 109(100), 108(6)
21	*o*-OCH_3	*o*-OCH_3	241(38), 240(8), 211(17), 196(5), 183(10), 154(7), 134(5), 132(5), 124(8), 123(100), 121(9), 120(19), 119(43), 118(5), 113(7), 109(5), 108(51), 104(5)

* Reprinted with permission, reference 44.

In the compounds derived from aliphatic aldehydes and ketones, α-cleavage was dominant and produced directly the base peaks in

$$CH_3CH{=}NCH_2CH_2CH_2CH_3$$

and

$$CH_3CH_2CH_2CH{=}NCH_2CH_2CH_2CH_3$$

TABLE 8. Mass spectra of some Schiff bases of pyridine-2-aldehyde.*

(pyridine ring with N)—CH=N—R

(Source: 2·2 × 10^{-5} torr)

R:	H	CH_3	C_2H_5	n-C_3H_7	i-C_3H_7	n-C_4H_9
Mol. Wt.:	106	120	134	148	148	162
m/e						
28	82·3	44·1	74·1	47·8	26·7	29·7
29	33·1	13·0	53·6	34·0	46·4	33·6
39	15·1	47·4	40·0	27·4	24·0	19·7
41					54·3	
42	1·6	81·0	22·7	18·7	42·1	6·3
51	55·3	48·2	55·2	22·3	44·7	14·6
52	100·0	78·5	79·8	43·0	85·9	30·0
65	7·7	46·6	61·1	41·4	43·9	22·0
78	36·8	41·5	52·0	23·5	52·1	22·0
79	91·0	36·4	45·5	27·0	55·4	24·5
92	7·1	49·3	69·0	63·0	71·8	59·3
105	8·0	25·5	33·5	13·2	18·3	15·2
106	16·9[a]	11·6	31·6	12·0	12·0	23·1
107	19·6[b]					
118		19·2	27·2	7·3	6·3	7·7
119		100·0	100·0	100·0	18·1	100·0
120		43·1[a]	41·5	9·5	3·5	14·8
121		22·7[b]				
133			19·0	6·6	100·0	16·5
134			14·6[a]		12·1	
135			30·1[b]			
148				2·3[a]	2·7[a]	
149				9·7[b]	0·7[b]	
162						5·6[a]
163						26·5[b]

[a] Molecular ion.
[b] Pressure-dependent $(M + 1)^+$ peak.

* The ion intensities of Tables 8 and 9 are standardized on 100 units for the base peak. (Tables 8 and 9 reprinted with permission, reference 46.)

by cleavage in the chain on the amine portion of the molecule. The compounds formed from aldehydes with methyl or ethyl amines gave weak peaks arising from β and γ cleavage, with the latter the more intense of the two. A four-membered ring was proposed to account for this as illustrated below:

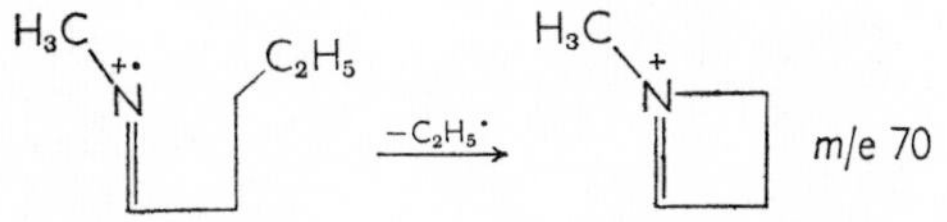

$$H_3C{-}\overset{+\cdot}{N}{=}CH{-}CH_2{-}CH_2{-}C_2H_5 \xrightarrow{-C_2H_5^{\cdot}} \text{(four-membered ring, } H_3C{-}\overset{+}{N}\text{)} \quad m/e\ 70 \qquad \text{(a)}$$

$$H_3C(H)\overset{+\cdot}{N}{-}CH{=}CH{-}CH_2{-}C_2H_5 \xrightarrow{-C_2H_5^{\cdot}} H_3C(H)\overset{+}{N}{-}CH{=}CH{-}\dot{C}H_2 \longrightarrow \text{(four-membered ring)} \quad m/e\ 70 \qquad \text{(b)}$$

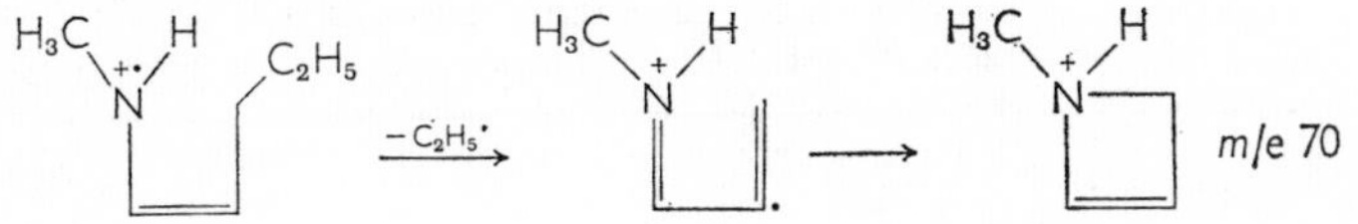

Route (b) represents allyl cleavage of a tautomeric enamine molecular ion. Another important peak was ascribed to a McLafferty rearrangement as follows:

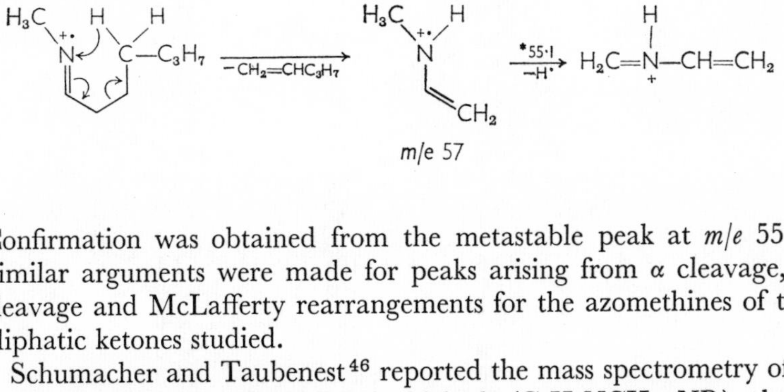

$$H_3C{-}\overset{+\cdot}{N}(H){-}CH{=}CH{-}CH(H){-}C_3H_7 \xrightarrow{-CH_2{=}CHC_3H_7} H_3C(H)\overset{+\cdot}{N}{-}CH{=}CH_2 \ (m/e\ 57) \xrightarrow[-H^{\cdot}]{*55\cdot 1} H_2C{=}\overset{+}{N}(H){-}CH{=}CH_2$$

Confirmation was obtained from the metastable peak at *m/e* 55·1. Similar arguments were made for peaks arising from α cleavage, γ cleavage and McLafferty rearrangements for the azomethines of the aliphatic ketones studied.

Schumacher and Taubenest[46] reported the mass spectrometry of a series of Schiff bases of pyridine-2-aldehyde ($C_5H_4NCH{=}NR$) where R was H, CH_3, C_2H_5, n-C_2H_7, i-C_3H_7, n-C_4H_9, cyclohexyl and phenyl. The base peak at $m/e = 119$ for R = CH_3, C_2H_5, n-C_3H_7 and n-C_4H_9 was explained as β cleavage of the aliphatic chain on the nitrogen atom. An intense peak at *m/e* 92 was assigned to the nitrogen analogue of the tropilium ion, $C_6H_6N^+$. A route to this ion was proposed from a cyclic four-membered transition state involving loss of a neutral molecule as shown on page 174. The spectra are shown in Tables 8 and 9.

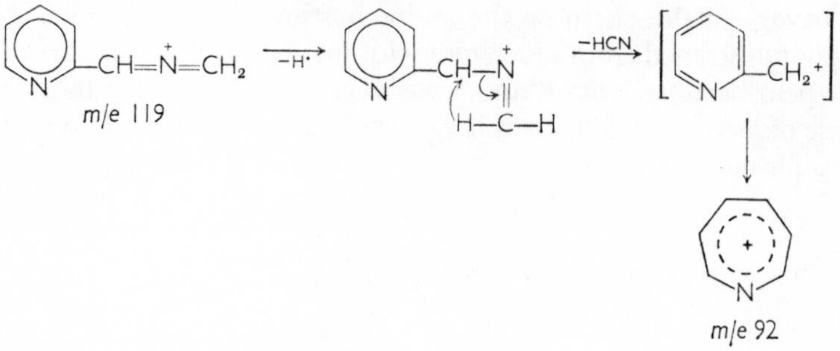

TABLE 9. Mass spectra of some Schiff bases of pyridine-2-aldehyde (see Table 8) with R = cyclohexyl and R = phenyl.

R:	Cyclohexyl		Phenyl
Mol. Wt.:	188		182
m/e		*m/e*	
		27	39·5
28	61·7	28	38·4
29	52·1	29	2·6
		39	35·9
41	98·3	51	96·3
55	71·4	52	77·0
56	59·5	53	33·0
65	34·2	63	26·6
66	19·5	64	23·3
67	34·1	76	23·0
78	36·3	77	100·0
79	60·9	78	56·8
80	39·1	79	73·9
83	32·7		
92	49·6	104	32·4
93	41·2	105	36·2
105	43·7	154	32·0
106	36·8	155	43·6
107	64·0	180	9·8
118	43·3	181	80·5
119	47·2	182	83·5[a]
131	61·4	183	38·8[b]
132	64·0		
145	100·0		
159	30·8		
187	19·2		
188	57·5[a]		
189	26·5[b]		

[a,b] See Table 8.

Mass spectra of a series of double Schiff bases of *trans*-1,2-diamino-cyclopropane have been obtained by Staab and Wünsche[47]. Compounds investigated were of the type

$$\begin{array}{ccc} R & & CHN{=}CHAr \\ & \diagdown\!\!\diagup & | \\ & C & | \\ & \diagup\!\!\diagdown & | \\ R' & & CHN{=}CHAr \end{array}$$

and are shown in Table 10.

TABLE 10. Table of double Schiff bases investigated by mass spectrometry.

$$\begin{array}{ccc} R & & CHN{=}CHAr \\ & \diagdown\!\!\diagup & | \\ & C & | \\ & \diagup\!\!\diagdown & | \\ R' & & CHN{=}CHAr \end{array}$$

Compound	R	R′	Ar
22	H	H	*p*-$C_6H_4N(CH_3)_2$
23	H	H	C_6H_5
24	C_6H_5	H	C_6H_5
25	H	H	*o*-C_6H_4OH
26	C_6H_5	C_6H_5	C_6H_5
27	C_6H_5	C_6H_5	*o*-C_6H_4OH

Molecular ion peaks were obtained for all of the compounds and the relative intensities varied from 6·3% for **25** to 91·3% for **26**. The low intensity of the peak in **25** was explained as being due to ring formation involving the *ortho* hydroxy group. Base peaks arose from the parent ion according to:

$$\begin{array}{ccc} R & & CH{-}N{=}CHAr \\ & \diagdown\!\!\diagup & | \\ & C & | \\ & \diagup\!\!\diagdown & | \\ R' & & CH{-}\overset{+\cdot}{N}{=}CHAr \end{array} \longrightarrow [ArC_2NH]^{\cdot} + \underset{\text{base peak}}{[ArC_3RR'NH_3]^{+}}$$

Further details should be obtained from the original article where the spectra are presented.

These four reports cover widely different types of azomethine compounds and indicate the power of the mass spectrometric method. The spectra are distinctive in every case both between types of Schiff bases and within a given type. *Ortho*-hydroxy substitution on a phenyl ring is especially noteworthy and could be recognized almost by inspection of the mass spectrogram.

VII. NUCLEAR MAGNETIC RESONANCE

N.m.r. has been applied to the analysis of azomethines predominantly from a structure standpoint. The discussion below describes the areas of azomethine chemistry which have been elucidated by n.m.r.

McCarthy and Martell[48] studied the n.m.r. (proton) of acetylacetone and 14 other β-diketone diimine Schiff bases. Chemical shifts and coupling constants are given.

Slomp and Lindberg[49] studied chemical shifts of protons in nitrogen containing compounds. Among the categories studied were azomethines. A chart is given depicting the chemical shift of the proton as related to the configurations around the $>C{=}N-$ grouping.

McDonagh and Smith[50] obtained n.m.r. spectra to study the reaction products of several substituted benzaldehydes with *o*-hydroxybenzylamine. The purpose of the study was to elucidate the tautomerism of these compounds. Nelson and Worman[51] also studied tautomerism of azomethines via n.m.r.

Rieker and Kessler[52] studied steric hindrance and isomerization of the $>C{=}N-$ double bond in fifteen quinone anils. Chemical shifts and their solvent dependency were studied. Ultraviolet measurements were also made and absorption shifts were correlated with ring distortions.

Tori et al.[53] studied the relation between allylic or homoallylic coupling constants and electron localization on C=N double bonds in some Schiff bases and their *N*-oxides. Coupling constants are also reported.

Staab et al.[54] used proton n.m.r. to elucidate the *syn*- and *anti*-isomerism of Schiff bases.

Binsch et al.[55] used ^{15}N n.m.r. to determine coupling constants in nitrogen compounds among which the azomethine group is included. The coupling of ^{15}N to directly bonded protons, to ^{13}C, to ^{15}N, and to protons separated by two and three bonds is discussed.

Dudeck et al.[56–59] studied the keto–enol equilibria in a variety of Schiff bases derived from β-diketones, *o*-hydroxyacetophenones, and *o*-hydroxyacetonaphthones.

Shapiro et al.[60] obtained the coupling constants of the *geminal* protons on $CH_2{=}N-$ azomethines. These are listed for eight compounds.

VIII. FLUORESCENCE AND OTHER PHOTOCHEMICAL PROPERTIES

Fluorescence of materials upon irradiation often provides a convenient approach to qualitative and quantitative measurements. Azomethines, especially the aromatic members, tend to fluoresce and hence a means toward analysis is provided. In fact, the fluorescence of the anils provides a means to determine some amines[61,62]. It must be kept in mind, however, that the fluorescence process is a delicate one and is readily influenced by impurities and conditions. Quenching or enhancement of fluorescence is very common and must be watched for when this analytical approach is used.

Nurmukhametov et al.[63] studied the luminescence spectra of 19 azomethines in powder form and in hexane solution. The azomethines were of the general formula $RC_6H_4CH{=}NC_6H_4R'$. The results were correlated with molecular structure.

A comprehensive study of azomethines of aromatic aldehydes was made[64,65], relating the photoluminescent process to concentration, temperature and structure. Thermochromic effects were also noted. Quenching effects and bathochromic shifts of spectra were also reported.

Terent'ev et al.[66] studied some azomethines of various vanillins and metal chelates of these compounds. The luminescence is attributed to the azomethine grouping.

The fluorescence of the aromatic azomethines of general structure

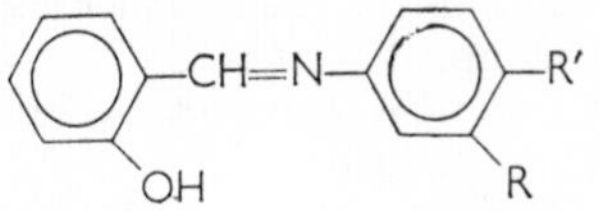

—CH=N— —R′
OH R

is so strong that they are proposed as fluorescent dyes[67].

In addition to fluorescence, some azomethines undergo isomerization or other photolytic reactions which either produce colour, or produce a change in colour. These reactions also have utility as analytical approaches.

Salicylideneaniline shows a phototropy[68], changing from yellow to red when exposed to ultraviolet light; the infrared spectrum, however, remains the same. This property is amenable to the analysis of this azomethine. Mixed crystals with benzylideneaniline exhibited similar behaviour. Anderson and Wettermark[69] studied the same phenomenon with *N*-(*o*-hydroxybenzylidene)aniline; *N*-(*o*-hydroxybenzylidene)-β-naphthylamine; *N*-(*p*-hydroxybenzylidene)aniline; *N*-benzylideneaniline and *N*-benzohydrylideneaniline. The dependence of the process on solvent and pH was also studied. The kinetics of the photo-induced isomerizations were also followed.

Becker and Richey[70] studied the photochromic properties of the nitrosalicylidene anils and a few substituted salicylidene-*o*-toluidines. The relationship of structural transformation to colour formation is described. In some instances, fluorescence was also observed.

IX. REFERENCES

1. N. Tarugi and F. Lenci, *J. Chem. Soc.*, **102**, 397 (1912).
2. C. G. Marcarovici and M. Marcarovici, *Acad. Rep. Populare Romine*, **4**, Nos. 1 and 2, 169 (1952); also *Chem. Abstr.*, **50**, 13665b (1956).
3. F. Feigl and E. Liebergott, *Anal. Chem.*, **36**, 132 (1964).
4. C. D. Wagner, R. H. Brown and E. D. Peters, *J. Am. Chem. Soc.*, **69**, 2609 (1947).
5. S. Siggia, J. G. Hanna and I. R. Kervenski, *Anal. Chem.*, **22**, 1295 (1950).
6. S. Siggia and E. Segal, *Anal. Chem.*, **25**, 830 (1953).
7. J. S. Fritz, *Anal. Chem.*, **25**, 578 (1953).
8. S. Freeman, *Anal. Chem.*, **25**, 1750 (1953).
9. J. Fritz, *Anal. Chem.*, **22**, 1028 (1950).
10. (a) P. A. Pichard and F. A. Iddings, *Anal. Chem.*, **31**, 1228 (1959).
    (b) R. Pohloudek-Fabini, B. Goeckeritz and H. Brueckner, *Pharm. Zentralhalle*, **104**, 315 (1965).
    (c) R. Hara and P. W. West, *Anal. Chim. Acta*, **15**, 193 (1956).
11. E. F. Hillenbrand, Jr., and C. A. Pentz, *Organic Analysis*, Vol. 3 (Ed. J. Mitchell, Jr., I. M. Kolthoff, E. S. Proskauer and A. Weissberger), Interscience, New York, 1956, pp. 194–196.
12. P. Zuman, *Collection Czech. Chem. Commun.*, **15**, 839 (1950).
13. P. Zuman, *Nature*, **165**, 485 (1950).
14. B. von Kastening, L. Holleck and G. A. Melkonian, *Z. Electrochem.*, **60**, 130 (1956).
15. M. Dezelic and K. Durzun, *Polarography 1964*, Vol. 2 (Ed. G. J. Hills), Macmillan, London, 1966, p. 879.
16. V. D. Bezuglyi, V. N. Dmitrieva and L. U. Skvortosova, *Kinetika i Kataliz.*, **6**, 737 (1965); *Chem. Abstr.*, **64**, 3332d (1966).
17. P. Zuman and M. Brezina, *Chem. Listy*, **46**, 516; 599 (1952).
18. M. Brezina and P. Zuman, *Chem. Listy*, **47**, 975 (1953).
19. B. Holleck and B. von Kastening, *Z. Electrochem.*, **60**, 127 (1956).
20. H. Lund, *Acta Chem. Scand.*, **13**, 249 (1959).
21. Yu. I. Vainshtein and Yu. A. Davydovskaya, *Tr. Vses. Nauchn.-Issled. Inst. Khim. Reaktivov*, **27**, 317 (1965); *Chem. Abstr.*, **65**, 5022e (1966).
22. Yu. I. Vainshtein and Yu. A. Davydovskaya, *Tr. Vses. Nauchn.-Issled. Inst. Khim. Reaktivov i Osobo Chist. Khim. Veshchestv*, **28**, 238 (1966); *Chem. Abstr.*, **67**, 28734z (1967).

23. R. E. Van Atta and D. R. Jamieson, *Anal. Chem.*, **31**, 1217 (1959).
24. M. E. Hall, *Anal. Chem.*, **31**, 2007 (1959).
25. E. N. Zil'berman A. A. Kalugin and E. M. Perepletchikova, *Russian Pat.* 143, 593 (1962); *Chem. Abstr.*, **57**, 3203c (1962).
26. V. N. Dmitrieva, V. B. Smelyakova, B. M. Krasovitskii and V. D. Bezuglyi, *J. Gen. Chem. U.S.S.R.*, **36**, 421 (1966).
27. N. F. Levchenko, L. Sh. Afanasiadi and V. D. Bezuglyi, *Zh. Obshch. Khim.*, **37**, 666 (1967).
28. M. Uehara, *Nippon Kagaku Zasshi*, **89**, 901 (1965).
29. V. N. Dmitrieva, L. Kononenko and V. D. Bezuglyi, *Theoret. Exp. Chem.*, **1**, 297 (1965).
30. R. M. Silverstein and G. C. Bassler, *Spectrometric Identification of Organic Compounds*, 2nd Ed., John Wiley and Sons, New York, 1967, p. 97.
31. J. Fabian, M. Legrand and P. Poirer, *Bull. Soc. Chim. France*, 1499 (1956).
32. F. H. Suydam, *Anal. Chem.*, **35**, 193 (1963).
33. W. L. Steele, *Dissertation Abstr.*, **25**, 61 (1964).
34. J. Fabian and M. Legrand, *Bull. Soc. Chim. France*, 1461 (1956).
35. L. E. Clougherty, J. A. Sousa and G. M. Wyman, *J. Org. Chem.*, **22**, 462 (1957).
36. H. H. Freedman, *J. Am. Chem. Soc.*, **83**, 2900 (1961).
37. V. I. Minkin, O. A. Osipov, V. A. Kogan, R. R. Shagidullin, R. L. Terent'ev and O. A. Raevskii, *Russian J. Phys. Chem.*, **38**, 938 (1964); *Zh. Fiz. Khim.*, **38**, 1718 (1964).
38. V. I. Minkin, Yu. A. Zhdanov, A. D. Garnovakii and I. D. Sadekov, *Zh. Fiz. Khim.*, **40**, 657 (1966); *Chem. Abstr.*, **65**, 4784a (1966).
39. D. N. Shigorin, I. Ya. Pavlenishvili, G. V. Panova, B. N. Bolotin and N. N. Shapet'ko, *Russian J. Phys. Chem.*, **40**, 822 (1966); *Zh. Fiz. Khim.*, **40**, 1516 (1966).
40. D. Heinert and A. E. Martell, *J. Am. Chem. Soc.*, **84**, 3257 (1962).
41. J. W. Ledbetter, D. N. Kramer and F. N. Miller, *J. Org. Chem.*, **32**, 1165 (1967).
42. E. Feytmans-de Medicis, *Bull. Soc. Chim. Belges*, **75**, 426 (1966).
43. H. A. Staab and F. Voegtle, *Chem. Ber.*, **98**, 2681 (1965).
44. D. J. Elias and R. G. Gillis, *Australian J. Chem.*, **19**, 251 (1966).
45. M. Fischer and C. Djerassi, *Chem. Ber.*, **99**, 1541 (1966).
46. E. Schumacher and R. Taubenest, *Helv. Chim. Acta*, **49**, 1455 (1966).
47. R. A. Staab and C. Wünsche, *Chem. Ber.*, **98**, 3479 (1965).
48. P. J. McCarthy and A. E. Martell, *Inorg. Chem.*, **6**, 781 (1967).
49. G. Slomp and J. G. Lindberg, *Anal. Chem.*, **39**, 60 (1967).
50. A. F. McDonagh and H. E. Smith, *Chem. Commun.*, 374 (1966).
51. D. A. Nelson and J. J. Worman, *Chem. Commun.*, 487 (1966).
52. A. Rieker and A. Kessler, *Z. Naturforsch.*, **b21**, 939 (1966).
53. K. Tori, M. Otsuru and T. Kubota, *Bull. Chem. Soc. Japan*, **39**, 1089 (1966).
54. H. A. Staab, F. Vogtle and A. Mannschreck, *Tetrahedron Letters*, 697 (1965).
55. G. Binsch, J. B. Lambert, B. W. Roberts and J. D. Roberts, *J. Am. Chem. Soc.*, **86**, 5564 (1964).
56. G. O. Dudeck and E. P. Dudek, *J. Am. Chem. Soc.*, **86**, 4283 (1964).
57. G. O. Dudeck and R. H. Holm, *J. Am. Chem. Soc.*, **84**, 2691 (1962).
58. G. O. Dudeck, *J. Am. Chem. Soc.*, **85**, 694 (1963).

59. G. O. Dudek and G. Volpp, *J. Am. Chem. Soc.*, **85**, 2697 (1963).
60. B. L. Shapiro, S. J. Ebersole, G. J. Karabatsos, F. M. Vane and S. L. Manatt. *J. Am. Chem. Soc.*, **85**, 4041 (1963).
61. L. Sassi, *Arch. Inst. Pasteur Tunis*, **33**, 451 (1956).
62. L. Juhlin and W. B. Shelley, *J. Histochem. Cytochem.*, **14**, 525 (1966).
63. R. N. Nurmukhametov, Y. I. Koslov, D. N. Shigorin and V. A. Puchkov, *Dokl. Akad. Nauk SSSR*, **143**, 1145 (1962).
64. O. A. Osipov, Y. A. Zhdanov, M. I. Knyazhanskii, V. I. Minkin, A. D. Garnovskii and I. D. Sadekov, *Zh. Fiz. Khim.*, **41** (3), 641 (1967).
65. M. I. Knyazhanskii, V. I. Minkin and O. A. Osipov, *Zh. Fiz. Khim.*, **41** (3), 649 (1967).
66. A. P. Terent'ev, E. G. Rukhadze, G. P. Talyzenkova and G. V. Panova, *Zh. Obshch. Khim.*, **36** (9), 1590 (1966).
67. All Union of Scientific Research Institute of Chemical Reagents and Pure Chemical Substances, *French Pat.* 1,427,102, (1966); *Chem. Abstr.*, **65**, 9069 (1966).
68. N. Ehara, *Nippon Kagaku Zasshi*, **82**, 941 (1961).
69. D. G. Anderson and G. Wettermark, *J. Am. Chem. Soc.*, **87**, 1433 (1965).
70. R. S. Becker and W. F. Richey, *J. Am. Chem. Soc.*, **89**, 1298 (1967).

CHAPTER 4

The optical rotatory dispersion and circular dichroism of azomethines

R. Bonnett

Queen Mary College, London, England

In this chapter it is proposed to discuss the optical dissymmetry effects (a term that will be used here to mean optical rotatory dispersion and circular dichroism) of compounds containing the azomethine group. Metal complexes will not be discussed. Systems such as aliphatic unconjugated azomethines in which the azomethine chromophore itself is insulated will be of especial interest, but other substituted, conjugated and otherwise related compounds will also be considered.

I. ULTRAVIOLET ABSORPTION OF AZOMETHINES

A. Unconjugated Azomethines

The chromophoric properties of the azomethine group in an aliphatic environment went largely unnoticed until as recently as 1963. In that year Bonnett and his colleagues, who had noted broad bands in the spectra of certain alkyl-1-pyrrolines at ~230 mμ, showed that *N*-neopentylidenealkylamines had a weak absorption band in a similar position[1,2]. Although standard texts seldom refer to it, there had in fact been earlier mention of the absorption spectra of aliphatic azomethines. For instance, Hires and Balog[3] had reported the spectrum of *N*-butylidenebutylamine (**1**), and spectra had also been presented for compounds **2**[4], **3**[5] and **4**[6].

n-PrCH=NBu-n

(**1**)

λ_{max} in EtOH 233 mμ, ε155

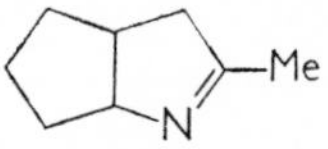

(**2**)

λ_{max} 229 mμ, ε195

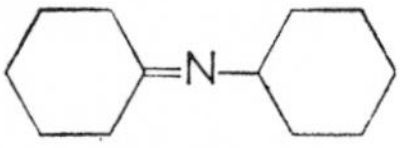

(**3**)

λ_{max} in isooctane (245 mμ), ε180; 300 mμ, ε60

(values read from published curve*)

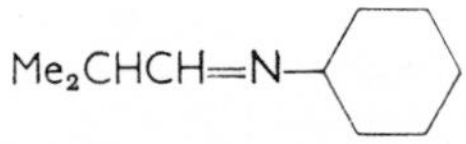

Me_2CHCH=N–

(**4**)

λ_{max} in cyclohexane 245 mμ, ε74

However, all these examples have hydrogen atoms on the carbon atom α to the carbon of the azomethine group: consequently the observed absorption might well have been ascribable to (a) products (such as **5**)[7]

n-PrCH=NBu-n (**1**) —150°, 3 h→ n-PrCH=C(Et)CH=NBu-n (**5**) 65%

which are known to absorb strongly in the 220 mμ region (see Section I.B) and which could arise by an autocatalysed aldol-type condensation and/or to (b) a small proportion of the tautomeric enamine (**6**), which would also be expected to absorb strongly in the 220 mμ region

n-PrCH=NBu-n (**1**) ⇄ EtCH=CH—NHBu-n (**6**)

* The 300 mμ band may be due to carbonyl contamination (e.g. from hydrolysis).

by virtue of its analogy with the tertiary enamine system (e.g.[8]

$Me_2C{=}CH{-}N$ (morpholino ring, O)

λ_{max}217 mμ, ε6600 in 0·001N NaOH). These possible complications are avoided by choosing an aliphatic aldo-azomethine with a quaternary carbon adjacent to the carbon of the functional group. Thus from pivalaldehyde (under nitrogen) *N*-neopentylidenealkylamines, which are colourless, rather sweet-smelling liquids of reasonable stability, may be prepared. These show weak absorption in the 240 mμ region, and such absorption is thus to be regarded as an inherent property of the unperturbed azomethine system. In Table 1 the spectra of various azomethines of this general type are summarized.

In those systems where formation of an enamine is possible the 240 mμ band is still seen, but under certain conditions an inflection (at ~ 220 mμ) can be detected, and this has been ascribed by Nelson and Worman[10] to the enamine chromophore. Indeed this provides a spectroscopic method for estimating the tautomeric ratio: in dilute solution in cyclohexane the enamine content may extend up to a few percent on this criterion (e.g. *N*-cyclohexylidene-n-butylamine 5%; *N*-cyclohexylidenecyclohexylamine < 1%) but in more polar solvents

TABLE 1. Electronic spectra of some unconjugated aliphatic azomethines.

	Solvent	λ_{max} (mμ)	ϵ_{max}	Reference
(**7**) $Me_3CCH{=}NBu$-n	EtOH	235	107	1
	Hexane	244	87	1
(**8**) $Me_3CCH{=}NBu$-*s*	MeOH	233	100	1
	EtOH	236	93	1
	t-BuOH	239	83	1
	Hexane	243	85	1
(**9**) $Me_3CCH{=}NBu$-*t*	EtOH	242	83	1
	Hexane	250	80	1
(**10**) 2,2,4,4-tetramethyl-1-pyrroline (structure: Me, Me, Me, Me, N)	EtOH	226	83	2
(**11**) $CH_2{=}NBu$-*t*[a]	Hexane	272	—	9

[a] In equilibrium with the cyclic trimer (1,3,5-tri-*t*-butylhexahydro-*s*-triazine).

(ethanol) the enamine tautomer is often not detected. Nevertheless the possibility of a contribution from the enamine tautomer must be borne in mind whenever the electronic spectra (Table 2)—and the optical dissymmetry effects—of azomethines which can form enamines are being considered.

TABLE 2. Electronic spectra of some enaminizable azomethines.

Compound	Solvent	λ_{max} (mμ)	ϵ_{max}	Reference
(**12**) $Me_2C{=}NBu\text{-}n$	Heptane	179	8900	11
	Cyclohexane	246	140	11
	Trimethyl phosphate	183	—[a]	11
		236	—[a]	11
	CH_3CN	235	192	12
	EtOH	232	200	11
(**13**) 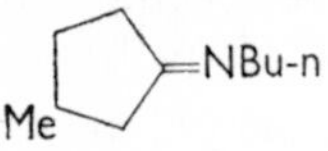	Cyclohexane	253	197	14
(**14**) 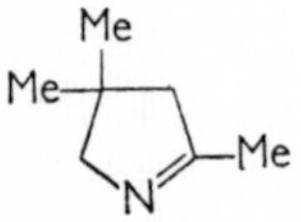	Hexane	230	195	2
	EtOH	221	210	2
(**15**) 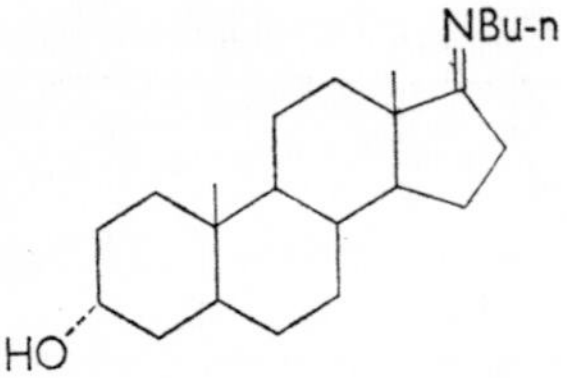	Hexane	248	175	15
	EtOH	238	220	15
(**16**) 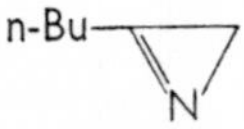	Cyclohexane	229	112	16
(**17**) 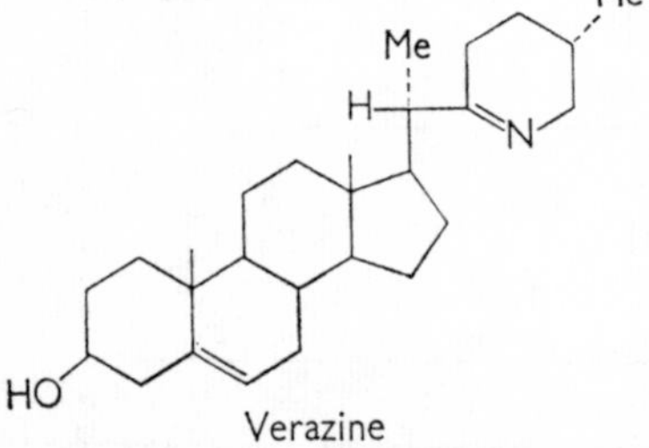 Verazine	EtOH	243	630	89

[a] Value not quoted in the preliminary communication.

Since the azomethine group is a weak chromophore and is located rather far into the quartz region, its effect is easily submerged if other chromophoric groups are present in the molecule. Thus in some of the steroidal azomethines (see Section II.C) the band is partially submerged in end-absorption and appears merely as an inflection. In *N*-neopentylidene-α-amino acid esters[17] (see Section II.B) the band is not clearly distinguished from absorption due to the ester chromophore.

The absorption band at 240 mμ has been attributed[1,6] to an $n \rightarrow \pi\dagger$ transition involving promotion of a non-bonded electron associated with nitrogen. Such electrons are formally sp^2 hybridized,

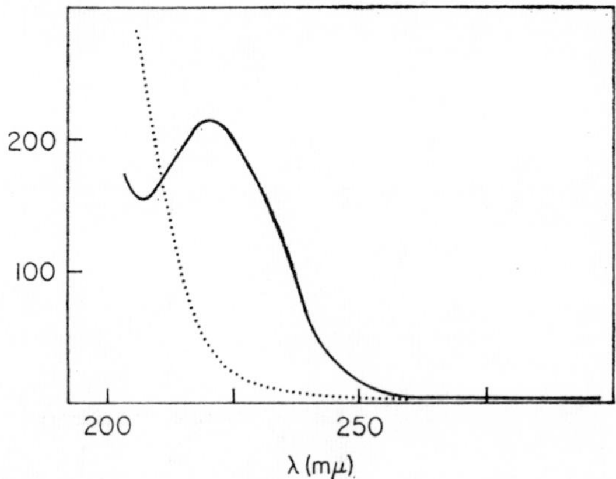

FIGURE 1. Electronic spectra of 2,4,4-trimethyl-1-pyrroline (**14**) (——) and the corresponding pyrrolidine (· · · · ·) in ethanol.

although Lipscomb and his colleagues[18] have pointed out that the nitrogen lone pair has less 2*s* character than would be predicted on this basis. That the double bond is involved is evident from the observation that the 240 mμ band is absent in the corresponding secondary amine (e.g. 2,4,4-trimethyl-1-pyrroline (**14**) → 2,4,4-trimethylpyrrolidine; Figure 1). The observation that acidification of the solution leads to the disappearance of the band accords with the postulate that lone-pair electrons are involved in the transition†. With a suitable system

† The analogous quaternary imminium ion appears to absorb with $\lambda_{max} \sim$ 220 mμ (e.g.[19] $Et_2CHCH{=}N^+$(cyclohexane ring) Cl^- λ_{max} 222 mμ, ϵ2250 in CH_3CN) although this location has been questioned[26] and somewhat lower values have been re-

such as 1-pyrroline, it may be shown (Figure 2) that protonation is a readily reversible process, i.e.

$$\underset{(\mathbf{14})}{\text{2,4,4-trimethyl-1-pyrroline}} \underset{NH_3}{\overset{H^+}{\rightleftharpoons}} \underset{(\mathbf{14a})}{\text{protonated (N}^+\text{–H)}}$$

(14) λ_{max} 221 mμ in EtOH ($n \rightarrow \pi^*$)

(14a) λ_{max} <220 mμ in EtOH ($\pi \rightarrow \pi^*$)

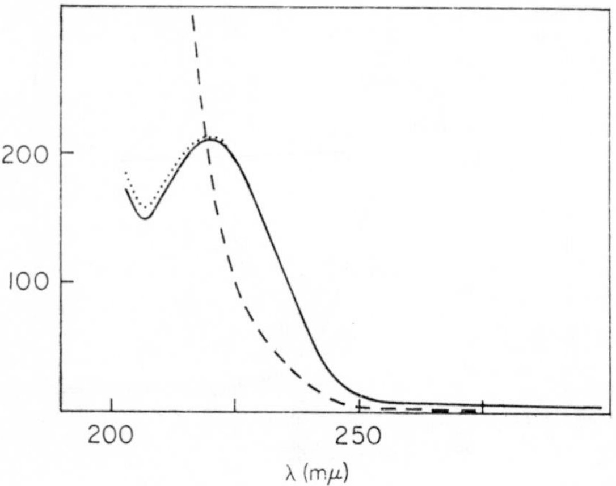

FIGURE 2. Electronic spectrum of 2,4,4-trimethyl-1-pyrroline (**14**) in ethanol (——); on treating with a trace of 6N HCl (– – – –); on further treating the acidified solution with a trace of aqueous ammonia (· · · · ·).

---

corded (e.g.[8] $Me_2C{=}CH{-}N$(morpholino) λ_{max} 196 mμ, ϵ2650 in 0·01N HCl).

An additional complication arises because the chemical stability of this system, at least under certain circumstances, appears to be open to doubt[20]. On the basis of the analogy of the spectra of polyalkylethylenes[21] the $R_2C{=}\overset{+}{N}HR$ ion would be expected to show a $\pi \rightarrow \pi^*$ transition at somewhat higher energy than that of the ternary immonium ion. It may be noted that nucleophilic attack of solvent alcohol on the ternary immonium ion would also remove the double-bond chromophore. However, the amino ketal so formed would be unlikely to regenerate the azomethine group instantaneously under basic conditions (cf. **14a** → **14**).

Moreover, the band, which is of rather low intensity, shows the solvent sensitivity (hypsochromic shift with more polar solvent) characteristic of $n \rightarrow \pi^*$ transitions, and the effect of substitution generally parallels that of the $n \rightarrow \pi^*$ band of the carbonyl chromophore[1,2]. It is interesting, however, that the transition is of considerably higher energy than has been predicted[22]. The low intensity of the band is possibly to be accounted for in terms similar to those applied to the pyridine $n \rightarrow \pi^*$ transition[23].

The strong band in the spectra of azomethines at ~180 mμ shows a small bathochromic shift with increasing solvent polarity (**12**, Table 2) and is attributed to a $\pi \rightarrow \pi^*$ transition[11].

B. Conjugated Azomethines

The conjugation of the azomethine chromophore with olefinic or aryl groups changes the spectrum considerably, since rather weak bands due to $n \rightarrow \pi^*$ transitions are now submerged by strong absorption associated with $\pi \rightarrow \pi^*$ transitions. Often bands still appear in the 230 mμ region, but their intensities ($\epsilon \sim$ 10,000) leave little room for confusion. Further bands or inflections, generally weaker, may be observed at longer wavelengths, and occasionally these have been identified as $n \rightarrow \pi^*$ components, including the special case[23] where the conjugated azomethine is part of an aromatic ring. Thus[6,23] pyridine in cyclohexane shows a $\pi \rightarrow \pi^*$ band at 251 mμ (ϵ2000) with an inflection (270 mμ) on the short-wavelength side attributed to a $n \rightarrow \pi^*$ transition. Similarly pyrimidine has λ_{max} 243 mμ (ϵ2030) and 298 mμ (ϵ326) in cyclohexane. The long-wavelength absorption of *N*-benzylidenealkylamines is the subject of conflicting reports. In some cases[3,5] (e.g. **24**, Table 3) inflections in the 280 mμ region have been observed; in others[30], including *N*-benzylidenemethylamine[24], these have not been noted. Benzophenone imine ($Ph_2C{=}NH$) is reported to have a strong absorption at 260 mμ (ϵ10,000) and a weaker one at 340 mμ (ϵ125) in absolute alcohol, although the curve presented[47] resembles that of the parent ketone rather closely. Several factors (weak or unrecognized inflections, stereochemical considerations, hydrolysis to carbonyl compound) may be involved here, and evidently further experimental work with carefully purified compounds is desirable. Although an aryl group appears to conjugate effectively when substituted at the carbon of the azomethine group, it may not do so when on the nitrogen. Thus the spectrum of *N*-cyclohexylideneaniline is rather reminiscent of that of aniline itself and there is evidence that in benzylideneaniline, the spectrum of which differs in detail from that of

TABLE 3. Electronic spectra of conjugated azomethines.

Compound	Solvent	λ_{max} (mμ)	ϵ_{max}	Reference
Unsaturated substituent on nitrogen				
(**18**) cyclohexylidene=N—Ph	Isooctane	225·1	6,500	25
		281·0	1,240	
(**19**) PrCH=NPh	EtOH	235	10,700	3
		287	2,630	
(**20**)[d] 2,5-dimethyl-3H-pyrrolium (Me, $\overset{+}{N}$H, Me)	5N H_2SO_4	(~275)[a]	2,500	26
Unsaturated substituent on carbon				
(**21**) CH_2=CH—CH=NEt	Cyclohexane[b]	213	10,000	27
(**22**) MeCH=CH—CH=NBu-n	CH_3CN	220	23,300	28
	CH_3CN/H^+	247·5	23,900	28
(**23**) Me_2C=CHCH=CHCH=NBu-n	MeOH	281	37,700	28
	CH_3CN	273·5	40,000	28
	CH_3CN/H^+	330	40,000	28
	Isooctane	270·8	39,100	28
(**24**) PhCH=NBu-n	EtOH	246	14,500	3
		(280)	1,520	
		(288)	1,000	
(**25**) PhMeC=N—cyclohexyl	EtOH[b]	240	10,000	5

Compound	Solvent			Ref.
(**26**) o-$MeC_6H_4MeC{=}N$–(cyclohexyl)	EtOH[b]	no maximum observed		5
		(~265)	~400	
	i-PrOH	244	10,800	29
(**27**) $Ph_2C{=}NMe$				
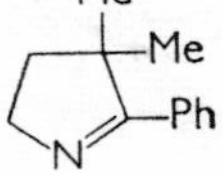 (**28**) Me, Me, Ph, N	EtOH	239	8,900	2
	EtOH/H^+	267		2
	Hexane	239	11,500	2
(**29**)[e] 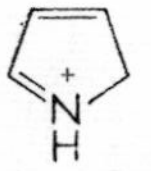 $\overset{+}{N}$, H	H_2SO_4	247	4,100	26
(**30**) 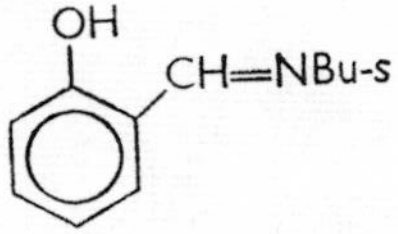 OH, CH=NBu-s	EtOH	253	12,300	30
		(278)	2,500	
		312	3,900	
		401	850	
	Dioxan	254	12,300	30
		(258)	11,500	
		316	4,470	
		(392)	32	
	Hexane	254	10,500	30
		(260)	9,800	
		317	4,700	

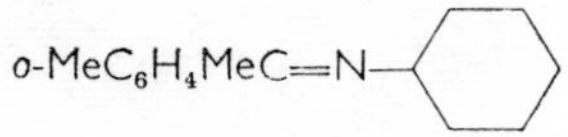

TABLE 3. Electronic spectra of conjugated azomethines (*continued*).

Compound	Solvent	λ_{max} (mμ)	ϵ_{max}	Reference
(31) OMe, CH=NBu-s	EtOH	250 304	25,100 10,000	30
(32) OH, CH=NCHMePh	EtOH	256 (283) 315 404	13,800 2,200 4,100 600	30
(33) CH=N—CH(Pr-i)COOK, OH, N	KBr disc[c]	322 422	86 106	31
	MeOH	250 323 414	8,600 3,400 2,000	31
	Dioxan	(250) (270) 324 425	6,600 5,500 2,400 4,200	33

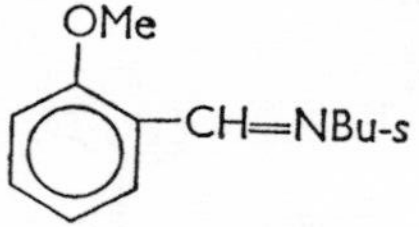

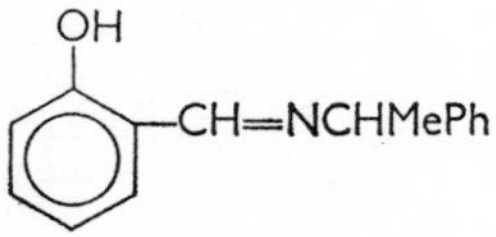

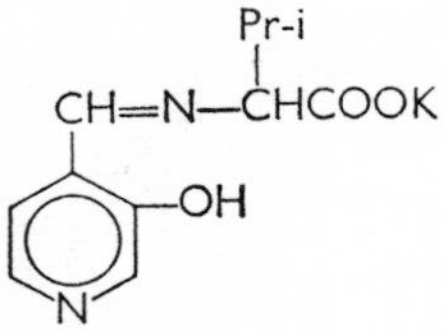

Compound	Solvent	λ	ε	Ref.
	Dioxan–0·05N HOAc	243	11,100	33
		328	4,200	
(34) PhCH=NPh	EtOH	256	16,200	24
		(310)	8,200	
	H_2SO_4	335	19,500	24
	Cyclohexane	227	12,000	32
		236	9,900	
		262	17,300	
		314	6,940	
(35) 2,4,6-Me₃C₆H₂–CH=NPh	Cyclohexane	273	17,500	32
		326	6,300	
(36) PhCH=N–C₆H₃-2,6-Me₂	Cyclohexane	251	22,100	32
		331	1,740	

[a] Wavelength values in parentheses denote inflections.
[b] Values read from published curve.
[c] Extinctions are relative values.
[d] Formed by protonation of 2,5-dimethylpyrrole: α-protonated species also present (band at ~237 mμ).
[e] Formed by protonation of pyrrole: in this case the α-protonated species predominates in the equilibrium.

stilbene[24], the *N*-phenyl group is twisted out of the plane of the rest of the chromophore[24,32]. That the *N*-substituent is twisted out of plane more readily than the *C*-substituent may be explained in terms of (a) the non-bonded distances[32] involved, thus:

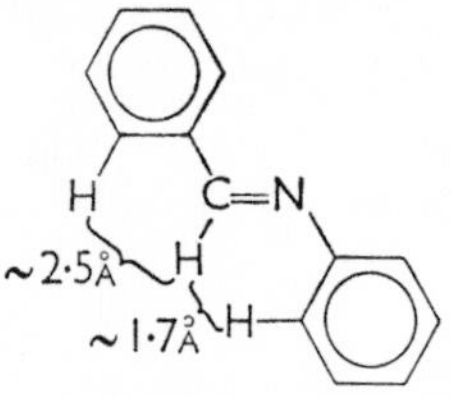

and (b) the development of orbital overlap between the lone pair and the *N*-aryl system in the twisted conformation. However, steric effects may also interfere with orbital overlap between the *C*-aryl and azomethine systems as is illustrated in the series **24**→**25**→**26** (Table 3).

In dramatic contrast to the behaviour of unconjugated azomethines, protonation of a *C*-conjugated azomethine causes a bathochromic shift of the absorption band. Table 3 illustrates this effect (**22**, **23**, **28**, **34**), and summarizes typical spectra for various kinds of conjugated azomethine.

Braude and his colleagues[13] have compared the spectra of conjugated azomethines with those of the corresponding polyenes and have examined the effect of other conjugating groups such as C═N itself. Conjugation can formally occur here in three ways, of which two, the azine (C═N—N═C) and diimine (—N═C—C═N—) systems, are familiar. With alkyl substituents both show intense maxima in the 205–210 mμ range, although a weak absorption at higher wavelength has been reported[6] for a tetrasubstituted azine [$(CH_2)_5C{=}N{-}N{=}C(CH_2)_5$, λ_{max} in cyclohexane 308 mμ, ϵ89]. Typical aryl derivatives on the other hand show maxima—often multiple maxima—in the 300 mμ region, e.g.[13] in ethanol:

n-BuN═CMeCMe═NBu-n	λ206 mμ	ε17,000
	209	18,500
n-PrCH═N—N═CHPr-n	205	13,000
	208	11,500
PhN═CMe—CMe═NPh	300	2,000
	325	2,000
PhCH═N—N═CHPh	300	36,000
	308	35,000

Two areas involving conjugated azomethines deserve special mention because of their relevance to biologically important processes. The first concerns the chemistry of vision[100]. The pigments ('visual purple') of the rod and cone structures of the retina are derived from vitamin A aldehydes and proteins (opsins). A characteristic compound is rhodopsin, from opsin and 11-*cis*-retinal (one of the thermodynamically less stable *cis*-configurations). It has $\lambda_{max} \sim 500$ mμ. There is suggestive evidence (but see below) that rhodopsin contains an azomethine linkage. On exposure to light rhodopsin is 'bleached', giving metarhodopsin II (λ_{max} 380 mμ) via various intermediates which can be detected at low temperatures. On hydrolysis metarhodopsin II gives opsin and all-*trans*-retinal: hence one of the consequences of the initial photo-reaction is *cis–trans* isomerization (another, presumably, is nerve stimulation). The process may be represented as follows:

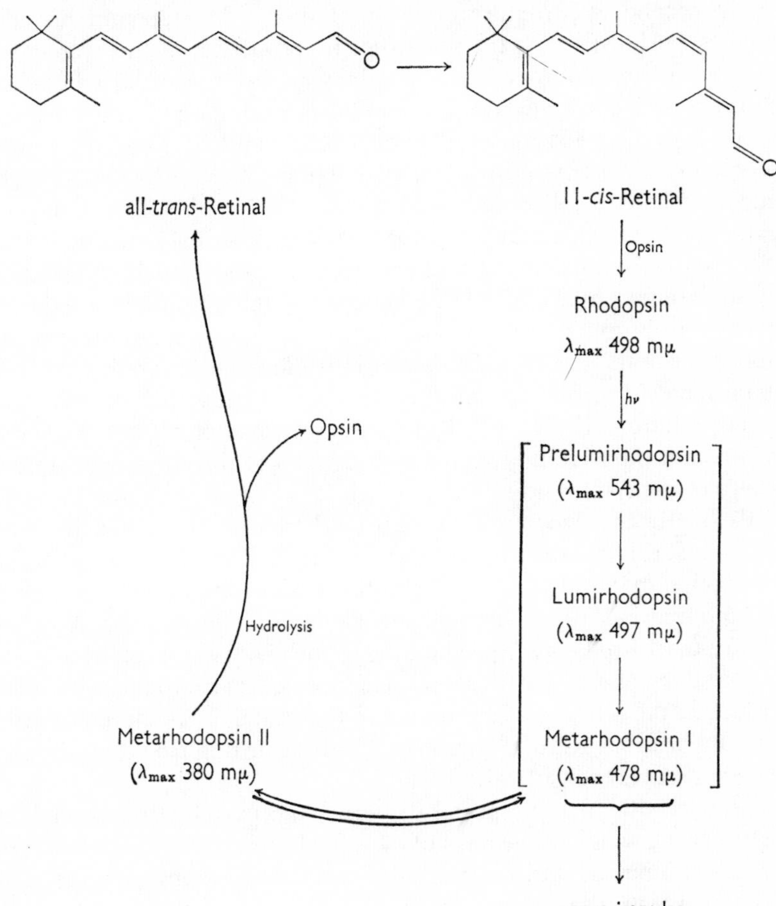

The intermediates in square brackets have been detected spectroscopically at low temperatures and are presumably to be regarded as short-lived intermediates in the process at ambient temperatures. Their chemical nature is not clear, but it is believed that they are derivatives of all-*trans*-retinal and that the differences are conformational ones in the protein moiety. Amongst the further complications are (i) isomerization to 9-*cis*-retinal and its derivatives (e.g. isorhodopsin) and (ii) the occurrence of pigments derived from another aldehyde (vitamin A_2 aldehyde) and from other opsins.

The evidence for the azomethine linkage is as follows. When rhodopsin is denatured with acid, another pigment, indicator yellow, is formed (indicator yellow appears to differ from metarhodopsin II in the secondary structure of the protein)[102]. Morton and his colleagues[101] showed that the spectroscopic properties of this were very similar to those of *N*-retinylidenemethylamine, the expected bathochromic shift occurring on protonation in each case. Thus indicator yellow had λ_{max} 365 mμ in alkaline solution and λ_{max} 440 mμ in the protonated form: *N*-retinylidenemethylamine (λ_{max} 355 mμ, ϵ57,400 in petrol) had λ_{max} 439 mμ in acid and λ_{max} 362 mμ in alkali. This provides good evidence that indicator yellow is *N*-retinylideneopsin, and is supported by the observed reduction of this (and model) compounds by sodium borohydride[102,103]. Rhodopsin cannot be directly reduced by sodium borohydride, but such reduction does proceed on irradiation. It appears that metarhodopsin II is much more rapidly reduced than metarhodopsin I, and the product is regarded as dihydrometarhodospsin II (λ_{max} 333 mμ); hydrolysis of this gives *N*-retinyl lysine, and the *N*-retinyl group is thought to be linked to the ϵ-amino group of the lysyl residue[100].

Evidently the azomethine group in rhodopsin is masked in some way: it is still necessary to account for this, and for the position of the absorption maxima with respect to the model compounds, since even on protonation the *N*-retinylidenealkylamine system has λ_{max} no higher than ~440 mμ. Specific interactions with the protein may be responsible, and there is some evidence for participation by thiol groups in the binding of the retinal to the protein[107]. Charge transfer has also been suggested[103], and the possible role of a *retro*-double bond system deserves consideration.

The second area of special interest concerns *o*-hydroxyarylidene Schiff bases, which in the form of pyridoxal derivatives[106] are involved in transamination and racemization reactions of α-amino acids. The *N*-salicylidene alkylamines in general give four bands in the near ultraviolet region (**30**, Table 3); the spectra are solvent sensitive and the band at ~400 mμ, while pronounced in ethanol solution, is almost entirely absent in hexane (Figure 3). Hires and Hackl suggested that this band is due to a solvated complex of the phenolic azomethine[34], and some other workers[92] have supported this interpretation.

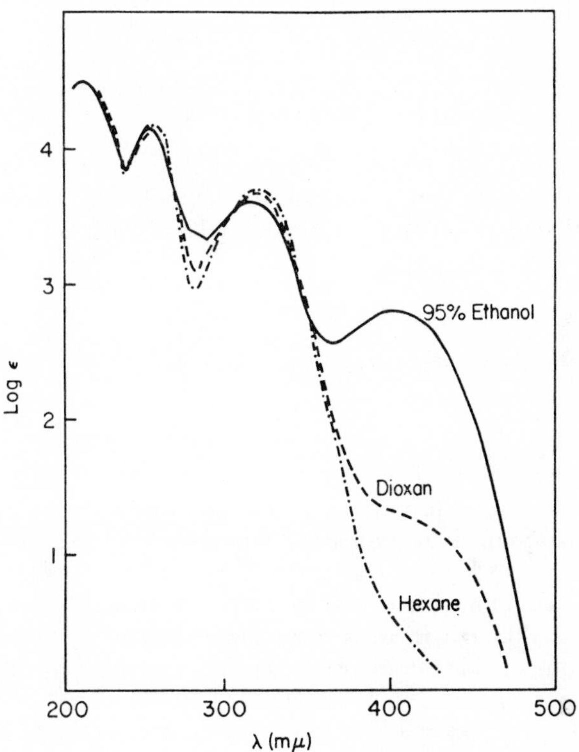

FIGURE 3. Electronic spectrum of (*S*)-(+)-*N*-salicylidene-α-phenyl ethylamine (**32**) in various solvents. (Reproduced from Smith, Cook & Warren, *J. Org. Chem.*, **29**, 2265 (1964) by permission of the authors and the editor.)

Earlier workers[93–95] had noticed that compounds in which intramolecular hydrogen bonding was possible showed the band at ~400 mμ, while if such bonding was prevented (e.g. *m*-hydroxybenzylideneaniline, *o*-methoxybenzylideneaniline, cf. **30** → **31**) this band was absent. The extra absorption was therefore attributed, rather vaguely, to intramolecular hydrogen bonding.

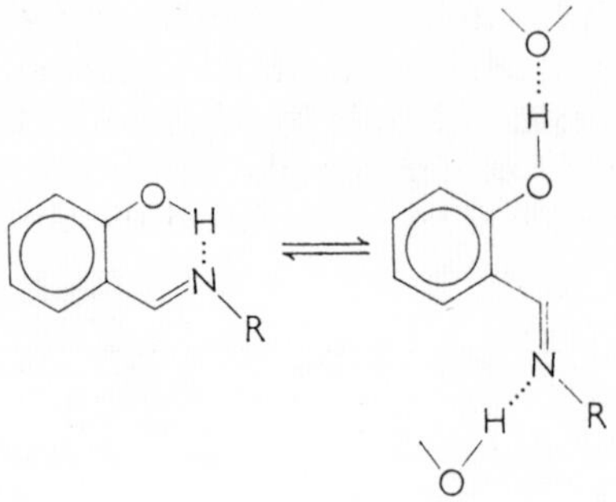

As early as 1941, however, Kiss and Auer[94] had considered an alternative explanation: a tautomeric equilibrium involving enol-imine and keto-amine forms, which may be represented as:

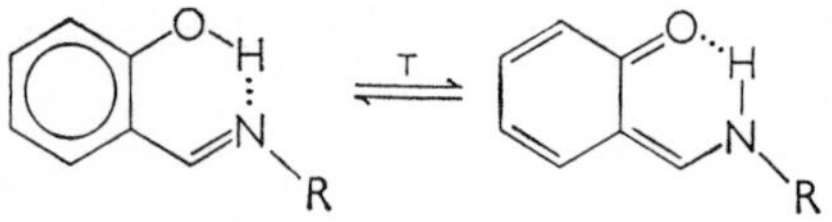

Enol-imine cis-Keto-amine

Other workers have favoured this interpretation[30,31,79,96] and bands have been assigned to individual tautomers[31,97] (e.g. for **33** ~250 and ~320 mμ belong to enol-imine form and ~270 and ~420 mμ belong to keto-amine; note that in this case both species are present in the solid). Although there is some disagreement[105], on balance the phenol-*o*-quinonoid tautomerism is strongly supported by the available evidence, which may be summarized as follows:

(a) *m*-Hydroxybenzylideneaniline does not show the 400 mμ band in ethanol, whereas the *o*-and *p*-derivatives do so (at 435 mμ and 427 mμ respectively)[94]. *m*- and *p*- derivatives can hydrogen bond intermolecularly, but neither can do so intramolecularly. The *p*-isomer can give a quinonoid tautomer, the *m*-isomer cannot. Hence it appears that hydrogen bonding *per se* is not responsible for the ~400 mμ band, but that the capacity for quinonoid tautomerization may be so.

(b) Hydrogen bonding generally results in a rather small red shift for a $\pi \rightarrow \pi^*$ band[31,96]; the ~400 mμ band is considerably displaced, however, and attribution to hydrogen bonding is therefore is not favoured.

(c) Increasing the polarity of the solvent increases the ϵ_{max} of the $\sim$400 mμ band in a uniform fashion, although the wavelength of the absorption maximum does not change very much, e.g.[97]

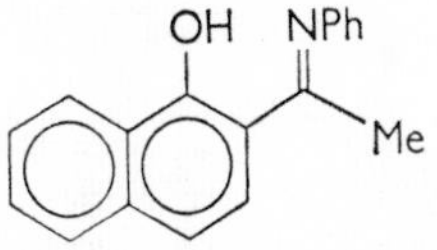

Solvent	Cyclohexane	CCl_4	$CHCl_3$	CH_3CN	MeOH
λ_{max}(mμ)	425	425	428	421	425
ϵ_{max}	1400	2000	5400	4600	7000

(d) Systems which may more readily assume quinonoid forms show a more prominent $\sim$400 mμ band. Thus in general naphthol derivatives show such a band (sometimes a doublet, and at longer wavelengths)[97] even in hydrocarbon solvents, as the example in (c) above shows.

(e) Data from n.m.r. studies on ^{15}N-salicylideneanilines and related compounds provide independent evidence for an equilibrium between enol-imine and keto-amine tautomers[97]. O.r.d. studies support these results[110].

An analogous tautomerization is thought to be implicated in the thermochromic and photochromic transformations of certain *N*-salicylideneanilines[99]. A recent study[104] of the absorption, excitation and fluorescence spectra of these compounds has led to the conclusion that the photo-reaction produces the *trans*-keto-amine from the enol-imine. The *trans*-keto-amine, which has λ_{max} at longer wavelength than the *cis*-keto-amine, isomerizes thermally to the *cis*-keto-amine–enol-imine system; thus for *N*-salicylidene-*o*-toluidine:

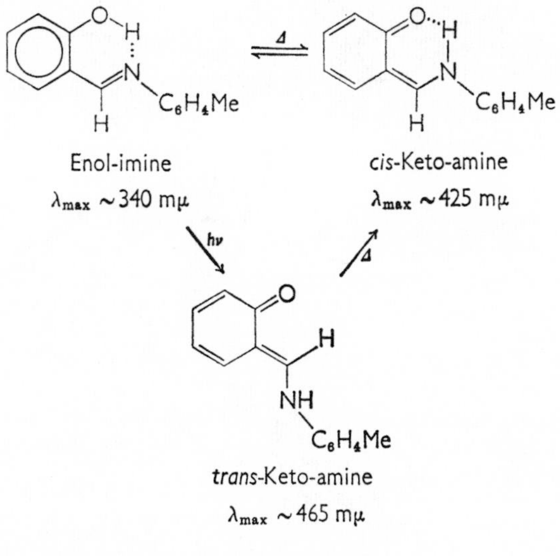

Enol-imine λ_{max} $\sim$340 mμ

cis-Keto-amine λ_{max} $\sim$425 mμ

trans-Keto-amine λ_{max} $\sim$465 mμ

TABLE 4. Electronic spectra of some substituted azomethines.

Compound type	Compound	Solvent	λ_{max}	ϵ_{max}	Reference
Oxime[c]	$Me_2C{=}NOH$[a]	EtOH	190	5,000	36
	$Me_2C{=}CHCMe{=}NOH$	EtOH[b]	236·5	12,600	44
	MeON= (steroid, C_8H_{17})	—[g]	206	8,300	37
Nitrone	5,5-Me, Me pyrroline N→O	EtOH Hexane	234 247	7,700 8,700	38 39
	4,4-Me, Me-2-Me pyrroline N→O	EtOH	229	9,000	38
	dihydroisoquinoline N→O	MeOH	211·5 228·5 304	6,400 12,100 15,900	40

	(pyrroline ring, N→O, Ph)	MeOH	205·5 221·5 14,300	8,500 8,000 288	40
	PhCH=NMe → O	MeOH	206·5 221·5 288	7,500 6,900 16,500	40
	PhCH=NPh → O	EtOH	238[e] 316	11,000 20,000	24
	(ring: Me, Me, N→O, Me / Me, Me, N→O, Me)	EtOH	218 347	5,400 12,300	90
Hydrazone	RCH=NNHMe R_2C=NNHMe	EtOH	228	4,700	41
	t-BuCH=$NNMe_2$	MeOH	236	5,600	9
		Hexane	239	7,100	9
		EtOH/H^+	no max. in 240 mμ region		9
	MeEtC=$NNMe_2$	H_2O	253	780	41
		EtOH	268	820	41
		Cyclohexane	274	820	41

TABLE 4. Electronic spectra of some subsituted azomethines (*continued*).

Compound type	Compound	Solvent	λ_{max}	ϵ_{max}	Reference
	PhCH=NNH$_2$	MeOH	212	12,000	42
			271	14,500	
	PhCH=NNMe$_2$	MeOH^f	222	8,500	42
			295	16,600	
		MeOH/H^{+f}	254	—	42
	PhCH=CHCH=NNMe$_2$	MeOH^f	237	7,250	42
			324	32,400	
Semicarbazone[c]	Me$_2$C=NNHCONH$_2$	EtOH	225·5	11,000	44
	Me$_2$C=CHCMe(=NNHCONH$_2$)	EtOH	260	12,000	46
Thiosemicarbazone[c]	Me$_2$C=NNHCSNH$_2$	EtOH	228·5	7,080	43
			271	21,200	
	PrCH=CHCMe(=NNHCSNH$_2$)	EtOH	246	10,150	44
			301·5	35,900	
Amidine	MeC(=NH)NH$_2$	Water	224	4,000	45
		Aq. acid	<220	—	45
	MeC(=NH)NHC$_6$H$_4$Cl-*p*	Water	236	8,100	45
		Aq. acid	228	7,000	45
	PhC)=NH)NH$_2$	Water	226	9,300	91
		Aq. acid	229	11,300	91
			270	900	

Guanidine	$NH_2C(=NH)NHC_6H_4Cl$-*p*	Water	243	9,000	45
		Aq. acid	234	9,750	45
	$NH_2C(=NH)MeNC_6H_4Cl$-*p*	Water	247	6,150	45
		Aq. acid	224	7,850	45
Imino ether	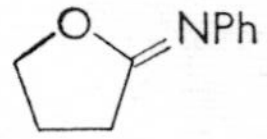 O, NPh	Isooctane	238·4	3,400	25
	PhC(=NH)OEt	Water	230	9,500	91
			270	620	
		Aq. acid	243	12,800	91
			(275)	1,400	
Nitrimine	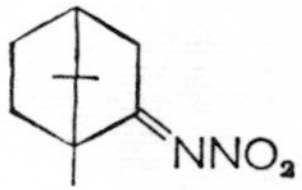 NNO_2	—[g]	270	500	48
Carbodiimide	EtN=C=NEt[d]	Hexane	230	200	44
			270	25	

[a] H. Saito, K. Nukada and M. Ohno [*Tetrahedron Letters*, 2124 (1964)[have reported a weak band at 310 mμ (ϵ15) in the spectrum of cyclohexanone oxime in isooctane and have attributed it to an $n \to \pi^*$ transition. However, our measurements with carefully purified cyclohexanone oxime in this solvent have failed to confirm the presence of this absorption band[9].

[b] Reference 36 shows λ_{max} 235 mμ, ϵ7100 and a weak shoulder at $\sim$290 mμ, $\epsilon \sim$ 1000 in water.

[c] See references 44 and 108 for other examples of this class in tabulated form.

[d] However, in the spectrum of dicyclohexylcarbodiimide (Sadtler Standard Spectra no. 9338) in cyclohexane no band or inflexion is observed down to 215 mμ ($\epsilon_{215} \approx$ 1100).

[e] Value read from published curve.

[f] Bands also reported[42] at 206–208 mμ.

[g] Not stated.

C. Other Substituted Azomethines

The ultraviolet spectra of compounds containing an azomethine group substituted in various ways are summarized in Table 4. In some cases further examples would be desirable to confirm the recorded data, and it is important to note that sometimes serious difficulties can arise because of autoxidation. For example[35] phenylhydrazones at spectroscopic concentrations are readily autoxidized, thus:

$$\text{(cyclohexylidene)}{=}NNHPh \xrightarrow[\text{rapid}]{\text{hexane},\ O_2} \text{1-(N{=}NPh)-1-OOH-cyclohexane}$$

λ_{max} in hexane	ϵ
268 mμ	10,400
413 mμ	140

In this series as a whole it is always desirable, and in some cases essential, that both the compound *and the solvent* should be freshly purified under nitrogen.

II. OPTICAL DISSYMMETRY EFFECTS OF UNCONJUGATED AZOMETHINES

The band in the 240 mμ region ascribed to the $n \rightarrow \pi^*$ transition of the azomethine chromophore is an 'optically active' one: in open-chain compounds the chromophore behaves as a symmetrical system which is dissymetrically perturbed by appropriate neighbouring substituents (generally those bearing asymmetric carbon atoms). Such azomethines generally have rather low rotational strengths ($a < 20$). In more rigid systems, however, such as 1-piperideines, the chromophore may be twisted in some way and high rotational strengths are then observed.

Glossary of terms

The abbreviations and conventions used here are those in common use and are as follows. Optical rotatory dispersion (o.r.d.) curves represent a plot of molecular rotation $[\Phi]$ in degrees against wavelength (λ, mμ). Circular dichroism (c.d.) spectra are plots of the difference $\Delta\epsilon$ between the molecular extinction coefficients for left- and right-handed circularly polarized light (i.e. $\Delta\epsilon = \epsilon_L - \epsilon_R$) against wavelength. Another mode of presentation (e.g. Figure 11) keeps to the degree unit by plotting molecular ellipticity $[\theta]$ in degrees against wavelength. The differential absorption of the left-handed and right-handed circularly polarized light beams results in the elliptical polarization of the emergent beam ($\Delta\epsilon$ and $[\theta]$ are linearly related by the equation $[\theta] = 3300\Delta\epsilon$ and these are merely alternative ways of representing the same information). To preserve uniformity of presentation in the tables the literature values of molecular ellipticity (which turn out to be less common in this series) have been divided by 3300 to convert them to $\Delta\epsilon$.

An o.r.d. curve may be *plain* (i.e. it shows no point of inflection) or *anomalous*. Anomalous curves are of most interest to chemists and are observed in the vicinity of the λ_{max} of an optically active chromophore. The relationship between the electronic spectrum, the optical rotatory dispersion and the circular dichroism associated with a single idealized optically active chromophore X (A, B and C are non-chromophoric) is shown on page 204.

The o.r.d. curve for such a system is sigmoid in shape as shown, and in order to avoid confusion with light absorption data the terms maximum and minimum are not used, but such points are referred to as *extrema*, each being either a *peak* (pk.) or a *trough* (tr.). If on going to lower wavelengths the first extremum is positive, the curve is said to show a *positive Cotton effect*. The circular dichroism in such a case also shows a positive Cotton effect, $\Delta\epsilon$ being positive as shown, the largest value, called a *positive maximum*, being reached at approximately the λ_{max} value of the conventional light absorption curve. The enantiomer of this hypothetical species will show the same light absorption, but the o.r.d. and c.d. curves will display negative Cotton effects, and will be enantiomeric with the two curves obtained in the first case as shown. The highest negative value of $\Delta\epsilon$ in the second c.d. spectrum is termed a *negative c.d. maximum*. In real systems several chromophores will be present, and this complicates the situation especially with respect to o.r.d. work: thus it is not unusual to observe a positive Cotton effect in which the first extremum has in fact a negative rotation value. This will happen, for example, when a weak positive Cotton effect in the accessible region (roughly down to 200 mμ) is superimposed on a strong negative plain curve due to a chromophore (an 'invisible giant') in the far ultraviolet.

In the tables the molecular rotations at the extrema are reported in the form $[\Phi]_{234} = +7000$, while the c.d. maxima are reported as $\Delta\epsilon_{217} = -1{\cdot}7$. By convention the values are listed starting at high wavelengths. An exclamation sign denotes a rotation at the lowest wavelength at which measurements could be (or were) made. As far as the *magnitudes* of the o.r.d. and c.d. effects are concerned, both

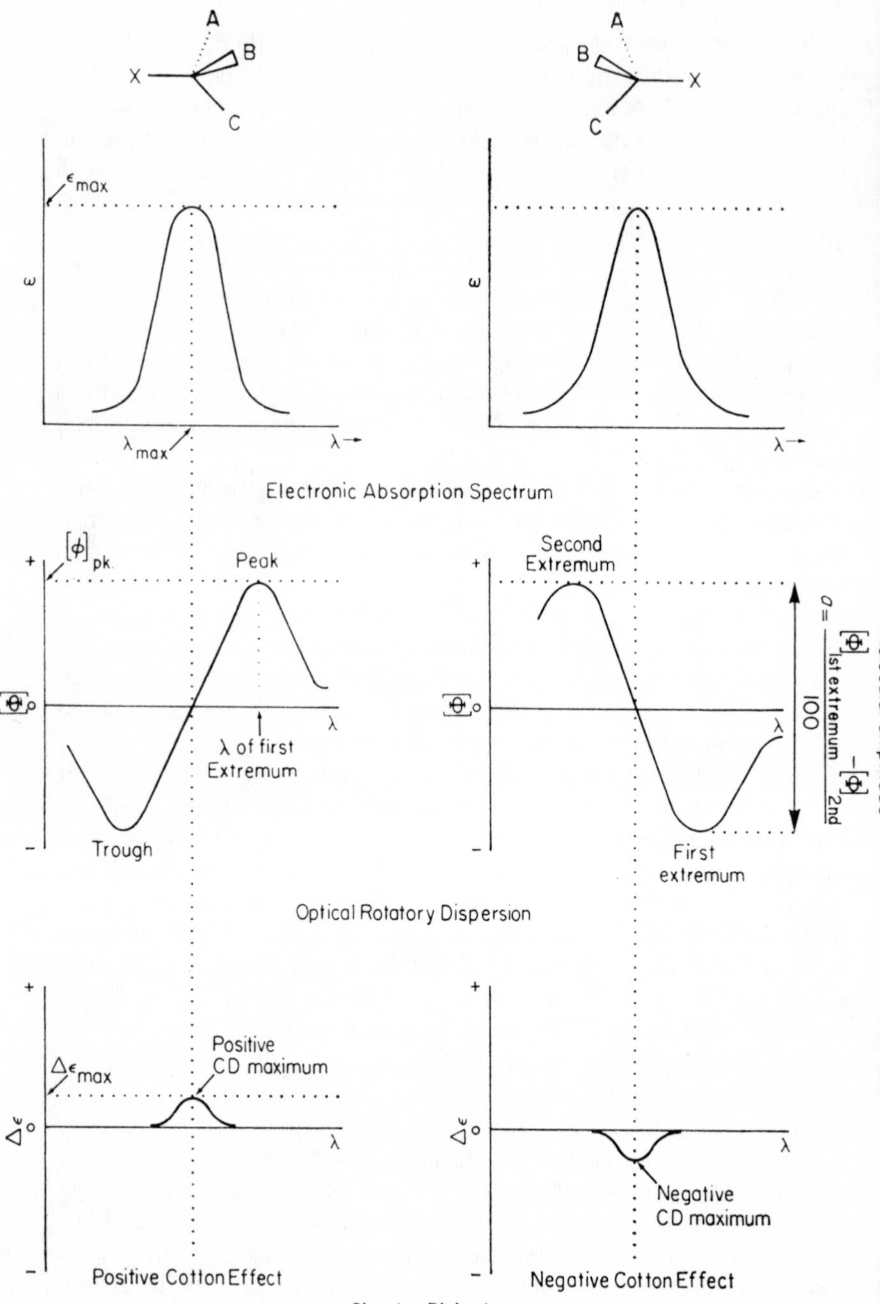

A
B
X
C
ϵ_{max}
ε
λ_{max}
λ→
Electronic Absorption Spectrum
$[\phi]_{pk.}$
Peak
Second
Extremum
$[\Phi]_0$
λ
λ of first
Extremum
Trough
First
extremum
Molecular amplitude
$a = \frac{[\Phi]_{1st\ extremum} - [\Phi]_{2nd}}{100}$
Optical Rotatory Dispersion
Positive
CD maximum
$\Delta\epsilon_{max}$
Δε
Negative
CD maximum
Positive Cotton Effect
Negative Cotton Effect
Circular Dichroism

depend on a quantity termed rotational strength which may be formulated in theoretical terms, but which is obtained experimentally from the area under the c.d. band. In o.r.d. work the rotational strength may be regarded as being roughly measured by the vertical distance between extrema, and is reported as the *molecular amplitude* *a* which may be represented as

$$\frac{[\Phi]_{\text{1st extremum}} - [\Phi]_{\text{2nd extremum}}}{100}$$

(the division by 100 is simply for convenience). For further details regarding nomenclature and general applications the reader is referred to *Optical Rotatory Dispersion and Circular Dichroism in Organic Chemistry* by P. Crabbé (Holden-Day, San Francisco, 1965) and to *Optical Rotatory Dispersion* by C. Djerassi (McGraw-Hill, New York, 1960). Chapter 12 of the latter book (by A. Moscowitz) and S. F. Mason [in *Quarterly Reviews*, **17**, 20 (1963)] give valuable theoretical introductions to the subject.

A. Dissymmetric Substituent at Carbon

Optically active ketones have been widely studied, but it was not until 1965 that optical dissymmetric effects of Schiff bases were compared with those of the parent ketones. Mason and Vane[14] examined the circular dichroism of (+)-3-methylcyclopentanone (**37**, X=O) and various analogues including the *N*-butylamine derivative (**37**, X=N—Bu-n). Although both the olefin (**37**, $X=CMe_2$ λ_{max} ~ 220 mμ) and the triphenyl phosphorane (**37**, $X=PPh_3$, λ_{max} 416·5 mμ in

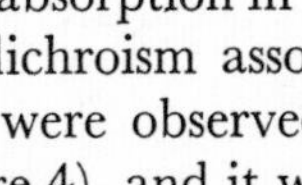

Me X

X=NBu-n; $\Delta\varepsilon_{255}$ = +0·89 (cyclohexane)

(**37**)

isooctane) showed specific absorption in the region examined, neither had measurable circular dichroism associated with this absorption. Circular dichroism bands were observed, however, with the ketone and the azomethine (Figure 4), and it was concluded that only those chromophores with classical non-bonding electrons (e.g. C=S and C=N) give $n \rightarrow \pi^*$ transitions which are analogous to that of the carbonyl group. The azomethine maxima are at lower wavelength than those for the carbonyl chromophore, but although the molecular extinction of the azomethine is higher than that of the ketone,

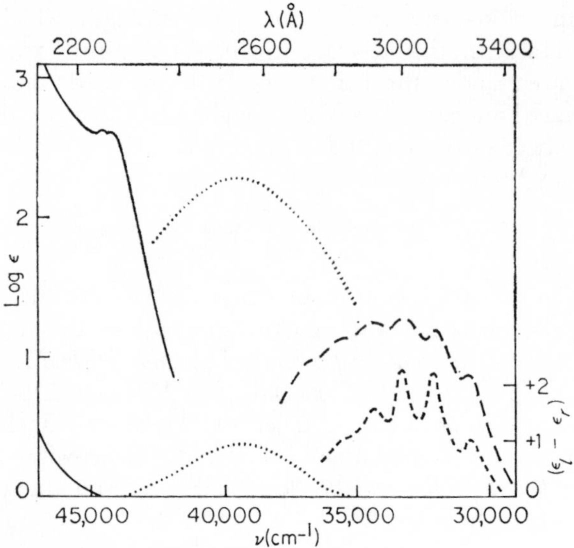

FIGURE 4. The circular dichroism (lower curves) and absorption spectra (upper curves) of the olefin (**37**, X = CH_3) (——), the imine (**37**, X = NBu-n) (· · · ·) and the ketone (**37**, X = O) (– – – –). (Reproduced from Mason and Vane, *Chem. Commun.*, 540 (1965) by permission of the authors and the editor.)

the rotational strength, albeit of the same sign, is lower (Figure 4). The magnetic moment of the $n \rightarrow \pi^*$ transition of the carbonyl chromophore has a value of 1 Bohr magneton, whereas that for the azomethine group is about 0·5 B.M.

Similar results have come from rotatory dispersion studies on azomethines derived from steroidal ketones[15], as is shown in Table 5. The ultraviolet maxima for both carbonyl and azomethine chromophores suffer typical bathochromic shifts in less polar solvents and the optical rotatory dispersion curves move in a similar fashion.

Two points require further comment. First, in all the azomethines referred to above (**37**–**39**) the possibility of a contribution from the

X HO H **(38)**

X HO H **(39)**

TABLE 5. Optical rotatory dispersion of some steroidal ketones and related azomethines[15].

Compound	Solvent	1st Extremum λ mμ	1st Extremum [Φ][a]	2nd Extremum λ mμ	2nd Extremum [Φ][a]	Amplitude *a*
(**38**) X=O (androsterone)	Hexane	320	+5910	275	−6250	+122
	MeOH	313	+7350	276	−7080	+144
(**38**) X=NBu-n	Hexane[b]	262	+3100	228	−2090	+52
	MeOH	246	+4880	220	−1420	+63
(**39**) X=O (3β-hydroxy-5α-androstan-16-one)	Hexane	323	−8550	266	+9400	−180
	MeOH	316	−11,900	282	+11,400	−233
(**39**) X=NBu-n	Hexane	269	−5200	230	+3630	−88
	MeOH[c]	254	−6570	219	+5200	−118

[a] A positive sign refers to a peak and a negative one to a trough.
[b] λ_{max} 248 mμ, ε175.
[c] λ_{max} 235 mμ, ε225.

enamine form must be borne in mind. While the proportion of enamine tautomer may be low, it need not be negligible: *N*-cyclopentylidene-*n*-butylamine (dilute solution in cyclohexane) has been estimated to contain 5% enamine, for example[10]. However, the proportion of enamine in alcoholic solution is thought to be low[10] and since such solutions show a marked Cotton effect the assignment of this to the azomethine chromophore appears to be satisfactory. Secondly, the azomethines can exist in two geometrically isomeric forms. However, the length of the alkyl chain of the nitrogen substituent appears to have only a small effect on rotational strength[15], which is dominated, for the systems considered, by the stereochemistry of the *C*-substituent, i.e. the cyclopentane ring. Differences have been noted with geometrically isomeric oximes, however (see Section III.B).

B. Dissymmetric Substituent at Nitrogen

1. Alkyl derivatives

The first example[1] of a pair of enantiomeric Cotton effect curves (Figure 5) associated with the azomethine chromophore was that of *N*-neopentylidene-*s*-butylamine (**40**, absolute configuration[49] for azomethine from (+)-*s*-butylamine shown). Extrema appear at about 263 mμ and 227 mμ with an amplitude of ~17 in hexane solution. Here not only is the formation of an enamine prevented, but the problem of geometrical isomerism is avoided, since for steric reasons formation of the *syn*-azomethine (e.g. **40**) will be much preferred. An alternative way of avoiding this problem is to use a symmetrical ketone, for example, to produce an azomethine such as the

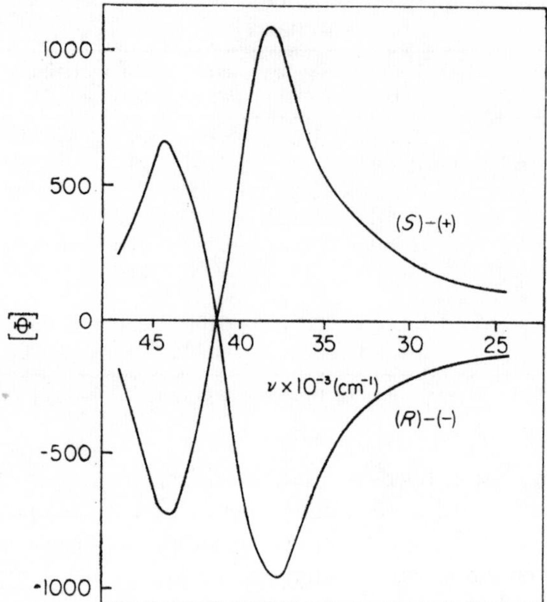

FIGURE 5. Optical rotatory dispersion of (*S*)-(+)- and (*R*)-(−)-*N*-neopentylidene-*s*-butylamine in hexane.

17β-aminosteroid derivative (**41**), which shows[50] a positive Cotton effect with an amplitude of 18 in dioxan ($[\Phi]_{272}$ + 970 pk., $[\Phi]_{234}$ − 800 tr.; λ_{max} 247 mμ, ε140). In both cases the (*S*)-configuration at C_1 of the aliphatic amine leads to a positive Cotton effect in the Schiff base; many more examples are needed before a correlation can be established here, however.

Me₃C–CH=N–C(H)(Me)(Et)   N=CMe₂ (steroid, HO, H)

(40)   **(41)**

2. α-Amino acid derivatives

During the course of a study on the interaction of carbonyl groups with biologically important substituents (e.g. NH_2, SH) Bergel and his collaborators[51,52] observed that α-amino acid esters suffered mutarotation (at the sodium D line) when dissolved in ketonic solvents: L-α-amino acid esters underwent a *laevo* mutarotation, while D-α-amino acid esters mutarotated in the opposite sense. With L-tyrosine ethyl ester in cyclopentanone the corresponding azomethine was isolated. More recently Bonnett, Klyne and their colleagues[17,53] have isolated a series of *N*-neopentylidene derivatives of the type **42**. These compounds show strong Cotton effects with the first extremum

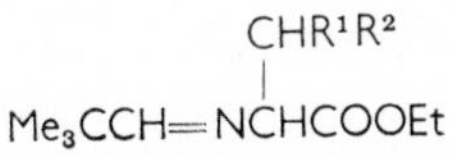

$$\mathrm{Me_3CCH{=}N\overset{\displaystyle CHR^1R^2}{\overset{|}{C}}HCOOEt}$$

(42)

[Structure (43): oxazolidine ring (O, NH) bearing COOEt on C-4 and Bu-*t* on C-2]

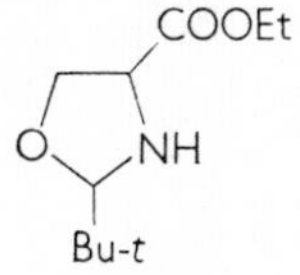

(43)

in the 250 mμ region (Figure 6): the compounds of L-configuration show a negative Cotton effect and those of D-configuration a positive one. This stands in contrast to the behaviour of the α-amino acids (or esters) themselves, where the Cotton effects are weaker, of opposite sign, and have the first extremum at lower wavelengths (~220 mμ). Generally when R^1 and R^2 are alkyl or hydrogen the second extremum has been reached, but in other cases (Table 6) this is usually not so. A good correlation has thus emerged between the sign of the Cotton effect of the *N*-neopentylidene derivative and the configuration of the α-amino acid ester. Certain predictable exceptions have been observed, however. For example, when R^1 contains a powerful chromophore this may interfere: *N*-neopentylidene-L-tryptophan ethyl ester shows a negative curve, but an extremum is not displayed in the 250 mμ region, though such extrema are observed for the derivatives of phenylalanine and tyrosine. The second small group of exceptions consists of those cases in which the azomethine chromophore is removed by further reaction. This happens with β-hydroxy-α-amino acid derivatives where cyclization to the corresponding oxazolidine occurs, e.g. **43** from serine ethyl ester. Although a Cotton effect of correct sign is observed it is quite weak (Figure 6), and possibly arises from a small proportion of the open-chain isomer (**42**, $R^1 = OH$, $R^2 = H$) present in equilibrium with the oxazolidine. It is noteworthy that in ketonic solvents, leucinol and similar 2-amino alcohols mutarotate in the opposite direction to that observed for the corresponding α-amino acid esters (e.g. L-(+)-leucinol $[\alpha]_D + 6{\cdot}9°$ in EtOH; max$[\alpha]_D + 49{\cdot}5°$ in Me_2CO): presumably oxazolidine–hydroxyazomethine equilibria are again involved[54], and these cases, and the related behaviour[54] of

TABLE 6. Optical rotatory dispersion of *N*-neopentylidene derivatives (**42**) of α-amino acid esters in methanol[17,53].

α-Amino acid	R^1	R^2	1st Extremum[a] λ (mμ)	1st Extremum[a] $[\Phi]$	2nd Extremum[a,b] λ (mμ)	2nd Extremum[a,b] $[\Phi]$	Amplitude *a*
D-Alanine	H	H	244	+6000	204	−11,000	+170
D-Alanine[c]	H	H	250	+7200	202	−13,000	+202
L-Valine	Me	Me	256	−5800	217	+9,000	−148
D-Valine	Me	Me	255	+6530	218	−10,400	+169
L-Isoleucine	Me	Et	256	−7860	219	+13,000	−209
L-Aspartic Acid	COOEt	H	244	−4650	213	+3,800!	—
L-Glutamic Acid	CH_2COOEt	H	250	−3880	217	+4,300!	—
L-Methionine	CH_2SMe	H	252	−2200	214	+1,800!	—
L-Lysine	$(CH_2)_3N{=}CHBu\text{-}t$	H	250	−5800	212	+7,320!	—
L-Phenyl alanine[c]	Ph	H	250	−8700	225	0!	—
L-Tyrosine[d]	*p*-C_6H_4OH	H	252	−6580	244	−5,050!	—
L-Tryptophan	β-indolyl	H	288	−1700 pk	244	−8,800!	—
L-Serine	—[e]		248	−1000	223	+655	−17

[a] Except where otherwise noted − refers to a trough and + to a peak.
[b] ! refers to rotation at lowest wavelength reached: not an extremum.
[c] Measured in hexane solution.
[d] Also $[\Phi]_{296}$ − 1460 tr., $[\Phi]_{289}$ − 1390 pk.
[e] Oxazolidine **43** formed.

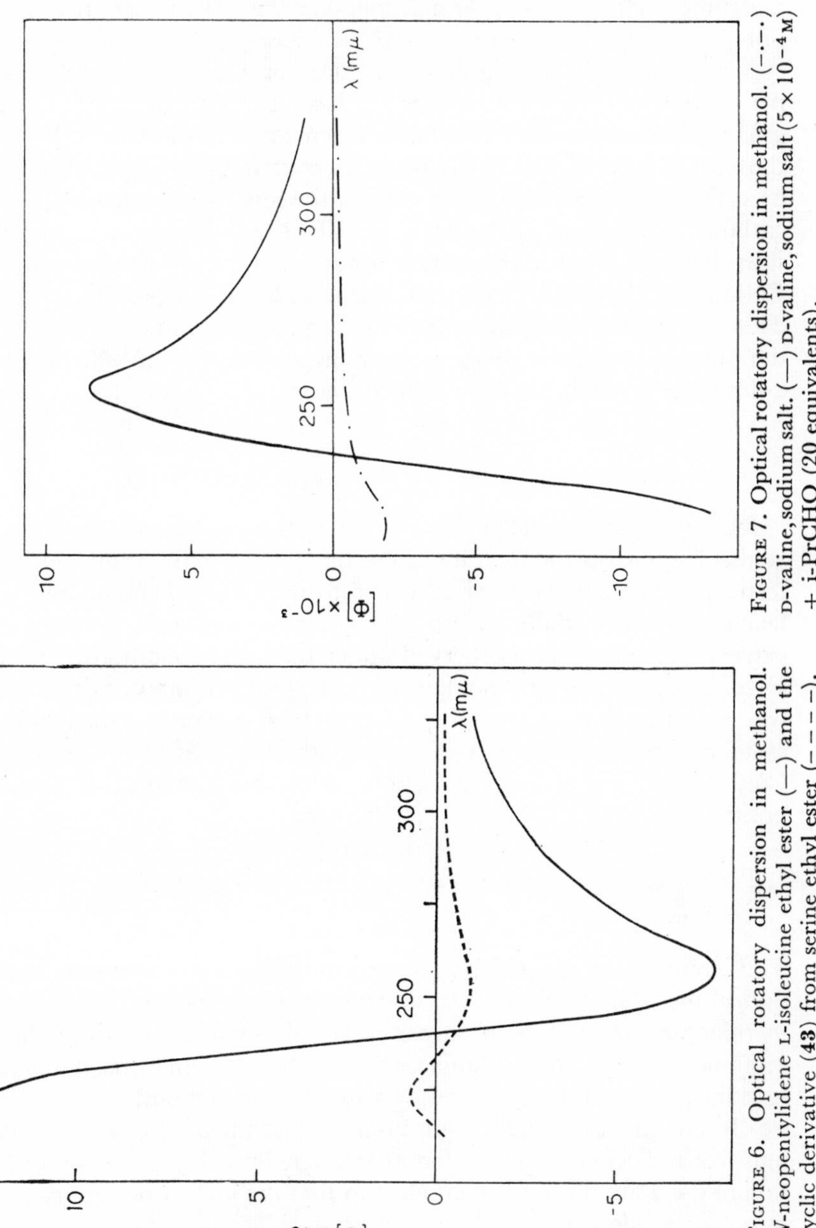

FIGURE 6. Optical rotatory dispersion in methanol. *N*-neopentylidene L-isoleucine ethyl ester (—) and the cyclic derivative (**43**) from serine ethyl ester (- - - -).

FIGURE 7. Optical rotatory dispersion in methanol. (-·-·-) D-valine, sodium salt. (—) D-valine, sodium salt (5×10^{-4} M) + i-PrCHO (20 equivalents).

L-alanine methylamide (→ imidazolidone?) merit further investigation.

For the purpose of assigning configuration it is rather simpler experimentally to work with the Schiff base derivatives of amino acid *anions*, which are not isolated but are generated in solution[53]. When the α-amino acid in anhydrous methanol containing two equivalents of base (Et_3N, NaOEt) is treated with an excess (~20 equivalents) of carbonyl component (*t*-BuCHO, i-PrCHO, cyclohexanone), a considerable and rapid mutarotation occurs with the development of a Cotton effect having its first extremum in the 250 mμ region (Figure 7). The similarity of such curves in both sign and magnitude to those of the isolated ester derivatives strongly suggests that the *N*-alkylidene-α-amino acid anion has been formed, thus:

$$\text{RCHO} + \text{NH}_2\underset{\text{R}}{\text{CH}}\text{COO}^- \rightleftharpoons \text{RCH{=}N}\underset{\text{R}}{\text{CH}}\text{COO}^- + \text{H}_2\text{O}$$

Figure 8 illustrates the application of this test to a variety of α-amino acids. The curves for D-valine show that either sodium alkoxide or tertiary amine can be used to generate the salt. D-Leucine and L-leucine give essentially enantiomeric curves, although the second extremum is not clearly observed under these test conditions. In the absence of powerful chromophores (e.g. tryptophan) and of situations where cyclization presumably occurs (e.g. serine → oxazolidine; cysteine → thiazolidine[55]), L-α-amino acids of type **44** give a negative Cotton effect under these conditions, while D-α-amino acids give a positive one.

$$\text{R}^1\text{R}^2\text{CH}\cdot\underset{\text{NH}_2}{\text{CH}}\text{COOH}$$

(**44**)

Thus configurations could be assigned to the following α-amino acids under the test conditions[53]: alanine, valine, leucine, isoleucine, methionine, aspartic acid, asparagine, glutamic acid, glutamine, arginine, lysine, phenylalanine, and tyrosine; while tryptophan, histidine, serine, threonine and cysteine did not respond characteristically. Within the stated limitations, this configuration test is both sensitive and convenient: it offers advantages over the direct examination of the α-amino acids themselves in that the first extremum is at a more accessible wavelength and, more significantly, shows a much higher rotation.

The origin of the high rotational strength is not altogether clear. The

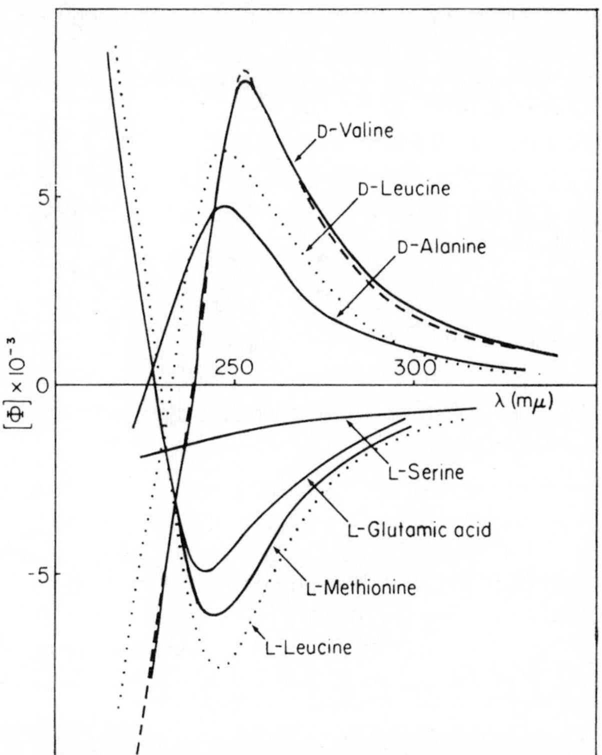

FIGURE 8. Configuration of α-amino acids using *N*-isobutylidene-α-amino acid anions. Test conditions: 5×10^{-4} M in anhydrous methanol +2 equivalents of NaOR or Et_3N +20 equivalents of isobutyraldehyde. The curves for D-valine using NaOH (– – – –) and Et_3N (——) are shown. Both enantiomers of leucine (· · · ·) are included.

electronic absorption spectra do not show separate bands for azomethine (~240 mμ) and ester carbonyl (~210 mμ) absorption: thus *N*-neopentylidene-L-alanine ethyl ester (**42**, $R^1 = R^2 = H$) has λ_{max} 217 mμ, ε290 in hexane. It is possible that the observed electronic and optical rotatory dispersion spectra represent the envelopes of, respectively, absorption and rotation associated with the individual chromophores. However, in the one case where circular dichroism has been recorded (*N*-neopentylidene-L-isoleucine ethyl ester in isooctane) a single negative maximum at 229 mμ has been found[56], and it seems likely that the two chromophores are coupled. It is interesting to note that electrostatic effects depicted in **45a, b** could well introduce an element of rigidity into the molecule, and confer upon the system the character of an inherently dissymmetric chromophore. In the form **45b** the system greatly resembles **45c**, which is the twisted form of the $\beta\gamma$-unsaturated carbonyl chromophore associated with a negative Cotton effect[83].

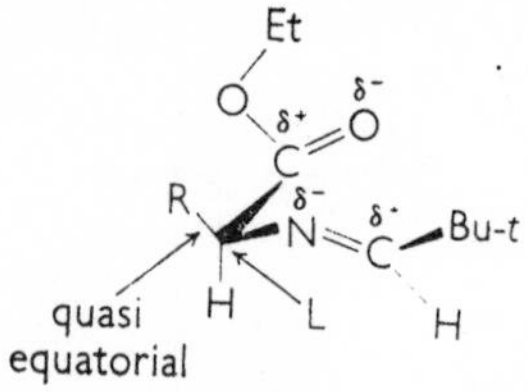

(**45a**) 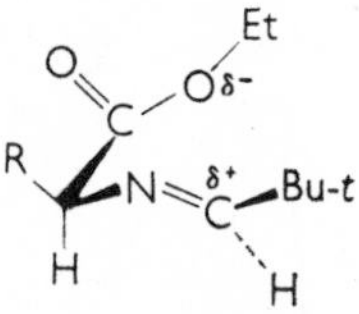 (**45b**) 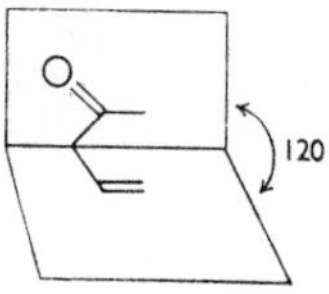 (**45c**)

C. Dissymmetric Substitution at Both Nitrogen and Carbon: Cyclic Azomethines

Compounds containing an azomethine system embedded in a rigid or fairly rigid molecular framework are of much interest, since their study should provide clear information on the effect of substituents upon Cotton effect. For carbonyl compounds the effect of substitution in rear octants has been thoroughly investigated[57], but the difficulty of introducing substituents into front octants has resulted in few examples, so that the complete rule has not been effectively tested. Indeed it is possible that a quadrant rule may be operative[58].

Since the azomethine group, unlike the carbonyl, has one bond directed towards the front octants, it is possible in principle to use azomethines to test front octant effects. A difficulty arises here, however, because of the lower symmetry of the azomethine function; the space around the carbonyl group can be divided up by the nodal surfaces of the orbitals concerned in the $n \rightarrow \pi^*$ transition and these can be regarded essentially as orthogonal planes. For the azomethine group, however, two of these surfaces are expected to be curved (Figure 9), and hence the division of space around the chromophore is not a straightforward operation[59]. Snatzke and his colleagues[59] have developed a treatment for the 1-piperideine system in the half-chair conformation: this suggests that the effect of the ring itself is dominant and leads to the following predicted correlation of Cotton effect with the chirality of the six-membered ring:

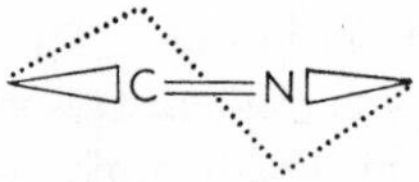

positive Cotton effect

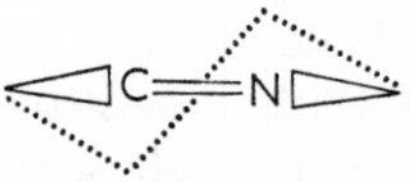

negative Cotton effect

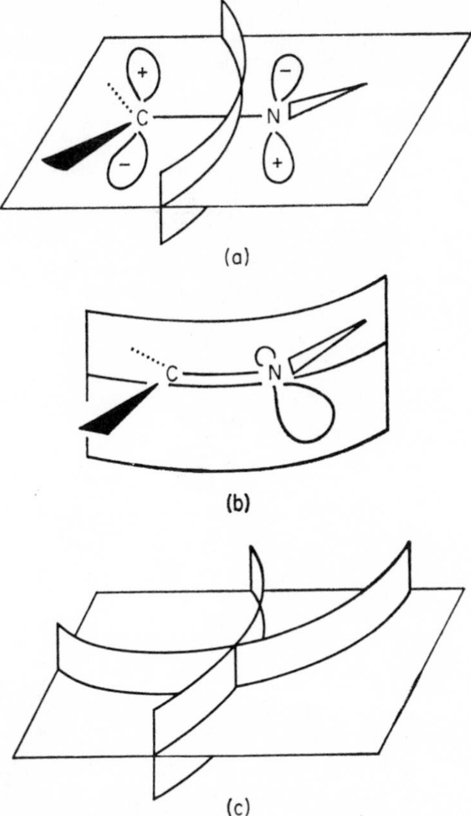

FIGURE 9. Schematic representation of the nodal surfaces of the orbitals involved in the $n\rightarrow\pi^*$ transition of the azomethine group. (a) π-antibonding orbital; (b) non-bonding orbital; Diagram (c) represents the unequal division of space about the azomethine group by these three surfaces. The yz surface is planar, but the xy one is probably slightly curved ,while the xz surface—pertaining to the non-bonded orbital, (b)—is likely to be much more so.

This 'rule' has not been sufficiently tested as yet, and evidently further substitution or ring fusion (especially at the 2,3 bond) in the vicinity of the azomethine function could make an important contribution. Some examples which illustrate the operation of the rule are given in Table 7.

It is noteworthy that the Cotton effect curves of compounds **50** and **51** are essentially mirror images (Figure 10); this reflects the enantiomeric nature of the environments (heavy outline) of the azomethine group in the two compounds. It has been reported[63] that the

TABLE 7. Correlation of Cotton effect with chirality of 1-piperideine ring.

Structure	Conformation of 1-piperideine ring	Predicted CE	Solvent	Observed CE	Reference
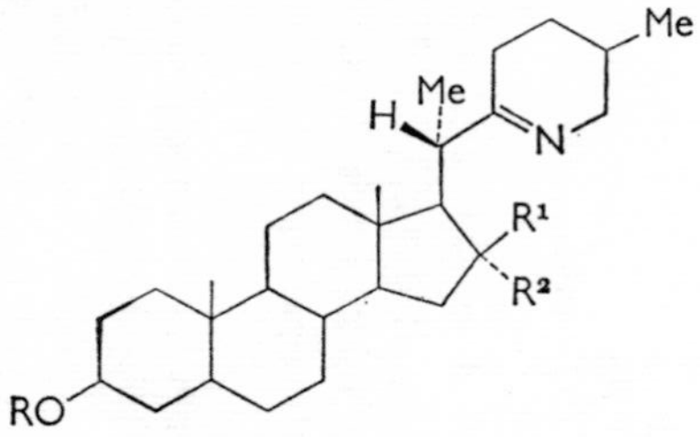 *Solanum* alkaloid series (25*R*)	*a* 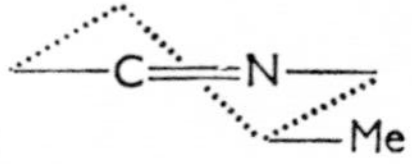	+ve			
(**46**) R = R¹ = R² = H			Dioxan	$\Delta\epsilon_{248.5}$ + 1·48	59
(**47**) R = R¹ = H, R² = OH			Dioxan	$\Delta\epsilon_{250}$ + 2·15	59
(**48**) R = Ac, R¹ = OAc, R² = H			Dioxan	$\Delta\epsilon_{298}$ − 0·07[b] $\Delta\epsilon_{248}$ + 2·24	59
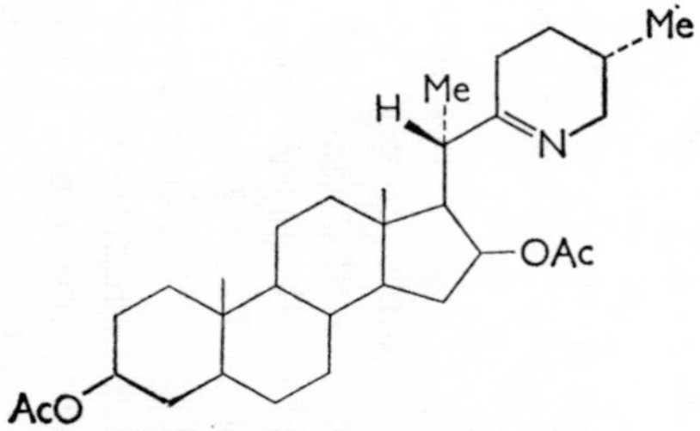 (25*S*)-3β,16β-Diacetoxy-22,26-imino-5α-cholest-22(*N*)-ene (**49**)	*a* 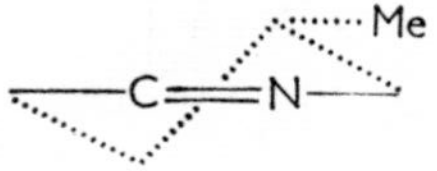	−ve	Dioxan	$\Delta\epsilon_{248}$ − 2·38	59

(**17**) Verazine (structure—see Table 2)		−ve	Dioxan	$\Delta\epsilon_{285}$ + 0·08 $\Delta\epsilon_{248}$ − 2·04	89
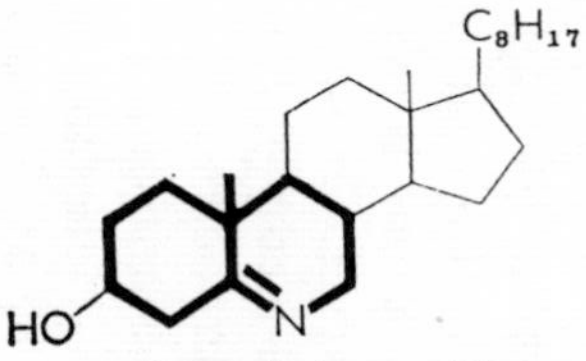 6-Azacholesterol[60] (**50**)	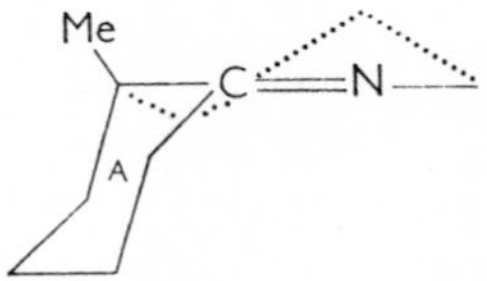 *c*	−ve	MeOH	$[\Phi]_{256}$ − 7500 tr $[\Phi]_{227}$ + 4870 pk	65
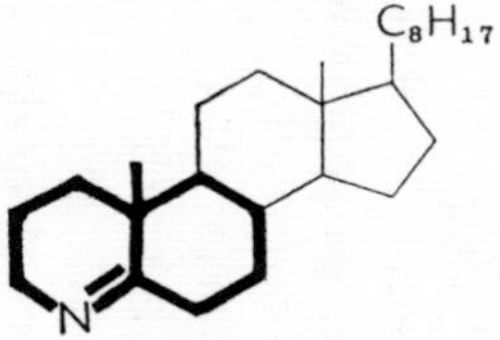 4-Azacholest-4-ene (**51**)	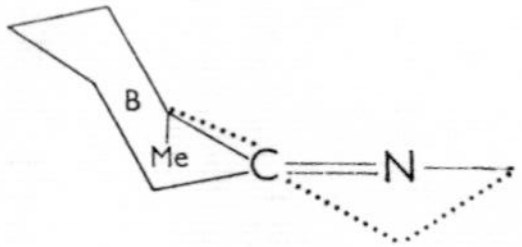 *c,d*	+ve	MeOH	$[\Phi]_{252}$ + 7100 pk $[\Phi]_{226}$ − 2300 tr	65

[a] Assuming ring methyl group adopts a quasi-equatorial position.
[b] The origin of this effect is not clear.
[c] Based on Dreiding models.
[d] Ring A can adopt another conformation (move $C_{(2)}$) which gives an incorrect prediction.

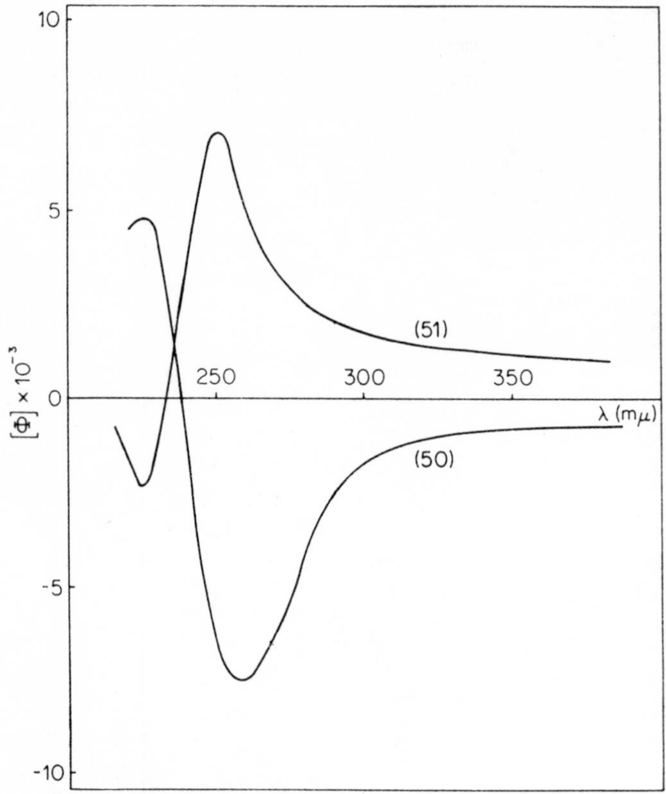

FIGURE 10. Optical rotatory dispersion in methanol of 4-azacholest-4-ene (**51**) and 6-azacholesterol (**50**).

azomethine (**51**) shows only end-absorption in the ultraviolet; Dr. Wood[9] has found, however, that this substance, purified via the picrate or by thin layer chromatography, has m.p. 103°, λ_{max} 241 mμ, ϵ148 in ethanol λ_{max} 256 mμ, ϵ137 in hexane. It is again difficult to eliminate entirely the possibility that the enamine is making some contribution; that this is not important, however, for 4-azacholest-4-ene is suggested by the observation[64] that 4-methyl-4-azaandrost-5-en-17-β-ol acetate (**52**) has (a) electronic absorption which is at lower wavelengths and which is much less solvent sensitive (λ_{max} 219 mμ, ϵ8700 in ethanol; λ_{max} 221 mμ, ϵ8500 in cyclohexane) and (b) a Cotton effect of opposite sign to that shown by **51**. The azomethines

(**55**, **56**) derived[61,62] from atisine (**53**) and veatchine (**54**) cannot form enamines for structural reasons, but the curves show o.r.d. Cotton effects in the 250 mμ region which disappear as expected in acid media or when the corresponding piperideine (e.g. from **56**) is examined[65]. However, preliminary circular dichroism results[9] indicate that the situation here may be more complex than the o.r.d. curves suggest: thus 'double-humped' curves, possibly associated with the solvation of the azomethine system, have been observed in methanol.

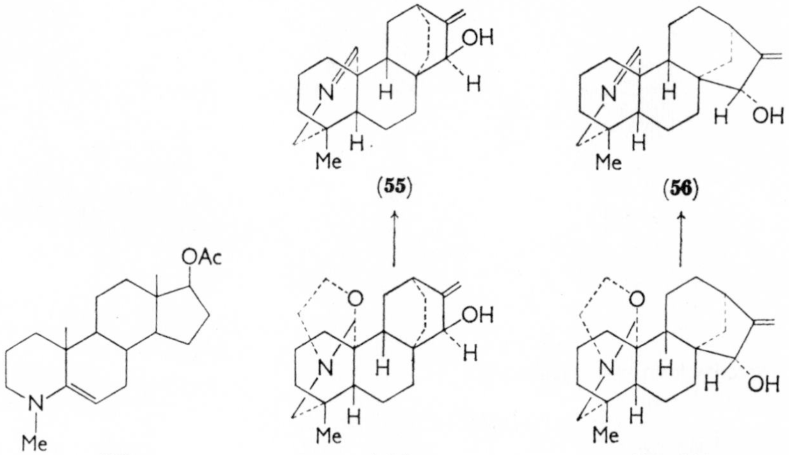

(**55**) (**56**)

(**52**) Atisine (**53**) Veatchine (**54**)

The analysis of the optical dissymmetric effects of cyclic azomethines possessing other ring sizes has not advanced sufficiently as yet to merit detailed discussion. Evidently synthesis of suitable models is now required. Table 8 summarizes various results that have been obtained. The azirine (**57**) is known to hydrolyse readily[67] and consequently in this case treatment with acid and then with base does not regenerate the original curve: a new Cotton effect appears, however, in the 300 mμ region due to the carbonyl chromophore. With concurchine and its derivatives the amplitudes observed are low, and probably partial rotations due to other chromophores (amino function, double bond in rings A and B) or to solvation are interfering to some extent. Circular dichroism studies—especially of possible solvent effects—are clearly desirable here, since preliminary circular dichroism measurements reveal 'double-humped' curves (e.g. **59**) for methanol solutions.

TABLE 8. Optical dissymmetry effects of some cyclic azomethines (azirine and pyrroline rings).

Compound	Solvent	Optical dissymmetry effects	Reference
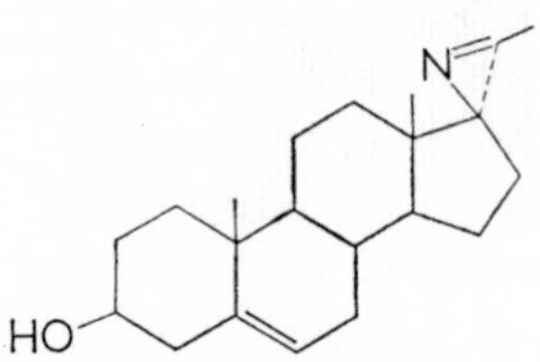 Azirine from pregnenolone (**57**)	MeOH	$[\Phi]_{259}$ − 4520 tr $[\Phi]_{228}$ + 230 pk	66, 67
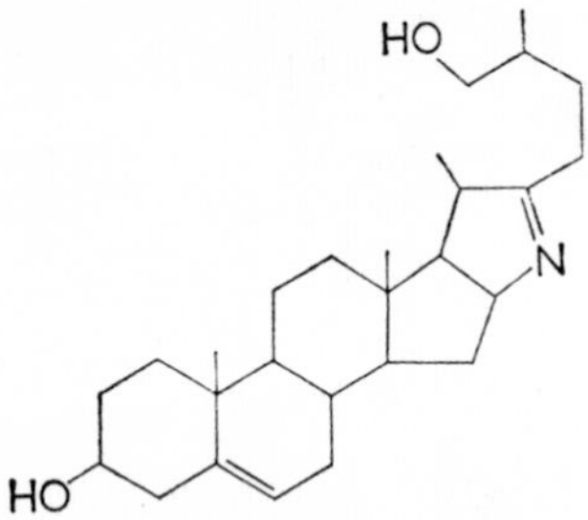 1-Pyrroline from kryptogenin (**58**)	MeOH	$[\Phi]_{241}$ − 13,100 tr	66, 68
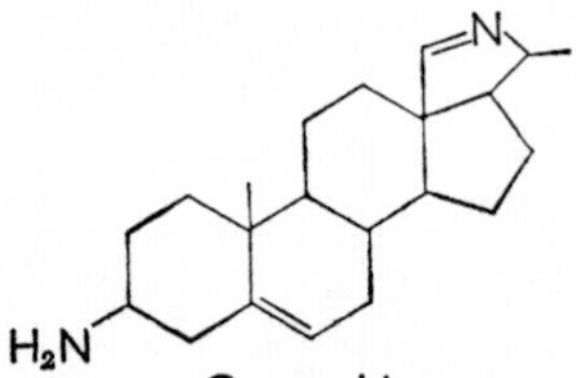 Concurchine (**59**)	MeOH	$[\Phi]_{258}$ − 1700 tr $[\Phi]_{243}$ − 1080 pk	66, 69, 70
	MeOH	$\Delta\epsilon_{258}$ − 0·26 $\Delta\epsilon_{229}$ + 0·72	76, 109
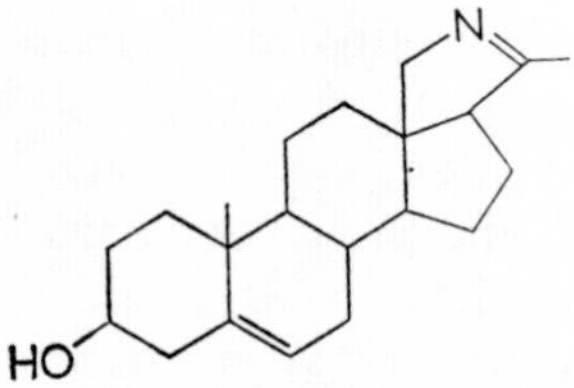 3β-Hydroxy-*N*-desmethyl cona-5,20(*N*)-dienine (**60**)	Dioxan	$\Delta\epsilon_{230}$ − 2·08	59, 71, 109

III. OPTICAL DISSYMMETRY EFFECTS OF CONJUGATED AND OTHER SUBSTITUTED AZOMETHINES

A. Aryl Azomethines

The high extinction coefficients of conjugated compounds frequently result in difficulties in measuring optical dissymmetry effects (unless these are *very* large) in the vicinity of the absorption band. Various authors[30,72,74] have recorded o.r.d. data for the benzylidene derivatives of optically active amines, and although the rotations observed are larger than those of the amines themselves, the measurements have often had to be confined to limited ranges. Thus the optical rotatory dispersion curves[30] for (*S*)-(+)-*N*-benzylidene-α-phenylethylamine (**61a**) and (*S*)-(+)-*N*-benzylidene-α-benzylethylamine (**61b**) are plain positive with rotations rapidly increasing down to about 290 mμ(!). The rotatory dispersion of several Schiff bases obtained from (*S*)-(−)-α-phenylethylamine and aromatic aldehydes (e.g. substituted benzaldehydes, pyridine aldehydes, furfural, 2-formylthiophen) have been measured in the range 486–656 mμ[72]. Potapov, Terentev and their coworkers, who made earlier studies[73] of such systems at the wavelength of the sodium D line, have advocated

$$\mathrm{PhCH{=}N{-}\overset{\overset{\displaystyle R}{\vdots}}{\underset{\underset{\displaystyle CH_3}{\vdots}}{C}}{-}H}$$

(**61a**) R=Ph

(**61b**) R=$PhCH_2$

the use of these plain curves for the assignment of configuration, in the sense that compounds of configuration **61**, where R > Me, are expected to possess plain positive curves[74]. The (*S*)-configuration of (+)-1-methyl-3-phenylpropylamine has been assigned in this way.

The signs of plain curves are clearly more satisfactory for configurational assignments than are $[\Phi]_D$ values: the comparison of complete Cotton effect curves is a sounder basis still. Cotton effects have indeed been observed in the circular dichroism of certain benzylidene derivatives of aliphatic amines: thus[80] *N*-benzylidene-20-α-(*S*)-amino-3β-hydroxy-5α-pregnane (i.e. the *N*-benzylidene derivative of the amine corresponding to **62a**, Table 9) has $\Delta\epsilon_{270}$ + 1·97 and $\Delta\epsilon_{285}$ + 1·9 in dioxan.

Much more attention has been paid to substituted benzylidene derivatives. The Russian authors[74] have shown, for example, that

TABLE 9. Optical dissymmetry effects of N-salicylidene derivatives of some aliphatic amines[c].

Compound and configuration	Solvent	Optical dissymmetry effects	Reference
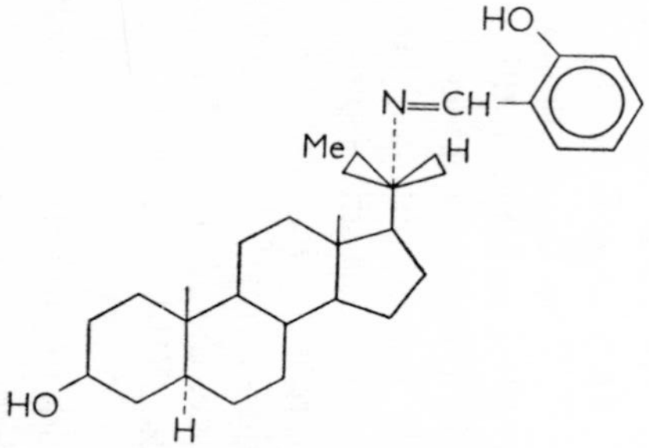 N-Salicylidene-20-α-(S)-amino-3β-hydroxy-5α-pregnane (**62a**)	Dioxan	$\Delta\epsilon_{315}$ + 2·91	80
(**62b**) N-Salicylidene-20β-(R)-amino-3β-hydroxy-5α-pregnane	Dioxan	$\Delta\epsilon_{315}$ − 4·13	80
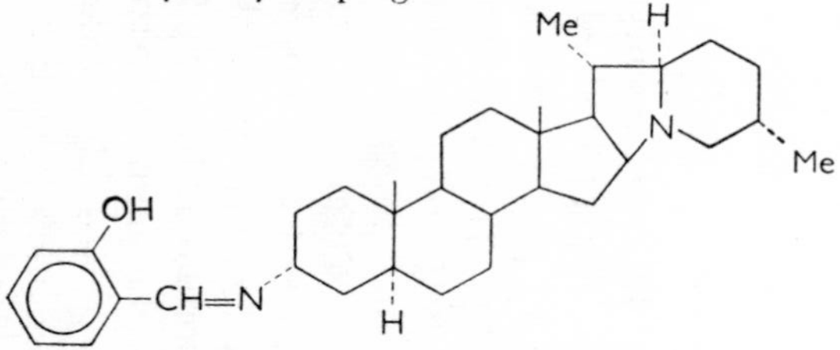 N-Salicylidene-3α-(R)-amino-5α-solanidane (**63a**)	Dioxan	$\Delta\epsilon_{317\cdot5}$ − 1·15	75

(**63b**) *N*-Salicylidene-3α-(*R*)-amino-5β-solanidane	Dioxan	$\Delta\epsilon_{323}$ − 0·69	75
(**63c**) *N*-Salicylidene-3β-(*S*)-amino-5α-solanidane	Dioxan	$\Delta\epsilon_{315}$ + 0·91	75
N-Salicylidene-17α-(*R*)-amino-3β-hydroxy-5α-androstane (**64a**)	Dioxan	$\Delta\epsilon_{315}$ − 4·57	80
(**64b**) *N*-Salicylidene-17β-(*S*)-amino-3β-hydroxy-5α-androstane	Dioxan	$\Delta\epsilon_{315}$ + 3·38 $\Delta\epsilon_{275}$ − 2·07	80
(**65**)			
a. R = Et; R′ = Me (*S*)	EtOH	$[\Phi]_{437}$ + 370 pk	79
b. R = Me; R′ = COOEt D[b]	EtOH[a]	Plain −ve $[\Phi]_{450}$ − 13!	79
c. R = Bu-i; R′ = COOEt D[b]	EtOH[a]	Plain + ve $[\Phi]_{360}$ + 890!	79
d. R = PhCH$_2$; R′ = COOMe D[b]	Dioxan	$\Delta\epsilon_{320}$ + 3·72	80

[a] Absolute ethanol.
[b] Measured on the enantiomer.
[c] For other examples of derivatives of steroidal amines with o.r.d./c.d. data in tabulated form, see reference 111.

p-dimethylaminobenzylidene derivatives give plain curves (down to about 300 mμ) of the same sign as those shown by the benzylidene derivatives (**61**), but with much higher rotations. Several groups have examined *N*-salicylidene derivatives (including those with further substitution, e.g. nitro-, chloro- and bromo-[30]) which, as might be expected from the solvent dependence of their electronic spectra (see Section I.B), give rather interesting results. Bertin and Legrand[77] studied the circular dichroism of some azomethines derived from 20-amino-steroids (e.g. **62**); the 20α derivative in dioxan showed a positive maximum at 315 mμ, and the 20β derivative a negative one at this wavelength. 3- and 17-Amino steroidal compounds (e.g. **63, 64**) have also been examined[75,80]. In the 3-amino system, for example, the *N*-salicylidene derivative of a 3α-amine shows a negative c.d. maximum in the 315 mμ region: in the examples studied (derivatives of *Solanum* steroidal alkaloids[75]) although the magnitude of the dissymmetry effect at ~315 mμ is sensitive to a change of configuration at position 5, its *sign* does not change (e.g. **63a**, **63b**).

The 400 mμ band has aroused a certain amount of discussion[78] since it was originally reported to be inactive[77]. It appears to be active, however, although this activity is often difficult to detect. Snatzke[75] has reported cases (e.g. **63a**) where the effect was not detected in circular dichroism spectra, but emerged in rotatory dispersion (both in dioxan). In the salicylidene derivatives of α- and β-arylalkyl amines (see below and Table 10) this effect is clearly demonstrable in ethanol, which is evidently a good solvent in which to look for it (cf. Figure 3). (*S*)-(+)-*N*-Salicylidene-*s*-butylamine (**65a**) shows a weak extremum $[\Phi]_{437}$ + 370 in ethanol, but in hexane displays a plain positive curve down to 375 mμ (!).

Thus although the 400 mμ band is not generally useful, the Cotton effect at 315 mμ appears to be a valuable means of assigning configuration to amines of the type $R_1R_2\overset{*}{C}HNH_2$, since *N*-salicylidene derivatives of (*S*)-configuration show a positive Cotton effect, those of (*R*)-configuration a negative one. In applying this sort of correlation to a situation where optical dissymmetry effects depend on rotamer populations, which in turn depend on the relative size of substituents, it must be recognized that the Cahn priority sequence[98] will not always be the order of decreasing steric requirement. Thus it has been suggested[80] that *N*-salicylidene-D-phenylalanine methyl ester (**65d**, D-(*R*) configuration, yet $\Delta\epsilon_{320}$ is positive) represents such a case. (See also below, where aralkylamine derivatives are considered more fully.)

Few α-amino acid ester derivatives have been studied: the derivatives of D-alanine (**65b**) and D-leucine (**65c**) show plain curves in the very limited range observed, apparently of opposite signs. Since the alanine derivative has a very low rotation it is possible that racemization has occurred here: it is conceivable, however, that hydrogen bonding in the salicylidene moiety interferes with the electrostatic interaction shown in **45** and that the sign of the curve is determined by the relative sizes of the groups at the asymmetric centre.

This work has been extended by Smith and his colleagues to derivatives of α- and β-arylalkylamines. Further penetration into the ultraviolet is possible with these compounds and large rotational strengths have been observed. It has been suggested[81] that the aryl group of the amine moiety interacts with the hydrogen-bonded salicylidenimine system to produce an inherently dissymmetric chromophore, although this is not attended by any marked change in the electronic spectrum (**30**, **32**, Table 3). For *N*-salicylidene derivatives of α-arylalkylamines having the (*S*) configuration (**66**), three positive circular dichroism maxima have been observed in *ethanol* at about 255, 315 and 405 mμ: these are assigned to transitions in the salicylidenimine system (tautomers, see Section I.B). In addition, a negative Cotton effect at ~280 mμ (not shown by β-arylalkylamino derivatives) is assigned to the $\pi \rightarrow \pi^*$ transition in the aryl group of the amine moiety (Figure 11).

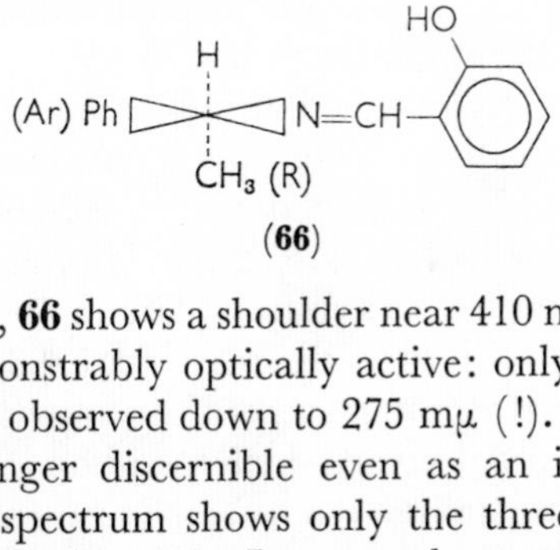

(Ar) Ph, H, CH_3 (R), N=CH–, HO

(**66**)

In *dioxan* solution, **66** shows a shoulder near 410 mμ (see Section I.B) but this is not demonstrably optically active: only one Cotton effect (+ve, ~315 mμ) is observed down to 275 mμ (!). In *isooctane* the 400 mμ band is no longer discernible even as an inflection, and the circular dichroism spectrum shows only the three Cotton effects at lower wavelength (Table 10). In general, compounds having the configuration **66*** again show positive Cotton effects at 255 mμ and 315 mμ: this correlation applies even when R is a very bulky alkyl group (*t*-Bu, **69**) and has been used to assign configurations[79] to

* When R is an alkyl group having a lower priority than phenyl this is (*S*): when R is alkoxycarbonyl, e.g. **65d**, **70**, it turns out to be (*R*). The configurational relationship of the aryl group, the salicylidenimino fragment and the hydrogen atom is the same in each case.

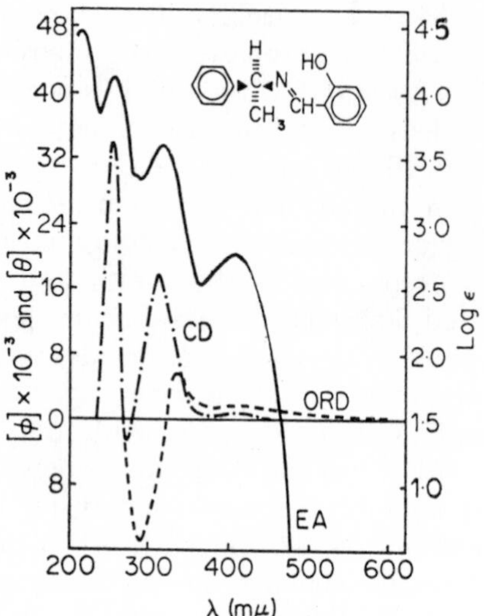

FIGURE 11. Electronic absorption (EA), rotatory dispersion (ORD) (both in 95% ethanol) and circular dichroism (CD) (in absolute ethanol) of (*S*)-(+)-*N*-salicylidene-α-phenylethylamine. (Reproduced from Smith and Records, *Tetrahedron*, **22**, 813 (1964) by permission of the authors and editor.)

(−)-α-phenyl-n-propylamine and (−)-α-phenylneopentylamine (both (*S*), the configuration of the latter being confirmed by chemical methods[82]). Table 10 summarizes some of the results obtained in this series, and gives examples of related β-aryl (e.g. **70**, **71**) and α,β-diaryl (e.g. **72**) derivatives. The latter type does not obey the rule given above, since (*S*)-*N*-salicylidenediphenylethylamine (**72**) does not show a Cotton effect at 315 mμ in ethanol, and the circular dichroism has a negative maximum at 262 mμ. In dioxan, however, an additional weak positive maximum is found at 319 mμ. These results have been interpreted[79,82] in terms of the summation of rotatory contributions from intramolecularly hydrogen-bonded conformers of **72**. In ether–pentane–ethanol at room temperature the long wavelength Cotton effect is not observed, but at −192° a negative maximum at 314 mμ appears: this presumably reflects a change in the conformational equilibrium of the system[79,82,83]. It is interesting to note that both (*S*)-*N*-salicylidene-α-phenylethylamine (**68**) and (*S*)-*N*-salicylidene-α-phenylneopentylamine (**69**) show increased rotational strengths at −192°, an observation[79] which illustrates the difficulty of interpretation of the Cotton effects of open-chain compounds in simple conformational terms.

TABLE 10. Optical dissymmetry effects of some *N*-salicylidene-aralkylamines[81].

Compound	R	R′	Solvent	Optical dissymmetry effect[a]
(**67**)	α-Naphthyl	Me	EtOH[c]	$\Delta\epsilon_{406} + 0{\cdot}4$, $\Delta\epsilon_{316} + 7{\cdot}9$, $\Delta\epsilon_{285} - 5{\cdot}5$, $\Delta\epsilon_{256} + 17$
(**68**)	Ph	Me	EtOH[b]	$[\Phi]_{428} + 1800$, $[\Phi]_{397} + 1400$, $[\Phi]_{337} + 5900$, $[\Phi]_{291} - 15{,}000$
			EtOH[c]	$\Delta\epsilon_{405} + 0{\cdot}33$, $\Delta\epsilon_{315} + 5{\cdot}5$, $\Delta\epsilon_{274} - 0{\cdot}91$, $\Delta\epsilon_{253} + 10{\cdot}3$
			Dioxan	$\Delta\epsilon_{317} + 5{\cdot}5$
			Isooctane	$\Delta\epsilon_{319} + 12$, $\Delta\epsilon_{274} - 1{\cdot}2$, $\Delta\epsilon_{253} + 17{\cdot}6$
(**69**)	Ph	*t*-Bu	EtOH[c]	$\Delta\epsilon_{396} + 0{\cdot}51$, $\Delta\epsilon_{314} + 5{\cdot}7$, $\Delta\epsilon_{254} + 7{\cdot}9$
(**70**)	$PhCH_2$	Me	EtOH[c]	$\Delta\epsilon_{399} + 0{\cdot}6$, $\Delta\epsilon_{312} + 4{\cdot}5$, $\Delta\epsilon_{251} + 11{\cdot}8$
			Isooctane	$\Delta\epsilon_{317} + 7{\cdot}3$, $\Delta\varepsilon_{253} + 9{\cdot}1$
(**65d**)	$PhCH_2$	COOMe[d,80]	Dioxan	$\Delta\epsilon_{320} + 3{\cdot}72$
(**71**)	*p*-$HOC_6H_4CH_2$	COOMe[d]	MeOH	$[\Phi]_{430}$ inflection, $[\Phi]_{337} + 12{,}000$, $[\Phi]_{300} - 8200$
			MeOH	$\Delta\epsilon_{415} + 0{\cdot}28$, $\Delta\epsilon_{319} + 5{\cdot}5$, $\Delta\epsilon_{262} + 13{\cdot}3$
(**72**)	Ph	$PhCH_2$	EtOH	$[\Phi]_{275} - 8400!$
			EtOH[c]	$\Delta\epsilon_{262} - 6{\cdot}4$
			Dioxan	$\Delta\epsilon_{319} + 1{\cdot}8$
			Isooctane	$\Delta\epsilon_{258} - 4{\cdot}5$

[a] Values quoted are for o.r.d. extrema (+ = pk, − = tr) and for c.d. maxima.
[b] For electronic spectrum see Table 3.
[c] Anhydrous ethanol.
[d] D-configuration (i.e. (*R*) on Cahn–Ingold–Prelog convention); measured on the enantiomer.

The reaction of β-diketones with amines leads to 1:1 condensation products which may be regarded as internally H-bonded tautomeric systems, e.g. **73**. Bergel and Butler[84] have examined the rotatory

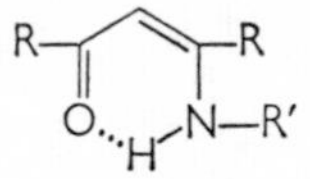

(73) 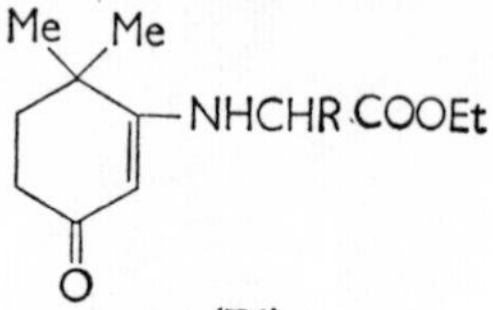 **(74)**

dispersion of the acetylacetone derivative of L-tyrosine ethyl ester down to 300 mμ (!): it showed a plain negative curve with $[\Phi]_{350} = -12{,}000$ (*N*-cyclopentylidene-L-tyrosine ethyl ester has $[\Phi] = 1000$ at this wavelength). The analogous derivatives from dimedone[85] exist in the conjugated carbonyl form (**74**) (and not as the tautomeric conjugated azomethine) and show strong Cotton effects (amplitudes of several hundreds) with the first extremum in the 300 mμ region.

B. Other Substituted Azomethines

1. Oximes

Lyle and Barrera[86] have observed plain curves for various oximes down to about 250 mμ. Plain negative curves are found for the two geometrically isomeric oximes derived from (+)-3-methylcyclohexanone: however, isomer **75a** shows a considerably larger rotation than **75b** (Figure 12), and the corresponding benzoates have Cotton effects of opposite sign and hence are readily distinguished. More recently Crabbé and Pinelo[87] have reached much shorter wavelengths using circular dichroism and have detected Cotton effects in the 200 mμ region (Table 11).

In certain cases the Cotton effect curve is solvent dependent. It may be recalled that a change in configuration at $C_{(5)}$ did not alter the sign of the Cotton effect for certain 3-*N*-salicylidenimino steroids (**63a, 63b**

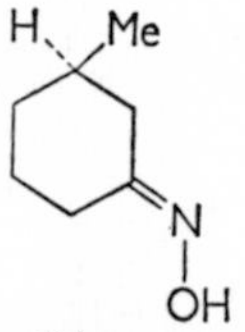

β-isomer
(75a)

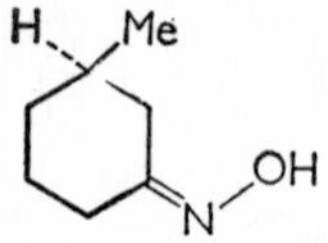

α-isomer
(75b)

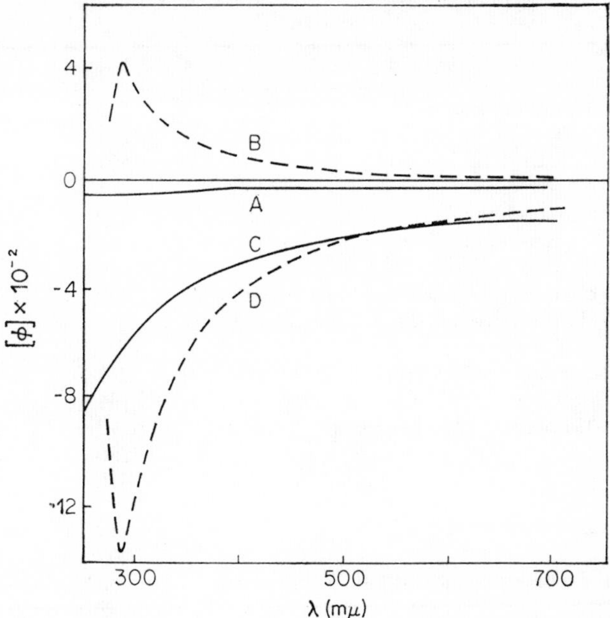

FIGURE 12. Rotatory dispersion in ethanol. A, 3-methylcyclohexanone oxime (α-form **75b**); B, the corresponding benzoate; C, 3-methylcyclohexanone oxime (β-form **75a**); D, the corresponding benzoate. (Reproduced from Lyle and Barrera, *J. Org. Chem.*, **29**, 3311 (1964) by permission of the authors and editor.)

in Table 9) although the magnitude of the effect did change. As is shown by examples **78** and **79**, the movement of the chromophore one step nearer does lead to a change in sign (although in the 5β-compound a turning point is not apparently observed). Again the Cotton effect of geometrically isomeric oximes (**80, 81**) differ in magnitude but not in sign (cf. **75a, 75b** above). The rotational strengths encountered are much larger than those found for the parent carbonyl compounds: this is especially so with conjugated systems (e.g. **85, 86,**

OH HON **(85)**

OAc OAc NOH **(86)**

TABLE 11. Circular dichroism of some oximes and related compounds in ethanol[87].

Compound	c.d.λ_{max}	$\Delta\epsilon$
(**76**) (+)-Camphor oxime	215	+0·58
	195[a]	−9·2
(**77**) (−)-Menthone oxime	197	−4·3
(**78**) 5α-Pregnan-3-one oxime	210	+2·2
(**79**) 5β-Pregnan-3-one oxime	195[a]	−0·82
(**80**) 17β-Hydroxy-5α-androstan-3-one α-oxime	210	+2·3
(**81**) 17β-Hydroxy-5α-androstan-3-one β-oxime	210	+3·2
(**82**) (−)-Methone oxime acetate	217	+1·45
	194[a]	−8·5
(**83**) (+)-Camphor oxime benzoate	273	+0·05
	230	−0·47
	210	−0·48
(**85**) Testosterone oxime	246	+14·7
	209	−4·1
(**86**) 3β,17β-Diacetoxyandrost-5-en-7-one oxime	235	−21·5

[a] Lowest wavelength reached: not a maximum.

in which the chromophores have opposite chirality); and the oxime appears to be a valuable and sensitive derivative with which to examine the stereochemistry of optically active dienone systems[87].

2. *N,N*-Dimethylhydrazones

The *N,N*-dimethylhydrazone of (+)-3-methylcyclohexanone has an absorption band in the 280 mμ region (cf. Table 4), and associated with it is found a Cotton effect curve which has the same sign as that of the parent ketone, but differs from it particularly with respect to the wavelength of extrema, as is shown in Table 12.

TABLE 12. Optical rotatory dispersion of the *N,N*-dimethylhydrazone of (+)-3-methylcyclohexanone[9].

Solvent	Rotatory dispersion	*a*
Hexane	$[\Phi]_{296}$ + 800 pk; $[\Phi]_{245}$ − 2650 tr; $[\Phi]_{227}$ − 2500 !	+34·5
MeOH	$[\Phi]_{292}$ + 375 pk; $[\Phi]_{250}$ − 1650 tr; $[\Phi]_{213}$ − 2380 !	+20
Parent ketone in MeOH[88]	$[\Phi]_{309}$ + 1090 pk; $[\Phi]_{270}$ − 1510 tr	+26

Clearly the study of the azomethine chromophore by optical dissymmetry effects has much further to go before a fund of information approaching that available on the carbonyl chromophore is attained. The very fact, however, that the azomethine system is somewhat—but not alarmingly—more complex than the carbonyl, and in particular, has a valence bond projecting towards the front, makes this a potential source of new information, especially concerning front octant effects. The examination of cyclic azomethines of known geometry, of solvent sensitivities, and of the host of possibilities suggested by Table 4—which has been barely touched—make this appear a promising area for further investigation.

ACKNOWLEDGEMENTS

Professors S. F. Mason and H. E. Smith are thanked for furnishing some unpublished results. This review was written during the tenure of a visiting Professorship at the University of British Columbia: the author is grateful to the Chemistry Department of that University for its warm hospitality during this period, to Professor W. Klyne and Dr. Emerson for measurements and discussions, to Dr. A. F. McDonagh for helpful suggestions and to colleagues in other laboratories for furnishing small quantities of steroidal azomethines.

REFERENCES

1. R. Bonnett, N. J. David, J. Hamlin and P. Smith, *Chem. Ind.* (*London*), 1836 (1963).
2. R. Bonnett, *J. Chem. Soc.*, 2313 (1965).
3. J. Hires and J. Balog, *Acta Univ. Szeged. Acta Phys. Chem.*, **2**, 87 (1956).
4. R. Griot and T. Wagner-Jauregg, *Helv. Chim. Acta*, **41**, 867 (1958).
5. E. D. Bergmann, Y. Hirschberg, S. Pinchas and E. Zimkin, *Rec. Trav. Chim.*, **71**, 192 (1952).
6. S. F. Mason, *Quart. Rev.*, **15**, 287 (1961).
7. W. S. Emerson, S. M. Hess and F. C. Uhle, *J. Am. Chem. Soc.*, **63**, 872 (1941).
8. E. J. Stamhuis, W. Maas and H. Wynberg, *J. Org. Chem.*, **30**, 2160 (1965).
9. R. Bonnett and J. Wood, unpublished results.
10. D. A. Nelson and J. J. Worman, *Chem. Commun.*, 487 (1966).
11. D. A. Nelson and J. J. Worman, *Tetrahedron Letters*, 507 (1966).
12. A. Williams and M. Bender, *J. Am. Chem. Soc.*, **88**, 2508 (1966).
13. H. C. Barany, E. A. Braude and M. Pianka, *J. Chem. Soc.*, 1898 (1949).
14. S. F. Mason and G. W. Vane, *Chem. Commun.*, 540 (1965). S. F. Mason, personal communication.
15. R. Bonnett and T. R. Emerson, *J. Chem. Soc.*, 4508 (1965).
16. G. Smolinsky, *J. Org. Chem.*, **27**, 3557 (1962).

17. Z. Badr, R. Bonnett, T. R. Emerson and W. Klyne, *J. Chem. Soc.*, 4503 (1965).
18. M. D. Newton, F. P. Boer and W. N. Lipscomb, *J. Am. Chem. Soc.*, **88**, 2367 (1966).
19. G. Opitz, H. Hellmann and H. W. Schubert, *Ann. Chem.*, **623**, 117 (1959). cf. N. J. Leonard and D. M. Locke, *J. Am. Chem. Soc.*, **77**, 437 (1955).
20. S. W. Pelletier and P. C. Parthasarathy, *J. Am. Chem. Soc.*, **87**, 777 (1965).
21. L. C. Jones and L. W. Taylor, *Anal. Chem.*, **27**, 228 (1955).
22. J. D. Roberts and M. C. Caserio *Supplement for Basic Principles of Organic Chemistry*. Benjamin, New York, 1964, p. 17, gives an estimate of $\sim$323 mμ.
23. S. F. Mason, *J. Chem. Soc.*, 1240 (1959). For reviews of *N*-heteroaromatic $n \rightarrow \pi^*$ transitions see M. Kasha in *Light and Life* (Ed. W. D. McElroy and B. Glass), John Hopkins Press, Baltimore, 1961, p. 31; L. Goodman, *J. Mol. Spectroscopy*, **6**, 109 (1961); S. F. Mason, *Physical Methods in Heterocyclic Chemistry* (Ed. A. R. Katritsky), Vol. 2, Academic Press, 1963, p. 7.
24. P. Brocklehurst, *Tetrahedron*, **18**, 299 (1962).
25. H. Saito and K. Nukada, *Tetrahedron*, **22**, 3313 (1966).
26. E. B. Whipple, Y. Chiang and R. L. Hinman, *J. Am. Chem. Soc.*, **85**, 26 (1963); Y. Chiang, R. L. Hinman, S. Theodoropulos and E. B. Whipple, *Tetrahedron*, **23**, 745 (1967).
27. C. B. Pollard and R. F. Parcell, *J. Am. Chem. Soc.*, **73**, 2925 (1951).
28. E. M. Kosower and T. S. Sorensen, *J. Org. Chem.*, **28**, 692 (1963).
29. M. Fischer, *Tetrahedron Letters*, 5273 (1966).
30. H. E. Smith, S. L. Cook and M. E. Warren, *J. Org. Chem.*, **29**, 2256 (1964).
31. D. Heinert and A. E. Martell, *J. Am. Chem. Soc.*, **85**, 183 (1963).
32. W. F. Smith, *Tetrahedron*, **19**, 445 (1963); V. I. Minkin, Y. A. Zhdanov, E. A. Medyantzeva and Y. A. Ostroumov, *Tetrahedron*, **23**, 3651 (1957).
33. D. Heinert and A. E. Martell, *J. Am. Chem. Soc.*, **85**, 188 (1963).
34. J. Hires and L. Hackl, *Acta Univ. Szeged. Acta Phys. Chem.*, **5**, 19 (1959).
35. A. J. Bellamy and R. D. Guthrie, *J. Chem. Soc.*, 2788 (1965).
36. H. Ley and H. Wingchen, *Ber.*, **67**, 501 (1934).
37. J. E. Baldwin and N. H. Rogers, *Chem. Commun.*, 524 (1965).
38. R. Bonnett, R. F. C. Brown, V. M. Clark, I. O. Sutherland and Sir Alexander Todd, *J. Chem. Soc.*, 2094 (1959); L. S. Kaminsky and M. Lamchen, *J. Chem. Soc.* (*B*), 1085 (1968).
39. R. Bonnett, S. C. Ho and J. A. Raleigh, *Can. J. Chem.*, **43**, 2717 (1965).
40. J. Thesing and W. Sirrenberg, *Chem. Ber.*, **91**, 1978 (1958).
41. G. J. Karabatsos, R. A. Taller and F. M. Vane, *Tetrahedron Letters*, 1081 (1964); *Tetrahedron*, **24**, 3557, 3923 (1968).
42. G. Adembri, P. Sarti-Fantoni and E. Belgodere, *Tetrahedron*, **22**, 3149 (1966).
43. L. K. Evans and A. E. Gillam, *J. Chem. Soc.*, 565 (1943).
44. A. E. Gillam and E. S. Stern, *Electronic Absorption Spectroscopy*, Arnold, London, 1957.
45. J. C. Gage, *J. Chem. Soc.*, 221 (1949).
46. W. Menschick, I. H. Page and K. Bossert, *Ann. Chem.*, **495**, 225 (1932).
47. J. Meisenheimer and O. Dorner, *Ann. Chem.*, **502**, 156 (1933).
48. J. P. Freeman, *Chem. Ind.* (*London*), 1624 (1960).

49. A. Kjaer and S. E. Hansen, *Acta Chem. Scand.*, **11**, 898 (1957).
50. H. E. Smith and A. F. McDonagh, unpublished results; A. F. McDonagh, *Thesis*, Vanderbilt Univ., 1967.
51. F. Bergel and G. E. Lewis, *Chem. Ind.* (*London*), 774 (1955).
52. F. Bergel, G. E. Lewis, S. F. D. Orr and J. Butler, *J. Chem. Soc.*, 1431 (1959) and references therein; cf. H. E. Smith, M. E. Warren and A. W. Ingersoll, *J. Am. Chem. Soc.*, **84**, 1513 (1962).
53. Z. Badr, R. Bonnett, W. Klyne, R. J. Swan and J. Wood, *J. Chem. Soc.* (*C*), 2047 (1966).
54. F. Bergel and M. A. Peutherer, *J. Chem. Soc.*, 3965 (1964).
55. F. Bergel and K. R. Harrap, *J. Chem. Soc.*, 4051 (1961) and references therein.
56. This spectrum was kindly recorded by the Japan Spectroscopic Co. Ltd.
57. W. Moffit, R. B. Woodward, A. Moscowitz, W. Klyne and C. Djerassi, *J. Am. Chem. Soc.*, **83**, 4013 (1961).
58. G. Wagnière, *J. Am. Chem. Soc.*, **88**, 3937 (1966). Cf. C. Djerassi and W. Klyne, *J. Chem. Soc.*, 2390 (1963).
59. H. Ripperger, K. Schreiber and G. Snatzke, *Tetrahedron*, **21**, 1027 (1965). Cf. E. Bianchi, C. Djerassi, H. Budzikiewicz and Y. Sato, *J. Org. Chem.*, **30**, 754 (1965).
60. H. Lettré and L. Knof, *Chem. Ber.*, **93**, 2860 (1960).
61. D. Dvornik and O. E. Edwards, *Chem. Ind.* (*London*), 623 (1958).
62. S. W. Pelletier, *J. Am. Chem. Soc.*, **82**, 2389 (1960). *Experientia*, **20**, 1 (1964).
63. C. W. Shoppee, R. W. Killick and G. Kruger, *J. Chem. Soc.*, 2275 (1962).
64. A. Yogev and Y. Mazur, *Tetrahedron*, **22**, 1317 (1966).
65. R. Bonnett, T. R. Emerson and J. Wood in *Some Newer Physical Methods in Structural Chemistry* (Eds. R. Bonnett and J. G. Davis), United Trade Press, London, 1967, p. 178.
66. R. Bonnett and T. R. Emerson, unpublished results.
67. D. F. Morrow and M. E. Butler, *J. Hetero. Chem.*, **1**, 53 (1964).
68. F. C. Uhle and F. Sallmann, *J. Am. Chem. Soc.*, **82**, 1190 (1960).
69. Q. Khuong-Huu, L. Lábler, M. Truong-Ho and R. Goutarel, *Bull. Soc. Chim. France*, 1564 (1964).
70. Q. Khuong-Huu, M. Truong-Ho, L. Lábler, R. Goutarel and F. Šorm, *Coll. Czech. Chem. Commun.*, **30**, 1016 (1965).
71. J. Hora and V. Černy, *Collection Czech. Chem. Commun.*, **26**, 2217 (1961).
72. F. Nerdel, K. Becker and G. Kresze, *Chem. Ber.*, **89**, 2862 (1962).
73. V. M. Potapov, A. P. Terentev and S. P. Spivak, *J. Gen. Chem. USSR*, **31**, 2251 (1961) and references therein; V. M. Potapov, V. M. Demyanovich, L. I. Lauztina and A. P. Terentev, *J. Gen. Chem. USSR*, **32**, 1161 (1962).
74. V. M. Potapov, V. M. Demyanovich and A. P. Terentev, *J. Gen. Chem. USSR*, **35**, 1541 (1965).
75. H. Ripperger, K. Schreiber, G. Snatzke and K. Heller, *Z. Chem.*, **5**, 62 (1965).
76. Dr. G. C. Barrett (West Ham College of Technology, London), unpublished results.
77. D. Bertin and M. Legrand, *Compt. Rend.*, **256**, 960 (1963).
78. A. Moscowitz, personal communication to P. Crabbé, *Tetrahedron*, **20**, 1211 (1964) and there reference 105.

79. M. E. Warren and H. E. Smith, *J. Am. Chem. Soc.*, **87**, 1757 (1965).
80. L. Velluz, M. Legrand and M. Grosjean, *Optical Circular Dichroism*, Academic Press, New York, 1965, p. 156 and appendix F.
81. H. E. Smith and R. Records, *Tetrahedron*, **22**, 813 (1966).
82. H. E. Smith and T. C. Willis, *J. Org. Chem.*, **30**, 2654 (1965).
83. A. Moscowitz, K. Mislow, M. A. W. Glass and C. Djerassi, *J. Am. Chem. Soc.*, **84**, 1945 (1962).
84. F. Bergel and J. Butler, *J. Chem. Soc.*, 4047 (1961).
85. B. Halpern and L. B. James, *Australian J. Chem.*, **17**, 1282 (1964); P. Crabbé and B. Halpern, *Chem. Ind.* (*London*), 346 (1965); P. Crabbé, B. Halpern and E. Santos, *Bull. Soc. Chim. France*, 1446 (1966).
86. G. G. Lyle and R. M. Barrera, *J. Org. Chem.*, **29**, 3311 (1964).
87. P. Crabbé and L. Pinelo, *Chem. Ind.* (*London*), 158 (1966).
88. C. Djerassi and G. W. Krakower, *J. Am. Chem. Soc.*, **81**, 237 (1959).
89. G. Adam, K. Schreiber, J. Tomko and A. Vassova, *Tetrahedron*, **23**, 167 (1967).
90. M. Lamchen and T. W. Mittag, *J. Chem. Soc.* (*C*), 2300 (1966).
91. E. S. Hand and W. P. Jencks, *J. Am. Chem. Soc.*, **84**, 3505 (1962).
92. J. Charette, G. Falthanse and P. Teyssie, *Spectrochim. Acta*, **20**, 597 (1964).
93. R. Tsuchida and T. Tsumaki, *Bull. Chem. Soc. Japan*, **13**, 527 (1938).
94. A. Kiss and G. Auer, *Z. Phys. Chem.* (*Leipzig*), **A189**, 344 (1941).
95. L. N. Ferguson and I. Kelly, *J. Am. Chem. Soc.*, **73**, 3707 (1951).
96. L. A. Kazitsyna, N. B. Kupletskaya, A. L. Polstyanko, B. S. Kikot, Y. A. Kolesnik and A. P. Terentev, *J. Gen. Chem. USSR*, **31**, 286 (1961).
97. G. O. Dudek and E. P. Dudek, *J. Am. Chem. Soc.*, **88**, 2407 (1966).
98. R. S. Cahn, C. K. Ingold and V. Prelog, *Experientia*, **12**, 81 (1956).
99. M. D. Cohen and G. M. J. Schmidt, *J. Phys. Chem.*, **66**, 2442 (1962); M. D. Cohen and S. Flavian, *J. Chem. Soc.* (*B*), 321 (1967).
100. R. Hubbard, D. Bownds and T. Yoshizawa, *Cold Spring Harbour Symp. Quant. Biol.*, **30**, 301 (1965); G. Wald, *Science*, **162**, 230 (1968); M. Akhtar, P. T. Blosse and P. B. Dewhurst, *Chem. Commun.* 631 (1967).
101. G. A. J. Pitt, F. D. Collins, R. A. Morton and P. Stok, *Biochem. J.*, **59**, 122 (1955).
102. D. Bownds and G. Wald, *Nature*, **205**, 254 (1965).
103. N. Akhtar, P. T. Blosse and P. B. Dewhurst, *Life Sciences*, **4**, 1221 (1965).
104. R. S. Becker and W. F. Richey, *J. Am. Chem. Soc.*, **89**, 1298 (1967).
105. V. I. Minkin, O. A. Osipov, V. A. Kogen, R. R. Shagidullin, R. L. Terentev and O. A. Raevskii, *Russ. J. Phys. Chem.*, **38**, 938 (1964).
106. Y. Matsushima and A. E. Martell, *J. Am. Chem. Soc.*, **89**, 1322 (1967).
107. G. Wald and P. K. Brown, *J. Gen. Physiol.*, **35**, 797 (1951).
108. P. J. Orenski and W. D. Closson, *Tetrahedron Letters*, 3629 (1967).
109. For the circular dichroism of some related nitrone derivatives see J. Parello and X. Lusinchi, *Tetrahedron*, **24**, 6747 (1968).
110. G. Dudek, *J. Org. Chem.*, **32**, 2016 (1967).
111. H. Ripperger, K. Schreiber, G. Snatzke and K. Ponsold, *Tetrahedron*, **25**, 827 (1969).

CHAPTER **5**

Basic and complex-forming properties

J. W. SMITH

Bedford College, London, England

I. INTRODUCTION

By virtue of the presence of a lone-pair of electrons on the nitrogen atom and of the general electron-donating character of the double bond, compounds containing the azomethine group should all possess basic properties. These are demonstrated by the acceptance of a proton from a Lowry-Brønsted acid to form the conjugate cation, by a tendency to form a hydrogen-bonded complex with a compound containing a hydrogen atom linked directly with an oxygen or a nitrogen atom, and by behaviour as a Lewis base in donating an electron pair to a metal atom in the formation of a coordination compound.

Although these properties should be very closely interlinked, there has as yet been very little attempt to correlate them, the relative behaviours of various groups of compounds containing the C=N bond having been studied in relation to these basic functions from quite independent points of view. They will therefore be discussed separately here.

II. BASIC STRENGTHS

The strength of a base is normally expressed in terms of the pK_a value, where pK_a = $-\log_{10} K_a$, K_a being the acid dissociation constant of the conjugate acid[1]. The stronger the base the higher is the pK_a value, which for amines ranges from about 10–11 for primary and secondary alkyl amines to about 4–5 for aryl amines. Ammonia has a pK_a value of about 9·2.

Due principally to the fact that many are relatively unstable under conditions suitable for measurement of their pK_a values, the basic strengths of only very few compounds of which the basic character depends on the presence of the azomethine group have been determined. The only series of compounds which have been studied in this respect are some derivatives of diphenylketimine ($Ph_2C{=}NH$) and of benzylidene-*t*-butylamine (PhCH=NBu-*t*). The pK_a values for the conjugate acids of these bases and of some related compounds are shown in Table 1.

It is very difficult to draw any inferences regarding the effects of substituents on the pK_a values of the derivatives of diphenylketimine, as anomalies abound amongst the data. Diphenylketimine itself is a relatively weak base, and the still lower pK_a values of 2-hydroxy- and 2,4,6-trihydroxydiphenylketimine suggest that the presence of a 2-hydroxy group produces a large decrease in base strength. On the other hand 2,4-dihydroxy-6-methyldiphenylketimine is almost as strong a base as 2-methyldiphenylketimine. Again, a 2-methoxy group seems to increase the base strength slightly, but 2-methoxy-4-hydroxydiphenylketimine is appreciably weaker as a base than 4-hydroxydiphenylketimine. Halogen substituents exert their expected effect in causing a withdrawal of electron density and hence in reducing the base strength. The substitution of one of the phenyl groups in diphenylketimine by a methoxy group to give iminomethoxylmethylbenzene appears to cause about the same reduction in base strength as the introduction of a halogen atom into the ring.

In view of the apparent inconsistencies in the results further studies in this series of compounds would obviously be of interest.

Simple aliphatic imino compounds cannot be studied, and the one isolated measurement recorded for a simple oxime serves only to underline the extemely weak basic character of the oxime group.

TABLE 1. pK_a Values of the conjugate acids of compounds containing the azomethine group.

Compound	pK_a	Reference
Diphenylketimine	7·18	2
2-Methyldiphenylketimine	6·79	2
2-Chlorodiphenylketimine	5·59	2
3-Chlorodiphenylketimine	5·69	2
2-Hydroxydiphenylketimine	5·00	2
3-Hydroxydiphenylketimine	7·08	2
4-Hydroxydiphenylketimine	6·45	2
2-Methoxydiphenylketimine	7·29	2
3-Methoxydiphenylketimine	6·59	2
2,4-Dimethyldiphenylketimine	6·79	2
2,5-Dimethyldiphenylketimine	6·79	2
3,5-Dimethyldiphenylketimine	7·18	2
2,6-Dimethyldiphenylketimine	6·29	2
2,4-Dihydroxydiphenylketimine	5·00	2
2-Methoxy-4-hydroxydiphenylketimine	5·99	2
2,4-Dimethoxydiphenylketimine	8·30	2
2,4,6-Trihydroxydiphenylketimine	5·20	2
2,4-Dihydroxy-6-methyldiphenylketimine	6·75	2
Iminomethoxylmethylbenzene	5·8	3
Acetoxime	0·99	4
Benzylidene-*t*-butylamine	6·7	5
p-Nitrobenzylidene-*t*-butylamine	5·4	5
m-Bromobenzylidene-*t*-butylamine	6·1	5
p-Chlorobenzylidene-*t*-butylamine	6·5	5
p-Methylbenzylidene-*t*-butylamine	7·4	5
p-Methoxybenzylidene-*t*-butylamine	7·7	5
p-Chlorobenzylideneaniline	2·80	6

Fairly systematic variations are observable in the pK_a values of benzylidene-*t*-butylamine and its derivatives. Here, as is to be expected, the basicity is increased by the presence of 4-methyl and 4-methoxy groups, which tend to increase the electron density at the nitrogen atom, and decreased by nitro or halogeno substituents which tend to withdraw electron density from it. These effects are very similar to, but with the apparent exception of the effect of the methyl group are less pronounced than, those observed for the derivatives of aniline[7]. Strong electron withdrawal also accounts for the very low basic strength of *p*-chlorobenzylideneaniline.

TABLE 2. Association constants and thermodynamic data for complex formation between derivatives of benzylideneaniline and *p*-nitrophenol.

Substituent[a] *p*	*p′*	$\sigma +$ [b]	$K \times 10^{-1}$ at 18°	27·5°	48°	ΔG(27·5°) (kcal/mole)	$\Delta H°$ (kcal/mole)	$\Delta S°$ (cal/deg mole)
NMe_2	H	−1·79	40	28	13	−3·4	−7·0	−12
OMe	H	−0·764	13	8·8	4·5	−2·7	−6·6	−13
Me	H	−0·306	8·4	6·1	3·1	−2·7	−6·0	−12
H	H	0·0	5·9	3·9	1·9	−2·2	−7·1	−16
Cl	H	0·122	4·2	2·9	1·4	−2·0	−6·5	−15
Br	H	0·148	3·8	2·5	1·3	−1·9	−6·6	−15
NO_2	H	0·777	2·3	1·2	0·66	−1·4	−7·2	−19
H	NMe_2	−0·609	19	14	7·3	−3·0	−5·9	−10
H	OMe	−0·268	9·4	6·3	3·5	−2·5	−6·0	−12
H	Me	−0·170	8·0	6·1	3·1	−2·4	−6·0	−12
H	Cl	0·227	3·2	2·5	1·1	−1·9	−6·7	−16
H	Br	0·232	3·2	2·1	1·3	−1·8	−5·5	−12
H	NO_2	0·778	2·2	1·3	0·56	−1·6	—	—
NMe_2	NMe_2		10·0	8·0	5·3	−4·0	−4·1	−0·28
NMe_2	Cl			1·5	0·99	−3·0	—	—
OMe	OMe		1·8	1·2	0·64	−2·8	−6·3	−11
OMe	Me		1·5	1·2	0·52	−2·9	−6·8	−13
NO_2	NMe_2		0·59	0·36	0·29	−2·2	−6·2	−13
Cl	Cl		0·28	0·15	0·085	−1·6	−7·0	−18

[a] *p* and *p′* refer to substituents in the rings linked to the carbon and nitrogen atoms, respectively, of the azomethine group.
[b] Values taken from data of Okamoto and Brown[9].

III. HYDROGEN-BONDING PROPERTIES

The basic character of the azomethine group is also revealed by the fact that in aprotic solvents containing it there is a tendency for the lone pair of electrons on the nitrogen atom to interact with hydroxylic compounds to yield hydrogen-bonded complexes. An example is furnished by the reaction between benzylideneaniline and *p*-nitrophenol in carbon tetrachloride solution to give the complex **1**.

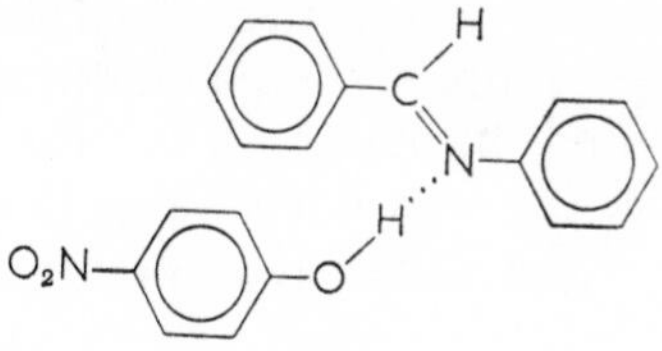

O_2N–C$_6$H$_4$–O–H···N(=CH–C$_6$H$_5$)–C$_6$H$_5$

(1)

Weinstein and McImrich have made an interesting study of the effects of 4- and 4′-substituents on the equilibrium constant for complex formation in this system[8]. The results, together with the values of the free energy, enthalpy and entropy changes of the process as derived from these figures, are shown in Table 2. In each series of monosubstituted derivatives, the association constants show a progressive decrease as the Hammett σ constants of the substituents are changed to lower negative and finally more positive values, thus showing a direct correlation with the change in electron density at the nitrogen atom brought about by the inductive and mesomeric effects of the substituents.

The values of $\Delta S°$ derived from the association constant and its temperature variation give an indication that it may tend towards higher negative values as the value of the association constant decreases. On the other hand $\Delta H°$ does not seem to vary greatly through the series. These results, however, must be taken as indicative only, as the method does not permit the evaluation of data suitable for precise and detailed analysis.

IV. COMPLEXES WITH METALS

A. Introduction

The most characteristic respect in which compounds containing the C═N bond show basic properties is in the formation of complexes with metals. These complexes provide some very characteristic series of coordination compounds, and consequently a large number of them have been prepared and their properties examined and compared.

The basic strength of the C=N group is insufficient by itself to permit of the formation of stable complexes by simple coordination of the lone pair to a metal ion. Therefore, in order that stable compounds should be formed it is necessary that there should also be present in the molecule a functional group with a replaceable hydrogen atom, preferably a hydroxyl group, near enough to the C=N group to permit the formation of a five or six membered ring by chelation to the metal atom. Among the simplest compounds which meet this requirement are salicylaldimine (**2**) and its *N*-hydroxy derivative salicylaldoxime (**3**), and by far the most intensive studies have been made on the metal complexes of the former and its derivatives.

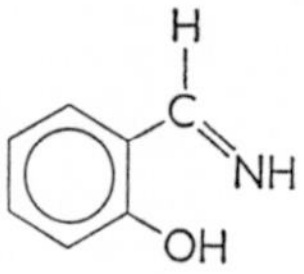

H C NH OH (**2**) H C NOH OH (**3**)

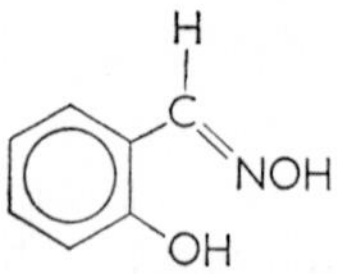

The compounds formed with metals by these and other Schiff's bases and known up to about 1964 have been listed and their properties discussed in detail by Holm, Everett and Chakravorthy[10], whilst some of the salicylaldimine complexes have formed the subject of reviews by Sacconi[11], and thus only a brief survey of the simpler groups of compounds is given here. Coordination compounds of this type, however, also include the well-known complexes of dimethylglyoxime with metals and other water-insoluble compounds, which have been used for some time in the detection and determination of certain metals.

B. Complexes of Salicylaldimine and Related Compounds

The salicylaldimine complexes of the general types **4** and **5** can often be made by the direct interaction between the metal ion and the appropriate Schiff's base in alcoholic or aqueous alcoholic solution and in the presence of a base such as sodium hydroxide or sodium acetate. Under these conditions the *N*-alkylsalicylaldimines tend to hydrolyse, however, so a more generally useful method of preparation is to reflux the salicylaldehyde complex of the metal with a slight excess of the primary amine in a non-aqueous solvent. This procedure was originally used by Schiff[12], who prepared a number of these compounds and proved their compositions.

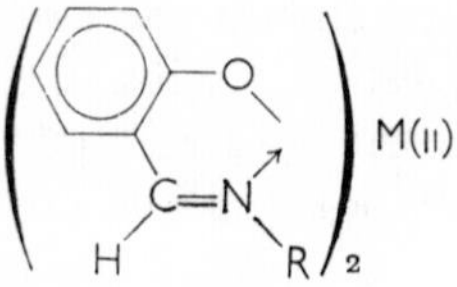

(4) 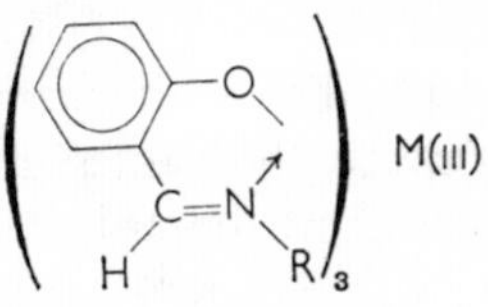 (5)

The coordination complexes formed with divalent metal ions (type **4**) are of considerable interest as they vary in structure from coplanar to tetrahedral, not only in dependence on the nature of the divalent metal involved, but also on the nature of the substituent on the nitrogen atom and the substituents, if any, on the aromatic ring of the salicylaldimine molecule.

An important method of studying the structures of these compounds is magnetic susceptibility measurement. The Ni^{2+}, Pd^{2+} and Pt^{2+} ions all have eight *d* electrons in their ultimate electronic shells. Hence in forming the Ni(II), Pd(II) and Pt(II) complexes, a process which can be envisaged as involving coordination of four electron pairs to these ions, two alternative structures may arise. If they use dsp^2 hybrid orbitals the bonds formed are effectively coplanar and, the *d* sub-shell being completed, there are no unpaired electrons, so the resultant spin S is zero and the compounds are diamagnetic. Alternatively, however, they can use sp^3 hybrid orbitals, in which case, following Hund's rule, there are two unpaired *d* electrons, so $S = 1$ and the complexes are paramagnetic.

The cobalt atom contains one electron less than the nickel atom and hence in the planar dsp^2 Co(II) complexes there is one unpaired electron whereas in the tetrahedral sp^3 complexes there are three. In the octahedral d^2sp^3 complexes of Co(III), however, there are no unpaired electrons and so they are diamagnetic.

The copper atom, on the contrary, has one electron more than the nickel atom and hence in both the planar and the tetrahedral complexes of Cu(II) there is one unpaired electron.

The Zn^{2+} ion has a complete group of 10 *d* electrons in its ultimate shell, so the tendency is for it to use sp^3 hybrid orbitals and form tetrahedral diamagnetic complexes.

The complexes of salicylaldimine itself (R = H) with Ni(II), Cu(II) and Pd(II) are all planar and crystallize in isomorphous forms*. They are diamagnetic and have the *trans* configuration[13–16]. The same is true of their *N*-hydroxy derivatives (complexes of salicylaldoxime) and of their annular-substituted derivatives. The unsubstituted complexes of salicylaldimine should therefore have zero dipole moments. The evidence[17] suggests that this is the case, the difference between the values of the molar polarizations and molar refractions being attributable to atom polarization, which may in certain cases be fairly high.

At the other end of the scale the complexes of Zn(II), which uses sp^3 hybrid orbitals, are all tetrahedral and have closed shell configurations. Their dipole moments range from 4·97 to 5·13 D, and these values have been used as standards for tetrahedral complexes in comparing the moments of other complexes of unknown stereochemistry[17].

In the Ni(II), Cu(II) and Pd(II) complexes with larger substituents on the nitrogen atoms, even methyl, ethyl or n-butyl groups, the alkyl substituents tend to be displaced out of the plane of the benzene rings in the solid state[18,19]. These displacements seem to arise from a tendency for the groups to attain a trigonal disposition around the nitrogen atom, this leading to steric interaction between the *N*-alkyl group and the oxygen atom of the second salicylaldimino group.

In pyridine solution a number of the Ni(II) complexes of the *N*-n-alkylsalicylaldimines are paramagnetic, with magnetic moments (μ) of about 3·1 Bohr magnetons (B.M.). This indicates the presence of unpaired electrons, and can be accounted for by the addition of two molecules of pyridine to each molecule of complex, as addition compounds of this composition have been isolated[20,21]. This addition of pyridine must be accompanied by a change from square planar (diamagnetic) to octahedral (paramagnetic) structure. Even in benzene or chloroform solution complexes of this type are slightly paramagnetic, with apparent magnetic moments of 0·2 to 1·0 B.M. This effect probably arises through the planar monomeric species being in equilibrium with dimeric or polymeric forms which are paramagnetic[22], through assumption of octahedral structure. The probability that polymeric species are involved is indicated by the fact that a paramagnetic form of the complex of *N*-methylsalicylaldimine with Ni(II) has been isolated in the solid state[23]. This is so

* In square planar complexes, the convention in nomenclature is that the *trans* isomer has the same groups of the ligand molecules occupying diametrically opposite corners of the square, whereas in the *cis* isomer they occupy adjacent corners.

exceptionally insoluble as to suggest that it has an octahedral polymeric structure. Further, Ferguson[24] has shown cryoscopically that this compound is partially associated in benzene solution.

These complexes also develop paramagnetism when fused[25], but this phenomenon cannot be ascribed solely to polymerization. It has been shown that at 180° the Ni(II) complex of *N*-methylsalicylaldimine is transformed into a tetrahedral form which is paramagnetic with a magnetic moment of 3·4 B.M. Even at room temperature there seems to be quite a close balance between the stabilities of the two forms, since in the presence of the corresponding complex of Zn(II), which is tetrahedral, the Ni(II) complex also crystallizes from chloroform solution in the tetrahedral form with magnetic moment 3·1 B.M.[26]. In the absence of foreign ions this Ni(II) complex even develops paramagnetism when heated in solution[23,27,28].

The essential planarity of these structures in solution at room temperature is shown by the fact that both for benzene and dioxan solutions the differences between the molar polarizations and molar refractions are only 22–44 c.c. for the *N*-methyl- and *N*-ethylsalicylaldimine complexes of Ni(II) and 8–24 c.c. for the *N*-ethyl-, *N*-n-propyl, *N*-n-butyl- and *N*-n-amylsalicylaldimine complexes of Pd(II)[47]. Although much greater than the differences commonly encountered for simple molecules, these figures are explainable in terms of high atom polarizations arising from vibrational motions of the rings. Theoretically the Pd–ligand bonds should be more homopolar than the Ni–ligand bonds, and therefore they should be associated with higher force constants. In accordance with this view it is found that if the molecules are treated as single unidimensional oscillators the force constants derived from these values of the atom polarization are $2–6 \times 10^{-12}$ erg/radian^2 for the Pd(II) complexes as compared with $0{\cdot}9–2{\cdot}1 \times 10^{-12}$ erg/radian^2 for the Ni(II) complexes.

The results of dielectric polarization measurements on these complexes contrast with those on the corresponding complexes of Cu(II), which will be discussed later. Here the differences (52–99 c.c.) are greater than can be accounted for on the basis of atom polarization alone.

When the *N*-alkyl substituent is still more bulky, as is the case with a *s*- or *t*-alkyl group, steric influences upon the structure become much more pronounced, with the result that the Ni(II) complexes tend to acquire tetrahedral configurations. The result is that the *N*-*s*-alkyl derivatives of the salicylaldimine complexes are much more paramagnetic in chloroform solution than are the *N*-n-alkyl derivatives[29].

Cryoscopic measurements on benzene solutions show that they are associated, but not to such an extent as to account for the whole of their paramagnetism. This must arise principally from the existence in chloroform solution of an equilibrium between the planar form for which the resultant spin is zero and the tetrahedral form for which it is 1 unit.

The existence of this equilibrium, which is much better developed than for the *N*-n-alkyl derivatives, has been confirmed by spectroscopic evidence, and it has been shown that it, too, tends to be shifted towards the tetrahedral form with rise of temperature[17,30]. The dipole moments of these compounds in dioxan solution also differ appreciably from zero, the values ranging from 2·34 D for the *N*-*s*-butyl- to 4·74 for the *N*-*t*-butylsalicylaldimine complex of Ni(II), the latter having almost the same moment as the corresponding complexes of Co(II) and Zn(II), which are known to be tetrahedral[17]. These observations were of considerable interest as they provided the first unambiguous evidence for the existence of tetrahedral chelate complexes of Ni(II).

This dependence of the structure upon the precise nature of the substituent of the nitrogen atom arises from the fact that the differences in free energy between the planar and tetrahedral forms are not large. From spectroscopic measurements the difference has been evaluated as just under 3 kcal/mole for the *N*-n-propylsalicylaldimine complex and its 5-substituted derivatives[31]. As a result the tetrahedral form is at too low concentration to be detected at room temperature. On the other hand the corresponding differences lie between +0·35 and −0·52 kcal/mole for the corresponding *N*-isopropyl derivatives[31,32], and similar values are observed in the presence of other secondary alkyl groups. Hence in solution in bibenzyl or in chloroform the two forms coexist at ordinary temperature, whilst at higher temperatures the tetrahedral form predominates.

The structures adopted in the solid state also depend on the relative lattice energies, and these may be determined by relatively minor structural factors. The latter become very significant when there are annular substituents in the salicylaldimine molecule. Thus the Ni(II) complexes of *N*-isopropylsalicylaldimine and of its 5-ethyl derivative are tetrahedral and paramagnetic, whereas the complexes of 5-methyl-, 5-n-propyl-, 5-chloro- and 5-nitro-*N*-isopropylsalicylaldimines are all planar and diamagnetic[31,32]. Further, 3-methyl-, 3-chloro-, and 3-bromo-*N*-isopropylsalicylaldimine all form planar complexes with Ni(II), but the corresponding 3-ethyl-, 3-isopropyl and 3-*t*-butyl derivatives are all tetrahedral and paramagnetic[30]. This difference of behaviour has been attributed in part to the influence of a statistical term arising from the different multiplicities of the fundamental electronic states, and in part to the greater facility with which the alkyl groups can rotate in a tetrahedral structure. Another entropy term arises, however, from the greater solvation of the planar forms[33]. Ring substitution in the 5,6-position of *N*-isopropylsalicylaldimine favours the planar form, whereas 3,4-substitution favours the tetrahedral structure. For the *N*-*t*-butyl-substituted complexes the large steric repulsions make the tetrahedral configurations the most stable even at very low temperatures.

x-Ray diffraction studies show that an *N*-phenyl group causes considerable distortion from overall coplanarity in the molecule[34–36]. In the solid state the Ni(II) complexes of *N*-aryl salicylaldimines are all either diamagnetic or wholly paramagnetic, the *N*-phenyl-, *N*-*o*-tolyl-, *N*-α-naphthyl-, *N*-2,4-dimethylphenyl- and *N*-2,5-dimethylphenyl derivatives all giving rise to diamagnetic solid complexes, whilst those produced by the *N*-*m*-tolyl- and *N*-*m*-chlorophenyl compounds are both paramagnetic. In solution in xylene or bibenzyl the complexes of *N*-phenyl- or *p*-substituted *N*-phenylsalicylaldimine are all paramagnetic and associated, but it is only at temperatures above about 70° that the spectra indicate the presence of tetrahedral forms. The Ni(II) complexes of *N*-*m*-tolyl- and *N*-*m*-chlorophenylsalicylaldimine, however, are more paramagnetic and more extensively associated in solution. The complexes of *o*-substituted *N*-phenylsalicylaldimine, on the contrary, are all either diamagnetic or only weakly paramagnetic in solution. It appears, therefore, that the presence of an *ortho* substituent prevents association by keeping the plane of the *N*-phenyl ring almost orthogonal to the plane of the chelate ring.

As the cobalt atom has one less electron than the nickel atom, the planar complexes of Co(II) contain one unpaired electron and hence they are paramagnetic. The magnetic susceptibility is therefore not such a sensitive tool in their study as it is for Ni(II) complexes, and more reliance has to be placed on other physical properties. The general tendency is for the tetrahedral structures to be more favoured than they are for Ni(II), Cu(II) or Pd(II) complexes. Thus whilst the complexes of Co(II) with salicylaldimine and salicylaldoxime both have ultraviolet absorption spectra which suggest that they have planar structures in solution[37], the *N*-alkyl derivatives have tetrahedral structures[38–40]. In benzene solution these complexes of *N*-alkyl salicylaldimines all have relatively large dipole moments, ranging from 4·62 D for the *N*-n-butyl-compound to 5·05 D for the *N*-*t*-butyl derivative[17]. Although slightly lower than those of the corresponding zinc complexes, these values are compatible only with tetrahedral structures. The reflectance spectra of the solids show that this structure persists in the solid state[17], and the *N*-n-butyl derivative is isomorphous with the corresponding complex of Zn(II), which has been definitely shown to be tetrahedral[38]. It is less surprising that the Co(II) complex of *N*-isopropylsalicylaldimine is isomorphous with the corresponding complex of Ni(II). Cryoscopic measurements in benzene solution have shown that both the *N*-alkyl and *N*-aryl complexes are monomeric in that solvent[41].

Radio-tracer investigations have shown that in its complexes with *N*-phenyl-, *N*-*m*-tolyl- and *N*-*o*-anisylsalicylaldimine, Co(II) exchanges rapidly with Co^{2+} ions in pyridine solution. The complexes are also broken down on an ion exchange column, whilst on electrolysis of their solutions cobalt migrates to the cathode. All these observations indicate that the compounds are relatively labile, as would be expected from their high-spin tetrahedral structures[42]. Such an arrangement of the ligands also accounts for the fact that when the aryl ring possesses an *ortho* substituent other than the methoxyl group the complexes are very difficult to isolate, presumably through steric hindrance to their formation.

The *N*-aryl salicylaldimine complexes of Co(II) have magnetic moments of 4·36 to 4·53 B.M. in the solid state and 4·22 to 4·56 B.M. in benzene solution. In pyridine solution, however, the values are somewhat higher, and unstable adducts with two molecules of pyridine have been isolated from the solutions. Hence, in spite of their tetrahedral structures, pyridine seems to be able to coordinate with these compounds.

Oxidation of the bis-salicylaldimine complexes of Co(II) in solution and in the presence of an excess of the imine leads to the formation of the tris complexes of Co(III), which have the general structure **5**. For the preparation of the *N*-n-alkyl salicylaldimine complexes, atmospheric oxidation suffices, but a stronger oxidizing agent such as hydrogen peroxide is needed to produce the *N*-phenyl derivatives, which tend to be reduced again to the Co(II) complexes on heating in solution[43].

As one salicylaldimine molecule is unable to span opposite corners of the octahedron, only two geometric isomers of these tris Co(III) complexes would be expected to exist. These are the *cis* form, in which the nitrogen atoms all occupy corners opposite to an oxygen atom, and the *trans* form in which one pair of nitrogen atoms occupy corners opposite to one another. In fact, only one form has been isolated in each case, and dipole moment[44] and nuclear magnetic resonance[45] evidence indicates that this is the *trans* form, as would be predicted from steric considerations.

Their crystal structures indicate that the Cu(II) complexes of salicylaldimine and of its *N*-n-alkyl derivatives have essentially *trans*-planar structures in the solid state, being isomorphous with the corresponding Ni(II) and Pd(II) complexes[46]. Spectroscopic evidence suggests that these complexes have the same structures in solution as in the solid phase, so it is rather surprising that the differences between the molar polarizations and molar refractions of these compounds have been reported to be very much larger than the corresponding differences for the Pd(II) complexes, which are also planar. If the differences are interpreted as arising from orientation polarization, they lead to dipole moments of 1·77 and 1·86 D for the *N*-n-propyl- and *N*-n-butylsalicylaldimine complexes, respectively. These values are independent of the solvent involved and of the concentrations of the solutions studied, this suggesting that the moments do not arise from the presence of associated species[47–49]. Their origin is as yet by no means clear. They may arise from a relatively slight departure from planarity in solution, not necessarily of the pseudo-tetrahedral type*. This would accord with the absence of the characteristic transition of the pseudo-tetrahedral form in the absorption spectra. Alternatively, however, such relatively low moments would arise if a low percentage of the pseudo-tetrahedral form were in equilibrium with the *trans* planar form. This had previously been suggested[50] to account for the large difference between the molar polarization and molar refraction of the Cu(II) complex of *N*-phenylsalicylaldimine.

The Cu(II) complexes of *N*-isopropyl- and *N*-*s*-butylsalicylaldimine, on the other hand, are isomorphous with the corresponding complexes of Ni(II), Zn(II) and Co(II), all of which are known to be tetrahedral[39,40]. Their magnetic moments, and that of the *N*-*t*-butylsalicylaldimine complex, lie in the range 1·89 to 1·92 B.M. They are therefore significantly higher than the values (1·83–1·86 B.M.) observed for the planar complexes with *N*-n-alkyl substituents, but are less than

* A pseudo-tetrahedral structure can be pictured as being attained when the bonds to the central atom in a planar complex are displaced alternately above and below the plane. In the limit if each becomes displaced by 45° a regular tetrahedral disposition of the bonds is attained, but in the pseudo-tetrahedral structures the displacements are much smaller.

the theoretical value of about 2·2 B.M. for a perfectly tetrahedral Cu(II) complex[51].

Spectroscopic evidence indicates that these complexes with branched chain alkyl substituents also have the same structures in solution as in the solid state, but their apparent dipole moments (2·72, 2·70 and 3·38 D for the *N*-isopropyl-, *N*-*s*-butyl- and *N*-*t*-butyl-derivatives, respectively) are still appreciably lower than the values for the corresponding Co(II) and Zn(II) complexes[48].

C. 2-Hydroxyacetophenonimine Complexes

Some studies have also been made of these 7-methyl derivatives of the salicylaldimine complexes. When the greenish-yellow complex of Ni(II) with 2-hydroxyacetophenone is heated in concentrated aqueous ammonia, it is converted into the deep red bis(2-hydroxyacetophenonimino)nickel (**6**). This complex and its *N*-hydroxy and *N*-methyl derivatives as initially formed are diamagnetic, but in refluxing in

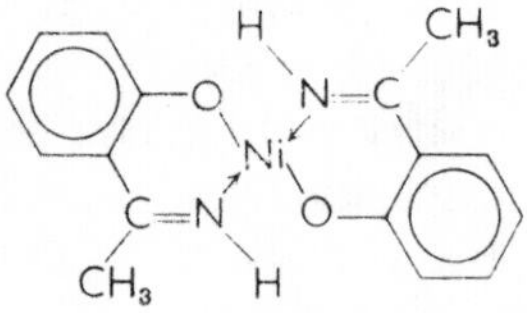

H CH$_3$ N=C O Ni O C=N CH$_3$ H

(6)

biphenyl at 254° they pass into paramagnetic isomers[53]. These isomers dissolve readily in chloroform, and on crystallizing from this solvent yield the diamagnetic forms again. Their magnetic susceptibilities do not follow the Curie law, but their apparent magnetic moments increase from 1·3–1·6 B.M. at about 80°K to 2·5–3·0 B.M. at 350°K and approximate to the behaviour expected for tetrahedral Ni(II). They differ, therefore, appreciably from the corresponding salicylaldimine complexes, but the reason for the decreased stability of the planar forms is not clear, as there should be no steric strain.

D. 2-Aminobenzylideneimine and Related Complexes

A few complexes have been prepared from the Schiff bases of 2-aminobenzaldehyde. The Ni(II) and Cu(II) derivatives have been reported to be slightly soluble and intensely coloured solids of the structure **7**, the Ni(II) compound being diamagnetic[54].

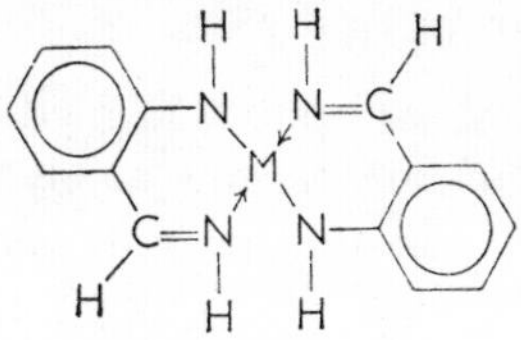

(7)

Compounds of an analogous type are formed by the self-condensation of 2-aminobenzaldehyde in absolute alcohol and in the presence of Ni^{2+} or Cu^{2+} ions. The products so obtained have been shown[55] to be mixtures from which complexes of the type **8** could be isolated as

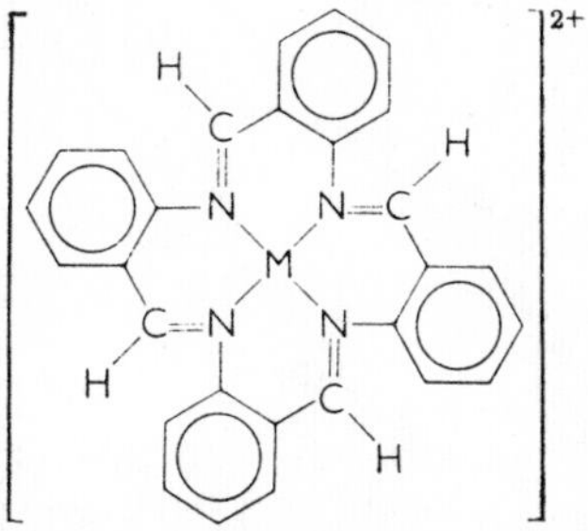

(8)

salts. Although attacked by bases, these complexes are unaffected by boiling mineral acids, even concentrated nitric acid, and the nickel is not precipitated by dimethylglyoxime. The perchlorate, tetrafluoroborate and tetraphenylborate of the nickel complex are all diamagnetic, whereas the iodide, nitrate and thiocyanate are all fully paramagnetic, with magnetic moments of 3·2 B.M. The chloride and bromide, on the other hand, have the intermediate magnetic moments of 1·68 and 1·47 B.M. respectively at room temperature. The magnetic susceptibilities of the two latter salts, however, do not obey the Curie law, and their variations with temperature have been interpreted in terms of thermal distribution between a singlet ground state, with no unpaired electrons, and a triplet excited state with two unpaired electrons. The enthalpy differences between these two states have been calculated to be 800 and 700 cal/mole for the chloride and bromide respectively[56]. The reasons why the nature of the anions should have such a profound effect upon the structure of the cation are still by no means clear, however.

E. Glyoxime Complexes

Probably the best known complexes formed by compounds containing the azomethine group are the metal glyoximates, including the deep red bis(dimethylglyoximato)nickel which has now been used for many years for detecting and determining nickel. After being first formulated with a very improbable seven-membered ring structure, later revised to a more reasonable six-membered ring, it is now known to have the five-membered ring structure **9** with hydrogen bonds linking the oxygen atoms of the two rings.

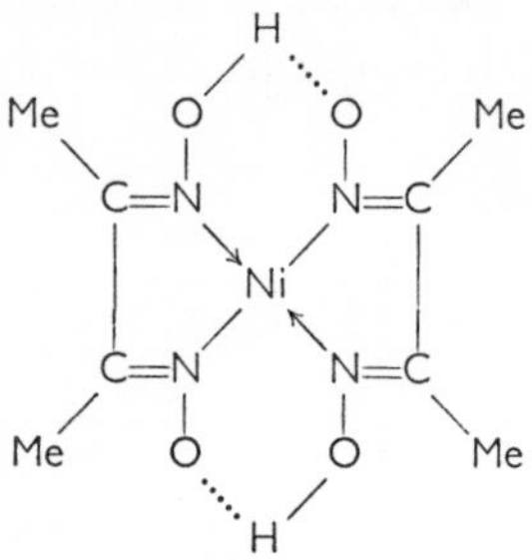

(**9**)

The five-membered ring structure was first suggested by Pfeiffer[57], on the basis of the observation that similar compounds to the glyoxime complexes were formed when one of the oxime groups was replaced by an imino or methylimino group.

The Ni(II) complex is insoluble in water and is only slightly soluble in chloroform. This is attributed to the presence of the polar hydrogen bridges, since the corresponding monomethyl ether of dimethylglyoxime leads to complexes insoluble in water but very soluble in chloroform[58]. This low solubility is undoubtedly a factor in causing these glyoxime complexes to be amongst the most stable coordination compounds formed by virtue of the presence of the azomethine group.

It was shown by x-ray diffraction measurements that the O—O distance in the Ni(II) complex with dimethylglyoxime is very short (2·44 Å), an observation which led Godycki and Rundle[59] to the suggestion that the hydrogen bond in its molecule may be symmetrical, with the hydrogen atom equidistant between the two oxygen atoms. This was in apparent agreement with the observations which had been made on the infrared spectrum of this compound. Rundle and Parasol[60] found that this contained a weak band at 1775 cm^{-1} which disappeared on deuteration, and which they therefore attributed to the O—H stretching mode. As the frequency attributable to this mode in the free glyoxime occurred at about 3100 cm^{-1}, they regarded this very large frequency shift as evidence for very strong intramolecular hydrogen bonding

This inference has been questioned, however, by Blinc and Hadzi[61], who have found another band at 2350 cm^{-1} in the infrared spectrum of the nickel dimethylglyoxime complex, and consider this to be attributable to the O—H stretching frequency, whilst the other band at 1780 cm^{-1} arises from the O—H bending mode. In the deuterated complex the corresponding bands were found at 1810 and 1910 cm^{-1} and at 1265 cm^{-1} respectively. Similarly the absorptions at 2340 and 1710 cm^{-1} for the dimethylglyoxime complex of Pd(II) were assigned to the O—H stretching and bending frequencies respectively. In sodium dimethylglyoximate, on the other hand, these bands occurred at 3020 and 1650 cm^{-1}, respectively, and were shifted to 2340 and 1237 cm^{-1} on deuteration. On this evidence Blinc and Hadzi suggested that the O—H···O bond may be bent, and that it is probably not symmetrical.

In contrast, the O—O distance in Pt(II) dimethylglyoximate has been found to be 3·03 Å and an infrared absorption at 3450 cm^{-1} is attributable to the O—H stretching frequency, so it appears that there is no hydrogen bonding in its molecule.

The tris(dimethylglyoximino) complex of Co(III) has an octahedral structure. Hence, as would be expected, its infrared spectrum does not include the band around 1780 cm^{-1} such as is observed with the Ni(II) and Pd(II) complexes, since it cannot form intramolecular hydrogen bonds[62]. On the other hand, the ions of the type Co(dimethylglyoxime)$_2$XY$^-$, where X and Y are Cl, Br or NO_2, all show absorptions in the region 1680–1770 cm^{-1} and so presumably contain such bonds[63]. The groups X and Y must therefore occupy *trans* positions in all these compounds.

Similar compounds are also formed by Rh(III) and Ir(III) and these have been very extensively investigated[64].

VI. REFERENCES

1. J. W. Smith in *The Chemistry of the Amino Group* (Ed. S. Patai), Interscience, London, 1968, p. 163.
2. J. B. Culbertson, *J. Am. Chem. Soc.*, **73**, 4818 (1951).
3. J. T. Edward and S. C. R. Meacock, *J. Chem. Soc.*, 2009 (1957).
4. J. K. Wood, *J. Chem. Soc.*, **83**, 568 (1903).
5. E. H. Cordes and W. P. Jencks, *J. Am. Chem. Soc.*, **85**, 2843 (1963).
6. E. H. Cordes and W. P. Jencks, *J. Am. Chem. Soc.*, **84**, 832 (1962).

7. J. W. Smith in *The Chemistry of the Amino Group* (Ed. S. Patai), Interscience, London, 1968, p. 185.
8. J. Weinstein and E. McImrich, *J. Am. Chem. Soc.*, **82**, 6064 (1960).
9. Y. Okamoto and H. C. Brown, *J. Org. Chem.*, **22**, 485 (1957).
10. R. H. Holm, G. W. Everett, Jr. and A. Chakravorthy, *Progress in Inorganic Chemistry* (Ed. F. A. Cotton), Vol. 7, Interscience, New York, 1966, pp. 83–214.
11. L. Sacconi, *Coord. Chem. Rev.*, **1**, 126; 192 (1966).
12. H. Schiff, *Ann. Chem.*, **150**, 193 (1869).
13. J. M. Stewart and E. C. Lingafelter, *Acta Cryst.*, **12**, 842 (1959).
14. M. v. Stackelberg, *Z. Anorg. Allgem. Chem.*, **253**, 136 (1947).
15. S. H. Simonsen and C. E. Pfluger, *Acta Cryst.*, **10**, 471 (1957).
16. G. N. Tischenko, P. M. Zorkii and M. A. Porai-Koshits, *Zh. Strukt. Khim.*, **2**, 434 (1961).
17. L. Sacconi, P. Paoletti and M. Ciampolini, *J. Am. Chem. Soc.*, **85**, 411 (1963).
18. E. Frasson, C. Panattoni and L. Sacconi, *J. Phys. Chem.*, **63**, 1908 (1959).
19. E. Frasson, C. Panattoni and L. Sacconi, *Acta Cryst.*, **17**, 85, 477 (1964).
20. H. E. Clark and A. L. Odell, *J. Chem. Soc.*, 3431 (1955).
21. F. Basolo and W. R. Matoush, *J. Am. Chem. Soc.*, **75**, 5663 (1953).
22. R. H. Holm, *J. Am. Chem. Soc.*, **83**, 4683 (1961).
23. L. Sacconi, P. Paoletti and R. Cini, *J. Am. Chem. Soc.*, **80**, 3583 (1958); *J. Inorg. Nucl. Chem.*, **8**, 492 (1958).
24. J. Ferguson, *Spectrochim. Acta*, **17**, 316 (1961).
25. L. Sacconi, R. Cini, M. Ciampolini and F. Maggio, *J. Am. Chem. Soc.*, **82**, 3487 (1960).
26. L. Sacconi, M. Ciampolini and G. P. Speroni, *J. Am. Chem. Soc.*, **87**, 3102 (1965).
27. C. M. Harris, S. L. Lenzer and R. L. Martin, *Australian J. Chem.*, **11**, 331 (1958).
28. H. C. Clark and R. J. O'Brien, *Can. J. Chem.*, **39**, 1030 (1961).
29. R. H. Holm and T. M. McKinney, *J. Am. Chem. Soc.*, **82**, 5506 (1960).
30. R. H. Holm and K. Swaminathan, *Inorg. Chem.*, **2**, 181 (1963).
31. L. Sacconi, M. Ciampolini and N. Narda, *J. Am. Chem. Soc.*, **86**, 819 (1964).
32. R. H. Holm, A. Chakravorty and G. O. Dudek, *J. Am. Chem. Soc.*, **86**, 379 (1964).
33. D. R. Eaton, W. D. Phillips and D. J. Caldwell, *J. Am. Chem. Soc.*, **85**, 397 (1963).
34. L. Wei, R. M. Stogsdill and E. C. Lingafelter, *Acta Cryst.*, **17**, 1058 (1964).
35. R. H. Holm and K. Swaminathan, *Inorg. Chem.*, **1**, 599 (1962).
36. L. Sacconi and M. Ciampolini, *J. Am. Chem. Soc.*, **85**, 1750 (1963).
37. H. Nishiwaka and S. Yamada, *Bull. Chem. Soc. Japan*, **37**, 8, 1154 (1964).
38. E. Frasson and C. Panattoni, *Z. Krist.*, **116**, 154 (1961).
39. L. Sacconi and P. L. Oriole, *Ric. Sci. Rend.*, **32**, 645 (1962).
40. L. Sacconi, P. L. Oriole, P. Paoletti and M. Ciampolini, *Proc. Chem. Soc.*, 256 (1962).
41. L. Sacconi, M. Ciampolini, F. Maggio and F. P. Cavasino, *J. Am. Chem. Soc.*, **84**, 3246 (1962).

42. B. O. West, *Nature*, **165**, 122 (1950); **173**, 1187 (1954); *J. Chem. Soc.*, 3115 (1952); 1374 (1962).
43. B. O. West, *J. Chem. Soc.*, 4944 (1960).
44. M. Ciampolini, F. Maggio and F. B. Cavasino, *Inorg. Chem.*, **3**, 1188 (1964).
45. A. Chakrovorthy and R. H. Holm, *Inorg. Chem.*, **3**, 1521 (1964).
46. E. Frasson, C. Panattoni and L. Sacconi, *Gazz. Chim. Ital.*, **92**, 1470 (1962).
47. R. G. Charles and H. Freiser, *J. Am. Chem. Soc.*, **73**, 5223 (1951).
48. V. A. Kogan, D. A. Osipov, V. I. Minkin and M. I. Gorelov, *Dokl. Akad. Nauk SSSR*, **153**, 594 (1963).
49. L. Sacconi, M. Ciampolini, F. Maggio and F. P. Cavasino, *J. Inorg. Nucl. Chem.*, **19**, 73 (1961).
50. J. Macqueen and J. W. Smith, *J. Chem. Soc.*, 1821 (1956).
51. B. N. Figgis, *Nature*, **182**, 1568 (1958).
52. L. Sacconi, P. Paoletti and G. Del Re, *J. Am. Chem. Soc.*, **79**, 4062 (1957).
53. C. M. Harris, S. L. Lenzer and R. L. Martin, *Australian J. Chem.*, **14**, 420 (1961).
54. P. Pfeiffer, T. Hesse, H. Pfitzinger, W. Scholl and H. Thielert, *J. Prakt. Chem.*, **150**, 261 (1938).
55. G. A. Milson and D. H. Busch, *J. Am. Chem. Soc.*, **86**, 4830, 4834 (1964).
56. S. L. Holt, Jr., R. C. Bouchard and R. L. Carlin, *J. Am. Chem. Soc.*, **86**, 519 (1964).
57. P. Pfeiffer, *Ber.*, **63**, 1881 (1930).
58. F. Feigl, *Chemistry of Specific Selective and Sensitive Reactions*, Academic Press, New York, 1949, p. 408.
59. L. E. Godycki and R. E. Rundle. *Acta Cryst.*, **6**, 487 (1953).
60. R. E. Rundle and M. Parasol, *J. Chem. Phys.*, **20**, 1487 (1952).
61. R. Blinc and D. Hadzi, *J. Chem. Soc.*, 4536 (1958); *Spectrochim. Acta*, **16**, 853 (1960).
62. A. Nakahara, J. Fujita and R. Tsuchda, *Bull. Chem. Soc. Japan*, **29**, 296 (1956).
63. J. Fujita, A. Nakahara and R. Tsuchda, *J. Chem. Phys.*, **23**, 1541 (1955).
64. F. P. Dwyer and R. N. Nyholm, *J. Proc. Roy. Soc. N.S.W.*, **75**, 127 (1941); **76**, 133, 275 (1942; **77**, 116 (1943); **78**, 266 (1944).

CHAPTER 6

Additions to the azomethine group

KAORU HARADA

University of Miami, Florida, U.S.A.

I. INTRODUCTION

Azomethines, which include aldimines, RCH=NR′, and ketimines, RR′C=NR″, are considered to be analogues of carbonyl compounds. Many of the chemical properties of azomethines are indeed similar to those of carbonyl compounds. The addition reactions of azomethines are mainly composed of reactions in which a variety of reagents add to the polarized >C=N— double bond ($>\overset{\delta+}{C}=\overset{\delta-}{N}—$). Therefore, nucleophilic reagents attack the carbon atom of the azomethine linkage. In the reaction of alkyl halides with azomethines, the alkyl group attaches to the nitrogen atom of the azomethine group. On the other hand, electrophilic reagents such as the Grignard reagent may react with the azomethines so that the alkyl or aryl group of the Grignard reagent attaches to the carbon atom of the azomethine linkage. Hydrogenation of the azomethines to the corresponding secondary amines may easily be carried out in several ways which are usually difficult in the carbonyl compounds.

So-called 'Schiff bases' (*N*-substituted imines) are treated in this article primarily. Some of the oximes and hydrazones are described but these are not emphasized. The stereochemistry of the azomethines in the hydrogenation reaction is described. Some of the addition reactions of immonium salts $>C=\overset{+}{N}<$ are also described. Cycloaddition reactions of the azomethines, such as the reactions with diazomethane, isocyanate, isothiocyanate, ketene, peroxy acid, etc., will be described in Chapter 7.

II. ADDITION REACTIONS

A. Addition Reactions Involving Organic and Inorganic Reagents

I. Hydrogen cyanide

Addition reactions of hydrogen cyanide to the carbon–nitrogen double bond of the azomethine linkage yield amino nitriles, in which the nitrile groups attach to the carbon atom of the azomethine linkage[1].

Hydrogen cyanide reacts with the trimer of methyleneaminoacetonitrile, $(CH_2{=}NCH_2CN)_3$, in the presence of hydrochloric acid to form iminodiacetonitrile (equation (1)[2]. The imine reacts with

$$(CH_2{=}NCH_2CN)_3 + 3\,HCN \longrightarrow 3\,NH(CH_2CN)_2 \tag{1}$$

hydrogen cyanide to form an α-amino nitrile[3-5]. Through the addition of hydrogen cyanide, benzophenoneimine gives α-aminodiphenylacetonitrile (equation 2)[3,5], and fluorenoneimine yields the corresponding amino nitrile (equation 3)[6]. Ethylenebis(*o*-hydroxyphenyl)azomethine is converted to *N*,*N'*-alkylenediaminephenyl-

$$(C_6H_5)_2C{=}NH \xrightarrow{HCN} (C_6H_5)_2CH(NH_2)CN \qquad (2)$$

$$\text{fluorenone imine } (C{=}NH) \longrightarrow \text{9-amino-9-cyanofluorene } (H_2N, CN) \qquad (3)$$

$$o\text{-}HOC_6H_4CH{=}N(CH_2)_2N{=}CHC_6H_4OH\text{-}o \longrightarrow o\text{-}HOC_6H_4CH(CN)NH(CH_2)_2NHCH(CN)C_6H_4OH\text{-}o \qquad (4)$$

acetonitrile[7] by addition of hydrogen cyanide (equation 4). These reactions are usually carried out in ether[6] or in benzene[5] under anhydrous conditions. However, an aqueous hydrogen cyanide–pyridine system has also been used successfully (equation 5)[8].

$$(CH_3)_2C{=}NOH \xrightarrow[\text{pyridine}]{HCN,H_2O} (CH_3)_2C(NHOH)CN \qquad (5)$$

Yields resulting from the hydrogen cyanide addition reactions to the azomethine compounds are improved by the use of a sodium cyanide–phosphate buffer system instead of liquid hydrogen cyanide[9,10]. One other advantage of this method is that of the lessened danger. Acetone oxime is converted to the corresponding α-hydroxylamino nitrile by the use of this system (equation 6)[11]. $\Delta^{5(10)}$Dehydroquinolizidinium salts (**1**) and $\Delta^{4(9)}$hexahydropyrrocolinium salts (**2**) react with

$$(CH_3)_2C{=}NOH + NaCN \xrightarrow{\text{phosphate buffer}} (CH_3)_2C(NHOH)CN \qquad (6)$$

potassium cyanide to yield 10-cyanoquinolizidine (**3**), (equation 7) and 9-cyanooctahydropyrrocoline (**4**), (equation 8)[12].

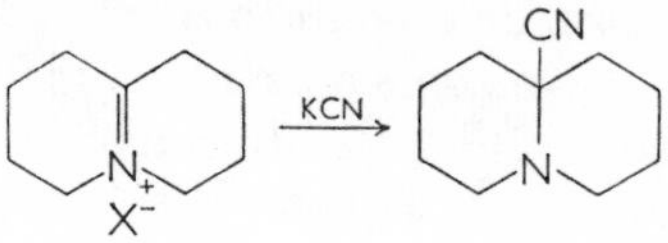

$$\text{(1)} \xrightarrow{KCN} \text{(3)} \tag{7}$$

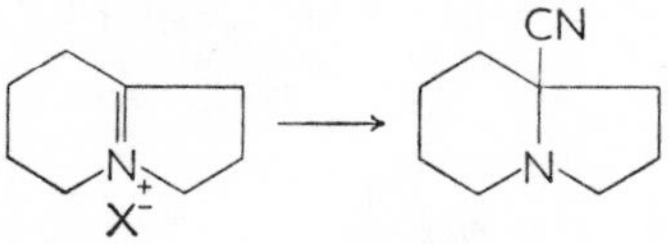

$$\text{(2)} \longrightarrow \text{(4)} \tag{8}$$

2. Sodium hydrogen sulphite

Sodium bisulphite reacts with azomethine compounds to yield the addition product **5**, a sodium α-aminosulphonate (equation 9)[13,14]. The adduct salt **5** is stable in aqueous solution, but it decomposes in

$$C_6H_5CH{=}NR' \xrightarrow{NaHSO_3} \underset{\substack{|\\ SO_3Na \\ (\mathbf{5})}}{C_6H_5CHNHR'} \tag{9}$$

boiling water, and is unstable in alkali solutions. By treatment with hydrochloric acid, salt **5** can be converted to the corresponding free α-aminosulphonic acid, which is stable under acidic conditions. The free aminosulphonic acid has an internal salt structure and is represented in equation (10):

$$\underset{\substack{|\\ SO_3H}}{ArCHNHR} \rightleftarrows \underset{\substack{|\\ SO_3^-}}{ArCH\overset{+}{N}H_2R} \tag{10}$$

Several α-aminosulphonic acids have thus been prepared (equations 11, 12)[13–15].

$$n\text{-}C_3H_7CH{=}NC_6H_5 \xrightarrow{NaHSO_3} \underset{\substack{|\\ NHC_6H_5}}{n\text{-}C_3H_7CHSO_3Na} \tag{11}$$

$$C_6H_5CH{=}NC_6H_5 \xrightarrow{NaHSO_3} \underset{\substack{|\\ NHC_6H_5}}{C_6H_5CHSO_3Na} \tag{12}$$

It seems worthwhile to describe here the chemical properties of the α-aminosulphonic acid as a reactive intermediate[15]. The sodium aminosulphonate (**5**) reacts with aniline to form the aryl amide of an α-arylaminoalkanesulphonic acid in good yield (equation 13)[15].

Potassium cyanide reacts with the resulting sulphonamide to form α-arylamino nitrile (equation 14)[15,16]. Diethyl malonate reacts with

$$\underset{\textbf{(5)}}{\underset{|}{\mathrm{RCHSO_3Na}}\atop\mathrm{NHAr}} \xrightarrow{\mathrm{ArNH_2}} \underset{|}{\mathrm{RCHSO_2NHAr}}\atop\mathrm{NHAr} \qquad (13)$$

$$\underset{|}{\mathrm{RCHSO_2NHAr}}\atop\mathrm{NHAr} \xrightarrow{\mathrm{KCN}} \underset{|}{\mathrm{RCHCN}}\atop\mathrm{NHAr} \qquad (14)$$

5 to form β-aryl-β-arylaminomethyl malonic esters (**6**), (equation 15)[15]. Similarly, **5** reacts with ethyl acetoacetate and acetylacetone to form β-aryl-β-arylaminomethyl acetoacetic ester (**7**), (equation 16), and β-aryl-β-arylaminomethyl acetylacetone (**8**), (equation 17)[15].

$$\mathrm{C_6H_5CH(NHAr)SO_3Na}\ \textbf{(5)} \xrightarrow{\mathrm{H_2C(COOC_2H_5)_2}} \mathrm{C_6H_5CH(NHAr)CH(COOC_2H_5)_2}\ \textbf{(6)} \qquad (15)$$

$$\textbf{(5)} \xrightarrow{\mathrm{CH_3COCH_2COOC_2H_5}} \mathrm{C_6H_5CH(NHAr)CH(COOC_2H_5)(COCH_3)}\ \textbf{(7)} \qquad (16)$$

$$\textbf{(5)} \xrightarrow{\mathrm{CH_3COCH_2COCH_3}} \mathrm{C_6H_5CH(NHAr)CH(COCH_3)_2}\ \textbf{(8)} \qquad (17)$$

3. Alkyl halide

The alkylation of Schiff bases with alkyl halides results in a quaternary immonium salt (**9**), which is converted to a secondary amine upon hydrolysis (equation 18)[17]. The alkylation is usually applied to the synthesis of secondary amines, and this method is known as the 'Decker alkylation method'. In this method of alkylation of primary amines, the addition products **9** are usually hydrolysed to secondary

$$\mathrm{C_6H_5CH{=}NR} \xrightarrow{\mathrm{R'X}} \underset{\textbf{(9)}}{\left[\mathrm{C_6H_5CH{=}\overset{+}{N}R(R')}\right]\mathrm{X^-}} \xrightarrow{\mathrm{H_2O}} \mathrm{RNHR'} + \mathrm{C_6H_5CHO} \qquad (18)$$

amines without isolation. Benzylidene derivatives are often used as the Schiff bases. Yields are satisfactory when R′ is a methyl group.

However, when larger alkyl groups are introduced, results are less satisfactory[18]. Intermediate immonium salts (**9**), which are reactive to various reagents, can also be prepared by various ways[19,20].

Allylamine is methylated by this method[21]. *N*-Benzylideneallylamine is heated with methyl iodine in a sealed tube. The resulting solid adduct is hydrolysed with water. Allylmethylamine is obtained in 71% yield (equation 19). *N*-Benzylidene-β-methyl-β-cyclohexylethylamine and methyl iodide yield *N*-β-dimethyl-β-cyclohexyl-ethylamine[22]. β-Phenylpropylalkylamines are prepared in the same way[18].

$$CH_2{=}CHCH_2N{=}CHC_6H_5 \xrightarrow{CH_3I} \left[CH_2{=}CHCH_2\overset{+}{N}(CH_3){=}CHC_6H_5\right]I^- \xrightarrow{H_2O} CH_2{=}CHCH_2NHCH_3 \quad (19)$$

4. Thiol, hydrogen sulphide

A Schiff base reacts with thioglycolic acid under refluxing with benzene to yield a 5-membered cyclized product, 4-thiazolidone (equation 20)[23-25]. An intermediate addition product, aminothioether, may be isolated. Benzophenone anil or benzaldehyde anil are reduced by *p*-thiocresol upon heating (equations 21, 22)[26]. Under suitable conditions benzaldehyde anil has been found to form addition products with thiol (equation 23)[27]. The addition products decompose easily by addition of dilute sodium hydroxide solution to form the

$$C_6H_5CH{=}N{-}C_6H_4(COOH) \xrightarrow{HSCH_2COOH} \left[C_6H_5CH(SCH_2COOH){-}NH{-}C_6H_4(COOH)\right] \longrightarrow C_6H_5\text{-}\underset{S-CH_2-C(=O)}{HC{-}N}\text{-}C_6H_4(COOH) \quad (20)$$

$$(C_6H_5)_2C{=}NC_6H_5 \xrightarrow{2\,p\text{-}CH_3C_6H_4SH} (C_6H_5)_2CH{-}NHC_6H_5 + (p\text{-}CH_3C_6H_4S{-})_2 \quad (21)$$

$$C_6H_5CH{=}NC_6H_5 \xrightarrow{2\,p\text{-}CH_3C_6H_4SH} C_6H_5CH{-}NHC_6H_5 + (p\text{-}CH_3C_6H_4S{-})_2 \quad (22)$$

original thiol and the Schiff base. The reduction of the addition product occurs as readily as does the direct reduction of anil with *p*-toluenethiol.

$$C_6H_5CH{=}NC_6H_5 \xrightarrow[\text{addition}]{p\text{-}CH_3C_6H_4SH} C_6H_5CH(SC_6H_4CH_3\text{-}p)NHC_6H_5$$

$$C_6H_5CH{=}NC_6H_5 \xrightarrow{2\,p\text{-}CH_3C_6H_4SH} C_6H_5CH_2NHC_6H_5 + (p\text{-}CH_3C_6H_4S)_2 \xleftarrow{p\text{-}CH_3C_6H_4SH} C_6H_5CH(SC_6H_4CH_3\text{-}p)NHC_6H_5 \quad (23)$$

Various addition reactions of thiols with *N*-benzylideneanthranilic acid have been studied[28]. Most of the thiols have given excellent yields of the corresponding addition products (Table 1).

$$C_6H_5CH{=}N\text{-}C_6H_4\text{-}COOH \xrightarrow{RSH} C_6H_5CH(SR)NH\text{-}C_6H_4\text{-}COOH \quad (24)$$

Carbon–nitrogen double bonds which are parts of aromatic systems such as pyridine, quinoline, isoquinoline and benzothiazole do not react with *p*-thiocresol[29]. Aryl thiols reduce the carbon–carbon double bond of benzalquinolidine to 2-(β-phenylethyl)quinoline (equation 25).

$$2\text{-}(C_6H_5CH{=}CH)C_9H_6N \xrightarrow{2\,p\text{-}CH_3C_6H_4SH} 2\text{-}(C_6H_5CH_2CH_2)C_9H_6N + (p\text{-}CH_3C_6H_4S)_2 \quad (25)$$

TABLE 1. The addition of thiols to *N*-benzylidene-anthranilic acid[28].

R	Yield of addition product (%)
CH_2COOH	92
CH_2CH_2COOH	99
C_6H_5	91
p-MeC_6H_5	86
$C_6H_5CH_2$	70
i-Pr	57
i-Bu	62
n-Bu	49
t-Bu	88

Hydrogen sulphide easily adds to ketimine in ether at relatively low temperatures ($-40°$ to $0°c$) to form *gem*-dithiol (equation 26)[30]. It could be assumed that an intermediate addition product might be

$$\begin{matrix} R \\ \quad\diagdown \\ \quad C{=}NR'' \\ \quad\diagup \\ R' \end{matrix} \xrightarrow{H_2S} \left(\begin{matrix} R \\ \quad\diagdown \\ \quad C{-}NHR'' \\ \quad\diagup\ | \\ R'\ \ SH \end{matrix} \right) \xrightarrow{H_2S} \begin{matrix} R \quad SH \\ \diagdown\ \diagup \\ C \\ \diagup\ \diagdown \\ R' \quad SH \end{matrix} + R''NH_2 \qquad (26)$$

formed during the reaction. The reaction is analogous to that of anil and thiol to form aniline and *gem*-dithioether (equation 27)[27]. $\Delta^{5(10)}$De-hydroquinolizidinium salt reacts with thiolates to yield thioethers (equation 28)[31].

$$C_6H_5CH{=}N{-}C_6H_4{-}COOH \xrightarrow{2\,p\text{-}CH_3C_6H_4SH} C_6H_5CH(SC_6H_4CH_3\text{-}p)_2 + H_2N{-}C_6H_4{-}COOH \qquad (27)$$

$$\Delta^{5(10)}\text{-dehydroquinolizidinium } X^- \xrightarrow{p\text{-}CH_3C_6H_4SK} \text{10-}(SC_6H_4CH_3\text{-}p)\text{quinolizidine} \qquad (28)$$

5. Acetophenone

N-Benzylidenemethylamine reacts with nitroacetophenone in ether in the presence of acetic anhydride to form a nitro compound (**10**) in 48% yield (equation 29)[32]. Schiff bases condense with acetophenone or its derivatives to yield amino ketones (**11**) with an amine hydro-

$$C_6H_5CH{=}NCH_3 + C_6H_5COCH_2NO_2 \xrightarrow[\text{ether}]{Ac_2O} \underset{\textbf{(10)}}{C_6H_5CO\underset{\displaystyle NO_2}{\underset{|}{C}}{=}CHC_6H_5} \qquad (29)$$

chloride as a catalyst (equation 30)[33–36]. The reaction may proceed in three ways depending on the chemical nature of the reactants (equation 31)[35].

$$p\text{-}C_6H_5C_6H_4N{=}CHC_6H_5 + CH_3COC_6H_5 \xrightarrow[p\text{-}C_6H_5C_6H_4N{=}CHC_6H_5\cdot HCl]{EtOH} \underset{\textbf{(11)}}{p\text{-}C_6H_5C_6H_4NH\underset{\displaystyle C_6H_5}{\underset{|}{C}}HCH_2COC_6H_5} \qquad (30)$$

When the aryl group contains a nitro group in the *m*- or *p*-position, the reaction results in an amino ketone (**11**) (equation 31a). When the

$$ArN{=}CHAr' + MeCOAr'' \xrightarrow[EtOH]{HCl} ArNHCH(Ar')CH_2COC_6H_5 \quad (11) \qquad (31a)$$

$$ArN{=}CHAr' + MeCOAr'' \longrightarrow Ar'CH{=}CHCOAr'' + ArNH_2 \quad (10) \qquad (31b)$$

$$ArN{=}CHAr' + MeCOAr'' \longrightarrow \text{Schiff base unchanged} \qquad (31c)$$

Ar′ contains a nitro group in the *m*- or *p*-position, the product is an α,β-unsaturated ketone (**10**) by elimination of aniline (equation 31b). When the Schiff base is benzal-*p*-anisidine or benzal-*p*-phenetidine, it is recovered unchanged (equation 31c). A recent study[37] of this type of reaction has shown that the uncatalysed addition of ketones to Schiff bases reported in the literature[38,39] could not be repeated. The addition reaction can be brought about by addition of a small amount of hydrochloric acid (equation 32). Adduct **11** decomposes to

$$C_6H_5CH{=}NC_6H_5 + CH_3COC_6H_5 \xrightarrow{H^+} \underset{(\mathbf{11})}{C_6H_5CH(NHC_6H_5)CH_2COC_6H_5} \xrightarrow[H_2SO_4]{AcOH} \underset{(\mathbf{10})}{C_6H_5CH{=}CHCOC_6H_5} + C_6H_5NH_2 \qquad (32)$$

liberate amine and forms an unsaturated ketone (**10**) in a solution of glacial acetic acid or concentrated sulphuric acid.

6. Phenylacetic ester

Reaction of Schiff bases with ethyl phenylacetate in the presence of anhydrous aluminium chloride yields an ester of α-phenyl-β-aryl-β-anilinopropionic acid (equation 33)[40]. Several similar studies of this

$$C_6H_5CH_2COOEt + C_6H_5CH{=}NC_6H_5 \xrightarrow[benzene]{AlCl_3} C_6H_5NHCH(C_6H_5)CH(C_6H_5)COOEt \qquad (33)$$

type of reaction have been made[41–44]. The method has possible application as a general way to synthesize α,β-diaryl-β-aminopropionic acid.

$$C_6H_4CH_2COOEt + C_6H_5CH{=}N{-}R \xrightarrow{AlCl_3} RNHCH(C_6H_5){-}CH(C_6H_4X)COOEt \quad \text{(refs. 2,5)} \qquad (34)$$

$$X = p\text{-}CH_3 \text{ or } p\text{-}NO_2$$

Without the use of phenylacetic acid ester, the Schiff base reacts with sodium lithium phenylacetate under similar conditions to form

β-arylaminopropionic acids in 74% yield[45]. The Schiff base reacts with free phenylacetic acid without any catalyst by heating at 100°c to yield α,β-diaryl-β-aminopropionic acid in 75% yield[46]. In a similar way α-lithiotoluenesulphonic acid reacts with a Schiff base to form 1,2-diphenyl-2-anilinoethanesulphonic acid (equation 35)[47].

$$C_6H_5CH_2SO_3Na + Li, PhBr \xrightarrow[\text{ether}]{C_6H_5CH{=}NC_6H_5} C_6H_5NHCH(C_6H_5){-}CH(C_6H_5)SO_3Na \quad (35)$$

7. Benzoyl cyanide

Benzophenoneimine reacts with benzoyl cyanide to form the addition product α-benzoylaminodiphenylacetonitrile (equation 36)[48].

$$(C_6H_5)_2C{=}NH \xrightarrow[\text{ether}]{C_6H_5COCN} (C_6H_5)_2C(CN)NHCOC_6H_5 \quad (36)$$

The reaction is analogous to that of hydrogen cyanide with a Schiff base. The Schiff base also forms the addition product with benzoyl cyanide[48].

8. Trihaloacetic acid

Trihaloacetic acids react with Schiff bases derived from ethylenediamine, aniline and cyclohexylamine in benzene or in toluene to yield trihaloaminoethanes by elimination of carbon dioxide[49,50]. A mechanism for this reaction has been proposed which consists of an intermediate formation and subsequent decarboxylation of the esters of α-amino alcohols (equation 37)[50].

$$-\overset{|}{C}{=}NH \xrightarrow{X_3CCOOH} -\overset{|}{C}(O{-}C({=}O){-}CX_3)NH- \longrightarrow -CH(CX_3)NH- + CO_2 \quad (37)$$

$$C_6H_5CH{=}NC_6H_5 \xrightarrow{Cl_3CCOOH} Cl_3CCH(C_6H_5)NHC_6H_5 \ (68\%) \quad (38)$$

$$(C_6H_5CH{=}NCH_2{-})_2 \xrightarrow{Cl_3CCOOH} (Cl_3CCH(C_6H_5)NHCH_2{-})_2 \ (60\%) \quad (39)$$

9. Aromatic aldehyde

When a Schiff base is mixed with an aromatic aldehyde in absolute alcohol or in absolute alcohol–toluene, the aldehyde component of the

Schiff base is exchanged. The ease of exchange of aromatic aldehyde to the Schiff base is in the order of [51]

$$o\text{-}HOC_6H_5CHO > p\text{-}HOC_6H_4CHO > C_6H_5CHO$$

When 2-methylpropanal is used, benzylideneaniline reacts with the aldehyde to yield the addition product 3-phenyl-3-anilino-2,2-dimethylpropylideneaniline (equation 40) [52].

$$C_6H_5CH{=}NC_6H_5 \xrightarrow[\text{alcohol}]{(CH_3)_2CHCHO} C_6H_5CH(NHC_6H_5)C(CH_3)_2CH{=}NC_6H_5 \quad (40)$$

10. Carboxylic acid chloride

Phthaloylglycyl chloride reacts easily with benzylideneaniline in benzene in the presence of triethylamine to form the β-lactam (equation 41) [53]. Formation of an intermediate acylamino aldoketene which

$$C_6H_4(CO)_2NCH_2COCl \xrightarrow{-HCl} \left[C_6H_4(CO)_2NCH{=}C{=}O\right] \xrightarrow{C_6H_5CH=NC_6H_5} C_6H_4(CO)_2N\text{-}\overset{|}{C}H\text{-}C(=O)\text{-}N(C_6H_5)\text{-}\overset{|}{C}H(C_6H_5) \quad (41)$$

adds to benzylideneaniline to yield the β-lactam is assumed to occur during the reaction. *N*-Benzylidenemethylamine, on the other hand, reacts with carboxylic acid chloride in ether to form the addition product *N*-α-haloalkyl carboxylic acid amide (**12**), (equation 42) [54]. Alcoholysis of **12** yields *N*-α-alkoxyalkylcarboxylic acid amide (**13**).

$$C_6H_5CH{=}NCH_3 \xrightarrow{RCOCl} \underset{(\mathbf{12})}{C_6H_5CH(Cl)N(CH_3)COR} \rightleftharpoons \left[C_6H_5\overset{+}{C}HN(CH_3)COR \longleftrightarrow C_6H_5CH{=}\overset{+}{N}(CH_3)COR\right]Cl^- \xrightarrow[CH_3OH]{N(C_2H_5)_3} \underset{(\mathbf{13})}{C_6H_5CH(OCH_3)N(CH_3)COR} \quad (42)$$

Similarly *N*-benzylidenemethylamine reacts with cyanoacetyl chloride to form addition product **14** which is converted with triethylamine by elimination of hydrogen chloride to the corresponding β-lactam (equation 43)[55].

$$C_6H_5CH{=}NCH_3 \xrightarrow{NCCH_2COCl} \underset{(\mathbf{14})}{C_6H_5\underset{Cl}{\underset{|}{C}}H{-}\underset{COCH_2CN}{\underset{|}{N}}CH_3} \xrightarrow[-HCl]{Et_3N} C_6H_5{-}\underset{NC-CH}{\underset{|}{C}}H{-}\underset{CO}{\underset{|}{N}}{-}CH_3 \quad (43)$$

11. Maleic anhydride

Anils react with maleic anhydride in the presence of water to form maleanilic acid (**15**) and aldehyde[56,57]. When an anil is heated with

$$RCH{=}NC_6H_5 + \begin{matrix} CH{-}CO \\ \| \quad\;\; \diagdown \\ \| \qquad\; O \\ \| \quad\;\; \diagup \\ CH{-}CO \end{matrix} \xrightarrow{H_2O} \left[\begin{matrix} RCHNC_6H_5 \\ | \quad | \\ HO \;\; COCH{=}CHCOOH \end{matrix} \right] \longrightarrow$$

$$\underset{(\mathbf{15})}{C_6H_5NHCOCH{=}CHCOOH} + RCHO \quad (44)$$

maleic anhydride in toluene, maleanilic acid is also obtained[58,59], whereas the formation of a condensation product has been reported when the mixture is heated without using the solvent[58]. Crotonaldehyde anil and cinnamic aldehyde anil react with maleic anhydride in xylene to form an addition product[60].

12. Grignard reagent

Grignard reagents react with azomethine compounds to form addition products (**16**) which on hydrolysis result in secondary amines (**17**), (equation 45)[61]. The reaction is usually applied to the Schiff

$$C_6H_5CH{=}NR \xrightarrow{R'MgX} \underset{(\mathbf{16})}{C_6H_5\underset{R'}{\underset{|}{C}}H\underset{MgX}{\underset{|}{N}}R} \xrightarrow{H_2O} \underset{(\mathbf{17})}{C_6H_5\underset{R'}{\underset{|}{C}}HNHR} \quad (45)$$

bases which are prepared from aryl aldehydes. In this addition reaction the alkyl group of the Grignard reagent is attached to the carbon atom of the azomethine compound. The reactions with Grignard reagents provide a general synthetic method for secondary amines of the type RR′CHNHR″.

Following the classical work of Busch[61], Moffett[62] and Campbell[63], various secondary amines are prepared by use of the Grignard reagents. A solution of the Schiff base is added to an excess of Grignard reagent and the addition complex is decomposed with ice and hydrochloric acid. In these reactions, intermediate addition products have not been identified.

$$C_6H_5CH{=}NCH_3 \xrightarrow{C_6H_5CH_2MgX} C_6H_5\underset{\displaystyle CH_2C_6H_5}{\underset{|}{C}}HNHCH_3 \quad (95\%) \quad \text{(ref. 62)} \quad (46)$$

$$C_6H_5CH{=}NC_6H_{11} \xrightarrow{C_6H_5CH_2MgX} C_6H_5\underset{\displaystyle CH_2C_6H_5}{\underset{|}{C}}HNHC_6H_{11} \quad (40\%) \quad \text{(ref. 62)} \quad (47)$$

Attempts have been made to isolate the addition compounds resulting from the action of ethylmagnesium bromide and phenylmagnesium bromide on quinoline in ether[64] (quinoline could be regarded as a cyclic azomethine compound). It was found that the products were composed of one mole of the organomagnesium compound and one mole of quinoline. However, the addition products could not be analysed due to their hygroscopic character. Methylmagnesium iodide–azomethine addition products have been isolated and analysed[65]. The analytical data show that equimolar quantities of the azomethine and Grignard reagent had reacted to form the addition compound. Alkenylmagnesium halides ($RCH{=}CHMgX$) also react normally with the azomethine linkage of Schiff bases[66]. Several unsaturated secondary amines

$$>C{=}C\underset{\displaystyle C_6H_5}{\underset{|}{C}}HNHR$$

have been prepared by this method.

$$C_6H_5CH{=}NC_6H_5 \xrightarrow{CH_3CH{=}CHMgBr} CH_3CH{=}CH\underset{\displaystyle C_6H_5}{\underset{|}{C}}HNHC_6H_5 \quad (64\%) \quad (48)$$

$$C_6H_5CH{=}NBu \xrightarrow{(CH_3)_2C{=}CHMgBr} (CH_3)_2C{=}CH\underset{\displaystyle C_6H_5}{\underset{|}{C}}HNHBu \quad (50\%) \quad (49)$$

The effect of manganous salt on the reaction of a Schiff base (*N*-benzylidene-n-butylamine) with a Grignard reagent (i-PrMgI) has been studied[67,68]. Without manganous salt, the reaction resulted in normal products, 1-phenyl-1-butylamino-2-methylpropane (**18**) and benzylbutylamine (**19**). In the presence of 5 mole % anhydrous manganous chloride the reaction results in **18** and **19**, as well as the 'dimer' **20** of the original Schiff base (*N,N'*-dibutyl α,α'-diphenylethylenediamine) and yielded a mixture of diastereoisomers of *meso*

$$C_6H_5CH{=}NBu + i\text{-}PrMgI \longrightarrow \begin{cases} CH_3CH(CH_3){-}CH(C_6H_5)NHBu \ (\mathbf{18}) \\ C_6H_5CH_2NHBu \ (\mathbf{19}) \end{cases} \qquad \xrightarrow{Mn^{2+}} \begin{cases} C_6H_5CHNHBu \\ | \\ C_6H_5CHNHBu \end{cases} (\mathbf{20}) + \mathbf{18} \text{ and } \mathbf{19} \tag{50}$$

and racemic isomers. The yield of **20** was strongly dependent on the type of halogen in the Grignard reagent, and increased according to the sequence of I, Br, Cl when 2·5 mole % of MnX_2 was used. The ratio of racemic to *meso* form increased in the same sequence from 1·1 to 1·4 to 8·3. The formation of a 'dimeric' product of Schiff base ($C_6H_5CH{=}NR$) is also dependent on the size of R and also R′ of the Grignard reagent (R′MgX)[69]. *N*-Benzylidenethylamine and *t*-butylmagnesium chloride result in *N*-*N*′-diethyl-α,α′-diphenylethylamine instead of the expected addition product (equation 51)[70].

$$C_6H_5CH{=}NC_2H_5 \xrightarrow[\text{ether}]{Me_3CMgCl} \begin{matrix} C_6H_5CHNHC_2H_5 \\ | \\ C_6H_5CHNHC_2H_5 \end{matrix} \tag{51}$$

Sterically hindered reactions of Grignard reagents with Schiff bases have been studied[71]. *N*-Benzylidene-*t*-butylamine reacts with allylmagnesium bromide; however, methylmagnesium iodide does not react even under forced conditions (equation 52). *N*-Benzylidenemethylamine, however, reacts with *t*-butylmagnesium chloride

$$C_6H_5CH{=}N\text{-}t\text{-}Bu \xrightarrow{CH_2=CH-CH_2MgBr} C_6H_5CH(CH_2{-}CH{=}CH_2)NH\text{-}t\text{-}Bu \tag{52}$$

$$C_6H_5CH{=}NCH_3 \xrightarrow{t\text{-}BuMgCl} C_6H_5CH(t\text{-}Bu)NHCH_3 \tag{53}$$

normally to give *N*-methyl α-*t*-butylbenzylamine (equation 53).

In the forced reaction of phenylmagnesium bromide with benzophenone anil, it was found that addition of the Grignard reagent results in a lateralnuclear 1,4 addition to the conjugated system consisting of the azomethine linkage and the adjacent carbon–carbon linkage of the phenyl group[72]. In the same way, benzophenone β-naphthylimine undergoes a 1,4 addition with phenylmagnesium bromide under forced conditions, yielding *o*-phenylbenzohydryl-β-naphthylamine (equation 54)[73].

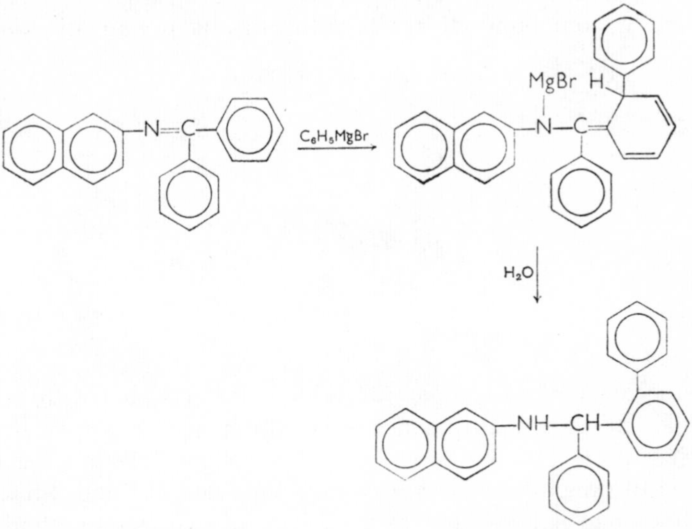

(54)

According to the literature[74] the yields of secondary amine obtained in the reaction of azomethines and the Grignard reagents are always less than 50% when a 1:1 ratio of Schiff base/RMgX is employed, although quantitative yields may be obtained when a 1:2 ratio is employed. This suggests that the Grignard reagent could have a structure R_2MgMgX_2 rather than RMgX. The addition reaction of the Grignard reagent to an azomethine linkage could be described as follows (equation 55):

$$RCH{=}NR' + R''_2MgMgX_2 \longrightarrow (R)(R'')CH{-}N(Mg{-}R''\cdot MgX_2)(R') \tag{55}$$

When the rate is expressed as rate = $k\,[R_2MgMgBr_2]$ [Schiff base], the observed data can be reasonably understood. The authors suggest a four-centre mechanism for the reaction (equation 56)[75].

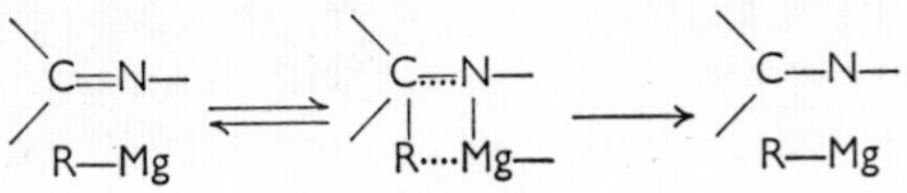

$$\begin{matrix} >C{=}N{-} \\ R{-}Mg \end{matrix} \rightleftharpoons \begin{matrix} >C{\cdots}N{-} \\ R{\cdots}Mg{-} \end{matrix} \longrightarrow \begin{matrix} >C{-}N{-} \\ R{-}Mg \end{matrix} \tag{56}$$

Various imines derived from aliphatic primary amines and enolizable aldehydes or ketones can be made to undergo complete enolization by refluxing with one equivalent of ethylmagnesium bromide in tetrahydrofuran. These readily prepared magnesium compounds react with alkyl halides to yield addition products which upon hydrolysis result in alkylated compounds in high yield[76]. The azomethine compound **21** reacts with ethylmagnesium bromide to

$$\underset{(21)}{CH_3\overset{CH_3}{\underset{H}{C}}CH{=}N\text{-}t\text{-}Bu} \xrightarrow[THF]{C_2H_5MgBr} \underset{(22)}{(CH_3)_2C{=}CHN(MgX)(t\text{-}Bu)} \xrightarrow{C_6H_5CH_2Cl}$$

$$CH_3\overset{CH_3}{\underset{CH_2C_6H_5}{C}}CH{=}N\text{-}t\text{-}Bu \xrightarrow{H_2O} \underset{(23)}{CH_3\overset{CH_3}{\underset{CH_2C_6H_5}{C}}CHO} \qquad (57)$$

form the magnesium complex **22**, which was alkylated with benzyl chloride and gave 2,2-dimethyl-3-phenylpropanal (**23**) upon hydrolysis (equation 57). Similarly by the use of the reaction the *ortho* position of cyclohexanone is alkylated (equation 58) and 3-methylcyclohexanone is converted to a mixture of DL-menthone and DL-isomenthone (equation 59)[76].

$$\text{cyclohexanone imine }(C_6H_{10}{=}NH) \xrightarrow{C_2H_5MgBr} \text{cyclohexenyl-}N(MgX)H \xrightarrow{BuI} \text{2-Bu-cyclohexylidene}{=}N(MgX)H \xrightarrow{H_2O} \text{2-butylcyclohexanone} \qquad (58)$$

$$\text{3-methylcyclohexanone imine }({=}NR) \xrightarrow{C_2H_5MgBr} \xrightarrow{i\text{-PrI}} \xrightarrow{H_2O} \underset{\text{DL-iso}}{\text{isomenthone}} + \underset{\text{DL}}{\text{menthone}} \qquad (59)$$

$\Delta^{1(10)}$-Dehydroquinolizidine salt reacts with methylmagnesium iodide to form 10-methylquinolizidine[12]. However, phenyl- and isopropylmagnesium iodide do not react with the dehydroquinolizidine salt because of steric hindrance (equation 60). Similarly, $\Delta^{4(9)}$

hexahydropyrrocolinium salt reacts with methylmagnesium iodide to yield 9-methyloctahydropyrrocoline (equation 61).

$$\xrightarrow{CH_3MgI} \quad (60)$$

$$\xrightarrow{CH_3MgI} \quad (61)$$

13. Alkyllithium

N-Benzylidene-*t*-butylamine reacts with methyllithium to yield *N*-α-methylbenzyl-*t*-butylamine (equation 62)[77]. Phenyllithium adds to benzophenone anil at the azomethine linkage only, resulting in

$$C_6H_4CH{=}N\text{-}t\text{-}Bu \xrightarrow{MeLi} C_6H_5CH(CH_3){-}NH\text{-}t\text{-}Bu \quad (62)$$

triphenylmethylaniline[78] (equation 63), whereas the Grignard reagent results in a 1,4 addition product. Similarly acetophenone anil

$$\xrightarrow{C_6H_5Li} \quad (63)$$

reacts with phenyllithium to yield a phenylated compound (equation 64)[79]. $\Delta^{5(10)}$ Dehydroquinolizidinium salt reacts with α-picolyl-

$$o\text{-}CH_3C_6H_4N{=}C(C_6H_5)(CH_3) \xrightarrow{C_6H_5Li} o\text{-}CH_3C_6H_4NHC(C_6H_5)(CH_3)C_6H_5 \quad (64)$$

lithium to form 10(α-picolyl)quinolizidine (equation 65)[12]. Pycolyllithium reacts with Schiff bases to form *N*,1-disubstituted 2-(2-pyridyl)ethylamine with a few exceptions (equation 66)[80]. Schiff bases derived from *N*-methylpyrrole-2-aldehyde and aniline

$$\text{quinolizinium } X^- \xrightarrow{\text{2-pyridyl-}CH_2Li} \text{9a-(2-pyridylmethyl)quinolizine} \qquad (65)$$

$$C_6H_5CH{=}NC_6H_5 \xrightarrow{\text{2-pyridyl-}CH_2Li} C_6H_5CH(NHC_6H_5)CH_2\text{-(2-pyridyl)} \qquad (66)$$

or 4-chloroaniline undergo addition with 2-picolyllithium followed by spontaneous deamination producing 2-{2'-[2''-(1''-methylpyrryl)] vinyl} pyridine (equation 67).

$$\text{1-methylpyrrol-2-yl-}CH{=}NC_6H_5 \xrightarrow{\text{2-pyridyl-}CH_2Li} \text{1-methylpyrrol-2-yl-}CHNHC_6H_5(CH_2\text{-2-pyridyl}) \xrightarrow{-C_6H_5NH_2} \text{1-methylpyrrol-2-yl-}CH{=}CH\text{-2-pyridyl} \qquad (67)$$

14. Metallic lithium, sodium

Anils react with metallic sodium or lithium in dry ether to form disodium or dilithium derivatives (equation 68)[81]. It is assumed that one metal atom is in the ionized state and the other atom is covalently

$$(C_6H_5)_2C{=}NC_6H_4Me\text{-}o \xrightarrow{Na} (C_6H_5)_2C(Na){-}N(Na){-}C_6H_4Me\text{-}o$$
$$(C_6H_5)_2C{=}NC_6H_4Me\text{-}o \xrightarrow{Li} (C_6H_5)_2C(Li){-}N(Li){-}C_6H_4Me\text{-}o \qquad (68)$$

bonded to the anil. These alkali metal derivatives of benzophenone anil react with methyl halides to give a mixture of di- and mono-methylated compounds (equation 69)[82]. In addition,

$o\text{-}CH_3C_6H_4NH_2$, $(C_6H_5)_2C{=}CH_2$, $(C_6H_5)_2{=}C(CH_3)_2$

and several small hydrocarbons have been identified from the reaction mixture.

$$o\text{-MeC}_6\text{H}_4\underset{\text{Li}}{\text{N}}\text{—}\underset{\text{Li}}{\text{C}}(\text{C}_6\text{H}_5)_2 \xrightarrow{\text{CH}_3\text{I}} o\text{-CH}_3\text{C}_6\text{H}_4\underset{\text{H}_3\text{C}}{\text{N}}\text{—}\underset{\text{CH}_3}{\text{C}}(\text{C}_6\text{H}_5)_2 + o\text{-CH}_3\text{C}_6\text{H}_4\text{NH}\underset{\text{H}_3\text{C}}{\text{C}}(\text{C}_6\text{H}_5)_2 \quad (69)$$

15. Magnesium–magnesium iodide

Ethylenediamine derivatives are prepared from benzylidenealkylamines or benzylidenearylamines with a magnesium–magnesium iodide mixture in ether or benzene solution (equations 70, 71)[83].

$$2\ \text{C}_6\text{H}_5\text{CH}{=}\text{NCH}_3 \xrightarrow{\text{Mg–MgI}_2} \text{C}_6\text{H}_5\text{CHNHCH}_3\text{—C}_6\text{H}_5\text{CHNHCH}_3 \quad (70)$$

$$2\ \text{C}_6\text{H}_5\text{CH}{=}\text{NC}_6\text{H}_5 \xrightarrow{\text{Mg–MgI}_2} \text{C}_6\text{H}_5\text{CHNHC}_6\text{H}_5\text{—C}_6\text{H}_5\text{CHNHC}_6\text{H}_5 \quad (71)$$

16. Nitroalkane

Nitroalkanes condense with anils under refluxing with alcohol[84,85]. Reaction of benzylideneaniline and nitromethane results in *N*-(2-nitro-1-phenylethyl)aniline[24]. Similarly, *N*-benzylideneaniline and nitroethane give *N*-(2-nitro-1-phenylpropyl)aniline[25]. However, under similar conditions, benzalazine, $C_6H_5CH{=}N{—}N{=}CHC_6H_5$, and nitromethane do not form any addition product. In ligroin or petroleum ether, ethyl α-nitroacetate reacts with various Schiff bases in the

$$\text{C}_6\text{H}_5\text{CH}{=}\text{NC}_6\text{H}_5 \xrightarrow{\text{CH}_3\text{NO}_2} \text{C}_6\text{H}_5\text{CH}(\text{NHC}_6\text{H}_5)\text{—CH}_2\text{NO}_2 \ \textbf{(24)}$$

$$\text{C}_6\text{H}_5\text{CH}{=}\text{NC}_6\text{H}_5 \xrightarrow{\text{C}_2\text{H}_5\text{NO}_2} \text{C}_6\text{H}_5\text{CH}(\text{NHC}_6\text{H}_5)\text{—CH}(\text{CH}_3)\text{NO}_2 \ \textbf{(25)} \quad (72)$$

presence of diethylamine to yield unstable addition products $[R'NHCH(R)\text{—}CH(NO_2)\text{—}COOC_2H_5]NH(C_2H_5)_2$ (**26**), in which R and R′ are usually aryl groups[86]. The procedure to precipitate the addition product **26** in chloroform with petroleum ether results in the decomposition of **26** and its conversion to the corresponding ethylamine salt of $R\text{—}CH[CH(NO_2)COOC_2H_5]_2$ (**27**) in good yield.

$$O_2NCH_2COOC_2H_5 \xrightarrow[Et_2NH]{C_6H_5N{=}CHC_6H_5} \left(C_6H_5NHCH(H_5C_6)\text{—}CH(NO_2)COOC_2H_5\right)NHEt_2 \;(99\%)$$

$$O_2NCH_2COOC_2H_5 \xrightarrow[Et_2NH]{CH_3N{=}CHC_6H_5} \left(CH_3NHCH(H_5C_6)\text{—}CH(NO_2)COOC_2H_5\right)NHEt_2 \;(97\%) \qquad (73)$$

17. Phosphorane

The reaction of ethylenetriphenylphosphorane (**28**) with a Schiff base (**29**) results in phenylallene in 62% yield (equation 74)[87]. The mechanism of the reaction is explained as follows (equation 75)[88].

$$\underset{(\mathbf{28})}{CH_3CH{=}P(C_6H_5)_3} + \underset{(\mathbf{29})}{C_6H_5CH{=}NC_6H_5} \longrightarrow CH_2{=}C{=}CHC_6H_5 + P(C_6H_5)_3 + C_6H_5NH_2 \qquad (74)$$

$$RCH_2\overset{-}{C}H\overset{+}{P}(C_6H_5)_3 + C_6H_5CH{=}NC_6H_5 \longrightarrow C_6H_5CH\text{—}NH\text{—}C_6H_5\,[\text{—}CH(CH_2R)\text{—}\overset{+}{P}(C_6H_5)_3] \longrightarrow$$

$$C_6H_5\text{—}CH(\text{—}NH\text{—}C_6H_5)\text{—}\overset{-}{C}(\overset{+}{P}(C_6H_5)_3)\text{—}CH(H)R \longrightarrow C_6H_5CH{=}C{=}CHR + P(C_6H_5)_3 + C_6H_5NH_2 \qquad (75)$$

Alkylenephosphoranes with no methylene group in the β-position to the phosphorus atom react with Schiff base to give olefins and phenyliminotriphenylphosphorane[87]. The mechanism of the reaction is explained below (equation 76)[88]. The addition product **30** cyclizes to form cyclic structure **31** which decomposes to olefin **32** and triphenylphosphinarylimine (**33**).

$$\underset{(28)}{\mathrm{R{-}\overset{H}{\underset{(C_6H_5)_3P^+}{C^-}}}} + \underset{(29)}{\mathrm{\overset{H}{\underset{NAr}{C}}{-}R'}} \xrightarrow{150\text{–}200°} \underset{(30)}{\mathrm{R{-}\overset{H}{\underset{(C_6H_5)_3P^+}{C}}{-}\overset{H}{\underset{N^-{-}Ar}{C}}{-}R'}} \longrightarrow$$

$$\underset{(31)}{\mathrm{R{-}\overset{H}{\underset{(C_6H_5)_3P}{C}}{-}\overset{H}{\underset{N{-}Ar}{C}}{-}R'}} \longrightarrow \underset{(32)}{\mathrm{R\overset{H}{C}{=}\underset{H}{C}R}} + \underset{(33)}{\mathrm{(C_6H_5)_3P{=}NAr}} \quad (76)$$

Alkali metal phosphides, MPR_2, react with Schiff bases in dioxan or benzene to yield addition products of the general formula $(C_6H_5{-}\underset{M}{N}{-}\underset{PR_2}{CH}{-}C_6H_5)$[89]. Lithium or potassium phosphides are used in this reaction. Schiff bases also react with diethylphosphite to form addition products (**34**), (equation 77)[90].

$$\mathrm{C_6H_5CH{=}NC_6H_5} \xrightarrow{\mathrm{H{-}P({=}O)(OEt){-}OEt}} \underset{(34)}{\mathrm{C_6H_5\underset{NHC_6H_5}{C}HP(O)(OEt)_2}} \quad (77)$$

18. Carbon monoxide

Carbon monoxide reacts with Schiff bases or ketoximes to form addition products. These addition products are often cyclic compounds (see cycloaddition), although a few non-cyclic compounds are reported.

Carbon monoxide and hydrogen react with methylphenylketoxime to give probably 3,4-dimethyl-3,4-diphenyl-2-azetidinone (**37**), (equation 78)[91].

$$2\,\mathrm{C_6H_5\overset{CH_3}{C}{=}NOH} \xrightarrow[\mathrm{H_2}]{\mathrm{CO}} \left(\underset{(35)}{\mathrm{(CH_3)(C_6H_5)C(NH{-}OH){-}C({=}O){-}C(CH_3)(C_6H_5)(NH{-}OH)}} \longrightarrow \underset{(36)}{\text{4-membered ring: } \mathrm{C({=}O)},\ \mathrm{C(CH_3)(C_6H_5)},\ \mathrm{NH},\ \mathrm{C(CH_3)(C_6H_5)}} \right) \longrightarrow \quad (78)$$

$$\underset{(37)}{\mathrm{C_6H_5{-}\overset{CH_3}{C}{-}C{=}O \;/\; C_6H_5{-}\underset{CH_3}{C}{-}N{-}H}}$$

Chemical analyses of the reaction products indicate that two moles of methylphenylketoxime condense with one mole each of carbon monoxide and hydrogen to yield **35**, which is then cyclized to compound **36** which rearranges to 3,4-dimethyl, 3,4-diphenyl 2-azetidinone (**37**).

When aromatic or aralkyl ketone oximes are treated with carbon monoxide and hydrogen using dicobalt octacarbonyl as a catalyst, they yield formamides and secondary amines (equation 79)[92]. A

$$(C_6H_5CH_2)_2C{=}NOH \xrightarrow[Co_2(CO)_8]{CO,H_2} \begin{cases}(C_6H_5CH_2)_2CHNHCHO \\ [(C_6H_5CH_2)_2CH]_2NH\end{cases} \quad (79)$$

proposed mechanism of the reaction is shown below (equation 80).

$$(C_6H_5CH_2)_2C{=}NOH \xrightarrow[-H_2O]{H_2} \underset{(\mathbf{38})}{(C_6H_5CH_2)_2C{=}NH} \longrightarrow \underset{(\mathbf{39})}{(C_6H_5CH_2)_2CHNH_2}$$

$$\underset{(\mathbf{39})}{(C_6H_5CH_2)_2CHNH_2} \xrightarrow{CO} \underset{(\mathbf{40})}{(C_6H_5CH_2)_2CHNHCHO}$$

$$\mathbf{38} + \mathbf{39} \xrightarrow[-NH_3]{} [(C_6H_5CH_2)_2C{=}NCH(CH_2C_6H_5)_2] \xrightarrow{H_2} [(C_6H_5CH_2)_2CH]_2NH \quad (80)$$

B. Hydrogenation Reactions

I. Catalytic hydrogenation

Aldimine or ketimine or a mixture of a carbonyl compound and amine (or ammonia) are readily converted to the corresponding secondary amine by catalytic hydrogenation. This method has been used for the preparation of secondary amines or for the alkylation of primary amines. Many studies on this subject were reviewed by Emerson[93]. In many preparations of secondary amines, the mixture of primary amines and carbonyl compounds are hydrogenated without isolating intermediate azomethine compounds. Raney nickel, platinum and palladium catalysts are usually used for hydrogenation. In some cases, intermediate aldimines and ketimines are isolated and then hydrogenated; however, in several cases these azomethine compounds cannot be isolated.

$$CH_3CH_2NH_2 + CH_3CHO \xrightarrow[Ni]{H_2} \begin{cases}(CH_3CH_2)_2NH & 55\% \\ (CH_3CH_2)_3N & 19\%\end{cases} \quad \text{(ref. 94)} \quad (81)$$

$$C_6H_{11}NH_2 + CH_3(CH_2)_2CHO \xrightarrow[Ni]{H_2} C_6H_{11}NH(CH_2)_3CH_3 \quad 91\% \text{ (ref. 95)} \qquad (82)$$

$$CH_3NH_2 + CH_3COCH_2CH_3 \xrightarrow[Pt]{H_2} CH_3CH_2CH(CH_3)NHCH_3 \quad 68\% \text{ (ref. 96)} \qquad (83)$$

$$CH_3CH(OH)CH_2NH_2 + CH_3\text{-}C_6H_9{=}O \xrightarrow[Pt]{H_2} CH_3CH(OH)CH_2NH\text{-}C_6H_{10}\text{-}CH_3 \quad 93\% \text{ (ref. 97)} \qquad (84)$$

$$\alpha\text{-}C_{10}H_7NH_2 + CH_3CHO \xrightarrow[Ni]{H_2} \alpha\text{-}C_{10}H_7NHCH_2CH_3 \quad 88\% \text{ (ref. 98)} \qquad (85)$$

$$CH_3CH_2N{=}CHCH_2CH_3 \xrightarrow[Pt]{H_2} CH_3CH_2NH(CH_2)_2CH_3 \quad 45\% \text{ (ref. 99)} \qquad (86)$$

$$CH_3N{=}CHC_6H_5 \xrightarrow[Ni]{H_2} CH_3NHCH_2C_6H_5 \quad 100\% \text{ (ref. 100)} \qquad (87)$$

$$C_5H_5(CH_2)_2N{=}CHC_6H_5 \xrightarrow[Ni]{H_2} C_6H_5(CH_2)_2NHCH_2C_6H_5 \quad 96\% \text{ (ref. 101)} \qquad (88)$$

Catalytic reductive amination of carbonyl compounds with ammonia could be classified in the category of catalytic hydrogenation of azomethine compounds. The reaction is used for preparation of primary amines[93].

$$RCOR' \xrightarrow[H_2]{NH_3} RCH(NH_2)R' \qquad (89)$$

Reductive amination of aliphatic and aromatic aldehydes with ammonia yields the corresponding primary amine. The reaction

$$C_6H_5CH_2CHO \xrightarrow[H_2]{NH_3} C_6H_5CH_2CH_2NH_2 \quad 64\% \text{ (ref. 102)} \qquad (90)$$

$$C_6H_5CHO \xrightarrow[H_2]{NH_3} C_6H_5CH_2NH_2 \quad 89\% \text{ (ref. 103)} \qquad (91)$$

$$C_4H_3O\text{-}CHO \longrightarrow C_4H_3O\text{-}CH_2NH_2 \quad 79\% \text{ (ref. 102)} \qquad (92)$$

$$C_6H_{10}{=}O \longrightarrow C_6H_{11}\text{-}NH_2 \quad 80\% \text{ (ref. 104)} \qquad (93)$$

$$C_6H_5COCH_3 \longrightarrow C_6H_5CH(NH_2)CH_3 \quad 44\text{–}52\% \qquad \text{(ref. 105)} \quad (94)$$

$$C_6H_5CH_2COCH_3 \longrightarrow C_6H_5CH_2CH(NH_2)CH_3 \quad \text{quantitative} \qquad \text{(ref. 106)} \quad (95)$$

products of reductive amination contain secondary amines and other more complex compounds. To minimize the formation of secondary amine, an excess of ammonia should be used. On the other hand, symmetrical secondary amines could be prepared by reductive amination by using an excess of carbonyl compound. For example, dibenzylamine is prepared from benzaldehyde and ammonia[107].

$$2\,C_6H_5CHO + NH_3 \xrightarrow[Ni]{H_2} (C_6H_5CH_2)_2NH \quad 81\% \qquad (96)$$
$$12\text{–}17\% \text{ primary amine}$$

The reductive amination of carbonyl compounds has been applied to the syntheses of α-amino acids. α-Keto acids are converted easily to the corresponding α-amino acids by the use of platinum or palladium catalysts in the presence of ammonia. Many of the natural α-amino acids are synthesized by use of these simple and mild reaction conditions.

$$CH_3COCOOH \xrightarrow{NH_3,H_2} \text{alanine} \qquad \text{(ref. 108)} \quad (97)$$

$$C_2H_5COCOOH \xrightarrow{NH_3,H_2} \alpha\text{-amino-n-butyric acid} \qquad \text{(ref. 108)} \quad (98)$$

$$(CH_3)_2CHCH_2COCOOH \xrightarrow{NH_3,H_2} \text{leucine} \qquad \text{(ref. 109)} \quad (99)$$

$$C_6H_5CH_2COCOOH \xrightarrow{NH_3,H_2} \text{phenylalanine} \qquad \text{(refs. 108–110)} \quad (100)$$

$$HOOCCH_2CH_2COCOOH \xrightarrow{NH_3,H_2} \text{glutamic acid} \qquad \text{(refs. 108, 111)} \quad (101)$$

$$HOOCCH_2COCOOH \xrightarrow{NH_3,H_2} \text{aspartic acid} \qquad \text{(ref. 108)} \quad (102)$$

When (a) benzylamine, (b) α-phenylgycinate, (c) α-alkylbenzylamine and (d) α(1-naphthyl)alkyl amine are used as constituents of aldimine and ketimine, the azomethine bonds of these compounds are hydrogenated normally. However, the C—N single bonding which is formed by hydrogenation is hydrogenolysed by the use of specific catalysts. These hydrogenolysis reactions can be applied to the preparation of primary amines. When optically active amines (b, c and d) are used as constituents of the Schiff base, catalytic reduction may introduce a new asymmetric carbon and subsequent hydrogenolysis could produce an optically active amine. The application of these reactions for the asymmetric synthesis of optically active amines will be described in Section II.B.8.

$$
\left.\begin{array}{ll}
\text{(a)} & \text{H(N)}\text{—}CH_2C_6H_5 \\
\text{(b)} & \text{H(N)}\text{—}\overset{*}{C}H(COO^-)\text{—}C_6H_5
\end{array}\right\} \xrightarrow{Pd/C} \text{primary amine}
$$

$$
\left.\begin{array}{ll}
\text{(a), (b)} & \\
\text{(c)} & \text{H(N)}\text{—}\overset{*}{C}H(R)\text{—}C_6H_5 \\
\text{(d)} & \text{H(N)}\text{—}\overset{*}{C}H(CH_3)\text{—}C_{10}H_7
\end{array}\right\} \xrightarrow{Pd(OH)_2/C} \text{primary amine} \qquad (103)
$$

2. Reduction using metals

Azomethine compounds have been reduced to secondary amines by the use of various types of metal reduction systems, such as (Na, EtOH), (Na–Hg, EtOH), (Zn, alkali), (Zn, acid), (Al, NaOH), (Al, EtOH), (Mg, MeOH). The method gives good results in some reductions of azomethine compounds; however, the reduction procedure has not been widely applied for the syntheses of secondary amines.

N-Benzylideneethylamine[112] and *N*-*m*-hydroxybenzylideneaniline[113] are reduced to the corresponding amines in good yield by sodium amalgam and ethyl alcohol (equations 104, 105). A mixture of

$$CH_3CH_2N{=}CHC_6H_5 \xrightarrow{Na\text{–}Hg,\,EtOH} CH_3CH_2NHCH_2C_6H_5\ (70\%) \quad \text{(ref. 112)} \quad (104)$$

$$m\text{-}HOC_6H_4CH{=}NC_6H_5 \xrightarrow{Na\text{–}Hg,\,EtOH} m\text{-}HOC_6H_4CH_2NHC_6H_5\ (100\%) \quad \text{(ref. 113)} \quad (105)$$

methylamine and cyclohexanone gives methylcyclohexylamine[114] by reduction with sodium and ethyl alcohol (equation 106). Similarly, *N*-benzylidenephenylethylamine gives the corresponding secondary amine[115] by the reduction of the sodium–alcohol system (equation 107).

$$CH_3NH_2 + O{=}C_6H_{10} \xrightarrow{Na,\,EtOH} CH_3NH\text{—}C_6H_{11} \quad \text{(ref. 114)} \quad (106)$$

$$C_6H_5(CH_2)_2N{=}CHC_6H_5 \xrightarrow{Na,\,EtOH} C_6H_5(CH_2)_2NHCH_2C_6H_5 \quad \text{(ref. 115)} \quad (107)$$

A mixture of 2,4,6-trimethylaniline and isobutyraldehyde gives 2,4,6-trimethylphenylisobutylamine[116] by reduction with zinc and hydrochloric acid (equation 108). Similarly, methyl *p*-aminobenzoate and butyraldehyde yield a secondary amine (equation 109)[117]. A mixture of methylamine and *m*-methoxybenzylmethyl ketone yields

$$2,4,6\text{-}(CH_3)_3C_6H_2NH_2 + (CH_3)_2CHCHO \xrightarrow{Zn,HCl} 2,4,6\text{-}(CH_3)_3C_6H_2NHCH_2CH(CH_3)_2 \quad 91\% \text{ (ref. 116)} \qquad (108)$$

$$CH_3(CH_2)_2CHO + CH_3OOC\text{-}C_6H_4\text{-}NH_2 \xrightarrow{Zn,\ HCOOH} CH_3OOC\text{-}C_6H_4\text{-}NH(CH_2)_3CH_3 \quad 47\% \text{ (ref. 117)} \qquad (109)$$

the corresponding *N*-methylamine by reduction with aluminium and water (equation 110)[118]. The Schiff base obtained by condensation of

$$m\text{-}CH_3OC_6H_4CH_2COCH_3 + CH_3NH_2 \xrightarrow{Al,\ H_2O} m\text{-}CH_3OC_6H_4CH_2\text{—}CH(NHCH_3)\text{—}CH_3 \quad \text{(ref. 118)} \qquad (110)$$

furfural with aniline is reduced by magnesium and methanol to a secondary amine (equation 111)[119,120]. When benzylidenealkylamines (**41**) and ethanol are treated with activated aluminium, a

$$C_4H_3O\text{—}CH{=}NC_6H_5 \xrightarrow{Mg,\ MeOH} C_4H_3O\text{—}CH_2NHC_6H_5 \qquad (111)$$

mixture of benzylalkylamines (**42**) and *NN'*-dialkyl-α-α'-diphenylethylenediamines (**43**) is obtained (equation 112)[121]. The ratio of **42**

$$\underset{(\mathbf{41})}{C_6H_5CH{=}NR} \xrightarrow{Al,EtOH} \underset{(\mathbf{42})}{C_6H_5CH_2NHR} + \underset{(\mathbf{43})}{C_6H_5CHNHR\text{—}C_6H_5CHNHR} \qquad (112)$$

and **43** does not depend on the concentration of **41** in inert solvents (toluene or ligroin). The use of dioxan and triethylamine favours the formation of **43**.

3. Lithium aluminium hydride

Lithium aluminium hydride reduces aromatic and aliphatic Schiff bases easily to yield secondary amines[122–124]. *N*-Phenyl-9-xanthydrylidenimine is not reduced by lithium aluminium hydride in ether. However, by using tetrahydrofuran as solvent and increasing the

$$(CH_3)_2CHCH{=}NCH(CH_3)_2 \xrightarrow{LiAlH_4} (CH_3)_2CHCH_2NHCH(CH_3)_2 \quad \text{(ref. 122)} \quad (113)$$

$$C_6H_5CH{=}NC_6H_5 \xrightarrow{LiAlH_4} C_6H_5CH_2NHC_6H_5 \quad \text{(ref. 123)} \quad (114)$$

$$(C_6H_5)_2C{=}NC_6H_5 \xrightarrow{LiAlH_4} (C_6H_5)_2CHNHC_6H_5 \quad \text{(ref. 124)} \quad (115)$$

$$\text{fluorenylidene}{=}NC_6H_5 \xrightarrow{LiAlH_4} \text{9-H-fluorenyl-}NHC_6H_5 \quad \text{(ref. 124)} \quad (116)$$

reaction temperature, the Schiff base is reduced to the corresponding secondary amine (equation 117)[124].

$$\text{xanthylidene}{=}NC_6H_5 \begin{cases} \xrightarrow[LiAlH_4]{ether} \not\rightarrow \\ \xrightarrow[LiAlH_4]{THF} \text{9-H-xanthyl-}NHC_6H_5 \end{cases} \quad (117)$$

4. Sodium borohydride

Sodium borohydride has been used for reduction of azomethine compounds. This reducing agent is more convenient than lithium aluminium hydride because a wide variety of solvents may be used.

Aromatic and aliphatic Schiff bases and also quaternary Schiff bases are reduced by sodium borohydride (equations 118, 119). Several

$$C_6H_5CH{=}NCH_3 \xrightarrow[\text{MeOH}]{\text{NaBH}_4} C_6H_5CH_2NHCH_3 \quad \text{(ref. 125)} \quad (118)$$

$$o\text{-}HOC_6H_4CH{=}N\text{-}2\text{-}C_{10}H_7 \xrightarrow{\text{NaBH}_4} o\text{-}HOC_6H_4CH_2NH\text{-}2\text{-}C_{10}H_7 \quad \text{(ref. 125)} \quad (119)$$

$$(R)C_6H_4\text{—}CH{=}N\text{—}C_6H_4(R') \xrightarrow{\text{NaBH}_4} (R)C_6H_4CH_2NH\,C_6H_4(R') \quad \text{(ref. 126)} \quad (120)$$

N-benzilidenaniline type compounds have been reduced by sodium borohydride to the corresponding secondary amines in 91–99% yield (equation 120)[126]. In this reduction, a reducible group such as nitro or chloro is not affected during the course of the reduction. A Schiff base, *N*-benzylidene-*p*-aminophenol, fails to yield the corresponding secondary amine under the conditions employed, probably due to a tautomerization involving a quinoid type structure[126]. Schiff bases prepared from aminopyridines are reduced to the corresponding *N*-alkyl aminopyridines (equation 121)[127].

$$R\text{—}N\langle C_5H_4 \rangle N{=}CHC_6H_5 \xrightarrow{\text{NaBH}_4} RN\langle C_5H_4 \rangle NHCH_2C_6H_5 \quad (121)$$

Sodium borohydride reduces Schiff bases derived from cysteamine and heteroaromatic aldehydes to yield *N*-substituted 2-aminoethanethiols (equation 122)[128]. *N*-Benzylidene-α-amino acid sodium salts

$$(ArCH{=}NCH_2CH_2S\text{-})_2 \xrightarrow{\text{NaBH}_4} 2\,ArCH_2NHCH_2CH_2SH \quad \text{(ref. 128)} \quad (122)$$

Ar = pyridyl or thienyl

can be reduced easily by sodium borohydride to *N*-benzyl-α-amino acids in aqueous solution. This is a very convenient method for obtaining *N*-benzylamino acid (equation 123) and therefore this compound is an intermediate in the synthesis of *N*-methyl-α-amino acid[129], which is usually troublesome to prepare.

$$C_6H_5CHO + \underset{\displaystyle NH_2}{RCHCOOH} \xrightarrow[H_2O]{\text{NaOH}} C_6H_5CH{=}N\underset{\displaystyle R}{CHCOONa} \xrightarrow{\text{NaBH}_4} C_6H_5CH_2NH\underset{\displaystyle R}{CHCOOH} \quad (123)$$

Quaternary Schiff bases containing a structure $—C{=}\overset{+}{N}{<}^{CH_3}$ are reduced by sodium borohydride to *N*-methyldihydro derivatives (equation 124)[130].

$$\text{Peganine methiodide} \xrightarrow{NaBH_4} \text{N-methyldihydropeganine} \tag{124}$$

5. Dimethylamine-borane

Schiff bases which have various functional groups are reduced to the respective secondary amines by using dimethylamine-borane in glacial acetic acid[131]. The method gives good yields (80–97%) and none of the functional groups of the Schiff base (—Cl, —NO_2, —OH, —OCH_3, —$COOC_2H_5$, —SO_2NH_2, —COOH) are affected under the reaction conditions (equations 125–128).

$$C_6H_5CH{=}NC_6H_5 \xrightarrow{BH_3NH(CH_3)_2} C_6H_5CH_2NHC_6H_5 \tag{125}$$

$$p\text{-}ClC_6H_4CH{=}NC_6H_4Cl\text{-}p \xrightarrow{BH_3NH(CH_3)_2} p\text{-}ClC_6H_4CH_2NHC_6H_4Cl\text{-}p \tag{126}$$

$$C_6H_5CH{=}NC_6H_4OH\text{-}p \xrightarrow{BH_3NH(CH_3)_2} C_6H_5CH_2NHC_6H_4OH\text{-}p \tag{127}$$

$$C_6H_5CH{=}NC_6H_4SO_2NH_2\text{-}p \xrightarrow{BH_3NH(CH_3)_2} C_6H_5CH_2NHC_6H_4SO_2NH_2\text{-}p \tag{128}$$

However, prolonged refluxing with an excess of the amine-borane in glacial acetic acid results in the acetyl derivative of the secondary amine[132]. When propionic acid and benzoic acid are used, propionamide and benzamide derivatives are obtained (equation 129).

$$C_6H_5CH{=}NC_6H_5 + BH_3\cdot NH(CH_3)_2 \begin{cases} \xrightarrow{AcOH} C_6H_5CH_2N(COMe)C_6H_5 \\ \xrightarrow{\text{propionic acid}} C_6H_5CH_2N(COEt)C_6H_5 \\ \xrightarrow{C_6H_5COOH} C_6H_5CH_2N(COPh)C_6H_5 \end{cases} \tag{129}$$

6. Formic acid

Formic acid is a well known reducing compound. Two common organic reactions to synthesize secondary amines by reduction with formic acid are known: the Leuckart reaction[133] and the Eschweiler reaction[134]. The Leuckart reaction is a convenient method for converting aldehydes and ketones to formyl derivatives of amines by heating with ammonium formate or formamide (equation 130)[135].

$$RCOR' + HCONH_2 \longrightarrow \begin{matrix} R \\ \quad\diagdown \\ \quad CHNHCHO \\ \quad\diagup \\ R' \end{matrix} \longrightarrow \begin{matrix} R \\ \quad\diagdown \\ \quad CHNH_2 \\ \quad\diagup \\ R' \end{matrix} \quad \text{Leuckart reaction} \quad (130)$$

The Eschweiler reaction is a methylation method of primary or secondary amines to the tertiary amine with formaldehyde and formic acid (equation 131).

$$RCH_2NH_2 + 2\,HCHO + 2\,HCOOH \longrightarrow RCH_2N(CH_3)_2 + 2\,CO_2 + 2\,H_2O \quad (131)$$

Eschweiler reaction

N-Benzylideneaniline is reduced to *N*-benzylaniline quantitatively by heating with triethylammonium formate at 140–160° (equation 132)[136]. Schiff bases which are composed of ethylenediamine and

$$C_6H_5CH{=}NC_6H_5 \xrightarrow[\text{heat}]{HCOO^-N^+Et_3} C_6H_5CH_2NHC_6H_5 \quad (132)$$

aromatic aldehydes are reduced by heating with formic acid (equation 133)[137]. Several Schiff bases are converted to the corresponding

$$(C_6H_5CH{=}NCH_2)_2 \xrightarrow[\text{heat}]{HCOOH} \left\{\begin{matrix} C_6H_5CHNHCH_2CH_2NH_2 \\ + \\ C_6H_5CH_2NHCH_2CH_2NHCH_2C_6H_5 \end{matrix}\right. \quad (133)$$

secondary amines by reduction with formic acid[138]. Hydrazones[139] and enamines[140] are also reduced by the use of formic acid. The formic acid reduction mechanism including the Leuckart reaction has been discussed[141]. Quinoline[142], which could be regarded as a Schiff base analogue, is also reduced to *N*-formyltetrahydroquinoline by formic acid (equation 134).

$$\text{Quinoline} \xrightarrow[\Delta]{HCOOH,\ HCOONa} \text{1,2,3,4-tetrahydroquinoline–N–CHO} \quad (134)$$

7. Electrolytic reduction

Several anils have been reduced to secondary amines by electrolytic reduction in sulphuric acid solution with lead or copper as cathode (equations 135,136)[143]. Similarly, oximes and phenylhydrazones were

$$C_6H_5CH{=}NC_6H_5 \xrightarrow[Pb]{H_2} C_6H_5CH_2NHC_6H_5 \quad (135)$$

$$p\text{-}CH_3C_6H_4CH{=}NC_6H_5 \xrightarrow[Pb]{H_2} p\text{-}CH_3C_6H_4CH_2NHC_6H_5 \quad (136)$$

converted to the corresponding amines by electrolytic reduction (equations 137–139)[144,145].

$$(CH_3)_2C{=}NOH \xrightarrow[Pb]{H_2} (CH_3)_2CHNH_2 \quad (137)$$

$$C_6H_5CH{=}NNHC_6H_5 \xrightarrow[Pb]{H_2} \begin{cases} C_6H_5NH_2 \ (43\%) \\ C_6H_5CH_2NHC_6H_5 \ (12\%) \end{cases} \quad (138)$$

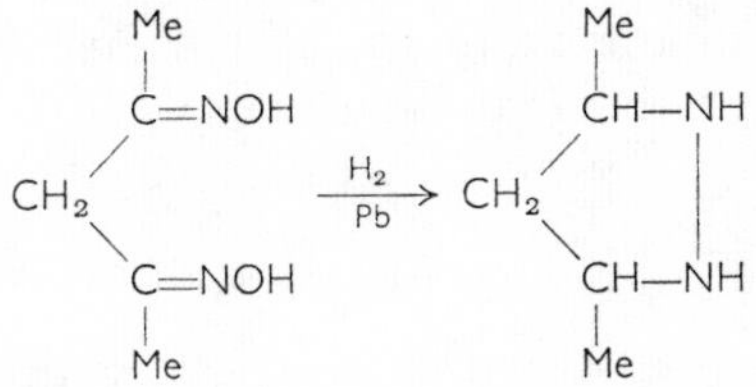

$$CH_2(C(Me){=}NOH)_2 \xrightarrow[Pb]{H_2} \text{cyclic } CH_2(CH(Me)NH)_2 \quad (139)$$

Glyoxylic oxime was reduced to glycine in yields higher than 70% (equation 140)[146].

$$HOOCCH{=}NOH \xrightarrow{H_2} HOOCCH_2NH_2 \quad (140)$$

8. Stereochemistry of hydrogenation of azomethine compounds

Benzil monooxime was hydrogenated by the use of palladium catalyst to form *erythro*-diphenylethanolamine (high m.p. isomer)[147] (equation 141). When the conformation of the substrate molecule in the reaction is a planar structure (**44**) as expected from steric and electronic considerations, the resulting hydrogenation product would be the *threo* isomer (equation 142). However, the resulting diphenyl-

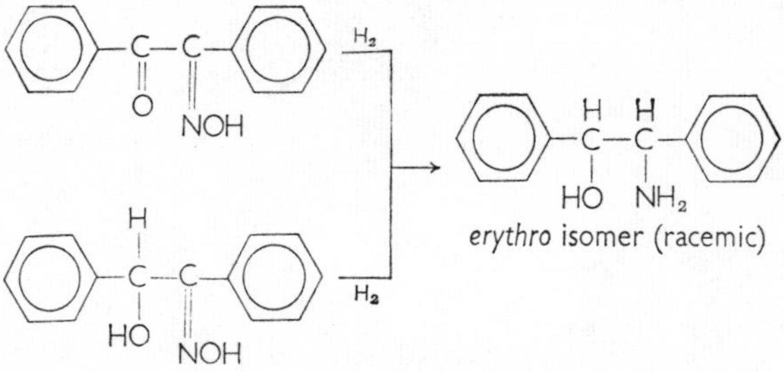

$$C_6H_5C(=O)C(=NOH)C_6H_5 \xrightarrow{H_2};\ C_6H_5CH(OH)C(=NOH)C_6H_5 \xrightarrow{H_2} \rightarrow C_6H_5CH(OH)CH(NH_2)C_6H_5 \quad (141)$$

erythro isomer (racemic)

$$\textbf{(44)} \xrightarrow{H_2} C_6H_5CH(OH)CH(NH_2)C_6H_5 \quad (142)$$

threo isomer (racemic)

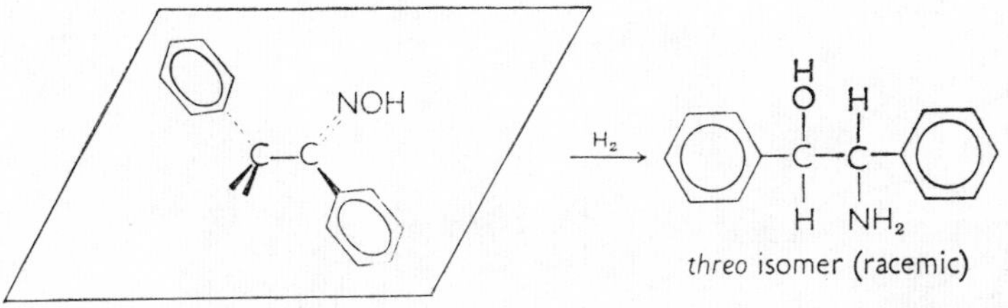

(**44**)

ethanolamine was found to be predominantly an *erythro* form. The catalytic reduction of benzoin oxime[148] also resulted in the *erythro* isomer. It has been considered that the catalytic hydrogenation proceeds as a *cis* addition to the double bond. Therefore, the conformation of the substrate molecule should be cisoidal, which is contrary to the accepted idea of the molecular conformation. Similar results have been reported in the synthesis of threonine (equations 143, 144)[149–152] and phenylserine (equation 145)[153].

$$CH_3COC(=NOH)COOEt \xrightarrow[PtO_2,Ni]{H_2} \longrightarrow CH_3CH(OH){-}CH(NH_2)COOH \quad \textit{erythro} \tag{143}$$

$$CH_3COCH(NHAc)COOEt \xrightarrow[Ni]{H_2} \longrightarrow CH_3CH(OH){-}CH(NH_2)COOH \quad \textit{erythro} \tag{144}$$

$$C_6H_5{-}COCH(NHAc)COOEt \xrightarrow[Pd]{H_2} \longrightarrow C_6H_5{-}CH(OH){-}CH(NH_2)COOH \quad \textit{erythro} \tag{145}$$

Chang and Hartung[154] proposed a mechanism for the hydrogenation of α-oximino ketones which explains the stereospecificity of

$$(Pd)_n \cdots HO{-}N{=}C(R){-}C(Ar){=}O \xrightarrow[\text{1,4 addition}]{H_2} (Pd)_n \cdots HO{-}N(H){-}C(R){=}C(Ar){-}O{-}H \xrightarrow{H_2} \begin{cases} \text{R; H–C–}NH_2\text{; H–C–OH; Ar} & \text{D-}\textit{erythro} \\ \text{R; }H_2N\text{–C–H; HO–C–H; Ar} & \text{L-}\textit{erythro} \end{cases}$$

$$\xrightarrow{H_2 \text{ 1,2 addition}} \begin{cases} (Pd)_n \cdots H_3\overset{+}{N}{-}C(H)(R){-}C(Ar){=}O \xrightarrow{H_2} \text{R; }H_2N\text{–C–H; HO–C–H; Ar} \quad \text{L-}\textit{erythro} \\ (Pd)_n \cdots H_3\overset{+}{N}{-}C(R)(H){-}C(Ar){=}O \xrightarrow{H_2} \text{R; H–C–}NH_2\text{; H–C–OH; Ar} \quad \text{D-}\textit{erythro} \end{cases} \tag{146}$$

the reaction whereby a single racemic modification (*erythro* form) is produced. The polar oxygen and nitrogen atoms of the substrate molecule absorb on the metal catalyst surface to form a rigid ring-like structure. The mechanism proposed by these authors is shown in equation (146).

The intermediate cyclic metal chelate compound could be shown in structure **45**. This intermediate thus formed is then absorbed on the catalyst surface and the *cis* addition of hydrogen results in predominantly *erythro* isomer.

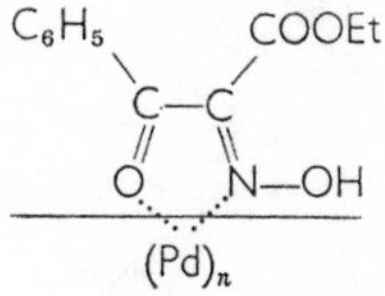

C_6H_5 COOEt

C—C

O N—OH

$(Pd)_n$

(45)

Overberger et al.[155] synthesized α,α'-dimethyldibenzylamine by catalytic hydrogenation of the Schiff base prepared from acetophenone and (*S*)-(−)-α-methylbenzylamine. The product is composed mainly of (*S,S*) amine (88%) with *meso* amine (*S,R*) as a minor component (12%)* (equation 147).

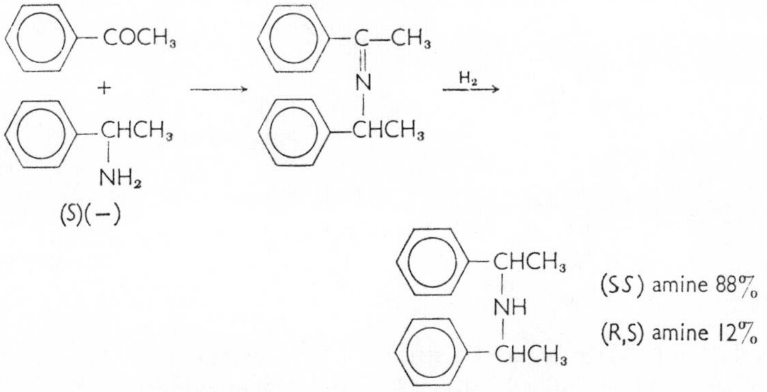

$C_6H_5COCH_3$ + $C_6H_5CH(NH_2)CH_3$ (S)(−) ⟶ $C_6H_5C(CH_3){=}N{-}CH(CH_3)C_6H_5$ $\xrightarrow{H_2}$ $C_6H_5CH(CH_3){-}NH{-}CH(CH_3)C_6H_5$

(*S,S*) amine 88%

(R,S) amine 12% (147)

Hiskey and Northrop successfully applied the sterically controlled reaction to the syntheses of α-amino acids (equation 148)[156]. They demonstrated the synthesis of 12–80% optically active amino acids by catalytic hydrogenation and subsequent hydrogenolysis of the

* The authors of the literature[155] use the (*R*)-configuration for (−)-α-methylbenzylamine which has been assigned an (*S*)-configuration.

Schiff bases (**46**) prepared from α-keto acids and optically active α-methylbenzylamine. When (*S*)-(−) amine was used, (*S*)-α-amino acid resulted. Kanai and Mitsui[157] reported the phenylglycine synthesis by the Hiskey reaction and proposed a steric course for the asymmetric

RCOCOOH + C_6H_5–CH(NH_2)CH$_3$ ⟶ RC(=N–C*H(CH$_3$)C_6H_5)COOH (**46**) $\xrightarrow[Pd/C]{H_2}$ RC*H(COOH)–NH–C*H(CH$_3$)C_6H_5 $\xrightarrow[Pd(OH)_2/C]{H_2}$ RC*HCOOH(NH$_2$) (148)

synthesis as illustrated in equation (149).

(**47a**) ⇌ (**47b**) ⟶ (**48**) ⟶ H_2N–C(COOH)(H)(Ph) (*S*) (149)

Harada and Matsumoto[158] studied the steric course by using several α-keto acids and (*S*)- and (*R*)-α-methylbenzylamine and (*S*)- and (*R*)-α-ethylbenzylamine. Accumulated data indicate that the conformation of the substrate molecule could be as illustrated by structure **48** in equation (150). The Schiff base (**48**) might form a five-membered cyclic structure **49** with the catalyst. Then the structure would be absorbed at the less bulky side of the molecule and the hydrogenation reaction would result in the formation of (*S*)-amino acid. The reduction of the Schiff bases (**48**) by the use of sodium borohydride

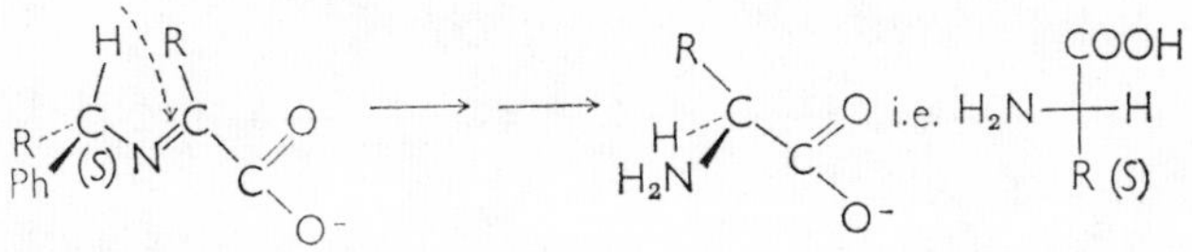

(150)

results in amino acids which have the same configuration as in catalytic hydrogenation; however, optical purities are much lower (10–20%)[159].

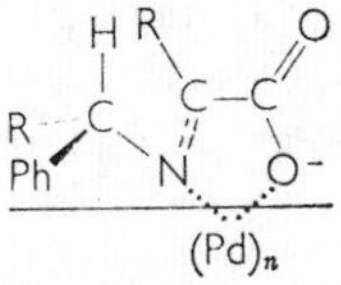

(49)

Harada reported the synthesis of optically active α-amino acids by hydrogenolytic asymmetric transamination[160]. Schiff bases of α-keto acids with (*S*)- and (*R*)-α-phenylglycine in aqueous alkaline solution were hydrogenated and hydrogenolysed to form optically active α-amino acids. When (*R*)-phenylglycine was employed, (*R*)-amino acids were obtained. Optical purity of the product is in the range of 40–60% (equation 151). These reactions are interesting because they are

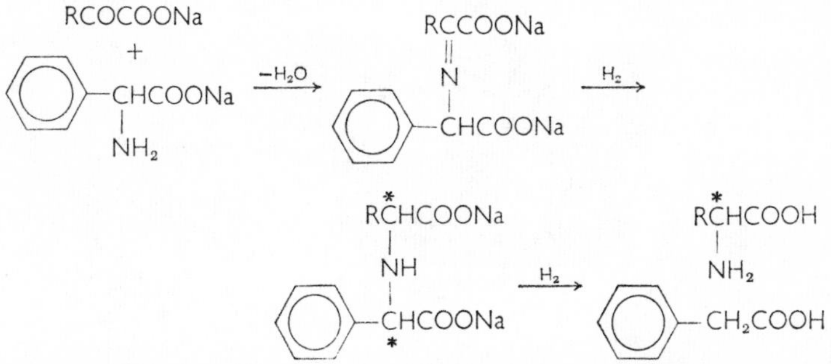

(151)

essentially a kind of asymmetric transamination reaction performed by catalytic hydrogenation and hydrogenolysis. These results throw light on the synthesis of isooctopine[161,162], (*S*)-arginine-(*S*)-alanine, from the Schiff base of L-arginine and pyruvic acid by catalytic hydrogenation in alkaline solution (equation 152).

Catalytic hydrogenation of oximes and benzylamine Schiff bases of menthyl esters of pyruvic acid, α-ketobutyric acid, and phenylglyoxylic acid were studied[163]. Optically active (*R*)-alanine (optical purity 16–25%), (*R*)-α-aminobutyric acid (8–21%), and (*R*)-phenylglycine

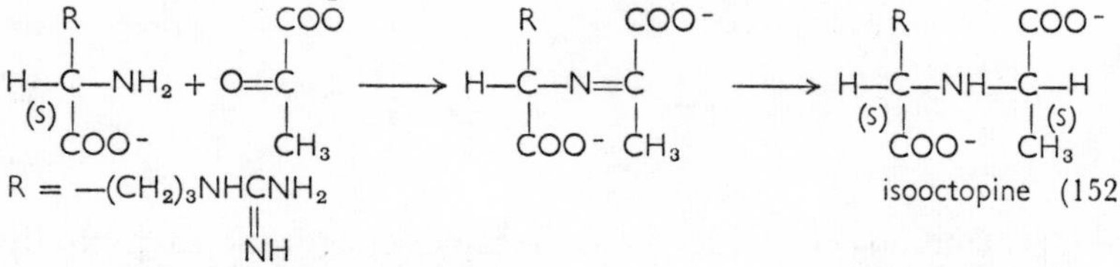

$$\underset{(S)}{H{-}\overset{R}{\underset{COO^-}{C}}{-}NH_2} + O{=}\overset{COO^-}{\underset{CH_3}{C}} \longrightarrow H{-}\overset{R}{\underset{COO^-}{C}}{-}N{=}\overset{COO^-}{\underset{CH_3}{C}} \longrightarrow \underset{(S)}{H{-}\overset{R}{\underset{COO^-}{C}}}{-}NH{-}\underset{(S)}{\overset{COO^-}{\underset{CH_3}{C}}{-}H}$$

$R = -(CH_2)_3NHCNH_2$ (C=NH)

isooctopine (152)

(44–49%) were obtained. The steric course of the reaction could be shown in equation (153). The most stable conformation might be structure **50**, since the C=O and C=N groups repel each other because of their electric dipoles. The menthyl residue is considered to

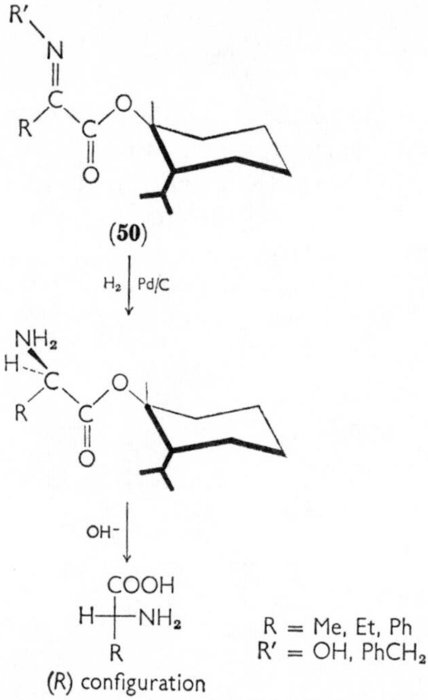

(50)

H_2 | Pd/C

NH_2, H, R, C, O

(153)

OH^-

$$H{-}\overset{COOH}{\underset{R}{C}}{-}NH_2$$

(*R*) configuration

R = Me, Et, Ph
R′ = OH, $PhCH_2$

take a conformation as proposed by Prelog[164]. The molecules would be absorbed with the less bulky side on a catalyst, and the hydrogen atoms would attack the C=N double bond from the backside of the plane of the paper. The resulting α-amino acids, after hydrogenation and hydrolysis, have an (*R*)-configuration. It seems likely that the substrate does not form a five-membered chelate intermediate with the catalyst, probably because the negative character of oxygen in the carbonyl group is not strong enough to combine with the metal catalyst.

To confirm the steric course shown in equation (153), catalytic hydrogenation of the Schiff bases of l-menthyl pyruvate with (*S*) and (*R*)-α-methylbenzylamine and with (*S*) and (*R*)-α-ethylbenzylamine was studied[158]. Observed results are shown in Table 2. These results

TABLE 2. Asymmetric syntheses of α-amino acids from menthyl pyruvate by using optically active α-alkylbenzyl amines[158].

Alkyl group	Configuration of α-alkylbenzyl amine	Yield of alanine (%)	Configuration of alanine	Optical purity of alanine (%)
Me	(*S*)-(−)	57	(*S*)-(+)	19
Et	(*S*)-(−)	56	(*S*)-(+)	15
Me	(*R*)-(+)	61	(*R*)-(−)	60
Et	(*R*)-(+)	55	(*R*)-(−)	36

suggest that the most preferred conformation of the substrate could be structures **51** and **52** in equations (154) and (155). In structure **51**, (*R*)-amine and the menthyl group cooperate with each other to yield a higher optical purity of the product; in structure **52**, the steric effects of (*S*)-amine and the menthyl group are reversed to give a lower optical purity.

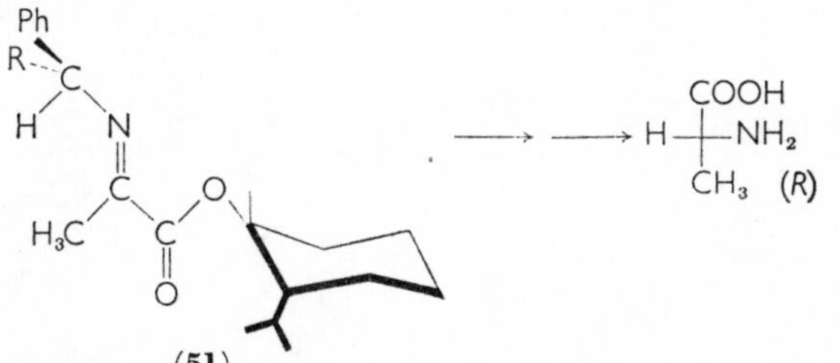

(**51**) ⟶ ⟶ COOH, H–C–NH_2, CH_3 (*R*) (154)

(**52**) ⟶ ⟶ COOH, H_2N–C–H, CH_3 (*S*) (155)

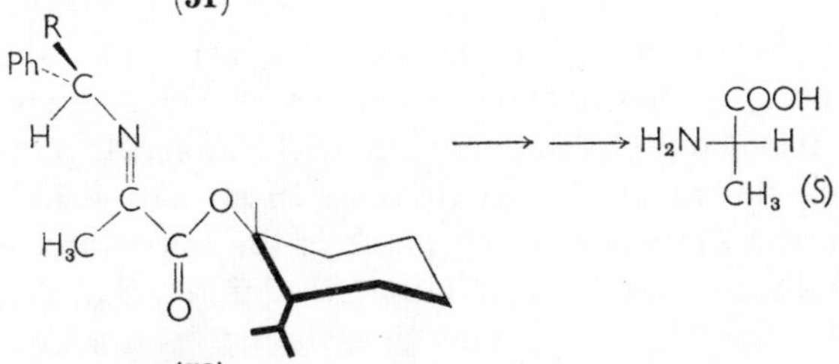

Hiskey and Northrop[165] reported alanylalanine formation from the Schiff base of benzylamine and pyruvyl-(*S*)-alanine by catalytic hydrogenation. They obtained (*R*)-alanyl-(*S*)-alanine and (*S*)-alanyl-(*S*)-alanine in the ratio of 2:1, which differed from the results expected by the application of the Prelog rule[164]. Kanai and Mitsu[157] suggested that the two carbonyl groups of the substrate molecule might be in the cisoidal conformation. Harada and Matsumoto[158] studied the catalytic hydrogenation of oximes of *N*-(*S*)-(−) and (*R*)-(+)-α-methylbenzylbenzoylformamide (**53**) and *N*-(*S*)-(−) and (*R*)-(+)-α-ethylbenzylbenzoylformamide (**54**). In these studies, (*R*)-(−)-phenylglycine was obtained when (*S*)-(−)-α-methylbenzylamine was used. However, when (*S*)-(−)-α-ethylbenzylamine was used, (*S*)-(+)-phenylglycine was synthesized.

The steric course of this asymmetric synthesis could be explained as shown in equations (156) and (157)[158]. It seems reasonable to assume that both structures **53** and **54** could take a cisoidal conformation. The carbonyl group of the amide bond is partially charged, because the amide bond could be regarded as a resonance hybrid of the lactam and dipolar structure[166]. Therefore the carbonyl group and hydroxyimino

R, H, N, Ph, C—C, C, CH_3, HO—N, O, H, $(Pd)_n$ ⟶ R, H, N, Ph, H_2N, C—C, C, H, H, O, CH_3 ⟶ COOH, H—+—NH_2, R (*R*)

(**53**) (156)

R, H, N, C_2H_5 (i-Pr), C—C, C, Ph (COOBu-i), HO—N, O, H, $(Pd)_n$ ⟶ ⟶ COOH, H_2N—+—H, R (*S*)

(**54**) (157)

group might be absorbed on the catalyst surface to form a five-membered chelate intermediate. Then the cyclic complex molecule could be absorbed on the less bulky side of the molecule and hydrogen would attack the molecule. In the case of structure **53**, hydrogen attacks from the front side of the plane of the paper and the *cis* addition of hydrogen would result in (*R*)-amino acid. However, in structure **54**, when (*S*)-(−)-α-ethylbenzylamine was used, the ethyl group could not occupy the same limited space between the substrate and catalyst as the menthyl group of (*S*)-(−)-α-methylbenzylamine did. Because of the steric hindrance, the ethyl group might rotate upward and a terminal hydrogen atom might be at the closest position to the catalyst. The clear difference of the configuration of the product by the use of α-methyl- and α-ethylbenzylamine might be due to the strong interaction and steric hindrance between the molecule and the catalyst. Assumed conformations of structures **53** and **54** in the form of the molecular model seem to be reasonable to explain the configuration of the reaction products. Further studies of this type of reaction also support the assumed steric course of the asymmetric reaction. *N*-Pyruvyl-(*S*)-(+)-alanine isobutyl ester and benzylamine resulted in (*R*)-alanyl-(*S*)-alanine. However, the benzylamine Schiff base of *N*-pyruvyl-(*S*)-(+)-valine isobutyl ester resulted in (*S*)-alanyl-(*S*)-valine[158].

III. ACKNOWLEDGEMENT

This work was aided by Grant NsG–689 of the National Aeronautics and Space Administration, U.S.A.

IV. REFERENCES

1. W. Miller and J. Plöchl, *Ber.*, **25**, 2020 (1892); **26**, 1545 (1893); **31**, 2699 (1899).
2. J. R. Bailey and H. L. Lochte, *J. Am. Chem. Soc.*, **39**, 2443 (1917).
3. G. E. P. Smith, Jr. and F. W. Bergstrom, *J. Am. Chem. Soc.*, **56**, 2095 (1934).
4. R. Tiollais, *Bull. Soc. Chim. France*, **14**, 959 (1947).
5. A. Dornow and S. Lüpfert., *Chem. Ber.*, **89**, 2718 (1956).
6. G. H. Harris, B. R. Harriman and K. W. Wheeler, *J. Am. Chem. Soc.*, **68**, 846 (1946).
7. J. Collazos, *Chim. Ind.* (*Paris*), **86**, 47 (1961).
8. F. Adickes, *J. Prakt. Chem.*, **161**, 271 (1943).
9. C. C. Porter and L. Hellerman, *J. Am. Chem. Soc.*, **61**, 754 (1939).
10. H. A. Lillevik, R. L. Hossfeld, H. V. Lindstrom, R. T. Arnold and R. A. Gortner, *J. Org. Chem.*, **7**, 164 (1942).
11. C. C. Porter and L. Hellerman, *J. Am. Chem. Soc.*, **66**, 1652 (1944).
12. N. J. Leonard and A. S. Hay, *J. Am. Chem. Soc.*, **78**, 1984 (1956).
13. E. Knoevenagel, *Ber.*, **37**, 4087 (1904).
14. H. Bucherer and A. Schwalbe, *Ber.*, **39**, 2810 (1906).
15. L. Neelakantan and W. H. Hartung, *J. Org. Chem.*, **24**, 1943 (1959).
16. W. V. Miller and J. Plöchl, *Ber.*, **25**, 2032 (1892).
17. H. Decker and P. Becker, *Ann. Chem.*, **395**, 362 (1913).
18. E. H. Woodruff, J. P. Lambooy and W. E. Burt, *J. Am. Chem. Soc.*, **62**, 922 (1940).
19. H. G. Reiber and T. D. Stewart, *J. Am. Chem. Soc.*, **62**, 3026 (1940).
20. N. J. Leonard and J. V. Paukstelis, *J. Org. Chem.*, **28**, 3021 (1963).
21. A. L. Morrison and H. Rinderknecht, *J. Chem. Soc.*, 1478 (1950).
22. B. I. Zenitz, E. B. Macks and M. L. Moore, *J. Am. Chem. Soc.*, **62**, 1117 (1947).

23. A. R. Surrey, *J. Am. Chem. Soc.*, **69**, 2911 (1947).
24. H. Erlenmeyer and V. Oberlin, *Helv. Chim. Acta*, **30**, 1329 (1947).
25. G. W. Stacy and R. J. Morath, *J. Am. Chem. Soc.*, **74**, 3885 (1952).
26. H. Gilman and J. B. Dichey, *J. Am. Chem. Soc.*, **52**, 4574 (1930).
27. G. W. Stacy, R. I. Day and R. J. Morath, *J. Am. Chem. Soc.*, **77**, 3869 (1955).
28. G. W. Stacy and R. J. Morath, *J. Am. Chem. Soc.*, **74**, 3885 (1952).
29. H. Gilman, J. L. Towle and R. K. Ingham, *J. Am. Chem. Soc.*, **76**, 2920 (1954).
30. B. Magnusson, *Acta Chem. Scand.*, **16**, 1536 (1962).
31. N. J. Leonard and A. S. Hay, *J. Am. Chem. Soc.*, **78**, 1984 (1956).
32. A. Dornow, A. Müller and S. Lüpfert, *Ann. Chem.*, **594**, 191 (1955).
33. N. S. Kozlov and Z. A. Ahramova, *Dokl. Akad. Nauk USSR*, **132**, 839 (1960); *Chem. Abstr.*, **54**, 20977 (1960).
34. N. S. Kozlov and I. A. Shur, *Zh. Obshch. Khim.*, **30**, 2492 (1960); *Chem. Abstr.*, **55**, 14459 (1961).
35. N. S. Kozlov, S. Ya Chumakov, L. Yu. Pinegine and I. A. Shur, *Uch. Zap. Permsk, Gos. Pedagog Inst.*, 269 (1961); *Chem. Abstr.*, **58**, 5558 (1963).
36. N. S. Kozlov, E. A. Britan and N. D. Zneva, *Zh. Obshch. Khim.*, **34**, 298 (1964); *Chem. Abstr.*, **60**, 15765 (1964).
37. A. H. Blatt and N. Gross, *J. Org. Chem.*, **29**, 3306 (1964).
38. F. E. Francis, *J. Chem. Soc.*, **75**, 865 (1899); **77**, 1191 (1900); F. E. Francis and E. B. Ludlam, *J. Chem. Soc.*, **81**, 956 (1902).
39. C. Mayer, *Bull. Soc. Chim. France*, **31**, 985 (1904).
40. B. I. Kurtev and N. M. Mollov, *Dokl. Akad. Nauk USSR*, **101**, 1069 (1955); *Chem. Abstr.*, **50**, 3416 (1956).
41. N. M. Mollov, N. Kh. Spasovska, *Compt. Rend. Acad. Bulgare Sci.*, **9**, 45 (1956); *Chem. Abstr.*, **51**, 17847 (1957).
42. B. I. Kurtev and N. Mollov, *Acta Chim. Acad. Sci. Hung.*, **18**, 429 (1959); *Chem. Abstr.*, **53**, 21805 (1959).
43. N. M. Mollov, M. V. Bozhilova and V. I. Baeva, *Compt. Rend. Acad. Bulgare Sci.*, **13**, 307 (1960); *Chem. Abstr.*, **55**, 23431 (1961).
44. N. M. Mollov and P. Ch. Petrova, *Izv. Inst. Org. Khim. Bulgar Akad. Nauk*, **1**, 127 (1964); *Chem. Abstr.*, **62**, 5223 (1965).
45. N. Marecov, G. Vasilev and V. Aleksieva, *Compt. Rend. Acad. Bulgare Sci.*, **10**, 217 (1957); *Chem. Abstr.*, **52**, 7243 (1958).
46. T. I. Bieber, R. Sites and Y. Chiang, *J. Org. Chem.*, **23**, 300 (1958).
47. N. Marekov and N. Petsev, *Compt. Rend. Acad. Bulgare Sci.*, **10**, 473 (1957); *Chem. Abstr.*, **52**, 12812 (1958).
48. A. Dornow and S. Lüpfert, *Chem. Ber.*, **89**, 2718 (1956).
49. A. Lukasiewicz, *Bull. Acad. Polon. Sci. Ser. Chim.*, **11**, 15 (1963); *Chem. Abstr.*, **60**, 9172 (1964).
50. A. Lukasiewicz, *Tetrahedron*, **20**, 1 (1964).
51. P. Nagy, *Szegedi Pedagog. Foiskola Evkonyve*, Part II, 205 (1960); *Chem. Abstr.*, **56**, 3394 (1962).
52. C. Mayer, *Bull. Soc. Chim. France*, **7**, 481 (1940).
53. J. C. Sheehan and J. J. Ryan, *J. Am. Chem. Soc.*, **73**, 1204 (1951).
54. H. Böhme and K. Hartke, *Chem. Ber.*, **96**, 600 (1963).
55. H. Böhme, S. Ebel and K. Hartke, *Chem. Ber.*, **98**, 1463 (1965).

56. F. Bergman, *J. Am. Chem. Soc.*, **60**, 2811 (1938).
57. M. R. Snyder, R. B. Hasbrouck and J. F. Richardson, *J. Am. Chem. Soc.*, **61**, 3558 (1939).
58. L. Tamayo and J. F. Vanes, *Anales Fis. Quim.* (*Madrid*), **43**, 777 (1947); *Chem. Abstr.*, **42**, 1571 (1948).
59. G. Caronna, *Gazz. Chim. Ital.*, **78**, 38 (1948); *Chem. Abstr.*, **42**, 6760 (1948).
60. B. I. Ardashev and Z. D. Markova, *Zh. Obshch. Khim.*, **21**, 1505 (1951); *Chem. Abstr.*, **46**, 5005 (1952).
61. M. Busch, *Ber.*, **37**, 2691 (1904); **38**, 1761 (1905).
62. R. B. Moffett and W. M. Hoehn, *J. Am. Chem. Soc.*, **69**, 1792 (1947).
63. K. N. Campbell, C. H. Helbing, M. P. Florkowski and B. K. Campbell, *J. Am. Chem. Soc.*, **70**, 3868 (1948).
64. F. Sachs and L. Sachs, *Ber.*, **37**, 3088 (1904).
65. P. M. Maginnity and T. J. Gair, *J. Am. Chem. Soc.*, **74**, 4958 (1952).
66. J. Ficini and H. Normant, *Bull. Soc. Chim. France*, 1454 (1957).
67. H. Thies, H. Schönenberger and K. Borah, *Naturwissenschaften*, **46**, 378 (1959).
68. H. Schönenberger, H. Thies, A. Zeller and K. Borah, *Naturwissenschaften* **48**, 129 (1961).
69. H. Thies and H. Schönenberger, *Arch. Pharm.*, **26**, 408 (1956).
70. H. Thies and H. Schönenberger, *Chem. Ber.*, **89**, 1918 (1956).
71. B. L. Emling, R. J. Horvath, A. J. Saraceno, E. F. Ellermeyer, L. Haile and L. D. Hudac, *J. Org. Chem.*, **27**, 657 (1959).
72. H. Gilman, J. E. Kirby and C. R. Kinney, *J. Am. Chem. Soc.*, **51**, 2252 (1929).
73. H. Gilman and J. Morton, *J. Am. Chem. Soc.*, **70**, 2514 (1948).
74. M. S. Kharasch and O. Reinmuth, *Grignard Reaction of Nonmetallic Substances*, Prentice-Hall, New York, N.Y., 1954.
75. R. E. Dessy and R. M. Salinger, *J. Am. Chem. Soc.*, **83**, 3530 (1961).
76. G. Stork and S. R. Dowd, *J. Am. Chem. Soc.*, **85**, 2178 (1963).
77. B. L. Embing, R. J. Horvath, A. J. Saraceno, E. F. Ellermeyer, L. Haile and L. D. Hudac, *J. Org. Chem.*, **24**, 657 (1959).
78. H. Gilman and J. E. Kirby, *J. Am. Chem. Soc.*, **55**, 1265 (1933).
79. B. M. Mikhailov and K. N. Kurdyumova, *Zh. Obshch. Khim.*, **28**, 355 (1958); *Chem. Abstr.*, **52**, 13685 (1958).
80. R. F. Shuman and E. D. Amstutz, *Rec. Trav. Chim.*, **84**, 441 (1965).
81. B. M. Mikhailov and K. N. Kurdyumova, *Zh. Obshch. Khim.*, **25**, 1734 (1955); *Chem. Abstr.*, **50**, 5591 (1965).
82. B. M. Mikhailov and K. N. Kurdyumova, *Zh. Obshch. Khim.*, **28**, 355 (1958); *Chem. Abstr.*, **52**, 13685 (1958).
83. H. Thies, H. Schönenberger and K. H. Bauer, *Arch. Pharm.*, **291/63**, 248 (1958).
84. C. D. Hund and J. S. Strong, *J. Am. Chem. Soc.*, **72**, 4813 (1950).
85. L. M. Kozlov and E. F. Fink, *Tr. Kazan. Khim. Tekhnol. Inst. im. S. M. Kirova*, **21**, 147 (1956); *Chem. Abstr.*, **51**, 11983 (1957).
86. A. Dornow and A. Frese, *Ann. Chem.*, **578**, 122 (1952).
87. H. J. Bestmann and F. Seng, *Angew. Chem.* (*Intern. Ed. Engl.*), **2**, 393 (1963).
88. H. J. Bestmann and F. Seng, *Tetrahedron*, **21**, 1373 (1965).

89. K. Issleib and R. D. Bleck, *Z. Anorg. Allgem. Chem.*, **336**, 234 (1965).
90. E. C. Ladd and M. P. Harvey, *Can. Pat.* 509,034 (1955); *Chem. Abstr.*, **50**, 10760 (1956).
91. A. Rosenthal, R. F. Astbury and A. Hubscher, *J. Org. Chem.*, **23**, 1037 (1958).
92. A. Rosenthal and M. Yalpani, *Can. J. Chem.*, **43**, 711 (1965).
93. W. S. Emerson, 'The Preparation of Amines by Reductive Alkylation', *Org. Reactions*, Vol. 4, John Wiley and Sons, New York, 1948, p. 174.
94. B. M. Vanderbilt, *U.S. Pat.* 2,219,879; *Chem. Abstr.*, **35**, 1065 (1941).
95. H. Adkins and C. F. Winans, *U.S. Pat.* 2,045,574; *Chem. Abstr.*, **30**, 5589 (1936).
96. A. Skita, F. Keil and H. Havemann, *Ber.*, **66**, 1400 (1933).
97. A. C. Cope and E. M. Hancock, *J. Am. Chem. Soc.*, **66**, 1453 (1944).
98. W. S. Emerson and W. D. Robb, *J. Am. Chem. Soc.*, **61**, 3145 (1939).
99. K. N. Campbell, A. H. Sommers and B. K. Campbell, *J. Am. Chem. Soc.*, **66**, 82 (1944).
100. J. W. Magee and H. R. Henze, *J. Am. Chem. Soc.*, **62**, 910 (1940).
101. K. Kindler, *Ann. Chem.*, **485**, 113 (1931).
102. E. J. Schwoegler and H. Adkins, *J. Am. Chem. Soc.*, **61**, 3499 (1939).
103. C. F. Winans, *J. Am. Chem. Soc.*, **61**, 3566 (1939).
104. M. R. Cantarel, *Compt. Rend.*, **210**, 403 (1940).
105. J. C. Robinson, Jr. and H. R. Snyder, *Org. Synth.*, **23**, 68 (1943).
106. P. Couturier, *Compt. Rend.*, **207**, 345 (1938); *Chem. Abstr.*, **32**, 8389 (1938).
107. C. F. Winans, *J. Am. Chem. Soc.*, **61**, 3566 (1939).
108. F. Knoop, H. Oesterlin, *Z. Physiol. Chem.*, **148**, 294 (1925).
109. A. Darapsky, J. Garmscheid, C. Kreuter, E. Engelmann, W. Eengels and W. Trinius, *J. Prakt. Chem.*, **146**, 219 (1936).
110. F. Knoop, H. Oesterlin, *Z. Physiol. Chem.*, **170**, 186 (1927).
111. F. Kögl, J. Halberstadt, T. J. Basendregt, *Rec. Trav. Chim.*, **68**, 387 (1949).
112. H. Zaunschirm, *Ann. Chem.*, **245**, 279 (1888).
113. E. Bamberger and J. Müller, *Ann. Chem.*, **313**, 97 (1900).
114. G. H. Coleman and J. J. Carnes, *Proc. Iowa Acad. Sci.*, **119**, 288 (1942); *Chem. Abstr.*, **37**, 5703 (1943).
115. N. A. Shepard and A. A. Ticknor, *J. Am. Chem. Soc.*, **38**, 381 (1916).
116. W. S. Emerson, F. W. Neumann and T. P. Moundres, *J. Am. Chem. Soc.*, **63**, 972 (1941).
117. A. R. Surrey and H. F. Hammer, *J. Am. Chem. Soc.*, **66**, 2127 (1944).
118. G. Hildebrandt, *U.S. Pat.* 2,344,356 (1944); *Chem. Abstr.*, **38**, 3421 (1944).
119. V. Hahn, R. Hansal, I. Markovcic and D. Vargazon, *Arhiv Kem.*, **26**, 21 (1954); *Chem. Abstr.*, **50**, 292 (1956).
120. R. Hansal, D. Vargazon and V. Hahn, *Arhiv Kem.*, **27**, 33 (1955); *Chem. Abstr.*, **50**, 4894 (1956).
121. H. Schönenberger, H. Thies and A. Rapp, *Arch. Pharm.*, **298**, 635 (1965) *Chem. Abstr.*, **63**, 16236 (1965).
122. A. H. Sommers and S. E. Aaland, *J. Org. Chem.*, **21**, 484 (1956).
123. R. F. Nystrom and W. G. Brown, *J. Am. Chem. Soc.*, **70**, 3738 (1948).
124. J. H. Billman and K. M. Tai, *J. Org. Chem.*, **23**, 535 (1958).

125. Z. Horii, T. Sakai and T. Inoi, *J. Pharm. Soc. Japan*, **75**, 1161 (1955); *Chem. Abstr.*, **50**, 7756 (1956).
126. J. H. Billman and A. C. Diesing, *J. Org. Chem.*, **22**, 1068 (1957).
127. G. N. Walker, M. A. Moor and B. N. Weaver, *J. Org. Chem.*, **26**, 2740 (1961).
128. T. P. Johnston and A. Gallagher, *J. Org. Chem.*, **29**, 2452 (1962).
129. P. Quitt, J. Hellerbach and K. Vogler, *Helv. Chim. Acta*, **46**, 327 (1963).
130. B. Witkop and J. B. Patrick, *J. Am. Chem. Soc.*, **75**, 4474 (1953).
131. J. H. Billman and J. W. McDowell, *J. Org. Chem.*, **26**, 1437 (1961).
132. J. H. Billman and J. W. McDowell, *J. Org. Chem.*, **27**, 2640 (1962).
133. R. Leuckart, *Ber.*, **18**, 2341 (1885).
134. W. Eschweiler, *Ber.*, **38**, 880 (1905).
135. M. L. Moore, *Organic Reactions*, Vol. 5, John Wiley and Sons, New York, 1949, p. 301.
136. E. R. Alexander and R. B. Wildman, *J. Am. Chem. Soc.*, **70**, 1187 (1948).
137. Z. Eckstein and A. Lukasiewicz, *Bull. Acad. Polon. Sci., Ser. Sci. Chim., Geol. Geograph.*, **7**, 789 (1959); *Chem. Abstr.*, **54**, 24679 (1960).
138. R. Baltzly and O. Kauder, *J. Org. Chem.*, **16**, 173 (1951).
139. A. N. Kost and I. I. Grandberg, *Zh. Obshch. Khim.*, **25**, 1719 (1955); *Chem. Abstr.*, **50**, 5544 (1956).
140. P. L. de Benneville and J. H. Macartney, *J. Am. Chem. Soc.*, **72**, 3073 (1950).
141. A. Lukasiewicz, *Tetrahedron*, **19**, 1789 (1963).
142. A. N. Kost and L. G. Yudin, *Zh. Obshch. Khim.*, **25**, 1947 (1955); *Chem. Abstr.*, **50**, 8644 (1956).
143. H. D. Law, *J. Chem. Soc.*, **101**, 154 (1912).
144. J. Tafel and E. Pfeffermann, *Ber.*, **35**, 1510 (1902).
145. J. Tafel and E. Pfeffermann, *Ber.*, **36**, 219 (1903).
146. H. D. C. Rapson and A. E. Bird, *J. Appl. Chem. (London)*, **13**, 233 (1963).
147. T. Ishimaru, *J. Chem. Soc. Japan (Pure Chem. Sec.)*, **81**, 643 (1960).
148. J. Weijland, K. Pfister, E. F. Swanezy, C. A. Robinson and M. Tishler, *J. Am. Chem. Soc.*, **73**, 1216 (1951).
149. H. Adkins and E. W. Reeve, *J. Am. Chem. Soc.*, **60**, 1328 (1938).
150. N. F. Albertson, B. F. Tullar and J. A. King, *J. Am. Chem. Soc.*, **70**, 1150 (1948).
151. K. Pfister, C. A. Robinson, A. C. Shabica and M. Tishler, *J. Am. Chem. Soc.*, **70**, 2297 (1948).
152. K. Pfister, C. A. Robinson, A. C. Shabica and M. Tishler, *J. Am. Chem. Soc.*, **71**, 1101 (1949).
153. W. A. Bolhofer, *J. Am. Chem. Soc.*, **74**, 5459 (1952).
154. Y. Chang and W. H. Hartung, *J. Am. Chem. Soc.*, **75**, 89 (1953).
155. C. G. Overberger, N. P. Marullo and R. G. Hiskey, *J. Am. Chem. Soc.*, **83**, 1374 (1960).
156. R. G. Hiskey and R. C. Northrop, *J. Am. Chem. Soc.*, **83**, 4798 (1961).
157. A. Kanai and S. Mitsui, *J. Chem. Soc. Japan (Pure Chem. Sec.)*, **89**, 183 (1966).
158. K. Harada and K. Matsumoto, *J. Org. Chem.*, **32**, 1794 (1967).
159. K. Harada and J. Oh-hashi, unpublished results.
160. K. Harada, *Nature*, **212**, 1571 (1966); *J. Org. Chem.*, **32**, 1799 (1967).

161. F. Knoop and C. Martius, *Z. Physiol. Chem.*, **258**, 238 (1939).
162. R. M. Herbst and E. A. Swart, *J. Org. Chem.*, **11**, 368 (1946).
163. K. Matsumoto and K. Harada, *J. Org. Chem.*, **31**, 1956 (1966).
164. V. Prelog, *Helv. Chim. Acta*, **36**, 308 (1953).
165. R. G. Hiskey and R. C. Northrop, *J. Am. Chem. Soc.*, **87**, 1753 (1965).
166. L. Pauling, *The Nature of the Chemical Bond*, 3rd Ed., Cornell University Press, Ithaca, New York, 1960, p. 281.

CHAPTER 7

Cycloaddition reactions of carbon–nitrogen double bonds

JEAN-PIERRE ANSELME

University of Massachusetts, Boston, U.S.A.

I. INTRODUCTION

For several years, the Diels–Alder reaction[1] was the only widely useful example of the so-called cycloaddition reactions. The extensive generalization by Huisgen and his school[2] of the concept of 1,3-dipolar cycloadditions, first recognized by Smith[3], has opened new avenues for investigations. The dimerization of olefins[4], as well as the addition of carbenes and nitrenes to unsaturated centres has extended the series to include three-, four-, five- and six-membered ring systems. Huisgen, Grashey and Sauer[2] in a previous volume in this series have reviewed cycloaddition reactions of alkenes, and the present chapter will deal with the various cycloadditions of the $\rangle$C=N— bond. Mechanistic aspects will be discussed only when solid evidence warrants such discussions, since even some extensive investigations in this field have yielded conflicting conclusions.

Conceptually, cycloadditions constitute one of the simplest reactions of organic chemistry. In theory, a single addition product (excluding the formation of isomers) is obtained from two reactants without elemental loss or gain, or without the fundamental alteration of either reactants. For example, 1,3-butadiene could react with ethylene to give 100% cyclohexene. Huisgen, Grashey and Sauer[2] have discussed the concept of cycloaddition in great detail, although the definition of a cycloaddition reaction remains the subject of wide discussion[5]. One point is quite certain, that is with such a broad spectrum of variations possible in both reactants undergoing cycloaddition, it would indeed be surprising that one inviolable mechanism should account for all the reactions, even those of the same type[6]. The extent to which one bond is formed with respect to the second bond will probably remain a moot point since even kinetic evidence can only tell us that the two bonds do not form exactly at the same time[7]. The term 'concerted' ought to be taken in the sense that no intermediate which would be capable of being isolated or trapped is formed.

In contrast with the tremendous number of investigations which have characterized cycloadditions to olefins[2], cycloadditions to imines seem almost insignificant. This state of affairs can be traced to several factors.

The greatly reduced stability of imines makes their use as 'shelf reagents' less rewarding since upon standing, many of them deteriorate. Thus, the more stable partners of cycloadditions such as alkenes, alkynes, nitriles and the like are chosen in preference to imines, particularly since investigation in the general field of cycloadditions to these systems is still not complete despite the enormous volume of publications. Furthermore, the possibility of tautomerism in *C* and/or *N*-alkyl substituted imines creates additional difficulties[8].

II. THREE-MEMBERED RINGS

Although potentially a very useful method for the preparation of aziridines (**1**) and diaziridines (**2**), the cycloaddition of carbenes and

$$>C{=}N{-} + {-}\ddot{C}{-} \longrightarrow \text{aziridine } (\mathbf{1}) \qquad (1)$$

$$>C{=}N{-} + {-}\ddot{N}\colon \longrightarrow \text{aziridine } (\mathbf{2}) \qquad (2)$$

nitrenes to imines has not been investigated very extensively. Furthermore, despite its explosive growth, the chemistry of carbenes is very young and that of nitrenes is of even more recent interest. The first report of the addition of a carbene to an imine was that of Fields and Sandri[9] (**3**, Ar = Ph) who showed that dichlorocarbene added to

$$Ph{-}CH{=}N{-}Ar + {:}CCl_2 \longrightarrow Ph{-}CH{-}N{-}Ar \text{ (ring closed by } CCl_2\text{)} \quad (\mathbf{3}) \qquad (3)$$

benzalaniline to give the corresponding dichloroaziridine. Although Cook and Fields[10] had applied the reaction to other anils (Ar = p-ClC_6H_4, p-$C_2H_5OC_6H_4$), the reaction has not been investigated to any large extent. Deyrup and Grunewald[11] have shown that the reaction of diphenylmethylene aniline (**4**) with chloroform and potassium *t*-butoxide gave 3,3-dichloro-1,2,2-triphenylaziridine (**5**).

$$Ph_2C{=}N{-}Ph \; (\mathbf{4}) + HCCl_3 \xrightarrow{t\text{-}BuO^-} Ph_2C{-}N{-}Ph \text{ (ring closed by } CCl_2\text{)} \; (\mathbf{5}) \qquad (4)$$

The possibility of a nucleophilic attack of trichloromethyl anion on the imine, followed by a displacement of chloride ion to give the aziridine (reaction 5) was eliminated on the basis of the following

$$Ph_2C{=}N{-}Ph + \bar{C}Cl_3 \longrightarrow Ph_2C(CCl_3){-}\bar{N}{-}Ph \longrightarrow Ph_2C{-}N(Ph) \text{ (ring closed by } CCl_2\text{)} + Cl^- \qquad (5)$$

experimental evidence. The reaction of 2,2-diphenyl-2-anilino-1,1,1-trichloroethane (reaction 6) with potassium *t*-butoxide gave **4** and **5** in

$$Ph_2C(CCl_3){-}NH{-}Ph + t\text{-}BuO^- \longrightarrow \mathbf{4} \text{ and } \mathbf{5} \qquad (6)$$

a ratio of 75:25. Furthermore, it was shown that less than 5% of the aziridine was generated if the reaction was carried out in the presence of tetramethylethylene. This indicated that at least 80% (and probably more) of the aziridine arose via the direct addition of the dichlorocarbene to the imine.

$$Ph_2C(CCl_3){-}NHPh \xrightarrow{t\text{-BuO}^-} Ph_2C(CCl_3){-}\bar{N}{-}Ph \longrightarrow Ph_2C{=}N{-}Ph + CCl_3^- \qquad (7)$$

$$CCl_3^- \xrightarrow{-Cl^-} :CCl_2 \xrightarrow{Ph_2C{=}NPh} \mathbf{5}$$

Further evidence was forthcoming when the same authors extended the reaction to prepare the monochloroaziridine; the reaction of benzalaniline with lithium dichloromethane gave the *cis*-aziridine (**6**).

$$PhCH{=}N{-}Ph + LiCHCl_2 \xrightarrow{<70^{\circ}} \text{cis-2-chloro-3-phenyl-1-phenylaziridine (Cl, H on C; Ph, H on C; N–Ph)} \qquad (8)$$

(**6**)

A mechanism involving addition of dichloromethide anion could not account for the observed stereoselectivity of the addition. Klaman, Wache, Ulm and Nerdel[12] also reported the reaction of anils with dichlorocarbene generated from chloroform by means of the ethylene oxide and bromide ion[13]. The ethylene oxide present in the reaction opened the aziridine ring to give the *N*-substituted oxindoles **7**.

$$R{-}C_6H_4{-}N{=}CPh_2 + \text{ethylene oxide} + HCCl_3 + Et_4\overset{+}{N}\overset{-}{Br} \longrightarrow R{-}C_6H_4{-}N(CPh_2)(CCl_2)\ \text{(aziridine)}$$

$R = H, CH_3$

$$R{-}C_6H_4{-}N(CPh_2)(CCl_2) + \text{ethylene oxide} \longrightarrow \text{5-R-3,3-diphenyl-1-}(CH_2CH_2Cl)\text{oxindole} \qquad (9)$$

(**7**)

This is somewhat related to the ring-opening reaction observed by Cook and Fields[10]. When the pyrrolidine immonium salt of cyclohexanone (**8**) was treated with trichloroacetate anion, the intermediate aziridinium salt underwent ring opening with the anion to give the observed product **9**.

(8) + $Cl_3CC(=O)\bar{O}$ ⟶ [aziridinium, CCl_2] $\xrightarrow{CCl_3CO_2^-}$ $OCCCl_3$ / CCl_2—N ⟶ $OCCCl_3$ / C(=O)—N (9) (10)

Leonard and his group[14] first made extensive use of the principle involved. The reaction of diazomethane with immonium salts gave the corresponding aziridinium salts which, with alcohol, underwent ring expansion. However, even more so than with imines themselves, these reactions probably do not involve a true cycloaddition process. The same would also be true for the reaction of dichlorocarbene with immonium salts[10] (reaction 9).

Speziale and coworkers[15] briefly mention the copper-catalysed decomposition of ethyl diazoacetate in the presence of benzalaniline. The product isolated, ethyl 3-anilinocinnamate, probably arose via tautomerization of the aziridine **10**.

$$PhCH{=}NPh + N_2CHCO_2Et \xrightarrow{Cu, 50^\circ} PhCH{-}CHCO_2Et\ (N{-}Ph)\ \textbf{(10)} \longrightarrow PhC(NHPh){=}CHCO_2Et \quad (11)$$

The reaction of dicarbene **11** with imines has been reported to give the corresponding allenes[16] **12**.

$$2\ R^1R^2C{=}N{-}R + {:}C{=}C{=}C{:}\ \textbf{(11)} \longrightarrow \textbf{(12)} \quad (12)$$

R = i-Pr, R¹ = Ph, R² = H
R = i-Pr, R¹ = R² = Ph

The formation of diaziridines from the reaction of imines with potential nitrene precursors is *not* a cycloaddition reaction in the majority of cases. This point has been emphasized by Schmitz[17,18], the main contributor to this field. Rather, a nucleophilic addition to the imine followed by displacement is indicated by the results obtained by Schmitz[19].

$$R_2C{=}N{-}R^1 + R^2\bar{N}Cl \longrightarrow R_2C{-}\bar{N}{-}R^1(N(R^2){-}Cl) \longrightarrow R_2C\langle N{-}R^1, N{-}R^2 \rangle \qquad (13)$$

Perhaps the only true cases involving nitrene insertions into imines are those reported by Graham. His initial communication[20] describing preparation of diaziridine **14** was later amplified by a more detailed report[21].

$$CH_2{=}N{-}t\text{-alkyl} + HNF_2 \xrightarrow{-HF} \left[H_2C\langle N{-}t\text{-alkyl}, N{-}F \rangle \right] \longrightarrow H_2C\langle N{=}N \rangle \qquad (14)$$

(13) **(14)**

The formation of oxaziranes by the reaction of imines with peroxy acids[22] is not to be classified as a cycloaddition, even though formally it may be regarded as a transfer of an oxygen atom to the imine. Recent mechanistic discussions[23] have suggested that the epoxidation of olefins may occur via a cyclic process.

$$>C{=}C< + HOOC(=O){-} \longrightarrow [\text{cyclic transition state: } C{-}C \cdots O \cdots O \cdots C{-},\ H \cdots O] \longrightarrow >C{-}C<(O) + {-}C(=O){-}OH \qquad (15)$$

The presence of the nitrogen atom in the imine bond makes this type of mechanism highly improbable. The protonation of nitrogen is the more likely possibility in this case followed by addition of the peracetate ion. The decomposition of the 1,2-addition product could take place in a concerted process or by the elimination of carboxylate ion, followed by deprotonation.

$$\text{>C=N-} + \text{HOOC-} \longrightarrow \text{>C=}\overset{+}{\text{N}}\text{-H} \longrightarrow \text{oxaziridine formation} \tag{16}$$

III. FOUR-MEMBERED RINGS

The formation of β-lactams from *N*-substituted imines and ketenes is the most widely known and studied example of the cycloaddition reactions of imines. In 1907, Staudinger reported the formation of 3,3,4,4-tetraphenyl-2-azetidinone from the reaction of diphenylketene with benzalaniline[24]. The β-lactam was formed in better than 70% yield and later reports by Staudinger and his group gave further

$$Ph_2C{=}C{=}O + PhN{=}CHPh \longrightarrow \text{3,3,4,4-tetraphenyl-N-phenyl-2-azetidinone} \tag{17}$$

examples of the cycloaddition of ketoketenes and of ketene itself to a number of imines[25]. Diphenylketene and biphenyleneketene were found to be the most reactive members. A major complication, particularly with alkyl ketenes, is the reaction of the ketene dimers with the imines to give piperidinediones[26].

It is apparent that the scope of the reaction as reported by Staudinger would be severely limited if ketoketenes were the only ketenes (in addition to ketene itself) which could be utilized. Of course, this limitation is not intrinsic to the reaction but rather is the result of the great tendency of aldoketenes to polymerize. For example, in 1956 Kirmse and Horner[27] prepared a number of 3-monosubstituted β-lactams by the photolytic decomposition of diazoketones in the presence of excess imine. Thus, monoaryl ketenes generated by the light-induced Wolff rearrangement of ω-diazoacetophenones gave good yields of the corresponding β-lactams. When the reaction was conducted in this fashion, the formation of piperidinediones was suppressed.

$$\mathrm{ArCOCHN_2} \xrightarrow[-\mathrm{N_2}]{h\nu} \mathrm{ArCH{=}C{=}O} \xrightarrow{\;\backslash\mathrm{C{=}N}-\;} \begin{array}{c} \mathrm{ArHC{-}C{=}O} \\ | \quad\;\; | \\ \mathrm{\backslash C{-}N\backslash} \end{array} \tag{18}$$

It was noted that electron-withdrawing groups rendered the ketenes less reactive. In contrast to phenylketene, *p*-chlorophenylketene gave a lower yield of adduct while *p*-nitrophenylketene did not react at all with benzalaniline.

A year later, Pfleger and Jäger[28] investigated the reaction of diphenylketene, phenylketene (generated by the silver oxide-catalysed decomposition of ω-diazoacetophenone) and ketene with a number of imines. Despite the fact that sometimes conflicting results are reported by different groups of investigators, the following conclusions appear to be upheld by the results of Staudinger, of Kirmse and Horner and of Pfleger and Jäger:

(a) Electron-withdrawing groups on the ketene moiety will hinder the addition.
(b) Electron-donating groups on the imine will facilitate the addition. (However, with increasing electron density of the imine, 2:1 adduct formation may become competitive.)

However, the rather shaky nature of these conclusions is made flagrantly obvious by the work of Zeifman and Knunyants[29], who recently reported that the reaction of ketene with *N*-tosyl hexafluoroacetone imine gave the adduct in 87% yield!

It would be dangerous to go beyond these very general conclusions in the absence of any systematic and careful investigation of this reaction. Kinetic and stereochemical data would be particularly useful in this respect. Steric factors seem to be of minor importance.

An unusual route to alkyl aldoketenes from pyrolysis of ethoxyalkynes has been reported by van Leusen and Arens[30]. In the presence of imines such as benzalaniline and benzophenone anil, the corresponding β-lactams could be obtained.

$$\mathrm{RC{\equiv}C{-}OEt} \xrightarrow{\Delta} \left[\mathrm{H}\cdots\mathrm{R{-}C{\equiv}C{-}O{-}CH_2{-}H_2C}\right] \longrightarrow \mathrm{RCH{=}C{=}O} \xrightarrow{\;\backslash\mathrm{C{=}N}-\;} \begin{array}{c} \mathrm{RHC{-}C{=}O} \\ | \quad\;\; | \\ \mathrm{\backslash C{-}NH} \end{array} \tag{19}$$

As a result of the research on penicillin, the addition of diphenylketene to 2-thiazoline was shown to give the corresponding azetidinone containing the penicillin skeleton[31]. Holley and Holley[32] reported the addition of dimethylketene to α-methylmercaptobenzalaniline.

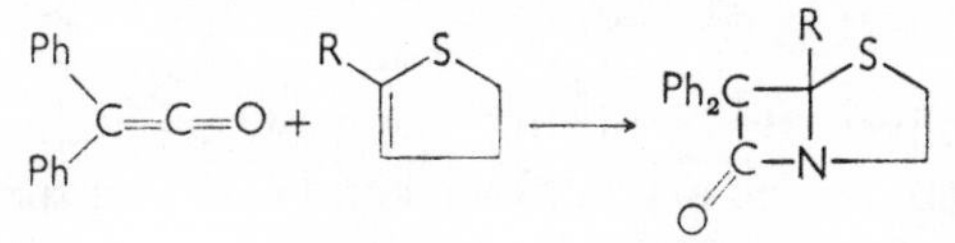

(20)

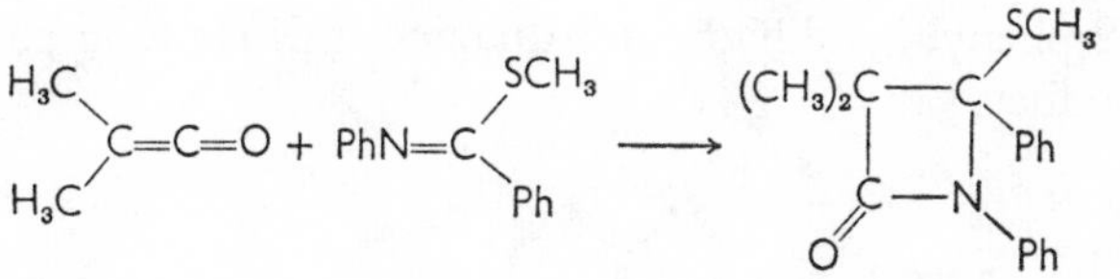

(21)

The reaction of diphenylketene with *p*-benzoquinone monoanil[33] gave an adduct as an oil having absorptions at 1670 cm^{-1} (α,β unsaturated carbonyl) and 1760 cm^{-1} (β-lactam carbonyl). The oil, upon standing, rearranged to **15**.

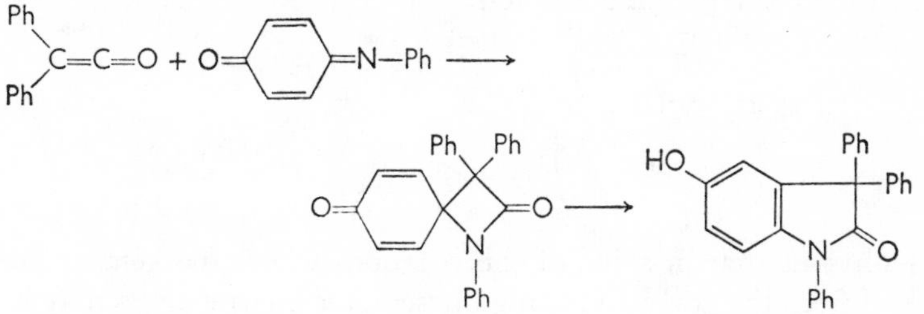

(22)

(15)

The reaction of mesoionic oxazolones with imines gave the corresponding β-lactams and this has been taken as evidence for the intermediate formation of ketenes[34].

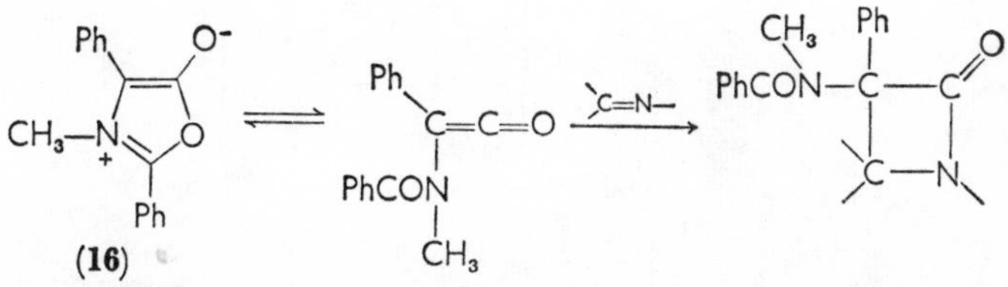

(23)

(16)

The penicillin project spurred the development of a specific method to prepare aminoazetidinones. Sheehan and Ryan[35] generated *N*,*N*-diacylaminoketenes from the corresponding acyl halides in the presence of imines.

$$(Acyl)_2NCH_2COCl + \rangle C{=}N{-} \longrightarrow \begin{array}{c} (Acyl)_2N{-}HC{-}C{=}O \\ | \quad\quad | \\ {-}C{-}N{<} \end{array} \tag{24}$$

With 2-thiazolines, the penicillin system was obtained. Difficulties were encountered when arylsulphonyl and carbobenzoxy groups were used as the *N*-acyl groups[36].

With 2-phenyl-5,6-dihydro-1,3-thiazine, this reaction gave the expected adducts[37].

$$R_2NCH_2COCl + \text{2-phenyl-5,6-dihydro-1,3-thiazine} \longrightarrow \text{adduct} \tag{25}$$

R = succinoyl, maleyl, phthaloyl

Bose and Kugajewsky[38a] have applied this method to generate diphenylketene in the presence of a number of amidines and were able to obtain the corresponding adducts.

$$Ph_2CHCOCl + R{-}\underset{NR_2}{C}{=}N{-}Ar \longrightarrow \begin{array}{c} Ph_2C{-}C{=}O \\ | \quad\quad | \\ R{-}\underset{NR_2}{C}{-}NAr \end{array} \tag{26}$$

It is obvious that in some of the reactions where the ketenes are generated from the acyl halide and a base, ionic intermediates may be involved and therefore the reactions are not true cycloadditions[38b].

$$\textbf{(17)} + PhN{=}C{=}S \longrightarrow [\text{spiro intermediate}] \longrightarrow \text{product} + CH_3N{=}C{=}S \tag{27}$$

The reaction of isocyanates with imines has been reported to give uretidinones[39]. However, the structural assignments are not certain. The product isolated from the reaction of **17** with phenylisothiocyanate has been rationalized as the result of the fragmentation of the initial uretidinethione[40].

The first examples of azetines (**17a**) have been recently reported by Effenberger and Maier[93]. The addition of ketene diaminals (**17b**) to *N*-arenesulphonyl benzaldimine gave the corresponding azetines in 35–70% yields.

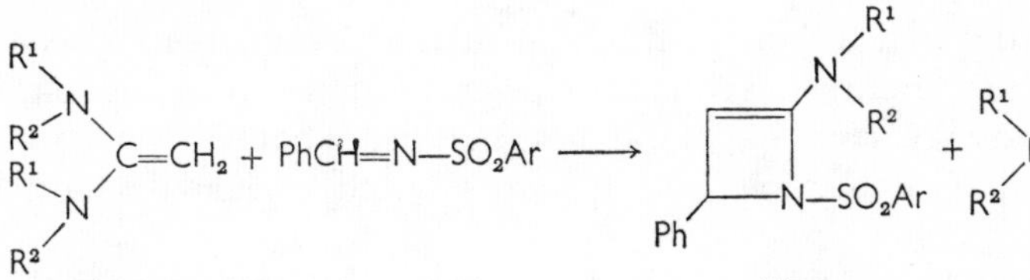

$$(R^1R^2N)_2C{=}CH_2 + PhCH{=}N{-}SO_2Ar \longrightarrow \text{azetine} + R^1R^2NH \qquad (28)$$

(**17b**) (**17a**)

Similarly Viehe and his group[94,95] have found the products resulting from the addition of ynamines (**17c**) to imines explainable in terms of an initial cycloadduct, i.e. an azetine which then opened to a larger ring. Subsequent hydrolysis left no doubt as to the validity of this interpretation.

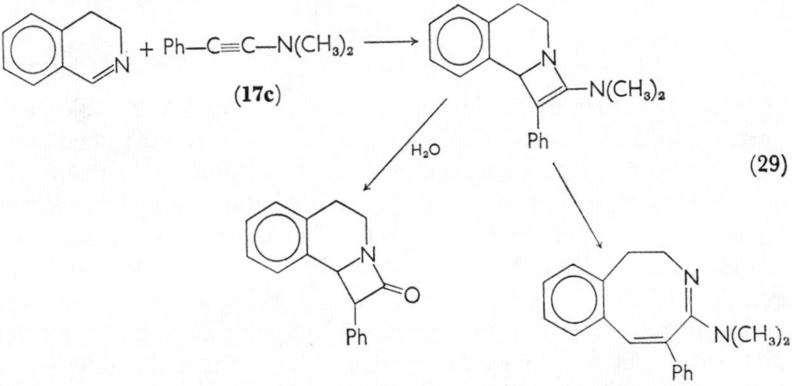

$$\text{3,4-dihydroisoquinoline} + Ph{-}C{\equiv}C{-}N(CH_3)_2 \longrightarrow \ldots \qquad (29)$$

(**17c**)

Finally, the head-to-head dimerization of some anils under the influence of light has recently been postulated to explain the formation of the observed product. No azetidines could be obtained from the irradiation of imines in the presence of olefins[41].

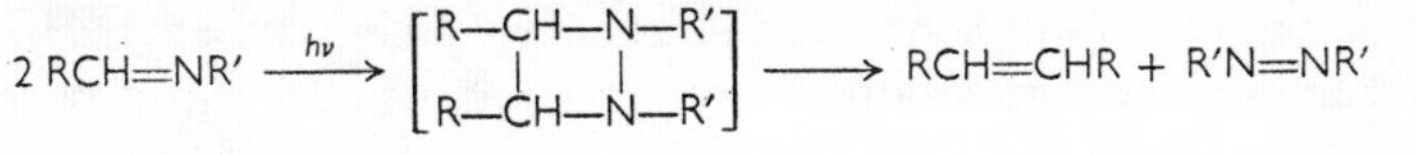

$$2\,RCH{=}NR' \xrightarrow{h\nu} \left[\begin{array}{c} R{-}CH{-}N{-}R' \\ | \quad\;\; | \\ R{-}CH{-}N{-}R' \end{array}\right] \longrightarrow RCH{=}CHR + R'N{=}NR' \qquad (30)$$

IV. FIVE-MEMBERED RINGS

Although Smith[3] as early as 1937 had recognized the concept of 1,3-dipolar cycloadditions, this type of reaction awaited the truly extraordinary contribution of Huisgen[2,42] and his school to gain full stature among organic reactions. So extensive has been their investigations, that 1,3-dipolar cycloadditions have reached, in the short span of ten years, a status almost equal to that of the Diels–Alder reaction.

A. *Diazoalkanes*

Unfortunately, as was pointed out in the introduction, 1,3-cycloadditions to imines have had to defer to those involving olefins, alkynes, nitriles, etc. One of the most significant contributions to this topic is that of Kadaba and Edwards[43] and of Kadaba and his group[44–46].

Although Meerwein[47] had earlier attempted to add diazomethane to Schiff bases, the credit for the first successful 1,3-cycloaddition to imines goes to Mustafa[48]. The reaction of nitro-substituted anils with diazomethane gave cyclic adducts to which structures such as **19** were assigned[48].

$$\mathrm{ArCH{=}NAr' + CH_2N_2} \longrightarrow \left[\begin{array}{c}\mathrm{ArHC{-}NAr'}\\ \mathrm{N{=}N{-}CH_2}\end{array}\right] \longrightarrow \begin{array}{c}\mathrm{Ar{-}HC{-}N{-}Ar'}\\ \mathrm{HN{-}N{=}CH}\end{array} \qquad (31)$$

(18) **(19)**

Mustafa postulated the initial formation of 1,2,4-triazolines (**18**) which tautomerized to the 2-isomers (**19**). His structural assignment was based on the fact that the adduct reverted to the anils and presumably diazomethane upon heating and upon hydrolysis. It seems unlikely in retrospect that structure **19** would qualify, particularly when examined in the context of Mustafa's data. Although the loss of molecular nitrogen is one of the most likely paths for the decomposition of 1-pyrazolines and 1-triazolines, examples of reverse 1,3-dipolar addition have been reported[49]. Buckley[50] reexamined the reaction of diazomethane with *p*-nitro-*N*-(*p*-nitrobenzylidene) aniline (**20**) and assigned the 1,2,3-triazoline structure to the product; he confirmed his assignment by showing that the adduct is identical with the product **21** formed from the reaction of *p*-nitrophenyl azide with *p*-nitrostyrene. The same author also first reported the catalytic effect of methanol on the formation of 1,5-diphenyl and 1-phenyl-5-*p*-chlorophenyl-Δ^2-1,2,3-triazolines. The cycloadducts are not obtained in the absence of

$$NO_2-C_6H_4-CH{=}N-C_6H_4-NO_2 + CH_2N_2 \longrightarrow$$

(20)

(32)

$$NO_2-C_6H_4-HC-N(C_6H_4-NO_2)\ (H_2C-N{=}N\ \text{ring}) \longleftarrow NO_2-C_6H_4-CH{=}CH_2 + N_3-C_6H_4-NO_2$$

(21)

methanol. In view of Buckley's findings, it is probable that the adduct of diphenyldiazomethane with **20** has the 1,2,3- rather than the 1,2,4-structure.

Kadaba and Edwards[43] reinvestigated the reaction of diazomethane with anils and confirmed the beneficial effect of methanol and water. They also showed that the reaction proceeded faster in dioxan than in diethyl ether. These authors found that the reaction followed second-order kinetics; it is to be pointed out, however, that the rate of reaction was measured by the consumption of diazomethane rather than by the formation of adduct. Perhaps more meaningful kinetic data on this reaction should be obtained. Electron-withdrawing groups on the anil moiety were found to favour the reaction whereas electron-donating groups hindered the addition.

A more extensive study of the addition of diazomethane to anils has been reported by Kadaba[44]. Unfortunately, only yields of adducts under varying time periods have been presented. Despite the lack of standardization, these data confirm the general view that electron-withdrawing groups favour the reaction. The results, when taken in conjunction with the reported catalytic effect of methanol or water, are interesting with regard to the possible mechanism of the reaction. An electron-withdrawing group on the amine ring is apparently more

TABLE 1. Reaction of diazomethane with anils[44].

Substituent		Yield	Time
C-Phenyl	*N*-Phenyl	(%)	(h)
p-NO_2	H	45·0	96
H	*p*-NO_2	75·0	42
p-Cl	H	41·0	168
H	*p*-Cl	64·1	96

effective than a similar group in the benzylidene portion (see Table 1). The dielectric constant of the solvent used seems to be rather uninfluential on the overall concerted nature of the reaction. This statement is to be taken with great caution since the question of the simultaniety of bond formation in cycloadditions is still unresolved. Diazomethane acts as a nucleophile, whilst water or methanol polarizes the $\rangle$C=N— bond much in the same fashion as acids catalyse the

$$\rangle\overset{\delta+}{C}=\overset{\delta-}{N}- + H-OR \longrightarrow \rangle\overset{\delta\delta+}{C}=\overset{\delta-}{N}\langle \cdots H-\overset{\delta-}{O}-R$$

addition of nucleophiles to the carbonyl group. The catalytic effect of Lewis acids during the addition of diazomethane to carbonyl groups is well known.

Very recently, Hoberg[51] reported the addition of diazomethane to benzalaniline to give the same triazoline as that previously obtained by Buckley and Kadaba. The reaction was performed at −78° with an equivalent of diethylaluminium chloride and an excess of benzalaniline. With diethylaluminium iodide, under essentially the same conditions (except that the anil was not in excess), excellent yields of the aziridine were obtained.

$$\text{Ph—HC—N—Ph (triazoline ring: } H_2C\text{, N=N)} \xleftarrow{Et_2AlCl} PhCH{=}NPh + CH_2N_2 \xrightarrow{Et_2AlI} PhCH{-}NPh \text{ (aziridine, } CH_2) \qquad (33)$$

However, it is highly improbable that either of these reactions is a true cycloaddition.

The greater sensitivity of the reaction to the presence of an electron-withdrawing group on the amine moiety indicates that the formation of the C—C bond in the transition state is relatively fast and much in advance of that of the N—N bond; the effect of the electron-withdrawing group would then be to stabilize the developing negative charge on the nitrogen. The following mechanism[44] has been suggested and seems consistent with the available data. The lack of solvent effect can be understood on the basis of this pseudo-concerted mechanism, Kadaba and Hannin[45,46] have noted the accelerating influence of *o*-substituents on the *C*-aryl group of the imines and have ascribed this effect to the steric inhibition of resonance in the imine;

$$ArCH{=}N{-}Ar' + H_2\bar{C}{-}\overset{+}{N}{=}N \longrightarrow \text{[transition state]} \longrightarrow \text{triazoline} \quad (34)$$

this would raise the ground-state energy and favour the addition. The inhibiting influence of a *p*-chloro group on the *C*-aryl portion has been ascribed to a mesomeric effect of the following type.

$$Cl{-}C_6H_4{-}CH{=}N{-}C_6H_5 \longleftrightarrow \overset{+}{Cl}{=}C_6H_4{=}CH{-}\bar{N}{-}C_6H_5$$

Knunyants and Zeifman[52] have reported the formation of 4,4-bis-(trifluoromethyl)-1,2,3-triazoline (**22**) in 92% yield from the reaction of hexafluoroacetone imine with diazomethane.

$$(CF_3)_2C{=}NH + CH_2N_2 \longrightarrow \underset{(\mathbf{22})}{(CF_3)_2C{-}NH{-}N{=}N{-}CH_2} \xrightarrow{h\nu} (CF_3)_2C{-}CH_2{-}N{-}H \quad (35)$$

The reaction of diazomethane with **23** is probably to be classified as a sequential-type reaction although formally it could be considered as a cycloaddition[53].

$$\underset{(\mathbf{23})}{SF_5N{=}C(Cl)CF_3} + CH_2N_2 \longrightarrow SF_5N{-}C(Cl)(CF_3){-}CH_2 \xrightarrow{CH_2N_2} SF_5N{-}C(CF_3)(CH_2Cl){-}CH_2 \quad (36)$$

With 3-(2,4-dinitrophenyl)-2-methylazirine, diazomethane gave 1-(2, 4-dinitrophenyl)-2-methyl-3-azido-1-propene possibly via the following path[53].

$$ArCH{-}N{=}C(CH_3) + CH_2N_2 \longrightarrow [ArCH{-}N{-}N{=}N{-}CH_2{-}C(CH_3)] \longrightarrow ArCH{=}C(CH_3){-}CH_2N_3 \quad (37)$$

$Ar = 2,4\text{-}(NO_2)_2C_6H_3{-}$

Backer and Bos[54] reported the reaction of diazomethane with a highly negatively substituted oxime **24** and isolated the *N*-methoxy-1,2,3-triazoline. Evidently, methylation of oxime occurred followed by the cycloaddition of diazomethane.

$$(CH_3SO_2)_2C{=}N{-}OH + CH_2N_2 \longrightarrow (CH_3SO_2)_2C{=}N{-}OCH_3 \longrightarrow$$

(**24**)

$$(CH_3SO_2)_2C\text{—}N(OCH_3)\text{—}N{=}N\text{—}CH_2 \quad \text{(ring)} \tag{38}$$

(**25**)

The triazoline **25** lost methane-sulphinic acid to give the corresponding triazole (**27**) under the influence of piperidine, and heat decomposed it to the aziridine **26**.

$$CH_3SO_2C\text{—}NOCH_3 \;(\text{HC}{=}N{-}N \text{ ring}) \xleftarrow{\text{piperidine}} (CH_3SO_2)_2C\text{—}NOCH_3 \;(H_2C\text{—}N{=}N \text{ ring}) \xrightarrow{\Delta} (CH_3SO_2)_2C\text{—}NOCH_3 \;(CH_2 \text{ ring}) \tag{39}$$

(**27**) + CH_3SO_2H ← (**25**) → (**26**) + N_2

B. *Nitrilimines*

The reaction of nitrilimines (**28**) with a number of aldimines has recently been reported by Huisgen and his group[55,56]. The reaction with aldimines gave fair to very good yields of the 1,2,4-triazolines (**29**); diphenyl nitrilimine was generated by two methods: thermal elimination of nitrogen from 1,5-diphenyltetrazole and base-catalysed elimination of hydrogen chloride from the hydrazidic chloride.

$$\text{R—N(tetrazole)C—R}^1 \xrightarrow[\Delta]{-N_2} \text{R—}\overset{-}{N}\text{—N}{=}\overset{+}{C}\text{—R}^1 \;(\mathbf{28}) \xleftarrow[Et_3N]{-HCl} \text{R—N(H)—N}{=}\text{C(Cl)—R}^1 \tag{40}$$

$$\mathbf{28} \xrightarrow{>C{=}N{-}} \text{R—N—N}{=}\text{C—R}^1,\ \text{—C—N— ring} \;(\mathbf{29})$$

The elimination of hydrogen chloride apparently was the preferred method for the generation of the nitrilimines. The presence of alkyl substituents did not have any detrimental effect on the course of the addition. A cyclic imine, 3,4-dihydroisoquinoline, was utilized and the adduct was obtained in 50% yield.

The addition of nitrilimines to other compounds containing the imine linkage has also been reported by Huisgen and his colleagues[56]. The initial adducts obtained from nitrilimines with aldoximes lost water to give the corresponding 1,2,4-triazoles. The yields were good to excellent in most cases.

$$\overset{+}{\text{R}}\text{C}{=}\text{N}{-}\overset{-}{\text{N}}{-}\text{R}^1 + \text{R}^2\text{CH}{=}\text{NOH} \longrightarrow \left[\text{RC}{=}\text{N}{-}\text{NR}^1{-}\text{CHR}^2{-}\text{N(OH)}{-}\right] \xrightarrow{-\text{H}_2\text{O}} \text{RC}{=}\text{N}{-}\text{N(R}^1){-}\text{C(R}^2){=}\text{N}{-} \quad (41)$$

It is interesting that, with *N*-phenylformaldoxime, *C*-carbethoxy-*N*-phenylnitrilimine (**30**) gave the 1,2,4-triazole-4-oxide (**31**) by the elimination of aniline from the initial adduct. Similarly, with diphenylnitrilimine, ethyl acetimidate gave the 1,2,4-triazole **32** as the product,

$$\underset{(\mathbf{30})}{\text{EtO}_2\text{CC(Cl)}{=}\text{NNHPh}} \xrightarrow[\text{Et}_3\text{N}]{-\text{HCl}} \text{EtO}_2\overset{+}{\text{C}}\text{C}{=}\text{N}{-}\overset{-}{\text{N}}{-}\text{Ph} \xrightarrow{\text{PhNHC(H)}{=}\text{NOH}}$$

$$\text{EtO}_2\text{CC}{=}\text{N}{-}\text{N(Ph)}{-}\text{CH(NHPh)}{-}\text{N(OH)}{-} \longrightarrow \underset{(\mathbf{31})}{\text{EtO}_2\text{CC}{=}\text{N}{-}\text{N(Ph)}{-}\text{CH}{=}\overset{+}{\text{N}}(\text{O}^-){-}} \quad (42)$$

aromatization having occurred by loss of ethanol.

$$\text{Ph}\overset{+}{\text{C}}{=}\text{N}{-}\overset{-}{\text{N}}\text{Ph} + \text{HN}{=}\text{C(OC}_2\text{H}_5)\text{CH}_3 \longrightarrow \underset{(\mathbf{32})}{\text{PhC}{=}\text{N}{-}\text{NPh}{-}\text{C(CH}_3){=}\text{N}{-}} \quad (43)$$

In light of our present understanding of these reactions, the first recorded example of the addition of a nitrilimine to an imine bond is that of Fusco and Musante[57]. They found that 1,3,5-triphenyl-1,2,4-triazole (**33**) was the product of the reaction of *N*-phenylbenzhydrazidic chloride with benzamidine. The benzamidine in this case also functioned as the dehydrohalogenating agent.

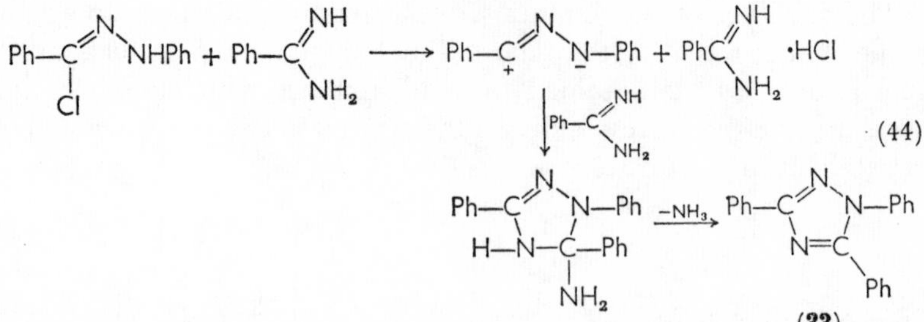

$$\text{Ph—C(Cl)=N—NHPh} + \text{Ph—C(=NH)NH}_2 \longrightarrow \text{Ph—}\overset{+}{\text{C}}\text{=N—}\overset{-}{\text{N}}\text{—Ph} + \text{PhC(=NH)NH}_2\cdot\text{HCl} \tag{44}$$

(33)

Bacchetti[58] reported a similar reaction of *N*-carbethoxybenzhydrazidic chloride (**34**) with benzamidine. Apparently, hydrolysis and decarboxylation took place in the last step.

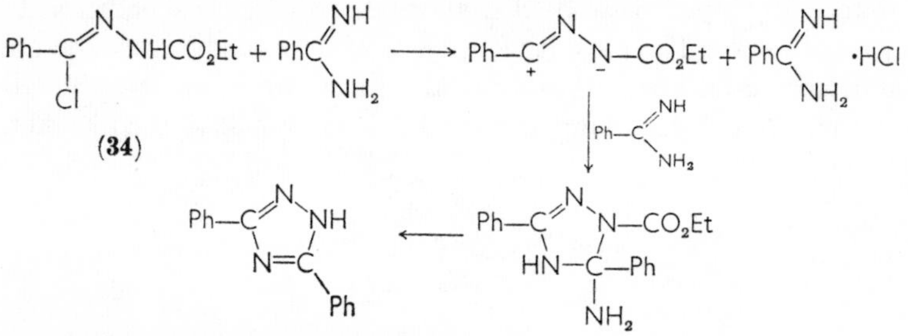

$$\text{Ph—C(Cl)=N—NHCO}_2\text{Et} + \text{PhC(=NH)NH}_2 \longrightarrow \text{Ph—}\overset{+}{\text{C}}\text{=N—}\overset{-}{\text{N}}\text{—CO}_2\text{Et} + \text{PhC(=NH)NH}_2\cdot\text{HCl} \tag{45}$$

(34)

C. Azomethine Imines and Ylids

Azomethine imines[59] (**35**) give good to excellent yields of adducts with a variety of imines, including *N*-aryl and *N*-alkyl imines of formaldehyde and aromatic aldehydes[60].

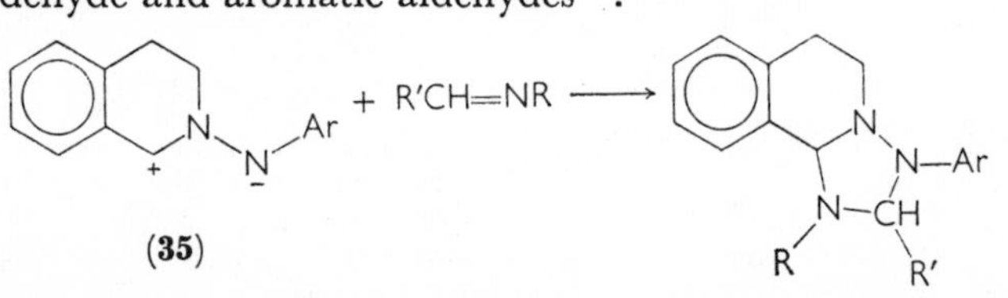

$$(35) + \text{R'CH=NR} \longrightarrow \tag{46}$$

(35)

Eicher, Hünig and Nikolaus[61] have interpreted the addition of *O*-alkyl nitrosimmonium salts (**36**) to *N*-aromatic heterocycles as a cycloaddition of the intermediate *N*-alkoxy azomethine imines (**37**). Pyridine, quinoline, isoquinoline, phenanthridine and benzothiazole all gave the corresponding adducts.

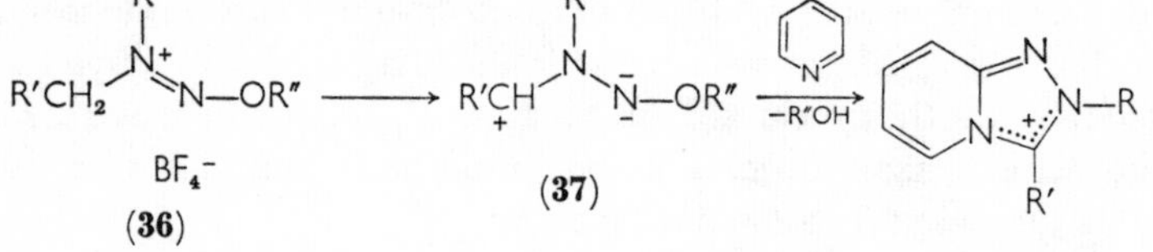

$$\text{R'CH}_2\text{—}\overset{+}{\text{N}}(\text{R})\text{=N—OR''}\ \text{BF}_4^- \longrightarrow \text{R'}\overset{+}{\text{C}}\text{H—N(R)—}\overset{-}{\text{N}}\text{—OR''} \xrightarrow[-\text{R''OH}]{\text{pyridine}} \tag{47}$$

(36) (37)

Similar adducts are obtained with normal aldimines. Azomethine ylid (**38**) gives fair yields of adducts with benzalaniline and *N*-methyl benzaldimine[62].

$$\mathbf{38} + RN{=}CHR' \longrightarrow \text{adduct} \qquad (48)$$

D. Nitrile Oxides

Nitrile oxides undergo 1,3-cycloadditions with imines. The early observation by Wieland[63] of the formation of **39** by the addition of hydrogen chloride to benzonitrile oxide can be rationalized as an addition to the α-chloro oxime.

$$Ph{-}C{\equiv}\overset{+}{N}{-}\bar{O} + HCl \longrightarrow Ph{-}C(Cl){=}N{-}OH \xrightarrow{PhC\equiv\overset{+}{N}-\bar{O}} \qquad (49)$$

(**39**)

In situ generation of nitrile oxides can also explain the formation of 1,2,4-oxadiazoles from the reaction of α-chloro oximes with imidates[64].

$$R{-}C(Cl){=}NOH + 2\,R'{-}C(OR''){=}NH \longrightarrow \text{(1,2,4-oxadiazole)} + R'C(OR''){=}NH\cdot HCl + R''OH \qquad (50)$$

Mukayama and Hoshino[65] demonstrated the formation of nitrile oxides from the reaction of some primary nitroalkanes with phenylisocyanate in the presence of base, by cycloaddition to unsaturated centres. Benzalaniline gave 10% and 18% yields of adducts with acetonitrile oxide and propionitrile oxide.

$$RCH_2NO_2 + R_3N + PhN{=}C{=}O \longrightarrow R{-}C{\equiv}\overset{+}{N}{-}\bar{O} \xrightarrow{PhCH=N-Ph} \text{adduct} \qquad (51)$$

Aromatic nitrile oxides have been reported to give good to excellent yields with a variety of aldimines[66,67]. Both *N*- and *C*-alkyl as well as aryl-substituted imines were used. The Italian group also reported the reaction of nitrile oxides with ketimine[66].

$$\mathrm{Ar{-}C{\equiv}\overset{+}{N}{-}\overset{-}{O}} + \mathrm{RN{=}CHR'} \longrightarrow \text{Ar–C–N–R (1,2,4-oxadiazoline ring: N, O, R')} \quad (52)$$

Phenanthraquinone imine (**40**) and chrysenequinone imine have been found to give high conversions of cycloadducts with benzonitrile oxide[68].

$$\mathrm{Ph{-}C{\equiv}\overset{+}{N}{-}\overset{-}{O}} + (\mathbf{40}) \longrightarrow \text{spiro adduct (Ph, N, NH, O, O)} \quad (53)$$

(**40**)

Although the ozonolysis of imines could be considered as 1,3-cycloaddition[69–71], the reputed instability of primary ozonides would make the formation of the trioxazolidines quite unlikely. The reaction probably proceeds via the following path:

$$\mathrm{{>}C{=}N{-}} + \mathrm{\overset{-}{O}{-}O{-}\overset{+}{O}} \longrightarrow \text{C–O–O–O}^{+}\text{, N} \longrightarrow \text{oxaziridine (C, O, N)} \quad (54)$$

V. SIX-MEMBERED RINGS

Diels–Alder reactions involving imines are not well known. The early examples include the mention by Alder[72] of the reaction of the imino form of dimethyl anilinomaleate with dienes to give tetrahydropyridines (equation 55). However, no experimental data referring to this reaction were reported. The influence of an electron-withdrawing group on the imine seemed to be the key factor, since many of the known examples reported (see below) have at least one and often two electron-withdrawing groups attached to the imine bond.

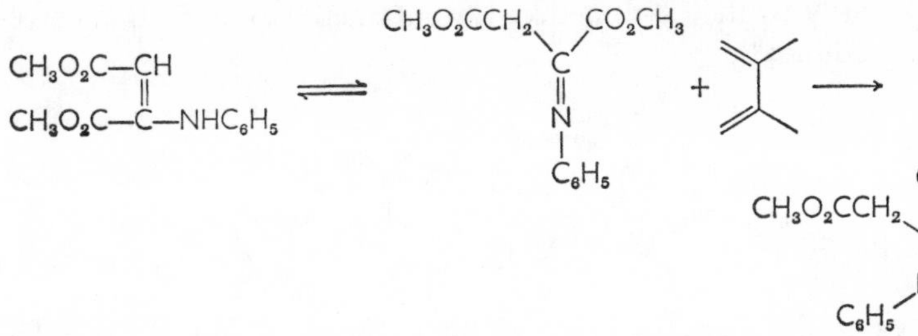

$$CH_3O_2C{-}CH{=}C(CO_2CH_3){-}NHC_6H_5 \rightleftharpoons CH_3O_2CCH_2(CO_2CH_3)C{=}N{-}C_6H_5 \quad (55)$$

A. Dienes

When imino chlorides are generated in the presence of dienes[73–77], products formally derived from the adducts are obtained.

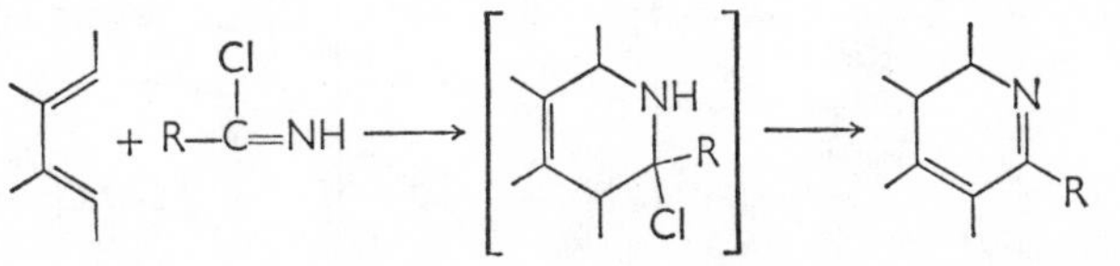

$$+\ R{-}C(Cl){=}NH \longrightarrow \quad (56)$$

Although a number of successful reactions had earlier been reported, the generality and reproducibility of these reactions had not been demonstrated. The results obtained with dienes where one of the double bonds was not part of an aromatic system eliminated the Houben–Hoesch reaction followed by ring closure as a possible alternative to the Diels–Alder reaction[77]. Similarly, the use of sulphuric acid instead of hydrochloric acid gave the imino sulphonates which also[78] underwent cycloadditions with dienes[78].

In the last few years, more conclusive examples of the Diels–Alder reaction of imines have been reported. Kresze and Albrecht[79,80] found that *N*-tosyl chloralimine underwent addition with 2,3-dimethyl-1,3-butadiene and cyclopentadiene but not with 1,3-cyclohexadiene. However, the more reactive *N*-tosyl fluoralimine did

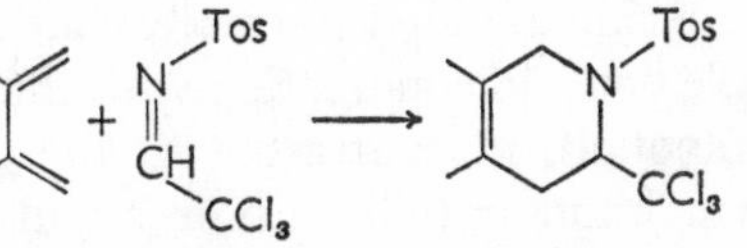

$$+\ Tos{-}N{=}CH{-}CCl_3 \longrightarrow \quad (57)$$

react with 1,3-cyclohexadiene. *N*-tosyl butyl glyoxylaldimine behaved in a similar fashion with a number of dienes. Hexafluoroacetone imine reacted easily with 2,3-dimethyl-1,3-butadiene to give the corresponding adduct[81].

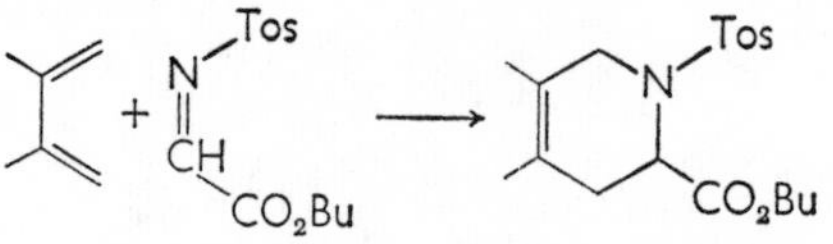

(58)

One example where the imine was not 'activated' was the reaction of 3,4-dihydroisoquinoline with ethyl 2,4-pentadienoate[82]. The final adducts in all cases possessed the double bond in conjugation with the carboxylate group. When R was methyl, the enamine tautomer

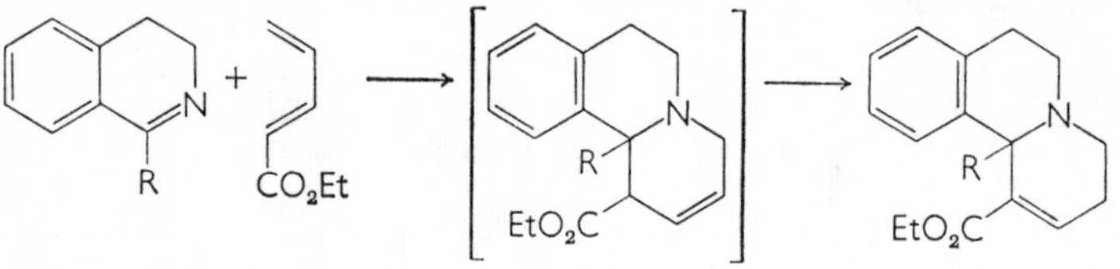

(59)

reacted to give two adducts **41** and **42** in a 4:1 ratio.

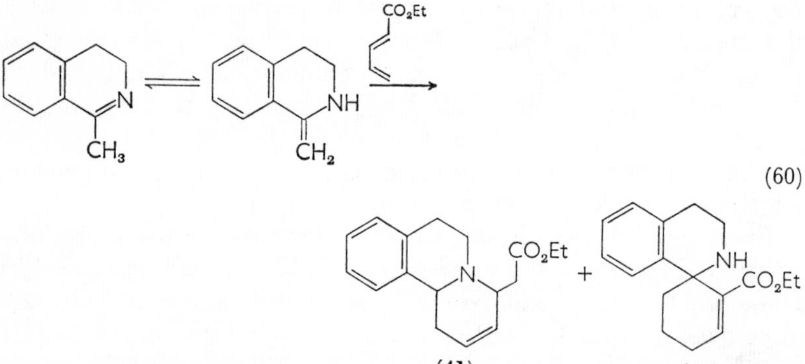

(60)

(41) **(42)**

3,4,5,6-Tetrahydropyridine added normally, but again the final product had the double bond in conjugation with the carboxylate group. The 2-methyl compound, in contrast with the dihydroisoquinoline case, gave the normal adduct (**43**) with an angular methyl group, in addition to the product of enamine addition. When 2-ethyl-1-piperideine was the dienophile, only enamine addition occurred to give **44**.

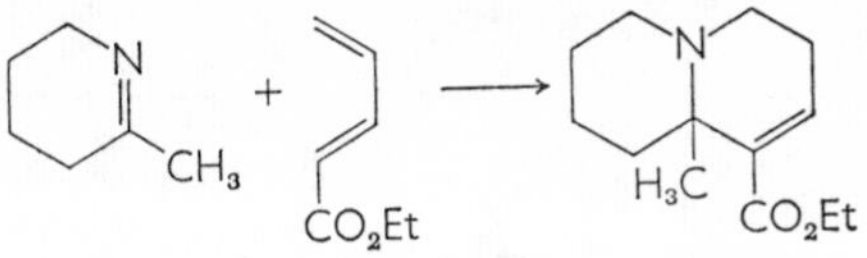

(43)

(61)

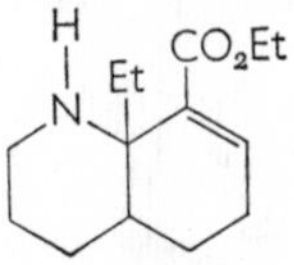

(44)

Böhme, Hartke and Miller[83] have reported the reaction of halomethyl amines with dienes. The reaction can be interpreted as a Diels–Alder reaction with the immonium salt. Very satisfactory yields were obtained under mild conditions.

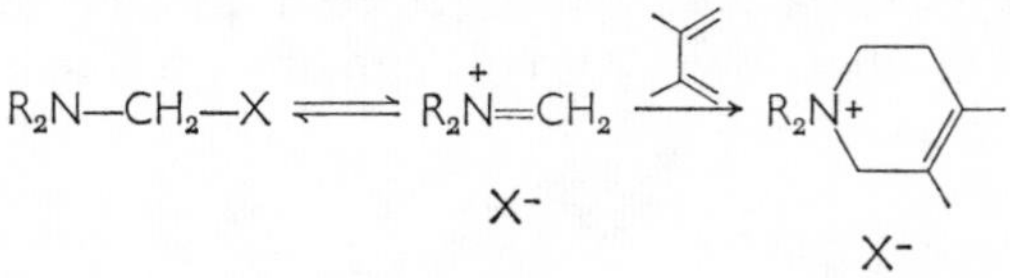

$$R_2N{-}CH_2{-}X \rightleftharpoons R_2\overset{+}{N}{=}CH_2\ X^- \longrightarrow R_2N^+ \ X^- \qquad (62)$$

Furan did not give an adduct. The reaction of carbalkoxy aminals with dienes constitutes another example of a Diels–Alder reaction of immonium salts[84]. The yields in this reaction are quite good.

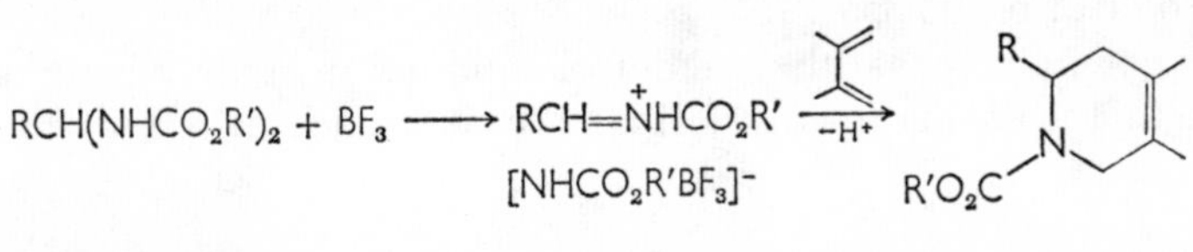

$$RCH(NHCO_2R')_2 + BF_3 \longrightarrow RCH{=}\overset{+}{N}HCO_2R'\ [NHCO_2R'BF_3]^- \xrightarrow{-H^+} \qquad (63)$$

B. *Heterodienes*

Conjugated dienes having one or more elements other than carbon in the diene portion are classified as heterodienes. In 1955, Lora-Tamayo[73,74] and his students reinterpreted the formation of *s*-triazines from the reaction of nitriles with hydrogen chloride as a Diels–Alder reaction. They viewed the initial formation of the heterodiene **45** and dienophile **46** as resulting from the reaction of two and one nitrile molecules respectively, with hydrogen chloride.

$$2\ R{-}C{\equiv}N + HCl \longrightarrow R{-}C({=}NH){-}N{=}C(Cl){-}R\ \textbf{(45)} + Cl(R)C{=}N{-}H\ \textbf{(46)} \longleftarrow R{-}C{\equiv}N + HCl \tag{64}$$

$\downarrow$ 2,4-dichloro-2,4,6-trialkyl-tetrahydrotriazine $\xrightarrow{-2\,HCl}$ 2,4,6-trialkyl-1,3,5-triazine

An analogous interpretation can be applied to the results of Weidinger and Kranz[85]. With sulphur trioxide, nitriles would react to give the diene system **47**. With added imine such as α-chlorochloraldimine, the triazine is obtained, elimination of sulphur trioxide and hydrogen chloride having obviously occurred. Amidines reacted in a similar

$$2\ R{-}C{\equiv}N + SO_3 \longrightarrow \textbf{(47)} \xrightarrow{Cl_3C(Cl)C{=}NH} \text{2,4-di-R-6-}CCl_3\text{-1,3,5-triazine} \tag{65}$$

fashion[14]. Imino ethers, amides, ureas and thioureas react to give the same type of products.

It is not unlikely that the reaction of amidines and guanidines[86] with the diethyl acetal of dimethyl formamide takes place via a similar path.

$$R{-}C({=}NH)NH_2 + (EtO)_2CHN(CH_3)_2 \longrightarrow R{-}C({=}NH){-}N{=}CH{-}N(CH_3)_2 \xrightarrow{R(H_2N)C{=}NH} \text{2,6-di-R-pyridine ring} \tag{66}$$

R = alkyl, aryl, dialkylamino

Goerdeler, Schenk and their colleagues[87–90] have recently reported the facile Diels–Alder reaction of a variety of imines with acyl and thioacyl isocyanates. Yields of adducts were very good in almost every case, for example, with R = phenyl or *p*-anisyl, 87% and 90% yields of compound **48** were obtained[87].

$$\text{RC(=S)–N=C=O} + \text{PhCH=NPh} \longrightarrow \textbf{(48)} \quad (67)$$

Excellent conversions were obtained when R was aryl- and alkyl-mercapto[88], alkyl- and arylamino[89], or alkoxy- and aryloxy[89].

Yields were somewhat more variable when compound **49** was used as the imine component[90].

$$\text{RC(=S)–N=C=O} + \textbf{(49)} \xrightarrow{\Delta} \textbf{(50)} \quad (68)$$

R = Et_2N, Ph_2N, PhO, Ph, cyclohexyl-O; R^1 = cyclohexyl, Ph.

Thiobenzoylisocyanate gave practically quantitative yields of the adduct with a number of *N*- and *C*-alkyl and aryl imines[91,92]. However, no cycloadducts were obtained with imidoyl chlorides, imidates,

$$\text{PhC(=S)–N=C=O} + \text{R–N=C}(R^1)(R^2) \longrightarrow \quad (69)$$

R, R^1 = alkyl, aryl
R^2 = H, alkyl, aryl

amidines, oxime ethers, hydrazones and *N*-phenyliminophosgene[91].

VI. REFERENCES

1. A. Wasserman, *Diels–Alder Reactions*, Elsevier, New York, 1965; J. Sauer. *Angew. Chem.*, **78**, 233 (1966); **79**, 76 (1967).
2. R. Huisgen, R. Grashey and J. Sauer, in *The Chemistry of Alkenes* (Ed. S. Patai), Interscience, New York, 1964, p. 739.
3. L. I. Smith, *Chem. Rev.*, **23**, 193 (1938).

4. J. D. Roberts and C. M. Sharts, *Org. Reactions*, **12**, 1 (1962).
5. J. E. Baldwin, *J. Org. Chem.*, **32**, 2438 (1967).
6. C. G. Overberger, N. Weinshenker, and J.-P. Anselme, *J. Am. Chem. Soc.*, **87**, 4119 (1965).
7. P. Beltrame, C. Veglio and M. Simonetta, *Chem. Commun.*, **996** (1967); *J. Chem. Soc.* (*B*), 867 (1967).
8. R. W. Layer, *Chem. Rev.*, **63**, 489 (1963).
9. E. K. Fields and S. Sandri, *Chem. Ind.*, 1216 (1959). P. K. Kadaba and J. O. Edwards, *J. Org. Chem.*, **25**, 1431 (1960).
10. A. G. Cook and E. K. Fields, *J. Org. Chem.*, **27**, 3686 (1962); see also reference 51.
11. D. H. Deyrup and R. B. Grunewald, *Tetrahedron Letters*, 32 (1965); see also *J. Am. Chem. Soc.*, **87**, 4538 (1965).
12. D. Klaman, H. Wache, K. Ulm and F. Nerdel, *Chem. Ber.*, **100**, 1870 (1967).
13. F. Nerdel and J. Buddrus, *Tetrahedron Letters*, 3585 (1965).
14. N. J. Leonard and K. Jahn, *J. Am. Chem. Soc.*, **84**, 4806 (1962); N. J. Leonard, K. Jahn, J. V. Pautkelis and C. K. Steinhardt, *J. Org. Chem.*, **28**, 1499 (1963).
15. A. J. Speziale, C. C. Tung, K. W. Ratts, A. Yao, *J. Am. Chem. Soc.*, **87**, 3460 (1965).
16. I. E. Den Besten and C. R. Wenger, *J. Am. Chem. Soc.*, **87**, 5500 (1965).
17. E. Schmitz, *Angew. Chem.*, **76**, 197 (1964).
18. E. Schmitz, in A. R. Katritzky, *Advances in Heterocyclic Chemistry*, Vol. 2, Academic Press, New York, 1964, p. 83.
19. E. Schmitz, *Chem. Ber.*, **97**, 2531 (1964); M. Anbar and G. Yagil, *J. Am. Chem. Soc.*, **84**, 1790 (1962).
20. W. H. Graham, *J. Am. Chem. Soc.*, **84**, 1063 (1962).
21. W. H. Graham, *J. Am. Chem. Soc.*, **88**, 4677 (1966).
22. W. D. Emmons, *J. Am. Chem. Soc.*, **79**, 5739 (1957); H. Krimm, *Chem. Ber.*, **91**, 1057 (1958); R. G. Pews, *J. Org. Chem.*, **32**, 1628 (1967).
23. H. Kwart, P. S. Starcher and S. Tinsley, *Chem. Commun.*, 335 (1967).
24. H. Staudinger, *Ann. Chem.*, **356**, 51 (1907); *Ber.*, **40**, 1145 (1907).
25. H. Staudinger, *Ber.*, **50**, 1035 (1917); H. Staudinger and J. Engle, *Ber.*, **50**, 1042 (1917); also additional papers.
26. H. Staudinger, H. W. Klever and P. Kober, *Ann. Chem.*, **374**, 1 (1910).
27. W. Kirmse and L. Horner, *Chem. Ber.*, **89**, 2759 (1956).
28. R. P. Pfleger and A. Jäger, *Chem. Ber.*, **90**, 2460 (1957).
29. Y. V. Zeifman and I. L. Knunyants, *Proc. Acad. Sci. USSR*, **173**, 354 (1967).
30. A. M. van Leusen and J. F. Arens, *Rec. Trav. Chem.*, **78**, 551 (1959).
31. H. T. Clarke, J. R. Johnson and R. Robinson, Eds., *The Chemistry of Penicillin*, Princeton University Press, Princeton, N. J., 1949, p. 849, 973.
32. A. D. Holley and R. W. Holley, *J. Am. Chem. Soc.*, **73**, 3172 (1951).
33. C. W. Bird, *J. Chem. Soc.*, 3016 (1965).
34. R. Huisgen, E. Funke, F. C. Schaefer and R. Knorr, *Angew. Chem.*, **79**, 321 (1967).
35. J. C. Sheehan and J. J. Ryan, *J. Am. Chem. Soc.*, **73**, 1204, 4367 (1951); see also A. K. Bose, B. Anjaneyulu, S. K. Bhattacharya and M. S. Manhas, *Tetrahedron*, **23**, 4769 (1967); L. Paul, A. Draeger, G. Hilgetag, *Chem. Ber.*, **99**, 1957 (1966).

36. J. C. Sheehan and E. J. Corey, *Org. Reactions*, **9**, 388 (1957).
37. L. Paul, P. Polczynski and G. Hilgetag, *Chem. Ber.*, **100**, 2761 (1967).
38. (a) A. K. Bose and I. Kugajewsky, *Tetrahedron*, **23**, 957 (1967).
(b) H. Böhme, S. Ebel and K. Hartke, *Chem. Ber.*, **98**, 1463 (1965).
39. A. Senier and F. G. Shepheard, *J. Chem. Soc.*, **95**, 494 (1909).
40. J. W. Hale and N. A. Lange, *J. Am. Chem. Soc.*, **41**, 379 (1919). N. A. Lange, *J. Am. Chem. Soc.*, **48**, 2440 (1926).
41. S. Searles and R. H. Clasen, *Tetrahedron Letters*, 1627 (1965).
42. R. Huisgen, *Angew. Chem.*, **75**, 604, 741 (1963).
43. P. K. Kadaba and J. O. Edwards, *J. Org. Chem.*, **26**, 2631 (1961).
44. P. K. Kadaba, *Tetrahedron*, **22**, 2453 (1966).
45. P. K. Kadaba and N. F. Hannin, *J. Heterocyclic Chem.*, **4**, 301 (1967).
46. P. K. Kadaba, Paper presented at the Second Mid-Atlantic Regional Meeting of the ACS, Feb. 7, 1967 in New York.
47. H. Meerwein, *Angew. Chem.*, **A60**, 78 (1948).
48. A. Mustafa, *J. Chem. Soc.*, 234 (1949).
49. K. L. Rinehart and T. V. van Auken, *J. Am. Chem. Soc.*, **84**, 3736 (1962); (1962); W. Kirmse, *Chem. Ber.*, **93**, 2357 (1960).
50. G. K. Buckley, *J. Chem. Soc.*, 1850 (1954).
51. H. Hoberg, *Ann. Chem.* **707**, 147 (1967).
52. I. L. Knunyants and Y. V. Zeifman, *Bull. Acad. Sci. USSR*, 711 (1967).
53. A. Logothetis, *J. Org. Chem.*, **29**, 3049 (1964).
54. H. J. Backer and H. Bos, *Rec. Trav. Chem.*, **69**, 1223 (1950).
55. R. Huisgen, R. Grashey, H. Knupfer, R. Kunz and M. Seidel, *Chem. Ber.*, **97**, 1085 (1964).
56. R. Huisgen, R. Grashey, E. Aufderhaar and R. Kunz, *Chem. Ber.*, **98**, 642 (1965).
57. R. Fusco and C. Musante, *Gazz. Chim. Ital.*, **68**, 147 (1938).
58. T. Bacchetti, *Gazz. Chim. Ital.*, **91**, 866 (1967).
59. R. Grashey and K. Adelsberger, *Angew. Chem.*, **74**, 292 (1962).
60. R. Grashey, H. Leiterman, R. Schmidt and K. Adelsberger, *Angew. Chem.*, **74**, 491 (1962).
61. T. Eicher, S. Hünig and P. Nikolaus, *Angew. Chem.*, **79**, 682 (1967).
62. E. Steingruber, *Ph.D. Thesis*, Munich, 1964, p. 34; see also A. H. Cook, *J. Chem. Soc.*, 502 (1941).
63. H. Wieland, *Ber.*, **40**, 1667 (1907).
64. F. Eloy and R. Lenaers, *Bull. Soc. Chim. Belges*, **72**, 719 (1963).
65. T. Mukayama and T. Hoshino, *J. Am. Chem. Soc.*, **82**, 5339 (1950).
66. F. Lauria, V. Vecchietti and G. Tosolini, *Gazz. Chim. Ital.*, **94**, 478 (1964).
67. R. Huisgen, W. Mack, K. Herbig and K. Bost, unpublished results.
68. W. I. Awad, S. M. Abdel, R. Omran and M. Sohby, *J. Org. Chem.*, **31**, 331 (1966).
69. J. S. Belew and J. T. Person, *Chem. Ind.*, 1246 (1959).
70. A. H. Riebel, R. E. Erickson, C. J. Abshire and P. S. Bailey, *J. Am. Chem. Soc.*, **82**, 1801 (1960).
71. R. E. Miller, *J. Org. Chem.*, **26**, 2327 (1961).
72. K. Alder, in *Newer Methods of Preparative Organic Chemistry*, Interscience, New York, 1948, p. 504.

73. R. Madronero, E. F. Alvarez, M. Lora-Tamayo, *Ann. Real. Soc. Espan.*, **B51**, 276 (1955).
74. R. Madronero, E. F. Alvarez, M. Lora-Tamayo, *Ann. Real. Soc. Espan.*, **B51**, 465 (1955).
75. G. G. Munoz, M. Lora-Tamayo and R. Madronero, *Ann. Real. Soc. Espan.*, **B57**, 465 (1961).
76. T. L. Aparicio, M. Lora-Tamayo and R. Madronero, *Ann. Real. Soc. Espan.*, **B57**, 173 (1961).
77. M. Lora-Tamayo, G. G. Munoz, and R. Madronero, *Bull. Soc. Chim. France*, 1331 (1958).
78. M. Lora-Tamayo, G. G. Munoz and R. Madronero, *Bull. Soc. Chim. France*, 1334 (1958).
79. G. Kresze and R. Albrecht, *Chem. Ber.*, **97**, 490 (1964).
80. R. Albrecht and G. Kresze, *Chem. Ber.*, **98**, 1431 (1965).
81. W. J. Middleton and C. G. Krespan, *J. Org. Chem.*, **30**, 1398 (1965).
82. F. Bohlman, D. Habeck, E. Poetsch and D. Schumann, *Chem. Ber.*, **100**, 2742 (1967).
83. H. Böhme, K. Hartke and A. Müller, *Chem. Ber.*, **96**, 607 (1963).
84. R. Merten and G. Miller, *Angew. Chem.*, **74**, 866 (1962); *Chem. Ber.*, **97**, 682 (1964).
85. H. Weidinger and J. Kranz, *Chem. Ber.*, **96**, 2070 (1963).
86. H. Bredereck, F. Effenberger and A. Hoffmann, *Chem. Ber.*, **96**, 3625 (1963); **97**, 61 (1964).
87. J. Goerdeler and H. Schenk, *Chem. Ber.*, **98**, 383 (1965).
88. A. Schenk, *Chem. Ber.*, **99**, 1258 (1966).
89. J. Goerdeler and K. Jonas, *Chem. Ber.*, **99**, 3572 (1966).
90. J. Goerdeler and R. Sappelt, *Chem. Ber.*, **100**, 2064 (1967).
91. J. Goerdeler and R. Weiss, *Chem. Ber.*, **100**, 1627 (1967).
92. O. Tsuge, M. Toshiro and R. Mizuguchi, *Chem. Pharm. Bull.*, **14**, 1055 (1966).
93. F. Effenberger and R. Maier, *Angew. Chem.*, **78**, 389 (1966).
94. R. Fuks and H. G. Viehe, unpublished results.
95. R. Fuks, R. Buijle and H. G. Viehe, *Angew. Chem.*, **78**, 594 (1966)

CHAPTER **8**

Substitution reactions at the azomethine carbon and nitrogen atoms

R. J. MORATH

College of St. Thomas, St. Paul, Minnesota, U.S.A.

and

GARDNER W. STACY

Washington State University, Pullman, Washington, U.S.A.

I. INTRODUCTION

Substitution reactions of azomethines are very conveniently considered as substitution at carbon or nitrogen respectively. Such a discussion by reaction type provides greater perspective than the more traditional account by compound class. An insight as to the large variety of displacements of azomethines and their utility in preparative organic chemistry is thus achieved.

II. SUBSTITUTIONS AT THE CARBON ATOM

A. *Displacements of a Chloro Substituent*

1. Imidyl chlorides, HN=C(R)Cl and RN=C(R¹)Cl

The imidyl chlorides (also called imino chlorides) (**1**) can be considered to be derivatives of the hypothetical imidic acid (**2**), in which

$$\underset{\textbf{(1)}}{\mathrm{RC(Cl){=}NH}} \qquad \underset{\textbf{(2)}}{\mathrm{RC(OH){=}NH}}$$

the hydroxyl group has been replaced by a chlorine. No claims have been made for the isolation of the free imidic acid[1–4]. The imidyl chlorides can be prepared from the reaction of an amide with phosphorus pentachloride.

$$\mathrm{RCONHR^1} \rightleftharpoons \mathrm{RC(OH){=}NR^1} \xrightarrow{\mathrm{PCl_5}} \mathrm{RC(Cl){=}NR^1} + \mathrm{HCl} + \mathrm{POCl_3}$$

Disubstituted amides on treatment with phosphorus pentachloride yield amido chlorides (**3**), from which the imidyl chlorides can be obtained on pyrolysis.

$$\mathrm{RCONR_2^1} \xrightarrow{\mathrm{PCl_5}} \underset{\textbf{(3)}}{[\mathrm{RC(Cl){=}NR_2^1}]^+\,\mathrm{Cl^-}} \longrightarrow \mathrm{RC(Cl){=}NR^1} + \mathrm{R^1Cl}$$

Although Walther and Grossman[5] prepared *N*-phenyl-α-(*p*-chlorophenyl)acetamidine (**4**) from *p*-chlorophenylacetamide by the reaction

$$p\text{-}\mathrm{ClC_6H_4CH_2CONH_2} \xrightarrow{\mathrm{PCl_5}} p\text{-}\mathrm{ClC_6H_4CH_2C(Cl){=}NH} \xrightarrow{\mathrm{PhNH_2}} \underset{\textbf{(4)}}{p\text{-}\mathrm{ClC_6H_4CH_2C(NH_2{\cdot}HCl){=}NPh}}$$

shown, the extension of this method for the preparation of unsubstituted amidines is unsatisfactory. Kirsanov[6,7] has shown that neither phosphorus oxychloride nor imidyl chlorides are formed from the reactions of unsubstituted amides with phosphorus pentachloride, but rather $ArCON{=}PCl_3$.

Janz and Danyluk[8] have shown that addition products from the reactions of acetonitrile and anhydrous hydrogen halides have the composition $CH_3CN \cdot 2HX$. From infrared studies, they conclude that $CH_3CN \cdot 2HX$ exists as the nitrilium salt **5** with X = Cl and as the iminohydrohalides **6** with X = Br, I.

$$[CH_3CNH]^+HCl_2^- \qquad CH_3C(X){=}NH \cdot HX \qquad (X = Br, I)$$

(5) **(6)**

The kinetics of hydrolysis of $RN{=}C(R')Cl$ in aqueous acetone leading to amides has been studied[9].

The chloro substituent in imidyl chlorides is readily displaced by nucleophilic reagents. *N*-Methylbutyramide reacts with phosphorus pentachloride to give *N*-methyl-α-chloroisobutyrimidyl chloride (**7**), from which the imidyl chlorine can be replaced by ethoxide in 84% yield to form ethyl *N*-methyl-α-chlorobutyrimidate (**8**)[10].

$$Me_2CHCONHMe \xrightarrow{PCl_5} \underset{(7)}{Me_2C(Cl){-}C(Cl){=}NMe} \xrightarrow{NaOEt} \underset{(8)}{Me_2C(Cl){-}C(OEt){=}NMe}$$

Hence the imidyl halide appears more reactive than the alkyl halide.

N-Phenylbenzimidyl chloride (**9**) reacts with sodium alkoxides in the corresponding alcohol to give alkyl *N*-phenylbenzimidates (**10**)[11].

$$PhCONHPh \xrightarrow{PCl_5} \underset{(9)}{PhC(Cl){=}NPh} \xrightarrow{RONa,\ ROH} \underset{(10)}{PhC(OR){=}NPh}$$

R = Me, Et, $PhCH_2$, i-Bu, i-Pr, i-Am

Imidyl chlorides will react with alkoxides and phenoxides to form imidates (sometimes called imino esters)[12–22]. The product is isolated either directly, by removal of the solvent *in vacuo* and trituration of the

$$PhC(Cl){=}NPh + PhONa \longrightarrow PhC(OPh){=}NPh + NaCl$$

residue with water, or as the hydrochloride by addition of hydrogen chloride in an anhydrous solvent. The *O*-methyl imidates are free from the *N*-substituted amides.

Another displacing agent is the thiophenoxide ion[15], which in this specific case gives phenyl *N*-phenylthiobenzimidate. Imidyl halides

$$\underset{\displaystyle Cl}{PhC{=}NPh} + PhSNa \longrightarrow \underset{\displaystyle SPh}{PhC{=}NPh} + NaCl$$

(11)

will also react with phenol in pyridine[23], as illustrated by 8-hydroxyquinoline displacing the chlorine in *N*-phenylbenzimidyl chloride to form **12**[24].

$$\underset{\displaystyle Cl}{PhC{=}NPh} + \text{8-hydroxyquinoline} \longrightarrow \text{8-(PhC=NPh)-O-quinoline} + HCl$$

(12)

Busch[25,26] prepared Schiff bases from the reactions of imidyl chlorides with Grignard reagents.

$$\underset{\displaystyle Cl}{PhC{=}NAr} + RMgX \longrightarrow \underset{\displaystyle R}{PhC{=}NAr} + MgXCl$$

In some reactions the imidyl chloride is presumed to be an intermediate but is usually not isolated. The reduction of an acid chloride to an aldehyde might proceed through an imidyl chloride intermediate which becomes reduced to the Schiff base prior to hydrolysis and steam distillation[27]. The Stephen reaction[28] for the conversion of a

$$ArCOCl + PhNH_2 \longrightarrow ArCONHPh \xrightarrow{PCl_5} \underset{\displaystyle Cl}{ArC{=}NPh} \xrightarrow{SnCl_2,\ HCl}$$

$$ArCH{=}NPh \xrightarrow{\text{hydrolysis}} ArCHO + PhNH_2$$

nitrile into an aldehyde may also proceed through an imidyl chloride.

$$RCN + 2HCl \xrightarrow{\text{dry } Et_2O} \left[\underset{\displaystyle Cl}{RC{=}NH_2^+}\ Cl^-\right] \xrightarrow{SnCl_2}$$

$$RCH{=}NH_2^+\ Cl^- \xrightarrow{\text{hydrolysis}} RCHO + NH_4Cl$$

N-(*p*-Tolyl)diphenylacetamide (**13**) on treatment with phosphorus pentachloride yields *N*-(*p*-tolyl)diphenylacetimidyl chloride (**14**), which, although not isolated, forms methyl *N*-(*p*-tolyl)diphenylacetimidate (**15**) on treatment with sodium methoxide. On reaction with hydrochloric acid, **15** was hydrolysed back to the starting material **13**. The acetimidate **15** could also be obtained directly by the action of methanol or diphenylketene-*p*-tolylimine (**16**)[29].

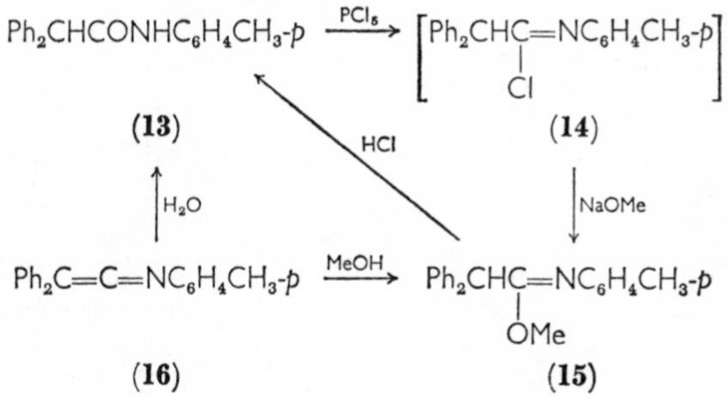

$$Ph_2CHCONHC_6H_4CH_3\text{-}p \xrightarrow{PCl_5} \left[Ph_2CHC(Cl){=}NC_6H_4CH_3\text{-}p\right]$$

(**13**) (**14**)

(**13**) ←HCl— (**15**); (**16**) —H_2O→ (**13**); (**14**) —NaOMe→ (**15**)

$$Ph_2C{=}C{=}NC_6H_4CH_3\text{-}p \xrightarrow{MeOH} Ph_2CHC(OMe){=}NC_6H_4CH_3\text{-}p$$

(**16**) (**15**)

The chlorine atom in imidyl chlorides can also be displaced by ammonia or by amines. Lossen[30] converted benzanilide into *N*-phenylbenzimidyl chloride, which on further treatment with ammonia was converted into *N*-phenylbenzamidine (**17**). *N,N'*-Diphenylbenzamidine (**18**) can be prepared by heating *N*-phenylbenzimidyl chloride

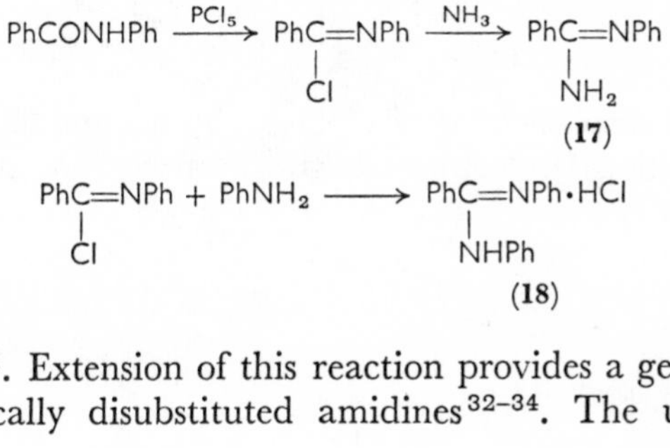

$$PhCONHPh \xrightarrow{PCl_5} PhC(Cl){=}NPh \xrightarrow{NH_3} PhC(NH_2){=}NPh$$

(**17**)

$$PhC(Cl){=}NPh + PhNH_2 \longrightarrow PhC(NHPh){=}NPh\cdot HCl$$

(**18**)

with aniline[31]. Extension of this reaction provides a general method for symmetrically disubstituted amidines[32–34]. The usual method

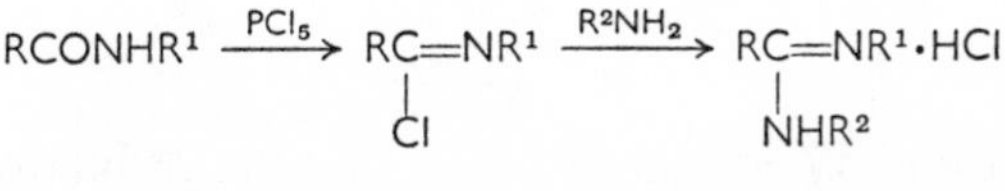

$$RCONHR^1 \xrightarrow{PCl_5} RC(Cl){=}NR^1 \xrightarrow{R^2NH_2} RC(NHR^2){=}NR^1\cdot HCl$$

for carrying out this reaction is first to form the imidyl chloride, to remove the excess reagent, and then to add the amine[33–36]. It is interesting to note that two different isomeric formulas for symmetrically disubstituted amidines can be written. von Pechmann[33,34,37] and later workers[38] demonstrated that attempts to prepare the two

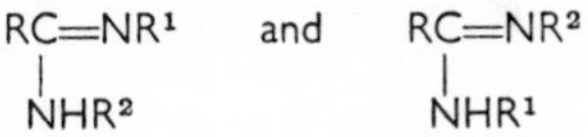

$$RC(NHR^2){=}NR^1 \quad \text{and} \quad RC(NHR^1){=}NR^2$$

forms resulted in the formation of only one compound. Although von Pechmann[39] reported the preparation of two isomeric *N*-(*p*-tolyl)-*N*′-

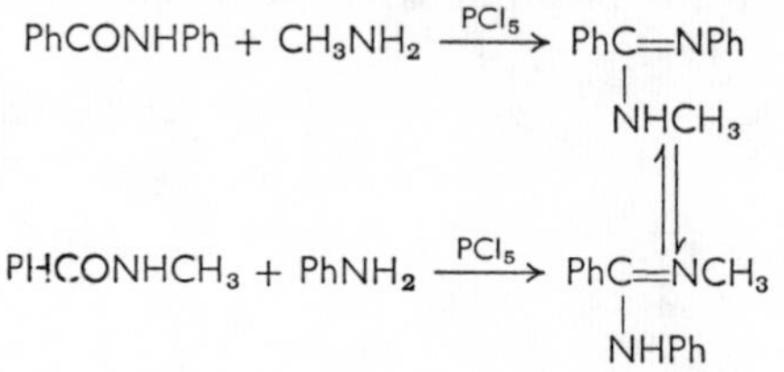

$$PhCONHPh + CH_3NH_2 \xrightarrow{PCl_5} PhC(=NPh)NHCH_3$$

$$\Updownarrow$$

$$PhCONHCH_3 + PhNH_2 \xrightarrow{PCl_5} PhC(=NCH_3)NHPh$$

phenylbenzamidines, it was later shown that a tautomeric equilibrium existed[37,40]. Cohen and Marshall[41] tried to prepare the two tautomers

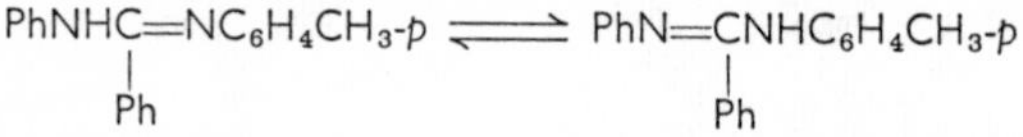

$$PhNHC(Ph)=NC_6H_4CH_3\text{-}p \rightleftharpoons PhN=C(Ph)NHC_6H_4CH_3\text{-}p$$

19 and **20** of a disubstituted amidine by the introduction of optically active bornyl groups into the molecule; however, they were unsuccessful.

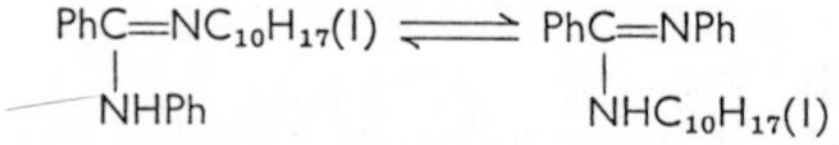

$$\underset{(\mathbf{19})}{PhC(NHPh)=NC_{10}H_{17}(l)} \rightleftharpoons \underset{(\mathbf{20})}{PhC(=NPh)NHC_{10}H_{17}(l)}$$

An unusual displacement of chlorine by an amide occurs in the following reactions[42], leading to the formation of *N*,*N*′-(*p*-tolyl)-acetamidine (**21**).

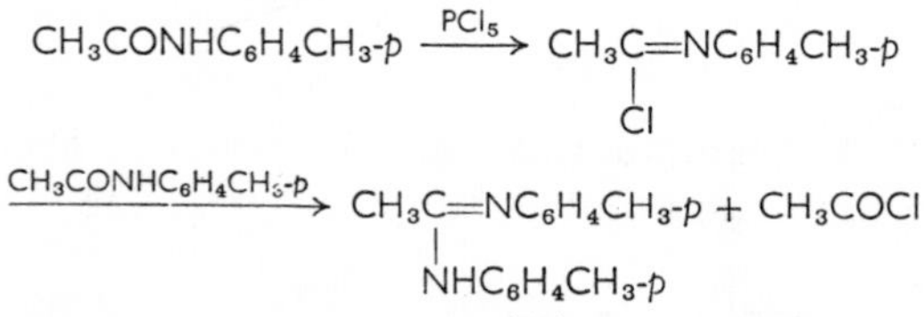

$$CH_3CONHC_6H_4CH_3\text{-}p \xrightarrow{PCl_5} CH_3C(Cl)=NC_6H_4CH_3\text{-}p$$

$$\xrightarrow{CH_3CONHC_6H_4CH_3\text{-}p} \underset{(\mathbf{21})}{CH_3C(NHC_6H_4CH_3\text{-}p)=NC_6H_4CH_3\text{-}p} + CH_3COCl$$

von Braun and Weissbach[43] noted that steric hindrance lowered the yields of **22** from 36% with X = H to ca. 10% with X = Cl.

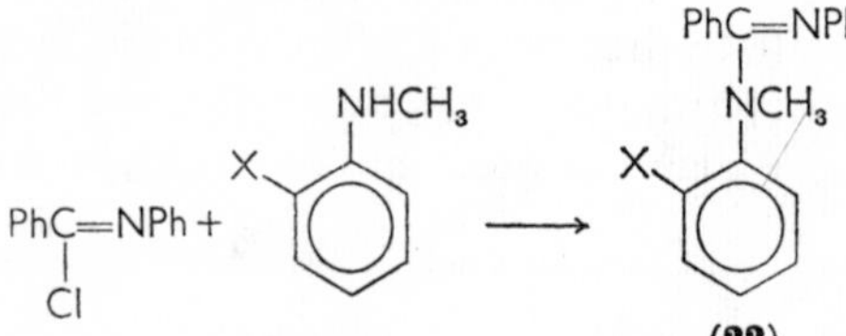

$$PhC(Cl)=NPh + \text{2-X-}C_6H_4NHCH_3 \longrightarrow \underset{(\mathbf{22})}{\text{2-X-}C_6H_4N(CH_3)C(Ph)=NPh}$$

The halogen can be displaced by substituted amines; for example, benzamidine (**23**) will displace the chlorine from *N*-phenylbenzimidyl chloride giving **24**[44].

$$\underset{\displaystyle Cl}{\overset{|}{PhC}}{=}NPh + \underset{\displaystyle NH_2}{\overset{|}{PhC}}{=}NH \longrightarrow \underset{\displaystyle NHC{=}NPh(Ph)}{\overset{|}{PhC}}{=}NH + HCl$$

(**23**) (**24**)

The chlorine in imidyl chlorides can be replaced by nucleophilic substitution to form Reissert compounds: *N*-phenylbenzimidyl chloride will react with isoquinoline and hydrogen cyanide in ether to give 2-(α-phenyliminobenzyl)-1,2-dihydroisoquinaldonitrile hydrochloride (**25**)[45].

$$PhC(Cl){=}NPh + HCN + \text{isoquinoline} \xrightarrow{Et_2O} \text{1-CN-1,2-dihydroisoquinoline-}N\text{-}C(Ph){=}NPh \cdot HCl$$

(**25**)

2. Hydroximic acid chlorides, HON=C(R)Cl and RON=C(R[1])Cl

The chlorine in benzohydroxamyl chloride (**26**) can be replaced by ammonia to give benzamidoxime (**27**)[46]. With hydroxylamine[47],

$$\underset{\displaystyle Cl}{\overset{|}{PhC}}{=}NOH + 2NH_3 \longrightarrow \underset{\displaystyle NH_2}{\overset{|}{PhC}}{=}NOH + NH_4Cl$$

(**26**) (**27**)

benzohydroxamyl chloride yields benzoxyamidoxime (**28**) which can be reduced with SO_2 to prepare benzamidoxime (**27**).

$$\underset{\displaystyle Cl}{\overset{|}{PhC}}{=}NOH + H_2NOH \longrightarrow \underset{\displaystyle NHOH}{\overset{|}{PhC}}{=}NOH + HCl$$

(**26**) (**28**)

$$\underset{\displaystyle NHOH}{\overset{|}{PhC}}{=}NOH \xrightarrow{SO_2} \underset{\displaystyle NH_2}{\overset{|}{PhC}}{=}NOH$$

(**28**) (**27**)

These chloroximes will also undergo nucleophilic substitutions with sodium azide[48] or with amines to yield *N*-substituted amidoximes.

$$\underset{\displaystyle Cl}{\overset{|}{RC}}{=}NOH + NaN_3 \longrightarrow \underset{\displaystyle N_3}{\overset{|}{RC}}{=}NOH + NaCl$$

$$\underset{\displaystyle Cl}{\overset{|}{RC}}{=}NOH + 2R^1NH_2 \longrightarrow \underset{\displaystyle NHR^1}{\overset{|}{RC}}{=}NOH + R^1NH_3^+\ Cl^-$$

They will also react with alkoxides[49,50].

$$\underset{\displaystyle Cl}{\overset{}{RC}}{=}NOH + R^1ONa \longrightarrow \underset{\displaystyle OR^1}{RC}{=}NOH + NaCl$$

The halide in ethyl benzohydroxamyl chloride (**29**) can be replaced by hydroxide in hydrolyses to give ethyl benzohydroxamate (**30**)[46,49,51]. On treatment with alcoholic solutions of ammonia for

$$\underset{\displaystyle Cl}{PhC}{=}NOEt + H_2O \longrightarrow \underset{\displaystyle OH}{PhC}{=}NOEt + HCl$$

(**29**) (**30**)

6–8 h at 160–180°, the halo group of the *O*-alkyl hydroximic acid halides is replaced by ammonia, leading to the formation of amidoximes[52].

$$\underset{\displaystyle X}{RC}{=}NOR + NH_3 \longrightarrow \underset{\displaystyle NH_2}{RC}{=}NOH$$

3. Heterocyclic chlorides

Reactive halogen on a nitrogen heterocyclic ring is well known in pyridines, pyrimidines and quinolines, among many others. Another case in point is the chloro substituent in pseudosaccharin chloride, which can be displaced by alcohols yielding solid ethers as alcohol derivatives[53,54].

[Pseudosaccharin chloride (3-chloro-1,2-benzisothiazole 1,1-dioxide; Cl, N, S, O_2) $\xrightarrow{RCH_2OH}$ 3-OCH_2R analogue (OCH_2R, N, S, O_2) + HCl]

B. *Displacements of Alkoxy Groups*

1. Alkyl imidates, HN=C(R)OR¹ and RN=C(R¹)OR²

Pinner[55,56] treated nitriles, dissolved or suspended in anhydrous alcohols, with an excess of dry hydrogen chloride, forming imidate hydrochlorides (also called imino esters or imino ethers). Acetonitrile with ethanol and hydrogen chloride gives ethyl acetimidate hydrochloride (**31**).

$$CH_3CN + EtOH + HCl \longrightarrow \underset{\displaystyle OEt}{CH_3C}{=}NH\cdot HCl$$

(**31**)

Normally when dry hydrogen chloride is passed into a mixture of an alkyl cyanide and an alcohol dissolved in ether, the hydrochloride of the alkyl imidate precipitates out of solution; presumably the imidyl chloride is the intermediate in the reaction. The hydrochlorides of the imidates are crystalline, water soluble compounds. If sodium carbonate

$$RCN + HCl \longrightarrow \left[\begin{matrix} RC{=}NH \\ | \\ Cl \end{matrix}\right] \xrightarrow{R^1OH} \begin{matrix} RC{=}NH \\ | \\ H\overset{+}{O}R^1 \end{matrix} + Cl^- \longrightarrow \begin{matrix} RC{=}NH \cdot HCl \\ | \\ OR^1 \end{matrix}$$

or potassium or sodium hydroxide[55] is added in the presence of ether, the free alkyl imidate that is formed passes into the ether layer. The imidic ester hydrochlorides, however, do have limited stability[55,57]. They decompose to the original nitrile and alcohol; they are converted into esters by hydrolysis, the reaction being catalysed by acids; or on being warmed with excess alcohol, they form ortho esters. Since

$$\begin{matrix} RC{=}NH \cdot HCl \\ | \\ OR \end{matrix} \longrightarrow RCN + ROH + HCl$$

$$\begin{matrix} RC{=}NH \cdot HCl \\ | \\ OR \end{matrix} + H_2O \longrightarrow RCOOR + NH_4Cl$$

$$\begin{matrix} RC{=}NH \cdot HCl \\ | \\ OR^1 \end{matrix} + 2R^1OH \longrightarrow RC(OR^1)_3 + NH_4Cl$$

the imidic esters are always obtained as hydrochlorides, subsequent reactions must be carried out in anhydrous media. An exception to this instability is ethyl nicotinimidate[58], which on being dissolved in aqueous alcohol containing ammonium chloride is converted into nicotinamidine.

Alkyl imidates readily react with ammonia dissolved in alcohol to yield amidine hydrochlorides[59–63]. The ethoxy group in ethyl acetimidate (**32**) is displaced by ammonia to give acetamidine hydrochloride (**33**). The nucleophile can even be an ammonium salt in aqueous alcohol[58,64–66] as solvent. Substituted acetonitriles also form

$$\begin{matrix} RC{=}NH \\ | \\ OR \end{matrix} + NH_4X \xrightarrow[50–70°]{EtOH\ (aq)} \begin{matrix} RC{=}NH \cdot HX \\ | \\ NH_2 \end{matrix} + ROH$$

imidates from which the alkoxy group can be displaced by ammonia[67,68] or amines[63,69].

$$PhNHCH_2CN \longrightarrow \begin{matrix} PhNHCH_2C{=}NH \cdot HCl \\ | \\ OEt \end{matrix} \longrightarrow \begin{matrix} PhNHCH_2C{=}NH \cdot HCl \\ | \\ NH_2 \end{matrix}$$

Tafel and Enoch[70,71] treated benzamide (**34**) with silver nitrate and sodium hydroxide in aqueous solution to prepare silver benzamide (**35**) which was converted into ethyl benzimidate (**36**) with ethyl iodide. Benzamidine (**37**) was obtained after displacement of the ethoxy

$$\underset{(\mathbf{34})}{PhCONH_2} \longrightarrow \underset{(\mathbf{35})}{\underset{|}{PhC{=}NH}\atop OAg} \longrightarrow \underset{(\mathbf{36})}{\underset{|}{PhC{=}NH}\atop OEt} \longrightarrow \underset{(\mathbf{37})}{\underset{|}{PhC{=}NH}\atop NH_2}$$

group with ammonia. Ethyl *N*-phenylacetimidate (**38**) was also prepared in a similar way[20,72,73].

$$\underset{|}{CH_3C{=}NPh}\atop OAg \; + EtI \longrightarrow \underset{(\mathbf{38})}{\underset{|}{CH_3C{=}NPh}\atop OEt}$$

Addition compounds of nitrile with hydrogen chloride do not produce amidines when treated with ammonia, but rather regenerate the nitrile[74]; this led to the suggestion that they were salt-like complexes of the nitrile with hydrogen chloride and not true imidyl chlorides (cf. Janz and Danyluk[8] in Section II.A.1).

The alkoxy group can be replaced from imidate salts with primary amines with the formation of monosubstituted amidines (**39**)[55,75]. The

$$\underset{|}{RC{=}NH\cdot HCl}\atop OR^1 \; + R^2NH_2 \longrightarrow \underset{(\mathbf{39})}{\underset{|}{RC{=}NH\cdot HCl}\atop NHR^2} + R^1OH$$

base strength of the nucleophile is important and the yields are best when R^2 is aliphatic[55,59,62,76–79]. With an excess of primary amine at a higher temperature for longer reaction periods, *N,N'*-disubstituted amidines (**40**) are produced[55,75,76,80].

$$\underset{|}{RC{=}NH\cdot HCl}\atop OEt \; + 2R^1NH_2 \longrightarrow \underset{(\mathbf{40})}{\underset{|}{RC{=}NR^1}\atop NHR^1} + NH_4Cl + EtOH$$

Secondary amines can be used as nucleophilic agents for alkyl imidates and lead to unsymmetrical disubstituted amidines (**41**)[55,75,77,81–84]. Imidates do not react with tertiary amines[55,79].

$$\underset{|}{RC{=}NH\cdot HCl}\atop OEt \; + R^1R^2NH \longrightarrow \underset{(\mathbf{41})}{\underset{|}{RC{=}NH\cdot HCl}\atop NR^1R^2} + EtOH$$

N-Phenyl-*N'*-(*p*-tolyl)formamidine (**44**) can be obtained from either ethyl *N*-phenylformimidate (**42**) and *p*-toluidine, or from ethyl *N*-(*p*-tolyl)formimidate (**43**) and aniline, provided there is rigid exclusion of all traces of acid[85].

$$\underset{\textbf{(42)}}{\underset{\text{OEt}}{\text{HC}{=}\text{NPh}}} + p\text{-MeC}_6\text{H}_4\text{NH}_2 \longrightarrow \underset{\textbf{(44)}}{\underset{\text{NH}p\text{-MeC}_6\text{H}_4}{\text{HC}{=}\text{NPh}}} + \text{EtOH}$$

$$\underset{\textbf{(43)}}{\underset{\text{OEt}}{\text{HC}{=}\text{N}p\text{-MeC}_6\text{H}_4}} + \text{PhNH}_2 \longrightarrow \textbf{(44)}$$

However, in the presence of an acid (e.g. the amine salt), a mixture of **44**, *N,N′*-diphenylformamidine (**45**), and *N,N′*-di(*p*-tolyl)formamidine (**46**) is obtained. This mixture has a sharp melting point (86°) and cannot be separated by crystallization. An explanation of this has been given[86].

$$\underset{\textbf{(45)}}{\underset{\text{NHPh}}{\text{HC}{=}\text{NPh}}} \qquad \underset{\textbf{(46)}}{\underset{\text{NHC}_7\text{H}_7}{\text{HC}{=}\text{NC}_7\text{H}_7}}$$

The imidate base in a suitable organic solvent, when shaken with an aqueous solution of the salt of an α-amino acid ester, yields an *N*-substituted imidate[87–95], but this is not a substitution reaction as we have defined.

$$\underset{\text{OR}^1}{\text{RC}{=}\text{NH}} + \text{HCl}\cdot\text{H}_2\text{NCH(CO}_2\text{Et)}_2 \longrightarrow \underset{\text{OR}^1}{\text{RC}{=}\text{NCH(CO}_2\text{Et)}_2}$$

The condensation of the free imidate with an α-aminocyanoacetic ester itself yields an imidazole[96]. However, ethyl phenylacetimidate (**47**) will react with glycine[97]. Further ethyl carbobenzoxyamino-

$$\underset{\text{OR}^1}{\text{RC}{=}\text{NH}} + \text{NH}_2\text{CH(CN)CO}_2\text{Et} \longrightarrow \text{2-R-4-CO}_2\text{Et-5-NH}_2\text{-imidazole (N–H)}$$

$$\underset{\textbf{(47)}}{\underset{\text{OEt}}{\text{PhCH}_2\text{C}{=}\text{NH}}} + \text{H}_2\text{NCH}_2\text{COOH} \longrightarrow \underset{\text{NHCH}_2\text{COOH}}{\text{PhCH}_2\text{C}{=}\text{NH}} + \text{EtOH}$$

acetimidate hydrochloride (**48**) will react with glycine to form an intermediate from which **49** can be obtained by cleavage with hydrobromic acid–acetic acid[97]. With model compounds it has been established that only the amino groups of proteins react with alkyl imidates in aqueous solutions[98].

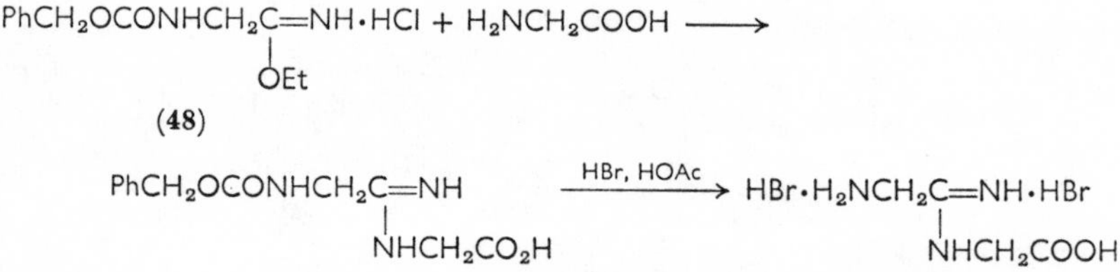

$$PhCH_2OCONHCH_2C(OEt){=}NH \cdot HCl + H_2NCH_2COOH \longrightarrow$$

(48)

$$PhCH_2OCONHCH_2C(NHCH_2CO_2H){=}NH \xrightarrow{HBr,\ HOAc} HBr \cdot H_2NCH_2C(NHCH_2COOH){=}NH \cdot HBr$$

(49)

The *N*-substituted imidate **50** can have its ethoxy group displaced by an amine to yield an intermediate which can cyclize to form hydroxyimidazoles (**51**)[95]. Analogously, aminoacetone displaces the ethoxy group and cyclizes to give 2-alkyl-4-methylimidazole (**52**)[99].

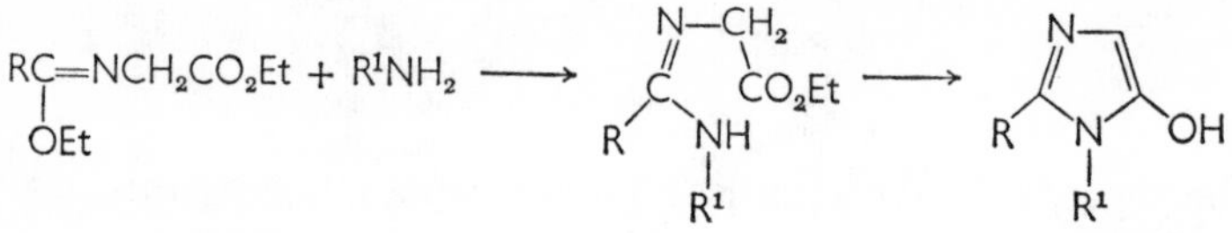

$$RC(OEt){=}NCH_2CO_2Et + R^1NH_2 \longrightarrow RC(NHR^1){=}N{-}CH_2CO_2Et \longrightarrow \text{(51)}$$

(50) (51)

$$RC(OEt){=}NH + H_2NCH_2COCH_3 \longrightarrow \text{(52)}$$

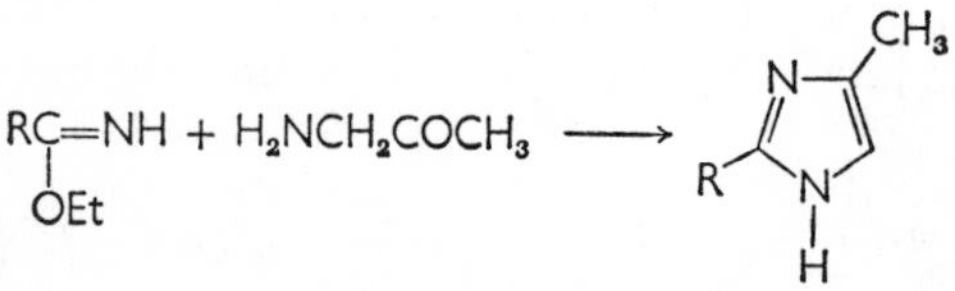

(52)

An interesting set of reactions is described by Roberts and Vogt[100], in which any one of three products can be isolated from the reaction of ethyl orthoformate (**53**) and aniline. *N,N'*-Diphenylformamidine (**55**)

$$\underset{(53)}{(EtO)_3CH} + PhNH_2 \xrightarrow[<140^\circ]{H^+} \underset{(54)}{PhN{=}CHOEt} + 2EtOH$$

$$\mathbf{54} + PhNH_2 \longrightarrow \underset{(55)}{PhN{=}CHNHPh} + EtOH$$

$$\mathbf{54} + H_2SO_4 \xrightarrow{140^\circ} \underset{(56)}{PhN(Et)CHO}$$

can be obtained if no acid catalyst is used or in the presence of catalyst with two moles of aniline per mole of ethyl orthoformate (**53**), since the intermediate, ethyl *N*-phenylformimidate (**54**) reacts with more aniline forming **55**. Ethyl *N*-phenylformimidate (**54**) can be made when aniline reacts with equimolar (or greater) amounts of ethyl orthoformate (**53**) in the presence of hydrochloric acid, sulphuric acid, *p*-toluenesulphonic acid or acetic acid at temperatures up to 140°. *N*-Ethylformanilide (**56**) can be obtained with sulphuric acid as catalyst with the temperature being raised to 140° after the ethanol initially produced is removed.

Substituted alkoxy groups can be replaced from imidates (**57**), prepared from propanesultone and acetamide, by amines and amino acids[101]. The group may also be displaced by substituted

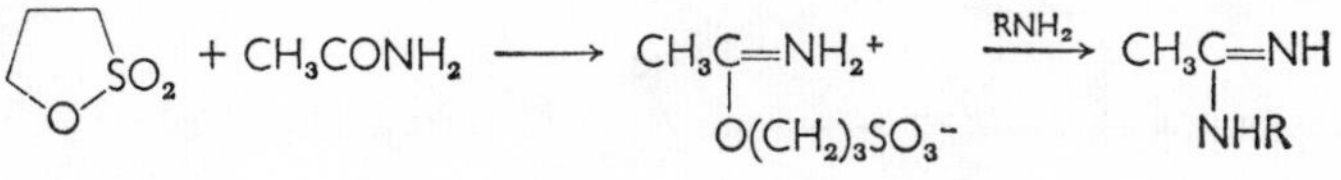

$$\text{(propanesultone, } SO_2\text{)} + CH_3CONH_2 \longrightarrow \underset{O(CH_2)_3SO_3^-}{CH_3C{=}NH_2^+} \xrightarrow{RNH_2} \underset{NHR}{CH_3C{=}NH}$$

(**57**)

hydrazine derivatives (**58**), yielding amidrazinoacetic acids (**59**)[102].

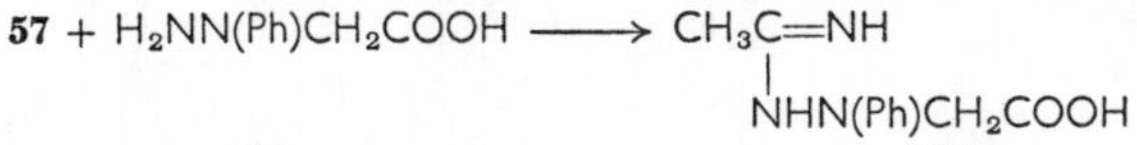

$$\mathbf{57} + H_2NN(Ph)CH_2COOH \longrightarrow \underset{NHN(Ph)CH_2COOH}{CH_3C{=}NH}$$

(**58**) (**59**)

N-Acyl derivatives of imino esters (**60**) react with hydrazine to yield 3,5-disubstituted-1,2,4(1H)-triazoles (**61**)[103].

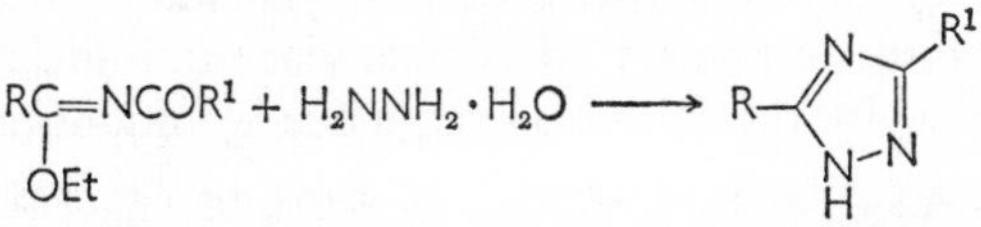

$$\underset{OEt}{RC{=}NCOR^1} + H_2NNH_2{\cdot}H_2O \longrightarrow \text{R-(1,2,4-triazole, N-H)-}R^1$$

(**60**) (**61**)

The displacements of the alkoxy groups from aromatic imidates by hydrazine leads to amidrazones (**62**)[55,104–108]. These substitutions can

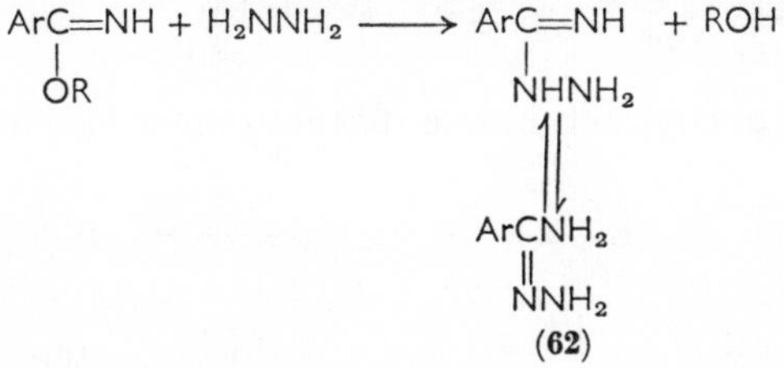

$$\underset{OR}{ArC{=}NH} + H_2NNH_2 \longrightarrow \underset{NHNH_2}{ArC{=}NH} + ROH \rightleftharpoons \underset{NNH_2}{ArC(NH_2){=}}$$

(**62**)

take one of three paths[109,110].

a. Equimolar reactants. If the temperature is maintained at 0°, equimolar quantities react to yield amidrazones.

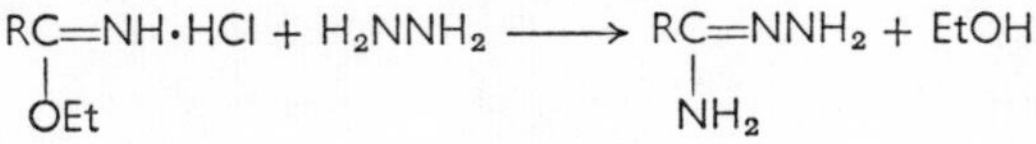

$$\underset{OEt}{RC{=}NH{\cdot}HCl} + H_2NNH_2 \longrightarrow \underset{NH_2}{RC{=}NNH_2} + EtOH$$

b. Two parts hydrazine to one part imidate. Again with careful maintenance of temperature at 0°, two parts of hydrazine react with one part imidate.

$$RC(OEt){=}NH\cdot HCl + 2H_2NNH_2 \longrightarrow RC(NHNH_2){=}NNH_2 + EtOH + NH_4Cl$$

c. At a temperature of 40–50°. When the temperature is raised to 40–50°, sunstituted tetrazines (**62a**) and *N*-aminotriazoles (**62b**) are formed.

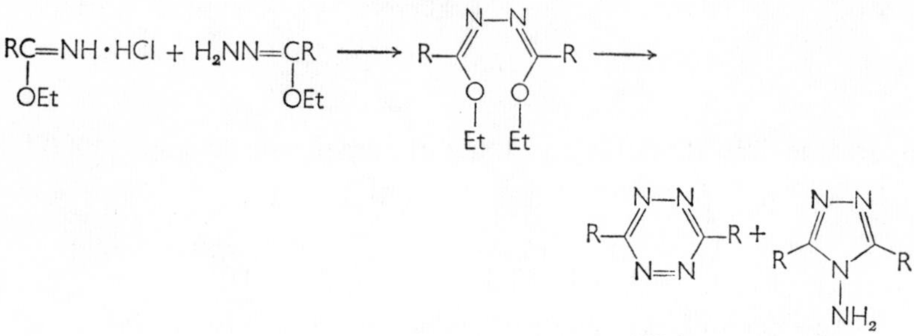

$$RC(OEt){=}NH\cdot HCl + H_2NN{=}CR(OEt) \longrightarrow \text{R–(dihydrotetrazine, OEt, OEt)–R} \longrightarrow \text{R–(tetrazine)–R} + \text{R–(4-amino-1,2,4-triazole, }N\text{-}NH_2\text{)–R}$$

(**62a**) (**62b**)

Bernton[111] formed the nitrite of phenylacetamidine (**64**) by the reaction of sodium nitrite on ethyl phenylacetimidate hydrochloride (**63**), presumably by the following sequence. Ammonium nitrate also

$$PhCH_2C(OEt){=}NH\cdot HCl + NaNO_2 + H_2O \longrightarrow PhCH_2CO_2Et + NH_4NO_2 + NaCl$$

(**63**)

$$PhCH_2C(OEt){=}NH + NH_4NO_2 \longrightarrow PhCH_2C(NH_2){=}NH\cdot HNO_2 + EtOH$$

(**64**)

converts ethyl phenylacetimidate (**65**) into phenylacetamidine nitrate (**66**)[111].

$$PhCH_2C(OEt){=}NH + NH_4NO_3 \longrightarrow PhCH_2C(NH_2){=}NH\cdot HNO_3$$

(**65**) (**66**)

An odd displacement was found by Rule[112], who noted that ethyl mandeloimidate (**67**) was converted into mandelamidine mandelate

$$2PhCHOHC(OEt){=}NH + 2H_2O \longrightarrow PhCHOHC(NH_2){=}NH\cdot PhCHOHCOOH + EtOH$$

(**67**) (**68**)

(**68**) by shaking **67** with water at room temperature for five days. Although this decomposition is rather unusual, it was ascertained that free imidic esters prepared from other cyanohydrins reacted with water analogously.

Simple *N*-substituted imidates when heated under reflux in benzene or alcohol with sulphonamides undergo displacement of the ethoxy group to form sulphonyl derivatives of the amidines (**69**)[113,114]. The

$$CH_3C(OEt){=}NR + R^1SO_2NH_2 \longrightarrow [CH_3C(NHSO_2R^1){=}NR] \longrightarrow CH_3C(NHR){=}NSO_2R^1 \quad (\mathbf{69})$$

amino element of a sulphonamide group is a better nucleophile than that of a simple amine; thus, *p*-aminobenzenesulphonamide reacts with imidates[115,116]. These *N*-sulphonyl derivatives of the imidates

$$RC(OR^1){=}NH + H_2NSO_2{-}C_6H_4{-}NH_2 \longrightarrow RC(NH_2){=}NSO_2{-}C_6H_4{-}NH_2$$

undergo nucleophilic attack with displacement of the alkoxy group with dimethylamine[117] and with water[118,119]. Barber[120] isolated two forms

$$R^1C(OEt){=}NSO_2R + Me_2NH \longrightarrow R^1C(NMe_2){=}NSO_2R + EtOH$$

$$R^1C(OEt){=}NSO_2R + H_2O \longrightarrow [R^1C(OH){=}NSO_2R] \longrightarrow R^1C(=O)NHSO_2R$$

of substituted sulphonamidines **70** and **71**; **70** was converted easily into the more stable **71**. However, Northey and coworkers[121] concluded

$$RC(OEt){=}NSO_2Ar + NH_3 \longrightarrow RC(NH_2){=}NSO_2Ar + EtOH \quad (\mathbf{70})$$

$$RC(NH_2){=}NH + ArSO_2Cl \longrightarrow RC(NHSO_2Ar){=}NH + HCl \quad (\mathbf{71})$$

that the sulphonamidines are better represented by **70**, since they do not form alkali salts and since hydrolysis products are consistent with such a structure.

$$RC(NH_2){=}NSO_2Ar \xrightarrow{H^+ \text{ or } OH^-} RCONH_2 + ArSO_2NH_2$$

When an alcoholic solution of ethyl *p*-sulphamidobenzimidate hydrochloride (**72**) is treated with *O*-methylhydroxylamine at 37° in a pressure bottle, ammonium chloride separates out and two products can be isolated[122]. The first product is the result of a nucleophilic displacement of the ethoxy group and is the *O*-methyl derivative of sulphamidobenzoxyamidoxime (**73**), which is soluble in dilute hydrochloric acid; the second product (**74**), formed by the nucleophilic displacement of the amino group in **73**, insoluble in dilute hydrochloric acid, is the dimethyl ether of the amidoxime.

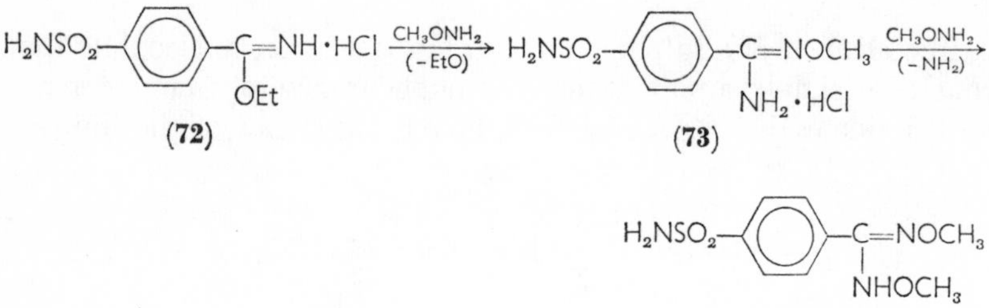

$$H_2NSO_2-C_6H_4-C(OEt){=}NH\cdot HCl \xrightarrow[(-EtO)]{CH_3ONH_2} H_2NSO_2-C_6H_4-C(NH_2\cdot HCl){=}NOCH_3 \xrightarrow[(-NH_2)]{CH_3ONH_2}$$

(**72**) (**73**)

$$H_2NSO_2-C_6H_4-C(NHOCH_3){=}NOCH_3$$

(**74**)

Nucleophilic displacements of imidates with monosubstituted hydrazines leads to N_1-substituted amidrazones[55,123–127], and a

$$RC(OEt){=}NH\cdot HCl + H_2NNHR^1 \longrightarrow RC(NH_2){=}NNHR^1 + EtOH$$

N_1,N_1-disubstituted amidrazone (**76**) can be obtained by the displacement reaction of ethyl mandeloimidate hydrochloride (**75**) with unsymmetrical diphenylhydrazine[128].

$$PhCHOHC(OEt){=}NH\cdot HCl + Ph_2NNH_2 \longrightarrow PhCHOHC(NH_2){=}NNPh_2$$

(**75**) (**76**)

The imidates also undergo nucleophilic substitution with hydrazoic acid[129–131].

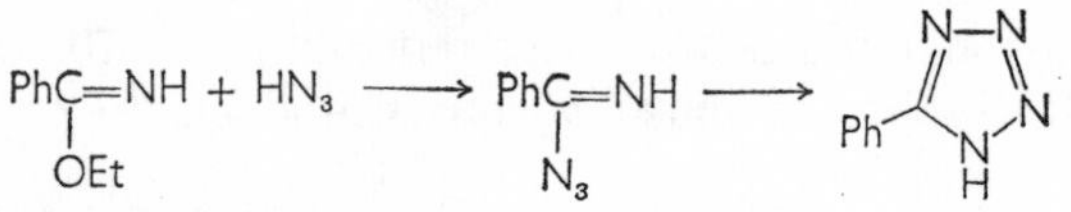

$$PhC(OEt){=}NH + HN_3 \longrightarrow PhC(N_3){=}NH \longrightarrow \text{5-phenyltetrazole (Ph-}CN_4H\text{)}$$

Ethyl benzimidate (**77**) undergoes displacement with hydroxylamine to form benzamidoxime (**78**)[132,133].

$$\underset{(77)}{\mathrm{PhC(OEt){=}NH}} + \mathrm{H_2NOH} \longrightarrow \left[\mathrm{PhC(NHOH){=}NH}\right] \longrightarrow \underset{(78)}{\mathrm{PhC(NH_2){=}NOH}}$$

Methyl *N*-phenylbenzimidate (**79**) reacts with phenylmagnesium bromide in toluene at 200° to give *N*-benzhydrylideneaniline (**80**)

$$\underset{(79)}{\mathrm{PhC(OCH_3){=}NPh}} + \mathrm{PhMgBr} \longrightarrow \underset{(80)}{\mathrm{Ph_2C{=}NPh}} \longrightarrow \mathrm{Ph_2C{=}O}$$

which, of course, can be hydrolysed to benzophenone. Imidates react with Grignard reagents to yield ketimines (**81**) which are also hydrolysed to ketones[128]. Ethyl *N*-phenylformimidate (**82**) reacts with phenylmagnesium bromide to form *N*-benzylideneaniline (**83**)[134–136].

$$\mathrm{RC(OEt){=}NH} + \mathrm{PhMgBr} \longrightarrow \underset{(81)}{\mathrm{RC(Ph){=}NH}} \longrightarrow \mathrm{RC(Ph){=}O}$$

$$\underset{(82)}{\mathrm{HC(OEt){=}NPh}} + \mathrm{PhMgBr} \longrightarrow \underset{(83)}{\mathrm{PhCH{=}NPh}}$$

Compounds such as **84** react with diazomethane to form *N*-methyl derivatives (**85**), which react with water and alcohols to form amides and imino esters respectively[137]; **86** reacts exothermically with amines to form amidines (**87**).

$$\underset{(84)}{\mathrm{(MeSO_2)_2CHCN}} + \mathrm{CH_2N_2} \longrightarrow \underset{(85)}{\mathrm{(MeSO_2)_2C{=}C{=}NMe}}$$

$$\mathbf{85} + \mathrm{MeOH} \longrightarrow \underset{(86)}{\mathrm{(MeSO_2)_2CHC(OMe){=}NMe}}$$

$$\mathbf{86} + \mathrm{PhNH_2} \longrightarrow \underset{(87)}{\mathrm{(MeSO_2)_2CHC(NHPh){=}NMe}}$$

The base-catalysed reaction of nitriles[138] with alcohols gives imidates which can be converted into amidines. This sequence of reactions sometimes offers advantages over the Pinner reaction with hydrogen chloride and alcohols.

$$\mathrm{RCN + R^1OH} \xrightarrow{\mathrm{NaOR^1}} \mathrm{RC(=NH)OR^1} \xrightarrow{\mathrm{NH_4Cl(alc)}} \mathrm{RC(=NH)NH_2 \cdot HCl}$$

$$\mathrm{RCN + R^1OH} \xrightarrow{\mathrm{HCl}} \mathrm{RC(=NH)OR^1 \cdot HCl} \xrightarrow{\mathrm{NH_3(alc)}} \mathrm{RC(=NH)NH_2 \cdot HCl}$$

$$\mathrm{RC(=NH)OR^1 \cdot HCl} \xrightarrow{\mathrm{K_2CO_3(aq)}} \mathrm{RC(=NH)OR^1}$$

Acetyl-L-phenylalanine nitrile (**88**) was made from the corresponding amide with phosphorus oxychloride in pyridine and converted into the imidate (**89**), with retention of configuration, and was then transformed into an amidine (**90**) by reaction with amines[139]. Imidates from the reactions of *N*-protected α-amino acid amides and Et_2O/BF_4 were reacted with amines to form amidines and with α-amino acids to form iminodipeptides[140]. Nucleophilic displacement of the ethoxy group also occurred when ethyl benzimidate was treated with β-, γ-, δ- or α-amino acids to form *N*-substituted amidines (**91**)[141].

$$\underset{(\mathbf{88})}{\mathrm{PhCH_2CH(NHAc)CN}} \longrightarrow \underset{(\mathbf{89})}{\mathrm{PhCH_2CH(NHAc)\text{—}C(=NH)OR}} \longrightarrow \underset{(\mathbf{90})}{\mathrm{PhCH_2CH(NHAc)\text{—}C(=NH)NH_2}}$$

$$\underset{(\mathbf{91})}{\mathrm{PhC(=NH_2^+)NHCH(R)(CH_2)_nCO_2^-}} \qquad (n = 0\text{–}4)$$

Perfluoroacyl imidates have also been made from perfluoronitriles[142]. These imidates will then undergo nucleophilic displacement with ammonia, hydroxylamine and hydrazine to give perfluoroacyl amidines, amidoximes and hydrazidines[143]. Methyl perfluorobutyrimidate (**92**) undergoes nucleophilic displacement by ammonia yielding **93**.

$$\underset{(\mathbf{92})}{\mathrm{C_3F_7C(=NH)OMe}} + \mathrm{NH_3} \longrightarrow \underset{(\mathbf{93})}{\mathrm{C_3F_7C(=NH)NH_2}} + \mathrm{MeOH}$$

An interesting intramolecular displacement of an alkoxy group occurs in the cyclic imidate, (3*R*,5*R*)-*N*-cyclopropyl-5-bromomethyl-3-(*o*-hydroxyphenyl)-3-phenyl-2-tetrahydrofuranoneimine (**94**), which on treatment with sodium methoxide in methanol gives *N*-cyclopropyl-3-(2′,3′-epoxypropyl)-3-phenyl-2-benzofuranoneimine (**95**)[144]. It is apparent that the phenolate ion is the nucleophilic species attacking the carbon atom of the azomethine group with displacement

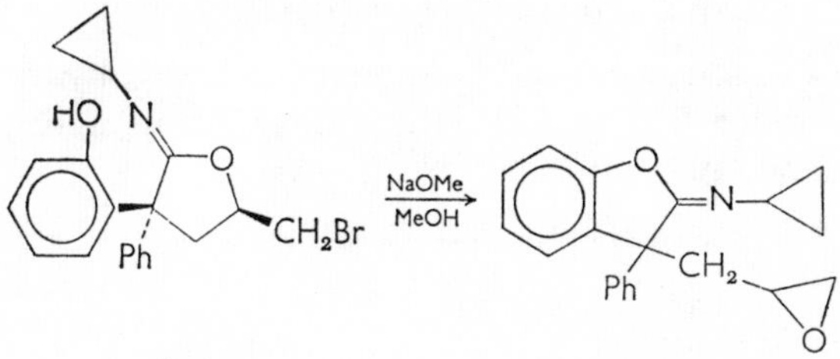

(**94**) (**95**)

of the alkoxy group which simultaneously displaces the bromo group, leading to the epoxy compound. Since the bromomethyl group is 'up' in **94**, models show that the phenolate ion can also attack that carbon with elimination of bromide, forming 4,5-benzo-7-cyclopropylimino-3,8-dioxa-6-phenylbicyclo[4.2.1]nonane (**96**). Of course, this latter

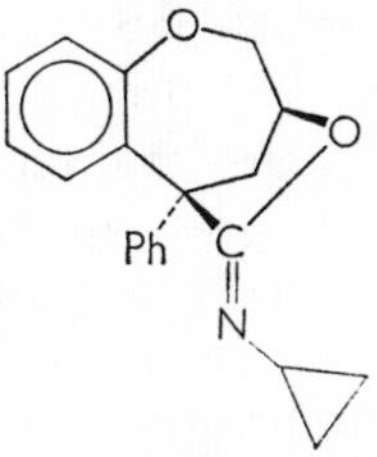

(**96**)

reaction is not a substitution reaction involving either carbon or nitrogen of the azomethine. However, the diastereoisomer of **94** in which the 5-bromomethyl group is in the (*S*) configuration gives only the compound analogous to **95** on treatment with base. It cannot give a compound like **96** because the bromomethyl group is too far removed from the nucleophilic phenolate ion[144].

2. O-Alkylisoureas, $HN{=}C(NH_2)OR$

The hydrazinolysis of *O*-alkylisoureas leads to aminoguanidine derivatives. *O*-Ethylisourea hydrogen sulphate (**97**) (prepared from urea and diethyl sulphate), when dissolved in water and treated with aqueous hydrazine at 60°, undergoes a nucleophilic substitution to yield aminoguanidine (**98**)[145].

$$\underset{\substack{|\\ \text{OEt}\\ \textbf{(97)}}}{\text{H}_2\text{NC}{=}\text{NH}} + \text{H}_2\text{NNH}_2 \longrightarrow \underset{\substack{|\\ \text{NHNH}_2\\ \textbf{(98)}}}{\text{H}_2\text{NC}{=}\text{NH}} + \text{EtOH}$$

3. Alkyl hydroxamic acids, HON=C(R)OR[1]

When an alcoholic solution of ammonia and ethyl hydroxamic acid (**99**) is heated for 8 h at 175°, the ethoxy group is displaced and benzamidoxime (**100**) is formed[146].

$$\underset{\substack{|\\ \text{OEt}\\ \textbf{(99)}}}{\text{PhC}{=}\text{NOH}} + \text{NH}_3 \longrightarrow \underset{\substack{|\\ \text{NH}_2\\ \textbf{(100)}}}{\text{PhC}{=}\text{NOH}} + \text{EtOH}$$

Good yields of hydroxamic esters are obtained when hydroxylamine in aqueous solution is shaken with ethereal solutions of imidates[147–149].

$$\underset{\substack{|\\ \text{OEt}}}{\text{RC}{=}\text{NH}} + \text{H}_2\text{NOH} \longrightarrow \underset{\substack{|\\ \text{OEt}}}{\text{RC}{=}\text{NOH}}$$

C. *Displacements of Thioalkoxy Groups*

1. Alkyl thioimidates, RN=C(R[1])SR[2]

Thioimidates can be prepared from the reactions of nitriles with thiols in the presence of hydrogen chloride[87,150–153].

$$\text{RCN} + \text{HCl} + \text{R}^1\text{SH} \longrightarrow \underset{\substack{|\\ \text{SR}^1}}{\text{RC}{=}\text{NH}\cdot\text{HCl}}$$

The thioalkoxy groups can be displaced with amines and ammonia in absolute alcohol leading to amidines[154].

$$\underset{\substack{|\\ \text{SMe}}}{\text{RC}{=}\text{NH}\cdot\text{HI}} + \text{R}^1\text{NH}_2 \longrightarrow \underset{\substack{|\\ \text{NHR}^1}}{\text{RC}{=}\text{NH}}$$

S-Alkyl isothioanilides will react with aromatic amines to produce symmetrically disubstituted amidines[155,156]. Ethyl *N*-phenylthioimidate [*S*-ethyl isothioacetanilide] (**101**) suffers nucleophilic displacemen with aromatic amines to give *N*-phenyl-*N*′-arylacetamidine (**102**).

$$\underset{\substack{|\\ \text{SEt}\\ \textbf{(101)}}}{\text{CH}_3\text{C}{=}\text{NPh}} + \text{ArNH}_2 \longrightarrow \underset{\substack{|\\ \text{NHAr}\\ \textbf{(102)}}}{\text{CH}_3\text{C}{=}\text{NPh}} + \text{EtSH}$$

Thioacetamide reacts with propanesultone in benzene to produce the thioimidate **103**, and these will undergo substitution with amines and amino acids leading to amidines[101,102].

$$\text{(tetrahydrofuran-SO}_2\text{ sultone)} + CH_3CSNH_2 \longrightarrow CH_3C(=NH_2^+)S(CH_2)_3SO_3^-$$
(103)

2. *S*-Alkylisothioureas, HN=C(NH$_2$)SR, and derivatives

Hydrazinolysis of *S*-alkylisothiouronium salts (**104**) or ammonolysis of *S*-alkylisothiosemicarbazide (**105**) produces aminoguanidine[157].

$$\underset{\textbf{(104)}}{HN{=}C(SR)NH_2} + H_2NNH_2 \longrightarrow HN{=}C(NHNH_2)NH_2 + RSH$$

$$\underset{\textbf{(105)}}{HN{=}C(SR)NHNH_2} + NH_3 \longrightarrow HN{=}C(NH_2)NHNH_2 + RSH$$

However, *N*-alkylaminoquanidines cannot be obtained from *S*-methylisothiourea and alkyl hydrazines[158]. 1-Amino-1-methylguanidine (**107**) is obtained from either the nucleophilic attack of methylhydrazine on *S*-methylisothiourea or from the attack of ammonia on *S*,2-dimethylisothiosemicarbazide (**106**).

$$HN{=}C(NH_2)SMe + MeNHNH_2 \longrightarrow \underset{\textbf{(107)}}{HN{=}C(NH_2)N(CH_3)NH_2}$$

$$\underset{\textbf{(106)}}{HN{=}C(SMe)N(CH_3)NH_2} + NH_3 \longrightarrow \textbf{107}$$

The nitrogen atom adjacent to the alkyl group in the alkyl hydrazine is the more nucleophilic.

The thiomethyl group of *S*-methylisothiosemicarbazide is readily displaced with amines in boiling ethanol, yielding *N*-alkyl-*N'*-aminoguanidines (**108**) which react with aldehydes to form guanyl hydrazones (**109**). This latter product can also be obtained by the displacement from *S*-methylisothiosemicarbazones (**110**) with amines, but this

$$HN{=}C(SMe)NHNH_2 \xrightarrow{RNH_2} \underset{\textbf{(108)}}{HN{=}C(NHR)NHNH_2} \xrightarrow{R^1CHO} \underset{\textbf{(109)}}{HN{=}C(NHR)NHN{=}CHR^1}$$

$$\underset{\textbf{(110)}}{HN{=}C(SMe)NHN{=}CHR^1} + RNH_2 \longrightarrow \textbf{109}$$

reaction proceeds more slowly and is accompanied by the production of considerable tar[159].

2,*S*-Dimethylisothiosemicarbazide (**111**) reacts with methylhydrazine to give 1,3-dimethyl-1,3-diaminoguanidine (**112**)[160]. Again it is the more nucleophilic nitrogen that does the attacking.

$$\underset{\textbf{(111)}}{H_2NN(CH_3)\underset{\displaystyle SMe}{\underset{|}{C}}{=}NH} + MeNHNH_2 \longrightarrow \underset{\textbf{(112)}}{H_2NN(CH_3)\underset{\displaystyle N(CH_3)NH_2}{\underset{|}{C}}{=}NH}$$

The *S*-methyl derivatives of isothiocarbohydrazide (**113**) behave towards amines and hydrazines as do the *S*-alkylisothioureas and isothiosemicarbazides[161]. **114** is a diaminoguanidine derivative and **115** is a triaminoguanidine.

$$\underset{\textbf{(113)}}{H_2NNH\underset{\displaystyle SMe}{\underset{|}{C}}{=}NNH_2} + RNH_2 \longrightarrow \underset{\textbf{(114)}}{H_2NNH\underset{\displaystyle NHR}{\underset{|}{C}}{=}NNH_2}$$

$$\textbf{113} + H_2NNH_2 \longrightarrow \underset{\textbf{(115)}}{H_2NNH\underset{\displaystyle NHNH_2}{\underset{|}{C}}{=}NNH_2}$$

D. Displacements of Hydrogen

1. Schiff bases, RCH=NR[1]

Hydrogen attached to the carbon of azomethines can be displaced with sodium amide in dry toluene at 120°. *N*-Benzylideneaniline (**116**) yields *N*-phenylbenzamidine (**117**) and ammonia[162].

$$\underset{\textbf{(116)}}{PhCH{=}NPh} \xrightarrow{NaNH_2} \underset{\textbf{(117)}}{Ph\underset{\displaystyle NH_2}{\underset{|}{C}}{=}NPh} + NH_3$$

N-Alkylidenecyclohexylamine (**118**) can have its hydrogen replaced with an alkyl group[163].

$$\underset{\textbf{(119)}}{C_6H_{11}N{=}CHR} \xrightarrow{i\text{-}Pr_2NLi} [C_6H_{11}N{=}CR^-] \xrightarrow{R^1X} C_6H_{11}N{=}CRR^1$$

t-Butyl hypochlorite will substitute a chlorine for the hydrogen in Schiff bases prepared from trimethylacetaldehyde or from benzaldehyde and amines. The intermediate imidyl chlorides react with alcohols to give *N*-substituted imidates (**119**) or with amines to form substituted amidines (**120**)[164].

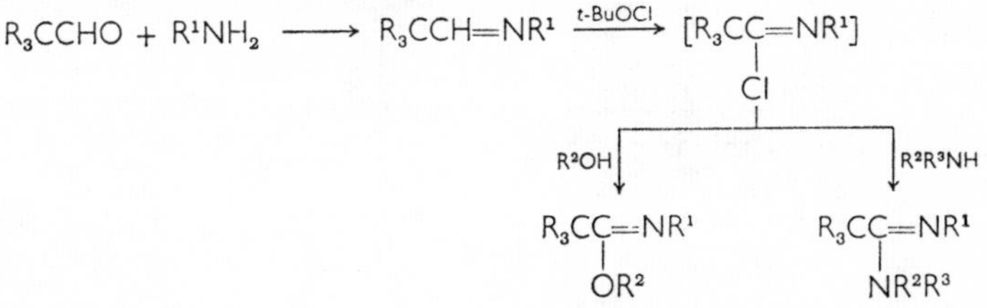

$$R_3CCHO + R^1NH_2 \longrightarrow R_3CCH{=}NR^1 \xrightarrow{t\text{-BuOCl}} [R_3C(Cl){=}NR^1]$$

$$[R_3C(Cl){=}NR^1] \xrightarrow{R^2OH} R_3C(OR^2){=}NR^1 \quad \textbf{(120)}$$

$$[R_3C(Cl){=}NR^1] \xrightarrow{R^2R^3NH} R_3C(NR^2R^3){=}NR^1 \quad \textbf{(119)}$$

The hydrogen can also be replaced using carbonyl-stabilized sulphonium ylids (**121**) which lead, after rearrangement, to 3-arylaminocinnamates and 3-arylaminocinnamamides (**122**)[165].

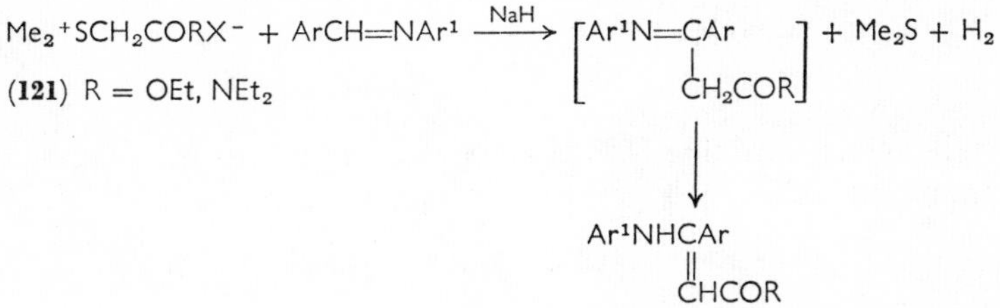

$$Me_2{}^+SCH_2CORX^- + ArCH{=}NAr^1 \xrightarrow{NaH} [Ar^1N{=}C(Ar)CH_2COR] + Me_2S + H_2$$

(**121**) R = OEt, NEt_2

$$\longrightarrow Ar^1NHC(Ar){=}CHCOR$$

(**122**) R = OEt, NEt_2

2. Aldoximes, RCH=NOH

The direct chlorination of aldoximes in either chloroform or dilute hydrochloric acid leads to hydroximic acid chlorides (chloroximes)[49,166–170]. The hydrogen atom is also substituted with nitrosyl chloride[170–173].

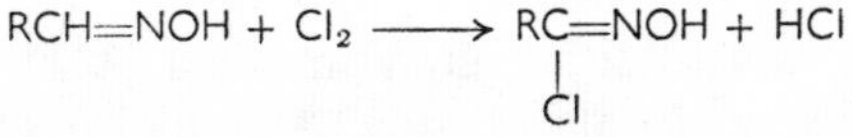

$$RCH{=}NOH + Cl_2 \longrightarrow RC(Cl){=}NOH + HCl$$

E. *Displacements of Amino Groups*

The action of nitrous acid on benzamidoxime (**123**) aided in its structural elucidation[174] by indicating the presence of the NOH grouping, which like hydroxylamine evolves nitrous oxide on treatment with one equivalent of nitrous acid. The *O*-alkylated benzamidoxime (**124**), when treated with nitrous acid[175], evolves nitrogen (proving the presence of the NH_2 group), but the intermediate hydroximic acid

$$H_2NOH + HONO \longrightarrow 2H_2O + N_2O$$

$$\underset{\substack{|\\ NH_2\cdot HCl \\ \mathbf{(123)}}}{PhC{=}NOH} + NaNO_2 \longrightarrow \underset{\substack{|\\ NH_2}}{PhC{=}O} + H_2O + N_2O + NaCl$$

(**125**) is not isolated. Yet with excess hydrochloric acid, the hydroximic acid chloride (**126**) is formed almost quantitatively[46,49,51,176–182].

$$\underset{\substack{|\\ NH_2 \\ \mathbf{(124)}}}{PhC{=}NOR} + HONO \longrightarrow N_2 + \underset{\substack{|\\ OH \\ \mathbf{(125)}}}{[PhC{=}NOR]} \xrightarrow{HCl} \underset{\substack{|\\ Cl \\ \mathbf{(126)}}}{PhC{=}NOR}$$

The *O*-alkyl hydroximic acid chlorides are stable, volatile with steam, and soluble in most organic solvents.

The amino group in amidoximes can be replaced by aromatic amines with the liberation of ammonia[175,183].

$$\underset{\substack{|\\ NH_2}}{RC{=}NOH} + ArNH_2 \longrightarrow \underset{\substack{|\\ NHAr}}{RC{=}NOH} + NH_3$$

N-Phenylbenzamidine (**127**) can be converted into *N,N'*-diphenylbenzamidine (**128**) by heating with excess aniline at 250°[184]. The

$$\underset{\substack{|\\ NH_2 \\ \mathbf{(127)}}}{PhC{=}NPh} + PhNH_2 \longrightarrow \underset{\substack{|\\ NHPh \\ \mathbf{(128)}}}{PhC{=}NPh} + NH_3$$

amino group in **127** can also be displaced with phenylhydrazine hydrochloride giving the tautomeric system **129a–129b**[5,33].

$$\underset{\substack{|\\ NH_2}}{PhC{=}NPh} + PhNHNH_2\cdot HCl \longrightarrow NH_4Cl + \underset{\substack{|\\ NHNHPh \\ \mathbf{(129a)}}}{PhC{=}NPh} \rightleftharpoons \underset{\substack{\|\\ NNHPh \\ \mathbf{(129b)}}}{PhCNHPh}$$

Amidines (**130**) and *N*-substituted amidines (**131**), in the presence of acids, undergo nucleophilic attack by hydroxylamine to form amidoximes[5,55,185–187].

$$\underset{\substack{|\\ NH_2 \\ \mathbf{(130)}}}{RC{=}NH} + H_2NOH + H^+ \longrightarrow \underset{\substack{|\\ NH_2}}{RC{=}NOH} + NH_4^+$$

$$\underset{\substack{|\\ NH_2 \\ \mathbf{(131)}}}{RC{=}NR^1} + H_2NOH + H^+ \longrightarrow \underset{\substack{|\\ NHR^1}}{RC{=}NOH} + NH_4^+$$

Excess ammonia will displace aniline from *N*,*N'*-diphenylbenzamidine (**132**) with the formation of benzamidine[188].

$$\underset{\text{NHPh}}{\overset{|}{\text{PhC}}}\text{=NPh} + 2\text{NH}_3 \longrightarrow \underset{\text{NH}_2}{\overset{|}{\text{PhC}}}\text{=NH} + 2\text{PhNH}_2$$

(**132**)

F. Displacements of Hydroxy Groups

The preparations of imidyl chlorides (Section II.A.1) were considered as the reaction of the iminol form of the amide with phosphorus pentachloride and thus could be considered in this section. However, the majority of evidence indicates that the position of equilibrium in the amide–iminol tautomerism favours the amide[189].

However, one can consider the reaction of ethyl benzohydroxamate (**133**) with phosphorus pentachloride as the replacement of a hydroxyl group by a chlorine[190].

$$\underset{\text{OH}}{\overset{|}{\text{PhC}}}\text{=NOEt} + \text{PCl}_5 \longrightarrow \underset{\text{Cl}}{\overset{|}{\text{PhC}}}\text{=NOEt} + \text{POCl}_3 + \text{HCl}$$

(**133**)

III. SUBSTITUTIONS AT THE NITROGEN ATOM

A. Displacements of Hydrogen

1. Aldimines and ketimines, RCH=NH and R_2C=NH

Aldimines and ketimines will react with Grignard reagents with the substitution of MgX for the hydrogen, and the grouping can be removed with ammonia under mild conditions; vigorous conditions will cause hydrolysis to aldehydes and ketones.

$$\text{R}_2\text{C=NH} + \text{R}^1\text{MgX} \longrightarrow \text{R}_2\text{C=NMgX} + \text{R}^1\text{H}$$

The reaction of ketimines with hypochlorous acid is reversible; hydrolysis of the *N*-chloroketimine liberates hypochlorous acid and not hydrogen chloride[191].

$$\text{R}_2\text{C=NH} + \text{HOCl} \rightleftharpoons \text{R}_2\text{C=NCl} + \text{H}_2\text{O}$$

N-Chloroketimines are considered to be intermediates in the reaction of ketimines leading to α-amino ketones and α-amino acids. The following reactions illustrate this[192], and the ensuing series is also to the point.

$$RCN + R^1CH_2MgBr \longrightarrow \underset{\displaystyle CH_2R^1}{RC{=}NMgBr} \xrightarrow{NH_3}$$

$$\underset{\displaystyle CH_2R^1}{RC{=}NH} \xrightarrow{t\text{-BuOCl}} \left[\underset{\displaystyle CH_2R^1}{RC{=}NCl}\right] \xrightarrow[2.\ HCl]{1.\ NaOMe} \underset{\displaystyle NH_2 \cdot HCl}{RCOCHR^1}$$

$$RCH_2CN + MeOH + HCl \longrightarrow \underset{\displaystyle OMe}{RCH_2C{=}NH \cdot HCl} \xrightarrow{HOCl}$$

$$\left[\underset{\displaystyle OMe}{RCH_2C{=}NCl}\right] \xrightarrow[2.\ H^+]{1.\ t\text{-BuOK}} \underset{\displaystyle NH_3^+}{RCHCO_2Me}$$

The methyl *N*-chloroimidate was not isolated but was rearranged immediately by adding it to a slight excess of potassium *t*-butoxide[192].

However, benzylphenyl-*N*-chloroketimine (**134**) was formed as a final product of the following set of reactions[193]; on treatment with

$$PhCN + PhCH_2MgX \longrightarrow \underset{\displaystyle CH_2Ph}{PhC{=}NMgX} \longrightarrow \underset{\displaystyle CH_2Ph}{PhC{=}NH} \xrightarrow{t\text{-BuOCl}} \underset{\displaystyle CH_2Ph}{PhC{=}NCl}$$

(134)

anhydrous hydrogen chloride in pentane, **134** formed the hydrochloride of benzylphenyl ketimine quantitatively[193]. The *N*-chloro- and *N*-bromobenzylphenyl ketimines have also been prepared from the ketimine with sodium hypobromite or sodium hypochlorite, respectively[194].

The hydrogen in aromatic ketimines can be replaced by the action of ketene or anhydrides[195].

$$Ar_2C{=}NH + CH_2{=}C{=}O \xrightarrow[Et_2O]{0°} Ar_2C{=}NCOCH_3$$

$$Ar_2C{=}NH + (RCO)_2O \xrightarrow{C_6H_6} Ar_2C{=}NCOR$$

2. Alkyl imidates, HN=C(R)OR^1

As discussed in Section II.B.1, imidates react with Grignard reagents with the formation of ketimines; however, the Grignard reaction fails with mandelamidine (**135**) and with its hydrochloride[128], with *N,N'*-diphenylformamidine (**136**)[136], and with N_1-phenylmandeloamidrazone (**137**)[128]. These observations suggest that the imino group in

$$\underset{\displaystyle NH_2}{PhCHOHC{=}NH} \qquad \underset{\displaystyle NH_2}{HC{=}NPh} \qquad \underset{\displaystyle NH_2}{PhCHOHC{=}NNHPh}$$

(135) **(136)** **(137)**

these compounds is unreactive to RMgX and therefore, by extension, the imino group in imidates is also unreactive.

Sulphonyl chlorides react with imidates leading to *N*-sulphonyl imidates[120]. Acid chlorides will react with imidates to give *N*-

$$2RC(OEt){=}NH + R^1SO_2Cl \longrightarrow RC(OEt){=}NSO_2R^1 + RC(OEt){=}NH \cdot HCl$$

acylimidates[196,197], and these undergo facile displacement of the ethoxy group to form the diacyl derivatives of ammonia.

$$RC(OEt){=}NCOR^1 + H_2O \longrightarrow \left[RC(OH){=}NCOR^1\right] \longrightarrow RCONHCOR^1$$

The reactions of hypobromous acid and hypochlorous acid on imidates give the corresponding *N*-halo derivatives[198–201].

$$RC(OEt){=}NH + HOX \longrightarrow RC(OEt){=}NX$$

$$X = Br, Cl$$

3. Amidines, $HN{=}C(R)NH_2$

Formamidine (**138**) reacts with acetic anhydride to yield a diacetyl derivative[202], while *N*-phenylbenzamidine (**139**) reacts with acetic anhydride to form **140**[202].

$$\underset{(138)}{HC(NH_2){=}NH} + Ac_2O \longrightarrow HC(NHCOCH_3){=}NCOCH_3$$

$$\underset{(139)}{PhC(NH_2){=}NPh} \rightleftharpoons PhC({=}NH)NHPh \xrightarrow{Ac_2O} \underset{(140)}{PhC({=}NCOCH_3)N(Ph)COCH_3}$$

Both aromatic and aliphatic amidines react with phenyl isocyanate to produce phenyl ureides (**141**)[55,203,83]. Benzamidine hydrochloride

$$RC(NH_2){=}NH + PhNCO \longrightarrow \underset{(141)}{RC(NHCONHPh){=}NCONHPh}$$

reacts with I_2/KI to form an iodoamidine[204,205] presumably with structure **142a** or **142b**, and sodium hypochlorite produces chloroamidines[205].

$$\underset{(142a)}{PhC(NH_2){=}NI} \qquad \underset{(142b)}{PhC(NHI){=}NH}$$

Amidines undergo the Hinsberg reaction with sulphonyl chlorides[120,121,206,207]. These sulphonyl derivatives are quite stable;

$$\underset{\displaystyle NH_2}{RC}{=}NH\cdot HCl + ArSO_2Cl + 2\,NaOH \longrightarrow \underset{\displaystyle NH_2}{RC}{=}NSO_2Ar + 2\,NaCl$$

the amide group of **143** is hydrolysed preferentially by alcoholic hydrogen chloride at 20°[120,207].

$$\underset{\displaystyle NH_2}{RC}{=}NSO_2{-}C_6H_4{-}NHCOCH_3 \xrightarrow{HCl,\ EtOH} \underset{\displaystyle NH_2}{RC}{=}NSO_2{-}C_6H_4{-}NH_2\cdot HCl$$

(**143**)

N,N-Dimethylbenzamidine (**144**) reacts with methyl iodide to give *N,N,N'*-trimethylbenzamidine hydroiodide (**145**)[208–211]. Alkylation

$$\underset{\displaystyle NMe_2}{PhC}{=}NH + MeI \longrightarrow \underset{\displaystyle NMe_2}{PhC}{=}NMe\cdot HI$$

(**144**) (**145**)

of a symmetrical disubstituted amidine leads to two products because of tautomerism of the system[209,212]. If R^1 and R^2 are greatly different

$$\underset{\displaystyle NHR^2}{RC}{=}NR^1 \rightleftharpoons \underset{\displaystyle \| \atop NR^2}{RC}NHR^1$$

$$\downarrow MeI \qquad\qquad \downarrow MeI$$

$$\underset{\displaystyle N(Me)R^2}{RC}{=}NR^1\cdot HI \qquad \underset{\displaystyle \| \atop NR^2\cdot HI}{RC}N(Me)R^1$$

in character, one product predominates; the alkyl group is attached to the less basic nitrogen.

N-Ethylbenzamidine (**146**) can be prepared by heating benzamidine with ethyl iodide at 100°[213], and continuation of the alkylation leads to disubstituted products[208,214].

$$\underset{\displaystyle NH_2}{PhC}{=}NH + EtI \longrightarrow \underset{\displaystyle NH_2}{PhC}{=}NEt$$

(**146**)

Unsubstituted amidines react in liquid ammonia with potassium or potassium amide with the replacement of only one hydrogen[188,215]. Copper or silver salts can be prepared from the potassium or sodium derivatives[215].

$$\underset{\substack{|\\NH_2}}{RC{=}NH} + K \xrightarrow{\text{liq. }NH_3} \underset{\substack{|\\NHK}}{RC{=}NH}$$

$$\underset{\substack{|\\NH_2}}{RC{=}NH} + KNH_2 \xrightarrow{\text{liq. }NH_3} \underset{\substack{|\\NHK}}{RC{=}NH} + NH_3$$

4. Imino carbonates, $(ArO)_2C{=}NH$

Cyanogen halides react with phenols to form imino carbonates, which then react with *t*-butylhypochlorite to yield *N*-chloro derivatives[216].

$$ArOH,\ ArONa + XCN \xrightarrow{0^\circ} (ArO)_2C{=}NH \xrightarrow{t\text{-}BuOCl} (ArO)_2C{=}NCl$$

B. *Displacements of Other Groups*

Chloro imines (**147**) can be prepared by the reaction of aldehydes with chloramine[217], and the chlorine atom can be replaced by treat-

$$RCHO + NH_2Cl \longrightarrow \underset{(\mathbf{147})}{RCH{=}NCl} + H_2O$$

ment with hydrogen chloride in ether[217]. These chloro imines will react with Grignard reagents to form Schiff bases[218].

$$RCH{=}NCl \xrightarrow{HCl(g),\ Et_2O} RCH{=}NH\cdot HCl + Cl_2$$

$$R_2C{=}NCl + R^1MgX \longrightarrow R_2C{=}NR^1$$

Amidoximes are reduced to amidines with $PhPCl_2$, as are oximes themselves[219]. Amidoximes can also be reduced to amidines with hydrogen and Raney nickel catalyst at 30 atm and 60–80°[220,221].

$$\underset{\substack{|\\NHPh}}{PhC{=}NOH} \xrightarrow{PhPCl_2} \underset{\substack{|\\NHPh}}{PhC{=}NH\cdot HCl}$$

$$Ph_2C{=}NOH \xrightarrow{PhPCl_2} Ph_2C{=}NH\cdot HCl$$

$$\underset{\substack{|\\NH_2}}{RC{=}NOH} + H_2 \longrightarrow \underset{\substack{|\\NH_2}}{RC{=}NH} + H_2O$$

Busch and Hobein[222] prepared amidines by the action of a Grignard reagent on phenylcyanamide (**148**). Both aromatic and alkali metal amides produce salts of amidines[215,223–225]. The reaction is best

$$\underset{(\mathbf{148})}{PhNHCN} + 2PhMgBr \longrightarrow \underset{\substack{|\\Ph}}{PhN(MgBr)C{=}NMgBr} \xrightarrow{H_2O} \underset{\substack{|\\Ph}}{PhNHC{=}NH}$$

carried out in anhydrous media (benzene, toluene, etc.)[215,223] with potassium amide being a particularly good reagent[215]. Conversion of

$$RCN + KNH_2 \longrightarrow RC(NH_2){=}NK$$

the alkali metal salts to the amidines is carried out by careful hydrolysis at low temperatures in order to avoid further hydrolysis to the amide[225].

$$RC(NH_2){=}NK + H_2O \longrightarrow RC(NH_2){=}NH + KOH$$

$$RC(NH_2){=}NK + HCl \xrightarrow{EtOH} RC(NH_2){=}NH \cdot HCl + KCl$$

The acyl group can be removed from *N*-acylketimines by treatment with gaseous HCl in benzene; hydrolysis with aqueous hydrochloric acid, however, leads to the carbonyl compound[195].

$$Ph_2C{=}NCOCH_3 \xrightarrow{HCl\ (g),\ PhH} Ph_2C{=}NH \cdot HCl$$

IV. REFERENCES

1. W. Eschweiler, *Ber.*, **30**, 998 (1897).
2. A. Hantzsch and E. Voegelen, *Ber.*, **34**, 3142 (1901).
3. J. E. MacKenzie, *J. Chem. Soc.*, **113**, 1 (1918).
4. H. G. Rule, *J. Chem. Soc.*, **113**, 3 (1918).
5. R. Walther and R. Grossmann, *J. Prakt. Chem.* [2], **78**, 478 (1908).
6. A. V. Kirsanov and R. G. Makitra, *Zh. Obshch. Khim.*, **26**, 907 (1956); *Chem. Abstr.*, **50**, 14633 (1956).
7. A. V. Kirsanov, *Khim. i Primenenie Fosfororgan. Akad. Nauk SSSR, Trudy 1-oi Konferents*, 99 (1955); *Chem. Abstr.*, **52**, 251 (1958).
8. G. J. Janz and S. S. Danyluk, *J. Am. Chem. Soc.*, **81**, 3850 (1959).
9. I. Ugi, F. Beck and U. Fetzer, *Chem. Ber.*, **95**, 126 (1962).
10. S. M. McElvain and C. L. Stevens, *J. Am. Chem. Soc.*, **69**, 2667 (1947).
11. A. E. Arbuzov and V. E. Shishkin, *Dokl. Akad. Nauk SSSR*, **141**, 349 (1961); *Chem. Abstr.*, **56**, 11491 (1962).
12. A. W. Chapman, *J. Chem. Soc.*, **121**, 1676 (1922).
13. A. W. Chapman, *J. Chem. Soc.*, **123**, 1150 (1923).
14. A. W. Chapman, *J. Chem. Soc.*, **127**, 1992 (1925).
15. A. W. Chapman, *J. Chem. Soc.*, 2296 (1926).
16. A. W. Chapman, *J. Chem. Soc.*, 1743 (1927).
17. J. Cymerman-Craig and J. W. Loder, *J. Chem. Soc.*, 4309 (1955).
18. A. Hantzsch, *Ber.*, **26**, 926 (1893).
19. E. R. H. Jones and F. G. Mann, *J. Chem. Soc.*, 786 (1956).
20. G. D. Lander, *J. Chem. Soc.*, **81**, 591 (1902).
21. G. D. Lander, *J. Chem. Soc.*, **83**, 320 (1903).

22. K. B. Wiberg and B. I. Rowland, *J. Am. Chem. Soc.*, **77**, 2205 (1955).
23. R. C. Cookson, *J. Chem. Soc.*, **643** (1953).
24. D. M. Hall, *J. Chem. Soc.*, 1603 (1948).
25. M. Busch and F. Falco, *Ber.*, **43**, 2557 (1910).
26. M. Busch and M. Fleischmann, *Ber.*, **43**, 2553 (1910).
27. A. Sonn and E. Müller, *Ber.*, **52B**, 1927 (1919).
28. H. Stephen, *J. Chem. Soc.*, 1874 (1925).
29. C. L. Stevens and J. C. French, *J. Am. Chem. Soc.*, **75**, 657 (1953).
30. W. Lossen, F. Mieurau, M. Kobbert, and G. Grabowski, *Ann. Chem.*, **265**, 129 (1891).
31. C. Gerhardt, *Ann. Chem.*, **108**, 219 (1858).
32. O. Wallach, *Ann. Chem.*, **184**, 1 (1877).
33. H. von Pechmann, *Ber.*, **28**, 2362 (1895).
34. H. von Pechmann, *Ber.*, **30**, 1779 (1897).
35. E. Bamberger and J. Lorenzen, *Ann. Chem.*, **273**, 269 (1893).
36. A. J. Hill and M. V. Cox, *J. Am. Chem. Soc.*, **48**, 3214 (1926).
37. H. von Pechmann, *Ber.*, **28**, 869 (1895).
38. E. E. Bures and M. Kundera, *Casopis Ceskoslov. Lekarnictva*, **14**, 272 (1934); *Chem. Abstr.*, **29**, 4750 (1935).
39. H. von Pechmann, *Ber.*, **27**, 1699 (1894).
40. W. Markwald, *Ann. Chem.*, **286**, 343 (1895).
41. J. B. Cohen and J. Marshall, *J. Chem. Soc.*, 328 (1910).
42. O. Wallach, *Ann. Chem.*, **214**, 202 (1882).
43. J. von Braun and K. Weissbach, *Ber.*, **56B**, 1574 (1932).
44. H. Ley and F. Muller, *Ber.*, **40**, 2950 (1907).
45. P. Davis and W. E. McEwen, *J. Org. Chem.*, **29**, 815 (1961).
46. A. Werner and C. Bloch, *Ber.*, **32**, 1975 (1899).
47. H. Ley and M. Ulrich, *Ber.*, **47**, 2941 (1914).
48. F. Eloy, *J. Org. Chem.*, **26**, 952 (1961).
49. A. Werner and H. Buss, *Ber.*, **27**, 2193 (1894).
50. A. Werner and A. Gemesens, *Ber.*, **29**, 1161 (1896).
51. G. Ponzio, *Gazz. Chim. Ital.*, **55**, 311 (1925).
52. F. Tiemann and P. Krüger, *Ber.*, **18**, 727 (1885).
53. J. R. Meadow and E. E. Reid, *J. Am. Chem. Soc.*, **65**, 457 (1943).
54. H. Böhme and H. Oper, *Z. Anal. Chem.*, **139**, 255 (1953).
55. A. Pinner, *Die Imidoäther und ihr Derivate*, Berlin, 1892.
56. A. Pinner and F. Klein, *Ber.*, **10**, 1889 (1877).
57. I. H. Derby, *Am. Chem. J.*, **39**, 437 (1908).
58. H. J. Barber and R. Slack, *J. Am. Chem. Soc.*, **66**, 1607 (1944).
59. R. E. Allen, E. L. Schumann, W. C. Day and M. G. Van Campen, *J. Am. Chem. Soc.*, **80**, 591 (1958).
60. J. N. Ashley, H. J. Barber, A. J. Ewins, G. Newbery and A. D. H. Self, *J. Chem. Soc.*, 103 (1942).
61. A. K. Bose, F. Greer, J. S. Gotts and C. C. Price, *J. Org. Chem.*, **24**, 1309 (1959).
62. N. W. Bristow, *J. Chem. Soc.*, 513 (1957).
63. M. Mangelberg, *Chem. Ber.*, **89**, 1185 (1956).
64. H. J. Barber, P. Z. Gregory, F. W. Major, R. Slack and A. M. Woolman, *J. Chem. Soc.*, 84 (1947).

65. H. J. Barber and R. Slack, *J. Chem. Soc.*, 82 (1947).
66. P. Z. Gregory, S. J. Holt and R. Slack, *J. Chem. Soc.*, 87 (1947).
67. Soc. pour l'ind. Chim. à Bâle, *Brit. Pat.* 529,055 (1940); *Chem. Abstr.*, **35**, 7976 (1941).
68. W. Steinkopf, *J. Prakt. Chem.*, **81**, 97, 193 (1910).
69. J. Vargha and I. Monduk, *Studia Univ. Babes-Bolyai Ser.*, **1**, 121 (1961); *Chem. Abstr.*, **58**, 4421 (1963).
70. J. Tafel and C. Enoch, *Ber.*, **23**, 103 (1890).
71. J. Tafel and C. Enoch, *Ber.*, **23**, 1550 (1890).
72. G. D. Lander, *J. Chem. Soc.*, **77**, 729 (1900).
73. G. D. Lander, *J. Chem. Soc.*, **79**, 690 (1901).
74. N. V. Sidgwick, *The Organic Chemistry of Nitrogen*, The Clarendon Press, Oxford, 1937, pp. 155–156.
75. R. L. Shriner and F. W. Neumann, *Chem. Rev.*, **35**, 351 (1944).
76. A. J. Hill and I. Rabinowitz, *J. Am. Chem. Soc.*, **48**, 732 (1926).
77. J. O. Jilek, M. Borovicka and M. Protiva, *Chem. Listy*, **43**, 211 (1949); *Chem. Abstr.*, **45**, 571 (1951).
78. H. M. Woodburn, A. B. Whitehouse and B. G. Pautler, *J. Org. Chem.*, **24**, 210 (1959).
79. M. Yamazaki, Y. Kitagawa, S. Hiraki and Y. Tsukamoto, *J. Pharm. Soc. Japan*, **73**, 294 (1953); *Chem. Abstr.*, **48**, 2003 (1954).
80. H. Bredereck, R. Gompper and H. Seiz, *Chem. Ber.*, **90**, 1837 (1957).
81. R. R. Burtner, *U.S. Pat.* 2,676,968 (1954); *Chem. Abstr.*, **49**, 6307 (1955).
82. A. Pinner, *Ber.*, **16**, 1643 (1883).
83. A. Pinner, *Ber.*, **23**, 2923 (1890).
84. G. Luckenbach, *Ber.*, **17**, 1423 (1884).
85. R. M. Roberts, *J. Am. Chem. Soc.*, **72**, 3603 (1950).
86. R. M. Roberts and R. H. DeWolfe, *J. Am. Chem. Soc.*, **76**, 2411 (1954).
87. H. Bader, J. D. Downer and P. Driver, *J. Chem. Soc.*, 2775 (1950).
88. J. W. Cornforth and R. H. Cornforth, *J. Chem. Soc.*, 96 (1947).
89. J. W. Cornforth and H. T. Huang, *J. Chem. Soc.*, 1969 (1948).
90. Chinoin R. T., *Hung. Pat.* 138,595 (1949); *Chem. Abstr.*, **48**, 10059 (1948).
91. A. Kjaer, *Acta Chem. Scand.*, **7**, 1024 (1953).
92. A. Salamon, *Müegyetemi Közlemények*, 72 (1949); *Chem. Abstr.*, **45**, 552 (1951).
93. E. Schmidt, *Ber.*, **47**, 2545 (1914).
94. G. Shaw and D. N. Butler, *J. Chem. Soc.*, 4040 (1959).
95. G. Shaw, R. N. Warrener, D. N. Butler and R. K. Ralph, *J. Chem. Soc.*, 1648 (1959).
96. A. H. Cook, A. C. Davis, I. Heilbron and G. H. Thomas, *J. Chem. Soc.*, 1071 (1949).
97. W. Reid, W. Stephan and W. von der Emden, *Chem. Ber.*, **95**, 728 (1962).
98. M. J. Hunter and M. L. Ludwig, *J. Am. Chem. Soc.*, **84**, 3491 (1962).
99. D. F. Elliot, *J. Chem. Soc.*, 589 (1949).
100. R. M. Roberts and P. J. Vogt, *J. Am. Chem. Soc.*, **78**, 4778 (1956).
101. W. Reid and E. Schmidt, *Ann. Chem.*, **676**, 114 (1964).
102. W. Reid and E. Schmidt, *Ann. Chem.*, **676**, 121 (1964).
103. B. G. Baccar and F. Mathis, *Compt. Rend.*, **261**, (1) (Group 8), 174 (1965); *Chem. Abstr.*, **63**, 11541 (1965).

104. A. Pinner, *Ber.*, **26**, 2126 (1893).
105. A. Pinner, *Ber.*, **27**, 984 (1894).
106. A. Pinner, *Ber.*, **30**, 1871 (1897).
107. A. Pinner, *Ann. Chem.*, **297**, 221 (1897).
108. A. Pinner, *Ann. Chem.*, **298**, 1 (1897).
109. W. Oberhummer, *Monatsh. Chem.*, **57**, 106 (1931).
110. W. Oberhummer, *Monatsh. Chem.*, **63**, 285 (1933).
111. A. Bernton, *Arkiv. Kemi Mineral. Geol.*, **7**, No. 13, 1 (1920).
112. H. G. Rule, *J. Chem. Soc.*, 3 (1918).
113. A. Heymons, *Ger. Pat.* 839,493 (1952); *Chem. Abstr.*, **47**, 1737 (1953).
114. E. B. Knott, *J. Chem. Soc.*, 686 (1945).
115. T. Fujisawa and C. Mizuno, *J. Pharm. Soc. Japan*, **72**, 698 (1952); *Chem. Abstr.*, **47**, 2726 (1953).
116. M. Miyazaki, *Japan. Pat.* 4477 (1952); *Chem. Abstr.*, **48**, 8259 (1954).
117. S. J. Angyal and W. K. Warburton, *Australian J. Sci. Research*, **4A**, 93 (1951).
118. P. Diedrich and M. Dohrn, *Ger. Pat.* 833,810 (1952); *Chem. Abstr.*, **47**, 1738 (1953).
119. P. Diedrich and M. Dohrn, *Ger. Pat.* 833,811 (1952); *Chem. Abstr.*, **50**, 1908 (1956).
120. H. J. Barber, *J. Chem. Soc.*, 101 (1943).
121. E. H. Northey, A. E. Pierce and D. J. Kertesz, *J. Am. Chem. Soc.*, **64**, 2763 (1942).
122. C. H. Andrews, H. King and J. Walker, *Proc. Royal Soc. (London), Ser. B.*, **133**, 20 (1946).
123. M. R. Atkinson and J. B. Polya, *J. Chem. Soc.*, 3319 (1954).
124. D. Jerchel and H. Fischer, *Ann. Chem.*, **574**, 85 (1951).
125. N. Kunimine and K. Itano, *J. Pharm. Soc. Japan*, **74**, 726 (1954); *Chem. Abstr.*, **49**, 11627 (1955).
126. A. Pinner, *Ber.*, **17**, 182 (1884).
127. H. Voswinckel, *Ber.*, **36**, 2483 (1903).
128. R. Roger and D. Neilson, *Chem. Rev.*, **61**, 179 (1961).
129. C. Ainsworth, *J. Am. Chem. Soc.*, **75**, 5728 (1953).
130. Chinoin R. T., *Hung. Pat.* 135,418 (1949); *Chem. Abstr.*, **48**, 10050 (1954).
131. A. G. Knoll, *Ger. Pat.* 521,870 (1929); *Chem. Abstr.*, **25**, 3364 (1931).
132. A. Pinner, *Ber.*, **17**, 184 (1884).
133. W. Lossen, *Ber.*, **17**, 1587 (1884).
134. L. Gatterman, *Ann. Chem.*, **393**, 215 (1912).
135. G. W. Monier-Williams, *J. Chem. Soc.*, **89**, 273 (1906).
136. L. I. Smith and J. Nichols, *J. Org. Chem.*, **6**, 489 (1941).
137. R. Dijkstra and H. J. Backer, *Proc. Kominkl. Ned. Akad. Wetenschap.*, **55B**, 382 (1952); *Chem. Abstr.*, **48**, 9900 (1954).
138. F. C. Schaefer and G. A. Peters, *J. Org. Chem.*, **29**, 815 (1961).
139. D. W. Wooley, J. W. B. Hershey and H. A. Jodlowski, *J. Org. Chem.*, **28**, 2012 (1963).
140. W. Reid and E. Schmidt, *Ann. Chem.*, **695**, 217 (1966).
141. W. Reid and D. Piechaczek, *Ann. Chem.*, **696**, 97 (1966).
142. H. C. Brown and C. R. Wetzel, *J. Org. Chem.*, **30**, 3724 (1965).
143. H. C. Brown and C. R. Wetzel, *J. Org. Chem.*, **30**, 3729 (1965).

144. H. E. Zaugg and R. J. Michaels, *J. Org. Chem.*, **28**, 1801 (1963).
145. E. Roberts, *Brit. Pat.* 817,749 (1959); *Chem. Abstr.*, **54**, 9775 (1960).
146. W. Lossen, *Ann. Chem.*, **252**, 214 (1889).
147. J. Houben and E. Pfankuch, *Ber.*, **59**, 2392 (1926).
148. J. Houben and E. Pfankuch, *Ber.*, **59**, 2397 (1926).
149. J. Houben and R. Zivadinovitsch, *Ber.*, **69**, 2352 (1936).
150. W. Autenrieth and A. Brüning, *Ber.*, **36**, 3464 (1903).
151. A. Bernthsen, *Ann. Chem.*, **197**, 341 (1879).
152. F. E. Condo, E. T. Kinkel, A. Fassero and R. L. Shriner, *J. Am. Chem. Soc.*, **59**, 230 (1937).
153. A. Pinner and F. Klein, *Ber.*, **11**, 1825 (1878).
154. P. Reynaud, R. C. Moreau and N. H. Thu, *Compt. Rend.*, **253**, 2540 (1961); *Chem. Abstr.*, **57**, 3356 (1962).
155. O. Wallach and H. Bleibtren, *Ber.*, **12**, 1061 (1879).
156. O. Wallach and M. Wusten, *Ber.*, **16**, 144 (1883).
157. G. W. Kirsten and G. B. L. Smith, *J. Am. Chem. Soc.*, **58**, 800 (1936).
158. W. G. Finnegan, R. A. Henry and G. B. L. Smith, *J. Am. Chem. Soc.*, **74**, 2981 (1952).
159. M. N. Shchukina and E. E. Mikhlina, *J. Gen. Chem.* (*Moscow*), **22**, 157 (1952).
160. W. R. McBride, W. G. Finnegan and R. A. Henry, *J. Org. Chem.*, **22**, 152 (1957).
161. E. S. Scott and L. F. Audrieth, *J. Org. Chem.*, **19**, 1231 (1954).
162. A. V. Kirsanov and Y. N. Ivaschchenko, *Bull. Soc. Chim. France* [5], **2**, 2109 (1935).
163. G. Wittig and H. Frommeld, *Chem. Ber.*, **97**, 3548 (1964).
164. H. Paul, A. Weise and R. Dettmer, *Chem. Ber.*, **98**, 1450 (1965).
165. A. J. Speziale, C. C. Tung, K. W. Ratts and A. Yao, *J. Am. Chem. Soc.*, **87**, 3460 (1965).
166. O. Piloty and H. Steinbock, *Ber.*, **35**, 3101 (1902).
167. A. Werner, *Ber.*, **27**, 2846 (1894).
168. H. Wieland, *Ann. Chem.*, **353**, 65 (1907).
169. H. Wieland, *Ber.*, **40**, 1676 (1907).
170. C. R. Kinney, E. W. Smith, B. L. Wooley and A. R. Willey, *J. Am. Chem. Soc.*, **55**, 3418 (1933).
171. H. Rheinboldt, *Ann. Chem.*, **451**, 161 (1927).
172. H. Rheinboldt and M. Dewald, *Ann. Chem.*, **451**, 273 (1927).
173. H. Rheinboldt, M. Dewald, F. Jansen and O. Schmitz-Dumont, *Ann. Chem.*, **451**, 161 (1927).
174. F. Tiemann and P. Krüger, *Ber.*, **17**, 1685 (1884).
175. E. Nordmann, *Ber.*, **17**, 2746 (1884).
176. O. L. Brady and F. Peakin, *J. Chem. Soc.*, 2267 (1929).
177. P. Knudsen, *Ber.*, **18**, 1068 (1885).
178. P. Krüger, *Ber.*, **18**, 1053 (1885).
179. L. H. Schubert, *Ber.*, **22**, 2433 (1889).
180. A. Spilker, *Ber.*, **22**, 2767 (1889).
181. A. Werner, *Ber.*, **25**, 27 (1892).
182. H. Wolff, *Ber.*, **22**, 2395 (1889).
183. J. Neff, *Ann. Chem.*, **280**, 294 (1891).

184. A. Bernthsen, *Ann. Chem.*, **184**, 321 (1876).
185. A. Lottermost, *J. Prakt. Chem.* [2], **54**, 116 (1897).
186. H. Muller, *Ber.*, **19**, 1669 (1886).
187. T. Kanazawa, E. Owada, M. Yoshida and T. Sato, *Nippon Kagaku Zasshi*, **76**, 654 (1955); *Chem. Abstr.*, **51**, 17814 (1957).
188. E. C. Franklin, *The Nitrogen System of Compounds*, ACS Monograph No. 68, Reinhold New York. 1935, p. 271.
189. L. Skulski, G. C. Palmer, and M. Calvin, *Tetrahedron Letters*, 1773 (1963).
190. W. Lossen, *Ber.*, **18**, 1198 (1885).
191. J. Steiglitz and P. P. Peterson, *Ber.*, **43**, 782 (1910).
192. H. E. Baumgarten, *J. Am. Chem. Soc.*, **59**, 2058 (1937).
193. H. E. Baumgarten, J. M. Petersen and D. C. Wolff, *J. Org. Chem.*, **28**, 2369 (1963).
194. K. N. Campbell, *J. Am. Chem. Soc.*, **59**, 2058 (1937).
195. J. E. Banfield, G. M. Brown, F. H. Davey, W. Davies and T. H. Ramsay, *Australian J. Sci. Res.*, **1A**, 330 (1948).
196. H. L. Wheeler and P. T. Walden, *Am. Chem. J.*, **19**, 129 (1897).
197. H. L. Wheeler, P. T. Walden and H. F. Metcalf, *Am. Chem. J.*, **20**, 64 (1898).
198. W. S. Hilbert, *Am. Chem. J.*, **40**, 150 (1908).
199. J. Steiglitz, *Am. Chem. J.*, **18**, 751 (1896).
200. J. Steiglitz, *Am. Chem. J.*, **29**, 49 (1903).
201. J. Steiglitz and R. B. Earle, *Am. Chem. J.*, **30**, 399 (1903).
202. A. Pinner, *Ber.*, **16**, 1655 (1883).
203. A. Pinner, *Ber.*, **22**, 1600 (1889).
204. J. Bougault and P. Robin, *Compt. Rend.*, **171**, 38 (1920); J. Bougault and P. Robin, *Compt. Rend.*, **172**, 452 (1921).
205. P. Robin, *Compt. Rend.*, **173**, 1085 (1921); P. Robin, *Compt. Rend.*, **177**, 1304 (1923).
206. Chinoin R. T., *Hung. Pat.* 127,837 (1941); *Chem. Abstr.*, **36**, 2271 (1942).
207. C. E. Kwarter and P. Lucas, *J. Am. Chem. Soc.*, **65**, 354 (1943).
208. F. L. Pyman, *J. Chem. Soc.*, 3359 (1923).
209. E. Beckmann and E. Fellrath, *Ann. Chem.*, **273**, 1 (1893).
210. J. von Braun, *Ber.*, **37**, 2678 (1904).
211. C. Chew and F. L. Pyman, *J. Chem. Soc.*, 2318 (1927).
212. H. von Pechmann and B. Heinze, *Ber.*, **30**, 1783 (1897).
213. A. Pinner and F. Klein, *Ber.*, **11**, 4 (1878).
214. F. L. Pyman, *J. Chem. Soc.*, 367 (1923).
215. E. F. Cornell, *J. Am. Chem. Soc.*, **50**, 3311 (1928).
216. M. Hedayatullah and L. Denivelle, *Compt. Rend.*, **256**, 4025 (1963).
217. C. R. Hauser, *J. Am. Chem. Soc.*, **52**, 1108 (1930).
218. J. W. LeMaistre, A. E. Rainsford and C. R. Hauser, *J. Org. Chem.*, **4**, 106 (1939).
219. A. Dornow and K. Fischer, *Chem. Ber.*, **99**, 68 (1966).
220. H. J. Barber and A. D. H. Self, *U.S. Pat.* 2,375,611 (1945); *Chem. Abstr.*, **39**, 3544 (1945).
221. N. Buu-Hoi, M. Welsch, N. Xuong and K. Thang, *Experimentia*, **10**, 169 (1954).
222. M. Busch and R. Hobein, *Ber.*, **40**, 4296 (1907).

223. A. J. Ewins, H. J. Barber, G. Newberry, J. N. Ashley and A. D. A. Self, *Brit. Pat.* 538,463 (1941); *Chem. Abstr.*, **36**, 3511 (1942).
224. A. V. Kirsanov and Y. N. Ivaschchenko, *Bull. Soc. Chim.* [5], **2**, 1944 (1935); A. V. Kirsanov and I. M. Polyakova, *Bull. Soc. Chim.* [5], **3**, 1600 (1936).
225. K. Ziegler, *U.S. Pat.* 2,049,582, *Chem. Abstr.*, **30**, 6389 (1936).

CHAPTER 9

syn–anti Isomerizations and rearrangements

C. G. McCarty

West Virginia University, Morgantown, U.S.A.

I. INTRODUCTION

The purpose of the first part of this chapter is to review some of the evidence establishing the existence of geometrical isomerism in the various classes of compounds containing the C=N double bond and to emphasize mainly some of the recent investigations of rates of isomerization in these compounds. Many of these recent studies are a result of the widespread use of modern instrumentation such as nuclear magnetic resonance spectroscopy.

In studies of rearrangements at the azomethine group one must often consider the existence of *syn–anti* isomerism and the configurational stability of such isomers under the conditions of the rearrangement reaction. For this reason, at least, it seems appropriate to consider both geometrical isomerizations and rearrangements at the C=N double bond in this one chapter.

The subject of rearrangements involving the azomethine group is so broad that it has been felt necessary to limit the discussion to only a few of the more common rearrangements in which the C=N double bond is contained in the starting material. Even in these cases, the emphasis is on the more recent studies in an attempt to provide the reader with a feeling of the current views on each subject.

II. *syn–anti* ISOMERIZATIONS

A. N-Aryl and N-Alkyl aldimines

The possibility of the existence of *syn–anti* (or *cis–trans*) isomerism in imines or Schiff bases derived from the reaction of an aldehyde with a primary amine has intrigued chemists for many decades. Only recently, however, has it been firmly established that such isomerism

$$\begin{array}{cc} R^1(H)C{=}N{-}R^2 & R^1(H)C{=}N{-}R^2 \\ \textit{syn or cis} & \textit{anti or trans} \end{array}$$

is indeed possible and that the rates of interconversion of such isomers can be measured quantitatively.

Most of the early investigations were with anils of benzaldehyde. The fact that hundreds of anils were reported to exhibit thermochromism[1] and dozens (mainly derivatives of salicylaldehyde) to display photochromism[2] led to many explanations of these phenomena in terms of *syn–anti* isomerization.

There have also been many reports of the isolation of anils in more than one crystalline form, usually of different colours and melting points[3]. This, quite naturally, was explained by some to be the isolation of stable *syn* and *anti* isomers. True *syn–anti* isomerism in these early cases was not unequivocally demonstrated and in many cases it has been later asserted that the two forms reported were just dimorphic forms. The danger of confusing polymorphism with stereoisomerism was pointed out by Curtin[4], with the illustration that benzophenone phenylimine (**1**) exists in two crystalline forms[5] even though no stereoisomerism is possible.

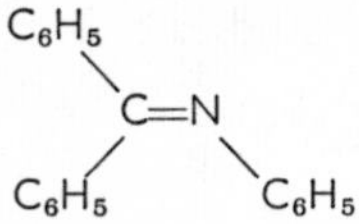

$(C_6H_5)_2C{=}N{-}C_6H_5$

(**1**)

$o\text{-}HOC_6H_4(H)C{=}N{-}C_6H_4CO_2R\text{-}p$

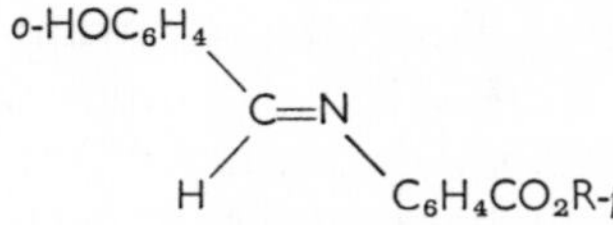

(**2**) $R = CH_3$
(**3**) $R = C_2H_5$

Schiff bases **2** and **3**, derivatives of salicylaldehyde, can be chosen to illustrate the course of events referred to in the previous paragraphs. Compound **2** can be isolated as yellow needles which melt at 145°[6]. Exposure of these needles to light gives orange-red needles of considerably higher melting point. By careful fractional crystallization from ethanol one can obtain two forms of the ethyl derivative (**3**) which have different melting points and are of different colour[7]. The results have been attributed by Manchot and Furlong[6,7] to the isolation and interconversion of the *syn* and *anti* isomers of these compounds.

It was claimed[8] that the results observed by Manchot and Furlong could be better explained in terms of polymorphism. That it is polymorphism in these cases was felt to be substantiated by the dipole moment studies of De Gaouck and Le Fevre[9] and by Jensen and Bang[10] who found identical dipole moments for the two forms of **3** in various solvents. What these authors failed to consider, though, is the possibility that the two crystalline forms of **3** were indeed *syn* and *anti* isomers which rapidly established the same equilibrium mixture of isomers upon being dissolved in the solvents. With the present knowledge of the rapid interconversion of isomers of *N*-aryl imines[11] they could not have ignored this possibility.

Most dipole moment studies are consistent with a planar *trans* structure for *N*-benzylideneanilines in solution. De Gaouck and Le Fevre[12] found that the dipole moment of *N*-benzylideneaniline (**4**) agreed well with that of *p*-chlorobenzylidene-*p*-chloroaniline (**5**). Spectral studies, however, have led to some disbelief in the planar

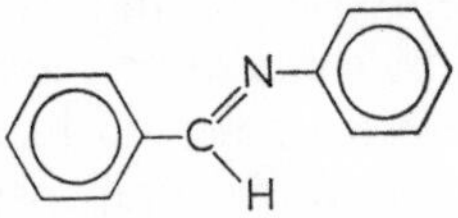

1·57 D
(**4**)

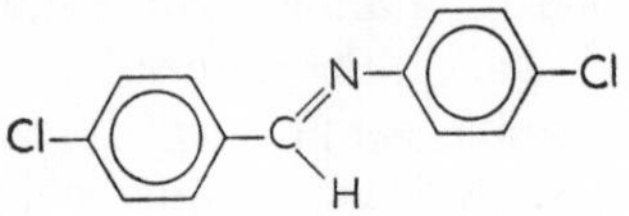

1·56 D
(**5**)

trans structure. Wiegand and Merkel[13] observed that the ultraviolet spectra of *trans*-stilbene (**6**), 2-phenylindene (**7**), and 2-phenylimidazole (**8**) were very similar and quite different from the spectrum of *N*-benzylideneaniline (**4**). Since dipole moment studies rule out a *cis* structure for **4** and the absorption spectrum is not consistent with a *trans* structure, they concluded that the preferred structure in solution is most likely linear at the nitrogen (**9**).

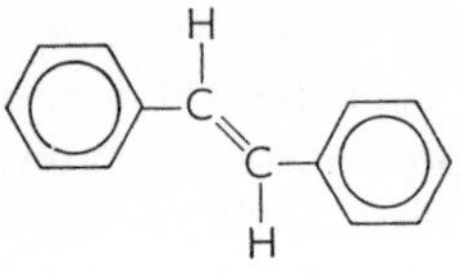

(**6**) 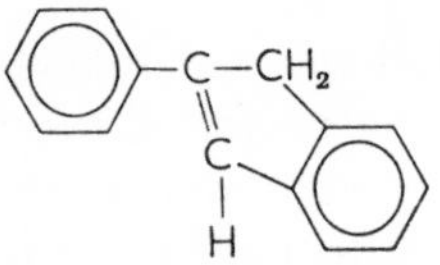 (**7**) 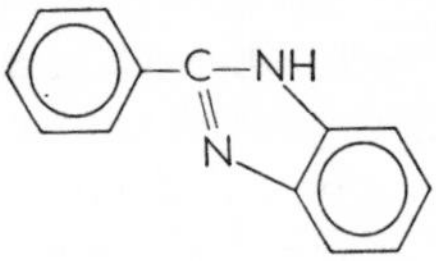 (**8**)

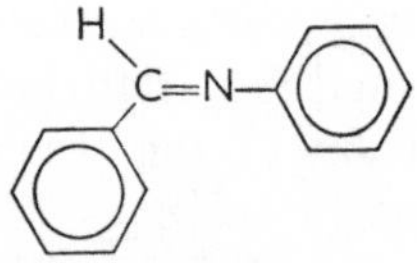

(**9**)

The anomalous absorption spectra of *N*-benzylideneanilines is, nevertheless, still consistent with a *trans* structure if one considers a deviation from coplanarity. The deviation from coplanarity as depicted in structure **10** can be expressed in terms of a dihedral angle, θ, between the plane of the aniline ring and the plane of the rest of the molecule. When $\theta = 0°$ there is no conjugation of the non-bonding electrons on nitrogen with the *N*-aryl group, whereas an angle of 90° represents optimal conditions for this type of conjugation. More

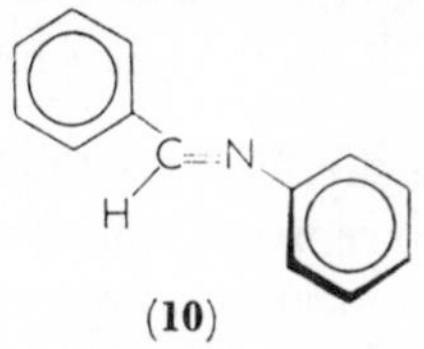

(**10**)

accurate measurements of both dipole moments and simple MO calculations[14] determine the angle θ to be approximately 60°, and *para* substituents in the aryl rings have little effect on this angle. A more complex MO method has recently been employed[15] to calculate the electronic spectra of *N*-benzylideneanilines assuming this deviation of about 60° from coplanarity, and the results are in quite good agreement with experimental results.

Recent x-ray crystallographic studies[16,17] have shown that some derivatives of *N*-benzylideneaniline may crystallize in a *trans* non-planar form. Of course, these observations may not be extrapolated to structures in solution.

In spite of the numerous studies on the structure of *N*-benzylidene-anilines there apparently still had not been a quantitative study of *syn–anti* isomerization in this type of compound at the time of a review on geometrical isomerization in conjugated molecules in 1955[18]. Seemingly, since *syn* isomers of *N*-benzylideneanilines apparently cannot be obtained from normal syntheses and are not favoured at equilibrium in solution, such a study would have to depend upon the *in situ* generation of the *syn* isomer as an unstable or transient species.

Fischer and coworkers found that at low temperatures photo-isomerizations could be carried out on systems which at room temperature would result in unstable, and therefore undetectable, isomers. In 1957 they reported[19] the results of their application of the low-temperature irradiation technique to *N*-benzylideneaniline and related compounds. When solutions of these Schiff bases were irradiated with light of various wavelengths at room temperature no short-lived or permanent change was noted. However, irradiation of the same solutions at $-100°$ or lower led to pronounced changes in the spectra. The actual extent of the changes depended on the wavelength of the light used for irradiation. Thermal reversion to the original absorption occurred at a convenient rate in the range of $-70°$ to $-40°$. The reversion process was found to be first order with an activation energy of 16–17 kcal/mole in methylcyclohexane. The spectral changes were completely reversible for all of the benzalaniline derivatives studied and non-existent for compounds such as **1**, which is unable to have two geometrical isomers.

Not many details were supplied by Fischer and Frei[19], but their communication appears to represent the first reported kinetic study of actual *syn–anti* isomerization at the azomethine group in *N*-benzylideneanilines (and may be the first kinetic study of isomerization at the carbon–nitrogen double bond in any system).

The development of high-energy flash photolysis units and their use, most notably by Wettermark and coworkers (see Chapter 12), allowed the conversions described by Fischer and Frei to be observed at temperatures closer to room temperature. Wettermark and Dogliotti[20] reported the observation of two photo-induced isomerizations in the photolysis of solutions of *N*-(*o*-hydroxybenzylidene)aniline (**11**) in the temperature range 15–70°. One process is associated with the formation of a new absorption band with a maximum around 470 mμ, which fades in the dark with a lifetime in the millisecond region. The fading of this band was found to be increased by the addition of small amounts of acid or base and the kinetic order of the process varied with the conditions.

Quinoid isomers have been proposed for the red form of **11** observed in the crystalline state and upon irradiation of solutions at low temperature[2,21]. The spectrum observed for the red form in a rigid glass and in the solid[22] is very similar to the spectrum observed by Wettermark upon irradiation of solutions at room temperature. Thus the quinonoid structure **12** has been assigned[20] to the species, giving rise to the short-lived band at 470 mμ. Probably many of the photochromic conversions noted for anils derived from salicylaldehyde[2] can be accounted for by equilibria between enol and keto structures. The role of *cis* and *trans* keto structures in photochromic anils is still the subject of active investigation[23].

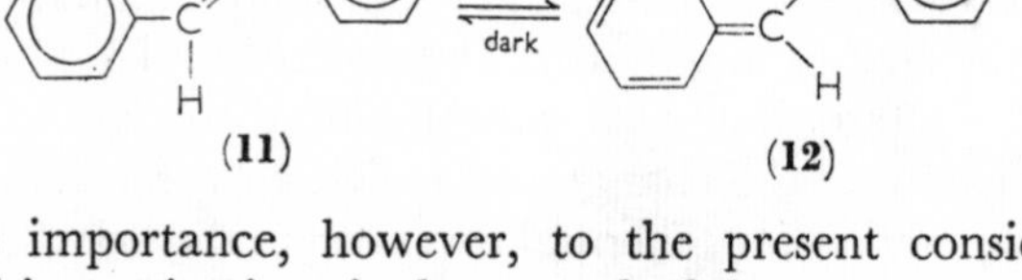

(**11**) $\underset{\text{dark}}{\overset{h\nu}{\rightleftharpoons}}$ (**12**) (1)

Of more importance, however, to the present consideration of geometrical isomerizations is the second of the processes observed by

TABLE 1. Kinetic parameters for the thermal relaxation of photo-isomerized solutions of some Schiff bases.[24]

Schiff base	$k_{30\cdot 0°}(s^{-1})$	$A \times 10^{11}(s^{-1})$	E_a(kcal/mole)
N-(*o*-hydroxybenzylidene) aniline (**11**)	1·67	0·3	14·2
N-(*o*-hydroxybenzylidene) -*β*-naphthylamine	1·93	3	15·4
N-(*p*-hydroxybenzylidene) aniline	0·72	2	15·7
N-benzylideneaniline (**4**)	1·46	12	16·5
N-benzhydrylidene-aniline (**1**)	No change		

Wettermark and Dogliotti[20] upon irradiation of solutions of **11**. This second change is associated with a decrease in optical density around 385 mμ upon photolysis and a restoration of absorption in the dark at a rate many times slower than the previously mentioned process. This slower process is independent of added acid or base and has an activation energy of 15 kcal/mole as derived from the first-order plots. It was assumed that the transient species with less absorption at 385 mμ than **11** is the *cis* or *syn* enol form of the imine and the process being measured is the *syn–anti* isomerization around the carbon–nitrogen double bond.

Some kinetic parameters for the thermal relaxation of the long-lived transformation (*syn–anti* isomerization) in several anils studied by Anderson and Wettermark[24] are summarized in Table 1. The anions of the first three anils in Table 1 in strong base also show *syn–anti* isomerization with an energy of activation comparable to that of the neutral species. The similarity in E_a values found for anils **11** and **4** provides convincing evidence that the transformations being observed for the salicylaldehyde derivatives do not involve the quinonoid tautomer. Furthermore the fact that compound **1**, for which *syn* and *anti* isomers do not exist, does not give rise to observable photo-induced transients also tends to support the proposition of *syn–anti* isomerization as the cause of the changes observed.

In order to determine the effect of substituents on the rate of isomerization, several substituted anils of the general structure **13** were

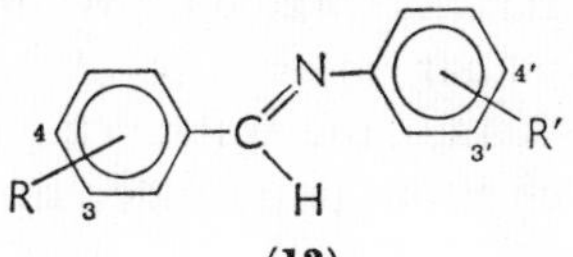

(**13**)

prepared and studied by the flash photolysis technique[25,26]. Each anil bore one substituent (NO_2, Br, Cl, CH_3, OH, $N(CH_3)_2$, or OCH_3) in the 3,3′,4, or 4′ position. All compounds except 4′-NO_2 showed changes in absorption upon photolysis which thermally reverted to the original absorption pattern by a first-order process assumed again to be *syn–anti* isomerization. The applicability of the Hammett equation (equation 2) was tested for the rate constants obtained and the results are shown in Table 2. The ρ values are

$$\log k = \rho\sigma + \log k_0 \tag{2}$$

positive, in agreement with the observation that electron-withdrawing groups facilitate the isomerization (Table 3).

TABLE 2. Applicability of the Hammett equation.[26]

	ρ	Correlation coefficient
Substitution in 3- or 4-position	0·35	0·90
Substitution in 3′- or 4′-position	1·85	0·98

TABLE 3. Rate constants for the thermal relaxation of photo-isomerized solutions of some *N*-benzylideneanilines.[a]

Substituent	Position	$k_{30°}(s^{-1})$
$N(CH_3)_2$	4	0·67
NO_2	4	2·17
$N(CH_3)_2$	4′	0·094
Br	4′	3·41
CH_3O	3	1·2 ± 0·2
NO_2	3	2·1 ± 0·1
$N(CH_3)_2$	3′	0·30 ± 0·01
NO_2	3′	16·6 ± 0·5

[a] Data are from references 25 and 26 and represent only a portion of those compounds reported.

From the magnitude of the ρ values in Table 2 it can be seen that the effect of a substituent is much greater when it is attached to the aniline ring than when it occupies the corresponding position on the benzaldehyde ring. This is interpreted by Wettermark[26] as being consistent with the non-planar model for *N*-benzilideneaniline (**10**), where the π-electron system of the aniline ring interacts more strongly with the non-bonding electrons on nitrogen than the π electrons from the rest of the system.

Whether the lack of observable isomerization in the 4′-NO_2 molecule is due to the process being too rapid to detect is not entirely clear. Wettermark[11,26] prefers an explanation based on a strong resonance interaction between the aniline nitrogen lone pair and the nitro group, thus leading to a maximum deviation from coplanarity and *sp* hybridization at the nitrogen. No geometrical isomers would then exist. This structure (**14**) is the same as that proposed by Ebara[27] but, as has been pointed out[11], this was on a misinterpretation of his experimental data.

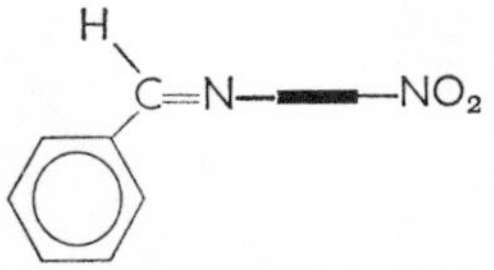

H
C=N—▬—NO_2

(**14**)

Most recently, Wettermark and Wallstrom[11] have found that the calculated C=N bond orders for a series of *N*-benzylidene type Schiff bases are inversely related to the rate constants for the thermal relaxation processes measured by them. It would be interesting to extend these MO calculations, if possible, to some of the compounds discussed in the following sections of this chapter to see if other rate data can be correlated with bond orders.

Reports of studies of the isomerization of aldimines derived from aliphatic amines are few and far between. The methylimines **15–18** were prepared by Curtin and coworkers[28]. Due to the method of synthesis, the lack of evidence for two isomers in the infrared and n.m.r. spectra, and the known preference for a *trans* configuration for benzene azomethane ($C_6H_5N{=}NCH_3$)[29], it was concluded that the *trans* or *anti* configuration predominates in solutions of these methylimines. Some similarities between these imines and some ketimines to be mentioned in the next section are also consistent with an *anti* assignment. There was no attempt by these workers[28], however, to generate the *syn* isomers and to study their reversion to the *anti* isomers, so no kinetic data are available.

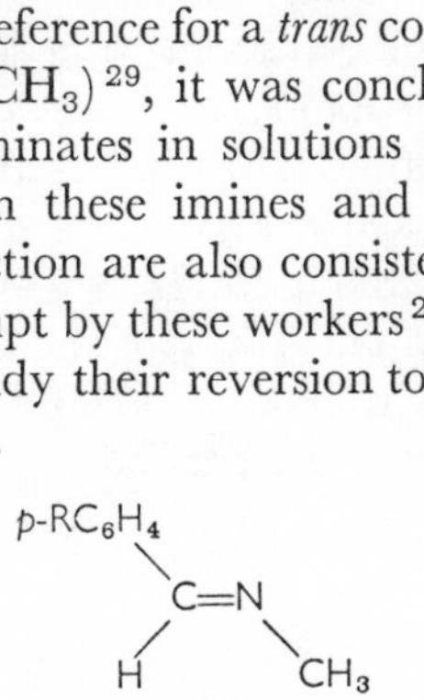

p-RC_6H_4
C=N
H CH_3

(**15**) R = H (**17**) R = CH_3O

(**16**) R = Cl (**18**) R = NO_2

Hine and Yeh[30] have recently determined the equilibrium constants for the formation of imines and water from isobutyraldehyde and several saturated amines ranging from methyl through *t*-butyl. A comparison of these imines with their olefinic analogues, the 4-methyl-2-pentenes, led to the conclusion that the *anti* form of the imines should be the more stable and was most likely the one being observed. More convincing, however, is the argument based on the constancy of the coupling constant (J_{AB}) between the α-hydrogen of the isopropyl group and the hydrogen attached to the sp^2 carbon as the group R is varied in **19** from methyl through *t*-butyl. It would certainly

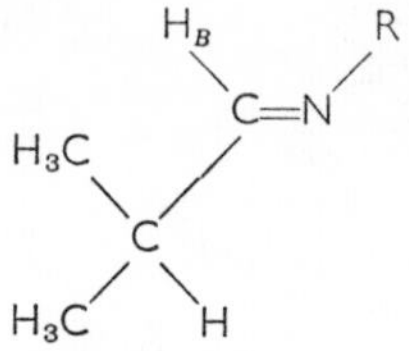

H_B, R, C=N, H_3C, C, H_3C, H

(19)

be expected that the relative amounts of contributing conformations and thus the net coupling constant would be changed if significant amounts of the *syn* isomers were present. Also, no extra bands were seen in the n.m.r. spectra which could be due to the *syn* isomers. Although this absence could be explained by a coincidence of chemical shifts in the two isomers or rapid establishment of a *syn–anti* equilibrium, these possibilities do not seem likely in view of the published evidence on the relatively slow isomerization of *N*-alkyl ketimines[28] and the chemical shift data for similar systems[31].

N-Alkylmethylene imines (**20–22**), the simplest of aldimines, have been prepared and their n.m.r. spectra examined[32,33]. The spectrum of *N*-methylmethyleneimine (**20**) appears as an ABX_3 system with the *A* and *B* hydrogens (each a doublet of quartets through spin–spin coupling) separated by about 25 Hz as would be expected for a non-linear structure[33]. An n.m.r. study of the configurational stability of these methylene imines would be theoretically possible through the observation of the temperature dependence of the separation between

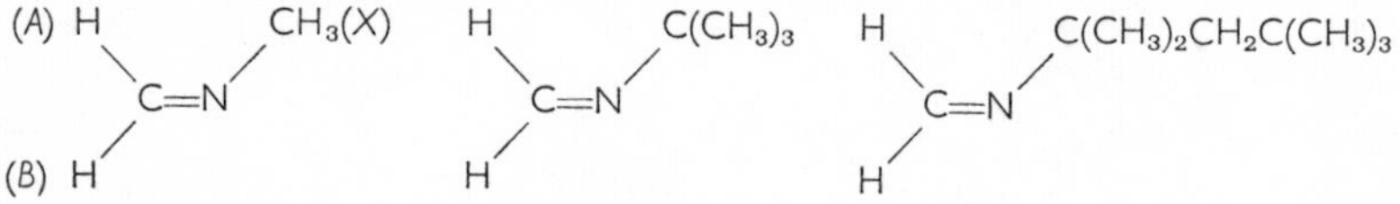

(A) H, $CH_3(X)$, C=N, (B) H — **(20)**

H, $C(CH_3)_3$, C=N, H — **(21)**

H, $C(CH_3)_2CH_2C(CH_3)_3$, C=N, H — **(22)**

the *A* and *B* signals. Apparently this has not been done and presumably this is due to the instability of methylene imines and the high temperatures that might be necessary to bring about rapid (on the n.m.r. time scale) inversion at nitrogen.

B. N-Alkyl and N-Aryl ketimines

The history of the study of isomerism of the *syn–anti* variety in ketimines (Schiff bases derived from ketones and aryl or alkyl amines) is much like that just described for aldimines. There have been several equivocal reports of the isolation of two isomers of ketimines in the older literature[3,18] and several of these have been later disputed[4].

For example, the reaction of *p,p'*-dichlorobenzalacetophenone with *p*-toluidine in benzene at room temperature reportedly gave yellow needles, m.p. 130°, when the reaction was terminated after 40 hours[34]. However, if the reaction mixture was allowed to stand for several weeks at room temperature the chief product was a colourless compound, m.p. 145°, which was formed together with the yellow needles. The two forms were separable by fractional crystallization, yielded different picrates and hydrochlorides, but had similar molecular weights. It was concluded that they represent the two geometrical isomers of the ketimine.

Ramart-Lucas and Hoch[35] obtained two forms of the anil of desoxybenzoin which had different melting points and different absorption spectra in the ultraviolet region. They assigned structure **23** to the higher melting isomer which possessed the longer wavelength absorption.

Taylor and Fletcher[36] felt that they had isolated geometrical isomers of imines in the 2-nitrofluorene series. The acid-catalysed condensation of 2-nitrofluorenone with *p*-toluidine resulted in a product, m.p. 193°, to which they assigned structure **24**. The base-catalysed condensation of 2-nitrofluorene with *p*-nitrosotoluene yielded a different isomer, m.p. 218°, which was assigned structure **25**. The structural assignments were made as a result of several considerations including differences in the ultraviolet absorption spectra of the two forms in ethanol. The

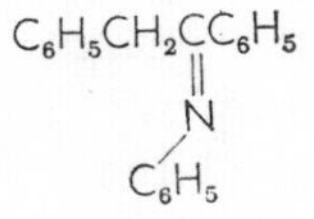

$C_6H_5CH_2CC_6H_5$ (=N–C_6H_5)

(23)

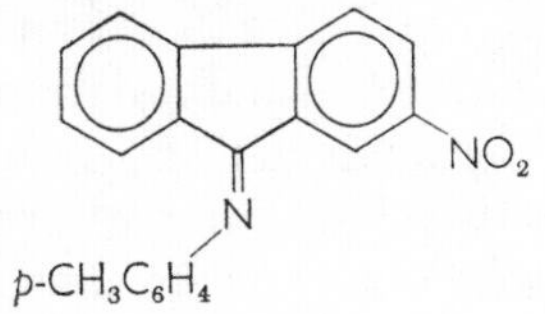

NO_2 ; N ; p-$CH_3C_6H_4$

(24)

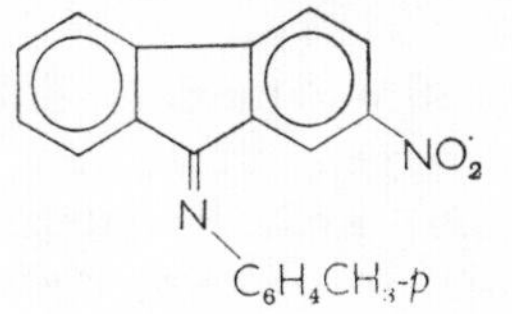

NO_2 ; N ; $C_6H_4CH_3$-p

(25)

spectral differences were quite small, however, and the unusual stereospecificity of the two reactions and configurational stability of the products led Curtin and Hausser[4] to attempt to reproduce this work of Taylor and Fletcher. This they could not do in entirety and they thus concluded that the structural assignments, **24** and **25**, should be open to question until further work is done on this system.

In 1962, Saucy and Sternbach[37] described the isolation of both forms of 2-methylamino-5-trifluoromethylbenzophenone methylimine (**26**, **27**). They assigned structure **26** to the lower melting isomer, mainly on the basis of the wide absorption band in its infrared spectrum at 3·3 μ attributed to a hydrogen bonded —N—H stretching mode. The other form, incapable of this hydrogen bonding, showed a sharper —N—H band at 2·8 μ.

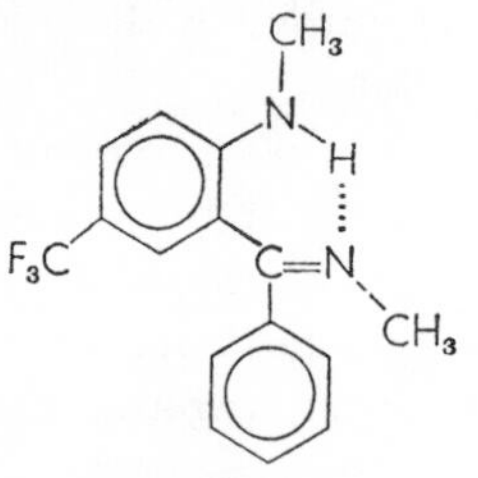

(**26**) 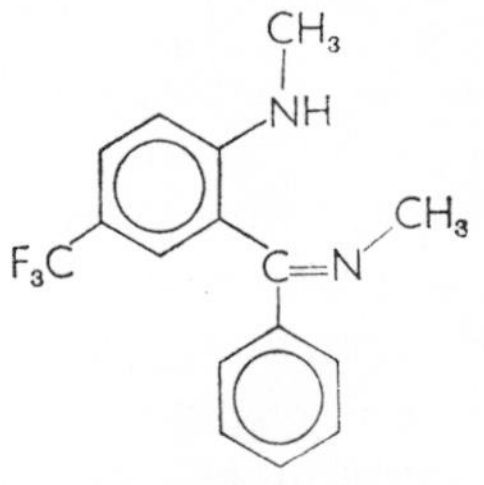 (**27**)

2-Amino-5-chlorobenzophenone and 4-(2-aminoethyl)morphine, when heated in xylene in the presence of zinc chloride, gave, upon fractional crystallization, two isomers designated α and β by Bell, Conklin and Childress[38,39]. The α-isomer (**28**) was white and melted at 140–142°, while the β-form (**29**) was pale yellow and melted in the range 112–114°. The melt of the β-isomer, when cooled after being held for a few minutes at 140–150°, yielded the α-isomer. The infrared and n.m.r. spectra of **28** and **29** were consistent with the ketimine structures but did not reveal the geometric orientation of the isomers. The assignments were made by relating the ultraviolet absorption spectra of the ketimine isomers with those of the corresponding oxime isomers (**30, 31**) of supposedly known configuration[40]. The spectra of **28** and **29** were quite different, but that of **28** was similar to the spectrum of **30**, and the curve of **29** was quite like the spectrum of **31**. The spectra observed in these sets of compounds may be explained by assuming that, because of steric crowding, the aromatic ring *cis* to the substituent on nitrogen is twisted out of the plane of the rest of the conjugated system[41] as depicted by structures **28–31**. The main

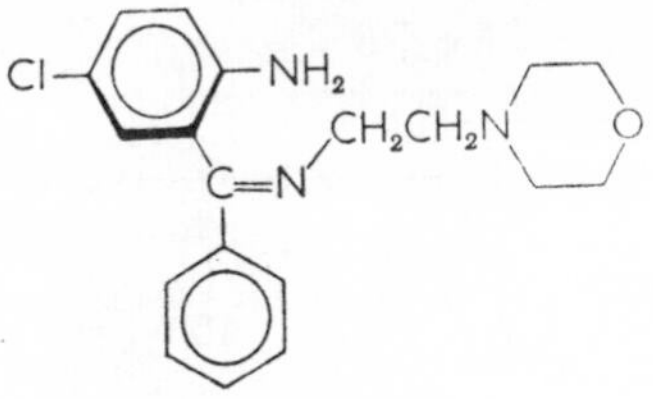

(28)

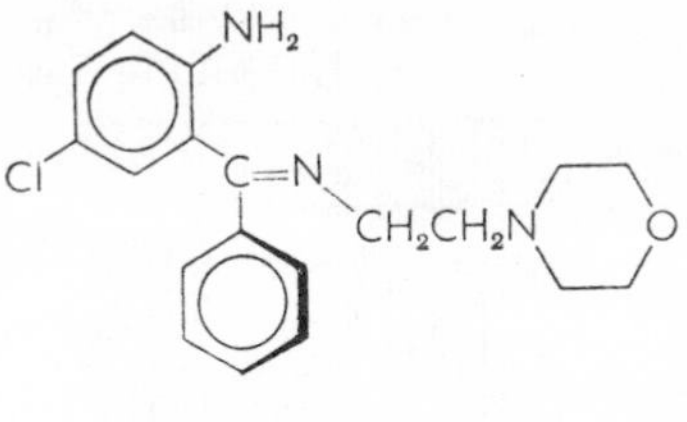

(29)

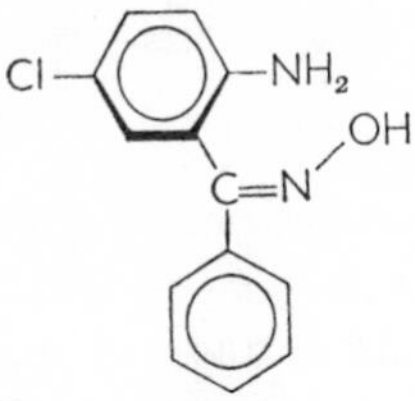

(30)

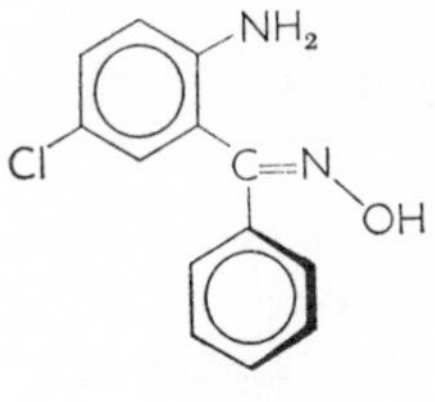

(31)

chromophore giving rise to the long wavelength absorption is thus the one ring and the carbon–nitrogen double bond which are coplanar. This ring twist argument could also have been used by Saucy and Sternbach[37] for assigning structures **26** and **27**, and the result would have been the same as their conclusion based on infrared evidence.

Geometric isomerism of simple aliphatic azomethines has been observed recently by n.m.r. spectroscopy[42–44]. The spectrum of the methylimine of acetone, for example, consists of three resonances of equal area at τ 8·20, 8·02 and 6·94, indicating that isomerization of the methyl group on nitrogen is slow on the n.m.r. time scale. A similar doubling of patterns was noted for other symmetrically *C*-substituted *N*-alkyl ketimines. For unsymmetrical ketimines it was possible through a consideration of peak areas to assign *syn–anti* ratios (**32**:**33**). Some of the results are summarized in Table 4.

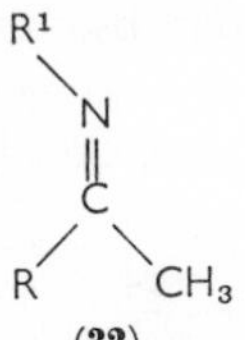

(32)

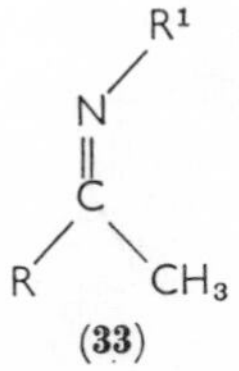

(33)

The first extensive study of the isomerization of ketimines is that of Curtin and Hausser[4]. Imines were chosen by these workers as model compounds to be used in place of vinyllithium compounds to study the effects of structural changes on the configurational stability of such

TABLE 4. Isomer ratios[42–44] of some unsymmetrically C-substituted azomethines $CH_3CR{=}NR^1$.

R	R^1	% *syn* (**32**)	% *anti* (**33**)
C_2H_5	CH_3	14	86
$(CH_3)_2CH$	CH_3	3	97
$(CH_3)_3C$	CH_3	0	100
C_3H_7	C_4H_9	23	77
$(CH_3)_3CCH_2$	C_4H_9	11	89
C_2H_5	$CH(CH_3)C_6H_5$	17	83
$(CH_3)_2CH$	$CH(CH_3)C_6H_5$	7	93
$(CH_3)_3C$	$CH(CH_3)C_6H_5$	0	100

species. Imines are isoelectronic with the corresponding vinyl carbanions thought to be intermediates in the *cis–trans* isomerization of vinyllithium compounds[45], and have the advantage of existing as

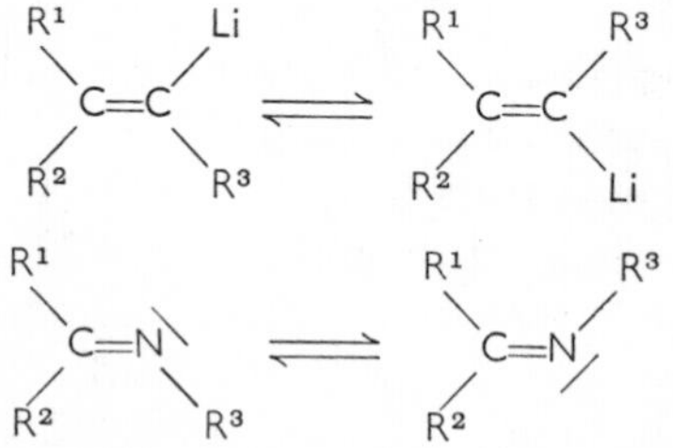

$$R^1R^2C{=}C(Li)R^3 \rightleftharpoons R^1R^2C{=}C(R^3)Li$$
$$R^1R^2C{=}\ddot{N}R^3 \rightleftharpoons R^1R^2C{=}N(R^3)\ddot{} \tag{3}$$

neutral molecules rather than as ions or ion aggregates. From the behaviour of vinyllithium compounds it was predicted that N-alkyl imines should possess a relatively high degree of configurational stability at room temperature.

Curtin and Hausser obtained by normal procedures the N-methylimines of p-chlorobenzophenone and p-nitrobenzophenone. The crystalline p-chlorobenzophenone methylimine was shown to be the pure *cis* isomer (**34**) (*cis* and *trans* were used in this study to define the relative positions of p-substituted phenyl and the substituent on nitrogen). It was found, by observation of bands unique to it in the

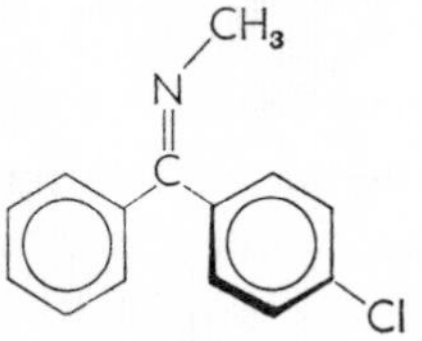

(**34**) 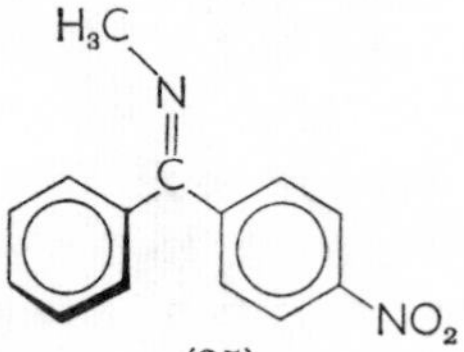 (**35**)

ultraviolet and infrared spectra, to isomerize in cyclohexane solution to an equilibrium mixture of 58% of the *trans* and 42% of the *cis* isomer at 100°. The first-order kinetics of this isomerization were measured over a temperature range of 40–60°. The resulting thermodynamic parameters are summarized in Table 5.

TABLE 5. Rates of uncatalysed *cis–trans* isomerization of methylimines **34** and **35** in cyclohexane.[4]

Compound	T (°C)	K_{eq}	$10^5\,k_{obs}$ (s^{-1})	E_a (kcal/mole)	ΔS^* (e.u.)
34	50·2	1·4	4·04	25 ± 0·3	−3·6 ± 1·0
35	50·2	1·6	4·26	27·1 ± 0·5	+3·4 ± 1·4

The *p*-nitrobenzophenone methylimine was shown to be the *trans* isomer (**35**). Its conversion in solution to an equilibrium mixture of 69·5% of the *trans* and 30·5% of the *cis* isomer (100°) was also conveniently followed spectrophotometrically over the range 40–60°, and the results are tabulated in Table 5. It is noteworthy that the effect of these substituents on the rate of isomerization is quite small.

The assignment of configurations of imines **34** and **35** was made with the aid of an empirical infrared correlation noted by Curtin and Hausser[4], and also through ultraviolet spectral analogies. Examination of infrared spectra of properly constituted olefins and azo compounds of known configurations revealed that the position of the phenyl hydrogen deformation frequency in the 700 cm^{-1} region is higher in that isomer with another atom or group *cis* to the phenyl ring, the difference between the two isomers generally being 5–9 cm^{-1}. No significant reversal of this trend has been found, and recently a very similar infrared method for structural assignments in olefins, imines and azo compounds has been communicated by Lüttke[46]. Use was made in this latter study of spectral differences in this same region of 700 cm^{-1}. Either of these infrared correlations should prove very useful for future configurational assignments.

Reports from other workers tend to confirm this high degree of configurational stability for common *N*-alkyl ketimines. In the previously mentioned cases where both isomers of one compound have been isolated[37–39] the compounds had an alkyl group on the nitrogen atom and no note was made of rapid spectral changes at room temperature. Recently Staab, Vögtle and Mannschreck[31] reported n.m.r. chemical shifts for the methyl groups in acetone *N*-benzylimine as shown in **36**. Presumably, as will be discussed in more detail later, the methyl group *cis* to the aromatic ring in **36** is the one with the higher field resonance as a result of a phenyl ring-current effect. The two methyl signals remained separated or uncoalesced even at 170° in quinoline; a fact which allows a calculation of a lower limit of 23 kcal/mole for the free energy of activation, ΔG^* [47].

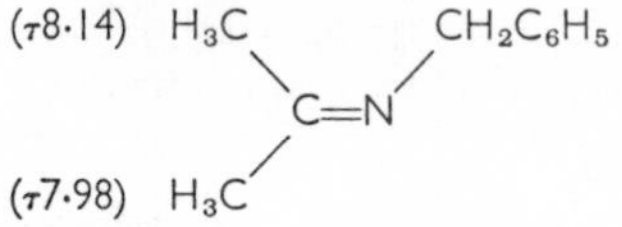

$$\begin{array}{l} (\tau 8{\cdot}14)\ H_3C \diagdown \qquad\qquad \diagup CH_2C_6H_5 \\ \qquad\qquad\qquad C{=}N \\ (\tau 7{\cdot}98)\ H_3C \diagup \end{array} \qquad \Delta G^* > 23 \text{ kcal/mole}$$

(**36**)

Some *N*-perfluoroalkyl ketimines seem to have much lower configurational stability for reasons not yet fully understood. The process shown in equation (4) is what is assumed to be responsible for the temperature dependence of the ^{19}F n.m.r. spectrum of this perfluoro imine[48]. It was observed that the two $=C—CF_3$ groups are

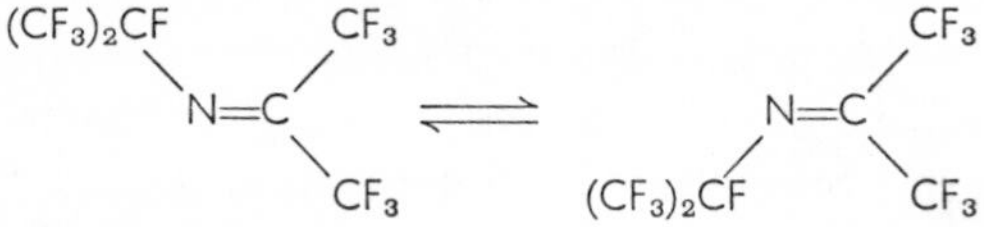

$$\begin{array}{l} (CF_3)_2CF \diagdown \qquad \diagup CF_3 \\ \qquad\qquad N{=}C \\ \qquad\qquad\qquad \diagdown CF_3 \end{array} \rightleftharpoons \begin{array}{l} \qquad\qquad\qquad \diagup CF_3 \\ \qquad\qquad N{=}C \\ (CF_3)_2CF \diagup \qquad \diagdown CF_3 \end{array} \qquad (4)$$

(**37**)

separated by more than 300 Hz at temperatures well below their coalescence temperature of 32°. By recording the peak separation as a function of temperature it was possible to calculate an energy of activation of 13 ± 3 kcal/mole for the alleged isomerization of **37**.

A temperature-dependent broadening of the *AB*-type pattern of the $=CF_2$ group in perfluoro-*N*-methyleneisopropylamine (**38**) has also

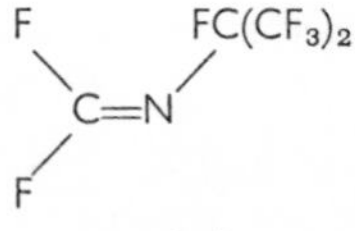

$$\begin{array}{l} F \diagdown \qquad \diagup FC(CF_3)_2 \\ \quad C{=}N \\ F \diagup \end{array}$$

(**38**)

been attributed[49] to isomerization about the C=N bond. In this case the energy of activation is apparently even lower, and is estimated to be about 11 kcal/mole. Since rapid isomerization is not observed at room temperature in perfluoroalkylidene derivatives of normal perfluoroalkyl amines[49], greater steric interactions in the ground state may partially explain the rapid isomerization of **37** and **38**, but this is yet to be established. It is doubtful that an argument based on steric factors alone could account for the difference of more than 10 kcal/mole in activation energy between **38** and **34** or **35**.

Curtin and McCarty[28,50] reported several lines of evidence to support the suggestion that *N*-aryl imines derived from unsymmetrically substituted benzophenones (**39–42**) are present in solution as a rapidly equilibrating mixture of *syn* and *anti* isomers but crystallize in a single stereoisomeric form. Each of the compounds **39–42** is a crystalline solid with a small melting point range ($\leq 1°$). Each of the first three (**39–41**)

$$(p\text{-}YC_6H_4)(C_6H_5)C{=}N{-}C_6H_4X\text{-}p$$

(**39**) $X = CH_3$, $Y = OCH_3$
(**40**) $X = N(CH_3)_2$, $Y = OCH_3$
(**41**) $X = Cl$, $Y = OCH_3$
(**42**) $X = OCH_3$, $Y = Cl$

in solution gave rise to two methoxyl proton peaks (of unequal area and 0·04–0·06 p.p.m. separation) in the n.m.r. spectra and to two infrared absorptions in the region near 700 cm^{-1} associated with the monosubstituted aromatic ring[4]. Each of the imines as the solid in a potassium bromide disk gave a single absorption in the 700 cm^{-1} region, but when the disk was heated for several minutes above the melting point of the particular imine, reground, and reformed, the monosubstituted aromatic absorption showed the development of a new band. Reference to the previously mentioned empirical infrared correlation in the 700 cm^{-1} region suggests that each of the imines **39–41** crystallizes largely in the configuration depicted with the $CH_3OC_6H_4$- ring *trans* to the *N*-aryl ring. The n.m.r. and infrared results are presented in Table 6. Similar results were found for *N*-aryl imines of other *p*-substituted benzophenones not shown in Table 6.

TABLE 6. N.m.r and infrared data for imines **39–41**.[28,50]

Compound	CH_3O (τ)	Solution infrared (cm^{-1})	KBr infrared (cm^{-1}) Before heat	After heat
(**39**)	6·22	700	703	703
	6·28	695		697
(**40**)	6·24	699	702	702
	6·28	695		696
(**41**)	6·23	698	701	701
	6·29	694		695

The ultraviolet spectrum of **40**, freshly dissolved in carbon tetrachloride at 0°, had a maximum at 275 mμ (ϵ 29,000) but on standing there developed a broad shoulder at 255 mμ. The approach to equilibrium was observed at various temperatures and the plots of $\ln(A_e - A_o)/(A_e - A)$ against t were linear, indicating a first-order process. The results are presented in Table 7. The unsymmetrical

TABLE 7. Rates of isomerization of imine **40** (4×10^{-4} M in carbon tetrachloride followed by observation of the absorption at 290 mμ).[28]

Temperature (°C ± 0·2°)	$10^4 k_{obs}$ (s^{-1})	E_a (kcal/mole)	ΔS^* (e.u.)
$-7·2$[a]	4·08 ± 0·05		
+2·8	15·0 ± 0·2	19·7 ± 0·4	−2·1 ± 1·5
+12·2	53·7 ± 0·5		
+62	($1·0 \times 10^4$)[b]		

[a] Duplicate runs made in each temperature range; only one shown.
[b] Extrapolated value.

tolyl imine (**39**), when dissolved in carbon tetrachloride, also underwent a change in the ultraviolet spectrum. This change was so rapid, however, even at −7·2°, that it was difficult to follow and only limited data were reported. One run at −7·2° gave $k_{obs} = 23·3 \pm 0·2 \times 10^{-4}$ s^{-1}. This substituent effect (admittedly very limited data) would be expected for a process with a Hammett ρ of +1·7.

The rate parameters shown in Table 7 and the methoxyl peak separation mentioned for the n.m.r. spectra of **39–41** suggest that this process might be more amenable to study by variable temperature n.m.r. techniques. With this objective in mind, Curtin and McCarty synthesized ketimines **43–46**. Each of the imines **43–45** showed, in

$$(p\text{-}CH_3OC_6H_4)_2C{=}N{-}C_6H_4\text{-}X\text{-}p$$

(**43**) X = H (τ6·20, 6·28)
(**44**) X = $N(CH_3)_2$ (6·25, 6·28)
(**45**) X = CH_3 (6·26, 6·31)
(**46**) X = $CO_2C_2H_5$ (6·26, broad)

carbon tetrachloride, the *p*-methoxyl proton absorption as two peaks of equal intensity with a separation of 0·03 to 0·08 p.p.m. The positions of these peaks are indicated next to the formulae. Imine **46** showed a broad peak at room temperature which also separated into a doublet when the sample was cooled below 40°. When **43–46** were heated to 70°, the methoxyl absorptions coalesced but reverted to their original position when the samples were cooled.

It was assumed that the process responsible for the temperature-dependent changes in the n.m.r. spectra of **43–46** was *syn–anti* isomerization which, when sufficiently rapid, made the environments of the two methoxyl groups equivalent. Measurement of the coalescence temperature for each compound allowed the rate of the process to be calculated at that temperature[47]. The results are shown in Table 8. The rates are shown extrapolated to a common temperature for convenience in comparison. The results are fairly insensitive to the value

TABLE 8. Isomerization rates of imines **43–46** in carbon tetrachloride.[28]

Compound, substituent	Coalescence temperature, (°C ±0·5°)	$\Delta\nu_{max}$, Hz at 60 MHz (± 0·15)	k_1, s^{-1} (± 0·3)	$k_1^{62\cdot2°}$, s^{-1} (extra-polated)
(**43**) H	69·5	3·92	8·7	4·4
(**44**) $N(CH_3)_2$	78·8	2·60	5·8	1·3[a]
(**45**) CH_3	62·2	4·90	10·9	11
(**46**) $CO_2C_2H_5$	29·8	5·99	12·4	180

[a] Extrapolated assuming $\Delta S^* = 0$.

assumed for ΔS^* but a value of $\Delta S^* = 0$ seems reasonable, considering the results obtained with an unsymmetrical analogue (compound **40**, Table 7). Plots of log $k_1^{62\cdot2°}$ versus Hammett σ^- were reasonably linear and gave a value of $\rho = +1\cdot67 \pm 0\cdot15$, in good agreement with that estimated from the limited data on the rate of isomerization of **39** and **40**. The correspondence of the data for **40** and **44** adds support to the hypothesis that it is the same process being measured by the two different instrumental techniques. Also it supports the conclusion that changes in substituents back on the benzophenone rings have little effect on the rate of isomerization[4].

A very similar study with similar results was carried out by Rieker and Kessler[51] on anils of 2,6-di-*t*-butyl-1,4-benzoquinone (**47**). In each of the anils prepared by them (different R groups) the *t*-butyl resonance in the n.m.r. spectrum appeared as two singlets at about

H² H¹ ⇌ H² H¹ (5)

(**47**)

τ 8·65 and 8·80. The hydrogens H¹ and H² appeared as an *AB*-system ($J_{AB} = 2\cdot6 \pm 0\cdot2$ Hz) in the region of τ 3·0 to 3·4.

The chemical shift difference between H¹ and H² was ascribed to the fact that the >C=N system and the aryl ring are probably not coplanar and the angle of twist may be as great as 90°. If so, this would be expected to lead to a shielding of the hydrogen *syn* to the aromatic ring relative to the other hydrogen due to the ring-current effect. N.m.r. measurements of the anisotropy effect of the phenylimino group have also been reported by Saito and Nukada[52], who found two peaks due to the α-methylene protons adjacent to the phenylimino group in the anils of cyclohexane and cyclopentane (**48, 49**). The latter authors

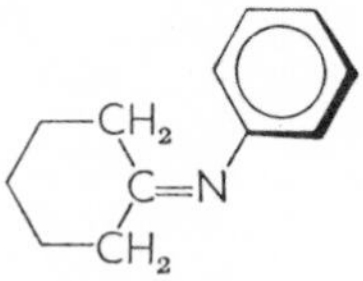

(**48**) 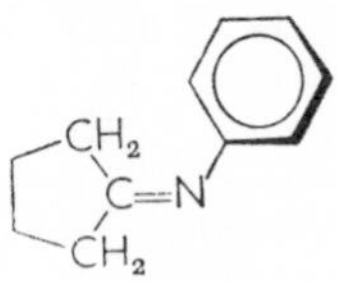 (**49**)

concluded from ultraviolet and infrared data in addition to the n.m.r. data that the $>$C=N plane is nearly perpendicular to the phenyl plane and that hybridization of the nitrogen is sp^2.

The temperature dependence of the *AB* pattern in the n.m.r. spectra of 12 benzoquinone anils, differing only in the substituted R, was determined by Rieker and Kessler[51]. These data allowed them to calculate the ΔG^* values for these compounds (some results are shown in Table 9). The ΔG^* values varied inversely with the Hammett σ_p values for the R groups and a plot of ΔG^* versus σ_p was found to be quite linear. The mechanistic implications of this are discussed in a later section.

TABLE 9. Temperature dependence of the *t*-butyl signal in the n.m.r. spectra of benzoquinone anils.[49]

Subst. R in **47**	$\Delta\nu_{max}$ in Hz ($\pm$ 0·4)	Coalescence temperature (°c $\pm$2°)	ΔG^* kcal/mole ($\pm$ 0·14)
$N(CH_3)_2$	4·7	144	22·80
OCH_3	6·8	152	22·94
H	9·2	140	22·03
Cl	8·4	128	21·42
$CO_2C_2H_5$	9·5	96	19·57

Curtin and coworkers[28] had also observed the anisotropic effect of the phenylimino group in the n.m.r. spectrum of the anil of acetone (**50**) which showed two methyl peaks of equal intensity at τ 7·96 and 8·33 in carbon tetrachloride at room temperature. In connection with an attempt to assign the peaks as shown next to the formula for **50**, the anil **51** was prepared. It was assumed to be the isomer with the bulky *t*-butyl and phenyl groups *trans* to one another as shown. With this assumption, which seems reasonable, it can be seen that the high field methyl peak in **50** is the one *syn* or *cis* to the phenyl as would be expected from the normal ring-current effect.

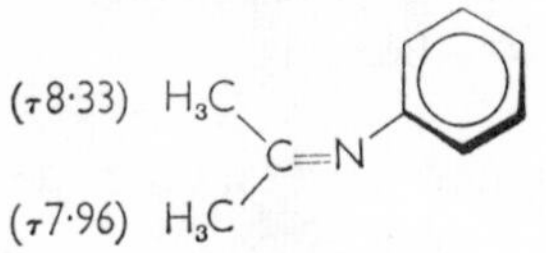

(τ8·33) H_3C—C=N—C_6H_5; (τ7·96) H_3C

(**50**)

(τ8·31) H_3C—C=N—C_6H_5; (τ8·81) $(CH_3)_3C$

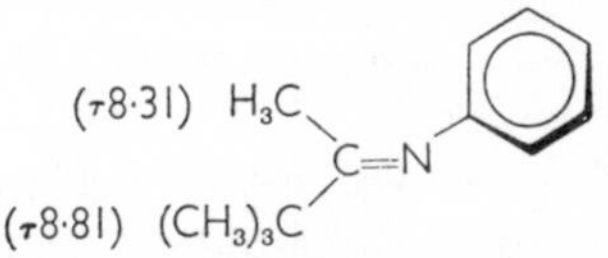

(**51**)

Compounds **50** and **51** were later synthesized and studied by Staab and coworkers[44], who obtained essentially the same chemical shift values as those shown. However, they also heated **50** in diphenyl ether as a solvent until the methyl peaks coalesced (126°) and calculated a ΔG^* of 20·3 kcal/mole for the temperature-dependent isomerization process assumed to be the reason for the collapse of the separate signals. This value is in quite good agreement with the values previously mentioned for other *N*-aryl ketimines[28,50].

Finally, it should be noted that *syn–anti* isomerism is observable in some parent ketimines. Lambert, Oliver and Roberts[53] have found evidence using n.m.r. spectroscopy that *s*-butylphenylketimine (**52**) in pentane at −60° exists as a mixture of *syn* and *anti* isomers. The equilibration process, studied by n.m.r. between −60 and −2°, is bimolecular and not unimolecular as it should be for a simple *syn–anti* isomerization. It was concluded that the process being observed probably involves a double proton exchange between two molecules (**53**).

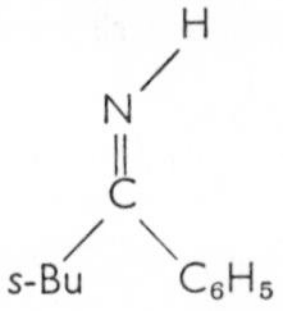

H—N=C(*s*-Bu)(C_6H_5)

(**52**)

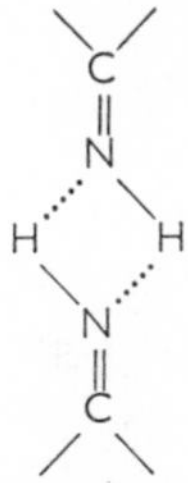

>C=N—H···N=C< (double H-bridged dimer)

(**53**)

C. Oximes and Oxime Ethers

The oximation of ketones and aldehydes is usually achieved through their reaction with hydroxylamine salts in the presence of a base. It has been known for many years that some aldehydes and unsymmetrical ketones yield two isomeric products which can sometimes be separated by fractional crystallization and are often interconvertible[3,54,55]. Although it was assumed by many that these isomers were the *cis* and *trans* or *syn* and *anti* forms of a structure containing the carbon–nitrogen double bond, this view was certainly not unanimous and could not really be proven until the advent of modern instrumentation.

It was noted soon after Beckmann's original work on the rearrangement of oximes to amides[56] that isomeric oximes give isomeric amides (equation 6). Thus the Beckmann rearrangement became the tool of

$$\mathrm{RCOR'} + \mathrm{NH_2OH} \rightleftharpoons \underset{\substack{\| \\ \mathrm{NOH}}}{\mathrm{RCR'}} \xrightarrow[\text{ether}]{\mathrm{PCl_5}} \mathrm{RNHCOR'} \quad \text{and/or} \quad \underset{\substack{\| \\ \mathrm{HON}}}{\mathrm{RCR'}} + \mathrm{H_2O} \xrightarrow[\text{ether}]{\mathrm{PCl_5}} \mathrm{R'NHCOR} \qquad (6)$$

chemists for assigning oxime configurations. It should be pointed out, however, that the assignments prior to the elegant work of Meisenheimer[57] in 1921 were wrong because the mechanism of the rearrangement was assumed to involve a *cis* exchange of groups rather than the presently accepted *trans* migration. Meisenheimer showed that benzil-*β*-monoxime (**54**) could be prepared from the ring opening of triphenylisoxazole (**55**) followed by hydrolysis of the benzoate ester and thus has the configuration shown. This monoxime gave phenylglyoxanilide (**56**) upon treatment with phosphorus pentachloride in ether which established the *trans* migration of groups.

$$\underset{(\mathbf{55})}{\text{triphenylisoxazole}} \xrightarrow{\mathrm{O_3}} \mathrm{C_6H_5COC(=N{-}OCOC_6H_5)C_6H_5} \xrightarrow[\text{ether}]{\mathrm{NaOH}}$$

$$\underset{(\mathbf{54})}{\mathrm{C_6H_5COC(=N{-}OH)C_6H_5}} \xrightarrow[\text{ether}]{\mathrm{PCl_5}} \underset{(\mathbf{56})}{\mathrm{C_6H_5NHCOCOC_6H_5}} \qquad (7)$$

The use of the Beckmann rearrangement for configurational assignments has to be viewed in light of the fact that oxime isomers are interconvertible and the isomerization is catalysed by the reagents often used to carry out the oxime to amide rearrangement; i.e. acids and bases[54]. Furthermore, the rate of the equilibration of isomers, the position of the equilibrium, and the migratory aptitude of the groups involved are all temperature dependent[41], and too often little effort has been expended in controlling the temperature of the Beckmann rearrangements, with the net result being that meaningful comparisons of experiments are difficult to come by in the older literature. For those cases where the oxime isomers are readily equilibrated under the reaction conditions, the Beckmann rearrangement often gives the same mixture of amides or the same amide for each of the two pure oxime isomers. This is true because the equilibration of isomers is faster than the rearrangement and the product composition becomes determined by the relative rates of migration of the two groups and is independent of the stereochemistry of the starting isomer[41].

With these complications in mind it is obvious that physical methods are far more desirable for confirming the existence of *syn–anti* isomerism in oximes and for assigning configurations. Dipole moment measurements have often been used for these purposes[58] but assignments resulting from such studies are little, if any, freer from ambiguities and criticisms than those from Beckmann rearrangement studies. Part of the trouble lies in the selection of the proper bond moments and bond angles and the rest is in the fact that oximes tend to exist as dimers or larger aggregates in solution.

x-Ray diffraction studies have established the non-linearity of the C=N—O group which gives rise to the isomerism in oximes, and a recent crystallographic analysis[59] of the *p*-bromobenzoate of benzil-α-monoxime shows clearly that the structure is **57**, with the *p*-bromo-

$$\begin{array}{l} C_6H_5COCC_6H_5 \\ \quad\quad\quad\;\; \| \\ \quad\quad\quad\;\; N \\ \quad\quad\quad\quad\; \diagdown \\ \quad\quad\quad\quad\;\; OCOC_6H_4Br\text{-}p \end{array}$$

(**57**)

benzoate group *trans* relative to the carbonyl group. This confirms Meisenheimer's original assignments based on his work with the β-monoxime (**54**). Jerslev[60] has determined the structure of both forms of *p*-chlorobenzaldoxime through an x-ray study, and his results also confirm earlier assumptions based on chemical reactivity of the two forms.

Differences in infrared[61] and ultraviolet[62] absorption spectra have also been used for structural assignments but spectroscopic measurements in neither of these regions have achieved the importance of nuclear magnetic resonance studies. Attesting to this is the large number of publications in this field confirming and extending the original n.m.r. work by Phillips[63] in 1958 on aldoximes and that of Lustig[64] in 1961 on ketoximes.

The now classic paper by Phillips was published at a time when the potential of n.m.r. as a tool for structure elucidation was just beginning to be appreciated by most organic chemists. He found that the n.m.r. spectra of several aliphatic aldoximes showed two multiplets, separated by 0·6 p.p.m., which were assignable to the aldehydic hydrogen atoms. The existence of two separate absorptions could be explained by the simultaneous existence of *syn* and *anti* isomers. Reasoning that the

$$\underset{syn}{\mathrm{R(H)C{=}N{-}OH}} \rightleftharpoons \underset{anti}{\mathrm{R(H)C{=}N{-}OH}} \tag{8}$$

proximity of oxygen induces a paramagnetic shift on the aldehyde proton in the *syn* form, he assigned the multiplet at lower field to the *syn* form. Area ratios allowed a determination of the equilibrium concentration of isomers of the aliphatic aldoximes. This previously could not be determined by classical means since the equilibria sometimes strongly favoured one form or rapid interconversion to one isomer occurred during crystallization. This, then, provided the first efficient determination of isomer ratios without disturbing the equilibrium.

In order to check Phillips' assignment of the low field —CH=NOH multiplet to the *syn* isomer, Lustig[64] recorded the spectra of the two *p*-chlorobenzaldoximes whose structures were known from x-ray studies[60]. Indeed, it was found that the aldehydic proton in the *anti* oxime lies at higher field than that of the *syn* from by about 0·7 p.p.m.

The location of the aldehyde proton absorption in some derivatives of *syn*- and *anti*-benzaldoxime and the effect of *m*- and *p*-substituents on the position of absorption has recently been described[65]. Nine pairs of isomeric benzaldoximes were examined and some of the results are shown in Table 10. It was noted that there is a good correlation between the chemical shift of the aldehyde proton and the Hammett σ-value of the substituent, R. The points for the *anti* isomers determined one line while the points for the *syn* isomers determined a different line.

Isomerism of the *syn–anti* variety was detected in several aliphatic ketoximes and ketoxime ethers by an n.m.r. method for the first time in 1961[64]. The isomerism was revealed by the fact that the resonances of protons on carbon atoms adjacent to the >C=NOH group appeared in the spectrum twice, with equal intensity. The separations, however, were as much as thirty times smaller than those observed by Phillips and others for the aldehyde proton in aldoxime isomers, and usually non-existent unless aromatic solvents were used or aromatic rings were part of the oxime molecule. For acetone oxime itself the largest separation was observed in benzene and it was only about 5 Hz.

TABLE 10. Chemical shifts of the aldehyde hydrogen in some *m*- and *p*-substituted benzaldoximes.[65]

Substituent (R)	Chemical shift[a] *syn*	*anti*
p-CH_3	2·00	2·82
p-NO_2	1·78	2·44
p-OCH_3	1·94	2·78
p-Cl	1·97	2·71
m-NO_2	1·74	2·48
m-Cl	1·92	2·57

[a] τ values for 6–15% solutions in tetrahydrofuran.

Karabatsos and coworkers[66,67] found that for oximes and oxime ethers of the general formula **58**, α-methyl hydrogens when *cis* to the —OX group usually resonate at higher field than when *trans* although

R(CH$_\beta$—CH$_\alpha$)C=N~OX

X = H, CH_3

(**58**)

NOH ‖ C, R, H, H_3C, CH_3

(**59**)

solvent changes may reverse this. α-Methylene and α-methine protons resonate at lower fields when *cis* than when *trans* with the difference being quite small for the methylene protons. This strange behaviour is explicable from considerations of preferred conformations and the geometrical dependence of the deshielding effect on each side of the oximido group[68]. Protons of freely rotating α-methyl groups and the relatively freely rotating α-methylene groups pass through different maxima and minima in deshielding when located on opposite sides of the oximido group, but the time–average chemical shift apparently becomes almost identical. On the other hand, the preferred conformation for an isopropyl group, for example, places the α-methine hydrogen in or near the C=NOX plane (**59**), thus making the behaviour of such protons much like that of aldehydic protons.

The geometry of the magnetic anisotropy associated with the oximido group is not known with certainty although there have been several recent studies on this subject. Saito and coworkers[69–71] feel that the main deshielding effect on the α-hydrogens arises from the proximity of the unshared pair of electrons on nitrogen, while Huitric and coworkers[68,72] have presented evidence for the greater effect being due to the proximity of the hydroxyl group of the oximido group.

Another n.m.r. correlation which is potentially useful for configuration assignments involves the chemical shift difference in the hydroxyl proton resonance of oxime isomers. Kleinspehn, Jung and Studniarz[73] studied 60 oximes in dimethyl sulphoxide and found that the —O$\underline{H}$ signals were fairly insensitive to variables such as concentration and temperature. For simple aldoximes the —O$\underline{H}$ signals for the *syn* isomers (—O$\underline{H}$ and =CH *cis*) ranged from 10·25 to 10·31 p.p.m. downfield from the TMS standard and the corresponding *anti* isomers were in the range of 10·60 to 10·68 p.p.m. The assignments made with the —O$\underline{H}$ resonance positions are consistent with those which could be made by noting the =C$\underline{H}$ absorption positions. For *syn*-methyl ketoximes the —O$\underline{H}$ range was 10·12 to 10·21 p.p.m. and

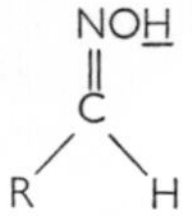

NO$\underline{H}$ (=C, R, H) 10·25–10·31 p.p.m.
(**60**)

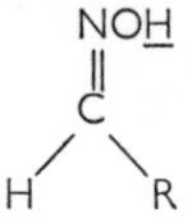

NO$\underline{H}$ (=C, H, R) 10·60–10·68 p.p.m.
(**61**)

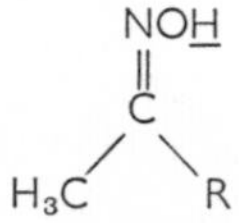

NO$\underline{H}$ (=C, H_3C, R) 10·05–10·08 p.p.m.
(**62**)

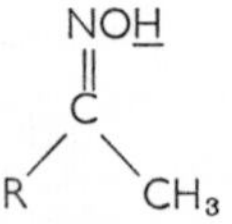

NO$\underline{H}$ (=C, R, CH_3) 10·12–10·21 p.p.m.
(**63**)

for the *anti* isomers, 10·05 to 10·08 p.p.m. downfield from TMS. Finally, of interest in those cases where only one isomer is obtainable, $\delta_{-OH}-\delta_{=CH}$ is approximately 3 p.p.m. for many *syn* aldoximes and about 4 p.p.m. for the corresponding *anti* aldoximes[73].

In addition to the aforementioned applications, these various empirical n.m.r. correlations, starting with the work of Lustig[64] and Phillips[63] have been used in recent years to assign configurations to α,β-unsaturated oximes (**64**[74], **65**[75], **66**[76]), to oximes of possible chemotherapeutic activity (**67**)[77], to the isomers of α-oximinoaryl acetonitrile (**68**)[78], and to the oxime ethers of benzaldehyde[79] and substituted benzaldehydes[80].

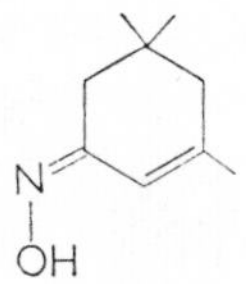

(**64**) 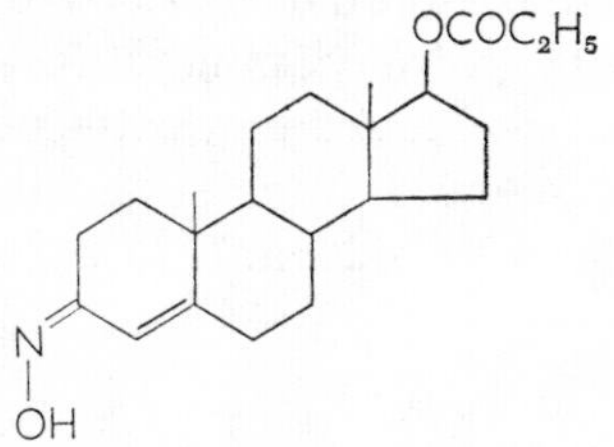 (**65**) 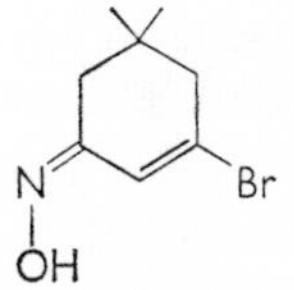 (**66**)

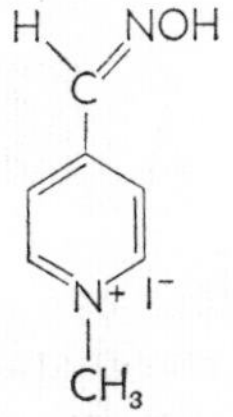

(**67**) 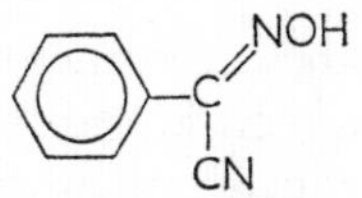 (**68**)

Much work on the isomerization of oximes or the equilibration of oxime isomers has been reported in the literature. As has been previously mentioned, isomerization sometimes accompanies the Beckmann rearrangement, especially when acid catalysts are employed. The relative stability of the isomers of an oxime depends upon steric as well as electronic effects[41]. The importance of steric factors is shown by the fact that *syn-t*-butyl phenyl ketoxime (**69**) isomerizes faster than *syn*-isopropyl phenyl ketoxime (**70**) under comparable acidic con-

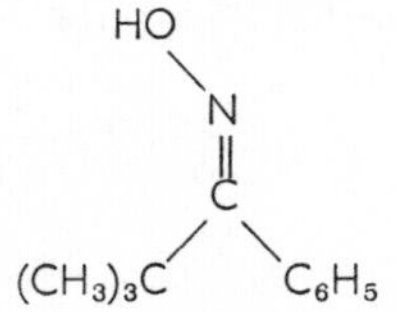

(**69**) 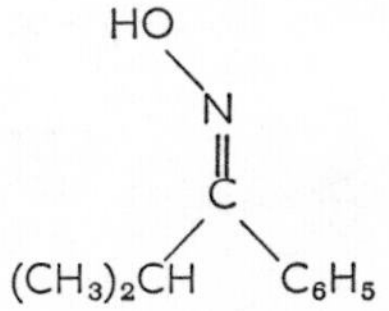 (**70**)

ditions[81]. Electrostatic effects are important in determining the relative stabilities of some *p*-substituted benzophenone oxime isomers in acidic media, and both electrostatic and steric effects must be considered when discussing relative stabilities of isomers of *o*-substituted benzophenone oximes[41].

It is well known that the favoured form of aromatic aldoximes is the *syn* (H and OH *cis*) configuration. These *syn* isomers, however, are readily converted to the *anti* isomers through the hydrochloride salt[3]. For example, 2,6-dimethyl-4-bromobenzaldehyde yields the *syn* oxime (**71**) upon reaction with hydroxylamine[80]. Rearrangement of **71** to the *anti* isomer (**72**) can be effected by treatment of **71** with hydrogen chloride in dry ether followed by neutralization with sodium carbonate. The conversion of *syn* aldoximes to *anti* aldoximes can also be

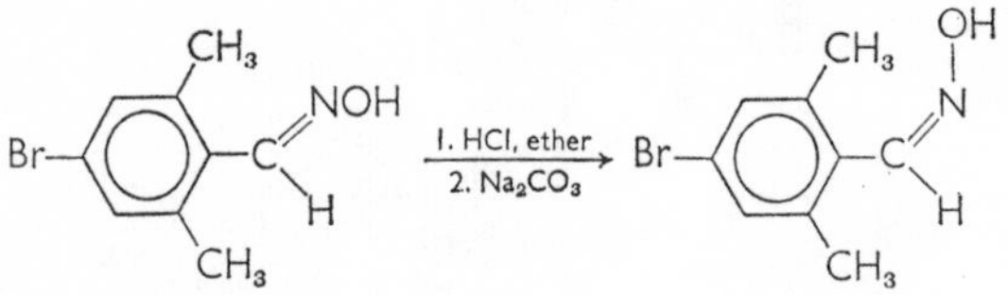

$$\text{(71)} \xrightarrow[\text{2. } Na_2CO_3]{\text{1. HCl, ether}} \text{(72)} \qquad (9)$$

(**71**) (**72**)

effected with protic acids other than hydrochloric and also with Lewis acids such as boron trifluoride[82].

The photochemical isomerization of oximes was first reported by Ciamician and Silber[83] in 1903 and has been frequently used since then as a means of obtaining the isomer not favoured under normal synthetic conditions. For example, it provides perhaps the simplest means of converting *syn*-isonicotinaldehyde oxime (**73**) to its *anti* analogue (**74**)[84]. In addition to leading to isomerization, irradiation

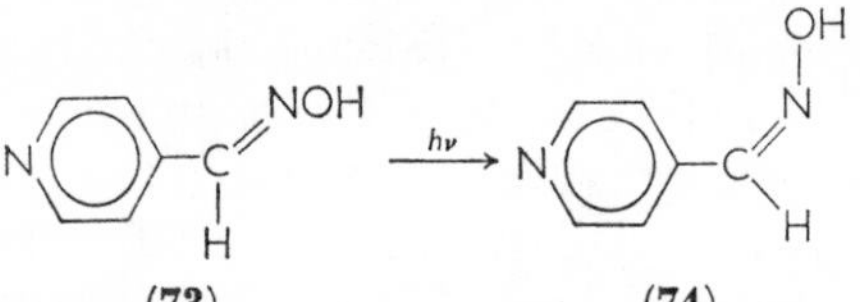

$$\text{(73)} \xrightarrow{h\nu} \text{(74)} \qquad (10)$$

(**73**) (**74**)

may also yield amides; *syn*-benzaldoxime (**75**), for example, gives benzanilide[85].

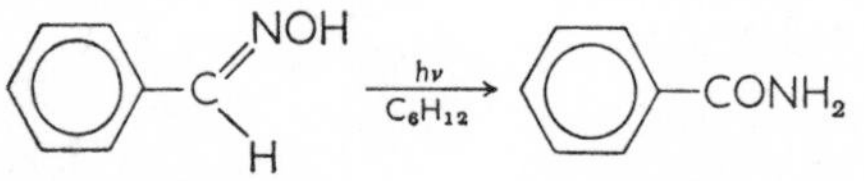

$$\text{(75)} \xrightarrow[C_6H_{12}]{h\nu} C_6H_5\text{—}CONH_2 \qquad (11)$$

(**75**) (**76**)

The uncatalysed, thermal isomerization of oximes has often been noted but, unfortunately, infrequently studied quantitatively. Among recent examples are the thermal conversion of **77** to **78**, a process which can be conveniently followed in inert solvents by n.m.r.[86], and the interconversion of *p*-benzoquinone monoximes (**79**, **80**), which also can be followed by n.m.r. and which probably involves the *p*-nitrosophenol (**81**) as shown[87–89]. Thermal isomerization of *anti*-benzaldoxime (**82**) during *O*-alkylation experiments probably explains the

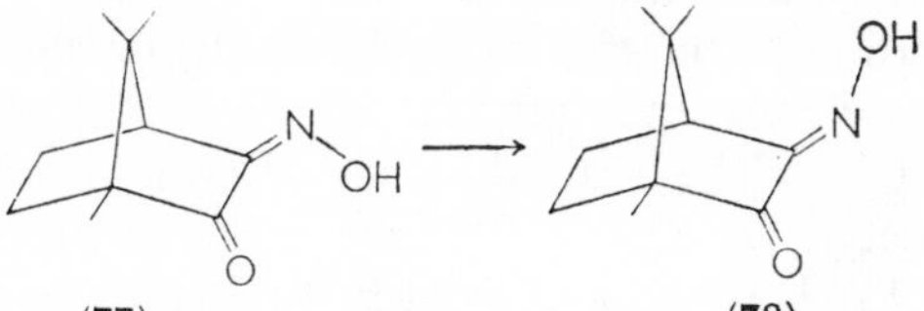

(**77**) → (**78**) (12)

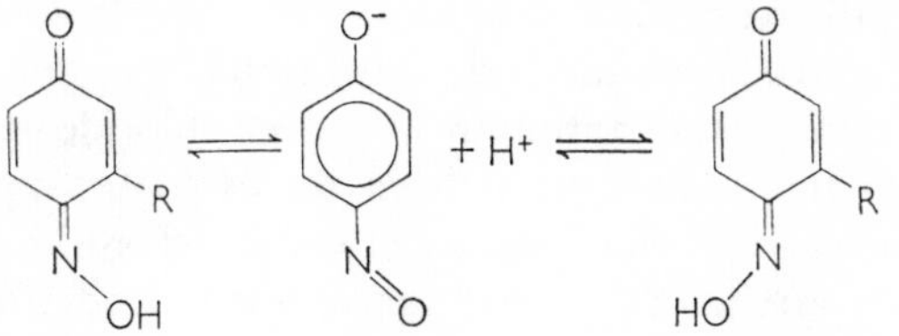

(**79**) ⇌ (**81**) + H⁺ ⇌ (**80**) (13)

mixture of oxime ethers noted by Buehler[79]. However, in this last case, a complex between silver ion and the oximes could be involved.

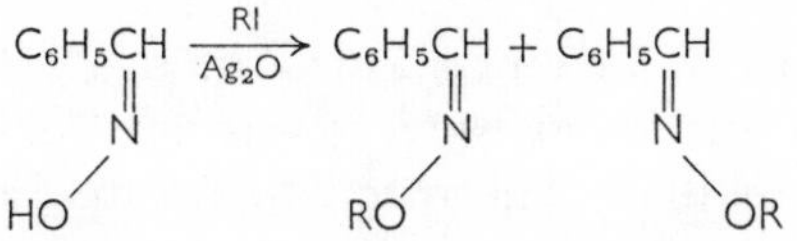

$$C_6H_5CH{=}N{-}OH \xrightarrow[Ag_2O]{RI} C_6H_5CH{=}N{-}OR + C_6H_5CH{=}N{-}OR \quad (14)$$

(**82**)

One of the first kinetic studies of the thermal *syn–anti* conversions of oxime isomers was carried out by Le Fevre and Northcott[90] on piperonaldoxime (**83**). The isomerization of the labile isomer (m.p. 144°) to the stable isomer (m.p. 112°) was followed at two temperatures in seven solvents. Energies of activation ranged from 26·6 kcal/mole in cyclohexanone to 22·6 kcal/mole in carbon tetrachloride and the rate constants could be roughly correlated with the dielectric constants of the solvents.

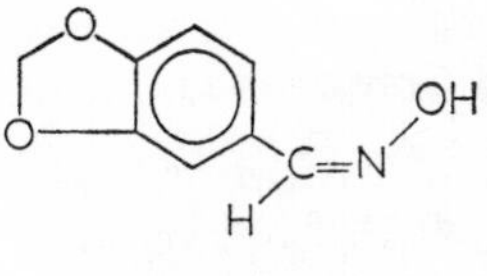

(**83**) 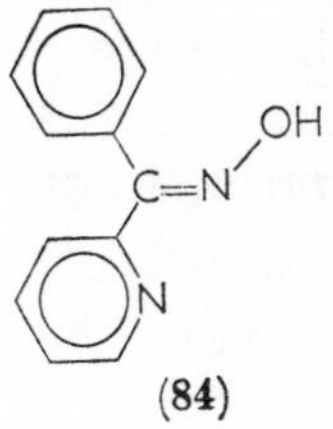 (**84**)

Vassian and Murmann[91] have studied the thermal *syn–anti* isomerization of phenyl-2-pyridyl ketoxime (**84**). Less than 1% rearrangement occurred on boiling of geometrical isomers for one hour in methyl or ethyl alcohol, water, chloroform or acetone. In the molten state at 175°, however, the two forms showed first-order behaviour for at least two half-lives and gave a value of $4{\cdot}22 \times 10^{-3}\ s^{-1}$ for k_{obs}. At 144° in cyclohexanol the rate of isomerization of the *syn* isomer (*syn*-phenyl) was $3{\cdot}4 \times 10^{-4}\ s^{-1}$ and that for the *anti* isomer was $1{\cdot}7 \times 10^{-4}\ s^{-1}$. Deviations from first-order behaviour suggested to the authors a complex mechanism possibly involving the solvent or impurities in the oximes.

Oxime ethers, like oximes, are susceptible to photochemical isomerization and to isomerization by hydrogen chloride in chloroform[92]. On the other hand, they have been found to be very resistant to thermal isomerization; the *β* isomer of *O*-methyl *m*-nitrobenzaldoxime being stable in the melt and remaining unchanged after 16 years at ambient temperatures[92]. The *O*-methyl ethers of *p*-chlorobenzophenone oxime (**85, 86**) were prepared by Curtin, Grubbs and McCarty[28]. The isomers were obtained in pure form and the structural assignments could be made with the infrared correlation mentioned earlier[28]. Ultraviolet and infrared spectra of the two forms were different enough to be used for a kinetic study. Very little change in the spectra of **86** was found when it was held for 528 h at 200° in degassed decane solution. The reaction was too slow to permit the calculation of a reliable rate constant but it seemed justified to set $10^{-7}\ s^{-1}$ as a maximum possible value at that temperature. At

$$(p\text{-ClC}_6H_4)(C_6H_5)C{=}N\text{-}OCH_3\ (\mathbf{85}) \rightleftharpoons (p\text{-ClC}_6H_4)(C_6H_5)C{=}N\text{-}OCH_3\ (\mathbf{86}) \qquad (15)$$

least it seems clear that oxime ethers are less readily isomerized thermally than are oximes and that oximes are much more stable in a configurational sense than Schiff bases, where the atom on nitrogen is carbon rather than oxygen.

D. Hydrazones, Azines and Related Compounds

Phenylhydrazine and 2,4-dinitrophenylhydrazine have been widely used as reagents for characterizing aldehydes and ketones[93] in spite of the variations in melting points which can be found in the literature for the resulting derivatives. The origin of these melting point discrepancies has usually been attributed to the existence of hydrazone isomers and in recent years the question of stereoisomerism of hydrazones and substituted hydrazones has received considerable attention.

Reports of the isolation of the two stereoisomers of a phenylhydrazone or 2,4-dinitrophenylhydrazone (DNP) are quite numerous[94–100]. Usually, however, these are cases where one of the isomers is intramolecularly hydrogen bonded. There have been fewer cases of the isolation of stereoisomers of simple DNP's[101–106]. Once again one must be very critical in reading the older reports since polymorphism, undoubtedly, has often been the reason for the isolation of two forms. The complexity of the situation can be seen from just reviewing the literature on the DNP of acetaldehyde[103]. Various investigators have isolated forms melting at 146° and 162°[104]; 168·5°, 156–157° and 149°[105]; 167–168° and 93–94°[106]. The confusion remained until the application of modern physical methods to the problem within the last decade.

Configurational assignments have been made by Ramirez and Kirby[107] using differences noted in the ultraviolet and infrared spectra of the two forms of several DNP's. The basis for their method was the observation of differences in the N—H stretching vibration of various related compounds having substituents on the α-carbon which could be hydrogen bonded to the N—H. The form of α-methoxypropiophenone DNP with the lower frequency and broader N—H band was assigned the *syn* configuration (**87**). Furthermore, the ultraviolet spectra of the two isomeric forms, **87** and **88**, were different with

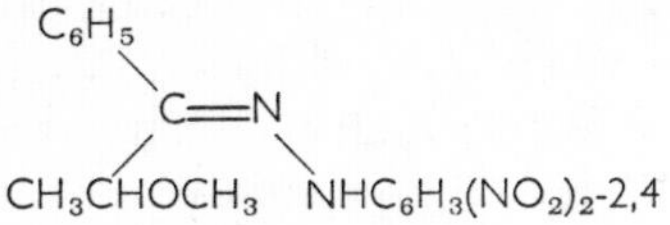

$C_6H_5(CH_3CHOCH_3)C{=}N{-}NHC_6H_3(NO_2)_2$-2,4

(**87**)

$C_6H_5(CH_3CHOCH_3)C{=}N{-}NHC_6H_3(NO_2)_2$-2,4

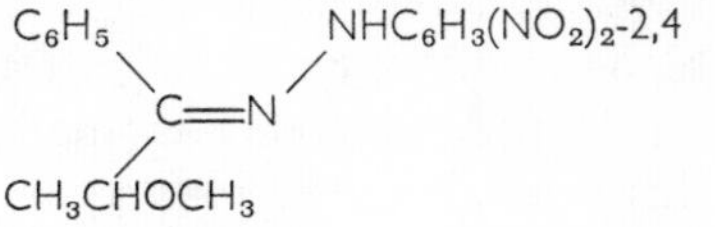

(**88**)

λ_{max} for **87** being about 21 mμ higher than that for **88**. The spectrum of the only isolable form of the parent propiophenone DNP most closely resembled that of **87** and thus was assigned the same *syn* configuration. This general type of argument was extended to many other DNP's.

Silverstein and Shoolery[108] were among the first to apply n.m.r. techniques to studies of DNP's. They noted differences in the aromatic hydrogen resonances of the two DNP's of ethyl benzoylacetate and explained the results in terms of a coplanar *anti* form (*trans* aryls, **89**) and a non-coplanar or sterically hindered *syn* form (**90**).

C_6H_5—C(CH_2CO_2Et)=N—$NHC_6H_3(NO_2)_2$-2,4

(89)

C_6H_5—C(CH_2CO_2Et)=N—$NHC_6H_3(NO_2)_2$-2,4

(90)

From consideration of the n.m.r. spectra of many ring-substituted phenylhydrazones, thiosemicarbazones and semicarbazones of aldehydes and ketones, Karabatsos and coworkers[66,103,109–111] have been able to draw the following conclusions pertinent to structural and configurational assignments: (a) All compounds neat or in solution exist in the imine form (**91**) with no azo (**92**) or enamine (**93**) forms detectable. (b) H_1 (in derivatives of aldehydes where $R^1 = H$, in **94**)

$(R^1CH_2)(R^2)C{=}NNHX$

(91)

$(R^1CH_2)(R^2)CHN{=}NX$

(92)

$(R^1CH{=})(R^2)C{-}NHNHX$

(93)

$(R^1(H_1))(R^2{-}CH_\beta{-}CH_\alpha)C{=}N{\sim}NHX$

(94)

resonates at lower magnetic field by 30–40 Hz when *cis* to —NHX than when *trans* regardless of solvent. (c) $H_\alpha(CH_3)$ generally resonates at higher field when *cis* to —NHX than when *trans*, although in some solvents the difference is small and in a few acetaldehyde derivatives this generalization is reversed. (d) $H_\alpha(CH_2)$ also generally resonates at higher field when *cis* to —NHX than when *trans* and the difference is around 20 Hz. (f) The position of H_β is very sensitive to both —NHX and the solvent and does not allow any generalizations to be made.

The differences, $\Delta\nu$ $(\nu_{cis} - \nu_{trans})$, for H_α(CH) and H_1 are large enough and consistent enough to be used with confidence for configurational assignments. Also reliable is the difference in the effect of aromatic solvents on the resonance positions of *cis* and *trans* hydrogens. Both are shifted upfield relative to their positions in aliphatic solvents, but the upfield shift of *cis* hydrogens is generally two to six times larger than that of the corresponding *trans* hydrogens. A specific hydrogen-bonded complex similar to **95** accommodates all of the data observed by Karabatsos and coworkers[109] for this effect of aromatic solvents.

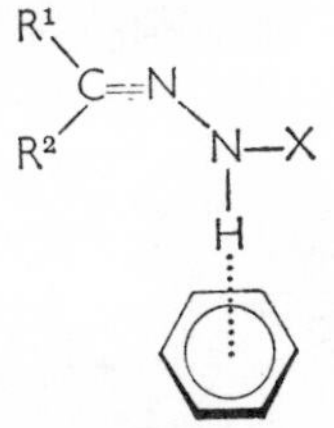

$$R^1R^2C{=}N{-}NH{-}X \cdots C_6H_6$$

(95)

Arbuzov, Samitov and Kitaev[112] have observed similar differences ($\nu_{cis} - \nu_{trans}$) in chemical shifts of protons of phenylhydrazones of some ketones and aldehydes. However, they do not entirely agree with Karabatsos on the analysis of the anisotropic effects of the benzene ring on the *cis* and *trans* hydrogens. An infrared study by Hadzi and Jan[113] has revealed a consistent difference of about 30 cm^{-1} between the —N—H stretching vibration in the spectra of *syn* and *anti* phenylhydrazones of aldehydes. The band in the spectrum of the *syn* isomers (aldehydic H and NHAr *cis*) is always the lower of the two and usually in the range of 3340–3350 cm^{-1} (0·02M in $CHCl_3$).

During their studies of phenylhydrazones and related derivatives of aldehydes and ketones, Karabatsos and coworkers learned much about the configurational stability of such compounds. The reaction of aliphatic aldehydes with 2,4-dinitrophenylhydrazine is apparently kinetically controlled, leading to the *syn* DNP. The rate of *syn–anti* isomerization is quite slow at room temperature in the absence of acid. Acid-free acetaldehyde DNP does not reach equilibrium after standing at room temperature for 10 days, whereas a trace of sulphuric acid effects the equilibration in less than a day[103]. The final *syn–anti* ratio is always found to favour the *syn* isomer but is quite solvent dependent. The corresponding ketone derivatives are often obtained as single isomers or as a mixture of isomers but isomerize in solution, in the absence of acid, faster than the aldehyde derivatives. The *syn* isomer of phenylacetone phenylhydrazone (**96**), for example, equilibrates with the *anti* isomer within a day at room temperature[103].

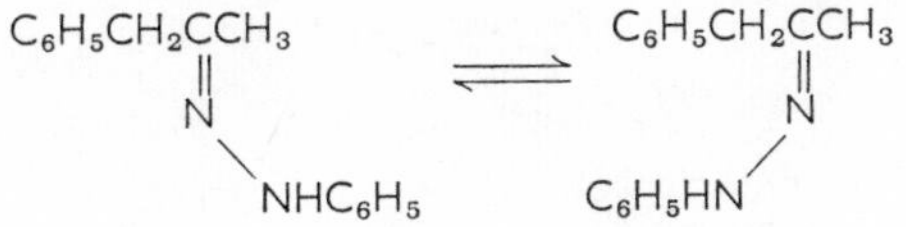

$$\underset{\textbf{(96)}}{C_6H_5CH_2C(CH_3){=}N{-}NHC_6H_5} \rightleftharpoons \underset{\textbf{(97)}}{C_6H_5CH_2C(CH_3){=}N{-}NHC_6H_5} \qquad (16)$$

Unlike derivatives where X is NHY in **98**, the compounds which have X being $N(CH_3)_2$ or $N(CH_3)C_6H_5$ exist only as a single isomer (*cis* H and X) that does not isomerize upon heating or adding acid[114,115]. Consideration of preferred conformations offers the

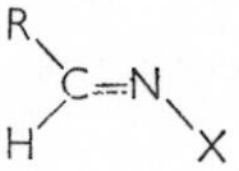

(**98**) 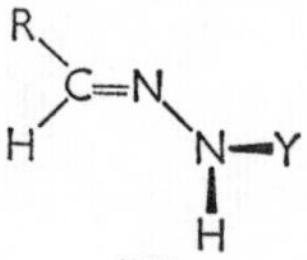 (**99**) 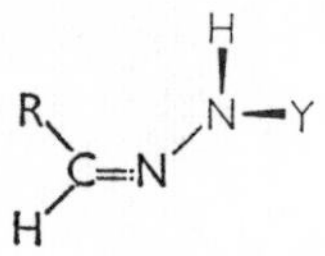 (**100**)

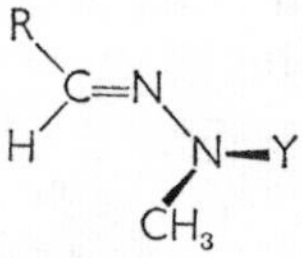

(**101**) 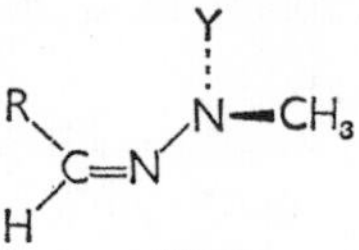 (**102**)

answer to this according to Karabatsos. When X is NHY both *syn* and *anti* isomers (**99, 100**) can assume conformations which allow for considerable overlap of the unshared pair of electrons on nitrogen with the π-electron system. When X is $N(CH_3)_2$ or $N(CH_3)C_6H_5$ this is possible only for the *syn* isomer (**101**) and thus it is heavily favoured even to the essential exclusion of the *anti* isomer.

Rates of *syn–anti* isomerizations of DNP's have occasionally been recorded. In their n.m.r. study of the DNP's of ethyl benzoylacetate,

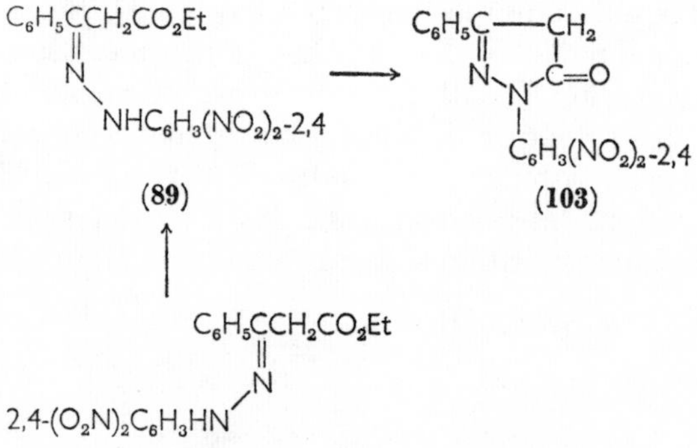

$C_6H_5C(=N-NHC_6H_3(NO_2)_2\text{-}2,4)CH_2CO_2Et$ (**89**) ⟶ C_6H_5C—CH_2 / N–N(C=O) / $C_6H_3(NO_2)_2\text{-}2,4$ (**103**) (17)

↑

$C_6H_5C(=N-NHC_6H_3(NO_2)_2\text{-}2,4)CH_2CO_2Et$ [$2,4\text{-}(O_2N)_2C_6H_3HN$] (**90**)

Silverstein and Shoolery[108] found that the rate of ethoxide-catalysed cyclization of the *syn* isomer (**90**) to give the pyrazolone (**103**) was identical to that for the *anti* isomer (**89**) ($9{\cdot}40 \times 10^{-3}$ s^{-1} at 25°). The unstable *syn* isomer must isomerize prior to cyclization. The clean, first-order reaction in both cases and the identity of rate constants showed the isomerization to be much faster than the cyclization under these conditions.

Hegarty and Scott[116] found that when eight alkylidene DNP's (**104**, R = alkyl) were brominated in 70% acetic acid containing potassium bromide, the process was independent of the concentration of bromine used. For the compound in which R = $CH(CH_3)_2$, the rate constant for bromination ($1{\cdot}6 \times 10^{-3}$ s^{-1}) was the same as for

R(H)C=N–$NHC_6H_3(NO_2)_2$-2,4 (**104**) ⇌ R(H)C=N–$NHC_6H_3(NO_2)_2$-2,4 (**105**) —Br_2→ R(Br)C=N–$NHC_6H_3(NO_2)_2$-2,4 (**106**) (18)

chlorination and did not vary when the bromine/tribromide ratio in solution was varied over a wide range. For the eight alkylidene DNP's they studied the rate constants were excellently correlated by the Taft equation[117] with $\delta = +0{\cdot}49$. This indicates the reaction being measured is sensitive only to the size of the group R, so it was concluded that the data are most consistent with a rate-determining conversion of the *syn* isomer (**104**, initially present in excess) to the more reactive *anti* isomer (**105**).

When considering possible hindrance to free rotation about the =N—N bond in dimethylhydrazones, an analogy can be drawn with the hindered rotation about the C—N bond in amides. In compound **111** Mannschreck and Koelle[118] noted no separation of the resonance peak for the methyl hydrogens at temperatures down to −65°, indicating a low barrier to rotation or, in other words, a small contribution of resonance form **110**.

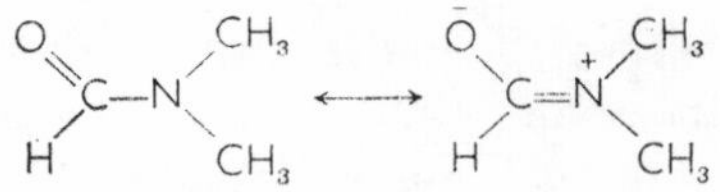

(107) (108)

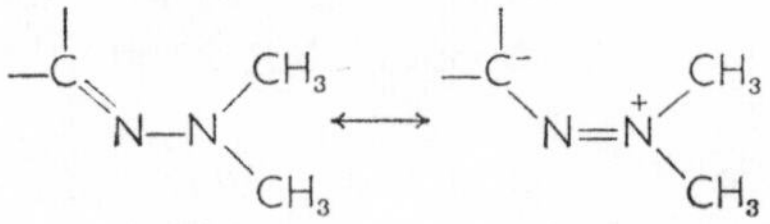

(109) (110) (19)

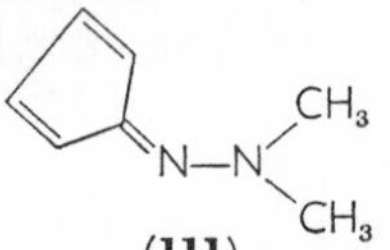

(111)

On the other hand, the ring protons of **111** remained an *ABCD* pattern up to 150°, which means that ΔG^* for *syn–anti* isomerization in this system is greater than 22 kcal/mole.

There are three azine isomers (**112–114**) theoretically possible from the reaction between hydrazine and an aldehyde or unsymmetrical ketone. Fleming and Mason[119] were the first to report the isolation and characterization of all three isomers of one azine (**112–114**,

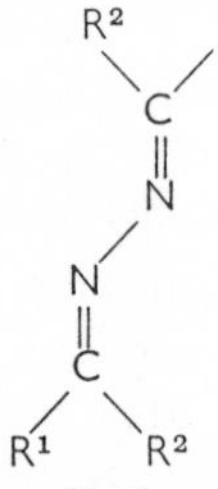

(112) 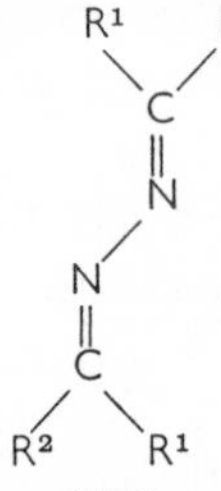 (113) 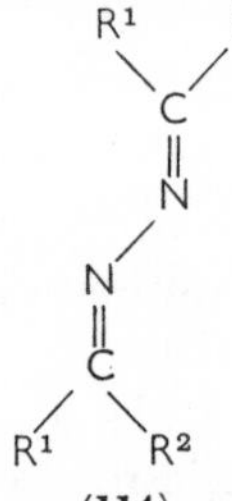 (114)

R^1 = *o*-$O_2NC_6H_4$, R^2 = CH_3). Differences in the ultraviolet and n.m.r. spectra allowed reliable assignments to be made. Isomer **112** was the most abundant and had the strongest ultraviolet absorption, consistent with the idea that it is the only isomer which involves little steric strain in approaching coplanarity. Isomer **112** is the only form observed when R^2 = H and R^1 = alkyl[120].

The various forms of the azine studied by Fleming and Mason were interconvertible to a certain extent upon irradiation of dioxan solutions of the pure isomers with ultraviolet light[119]. Thus far, however, compounds **115** and **116** have not yielded to variable temperature n.m.r. techniques; the methyl proton resonances remaining uncoalesced at 170°[44].

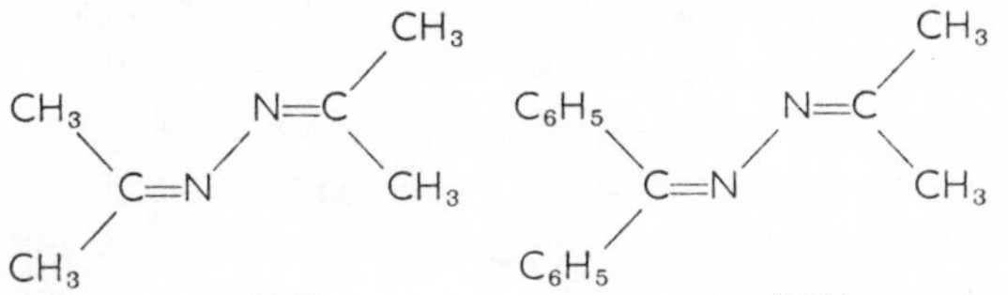

$$(CH_3)_2C{=}N{-}N{=}C(CH_3)_2 \quad \textbf{(115)} \qquad (C_6H_5)_2C{=}N{-}N{=}C(CH_3)_2 \quad \textbf{(116)}$$

E. N-Halimines

The *N*-chlorimines of substituted benzaldehydes (**117**) have been prepared and studied by Hauser and coworkers[121]. These were unstable compounds which spontaneously decomposed at varying rates into hydrogen chloride and a nitrile. Fractional crystallization yielded no indication of isomers of these *N*-chlorimines.

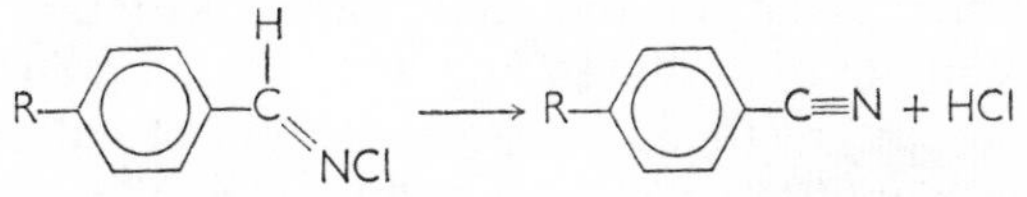

$$R{-}C_6H_4{-}CH{=}NCl \ \textbf{(117)} \longrightarrow R{-}C_6H_4{-}C{\equiv}N + HCl \ \textbf{(118)} \qquad (20)$$

On the other hand, the *N*-halimines of ketones (**120**) are fairly stable under ordinary conditions and the stereoisomers, where possible, have frequently been separated. Much of the early interest in these compounds was due to the fact that they had been proposed as intermediates in the Beckmann rearrangements of oximes when carried out with phosphorus pentachloride (equation 21)[122]. This was later

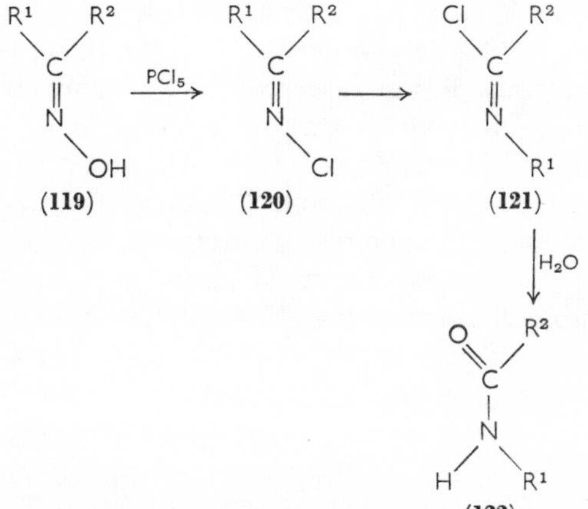

$$R^1R^2C{=}N{-}OH \ \textbf{(119)} \xrightarrow{PCl_5} R^1R^2C{=}N{-}Cl \ \textbf{(120)} \longrightarrow ClR^2C{=}N{-}R^1 \ \textbf{(121)} \xrightarrow{H_2O} O{=}C(R^2){-}NH{-}R^1 \ \textbf{(122)} \qquad (21)$$

proven not to be the case[123,124], but Beckmann type rearrangements of *N*-halimines can be effected by metal ion catalysts[125,126].

Peterson[123] separated the stereoisomers of several substituted benzophenone *N*-chlorimines and observed that each form could be recovered unchanged from solution, melt or vapour phase even in the presence of a few added crystals of the other form. Treatment of some of the isomers with dry chlorine or heating up to three hours at 100° were without effect on the melting points of the pure forms.

Curtin, Grubbs and McCarty[28] prepared and separated the two *N*-chlorimines of *p*-chlorobenzophenone (**123, 124**). A configurational

$$\underset{\textbf{(123)}}{(C_6H_5)(p\text{-}ClC_6H_4)C{=}N\text{—}Cl} \rightleftharpoons \underset{\textbf{(124)}}{(C_6H_5)(p\text{-}ClC_6H_4)C{=}N\text{—}Cl} \quad (22)$$

assignment based on infrared and ultraviolet spectral data was the opposite of that given in the older literature[127]. The thermal, uncatalysed equilibration of **123** and **124** was followed by an ultraviolet spectroscopic method. No isomerization of either isomer was detected in runs at 60° in cyclohexane for 500 h. The assumption of a reasonable error in the analytical method gave a value of $3{\cdot}2 \times 10^{-8}$ s^{-1} as the maximum limit on the rate of approach to equilibrium at this temperature. Because the *N*-chlorimines halogenate cyclohexane at temperatures above 60°[128] it was necessary to switch to other solvents for kinetic studies at higher temperatures. The approach of each isomer towards equilibrium was noted in several 60 h runs at 120° and 140° using benzene as solvent. The results indicated a half-life of about 500 h at 120° and 80 h at 140°. It seems quite likely, though, that homolysis of the N—Cl bond was occurring at these higher temperatures so the values probably do not represent rates of pure, thermal *syn–anti* isomerization. Similar results were obtained for the *N*-bromimines of *p*-chlorobenzophenone. However, these were very sensitive to traces of acid and underwent N—Br bond homolysis at lower temperatures.

Recent interest in the chemistry of tetrafluorohydrazine has led to the isolation of *syn* and *anti* isomers of numerous *N*-fluorimines. The reaction of tribromofluoromethane (**125**) and tetrafluorohydrazine (**126**)

$$CFBr_3 + N_2F_4 \xrightarrow{h\nu} (Br)(F)C{=}N\text{—}F + \underset{\textbf{(128)}}{(Br)(F)C{=}N\text{—}F} \quad (23)$$

initiated by ultraviolet irradiation produced the *syn* and *anti* isomers of *C*-bromodifluoromethyleneimine (**127, 128**) among other products[129]. The isomers were readily separated in pure form by gas chromatography. Configurational assignments were made by considering the differences in the ^{19}F coupling constants and drawing an analogy with the differences observed in *cis* and *trans* fluorinated olefins. There was no evidence for isomerization on the chromatography columns or upon standing in glass bulbs for several months. Irradiation of the *syn* isomer (**127**) gave measurable amounts of the *anti* (**128**) along with trifluoromethyleneimine, but there was no evidence for generation of the *syn* from the *anti*.

Reactions of tetrafluorohydrazine with olefins, acetylenes and allene have led to interesting products which often have been obtained as a mixture of *syn* and *anti* forms as evidenced by ^{19}F n.m.r. spectra[130–132]. Some of the products are shown in equations (24–29). As can be seen from equations (24), (28) and (29), sodium fluoride has been found to

$$CH_2{=}CHR \xrightarrow[2.\ NaF]{1.\ N_2F_4} \underset{\overset{\|}{FN}}{NCCR} + \underset{\overset{\|}{NF}}{NCCR} \qquad R = F,\ CH_3,\ CH_2OCOCH_3 \tag{24}$$

$$NCC{\equiv}CCN \xrightarrow[2.\ \Delta]{1.\ N_2F_4} NC\underset{\underset{NF}{\|}}{C}{-}\underset{\underset{NF_2}{|}}{C}FCN \tag{25}$$

$$F_3CC{\equiv}CC_4F_9 \xrightarrow[2.\ \Delta]{1.\ N_2F_4} F_3C\underset{\underset{NF}{\|}}{C}{-}\underset{\underset{NF_2}{|}}{C}FC_4F_9 + F_3C\underset{\underset{NF_2}{|}}{C}F{-}\underset{\underset{NF}{\|}}{C}C_4F_9 \tag{26}$$

$$CH_2{=}C{=}CH_2 + N_2F_4 \longrightarrow \underset{(\mathbf{130})}{CH_2F\underset{\underset{NF}{\|}}{C}CH_2NF_2} + \underset{(\mathbf{131})}{CH_2F\underset{\underset{FN}{\|}}{C}CH_2NF_2} \tag{27}$$

$$\mathbf{130} + NaF \xrightarrow{150°} CH_2F\underset{\underset{NF}{\|}}{C}CN \tag{28}$$

$$\mathbf{131} + NaF \xrightarrow{100°} CH_2F\underset{\underset{FN}{\|}}{C}CN \tag{29}$$

be a strong enough base to effect the elimination of hydrogen fluoride from some difluoroamino-*N*-fluorimino compounds to give the corresponding *N*-fluorimino nitriles. The stereospecificity indicated in the last two reactions gives some indication of the configurational stability of the *N*-fluorimines.

Dehydrofluorination of the adduct from N_2F_4 and *trans* stilbene leads to a mixture of all three of the possible *syn* and *anti* isomers (**132–134**)[133]. Similarly, the adduct from 2-butene and N_2F_4 yields **135–137**[134]. Irradiation of solutions of the pure isomers **135–137** with an

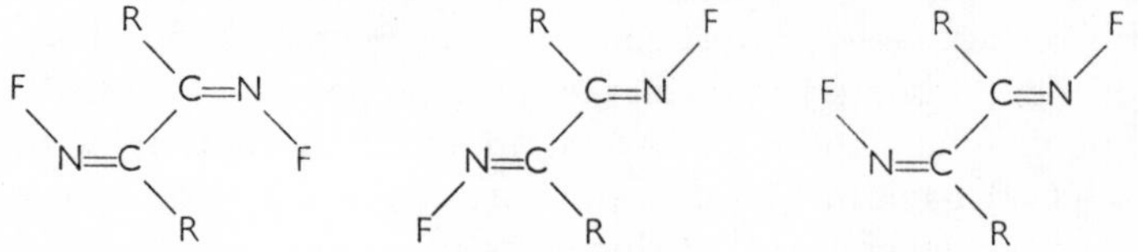

132–134, R = C_6H_5

135–137, R = CH_3

ultraviolet lamp resulted in production of the other two forms after only a few hours.

Related to the *N*-halimines of ketones are the chlorimido acid esters (**138**) studied by Stieglitz[135] and Hilpert[136]. Isomers presumed to be stereoisomers were separated by fractional crystallization. Heating the pure forms at 80° for 1 h or 140° for 5 min did not bring

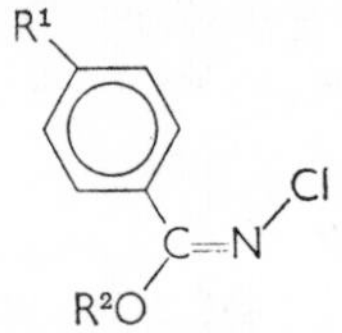

(**138**) 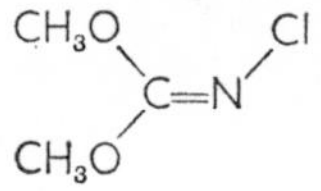 (**139**)

about an interconversion of isomers or a Beckmann rearrangement. However, isomerization could be brought about by the use of dry chlorine.

In studies of isomerization rates of iminocarbonates, the *N*-chloro derivative (**139**) has been found to exhibit a doublet in the n.m.r. spectrum obtained at room temperature[137]. There is no change in the spectrum when the sample is heated to 105° in methylcyclohexane or 170° in diphenyl ether.

F. Imidates, Iminocarbonates and Related Compounds

The thermal rearrangement of imidates (**140**) to disubstituted amides (Chapman rearrangement) has been reviewed recently[138]. There is little information available, however, on the rate of *syn–anti* isomerization at the azomethine linkage of imidates. Marullo and

$$R^1\text{—}C(OR^2)\text{=}NR^3 \longrightarrow R^1\text{—}C(\text{=}O)\text{—}N(R^2)R^3 \qquad (30)$$

(**140**)

coworkers[139] have reasoned that in the pyrolysis of imidates such as **141** to give amides and olefins the *syn–anti* isomerization most likely is

$$C_6H_5\text{—}C(OCH_2CH_2C_6H_5)\text{=}N\text{∼}C_6H_5 \xrightarrow{\Delta} C_6H_5\text{—}C(\text{=}O)\text{—}NHC_6H_5 + CH_2\text{=}CHC_6H_5 \qquad (31)$$

(**141**)

not the rate determining step since the pyrolysis reaction has an energy of activation of about 40 kcal/mole and *N*-aryl imines[28] and iminocarbonates[137] isomerize with energies of activation closer to 20 kcal/mole.

The complexity of the room temperature n.m.r. spectrum of phenyl *N*-methylacetimidate as prepared from phenol and acetoxime benzene sulphonate has been explained on the basis of the existence of *syn* (**142**) and *anti* (**143**) isomers in the ratio of 2:1, favouring the *anti* form[140].

$$CH_3\text{—}C(OC_6H_5)\text{=}N\text{—}CH_3 \rightleftharpoons CH_3\text{—}C(OC_6H_5)\text{=}N(CH_3) \qquad (32)$$

(**142**) (**143**)

At higher temperatures the peaks of the two forms begin to merge and have essentially coalesced by 70°.

Most hydroxamic acids can be isolated in two forms[141]. For example, ethyl benzohydroxamic acid (**144**) was found to exist in one form melting at 53·5° and another one melting at 67·5–68°[142]. Werner[143] suggested that these were *syn* and *anti* forms and demonstrated that they were not converted into one another by chemical reactions that interconvert oxime isomers. Also, the ethyl acetyl hydroxamate isomers (**145**) were stable up to 140°[144].

$$C_6H_5\text{—}C(OC_2H_5)\text{=}NOH \qquad C_6H_5\text{—}C(OC_2H_5)\text{=}NOCOCH_3$$

(**144**) (**145**)

A study by Curtin and Miller[145] of the rearrangement of benzimidoyl benzoates (**146**) to imides (**147**) via 1,3 aroyl migrations suggests that the imino compounds exist mainly in the *trans* aryl form as shown.

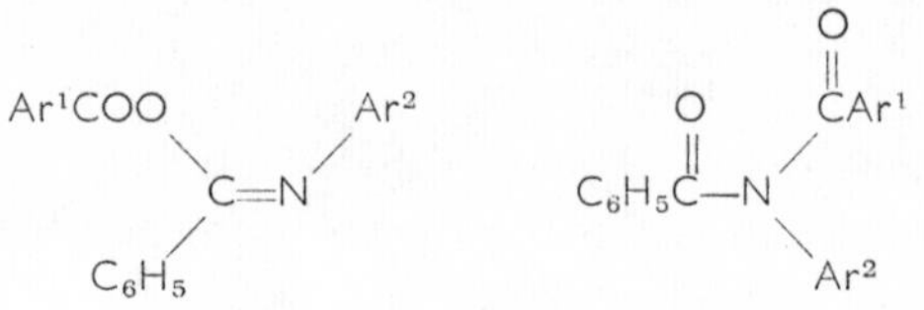

$$\underset{\textbf{(146)}}{\begin{array}{c} Ar^1COO \qquad\quad Ar^2 \\ \diagdown\ \ \diagup \\ C{=}N \\ \diagup \\ C_6H_5 \end{array}} \qquad \underset{\textbf{(147)}}{C_6H_5C({=}O){-}N(C({=}O)Ar^1)Ar^2}$$

That the rate of *syn–anti* isomerization of **146** is not the rate determining factor (the rearrangement probably proceeds through the *cis* aryl isomer) seems reasonable from a comparison of the observed rearrangement rates (10^{-4} to 10^{-5} s^{-1}) with those of isomerization of *N*-aryl ketimines[28] and *N*-aryl aldimines[24] (1–10 s^{-1}) at comparable temperatures.

The isomerization of iminocarbonates and iminothiocarbonates has been studied more quantitatively than the isomerization of imidates and other compounds just mentioned in this section. The n.m.r. spectra of compounds **148** all showed a single methoxyl resonance at

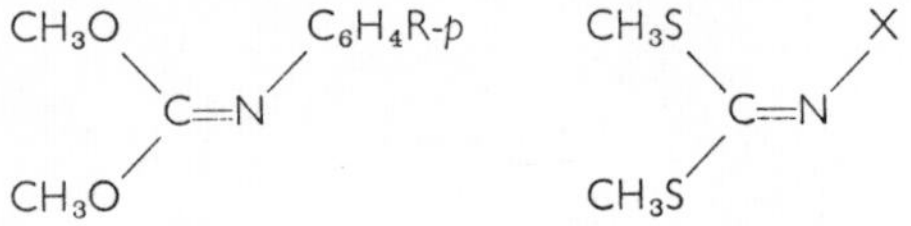

$$\underset{\textbf{(148)}}{(CH_3O)_2C{=}N{-}C_6H_4R\text{-}p} \qquad \underset{\textbf{(149)}}{(CH_3S)_2C{=}N{-}X}$$

room temperature, but cooling resulted in the formation of a distinct doublet[137]. Some of the data from a variable temperature n.m.r. study of these *N*-aryl iminocarbonates are summarized in Table 11. These values for **148** were confirmed in a more recent study by Wurmb-Gerlich, Vögtle, Mannschreck and Staab[44]. These latter workers also looked at the isomerization of other iminocarbonates and some iminothiocarbonates (**149**). Some of their data are also presented in Table 11. The most remarkable feature of the data in Table 11 is the low E_a or ΔG^* values compared with the values for *N*-aryl and *N*-alkyl ketimines[28]. Thus, substitution of CH_3O— or CH_3S— for alkyl or aryl groups on the carbon of the C=N double bond has a very pronounced effect on the rate of *syn–anti* isomerization at this multiple linkage.

The related unsymmetrical compounds (**150, 151**) apparently exist only in the form shown[44]. The chemical shifts shown by the formulae

TABLE 11. Isomerization data for compounds **148** and **149**.

Substituent	Solvent	$\Delta\nu$ (Hz[a])	T_c (°c[b])	E_a (kcal/mole)	ΔG_c^* (kcal/mole[c])
R in **148**:[d]					
H	Acetone	9·13	+0·8	13·4 ± 0.3	
Cl	Acetone	7·14	−1·6	16·6 ± 0.3	
CH_3	Acetone	8·03	+2·8	15·5 ± 0·3	
X in **149**:[e]					
C_6H_5	Acetone-d_6	2·5	−22		13·7 ± 0·3
CH_3	$(C_6H_5)_2O$	6	73		18·6 ± 0·3
CN	Acetone-d_6	13	1		14·0 ± 0·3

[a] Maximum separation observed at low temperature.
[b] Temperature of coalescence of the methyl peaks.
[c] Free energy of activation at T_c.
[d] Data from reference 138.
[e] Data from reference 44.

are from the room temperature spectra and cooling to −40 to −50° failed to broaden the peaks.

τ 7·70 CH_3S / τ 6·07 CH_3O — C=N — C_6H_5 **(150)**

τ 7·67 CH_3S / τ 7·12 CH_3O — C=N — CH_3 **(151)**

G. *syn–anti* Isomerization Mechanisms

Several possible mechanisms have been considered for geometrical isomerization at the C=N bond. When considering isomerization rates ranging over more than 10 powers of 10 and involving compounds with substituents of widely different resonance and inductive effects, as has been the case in the preceding sections, one cannot hope to explain all data by one isomerization mechanism. Quite possibly, as with most considerations of mechanisms, there must be considered a continuum of mechanisms between the extremes to be described.

First, one can consider a process involving the rupture of the π bond in a homolytic fashion to give a transition state resembling **152**, and further rotation about the axis through the carbon and nitrogen atoms to give the other isomer. It is generally agreed that many olefins isomerize by this kind of process, but these olefin isomerizations require high temperatures and have activation energies ranging from 36 to 60 kcal/mole. The barrier to the analogous rotation about the N=N double bond has been calculated[146,147] to be even higher than that for rotation about the C=C double bond, so it seems reasonable

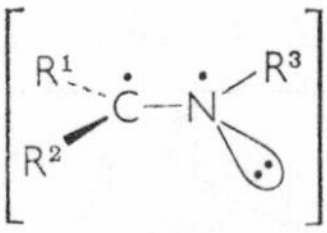

(152) 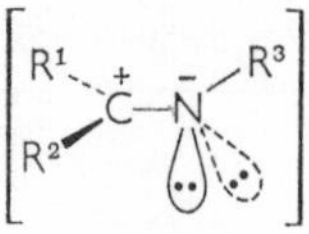 (153) 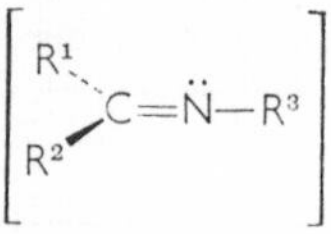 (154)

to assume that isomerizations through **152** would be associated with energies of activation of 50 kcal/mole or higher. It is obvious from the data presented thus far in this chapter that most imines isomerize with much lower E_a values, and thus this homolytic bond rupture path does not seem likely.

Another possible rotation process is one involving heterolytic rupture of the C=N bond and passage through a transition state resembling **153**. Of course, this structure is the extreme case of the normally considered polarization of the C=N bond in the ground state. As will be pointed out in more detail, destabilization of the double bond in the ground state via dipolar resonance structures resembling **153** may make such a rotation mechanism through **153** quite probable in some cases.

Unfortunately, many experimental results which can be cited in support of a mechanism proceeding through **153** are also consistent with the 'lateral shift'[28] mechanism, which is presumed to consist of the shift of the substituent attached to nitrogen from one side of the molecule through a linear transition state (**154**) to the other side. This resembles inversion through nitrogen in normal trisubstituted amines or ammonia itself except, in this case, the nitrogen adopts linear *sp* bonds, the π bond remains intact, and the unshared pair occupies the perpendicular *p* orbital on nitrogen. The motion assumed for this mechanism is shown in **155** and can be compared with **156** for the rotation mechanisms.

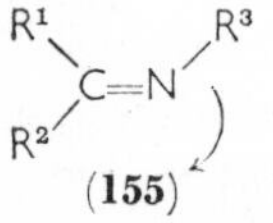

(155) 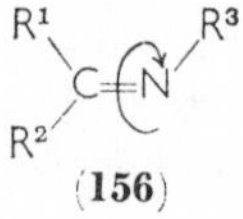 (156)

The results of studies of isomerization rates of *N*-aryl and *N*-alkyl ketimines and *N*-aryl aldimines are consistent with the lateral shift mechanism[44]. Curtin, Grubbs and McCarty[28] found that the effect of substituents on the aryl rings attached to the doubly bonded carbon can be correlated by a Hammett ρ of about $+0{\cdot}1$, whereas substituents on the phenyl ring bound to nitrogen have a much larger effect on the isomerization rate, being correlated by $\rho = +1{\cdot}5$. The accelerating effect of electron-withdrawing substituents in the *para* position of the *N*-aryl group can be understood by considering resonance structures such as **157** contributing to the linear transition state. Similarly, Wettermark[26] found the ρ-value for aniline ring

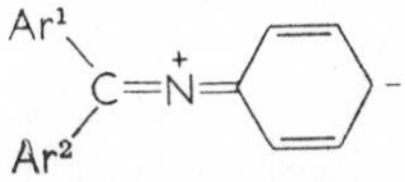

(**157**) 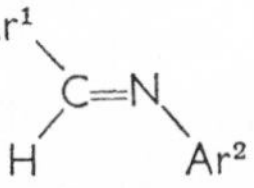 (**158**)

substitution in **158** to be $+1{\cdot}85$ while the ρ-value for substitution on the aldehyde ring is $+0{\cdot}41$.

These Hammett correlations for aldimines and ketimines seem to support a structure like **154** more than **153** where there should be appreciable substituent effects at both the carbon and the nitrogen rings. Also consistent with this picture is the fact that *N*-alkyl imines isomerize much faster than olefins but considerably slower than the *N*-aryl imines. Obviously, special resonance structures such as **157** are not available for stabilization of the linear transition state in the case of *N*-alkyl imines. Also, steric effects in the bent ground state are probably not so severe in the simple *N*-alkyl imines as in the *N*-aryl imines.

Also offered in support of the lateral shift mechanism is the rate enhancement observed when alkyl substituents are in the *ortho* positions of the *N*-aryl ring of *N*-aryl ketimines[44,148]. This enhancement increases with increasing size of the *ortho* substituents. From the Hammett correlation just mentioned for *para* substituents it would be predicted that alkyl groups should hinder the isomerization. Thus, the effect of *ortho* substituents is a steric one and it is felt that the relief of steric strain in going from the ground to the transition state should be greatest in the lateral shift mechanism.

One of the most striking results one can see when comparing the configurational stability at the C=N double bond in all the types of compounds considered in this chapter is the very great stability of compounds such as *N*-halimines and oxime ethers. In fact, in all cases where a heteroatom is directly bonded to the nitrogen of the C=N double bond, i.e., azines, hydrazones, oximes, etc., the configurational stability is much higher than in the usual *N*-alkyl and *N*-aryl imines. It is tempting to explain this through unusual stability of the double bond order in the bent ground state, since stabilization of the polarized structure with diminished electron density at nitrogen is not possible and just the opposite would be predicted. It is impossible at this point, however, to rule out the possibility that this high configurational stability may arise from unusual instability of the linear transition state (**154**) or the dipolar structure (**153**) leaving a path through **152** as the only mechanism for isomerization in some of these cases.

As was mentioned earlier, the cause of the surprisingly low barrier to isomerization of the *N*-perfluorimines in unclear. There may be special stabilization of a linear transition state by structures such as **159**[149]. The situation is probably more complex with difluoromethylene

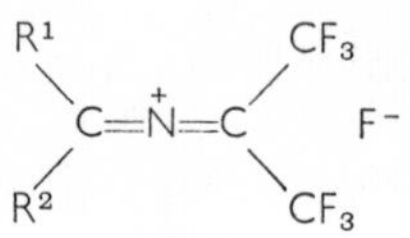

$$(R^1)(R^2)C{=}\overset{+}{N}{=}C(CF_3)_2 \quad F^- \qquad\qquad (R^1\overset{+}{O}{=})(R^2O)C{-}\overset{-}{N}R^3$$

(**159**) 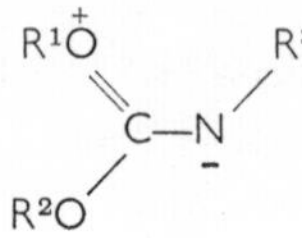 (**160**)

derivatives and not directly comparable to other imines, since recent evidence points to sp^3 hybridization at the carbon atom of the *gem*-difluoro group ($F_2C{=}$)[150], and thust he C=N bond in these compounds would be different from those in other imines.

Also uncertain is the mechanism of isomerization of compounds with oxygen or sulphur at the carbon of the C=N bond. The low barrier to *syn–anti* isomerization in the iminocarbonates and iminothiocarbonates may be rationalized as arising from a lowering of the carbon–nitrogen bond order relative to ketimines and aldimines as seen by considering structures like **160**. Approached from a different angle, it has been suggested that **160** represents the stabilization of the dipolar transition state (**153**) in the rotation mechanism[137].

III. REARRANGEMENTS

A. The Beckmann Rearrangement

1. Introduction

The most familiar and extensively investigated of the rearrangements involving the azomethine group is certainly the rearrangement of oximes to amides first reported by Beckmann in 1886[151] (equation 33). The catalyst is usually an acid which often serves also as the solvent. It is now recognized that a whole spectrum of mechanisms may exist for this rearrangement with the course of a given reaction

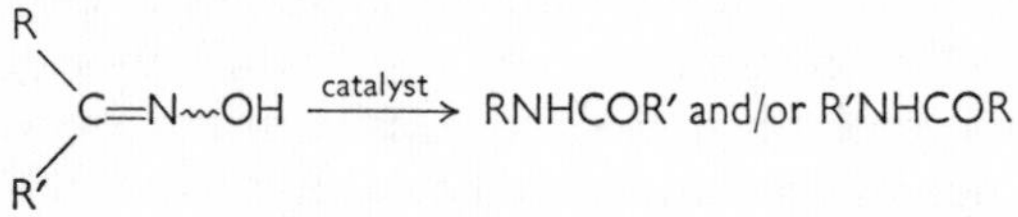

$$(R)(R')C{=}N{\sim}OH \xrightarrow{\text{catalyst}} \text{RNHCOR' and/or R'NHCOR} \tag{33}$$

being governed by the choice of catalyst and solvent among other things.

A very commonly proposed picture of the rearrangment is the one depicted in the following sequence (equation 34). The function of the

$$\begin{array}{l} \underset{R'}{\overset{R}{>}}C{=}N\diagup^{OH} \underset{}{\overset{H^+}{\rightleftharpoons}} \underset{R'}{\overset{R}{>}}C{=}N\diagup^{\overset{+}{O}H_2} \longrightarrow \left[\overset{R}{\diagdown}C{=}N\overset{\cdot\cdot OH_2}{}\ \underset{R'}{\cdot\cdot}\right]^+ \\ \downarrow \\ R{-}\overset{\overset{+}{O}H_2}{\overset{|}{C}}{=}N{-}R' \longleftarrow \left[R{-}C{\equiv}N{-}R' \ \ \overset{+}{O}H_2 \longleftrightarrow R{-}\overset{+}{C}{=}N{-}R' \ \ OH_2\right] \\ \downarrow {-H^+} \\ R{-}\overset{OH}{\overset{|}{C}}{=}N{-}R' \rightleftharpoons R{-}\overset{O}{\overset{\|}{C}}{-}NHR' \end{array} \qquad (34)$$

catalyst is to convert the —OH to a good leaving group. The stereospecificity which has been observed in so many reactions where the configuration of the starting oxime is known has led to the proposal of intramolecular migration of the group *trans* or *anti* to the departing group on the nitrogen in a synchronous fashion as shown by the pseudo three-membered ring transition state in equation (34). While many data can be correlated by such a picture of the Beckmann rearrangement, it will be shown in a later section that modifications in this scheme are required to account for the results of many recent studies.

Two excellent reviews of the Beckmann rearrangement have appeared in this decade[54,55]. The first, by Donaruma and Heldt[54], includes a comprehensive coverage of the synthetic utility of this reaction, while the studies of the mechanism of the reaction have been emphasized more by Smith[55]. With these two reviews being so recent and readily accessible it will only be necessary in this chapter to present some of the most recent applications and mechanistic studies. For convenience, the rearrangements of oxime esters and *N*-halimines are also included. No attempt has been made to compile an all-inclusive summary, but it is hoped that most of the pertinent references from the early 1960's usage most of 1967 have been included.

2. Experimental conditions

The Beckmann rearrangement proceeds under the catalytic action of any of a large number of reagents, most of which are protic or aprotic acids. Concentrated sulphuric acid is frequently employed, but phosphoric acid, phosphorus pentachloride, aryl sulphonyl chlorides, aluminium chloride, hydrogen bromide and hydrogen fluoride are also used as catalysts. Beckmann's mixture, which consists of a saturated solution of hydrogen chloride in a mixture of glacial acetic acid and acetic anhydride, is useful when the oxime reacting is insoluble in other media.

Polyphosphoric acid (PPA) has become a popular medium for carrying out Beckmann rearrangements. Pearson and Stone[152] have found that the rearrangement of substituted acetophenones to the isomeric amides proceeds 12–35 times as rapidly in PPA as in sulphuric acid. Their data also suggest that previous reports of the use of PPA often involve conditions that are too strenuous since, in most of the cases they studied, rearrangement occurred overnight at room temperature in PPA containing no more than the equivalent of 84% phosphorus pentoxide.

The use of PPA or other protic solvents as catalysts is known to often bring about geometical isomerization of oximes at a rate which may compete with or exceed the rate of rearrangement[41]. Thus, as has been mentioned earlier in this chapter, this possibility makes it necessary to view with caution many structural studies made via the Beckmann rearrangement and the assumption of a stereospecific migration of the group *trans* to the oxime —OH. The procedure of Craig and Naik[153], involving the rearrangement of the benzene- or *p*-toluenesulphonyl esters of oximes upon passage through a column of neutral alumina, has been shown to give good yields of Beckmann products with preservation of the configurational integrity of the oxime. An example of different results from the same starting oxime

NOH; PPA; O; NH; N H

(161) **(162)**

(35)

$NOSO_2C_7H_7$; $C_7H_7SO_2Cl$; C_5H_5N; Al_2O_3; H N; O

(163) **(164)**

obtained through the use of PPA and the Craig and Naik method is illustrated in equation (35)[154]. Assuming the oxime configuration to be as shown in **161** the results may be explained by a PPA-catalysed rearrangement of configuration prior to conversion to **162.**

Other studies on the selectivity of various catalysts include the one by Wiemann and Ham[155]. Treatment of oxime **165** with sulphuric acid yielded an oxazoline (**166**) while phosphorus pentachloride in ether afforded a good yield of the expected amide (**167**).

$$\textbf{(165)}\ \xrightarrow{H_2SO_4}\ \textbf{(166)};\qquad \textbf{(165)}\ \xrightarrow{Et_2O\ |\ PCl_5}\ \textbf{(167)}\ (\text{NHCOCH}_3) \qquad (36)$$

There have been a few reports of the use of formic acid as a rearrangement medium for certain *o*-hydroxy ketoximes[156], but the results in a recent paper by Van Es[157] would seem to indicate that it is an ideal medium for many other ketoximes as well; e.g. acetophenone oxime gives a 90% yield of acetanilide after six hours in formic acid.

In liquid sulphur dioxide, the Beckmann rearrangement has been reported to take place with unusual rapidity, at a low temperature, and without any side reactions, giving a product of extreme purity[158–161]. Cyclohexanone oxime-*p*-toluenesulphonate, for example, rearranges at 21·5° at a rate which is eighty times that in methanol, 480 times that in chloroform and 16,000 times that in carbon tetrachloride[161].

The oximes of cyclohexanone, benzophenone and acetophenone when added to a methylene chloride solution of iodine pentafluoride undergo a Beckmann rearrangement to give the expected amides (equation 37)[162].

$$(C_6H_5)_2C{=}N{-}OH \xrightarrow[CH_2Cl_2,\ 3^\circ]{IF_5} C_6H_5NHCOC_6H_5 \qquad (37)$$

An example of a Beckmann rearrangement accompanying dehydration is the conversion of *syn*-aldoximes into isonitriles, which has been effected by decomposing the addition compounds of the *syn*-aldoximes and methylketone diethylacetal (**168**) in the presence of a catalytic amount of boron trifluoride and mercuric oxide[163].

$$\underset{\textbf{(168)}}{CH_3CH_2C(OC_2H_5)_2-O-N{=}C(H)R} \xrightarrow[Et_2O]{BF_3/HgO} RNC + C_2H_5COOC_2H_5 + C_2H_5OH \qquad (38)$$

A novel rearrangement of 2-bromoacetophenones (**169**) in the presence of triphenylphosphine leads to *N*-arylacetimidoyl bromides (**170**) which can serve as precursors in the syntheses of ketenimines[164].

$$\underset{\textbf{(169)}}{Ar(BrCH_2)C{=}N-OH} \xrightarrow[CH_3CN]{(C_6H_5)_3P} \underset{\textbf{(170)}}{ArN{=}C(Br)CH_3} + (C_6H_5)_3P \rightarrow O \qquad (39)$$

$$\underset{\textbf{(171)}}{\text{1-Cl-1-}NO_2\text{-cyclohexane}} \xrightarrow[C_6H_6]{(C_6H_5)_3P} \underset{\textbf{(172)}}{\text{cyclohexylidene-}N(\rightarrow O)OP^+(C_6H_5)_3\ Cl^-} \xrightarrow{(C_6H_5)_3P} \underset{\textbf{(173)}}{\text{cyclohexylidene-}NOP^+(C_6H_5)_3\ Cl^-} + (C_6H_5)_3P{\rightarrow}O \qquad (40a)$$

$$\textbf{173} \longrightarrow \underset{\textbf{(174)}}{\text{2-chloro-azacycloheptene (C(Cl)=N)}} + (C_6H_5)_3P{\rightarrow}O \qquad (40b)$$

$$\textbf{174} \xrightarrow{H_2O} \underset{\textbf{(175)}}{\text{azepan-2-one (C(=O)NH)}} \qquad (40c)$$

The reaction is assumed to proceed through a quasiphosphonium salt analogous to the one postulated in the reaction of a *gem*-chloronitroalkane with triphenylphosphine (equation 40)[165].

Heterogeneous catalytic techniques for the Beckmann rearrangement have received further attention. Metallic copper is effective with some oximes although strenuous conditions are required[166]. Raney nickel has been used to convert acid-sensitive aromatic and aliphatic aldoximes to amides[167]. Even better, though, are some nickel compounds such as nickel acetate tetrahydrate. Acid zeolites (crystalline aluminosilicates) show catalytic activity in the rearrangement of common ketoximes to amides[168]. Optimum conditions include temperatures of 250–350°, a non-polar solvent, and an inert carrier gas.

3. Some recent novel applications

There is a continuing interest in the synthesis of steroids modified through the incorporation of heteroatoms in the polycyclic nucleus. Azasteroids are potentially available from normal ketosteroid oximes via the Beckmann rearrangement. The treatment of testosterone propionate oxime (**176**) or 17α-methyltestosterone acetate oxime (**177**) with thionyl chloride in dioxan gives only the 3-aza-Δ^{4a}-4-ketones (**178, 179**)[75]. This is in spite of the presence of some of the *anti* isomers of **176** and **177** in the starting mixtures. A more convenient method

OR R′ N OH $\xrightarrow[\text{dioxan}]{SOCl_2}$ OR R′ HN O (41)

(**176**) R = C_2H_5CO, R′ = H (**178**)

(**177**) R = CH_3CO, R′ = CH_3 (**179**)

for the rearrangement of such α,β-unsaturated steroidal ketoximes has been more recently reported by Kohen[169]. A dimethylformamide solution of **177** plus an equimolar quantity of *p*-toluenesulphonyl chloride gave, at room temperature, a 91% yield of the 3-azalactam (**179**) free of contamination by the unreactive *anti* oxime.

Through selective oximation of 5α-pregnane-3,20-dione and A-nor-5α-pregnane-2,20-dione, Nace and Watterson[170] have been able to prepare oximes **180** and **181** without protecting the free 20-carbonyl groups. Oxime **180** was rearranged by the Craig and Naik procedure[153] to give a 93% yield of the previously unknown A-homo-4-aza-5α-pregnane-3,20-dione (**182**). Similarly **181** gave a 1:1 mixture of **183** and **184** in 93% yield. Thus the Beckmann rearrangement provides an excellent route to A-azapregnane derivatives.

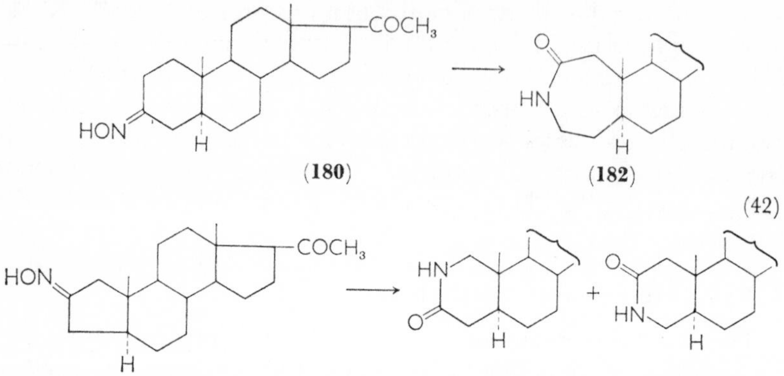

(**180**) (**182**)

(42)

(**181**) (**183**) (**184**)

A new synthesis of quinazoline-N^3-oxides (**186**) through the Beckmann rearrangement of *o*-acylaminoacetophenone oximes (**185**) has been reported by Kovendi and Kircz[171]. The quinazolinedione **188** is formed by a novel rearrangement of **187** at its melting point[172]. This appears to be a Beckmann rearrangement of **187** followed by an O → N migration of the carbonyl group.

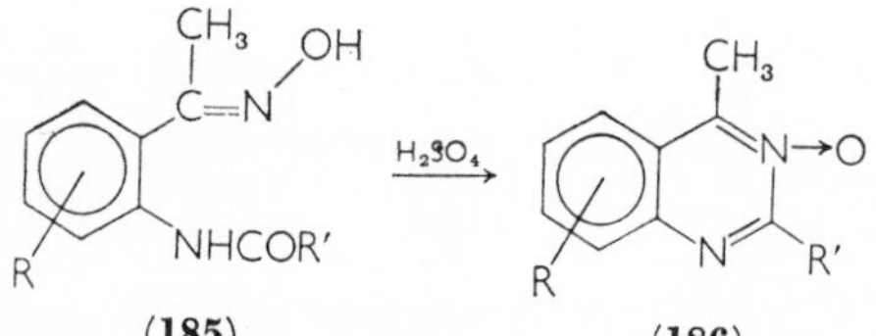

(**185**) $\xrightarrow{H_2SO_4}$ (**186**) (43)

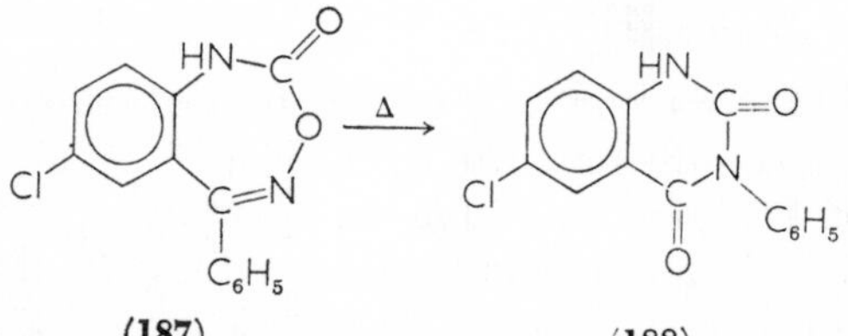

(**187**) $\xrightarrow{\Delta}$ (**188**) (44)

2-Isoxazolines may be regarded as cyclic ethers of oximes. Acetic anhydride–boron trifluoride etherate rearranges 3,5-diphenyl-2-isoxazoline (**189**) at room temperature, but other substituents in the isoxazoline ring alter the course of the reaction or result in no rearrangement[173].

$$\text{(189)} \xrightarrow[\text{BF}_3\text{-etherate}]{\text{Ac}_2\text{O}} C_6H_5NHCOCH{=}CHC_6H_5 + C_6H_5NHCOCH_2CH(OAc)C_6H_5 \qquad (45)$$

α-Oximinoaryl acetonitriles have been reported to be unreactive towards phosphorus pentachloride in ether[174]. Stevens[78], however, has found that some of these compounds do rearrange as shown in equation (46). Also, the *N*-fluorimine (**190**) readily undergoes a Beckmann rearrangement in concentrated sulphuric acid at 85° to give a high yield of *N*-(4-chlorophenyl)oxamide (**191**). Similar rearrangements of

$$Cl{-}C_6H_4{-}C(=NOH)CN \xrightarrow[\text{ether}]{PCl_5} Cl{-}C_6H_4{-}C(=NCl)CN + Cl{-}C_6H_4{-}N{=}C(Cl)CN + Cl{-}C_6H_4{-}C(=O)CN \qquad (46)$$

$$Cl{-}C_6H_4{-}C(=NF)CN \; (190) \xrightarrow{H_2SO_4} HOC(=O){-}C(=O)NH{-}C_6H_4{-}Cl \; (191) \qquad (47)$$

$$C_6H_5C(=NOTs)CN + RSH \xrightarrow{(C_2H_5)_3N} RSC(=NC_6H_5)CN \qquad (48)$$

$$R = C_2H_5 \text{ or } C_6H_5CH_2$$

the tosylate esters of these oximino nitriles occur in the presence of mercaptans and triethylamine (equation 48)[175].

The benzenesulphonyl esters of *syn-* and *anti*-7-oximino-1,3,5-cyclooctatriene give **192** and **193**, respectively, when treated with 60% aqueous acetone at 20°[176]. This arrangement probably proceeds through the monocyclic azacyclononatrienones which are not isolated.

$$C_8H_8{=}N{\sim}OSO_2C_6H_5 \xrightarrow[\text{water, } 20°]{\text{acetone}} \text{(192)} + \text{(193)} \qquad (49)$$

At one time it was believed that carbonyl derivatives of xanthen-9-one could not be prepared[177]. More recently, though, the synthesis of xanthen-9-one oximes has been reported[178] and the rearrangement of the oxime of 4-phenylxanthen-9-one (**194**) in PPA at elevated temperatures leads to a lactam which has been identified as having structure **195**[179].

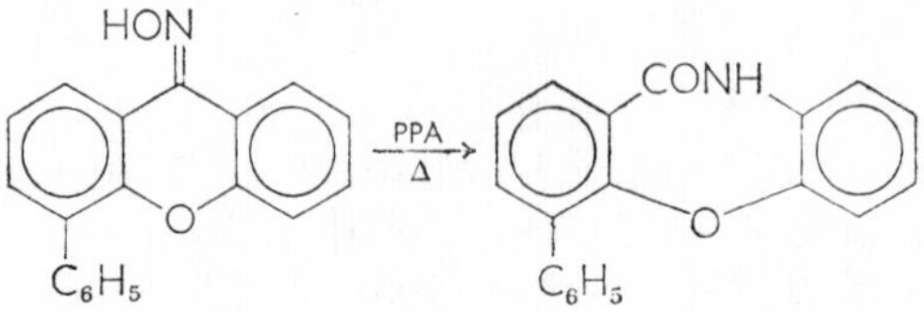

(50)

(**194**) (**195**)

Oximes have been conveniently prepared from keto derivatives of cyclopentadienylmanganese tricarbonyl complexes (**196**, R = CH_3, C_6H_5)[180]. In cases where *syn* and *anti* isomers were formed they were separated before being subjected to rearrangement conditions. Only the *syn* isomers (configuration shown) rearrange and all of the *anti* compounds are converted by phosphorus pentachloride in pyridine to $(C_6H_5N)_2Mn(CO)_3Cl$.

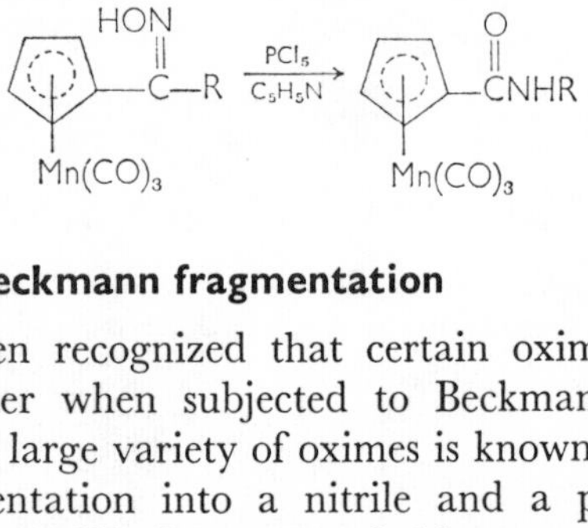

(51)

4. Examples of Beckmann fragmentation

It has long been recognized that certain oximes behave in an 'abnormal' manner when subjected to Beckmann rearrangement conditions. Now a large variety of oximes is known which undergo a heterolytic fragmentation into a nitrile and a positively charged species (equation 52)[181]. Processes of this type have been called

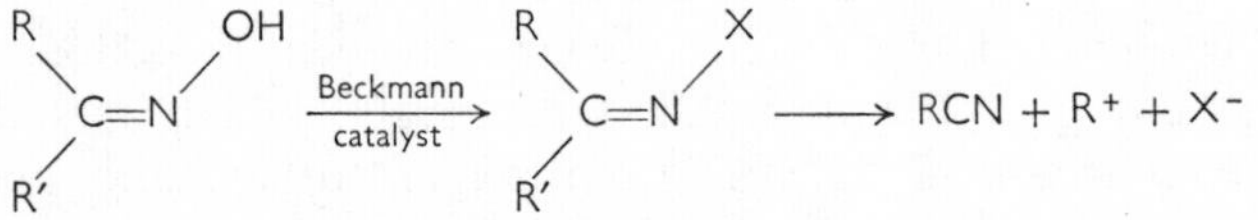

$$\text{RR}'\text{C=N-OH} \xrightarrow[\text{catalyst}]{\text{Beckmann}} \text{RR}'\text{C=N-X} \longrightarrow \text{RCN} + \text{R}^+ + \text{X}^- \quad (52)$$

second-order Beckmann, abnormal Beckmann, Beckmann fission and Beckmann fragmentation reactions. Conditions are especially favourable for this type of reaction when R^+ is relatively stable as in α-amino, α-hydroxy, α-alkoxy, α-keto, α-imino, α-diaryl, α-trialkyl or α-triaryl oximes. There is now evidence (Section III.A.5) that not all of these classes of oximes fragment directly to the nitrile. Some probably proceed through an intermediate common to both fragmentation and the normal Beckmann rearrangement. It is because of this relationship and the considerable recent interest in fragmentation reactions that several examples for a few of these types will be discussed.

a. α-Trisubstituted oximes. The literature on the Beckmann rearrangement contains several examples of oximes of bicyclic ketones undergoing fragmentation along with rearrangement to the expected amide when subjected to Beckmann catalysts[54]. The reaction of fenchone oxime (**197**) with *p*-toluenesulphonyl chloride in pyridine, with phosphorus pentachloride, or with sulphuric acid leads to the formation of a mixture of the lactam **198** and the unsaturated nitrile **199** or **200** (or perhaps a mixture of the two)[182]. Conversion of the nitrile to the lactam has been effected separately by the Ritter reaction with sulphuric acid, but the yield was poor so it was assumed that ion **201** is not involved in the major reaction path leading to the lactam.

NOH (**197**) → N≡C+ —ŌH→ NH, C=O (**198**)

(**201**) CN ⇌ (−H⁺ / H⁺) (**199**) CN and/or (**200**) CN

(53)

More recently Sato and Obase[183] have shown that the pyrolysis of camphor oxime (**202**) at 240° for 6–7 min affords a mixture of **203**, **204** and **205** in relative amounts 1:3·8:4·4. At 500° the principal product found was **206**. Interestingly, nitriles **204** and **205** were also found among the products from the photolysis of camphor oxime in methanol using a quartz high-pressure mercury lamp.

Early work on the Beckmann rearrangement of spiroketoximes showed that the products depend on the catalyst employed as well as the size of the rings[184,185]. For example, **207** gives **208** with phosphorus pentachloride as catalyst along with a small quantity of **209**[186]. On the other hand, **209** is the main product when phosphorus pentoxide

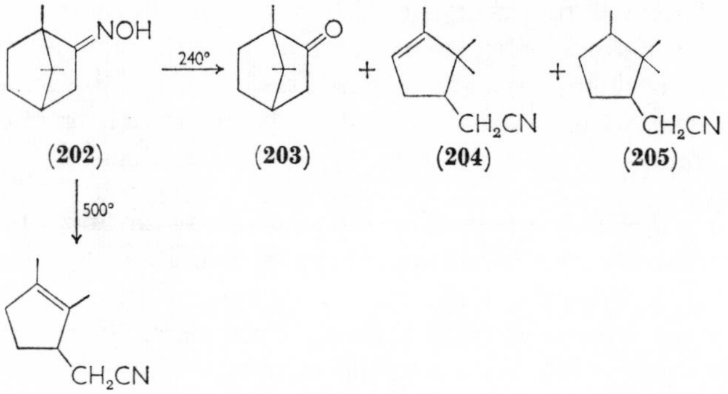

NOH, 240°, O, +, +, CH_2CN, CH_2CN
(202) (203) (204) (205)
500°
CH_2CN
(206)

(54)

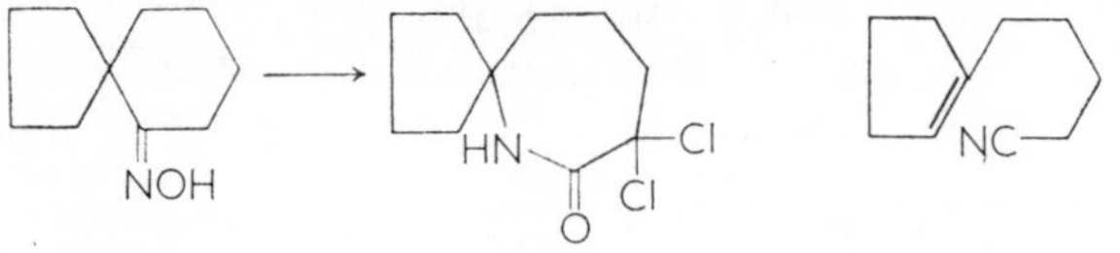

NOH, HN, Cl, Cl, O, NC
(207) (208) (209)

(55)

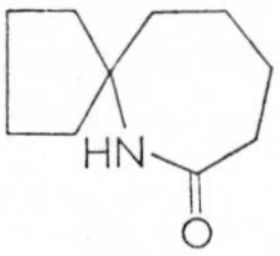

HN, O
(210)

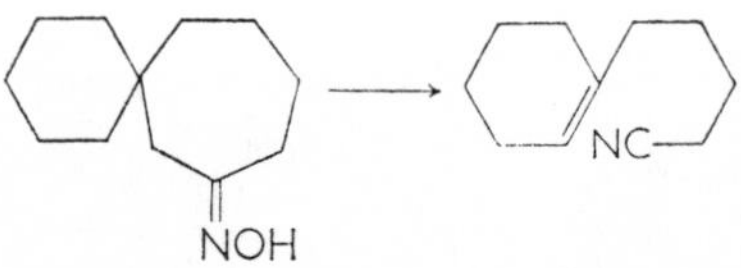

NOH, NC
(211) (212)

(56)

is used and **210** is the minor product. The spiroketoxime **211** gives solely nitrile **212** with phosphorus pentoxide as catalyst.

Conley and Nowak[187] investigated the rearrangement of several 2,2-dimethylcycloalkanone oximes (equation 57). With either phosphorus pentachloride or thionyl chloride as catalyst these oximes are cleaved mainly to unsaturated nitriles. With PPA, the oximes yield α,β-unsaturated ketones. These ketones were thought to be formed by an acid-catalysed Hoesch type cyclization of the originally formed unsaturated nitriles (equation 58). This possibility was confirmed by treatment of the unsaturated nitrile, obtained from the reaction of the

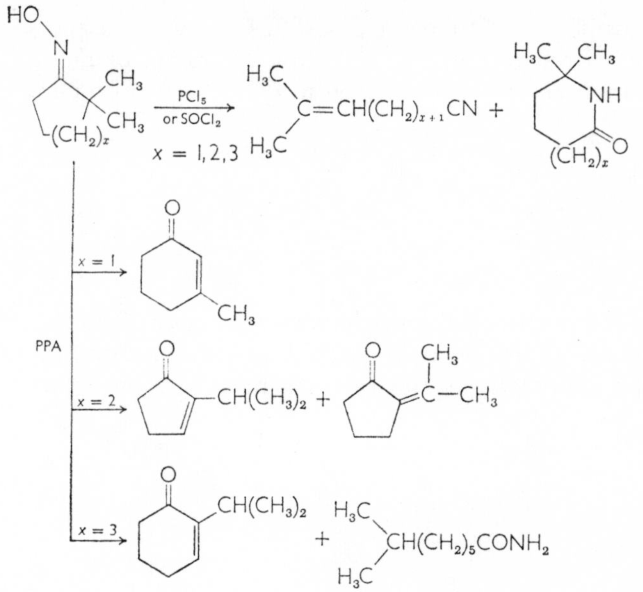

(57)

oxime with phosphorus pentachloride or thionyl chloride, with PPA to give the same α,β-unsaturated ketone as obtained directly from the reaction of the oxime with PPA. In those cases where the relationship

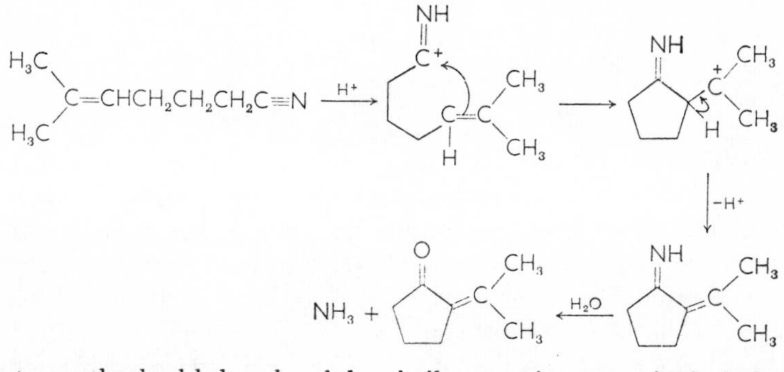

(58)

between the double bond and the nitrile group is not particularly favourable for cyclization then hydration to the open chain amide occurs. Somewhat similar results were obtained with the 2,2-diaryl cycloalkanone oximes[188]. The products were again dependent on ring size and on the catalyst.

Although 2,2-disubstituted-1-tetralone oximes (**213**) in PPA afford almost quantitative yields of the lactams expected from the normal Beckmann rearrangement (equation 59)[189] such is not the case with

1,1-disubstituted-2-tetralone oximes[190]. Conley and Lange[190] found that the rearrangement of 1,1-dimethyl-2-tetralone oxime (**214**) using phosphorus pentachloride resulted in a 93% yield of the unsaturated

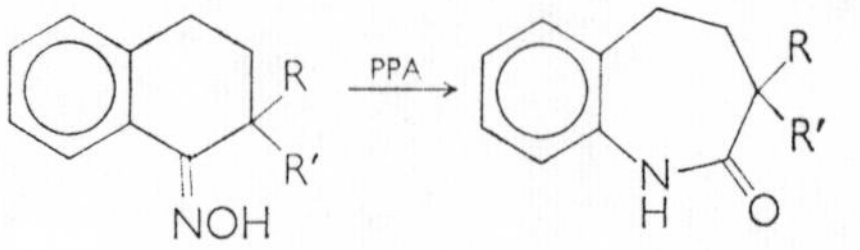

(**213**) —PPA→ (59)

nitrile (**215**) expected from oxime fragmentation (equation 60). Rearrangement of this oxime in hot PPA gave the lactam (**216**) expected from the normal Beckmann rearrangement and an α,β-unsaturated

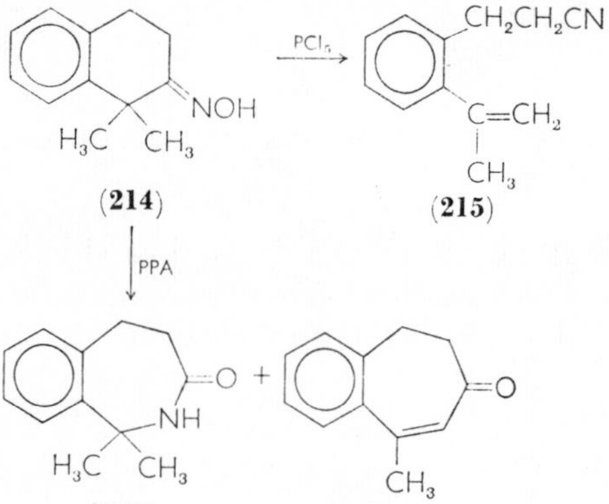

(**214**) —PCl_5→ (**215**)

(**214**) —PPA→ (**216**) + (**217**) (60)

ketone (**217**). The ratio of lactam to ketone ($\sim$3:1) was found to be identical to that obtained from independent nitrile cyclizations under comparable conditions of temperature and time. From these data, the authors concluded that the lactam, although it is the product expected from the normal rearrangements, is not produced in the initial step. Rather, they feel, the ketone and lactam are formed competitively from a common intermediate by Hoesch and Ritter cyclizations, respectively (equation 61).

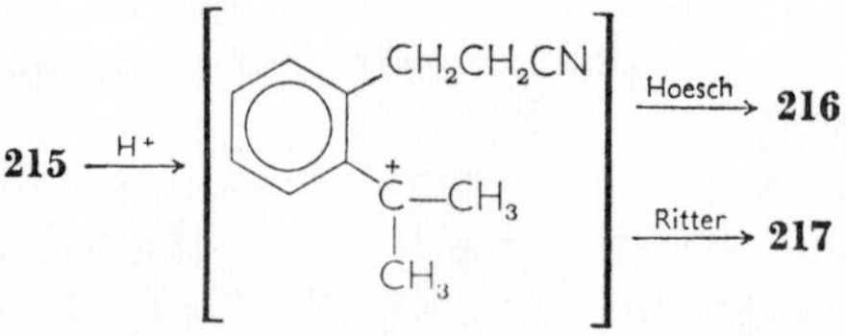

215 —H^+→ [$C_6H_4(CH_2CH_2CN)(\overset{+}{C}(CH_3)_2)$] —Hoesch→ **216**; —Ritter→ **217** (61)

This interesting possibility of a two-step, fragmentation–recombination sequence leading to so-called 'normal' Beckmann products from some oximes has received rather strong support in the form of experimental evidence from crossover and stereochemical studies[191]. Pinacolone oxime (**218**) and 2-methyl-2-phenylpropiophenone oxime (**219**) were employed in the crossover study. Treatment of **218** with PPA yielded a normal product (**220**) plus small amounts of acetamide. Since oximes like **218** normally fragment in PPA, it was believed that **218** rearranged by fragmentation and a Ritter type recombination (the validity of this supposition was shown by the synthesis of **220** from acetonitrile and *t*-butyl alcohol in acid). The reaction of **219** under identical conditions likewise led to a 'normal' product (**221**, also shown to result from the appropriate carbonium ion and nitrile in acid). Heating a mixture of oximes **218** and **219** in PPA led to the isolation of **220** and **221** plus the crossed products **222** and **223**. Heating **220** and **221** together in PPA resulted in their recovery unchanged.

$$(CH_3)_3C{-}\underset{\underset{NOH}{\|}}{C}{-}CH_3 \xrightarrow[80^\circ]{PPA} (CH_3)_3CNHCOCH_3 + CH_3CONH_2$$

(218) **(220)**

$$C_6H_5C(CH_3)_2\underset{\underset{NOH}{\|}}{C}C_6H_5 \xrightarrow[80^\circ]{PPA} C_6H_5C(CH_3)_2NHCOC_6H_5$$

(219) **(221)**

$$\mathbf{218} + \mathbf{219} \xrightarrow[80^\circ]{PPA} \mathbf{220} + \mathbf{221} + \underset{\mathbf{(222)}}{(CH_3)_3CNHCOC_6H_5} + \underset{\mathbf{(223)}}{C_6H_5C(CH_3)_2NHCOCH_3} \quad (62)$$

Kenyon and coworkers[192] have proven that the normal Beckmann rearrangement proceeds with retention of configuration of the migrating group. The intermolecularity shown by the results of the crossover study (equation 62) suggests that if the asymmetric carbon were trisubstituted, fission of the oxime would be expected and any original optical activity would be lost. To test this idea, a stereochemical study was carried out using the oxime of 9-acetyl-*cis*-decalin (**224**)[191]. Here a change in relative configuration might be observed instead of loss of optical activity. This oxime was rearranged under a variety of conditions as shown in equation (63). The formation of the stereoisomerized *trans* amide (**226**) from rearrangements in strong acid, along with the observation of the formation of acetonitrile when using phosphorus pentachloride as a catalyst, provides strong evidence for the fragmentation–recombination mechanism. Adding to this evidence is the fact that the *trans* amide (**226**) can be synthesized by the addition of acetonitrile to the 9-decalyl carbonium ion generated from β-decalol plus sulphuric acid. It may be that the *trans* amide is the kinetically controlled product in these reactions.

(**224**) → (**225**) (**226**) (63)

Reagent	*Products*
$C_6H_7SO_2Cl/C_5H_5N$	**225**, 92%
PPA, 25°	**226**, 61%
H_2SO_4, 85–90%	**226**, 40%
PCl_5	acetonitrile

Among other recent examples of fragmentation products from α-trisubstituted oximes is the work reported by Shoppee, Lack and Roy[193] on the rearrangement of the oximes of some aza-steroids. They found, for example, that thionyl chloride at −20° converts 5α-cholestan-1-one oxime (**227**) to approximately equal amounts of the lactam **228** and nitrile **229**.

(**227**) $\xrightarrow[-20°]{SOCl_2}$ (**228**) + (**229**) (64)

An interesting case where fragmentation is not observed is shown in equation (65). The oxime ester **230** in 85% ethanol at reflux temperature for 15 min gives practically a quantitative yield of the amide **231**[194].

(**230**) $\xrightarrow{85\%\ C_2H_5OH}$ (**231**) (65)

b. α-Disubstituted oximes. The possibility of a fragmentation–recombination reaction of α-disubstituted acyclic ketoximes has also been reported. When phenyl cyclohexyl ketoxime (**232**) is heated in PPA a mixture of amides is formed as shown in equation (66)[195]. Evidence in favour of the two-step mechanism comes from the fact that heating the expected fragmentation products, cyclohexene and benzonitrile, in PPA leads to *N*-cyclohexylbenzamide (**234**) in good yield along with some of the hydration product, benzamide.

$$(C_6H_5)(C_6H_{11})C{=}N{-}OH \xrightarrow[130°]{PPA} C_6H_5CONH_2 + C_6H_{11}{-}CONHC_6H_5 + C_6H_5CONH{-}C_6H_{11} \qquad (66)$$

(**232**) (**233**) (**234**)

Organic halides may also be formed as Beckmann cleavage products when phosphorus pentachloride is used as a catalyst. Hassner and Nash[196] found an extremely facile cleavage of 1,1-diaryl-2-propanone oximes (**235**) to diarylmethyl chlorides (**236**) in essentially quantitative yields. Little or no amide products were detected.

$$(C_6H_5)(Ar)CHC({=}NOH)CH_3 \xrightarrow[5\ min,\ 0°]{PCl_5/ether} (C_6H_5)(Ar)CHCl + CH_3CN \qquad (67)$$

(**235**) (**236**)

c. α-Oximino ketones. When α-oximino ketones are treated with strong acids or acid chlorides, or when they are dissolved in aqueous base and treated with acylating agents, they are cleaved to nitriles and carboxylic acids[197]. When alcohols are present, esters are formed (equation 68)[198] Ferris and coworkers have favoured a mechanism as shown in equation (69) to explain these abnormal Beckmann reactions. Another mechanism which has been considered is one which involves attack of the base on the carbonyl carbon with concerted fragmentation of the rest of the molecule (equation 70)[199]. The mechanism in these cleavage reactions has as yet not been unequivocally settled. Freeman[200] has concluded that, in the presence of a good nucleophile and no hindrance to attack at the carbonyl, the concerted mechanism may

$$\mathrm{R{-}\overset{O}{\overset{\|}{C}}{-}\overset{NOH}{\overset{\|}{C}}{-}R'} \xrightarrow[\mathrm{R''ONa\ +\ R''OH}]{\mathrm{Ac_2O}} \mathrm{R{-}\overset{O}{\overset{\|}{C}}{-}OR'' + R'CN} \tag{68}$$

$$\mathrm{R{-}\overset{O}{\overset{\|}{C}}{-}\overset{NOH}{\overset{\|}{C}}{-}R'} \xrightarrow[\mathrm{R''OH}]{\mathrm{NaOR''}} \mathrm{R{-}\overset{O}{\overset{\|}{C}}{-}\overset{NO^-}{\overset{\|}{C}}{-}R'\ \ Na^+} \xrightarrow{\mathrm{Ac_2O}} \mathrm{R{-}\overset{O}{\overset{\|}{C}}{-}\overset{N{-}OAc}{\overset{\|}{C}}{-}R''}$$

$$\downarrow$$

$$\mathrm{H^+ + R{-}\overset{O}{\overset{\|}{C}}{-}OR''} \xleftarrow{\mathrm{R''OH}} \mathrm{R{-}\overset{O}{\overset{\|}{C}}{}^+ + R'CN + AcO^-} \tag{69}$$

$$\downarrow \mathrm{R''O^-}$$

$$\mathrm{R{-}\overset{O}{\overset{\|}{C}}{-}OR''}$$

$$\mathrm{R{-}\underset{\nwarrow B^-}{\overset{O}{\overset{\|}{C}}}{-}\overset{N{-}OR'}{\overset{\|}{C}}{-}R''} \longrightarrow \mathrm{R{-}\overset{O}{\overset{\|}{C}}B + R''CN + {}^-OR'} \tag{70}$$

be operative and when this path cannot be followed the fragmentation process may occur.

Hassner and coworkers[201] have demonstrated that esters of steroidal α-oximino ketones undergo cleavage instead of Beckmann rearrangement under certain conditions. The corresponding free oximes with boiling acetic acid–acetic anhydride give mainly imides (equation 71) but it was shown[202] that the reaction proceeds first by cleavage to a nitrile acid (**238**) followed by ring closure to the imide (**239**). More recent work by Hassner and Wentworth[203] has provided evidence in

HO– … =N–OH, =O **(237)** $\xrightarrow[\mathrm{Ac_2O}]{\mathrm{HOAc}}$ AcO– … –COOH, –CN **(238)** $\longrightarrow$ AcO– … NH, =O, =O **(239)** (71)

favour of the concerted mechanism (equation 70). Treatment of the acetate of **237** (**237-OAc**) with primary amines at 25° yields amidonitriles (**240**). As the steric requirements of the amine increase the cleavage slows and more hydrolysis of the oxime results. When **237-OAc** is treated with methanol or ethanol, cleavage is complete within 24 hours, while with isopropyl or *t*-butyl alcohol cleavage is essentially absent. All of these results, they felt, are not compatible with a cleavage mechanism such as equation (69), and are more easily rationalized through equation (70).

$$(\textbf{237-OAc}) \xrightarrow{RNH_2} (\textbf{240}) \qquad (72)$$

d. α-Amino ketoximes. Grob and coworkers have extensively studied the fragmentation reactions of α-amino ketoximes[204–208]. In their early investigations they found that the *syn* and *anti* forms of the esters, e.g. tosylate, benzoate or *p*-nitrobenzoate or ethers, e.g. 2,4-dinitrophenyl or picryl of α-aminoacetophenone oximes undergo fragmentation to a carbimonium salt (**242**) and benzonitrile (equation 73)[204]. The rate of fragmentation in 80% ethanol depends markedly on the nature of the amino and ester or ether groups and is only slightly influenced by substituents in the benzene ring. Some of their results are shown in Tables 12 and 13.

$$\underset{(\textbf{241})}{R_2NCH_2C(Ar){=}N{-}X} \xrightarrow{80\%\ EtOH} \underset{(\textbf{242})}{R_2\overset{+}{N}{=}CH_2} + X^- + ArCN \qquad (73)$$

Three possible mechanisms may be considered for these reactions. First, there could be a synchronous process with simultaneous cleavage giving $R_2\overset{+}{N}{=}CH_2$, X^- and the nitrile as shown in **243**. Secondly, there could be fragmentation to $R_2\overset{+}{N}{=}CH_2$ and a carbanion (**244**) which eliminates X^- in a later step. Finally, one can consider loss of X^- to form an iminium ion (**245**) which subsequently decomposes to a carbimonium ion and nitrile.

A *p*-nitrophenyl group should stabilize **244** and enhance the rate relative to phenyl if the second mechanism holds and if **244** is formed

TABLE 12. Dependence of rate of reaction of compounds **241** ($Ar = C_6H_5$, $X = C_6H_5COO$) on the amino group[204].

R_2N—:	$C_6H_5(CH_3)N$—	$O(CH_2CH_2)_2N$—	$C_6H_5CH_2(CH_3)N$—	$(CH_3)_2N$—	$(CH_2)_5N$—	$(CH_2)_4N$—
$k_{rel}^{13°}$:	1	179	1360	1560	1920	3000

TABLE 13. Dependence of rate of reaction of compounds **241** [$Ar = C_6H_5$, $R_2N = C_6H_5(CH_3)N$] on the ester group[204]

X—:	C_6H_5COO—	*p*-$CH_3OC_6H_4COO$—	*p*-$NO_2C_6H_4COO$—	3,5$(NO_2)_2C_6H_3COO$—
$k_{rel}^{70°}$:	1	0·6	20	83

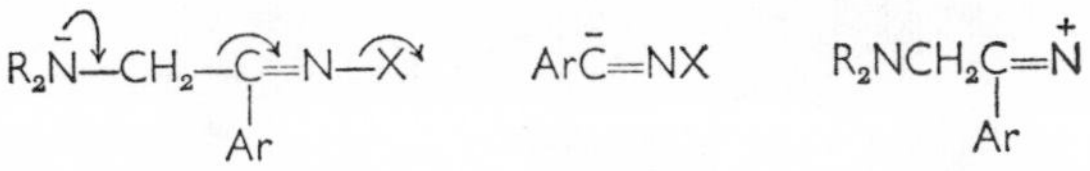

$R_2\ddot{N}{-}CH_2{-}C(Ar){=}N{-}X$ **(243)** $Ar\bar{C}{=}NX$ **(244)** $R_2NCH_2C(Ar){=}\overset{+}{N}$ **(245)**

in the rate-determining step. Similarly, one can argue that the *p*-nitrophenyl group should destabilize **245** relative to phenyl and decrease the rate. The failure of *p*-nitro or other substituents in the Ar group to greatly influence the rate along with the simultaneous and strong dependence of the rate on the nature of the amino group and the group X has been taken by the authors[204] as evidence for the synchronous mechanism of fragmentation. The process would thus amount to a coplanar *trans* elimination in the *anti* oxime series and an unusual *cis* elimination in the *syn* oxime series. The much lower reaction rates of the *syn* forms ($k_{anti}/k_{syn} \approx 2{,}000$) may reflect the stereoelectronically less-favourable transition state in these isomers. The rates of both forms, however, are greatly enhanced over those of comparable alkyl phenyl ketoxime derivatives.

Grob and coworkers[205,206] also found that the main reaction of the *p*-toluene sulphonate esters of a series of mono- and bicyclic α-amino ketoximes is fragmentation to carbimonium salts (equations 74 and 75). The products were isolated as the tosyl amides or amines. Compounds **246** and **247** give lactams which could be the products of a normal Beckmann rearrangement. However, these too could fragment to carbimonium ions which subsequently cyclize via a Ritter type reaction. This latter possibility was confirmed by the interception of the intermediate ions by cyanide ions (to give nitrile **248** in the case of compound **247**).

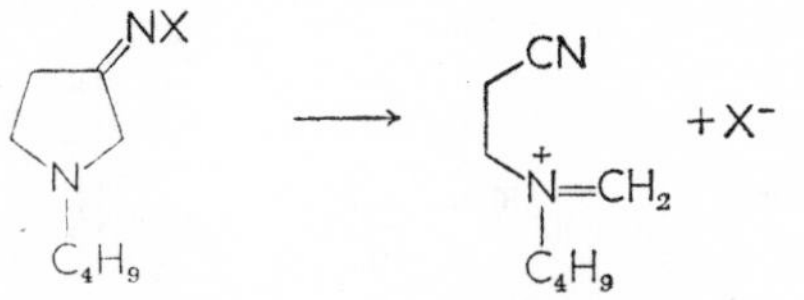

$$\text{1-}C_4H_9\text{-pyrrolidin-3-one oxime derivative (=NX)} \longrightarrow NC{-}CH_2CH_2{-}\overset{+}{N}(C_4H_9){=}CH_2 + X^- \quad (74)$$

$$\text{quinuclidin-3-one oxime derivative (=NX)} \longrightarrow \text{4-cyano-1-methylene-piperidinium} \ (\overset{+}{N}{=}CH_2) + X^- \quad (75)$$

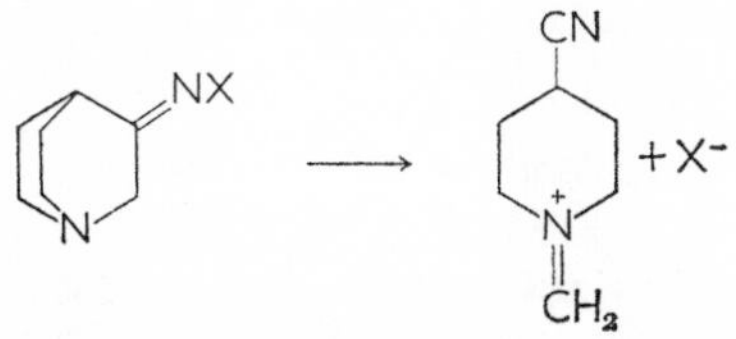

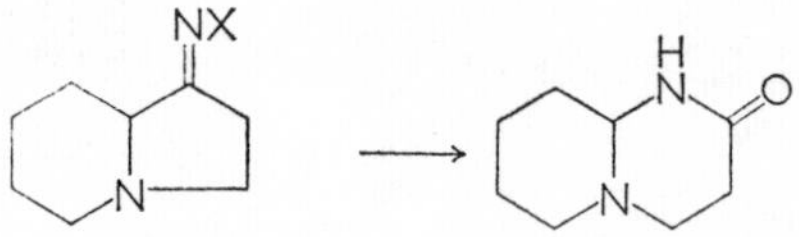

(246)

(76)

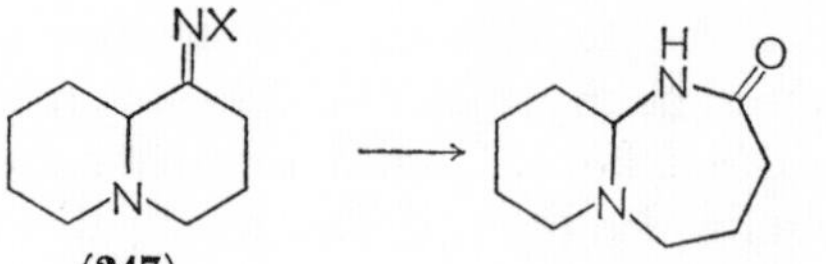

(247)

(77)

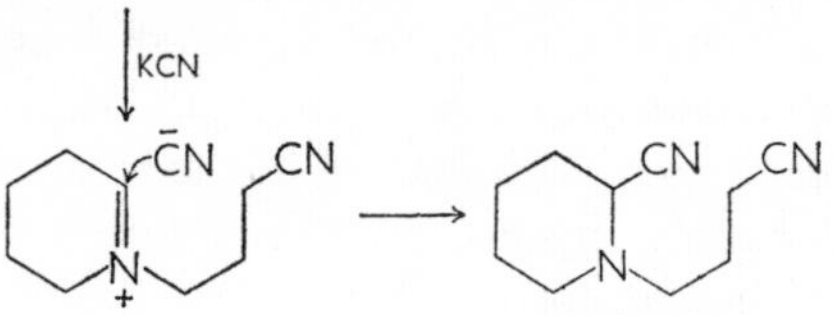

(248)

The rates of reaction of some α-amino ketoxime acetates and their homomorphous analogues under comparable conditions differ by as much as 10^8, favouring in all cases the α-amino ketoxime derivatives[207,208]. Acetates **249** and **251** quantitatively fragment to benzonitrile and cyclic immonium salts at rates 10^7 and 10^8, respectively, as high as those for the rearrangement of **251** and **252**[207]. Interestingly, even tosylate **253** undergoes fragmentation at a rate 2–4 times as high as that for the rearrangement of **254**[208]. In this last case the experimental results have been interpreted in terms of fragmentation of **253** to benzonitrile and the 2-quinuclidinyl cation (**255a, b**) which is the precursor of the other products. This suggests that there must be some delocalization of the positive charge on to the bridgehead nitrogen; i.e. some contribution from **255b.**

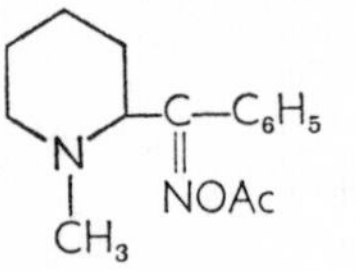

(249) 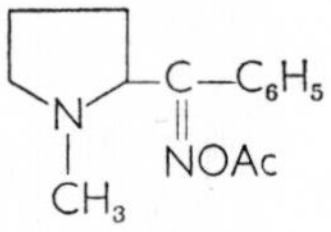 **(250)** 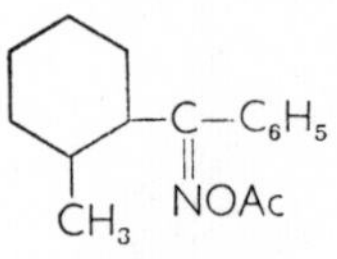 **(251)** 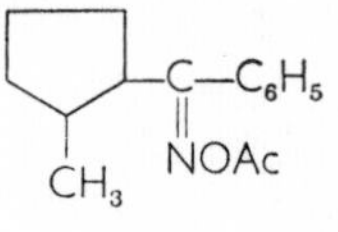 **(252)**

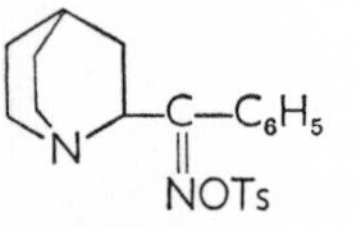

(253) 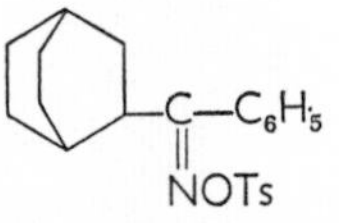 **(254)** 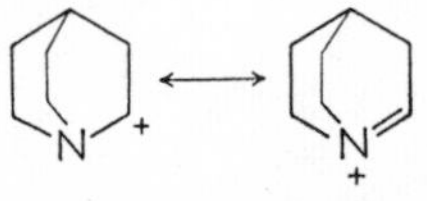 **(255a)** **(255b)**

e. Other examples. Among the many other examples of Beckmann fragmentation reactions reported during the last few years are the cleavage of α-hydroxy oximes to nitriles (equation 78)[209], the fragmentation of α-difluoroamino fluorimines to α-difluoroamino fluorides (equation 79)[210], and cleavage accompanied by rearragement in *N*-fluorimines **256**[133] and **257**[78].

$$\text{(HO, NOH decalin oxime)} \xrightarrow[-20°]{SOCl_2} \text{(keto nitrile, O, CN)} \tag{78}$$

$$\text{(2-Cl-2-}NF_2\text{-cyclohexanone N-fluorimine)} \xrightarrow[CH_2Cl_2]{BF_3} ClCF(NF_2)CH_2CH_2CH_2CN \tag{79}$$

$$\underset{\mathbf{(256)}}{C_6H_5C(=NF)-C(=NF)C_6H_5} \xrightarrow[EtOH]{NaOEt} C_6H_5CN + C_6H_5N{=}C(OEt)_2 \tag{80}$$

$$\underset{\mathbf{(257)}}{Cl-C_6H_4-C(=NF)CN} \xrightarrow[EtOH]{NaOEt} Cl-C_6H_4-CN + Cl-C_6H_4-N{=}C(OEt)_2 \tag{81}$$

$$Cl-C_6H_4-CN \longrightarrow Cl-C_6H_4-CONH_2$$

The fragmentation of β-keto ether or thioether oximes to nitriles and carbonyl compounds can be applied to structure elucidation of natural products and to the synthesis of compounds difficult to prepare by other routes. Several examples illustrating Beckmann cleavage in these oximes are shown in equations (82)[211], (83)[212] and (84)[213].

Ethyl-5-cyano-2-oximinopentanoate (**259**) has been obtained by the action of acetic anhydride on a NaOEt/EtOH solution of 1,2,3-cyclohexanetrione-1,3-dioxime (**258**)[214]. Reaction of 1-phenyl-1,2,3-butanetrione 2-oxime (**260**) with an excess of potassium cyanide in methanol results mainly in the formation of 2-hydroxy-2-cyanophenylacetamide (**261**) which can be rationalized as arising from the initial cleavage products[215]. Indolyl-3-pyruvic acid oxime (**262**) is probably

$$\text{dispiro oxime} \xrightarrow[C_6H_6]{PCl_5} \text{1-hydroxy-1-cyclohexyl-}CH_2CN + \text{cyclohexanone} \tag{82}$$

$$\text{MeO-tetralone oxime (SMe)} \xrightarrow[70^\circ]{Et_2NCF_2CHClF} \text{MeO-aryl(CN)-CH=CH-SMe} \xrightarrow[HClO_4]{15\%} \text{MeO-aryl(CN)-}CH_2CH_2CHO \tag{83}$$

$$\text{2-methoxycyclohexanone oxime} \xrightarrow[CCl_4,\ 0^\circ]{SOCl_2} NC(CH_2)_4CHClOCH_3 \xrightarrow{CH_3OH} NC(CH_2)_4CH(OCH_3)_2 \tag{84}$$

a precursor of the natural product indolyl-3-acetonitrile (**263**) since fragmentation of **262** to **263** can be achieved under conditions approximating the biochemical process[216].

$$\underset{\textbf{(258)}}{\text{2,6-bis(hydroxyimino)cyclohexanone}} \xrightarrow[2.\ Ac_2O]{1.\ EtOH/NaOEt} \underset{\textbf{(259)}}{EtO_2CC(=NOH)(CH_2)_3CN} \tag{85}$$

$$\underset{\textbf{(260)}}{C_6H_5C(=O)-C(=NOH)-C(=O)CH_3} \xrightarrow[\text{aq. MeOH}]{KCN,\ 50\%} \underset{\textbf{(261)}}{C_6H_5C(OH)(CN)-CONH_2} \tag{86}$$

$$\underset{\textbf{(262)}}{\text{indol-3-yl-}CH_2C(=NOH)CO_2H} \xrightarrow[41^\circ]{HCl/KCl} \underset{\textbf{(263)}}{\text{indol-3-yl-}CH_2CN} \tag{87}$$

5. The mechanism

Mechanistic interpretations offered by authors in attempts to explain 'abnormal' products, or normal products arising perhaps from abnormal routes, have been interspersed through the preceding consideration of Beckmann fragmentation reactions. It is necessary to consider here the results of some more intensive and quantitative studies which, in some cases, shed a more general light on the mechanism or spectrum of mechanisms for this reaction.

Much of our present understanding of the nature of the intermediates in the Beckmann rearrangement comes from the early work by Kuhara[217] and Chapman[218] and the more recent studies by Grob[219]. Kuhara and coworkers[217] showed that while oximes do not normally rearrange without an added catalyst, their benzenesulphonyl esters rearrange readily at room temperature, in the dark, and without any catalyst. Furthermore, they demonstrated that the initial product is an imidyl derivative (**264**) which undergoes further rearrangement by an O → N shift to finally give the amide which is usually isolated (equation 88).

$$\begin{matrix} R \\ \quad \diagdown \\ \quad C{=}N \\ \diagup \quad \diagdown \\ R' \qquad OSO_2C_6H_5 \end{matrix} \xrightarrow{\text{spontaneous}} \underset{(\mathbf{264})}{\begin{matrix} R'C{=}N{-}R \\ | \\ OSO_2C_6H_5 \end{matrix}} \longrightarrow \begin{matrix} R'C{-}N{-}R \\ \| \quad | \\ O \quad SO_2C_6H_5 \end{matrix} \tag{88}$$

Evidence for the imidyl intermediate comes also from its observed interception by nucleophiles such as alcohols and amines.

Among other things, Kuhara and his group[220] also found the ease of rearrangement in a series of oxime esters to be proportional to the strength of the esterifying acid; the order for several acids being $C_6H_5SO_3H > ClCH_2CO_2H > C_6H_5CO_2H > CH_3CO_2H$. Although acyl oximes derived from oximes plus sulphuric acid, PPA, or phosphorus pentachloride have not been isolated as intermediates, it is believed that they are formed from these catalysts and then rearrange rapidly in a manner analogous to the benzenesulphonyl esters.

Chapman and coworkers[218] established that in the rearrangement of the picryl ethers of *p*-substituted benzophenone oximes (**265**) electron-withdrawing substituents strongly decrease the rate and

$$\underset{(\mathbf{265})}{\begin{matrix} p\text{-}XC_6H_4 \\ \quad \diagdown \\ \quad C{=}N \\ \diagup \quad \diagdown \\ C_6H_5 \qquad OC_6H_2(NO_2)_3 \end{matrix}} \longrightarrow \begin{matrix} O \quad C_6H_2(NO_2)_3 \\ \| \quad \diagup \\ C_6H_5CN \\ \quad \diagdown \\ \qquad C_6H_4X\text{-}p \end{matrix} \tag{89}$$

electron-donating substituents strongly increase the rate when in the migrating group. The same substituents in the non-migrating phenyl group affect the rate in the same manner but to a much smaller degree. There also was observed an increase in rate as the ionizing ability of the solvent increased. Rates they determined followed the first-order rate law in every case and were lower than the rates of conversion of the corresponding imidyl esters to the same amides. These data led them to conclude that the transition state of the rate-determining step in these Beckmann rearrangements must involve partial ionization of the N—O bond with simultaneous migration of the aryl group *trans* to the picryl group.

The results of the studies by Kuhara and Chapman and their co-workers do contribute much towards the elucidation of the details of the mechanism of the rearrangement at least under the conditions they employed. However, the large variety of catalysts commonly used in the Beckmann rearrangement, along with the great variety in structures of oximes and oxime derivatives which are known to re-arrange, make it unfeasible to assign one mechanism to this reaction. Each set of reaction conditions may define a slightly different mechanism. All of the mechanisms, though, must have in common several basic steps; conversion of the oxime to the rearranging species, rearrangement and conversion of the intermediate to the isolated species.

Over the past few years Grob and his group[219] have provided many quantitative data on the mechanism of rearrangement of ketoxime tosylates in aqueous ethanol. Relative rates for the solvolysis of some of these tosylates in 80% ethanol (containing a two-molar equivalent of triethylamine to neutralize the toluenesulphonic acid formed) are given in Table 14. All of the reactions are first order in oxime ester concentration. As may be seen from the data, the rate increases with increased branching at the azomethine carbon, but the effect on rate is much less than the effect of the same branching on the rate of solvolysis of the corresponding chlorides. There is no relationship between the relative rates of rearrangement of the oxime tosylates and the corresponding alkyl chloride solvolyses. The rearrangement rates remain more constant over the series studied than the solvolysis rates, which increase rapidly with increased charge stabilization.

These data are consistent with partial ionization in the transition state to give an electron-deficient species. The electron deficiency, though, must be less concentrated than in an alkyl carbonium ion. Consistent with this idea is the observation that even 1-bicyclo-[2.2.2]octyl methyl ketoxime tosylate (**266**) reacts at an appreciable rate relative to *t*-butyl methyl ketoxime tosylate[219], in great contrast to the relative rates of solvolyses of the corresponding bromides. The lack of a consistent or linear increase in rate with increased branching has been attributed to a decrease in solvation of the transition state

TABLE 14. Rearrangement rates of some ketoxime tosylates (0·001M) in 80% ethanol at 23°.[219]

$$\begin{array}{c} R \\ \diagdown \\ C{=}N \\ \diagup \quad \diagdown \\ R' \qquad OTs \end{array}$$

R	R′	$k \times 10^5 (s^{-1})$	$k_{rel.}$	RCl Solvolysis, $k \times 10^5$ (25°)
CH_3	CH_3	1·07	1	
C_2H_5	CH_3	64·8	60	
$CH(CH_3)_2$	CH_3	868	810	$<1 \times 10^{-3}$
$CH(C_6H_5)_2$	CH_3	43·1	40	172
$C(CH_3)_3$	CH_3	931	870	0·924
$C(CH_3)(C_2H_5)C_6H_5$	CH_3	878	820	528
C_6H_5	CH_3	106	100	
$CH(CH_3)_2$	C_6H_5	45·8	1	
$C(CH_3)_3$	C_6H_5	55·3	1·2	
$C(CH_3)(C_2H_5)C_6H_5$	C_6H_5	247	5·4	
C_6H_5	C_6H_5	288	6·3	
Cyclopentanone oxime tosylate			1	
Cyclohexanone oxime tosylate			21	
Diethyl ketoxime tosylate			2·8	

due to steric crowding. A more recent study shows that steric hindrance to solvation is an effective factor contributing to relative rearrangement rates only for certain solvents or solvent mixtures[221].

$$\begin{array}{c} \text{(bicyclo[2.2.2]octyl)} \\ | \\ C{=}N \\ \diagup \quad \diagdown \\ H_3C \qquad OTs \end{array}$$

(266)

Since the constrained cyclohexyl and cyclopentyl systems cannot readily rearrange to a linear intermediate, it is of interest to compare the rates of cyclohexanone and cyclopentanone oxime tosylates to that of diethyl ketoxime tosylate. The relative rates observed by Grob and coworkers[219] (Table 14) show that the rate-determining step must not be migration to a linear structure as is required by the nitrilium ion (**267a, b**) but is more likely rearrangement to a structure such as **268** which is bent at the azomethine group. This bent intermediate can then ionize rapidly to the nitrilium ion, which serves as a common intermediate for the formation of rearrangement and fragmentation products. These ideas are incorporated in equation (90).

$$\text{RR'C=N–X} \xrightarrow{k_1} \underset{(\mathbf{268})}{\text{R'XC=N–R}} \underset{k_{-2}}{\overset{k_2}{\rightleftharpoons}} \underset{(\mathbf{267a})}{\text{R'–C}\equiv\overset{+}{\text{N}}\text{–R}} \longleftrightarrow \underset{(\mathbf{267b})}{\text{R'–}\overset{+}{\text{C}}\text{=N–R}} \;\; \text{X}^- \qquad (90)$$

$$(\mathbf{267a}) \underset{k_{-4}}{\overset{k_4}{\rightleftharpoons}} \text{R'CN} + \text{R'} \qquad (\mathbf{267b}) \xrightarrow[k_3]{\text{Y}^-} \text{R'YC=N–R}$$

$$\underset{(\mathbf{269})}{\text{R'–C}\overset{\delta+}{\underset{\delta-}{=}}\text{N (bridged by R and X)}}$$

The transition state of the isomerization step is pictured by Grob and coworkers[219] as a bridged intimate ion-pair complex (**269**) in which the relative positions of the groups around the azomethine double bond more closely resemble those in the starting material. This explains the acceleration of the reaction by polar solvents, the normal rates of the cyclic ketoxime tosylates, and avoids the formation of a concentrated charge at the oxime carbon. It is not to be implied that this two-step, migration–ionization mechanism applies to all groups R and R′. A one-step rate-determining rearrangement and ionization to the nitrilium ion might be applicable for aryl migration.

The direct or one-step ionization mechanism may also apply in cases where the group X leaves through assistance from complex formation. The nitrilium salt intermediates first described by Theilacker and Mohl[222] as resulting from the treatment of the *N*-chlorimine of benzophenone (**270**, R = C_6H_5) with antimony pentachloride have been isolated and characterized by Grob and coworkers[219]. The intermediates were isolated in good yield and were identified from their decomposition products and by infrared spectral comparison with authentic samples. Infrared spectra of the nitrilium salts indicated that they are linear as shown.

$$\underset{(\mathbf{270})}{\text{R(C}_6\text{H}_5\text{)C=N–Cl}} \xrightarrow{\text{SbCl}_5} \text{C}_6\text{H}_5\text{C}\equiv\overset{+}{\text{N}}\text{–R} + \text{SbCl}_6^- \qquad (91)$$

$$\text{R} = \text{C}_6\text{H}_5, \text{C(CH}_3\text{)}_3$$

The mechanistic scheme shown in equation (90) suggests that the ratio of amide formation to fragmentation depends upon the ratio of k_3 to k_4. Obviously, the nucleophilicity of Y^- and the nature of R^+ are determining factors in establishing this ratio. In solutions containing no good nucleophile, such as PPA, fragmentation may be favoured (assuming k_4/k_{-4} is large).

The question arises as to whether or not any Beckmann fragmentation reactions need be said to follow the mechanism in equation (90). As has been previously pointed out, the evidence for a one-step synchronous elimination mechanism is quite convincing for some cases where the substituents on the α-carbon have a large electron-donating ability; e.g. α-amino ketoximes. It seems now that oximes with groups R of lesser electron-donating ability which still fragment, e.g. α-trisubstituted and α-disubstituted oximes, do so through a normal rearrangement intermediate. Evidence for this is contained in Table 15[219].

TABLE 15. Amounts of fragmentation and relative rearrangement rates of some ketoxime tosylates[219]

R(H$_3$C)C=N–OTs

R	% Fragmentation	$k_{rel.}^{23°}$	$k_{rel.}^{25°}$ (RCl)
$CH(CH_3)_2$	—	1	< 0·001
$C(CH_3)_3$	10	1·1	1
$CH(C_6H_5)_2$	54	0·05	186
$C(CH_3)(C_2H_5)C_6H_5$	80	1·01	572

As seen from these data, there is no relationship between the amount of fragmentation and the rate of the rearrangement or between these rates and the rates of solvolyses of the corresponding chlorides. The fragmentation yields, though, do parallel relative solvolysis rates, indicating that the degree of fragmentation corresponds to the stability of the carbonium ion formed[219,223]. Thus, in these cases, it has been concluded that fragmentation follows the rate-determining step or, in other words, the step in which the nitrilium ion is formed. Further evidence for the nitrilium ion as the precursor to both fragmentation and rearrangement comes from the observation that **271** gives about the same ratio of amide and nitrile as is obtained under comparable conditions from **272**[219].

$[C_6H_5C{\equiv}\overset{+}{N}{-}C(CH_3)_3]\ FeCl_4^-$

(271)

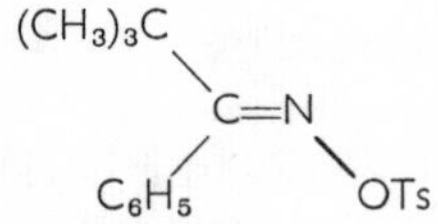

$(CH_3)_3C(C_6H_5)C{=}N{-}OTs$

(272)

N-Alkyl nitrilium salts have been found to be common intermediates for the rearrangement and fragmentation processes in the solvolyses of optically pure (−)-*anti*-2-methyl-2-phenylbutyrophenone oxime tosylate (**273**) and (−)-*anti*-3-methyl-3-phenyl-2-pentanone oxime tosylate (**274**) in methanol[224]. The products are the amide with retained configuration of the migrating group and 2-phenyl-2-methoxybutane (**275**) with predominant inversion.

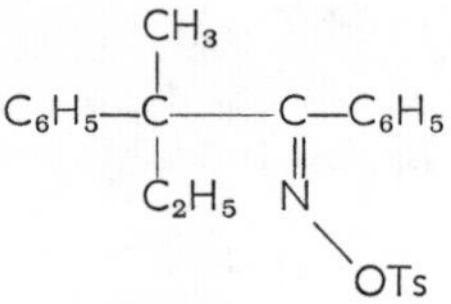

$C_6H_5{-}C(CH_3)(C_2H_5){-}C(C_6H_5){=}N{-}OTs$ **(273)**

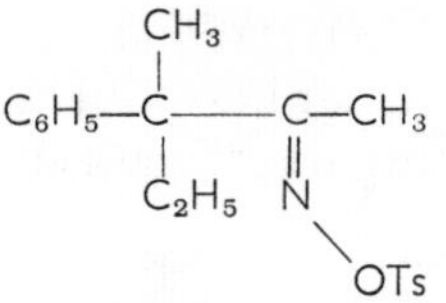

$C_6H_5{-}C(CH_3)(C_2H_5){-}C(CH_3){=}N{-}OTs$ **(274)**

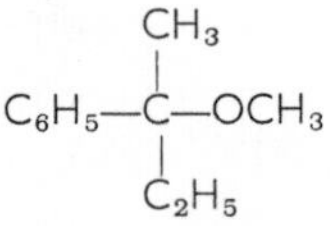

$C_6H_5{-}C(CH_3)(C_2H_5){-}OCH_3$ **(275)**

A 'univalent' nitrogen species has been proposed as a reaction intermediate for the Beckmann rearrangement but has, until recently, received little support since it cannot explain the stereospecificity normally observed without drastic modifications[55]. Recently, though, Lansbury and coworkers[225–227] have reported some examples of non-stereospecific rearrangements of 7-alkyl-1-indanone oximes. For example, they found that PPA converts 4-bromo-7-*t*-butyl-1-indanone-oxime (**276**) mainly into 1,8-ethano-7-bromo-4,4-dimethyl-3,4-dihydroisoquinoline (**277**) plus a mixture of lactams **278** and **279**.

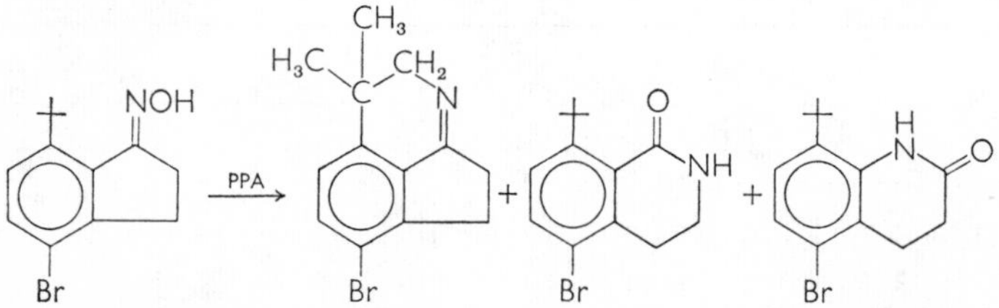

(276) $\xrightarrow{PPA}$ **(277)** + **(278)** + **(279)** (92)

Indanone oximes with less bulky substituents in the 7-position give more of the lactam from aryl migration and 8-substituted tetralone oximes give only Beckmann rearrangement with aryl migration[228].

The products in equation (92) are explained as arising from a combination of several steric effects[227]. Torsional strain in the transition state associated with aryl migration in **276**, combined with the steric hindrance offered by the *t*-butyl group, hinders aryl migration through the normal pseudo three-membered ring transition state. These effects are less severe in cases of smaller substituents or in the tetralone series where the ring is more flexible and normal stereospecific migration is then observed. Thus, Lansbury and coworkers have concluded that an iminium ion (**280**) is formed in the indanone series. This ion accounts for the alkyl migration and for the presence of **277** as an iminium insertion product.

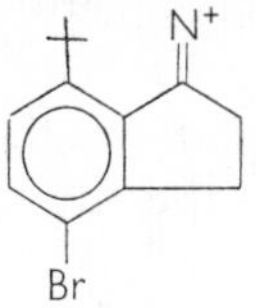

$$C_6H_5C{\equiv}\overset{+}{N}C_6H_5 \quad BF_4^- \qquad C_6H_5C(OC_2H_5){=}NC_6H_5$$

(**280**) 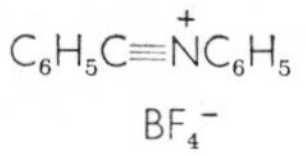 (**281**) 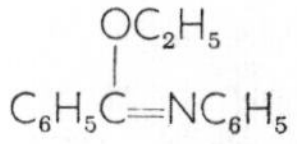 (**282**)

Loeppky and Rotman[126] hoped to trap an iminium ion from the treatment of *N*-chloro ketimines of benzophenone with silver tetrafluoroborate, reasoning that this reagent might satisfactorily replace intramolecular assistance to ionization. Using aqueous dioxan as solvent they found mainly the amides from normal Beckmann rearrangements. When aprotic solvents were employed, the products formed could be explained as arising from the nitrilium ion (**281**). The intermediacy of **281** was further substantiated by its trapping with sodium ethoxide in 1,2-dimethoxyethane to give ethyl *N*-phenylbenzimidate (**282**).

Pearson and McNulty[229] compared the rates of rearrangement of several substituted acetophenone oximes in concentrated sulphuric acid. Among their results was the observation that *p*-alkyl substituents show less electron-release ability (relative to hydrogen) than they do in most nucleophilic reactions. Also, the σ-constants of alkyl groups occupying positions *ortho* to each other were not additive in predicting rates. The correlation of log rate with the Hammett acidity constant, H_0, on the other hand, was good. In conclusion, structure **283** was offered as a picture of a 'transition complex' in these systems.

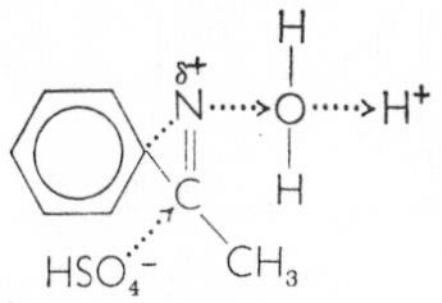

(**283**) 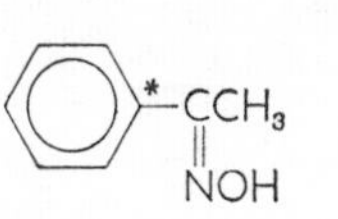 (**284**)

Yukawa and Kawakami[230] reported a reverse carbon-14 isotope effect ($k^{14}/k^{12} = 1{\cdot}12$) for the rearrangement of phenyl-1-labelled acetophenone oxime (**284**) which they interpreted as favouring a mechanism in which the migrating group participates in the N—O bond fission as in structure **283**. Glover and Raaen[231] have repeated this work and their results do not substantiate the previous findings of Yukawa and Kawakami. The discrepancy between these two studies is not clear at present.

To ascertain whether alkyl groups right on the azomethine carbon influence rates of rearrangement mainly through polar or through steric effects, Kawakami and Yukawa[232] have determined the rates of rearrangement of several alkyl methyl ketoximes (all in the *syn*-methyl configuration). The sequence found for the first-order rate constants in sulphuric acid was cycloheptyl ≈ cyclohexyl > 2-methylcyclopentyl > cyclopentyl > n-amyl. The effect of these migrating groups upon the reaction rates cannot be explained by their polar effect alone, since there is no correlation between the relative rates and Taft's σ^* values[177]. An explanation based on the steric effect of the migrating groups and relative changes in steric compression in the ground and transition states may be more satisfactory.

The kinetics and mechanism of the Beckmann rearrangement of alicyclic ketoximes in concentrated and fuming sulphuric acid have been extensively studied by Vinnik and Zarakhani. An excellent review of their work has recently been published[233]. On the basis of their results they propose that in all cases the alicyclic ketoximes are present as four equilibrated forms: the unionized form (**285**), the protonated form (**286**), the ion pair (**287**) and the dehydrated ion pair (**288**). The relationship of these forms is shown in equation (93). The observed

$$(CH_2)_n\,C{=}NOH + H_2SO_4 \rightleftharpoons (CH_2)_n\,C{=}\overset{+}{N}(H)OH + HSO_4^-$$

(285) **(286)**

$$\mathbf{286} + HSO_4^- \rightleftharpoons (CH_2)_n\,C{=}\overset{+}{N}(H)OH\cdot HSO_4^- \tag{93}$$

(287)

$$\mathbf{287} \rightleftharpoons (CH_2)_n\,C{=}\overset{+}{N}\cdot HSO_4^- + H_2O$$

(288)

$$\mathbf{288} \xrightarrow{\text{rate determining}} \text{lactam}$$

dependence of the rate constants on the acidity of the medium, the activity of water, and the activity of sulphuric acid is best explained by considering that the rate-determining step occurs from the dehydrated ion pairs as shown. At lower acid concentrations, such as were used by Pearson and coworkers[229], a step leading up to **288** may be rate-determining.

Although most of the aforementioned work by Grob and coworkers with ketoxime esters was carried out in 80% ethanol, there have been others who have studied the effect of solvent variation on rearrangement rates. Chapman[234] suggested that the rate enhancement provided by polar solvents could be correlated by the dipole moments of the solvents. Heldt[235] later asserted that rates could be reasonably well correlated by Kosower's *Z* values[236] or, that is, by the ionizing power of the solvents. These successful correlations may be due to the fortuitous choice of only a few solvents as pointed out by Tokura, Kawahara and Ikeda[161]. These authors used twelve solvents in their study of the rearrangement of cyclohexanone oxime tosylate. The rate constants decreased in the order: liquid sulphur dioxide > methanol > acetic acid > nitromethane > ethanol > acetonitrile > pyridine > chloroform > acetone > ethylene dichloride > methyl ethyl ketone > carbon tetrachloride. They could establish no good relationship between log *k* and the dielectric constants, or between log *k* and the dipole moments, or between log *k* and the log *Z* values of the solvents. An adequate explanation of the effect of such solvents on the rate of Beckmann rearrangement of oxime esters awaits further study.

B. The Chapman Rearrangement

1. Introduction

The thermal rearrangement of an aryl imidate (**289**) to an *N*-aroyldiaryl amine (**290**) was discovered by Mumm, Hesse and Vol-

$$\underset{\textbf{(289)}}{Ar^1\text{—}\overset{\displaystyle OAr^2}{\overset{|}{C}}\text{=N—}Ar^3} \xrightarrow{200\text{–}300°} \underset{\textbf{(290)}}{Ar^1\text{—}\overset{\displaystyle O}{\overset{\|}{C}}\text{—N}\begin{matrix}\diagup Ar^2\\ \diagdown Ar^3\end{matrix}} \qquad (94)$$

quartz[237] several years before the mechanism and scope of the reaction were studied by Chapman[238–241]. The reaction is now usually referred to as the Chapman rearrangement and less frequently as the

Chapman–Mumm rearrangement. Quite an extensive review of this reaction by Schulenberg and Archer[242] has been published recently. In the brief consideration of the reaction presented here, the broadest definition of the Chapman rearrangement will be used, some examples of the thermal rearrangement of alkyl imidates will be included along with examples of rearrangements brought about with the aid of added catalysts.

2. Mechanism

That the Chapman rearrangement is an intramolecular reaction was first suggested by Chapman himself[238] and later confirmed by Wiberg and Rowland[243]. The latter authors heated a mixture of compounds **291** and **292** and found both the x-ray powder pattern and the infrared spectrum of the mixture of amides formed to be identical to those of a mixture of amides obtained from heating the two imidates separately and mixing the products. In other words, no crossover product could be detected.

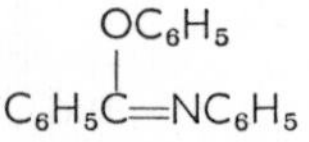

$$\underset{\textbf{(291)}}{C_6H_5C(OC_6H_5){=}NC_6H_5} \qquad \underset{\textbf{(292)}}{C_6H_5C(OC_6H_4Cl\text{-}p){=}NC_6H_4Cl\text{-}p}$$

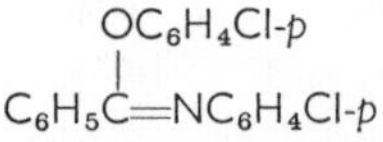

The rearrangement has been found to follow first-order kinetics with the rates being dependent on the nature and position of substituents in the aromatic rings. Chapman[240] found that electron-withdrawing substituents on the aryloxy ring (Ar^2 in **289**) of aryl imidates facilitate the reaction and he concluded that the rate of the reaction is associated with the acidity of the phenol from which the ester is derived. He also found that electron-withdrawing substituents in the arylimino ring (Ar^3) retard the reaction as they do to a lesser extent when in the *C*-aryl ring (Ar^1).

From their study of a large number of substituted imidates, Wiberg and Rowland[243] concluded that the Chapman rearrangement is essentially an intramolecular nucleophilic displacement on an aromatic ring. This description is consistent with the effects of the substituents in the various rings on the rate of rearrangement and can perhaps best be pictured as in equation (95). Additional support for the nucleophilic aromatic displacement nature of the rearrangement

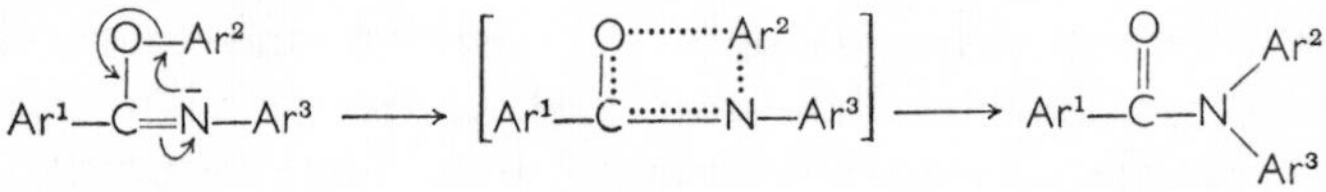

$$Ar^1{-}C(O{-}Ar^2){=}N{-}Ar^3 \longrightarrow \left[Ar^1{-}C(\overset{\cdots}{O}\cdots Ar^2)\cdots N{-}Ar^3\right] \longrightarrow Ar^1{-}C({=}O){-}N(Ar^2)(Ar^3) \qquad (95)$$

has come from the observation that if the log of the rate constants for the nucleophilic displacement reactions of a series of *p*-substituted *o*-nitrobromobenzenes by piperidine is plotted against the log of the rate constants for the Chapman rearrangement of a series of *p*-substituted (Ar^2) imidates, a very good correlation is obtained[243].

Chapman, as well as Wiberg and Rowland, found that imidates with *ortho* substituents in the aryloxy ring rearrange more readily than those with the same substituent in the *para* position of the aryloxy ring. Wiberg and Rowland[243] explained this enhanced reactivity of the *ortho*-substituted compounds as being due to an entropy effect; the argument being essentially that the *ortho* substituents restrict free rotation of the aryloxy ring in the reactant, and thus lessen the entropy decrease in going between the reactant and the cyclic transition state.

3. Synthetic utility

Chapman[241], in 1929, pointed out the fact that the thermal rearrangements of aryl imidates, which he had been studying, could be of value in synthesizing substituted diphenylamines via hydrolysis of the initially formed amides (equation 96). Many substituted diphenyl-

$$\mathrm{Ar^1{-}C(OAr^2){=}N{-}Ar^3} \xrightarrow{\Delta} \mathrm{Ar^1{-}C({=}O){-}N(Ar^2)(Ar^3)} \xrightarrow{\text{alc. KOH}} \mathrm{Ar^2NHAr^1} \qquad (96)$$

amines which can be obtained from the Chapman rearrangement of imidates are difficult to prepare by other methods while, admittedly, some are just as readily or more readily available from the Ullmann condensation[242].

A tabular summary of the reported applications of the Chapman rearrangement has been provided by Schulenberg and Archer[242]. Their literature coverage extends through the middle of 1963. A few selected examples will suffice here (equations 97–100).

Jamison and Turner[246] discovered that the Chapman rearrangement may be used to synthesize substituted acridones (equation 101). The process has been exploited by Singh and coworkers[248–250] in their syntheses of 9-amino acridines in search of antimalarial agents. In a similar fashion, Cymerman-Craig and Loder[251] used the Chapman rearrangement to prepare substituted benzacridones (equation 102) which can be reduced by sodium amalgam to benzacridines.

$$C_6H_5C(OC_6H_4Br\text{-}o)=NC_6H_4Br\text{-}o \xrightarrow[1\ h]{265°} C_6H_5C(=O)\text{—}N(C_6H_4Br\text{-}o)_2\ (86\%) \xrightarrow[100°,\ 8\ h]{aq.\ KOH} HN(C_6H_4Br\text{-}o)_2\ (87\%)$$ (97) (ref. 244)

$$C_6H_5C(O\text{-}C_6H_4\text{-}C(=O)C_6H_5)=NC_6H_5 \xrightarrow[40\ min]{270°} C_6H_5C(=O)\text{—}N(C_6H_5)\text{-}C_6H_4\text{-}C(=O)C_6H_5\ (90\%) \xrightarrow[1\ h]{alc.\ KOH} HN(C_6H_5)\text{-}C_6H_4\text{-}C(=O)C_6H_5\ (89\%)$$ (98) (ref. 245)

$$C_6H_5C(OC_6H_2Cl_3\text{-}2,4,6)=NC_6H_3Cl_2\text{-}2,4 \xrightarrow[2\ h]{270°} C_6H_5C(=O)\text{—}N(C_6H_2Cl_3\text{-}2,4,6)(C_6H_3Cl_2\text{-}2,4)\ (81\%) \xrightarrow{alc.\ KOH} HN(C_6H_2Cl_3\text{-}2,4,6)(C_6H_3Cl_2\text{-}2,4)\ (92\%)$$ (99) (ref. 246)

$$O_2S[C_6H_4\text{—}O\text{—}C(C_6H_5)=NC_6H_5]_2 \xrightarrow[4\ h]{220°} O_2S[C_6H_4\text{—}N(C_6H_5)\text{—}C(=O)C_6H_5]_2\ (97\%) \xrightarrow{aq.\ KOH} O_2S[C_6H_4\text{—}NH\text{—}C_6H_5]_2\ (85\%)$$ (100) (ref. 247)

$$C_6H_5C(O\text{-}C_6H_4\text{-}CO_2CH_3)=N\text{-}C_6H_4\text{-}Cl \xrightarrow{270°} C_6H_5C(=O)\text{—}N(C_6H_4CO_2CH_3)(C_6H_4Cl) \xrightarrow{320°} \text{2-chloroacridone (Cl, NH, C=O)}$$ (101)

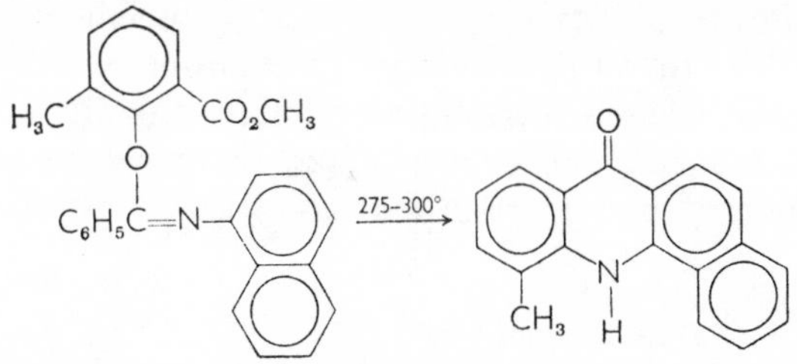

$$\text{(2-}CO_2CH_3\text{-6-}H_3C\text{-phenyl)}O\text{–}C(C_6H_5){=}N\text{–(1-naphthyl)} \xrightarrow{275\text{–}300^\circ} \text{(benzacridone: C=O, N–H, }CH_3\text{)} \quad (102)$$

4. Rearrangements of alkyl imidates

There have been several reports of the thermal rearrangement of alkyl imidates. Generally the temperatures required are higher than those needed for the rearrangement of aryl imidates and the yields are lower (equation 103)[252]. In contrast to the normal Chapman rearrangement, the rearrangement of alkyl imidates has been shown to

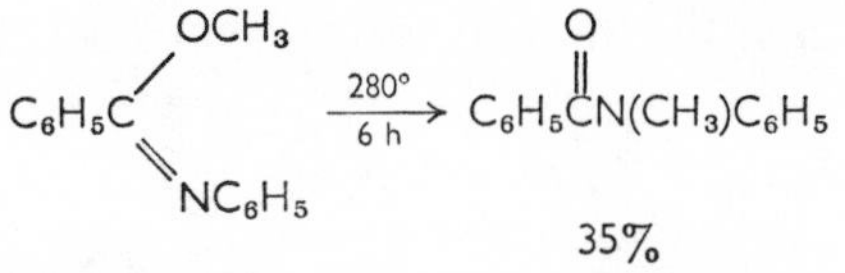

$$C_6H_5C(OCH_3){=}NC_6H_5 \xrightarrow[6\ h]{280^\circ} C_6H_5\overset{O}{\overset{\|}{C}}N(CH_3)C_6H_5 \quad (103)$$

35%

be intermolecular[253] and it has been proposed that it is a free-radical process[254]. A couple of recent and novel examples are shown in equations (104)[255] and (105)[256].

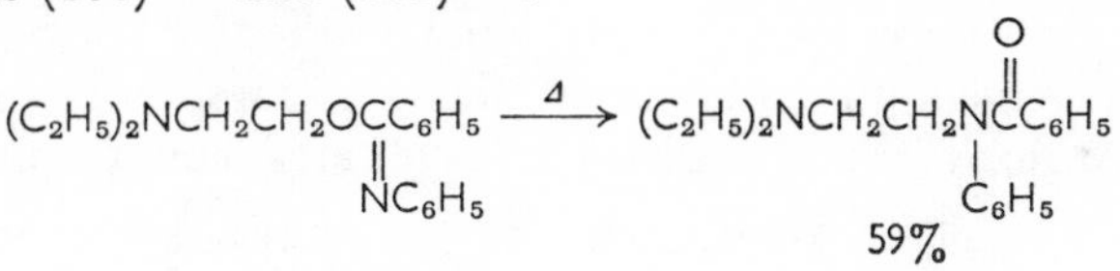

$$(C_2H_5)_2NCH_2CH_2OC({=}NC_6H_5)C_6H_5 \xrightarrow{\Delta} (C_2H_5)_2NCH_2CH_2N(C_6H_5)\overset{O}{\overset{\|}{C}}C_6H_5 \quad (104)$$

59%

$$\text{3-methoxy-1,2-benzisothiazole 1,1-dioxide} \xrightarrow{190^\circ} \text{2-methyl-1,2-benzisothiazol-3-one 1,1-dioxide }(C{=}O,\ NCH_3,\ SO_2) \quad (105)$$

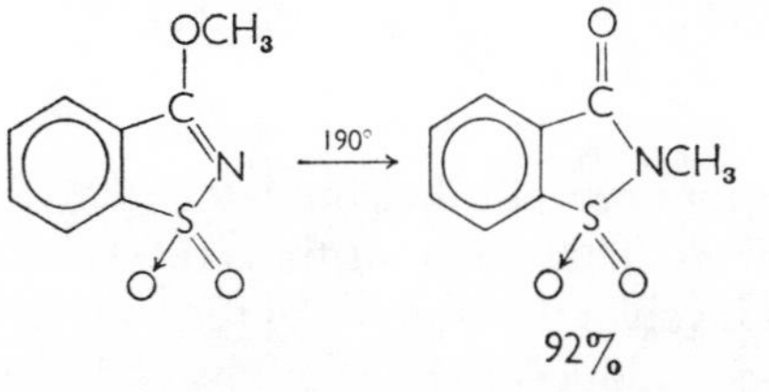

92%

Often, though, with *O*-alkyl groups larger than methyl, elimination by a unimolecular *cis* process (equation 106) is the normal result of pyrolysis of an alkyl imidate instead of rearrangement[139].

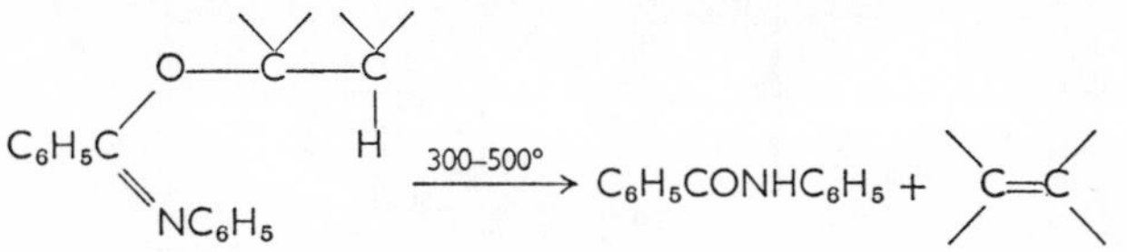

$$C_6H_5C(\text{–O–}\overset{\diagdown\diagup}{C}\text{–}\overset{\diagdown\diagup}{C}\text{–H}){=}NC_6H_5 \xrightarrow{300\text{–}500^\circ} C_6H_5CONHC_6H_5 + {>}C{=}C{<} \quad (106)$$

Allyl imidates readily undergo thermal rearrangement to *N*-allylic benzanilides by an intramolecular mechanism which must be analogous to the Claisen rearrangement, since the allylic group becomes inverted during the course of the rearrangement (equation 107)[257]. Roberts[258] and Hussein and coworkers[259,260] have studied

$$C_6H_5C(=NC_6H_5)OCH(R^1)CH{=}CHR^2 \xrightarrow{200^\circ} C_6H_5C(=O)N(C_6H_5)CH(R^2)CH{=}CHR^1 \qquad (107)$$

both the thermal and acid-catalysed rearrangements of allylic *N*-phenylformimidates (**293**). In general the acid-catalysed reactions proceed at much lower temperatures but give poorer yields of formanilides (**294**) than the thermal rearrangements (equation 108).

$$\underset{(\mathbf{293})}{HC(=NAr)OCH_2CH{=}CH_2} \xrightarrow[\text{or } H_2SO_4 \text{ at } 115^\circ]{200^\circ} \underset{(\mathbf{294})}{HC(=O)N(Ar)CH_2CH{=}CH_2} \qquad (108)$$

Alkyl halides also catalyse the conversion of alkyl imidates to amides at lower temperatures than those required for thermal rearrangement. The rearrangement is intermolecular, though, as is shown by equation (109)[261]. Arbuzov[262] and Shishkin[263] have carried out the rearrange-

$$CH_3C(=NC_6H_5)OC_2H_5 \xrightarrow[CH_3I]{150^\circ} CH_3C(=O)N(CH_3)C_6H_5 \qquad (109)$$

ment of a large number of alkyl imidates (**295**) in the presence of various alkyl halides. They feel that these reactions occur in two steps of which the first involves the addition of the alkyl halide to the imidate to form an ionic adduct (**296**)[264]. This adduct then undergoes cleavage to the final products as shown in equation (110).

$$\underset{(\mathbf{295})}{R^1C(=NR^3)OR^2} + R^4X \longrightarrow \underset{(\mathbf{296})}{\left[R^1{-}C(OR^2){=}\overset{+}{N}R^3R^4\right]X^-} \longrightarrow R^1{-}C(=O){-}NR^3R^4 + R^2X \qquad (110)$$

5. Rearrangements of acyl imidates

Numerous examples of rearrangements involving acyl migration from oxygen to the nitrogen of the carbon–nitrogen double bond can be found in the literature. In fact, it was the instability of acyl imidates towards rearrangement that led Mumm and coworkers[237] to study the more stable aryl imidates after several unsuccessful attempts to isolate the acyl derivatives. They had originally expected to obtain compound **298** when *N*-phenylbenzimidoyl chloride (**297**) was treated with sodium *m*-nitrobenzoate. Instead the imide **299** was obtained.

$$\underset{(\mathbf{297})}{C_6H_5CCl{=}NC_6H_5} + m\text{-}O_2NC_6H_4CO_2Na \longrightarrow \underset{(\mathbf{298})}{\left[m\text{-}O_2NC_6H_4CO_2\underset{C_6H_5}{\underset{|}{C}}{=}NC_6H_5\right]} \longrightarrow \underset{(\mathbf{299})}{m\text{-}O_2NC_6H_4C({=}O){-}N{-}C({=}O)C_6H_5} \quad (111)$$

More recent attempts to prepare and isolate simple acyl imidates have also been unsuccessful[265,266]. Curtin and Miller[145], reasoning that the rate of rearrangement of acyl imidates should be depressed by decreasing the nucleophilicity of the imido nitrogen (as in the normal Chapman rearrangement) were able to prepare and isolate several *N*-(2,4-dinitrophenyl)benzimidoyl benzoates (**300**). They found that these compounds rearrange to the corresponding imides (**301**) in

$$\underset{(\mathbf{300})}{p\text{-}XC_6H_4C({=}O){-}O{-}C(C_6H_5){=}NC_6H_3(NO_2)_2\text{-}2,4} \longrightarrow \underset{(\mathbf{301})}{p\text{-}XC_6H_4C({=}O){-}NC_6H_3(NO_2)_2\text{-}2,4{-}C({=}O)C_6H_5} \quad (112)$$

$X = H, CH_3O, Br, NO_2$

benzene or acetonitrile solution in the temperature range of 40–65°. The good first-order behaviour, insensitivity of the reaction to small amounts of acid, and the small positive ρ-value correlating the effect of substituents (X in **300**) led them to propose an intramolecular mechanism proceding via a normal carbonyl addition as shown in **302**.

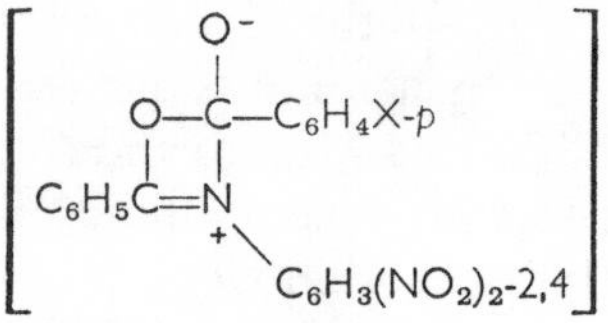

$$\left[\begin{array}{c} O^- \\ | \\ O\text{—}C\text{—}C_6H_4X\text{-}p \\ | \quad | \\ C_6H_5C{=}N^+\text{—}C_6H_3(NO_2)_2\text{-}2,4 \end{array}\right]$$

(302)

Cyclic acyl imidates (or 'isoimides') are much more stable than their open-chain analogues towards thermal rearrangement to imides. Presumably this is because they are sterically constrained in such a way that migration of the acyl group via a transition state similar to **302** is difficult or impossible[145]. However, the rearrangement of such cyclic compounds can be effected by high temperatures (equation 113)[145], or by basic solvents at moderately elevated temperatures, or by added nucleophilic catalysts at room temperature[267].

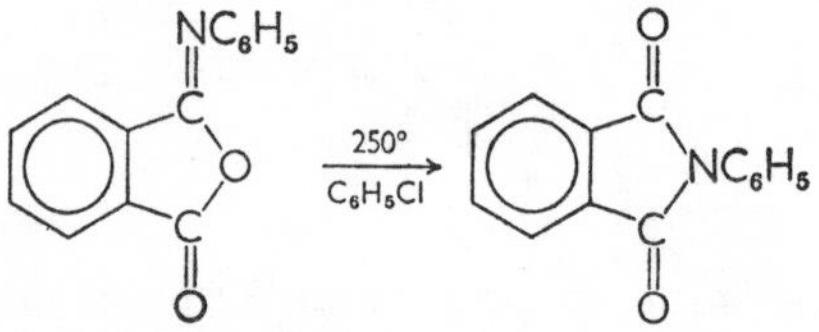

$$\text{(phthalisoimide, } C{=}NC_6H_5\text{)} \xrightarrow[C_6H_5Cl]{250^\circ} \text{(N-phenylphthalimide, } NC_6H_5\text{)} \qquad (113)$$

Hedaya, Hinman and Theodoropulos[268], in their study of the rearrangement of *N,N'*-biisoimides to *N,N'*-biimides, found the rate of rearrangement to be dependent on the nature of the solvent and the structure of the *N,N'*-biisoimide. More quantitative data are needed before one can understand how these factors lead to the qualitative order of rates they reported, but it seems certain that the mechanism of these rearrangements involves the basic solvents they employed as acyl group transfer agents. It is unlikely that unimolecular internal acyl migration is the mode of rearrangement. One of the reactions they observed is given in equation (114).

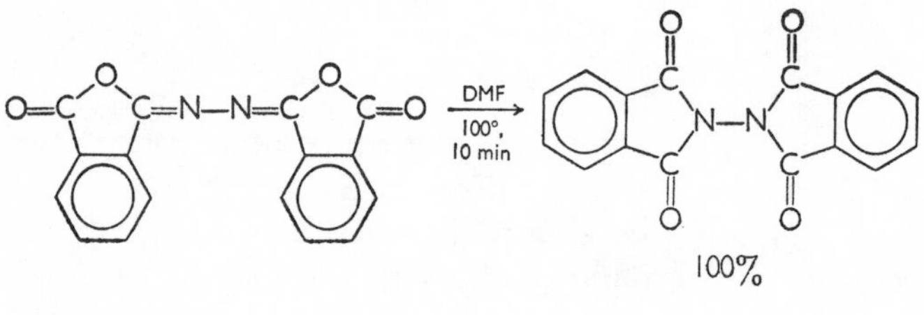

$$O{=}C\text{(isoimide)}C{=}N\text{—}N{=}C\text{(isoimide)}C{=}O \xrightarrow[100^\circ,\ 10\ \text{min}]{DMF} \text{(phthalimide)}N\text{—}N\text{(phthalimide)} \qquad (114)$$

100%

6. Related reactions

Related to the Chapman rearrangement, at least in the sense of involving O → N migrations, are the rearrangements of imidocarbonates. Meyers[269] obtained a high yield of ethyl ethyl (*p*-tolylsulphonyl)carbamate (**304**) by heating diethyl *N*-(*p*-tolylsulphonyl)imidocarbonate (**303**) for one hour at 200°. Similarly, diethyl

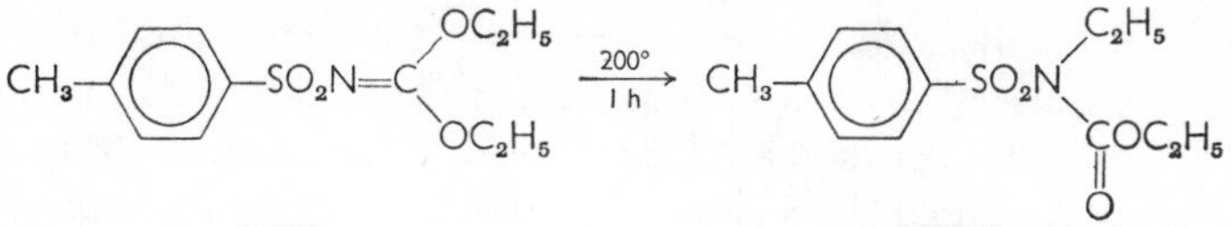

$$CH_3-C_6H_4-SO_2N{=}C(OC_2H_5)_2 \xrightarrow[1\,h]{200°} CH_3-C_6H_4-SO_2N(C_2H_5)C(=O)OC_2H_5 \qquad (115)$$

(**303**) (**304**)

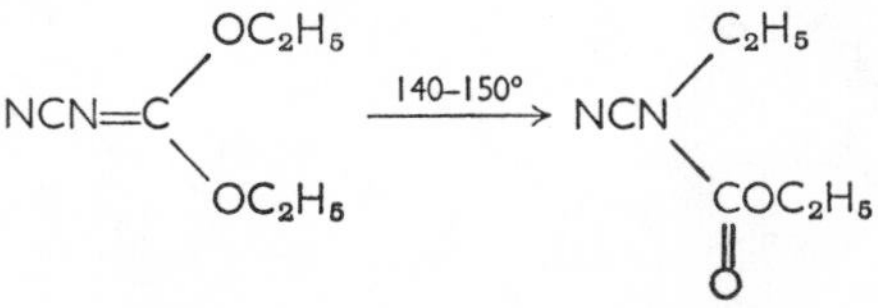

$$NCN{=}C(OC_2H_5)_2 \xrightarrow{140\text{–}150°} NCN(C_2H_5)C(=O)OC_2H_5 \qquad (116)$$

(**305**) (**306**)

N-cyanoimidocarbonate (**305**) rearranges quantitatively to **306** upon being heated for a short time at 140–150°[270]. Cyclic imidocarbonates rearrange to oxazolidones. In the case of ethylene *N*-phenylimidocarbonate (**307**), rearrangement to 3-phenyloxazolid-2-one (**308**) has been brought about by the action of lithium chloride at 200°[271] and by aluminium chloride at 100°[272]:

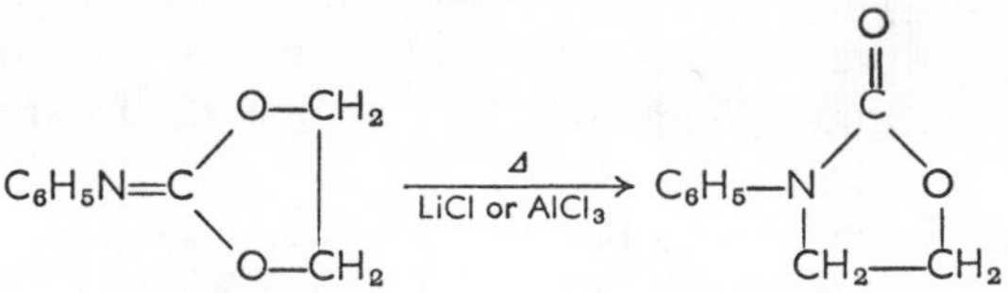

$$C_6H_5N{=}C\langle\begin{matrix}O-CH_2\\ O-CH_2\end{matrix} \xrightarrow[\text{LiCl or AlCl}_3]{\Delta} C_6H_5-N\langle\begin{matrix}C(=O)-O\\ CH_2-CH_2\end{matrix} \qquad (117)$$

(**307**) (**308**)

C. The Neber Rearrangement

1. Introduction

In 1926, Neber and Friedolsheim[273] reported that certain oxime tosylates (**309**) when treated with base followed by acid hydrolysis rearrange to α-amino ketones (**310**). Largely through the work of Neber and his associates[274–277] the versatility of this reaction in the

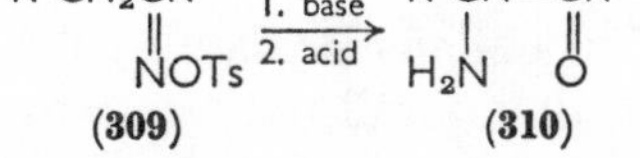

$$R^1CH_2C(=NOTs)R^2 \xrightarrow[2.\ \text{acid}]{1.\ \text{base}} R^1CH(NH_2)-C(=O)R^2 \qquad (118)$$

(**309**) (**310**)

synthesis of a variety of α-amino ketones has been well demonstrated. The tables in a review by O'Brien[278] should be consulted for specific examples. O'Brien's review is a survey of all of the information available on the Neber rearrangement up to the latter part of 1963.

2. Mechanism

Neber and Friedolsheim[273] first postulated a mechanism somewhat analogous to a Beckmann rearrangement for this base-catalysed reaction of oxime esters. Later, however, intermediates were isolated which Neber and coworkers[275,276] believed to be azirines (**311**) and they therefore modified their view of the mechanism to accommodate these cyclic species (equation 119).

$$2\text{-}NO_2\text{-}4\text{-}O_2N\text{-}C_6H_3\text{—}\underset{\diagdown N \diagup}{CH\text{—}C}\text{—}R \qquad R = CH_3,\ C_6H_5$$

(**311**)

$$R^1CH_2\underset{\underset{NOTs}{\|}}{C}R^2 \xrightarrow{\text{base}} R^1\underset{\diagdown N \diagup}{CH\text{—}C}R^2 \xrightarrow{\text{EtOH}} R^1\underset{\underset{H_2N}{|}}{CH}\text{—}\overset{\overset{OEt}{|}}{\underset{\underset{OEt}{|}}{C}}\text{—}R^2 \xrightarrow{H_3O^+} R^1\underset{\underset{H_3N^+}{|}}{CH}\text{—}\underset{\underset{O}{\|}}{C}R^2 \quad (119)$$

Cram and Hatch[279] felt that Neber's suggestion of the strained ring azirine as a reaction intermediate in this reaction warranted a critical reexamination. They repeated Neber's work, taking advantage of infrared spectral data not available, of course, to the earlier investigators, and confirmed the azirine structure as the most plausible for Neber's intermediate.

Cram and Hatch[280] reasoned that the systems containing the 2,4-dinitrophenyl group (the only systems which have led to the intermediates isolated by them and by Neber) may not be representative of the Neber rearrangement in general because of the influence of the nitro groups on the acidity of the benzylic hydrogens and on the resonance stabilization of the azirine ring system. They therefore submitted the *p*-toluenesulphonate of desoxybenzoin oxime (**312**) to the rearrangement conditions and succeeded in isolating the unstable 2,3-diphenyl-2-ethoxyethylenimine (**313**) which, when treated with lithium aluminium hydride, gave *cis*-diphenylethylenimine (**314**) and, when hydrolysed in aqueous acid, gave desylamine (**315**). In view of the isolation of **313** as an intermediate in this more representative Neber rearrangement, the overall reaction was formulated as occur-

$$C_6H_5CH_2C(=NOTs)C_6H_5 \xrightarrow[KOEt]{EtOH} C_6H_5CH\text{—}C(OEt)C_6H_5 \text{ (NH aziridine)}$$

(312) → **(313)**

(313) $\xrightarrow{LiAlH_4}$ $C_6H_5CH\text{—}CHC_6H_5$ (NH aziridine) **(314)**

(313) $\xrightarrow{HCl,\ H_2O}$ $C_6H_5CH(N^+H_3\ ^-Cl)\text{—}C(=O)C_6H_5$ **(315)**

(120)

ring via a mechanism as shown in equation (121), the unsaturated nitrene being included to explain the observed sterically indiscriminate character of the reaction.

$$B:^- \curvearrowright H\text{—}CH(R^1)\text{—}C(=NOTs)\text{—}R^2 \xrightarrow[-BH]{-OTs^-} R^1CH{=}C(\ddot{N}:)\text{—}R^2 \longrightarrow R^1CH\text{—}C\text{—}R^2 \text{ (2H-azirine, N)}$$

$$\xrightarrow{BH} R^1CH\text{—}C(B)\text{—}R^2 \text{ (NH aziridine)} \xrightarrow{H_3O^+} R^1CH(H_3N^+)\text{—}C(=O)\text{—}R^2$$

(121)

Another possible mechanism which must be considered is one in which the reaction is initiated by attack of the alkoxide ion on the carbon–nitrogen double bond followed by loss of the tosyloxy group. The resulting species would then be a saturated nitrene (**317**) which, by analogy with the behaviour of carbenes, should show little selectivity between insertion at either of two possible benzylic positions[281]. If selectivity were to be observed it should be at the less acidic C—H bond[282]. The experimental results reported by House and Berkowitz[282] lend no support to such a mechanism involving a saturated nitrene (equation 122). Instead, their data merely support the previous belief that when different α-protons are present, the more acidic will be removed. When in **316** $R^1 = p\text{-}O_2NC_6H_4$ and $R^2 = p\text{-}CH_3OC_6H_4$, the only Neber rearrangement product isolated was 1-amino-3-

(4-methoxyphenyl)-1-(4-nitrophenyl)propan-2-one (**318**). Furthermore, **318** was the product regardless of the stereochemistry of the starting oxime ester, although Beckmann rearrangement by-products proved the oxime esters not to be isomerized by the Neber rearrangement conditions.

$$EtO^- + R^1CH_2C(=NOTs)CH_2R^2 \;(\mathbf{316}) \longrightarrow R^1CH_2\text{—}C(OEt)(\bar{N}\text{—}OTs)\text{—}CH_2R^2 \xrightarrow{-OTs^-} R^1CH_2\text{—}C(OEt)(\ddot{N}{:})\text{—}CH_2R^2 \;(\mathbf{317}) \longrightarrow \text{C—H insertion} \quad (122)$$

$$O_2N\text{—}C_6H_4\text{—}CH(NH_2)\text{—}C(=O)\text{—}CH_2\text{—}C_6H_4\text{—}OCH_3 \quad (\mathbf{318})$$

In several recent studies[283–285], both azirines and alkoxyaziridines have been isolated and characterized from Neber and modified Neber reactions, and there seems little doubt now about their role as intermediates in the path from oxime esters to α-amino ketones. Also, the reported syntheses of azirines by a variety of different reactions[286–288] removes this ring system from the suspicion with which it was formerly viewed. Still unanswered, however, is the question of whether the separation of the tosyloxy group occurs prior to ring closure, as suggested by Cram[280] and House[282] and their associates, or in a concerted manner. The unsaturated nitrene may not really be needed to explain the lack of stereospecificity since the initial generation of the anion at the α-position should promote geometrical equilibration. If, furthermore, the subsequent reaction is fast relative to protonation, the stereochemical control of the competing Beckmann rearrangement observed by House and Berkowitz[282] would not be disturbed[289].

3. Related reactions

In a paper on the rearrangement of *N*,*N*-dichloro-*s*-alkyl amines (**319**) to α-amino ketones, Baumgarten and Bower[290] visualized a mechanism (equation 123) similar to that proposed for the Neber rearrangement of oxime tosylates[280]. Positive evidence for the *N*-chlorimines as intermediates in this rearrangement has been submitted by Alt and Knowles[291], who found that treatment of *N*-chlorocyclohexylimine (**323**) with sodium methoxide in methanol gave an excellent yield of 2-aminocyclohexanone (**324**). Evidence for the azirine (**321**) or alkoxyaziridine (**322**) structures comes from the

$$R^1CH(NH_2)CH_2R^2 \xrightarrow[H_2O]{Cl_2} R^1CH(NCl_2)CH_2R^2\ (\mathbf{319}) \xrightarrow{NaOMe} R^1C(=NCl)CH_2R^2\ (\mathbf{320}) \longrightarrow R^1C\overset{N}{=\!\!-}CHR^2\ (\mathbf{321}) \xrightarrow{MeOH} R^1C(OMe)\text{—}CHR^2\ (\text{NH bridge})\ (\mathbf{322}) \xrightarrow[H_2O]{HCl} R^1C(=O)\text{—}CHR^2(^+NH_3Cl^-) \quad (123)$$

$$\text{cyclohexylidene=N—Cl}\ (\mathbf{323}) \xrightarrow[MeOH]{NaOMe} \text{2-aminocyclohexanone (C=O, } NH_2)\ (\mathbf{324}) \quad (124)$$

observation[292] that in rearrangements proceeding through **320** (where $R^1 = R^2 = C_6H_5$) reduction of the reaction intermediate with lithium aluminium hydride leads to *cis*-2,3-diphenylethylenimine, just as in the Neber rearrangement of the oxime ester of desoxybenzoin[280].

In an extension of their work on rearrangements of *N*-chloro ketimines, Baumgarten and coworkers[293] studied the behaviour of *N*-chlorimino esters (**325**) under similar conditions. Again the products, α-amino acid esters or α-amino acids, could be rationalized as arising by way of a Neber-like rearrangement (equation 125).

Quaternary hydrazonium salts having α-hydrogens are also converted by a Neber-like rearrangement to α-amino ketones[294] and sometimes the intermediate azirines are isolable[283]. A recent attempt to extend this reaction to hydrazonium iodide acetals (**326**) did not, however, yield the expected results[295]. Instead of the expected 2,2-dialkoxyazirines (**327**) only α-imino ortho esters (**328**) were detected (equation 126).

$$RCH_2C(=NCl)Me \xrightarrow{t\text{-BuOK}} \left[RCH\text{—}COMe \text{ (azirine, N)} \right] \xrightarrow{t\text{-BuOH}} \left[RCH\text{—}C(OMe)(OBu\text{-}t) \text{ (aziridine, NH)} \right] \xrightarrow[HOH]{HCl} RCH(NH_3Cl)CO_2Me \xrightarrow[HOH]{HCl} RCH(NH_3Cl)CO_2H \quad (125)$$

(325)

$$R^1C(=N\text{—}\overset{+}{N}(CH_3)_3I^-)CH(OR^2)_2 \xrightarrow[(CH_3)_2CHOH]{(CH_3)_2CHONa} \begin{cases} \not\rightarrow R^1C\text{—}C(OR^2)_2 \text{ (azirine, N)} \\ \rightarrow R^1C(=NH)\text{—}C(OR^2)_2OCH(CH_3)_2 \end{cases} \quad (126)$$

(326) (327) (328)

It should be noted that structure **326** contains only an α-methinyl hydrogen and it has been suggested[280] that an α-methylene or α-methyl group is a necessary structural requirement for the Neber rearrangement. On the other hand, *N*-chloro ketimines with only an α-methinyl hydrogen available do rearrange[292]. It seems that the subject of structural limitations in Neber and Neber-like rearrangements needs to be reexamined with care taken to use the same substituents and reaction conditions for the oxime esters, the *N*-chloro ketimines, and the hydrazonium salts.

D. Other Rearrangements

1. Introduction

Some rearrangements involving the azomethine group, in addition to the aforementioned Beckmann, Chapman and Neber rearrangements, have been discussed by Smith[55]. A few not included in Smith's review are presented here.

2. α-Hydroxy imines

In 1962, Stevens and coworkers[296] reported a new thermal rearrangement of α-amino ketones. An example of this rearrangement is the conversion of 2-ethyl-2-methylaminobutyrophenone (**329**) to

3-methylamino-3-phenylhexan-4-one (**330**) in 35% yield at 240°. Similarly, α-methylaminocyclopentyl methyl ketone (**331**) is converted to 2-methyl-2-methylaminocyclohexanone (**332**) in a rearrangement involving ring enlargement. Stevens and coworkers proposed a mechanism involving two carbon skeleton migrations for these re-

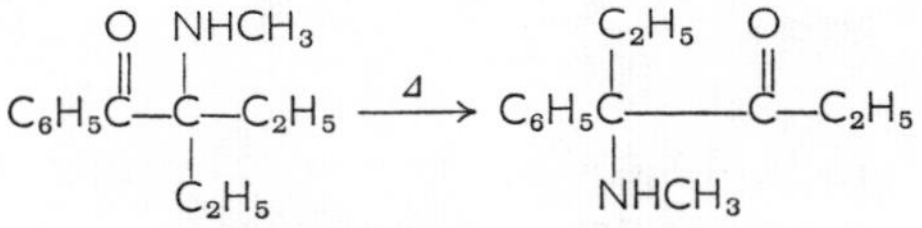

$$C_6H_5C(=O)-C(NHCH_3)(C_2H_5)-C_2H_5 \xrightarrow{\Delta} C_6H_5C(C_2H_5)(NHCH_3)-C(=O)-C_2H_5 \quad (127)$$

(**329**) (**330**)

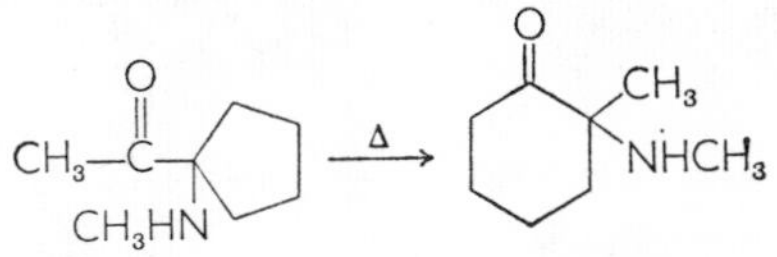

$$CH_3-C(=O)-C_5H_8(NHCH_3) \xrightarrow{\Delta} \text{2-}CH_3\text{-2-}NHCH_3\text{-cyclohexanone} \quad (128)$$

(**331**) (**332**)

arrangements. α-Hydroxy imines as possible intermediates were considered since these compounds had been proposed earlier as intermediates in other reactions leading to α-amino ketones[297]. The α-hydroxy imine **333**, for example, when placed under the conditions of the rearrangement reaction (equation 127), gave the product **330** in comparable yield. The isomeric α-hydroxy imine (**334**) was eliminated as a possible reaction intermediate since it did not give an appreciable yield of **330**.

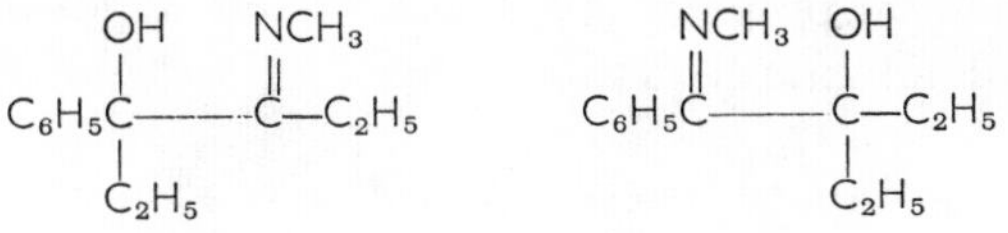

$$C_6H_5C(OH)(C_2H_5)-C(=NCH_3)-C_2H_5 \qquad C_6H_5C(=NCH_3)-C(OH)(C_2H_5)-C_2H_5$$

(**333**) (**334**)

As a result of extensive kinetic studies by Stevens and coworkers[298,299], along with stereochemical studies by Morrow[300] and Yamada[301] and their associates, the mechanism of this thermal rearrangement of α-hydroxy imines has been viewed as being an intramolecular concerted process. For the anil of 1-hydroxycyclopentyl phenyl ketone the mechanism has been pictured as shown in equation (129)[298].

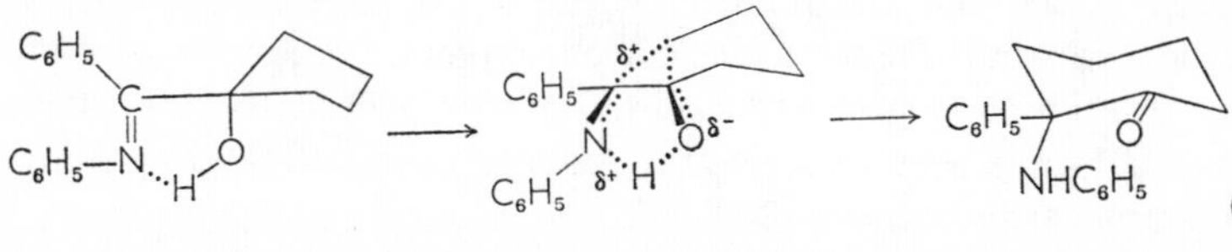

$$C_6H_5C(=NC_6H_5)-C_5H_8OH \longrightarrow [\text{transition state: } C_6H_5, N, C_6H_5, \delta^+, O^{\delta-}, \delta^+H] \longrightarrow 2\text{-}C_6H_5\text{-2-}NHC_6H_5\text{-cyclohexanone} \quad (129)$$

The acid salts of the α-hydroxy imines have been found to give better yields of α-amino ketones under less severe conditions than the α-hydroxy imines themselves[302]. Both the salts and the free imines have been used in the syntheses of a variety of interesting α-amino ketones, including steroidal α-amino ketones[300].

3. Hydrazones

Robev[303] found in 1954 that the phenylhydrazones of aromatic aldehydes (**335**) when treated with either sodium amide or phenyllithium in boiling xylene are converted to amidines (**336**). It was

$$\underset{(\mathbf{335})}{Ar^1CH{=}NNHAr^2} \xrightarrow[\text{or } C_6H_5Li,\ \text{xylene}]{NaNH_2} \underset{(\mathbf{336})}{Ar^1C({=}NAr^2)NH_2} \qquad (130)$$

quite logically thought that the hydrazones first decomposed to amines and nitriles since *N*-phenylamidines can be obtained from aromatic amines and nitriles under analogous conditions[304]. More recent studies, though, show that cleavage to nitriles and amines does not occur during the Robev rearrangement[305]. The reaction does, however, have an intermolecular character as shown by crossover studies[306] and does possess properties of a radical process since it is inhibited by added hydroquinone[305] and requires the presence of oxygen[307]. The radical mechanism proposed by Robev[308] still stands as being most consistent with all data (equation 131).

$$\begin{aligned} Ar^1CH{=}NNHAr^2 &\longrightarrow Ar^1CH{=}N\cdot + H\dot{N}Ar^2 \\ Ar^1CH{=}N\cdot &\longrightarrow Ar^1\dot{C}{=}NH \\ Ar^1\dot{C}{=}NH + H\dot{N}Ar^2 &\longrightarrow Ar^1C({=}NH)NHAr^2 \\ Ar^1C({=}NH)NHAr^2 &\rightleftharpoons Ar^1C(NH_2){=}NAr^2 \end{aligned} \qquad (131)$$

Imidyl azides (**337**) frequently cyclize to form tetrazoles (**338**)[309]. Both the azides and the tetrazoles have been found to rearrange under proper conditions to give carbodiimides[310,311]. Recently, hydrazidic azides (**339**) have been converted to semicarbazides (**341**), apparently via *N*-aminocarbodiimides (**340**)[312].

$$R^1-C(N_3)=N-R^2 \rightleftharpoons R^1-C\!=\!N-N\!=\!N-N(R^2) \text{ (tetrazole)} \quad (132)$$

(337) (338)

$$RC(N_3)=NNHAr \longrightarrow [R-N=C=NNHAr] \longrightarrow RNHC(=O)NHNHAr \quad (133)$$

(339) (340) (341)

4. Guanidines and amidines

N-Aminocarbodiimides have also been cited as intermediates in the base-catalysed rearrangements of *N*-haloguanidines (**342**)[313]. When no trapping agent is present, the carbodiimide (**343**) dimerizes to guanazine (**344**).

$$H_2N-C(=NR)-NHX \xrightarrow{OH^-} RN=C=N-NH_2 \longrightarrow RNHC\!\!\overset{N(NH_2)}{}\!\!CNHR \;(\text{ring } N-N) \quad (134)$$

(342) (343) (344)

A novel synthesis of carbodiimides from *N*-haloamidines (**345**) is shown in equation (135)[314]. Whether this rearrangement proceeds by way of migration of R^1 to a positively charged nitrogen (path A) or to a nitrene (path B) is not known.

$$R^1-C(=NR^2)-NHX \xrightarrow{Ag_2O} R^1C(=NR^2)-\overset{+}{N}H \xrightarrow{-H^+} R^1-C(=NR^2)-\ddot{N}: \quad (135)$$

(345)

Path A: $R^1C(=NR^2)-\overset{+}{N}H \longrightarrow {}^{+}C(=NR^2)-NHR^1 \xrightarrow{-H^+} R^2N=C=NR^1$

Path B: $R^1-C(=NR^2)-\ddot{N}: \longrightarrow R^2N=C=NR^1$

IV. REFERENCES

1. C. M. Brewster and L. H. Millam, *J. Am. Chem. Soc.*, **55**, 763 (1933).
2. G. H. Brown and W. G. Shaw, *Rev. Pure Appl. Chem.*, **11**, 1 (1961).
3. J. Meisenheimer and W. Theilacker in *Stereochemie*, Vol. 3 (Ed. K. Freudenberg), Franz Deuticke, Leipzig, 1932.
4. D. Y. Curtin and J. W. Hausser, *J. Am. Chem. Soc.*, **83**, 3474 (1961).
5. E. Knoevenagel, *J. Prakt. Chem.*, **89**, 38 (1914).

6. W. Manchot and J. R. Furlong, *Ber.*, **42**, 4383 (1909).
7. W. Manchot and J. R. Furlong, *Ber.*, **42**, 3030 (1909).
8. O. Anselmino, *Ber.*, **43**, 462 (1910).
9. V. De Gaouck and R. J. W. Le Fevre, *J. Chem. Soc.*, 1392 (1939).
10. K. A. Jensen and N. H. Bang, *Ann. Chem.*, **548**, 106 (1941).
11. G. Wettermark, *Svensk Kem. Tidskr.*, **79**, 249 (1967).
12. V. De Gaouck and R. J. W. Le Fevre, *J. Chem. Soc.*, 741 (1938).
13. C. Wiegand and E. Merkel, *Ann. Chem.*, **550**, 175 (1942).
14. V. I. Minkin, E. A. Medyantzeva and A. M. Simonov, *Dokl. Akad. Nauk SSSR*, **149**, 1347 (1963).
15. V. I. Minkin, Y. A. Zhdanov, E. A. Medyantzeva and Y. A. Ostroumov, *Tetrahedron*, **23**, 3651 (1967).
16. J. Bregman, L. Leiserowitz and G. M. J. Schmidt, *J. Chem. Soc.*, 2068 (1964).
17. J. Bregman, L. Leiserowitz and K. Osaki, *J. Chem. Soc.*, 2086 (1964).
18. G. Wyman, *Chem. Rev.*, **55**, 625 (1955).
19. E. Fischer and Y. Frei, *J. Chem. Phys.*, **27**, 808 (1957).
20. G. Wettermark and L. Dogliotti, *J. Chem. Phys.*, **40**, 1486 (1964).
21. M. D. Cohen and G. M. J. Schmidt, *J. Phys. Chem.*, **66**, 2442 (1962).
22. M. D. Cohen, Y. Hirshberg and G. M. J. Schmidt in *Hydrogen Bonding* (Ed. D. Hadzi), Pergamon Press, London, 1959, p. 293.
23. R. S. Becker and W. F. Richey, *J. Am. Chem. Soc.*, **89**, 1298 (1967).
24. D. G. Anderson and G. Wettermark, *J. Am. Chem. Soc.*, **87**, 1433 (1965).
25. G. Wettermark, J. Weinstein, J. Sousa and L. Dogliotti, *J. Phys. Chem.*, **69**, 1584 (1965).
26. G. Wettermark, *Arkiv Kemi*, **27**, 159 (1967).
27. N. Ebara, *Bull. Chem. Soc. Japan*, **33**, 534 (1960).
28. D. Y. Curtin, E. J. Grubbs and C. G. McCarty, *J. Am. Chem. Soc.*, **88**, 2775 (1966).
29. P. P. Birnbaum, J. H. Linford and D. W. G. Style, *Trans. Faraday Soc.*, **49**, 735 (1953).
30. J. Hine and C. Y. Yeh, *J. Am. Chem. Soc.*, **89**, 2669 (1967).
31. H. A. Staab, F. Vögtle and A. Mannschreck, *Tetrahedron Letters*, 697 (1965).
32. B. L. Shapiro, S. J. Ebersole, G. J. Karabatsos, F. M. Vane and S. L. Manatt, *J. Am. Chem. Soc.*, **85**, 4041 (1963).
33. C. F. Chang, B. J. Fairless and M. R. Willcott, *J. Mol. Spectr.*, **22**, 112 (1967).
34. F. Straus and A. Ackermann, *Ber.*, **43**, 596 (1910).
35. P. Ramart-Lucas and M. J. Hoch, *Bull. Soc. Chim. France* [5], **3**, 918 (1936).
36. M. E. Taylor and T. L. Fletcher, *J. Am. Chem. Soc.*, **80**, 2246 (1958).
37. G. Saucy and L. H. Sternbach, *Helv. Chim. Acta*, **45**, 2226 (1962).
38. S. C. Bell, G. L. Conklin and S. J. Childress, *J. Am. Chem. Soc.*, **85**, 2868 (1963).
39. S. C. Bell, G. L. Conklin and S. J. Childress, *J. Org. Chem.*, **29**, 2368 (1964).
40. T. S. Sulkowski and S. J. Childress, *J. Org. Chem.*, **27**, 4424 (1962).
41. P. A. S. Smith and E. P. Antoniades, *Tetrahedron*, **9**, 210 (1960).
42. D. A. Nelson and R. L. Atkins, *Tetrahedron Letters*, 5197 (1967).

43. S. S. Lande, *Diss. Abstr.*, **27**, 4313-B (1967).
44. D. Wurmb-Gerlich, F. Vögtle, A. Mannschreck and H. A. Staab, *Ann. Chem.*, **708**, 36 (1967).
45. D. Y. Curtin and W. J. Koehl, Jr., *J. Am. Chem. Soc.*, **84**, 1967 (1962).
46. W. Lüttke, *Ann. Chem.*, **668**, 184 (1963).
47. J. A. Pople, W. G. Schneider and H. J. Bernstein, *High-Resolution Nuclear Magnetic Resonance*, McGraw-Hill, New York, 1959, Chap. 13; J. M. Emsley, J. Feeney and L. H. Sutcliffe, *High Resolution Nuclear Magnetic Resonance Spectroscopy*, Pergamon Press, London 1965, Chap. 9.
48. S. Andreades, *J. Org. Chem.*, **27**, 4163 (1962).
49. P. H. Ogden and G. V. D. Tiers, *Chem. Commun.*, 527 (1967).
50. D. Y. Curtin and C. G. McCarty, *Tetrahedron Letters*, 1269 (1962).
51. A. Reiker and H. Kessler, *Tetrahedron*, **23**, 3723 (1967).
52. H. Saito and K. Nukada, *Tetrahedron*, **22**, 3313 (1966).
53. J. B. Lambert, W. L. Oliver and J. D. Roberts, *J. Am. Chem. Soc.*, **87**, 5085 (1965).
54. W. Z. Heldt and L. G. Donaruma, *Org. Reactions*, **11**, 1 (1960).
55. P. A. S. Smith in *Molecular Rearrangements*, Part I (Ed. P. de Mayo), Interscience, New York, 1962, p. 483 ff.
56. E. Beckmann, *Ber.*, **20**, 1507 (1887).
57. J. Meisenheimer, *Ber.*, **54**, 3206 (1921).
58. L. E. Sutton and T. W. J. Taylor, *J. Chem. Soc.*, 2190 (1931).
59. K. A. Kerr, J. M. Robertson, G. A. Sim and M. S. Newman, *Chem. Commun.*, 170 (1967).
60. B. Jerslev, *Nature*, **180**, 1410 (1958).
61. A. Palm and H. Werbin, *Can. J. Chem.*, **32**, 858 (1954).
62. O. L. Brady and H. J. Grayson, *J. Chem. Soc.*, 1037 (1933).
63. W. D. Phillips, *Ann. N. Y. Acad. Sci.*, **70**, 817 (1958).
64. E. Lustig, *J. Phys. Chem.*, **65**, 491 (1961).
65. I. Pejkovic-Tadic, M. Hranisavljevic-Jakovljevic, S. Nesic, C. Pascual and W. Simon, *Helv. Chim. Acta*, **48**, 1157 (1965).
66. G. J. Karabatsos, R. A. Taller and F. M. Vane, *J. Am. Chem. Soc.*, **85**, 2326, 2327 (1963).
67. G. J. Karabatsos and N. Hsi, *Tetrahedron*, **23**, 1079 (1967).
68. W. F. Trager and A. C. Huitric, *Tetrahedron Letters*, 825 (1966).
69. H. Saito and K. Nukada, *Tetrahedron Letters*, 2117 (1965).
70. H. Saito and K. Nukada, *J. Mol. Spectr.*, **18**, 1 (1965).
71. H. Saito, K. Nukada and M. Ohno, *Tetrahedron Letters*, 2124 (1964).
72. A. C. Huitric, D. B. Roll and J. R. DeBoer, *J. Org. Chem.*, **32**, 1661 (1967).
73. G. G. Kleinspehn, J. A. Jung and S. A. Studniarz, *J. Org. Chem.*, **32**, 460 (1967).
74. G. Slomp and W. J. Wechter, *Chem. Ind. (London)*, 41 (1962).
75. R. H. Mazur, *J. Org. Chem.*, **28**, 248 (1963).
76. W. R. Benson and A. E. Pohland, *J. Org. Chem.*, **30**, 1129 (1965).
77. E. J. Poziomek, D. N. Kramer, W. A. Mosher and H. O. Michel, *J. Am. Chem. Soc.*, **83**, 3916 (1961).
78. T. E. Stevens, *J. Org. Chem.*, **32**, 670 (1967).
79. E. Buehler, *J. Org. Chem.*, **32**, 261 (1967).

80. H. Hjeds, K. P. Hansen and B. Jerslev, *Acta Chem. Scand.*, **19**, 2166 (1965).
81. R. F. Brown, N. M. van Gulick and G. H. Schmid, *J. Am. Chem. Soc.*, **77**, 1094 (1955).
82. C. R. Hauser and D. S. Hoffenberg, *J. Org. Chem.*, **20**, 1491 (1955).
83. G. Ciamician and P. Silber, *Ber.*, **36**, 4266 (1903).
84. E. J. Poziomek, *J. Pharm. Sci.*, **54**, 333 (1965).
85. J. H. Amin and P. de Mayo, *Tetrahedron Letters*, 1585 (1963).
86. A. Daniel and A. Pavia, *C. R. Acad. Sci.*, *Paris*, *Ser. C*, 643 (1966).
87. R. K. Norris and S. Sternhell, *Tetrahedron Letters*, 97 (1967).
88. R. K. Norris and S. Sternhell, *Australian J. Chem.*, **19**, 841 (1966).
89. H. Uffmann, *Tetrahedron Letters*, 4631 (1966).
90. R. J. W. Le Fevre and J. Northcott, *J. Chem. Soc.*, 2235 (1949).
91. E. G. Vassian and R. K. Murmann, *J. Org. Chem.*, **27**, 4309 (1962).
92. O. L. Brady and L. Klein, *J. Chem. Soc.*, 874 (1927).
93. R. L. Shriner, R. C. Fuson, and D. Y. Curtin, *The Systematic Identification of Organic Compounds*, 5th Ed., John Wiley and Sons, New York, 1964.
94. G. W. Wheland, *Advanced Organic Chemistry*, John Wiley and Sons, New York, 1949, p. 346.
95. P. de Mayo and A. Stoessl, *Can. J. Chem.*, **38**, 950 (1960).
96. F. A. Isherwood and R. L. Jones, *Nature*, **175**, 419 (1955).
97. H. Van Duin, *Rec. Trav. Chim.*, **73**, 78 (1954).
98. D. Schulte-Frohlinde, R. Kuhn, W. Munzing and W. Otting, *Ann. Chem.*, **622**, 43 (1954).
99. M. N. Preobrazhenskaya, N. V. Uvarova, Y. N. Sheinker and N. N. Suvorov, *Dokl. Akad. Nauk SSSR*, **148**, 1088 (1963).
100. P. Hope and L. A. Wiles, *J. Chem. Soc.* (C), 283 (1966).
101. J. F. May, *Proc. S. Dakota Acad. Sci.*, **42**, 184 (1963).
102. L. Tschetter and E. M. Reiman, *Proc. S. Dakota Acad. Sci.*, **43**, 165 (1964).
103. G. J. Karabatsos, B. L. Shapiro, F. M. Vane, J. S. Fleming and J. S. Ratka, *J. Am. Chem. Soc.*, **85**, 2784 (1963).
104. C. K. Ingold, G. J. Pritchard and H. G. Smith, *J. Chem. Soc.*, 79 (1934).
105. W. M. D. Bryant, *J. Am. Chem. Soc.*, **60**, 2814 (1938).
106. H. Van Duin, *Thesis*, Free University of Amsterdam, 1961.
107. F. Ramirez and A. F. Kirby, *J. Am. Chem. Soc.*, **76**, 1037 (1954).
108. R. M. Silverstein and J. N. Schoolery, *J. Org. Chem.*, **25**, 1355 (1960).
109. G. J. Karabatsos, F. M. Vane, R. A. Taller and N. Hsi, *J. Am. Chem. Soc.*, **86**, 3351 (1964).
110. G. J. Karabatsos and R. A. Taller, *J. Am. Chem. Soc.*, **85**, 3624 (1963).
111. G. J. Karabatsos, J. D. Graham and F. M. Vane, *J. Am. Chem. Soc.*, **84**, 753 (1962).
112. A. Arbuzov, Y. Y. Samitov and Y. P. Kitaev, *Izv. Akad. Nauk SSSR Ser. Khim.*, 55 (1966).
113. D. Hadzi and J. Jan, *Rev. Roum. Chim.*, **10**, 1183 (1965).
114. G. J. Karabatsos, R. A. Taller and F. M. Vane, *Tetrahedron Letters*, 1081 (1964).
115. G. J. Karabatsos and K. L. Krumel, *Tetrahedron*, **23**, 1097 (1967).
116. A. F. Hegarty and F. L. Scott, *Chem. Commun.*, 521 (1967).
117. R. W. Taft in *Steric Effects in Organic Chemistry* (Ed. M. S. Newman), John Wiley and Sons, New York, 1956, p. 598.

118. A. Mannschreck and U. Koelle, *Tetrahedron Letters*, 863 (1967).
119. I. Fleming and J. Harley-Mason, *J. Chem. Soc.*, 5560 (1961).
120. E. Arnal, J. Elguero, R. Jacquier, C. Marzin and J. Wylde, *Bull. Soc. Chim. Fr.*, 877 (1965).
121. C. R. Hauser, A. G. Gillaspie and J. W. Le Maistre, *J. Am. Chem. Soc.*, **57**, 567 (1935).
122. A. Hantzch and F. Kraft, *Ber.*, **24**, 3511 (1891).
123. P. P. Peterson, *Am. Chem. J.*, **46**, 325 (1911).
124. S. L. Reid and D. B. Sharp. *J. Org. Chem.*, **26**, 2567 (1961).
125. W. Theilacker and H. Mohl, *Ann. Chem.*, **563**, 99 (1949).
126. R. N. Loeppky and M. Rotman, *J. Org. Chem.*, **32**, 4010 (1967).
127. W. Theilacker and K. Fauser, *Ann. Chem.*, **539**, 103 (1939).
128. D. Y. Curtin and C. G. McCarty, *J. Org. Chem.*, **32**, 223 (1967).
129. D. H. Dybvig, *Inorg. Chem.*, **5**, 1795 (1966).
130. A. L. Logothetis and G. N. Sausen, *J. Org. Chem.*, **31**, 3689 (1966).
131. G. N. Sausen and A. L. Logothetis, *J. Org. Chem.*, **32**, 2261 (1967).
132. T. E. Stevens, *J. Org. Chem.*, **32**, 670 (1967).
133. F. A. Johnson, C. Haney and T. E. Stevens, *J. Org. Chem.*, **32**, 466 (1967).
134. S. K. Brauman and M. E. Hill, *J. Am. Chem. Soc.*, **89**, 2127 (1967).
135. J. Stieglitz and R. B. Earle, *Am. Chem. J.*, **30**, 399 (1903).
136. W. S. Hilpert, *Am. Chem. J.*, **40**, 150 (1908).
137. N. P. Marullo and E. H. Wagener, *J. Am. Chem. Soc.*, **88**, 5034 (1966).
138. J. Schulenberg and S. Archer, *Org. Reactions*, **14**, 1 (1965).
139. N. P. Marullo, C. D. Smith and J. F. Terapane, *Tetrahedron Letters*, 6279 (1966).
140. M. Kandel and E. H. Cordes, *J. Org. Chem.*, **32**, 3061 (1967).
141. H. L. Yale, *Chem. Rev.*, **33**, 209 (1943).
142. W. Lossen and J. Zanni, *Ann. Chem.*, **182**, 220 (1876).
143. A. Werner, *Ber.*, **26**, 1561 (1893).
144. A. Werner and J. Subak, *Ber.*, **29**, 1153 (1896).
145. D. Y. Curtin and L. L. Miller, *J. Am. Chem. Soc.*, **89**, 637 (1967).
146. J. Binenboym, A. Burcat, A. Lifshitz and J. Shamir, *J. Am. Chem. Soc.*, **88**, 5039 (1966).
147. E. R. Talaty and J. C. Fargo, *Chem. Commun.*, 65 (1967).
148. H. Kessler, *Angew. Chem. Intern. Ed. Engl.*, **6**, 977 (1967).
149. J. Hine, *J. Am. Chem. Soc.*, **85**, 3239 (1963).
150. W. A. Bernett, preprint of a paper provided via personal communication with P. H. Ogden.
151. E. Beckmann, *Ber.*, **19**, 988 (1886).
152. D. E. Pearson and R. M. Stone, *J. Am. Chem. Soc.*, **83**, 1715 (1961).
153. J. C. Craig and A. R. Naik, *J. Am. Chem. Soc.*, **84**, 3410 (1962).
154. J. B. Hester, Jr., *J. Org. Chem.*, **32**, 3804 (1967).
155. J. Wiemann and P. Ham, *Bull. Soc. Chim. France*, 1005 (1961).
156. J. Meisenheimer, W. Theilacker and O. Beiszwenger, *Ann. Chem.*, **495**, 249 (1932).
157. T. Van Es, *J. Chem. Soc.*, 3881 (1965).
158. A. Striegler, *J. Prakt. Chem.*, **15**, 1 (1961).
159. R. Tada, Y. Masubuchi and N. Tokura, *Bull. Chem. Soc. Japan*, **34**, 209 (1961).

160. N. Tokura, K. Shiina and T. Terashima, *Bull. Chem. Soc. Japan*, **35**, 1986 (1962).
161. N. Tokura, T. Kawahara and S. Ikeda, *Bull. Chem. Soc. Japan*, **37**, 138 (1964).
162. T. E. Stevens, *J. Org. Chem.*, **26**, 2531 (1961).
163. T. Mukaiyama, K. Tonooka and K. Inoue, *J. Org. Chem.*, **26**, 2202 (1961).
164. M. Masaki, K. Fukui and M. Ohta, *J. Org. Chem.*, **32**, 3564 (1967).
165. M. Ohno and N. Kawabe, *Tetrahedron Letters*, 3935 (1966).
166. E. V. Genkina and M. N. Gorodisskaya, *Zh. Vses. Khim. Obshch.*, **9**, 709 (1964); *Chem. Abstr.*, **62**, 8967 (1965).
167. L. Field, P. B. Hughmark, S. H. Shumaker and W. S. Marshall, *J. Am. Chem. Soc.*, **83**, 1983 (1961).
168. P. S. Landis and P. B. Venuto, *J. Catal.*, **6**, 245 (1966).
169. F. Kohen, *Chem. Ind. (London)*, 1378 (1966).
170. H. R. Nace and A. C. Watterson, Jr., *J. Org. Chem.*, **31**, 2109 (1966).
171. A. Kovendi and M. Kircz, *Chem. Ber.*, **98**, 1049 (1965).
172. T. S. Sulkowski and S. J. Childress, *J. Org. Chem.*, **27**, 4424 (1962).
173. G. Cainelli, S. Morrocchi and A. Quilico, *Gazz. Chim. Ital.*, **95**, 1115 (1965).
174. M. R. Zimmerman, *J. Prakt. Chem.*, [2] **66**, 353 (1902).
175. A. Kaneda, M. Nagatsuka and R. Sudo, *Bull. Chem. Soc. Japan*, **40**, 2705 (1967).
176. M. Kröner, *Chem. Ber.*, **100**, 3162 (1967).
177. R. Fosse, *Ann. Chim.*, **6**, 13 (1916).
178. N. Campbell, S. R. McCallum and D. J. MacKenzie, *J. Chem. Soc.*, 1922 (1957).
179. A. T. Troshchenko and T. B. Lobanova, *Zh. Org. Khim.*, **3**, 501 (1967).
180. E. Cuingnet and M. Tarterat-Adalberon, *Bull. Soc. Chim. France*, 3728 (1965).
181. C. A. Grob and P. W. Schiess, *Angew. Chem. Intern. Ed. Engl.*, **6**, 1 (1967).
182. R. W. Cottingham, *J. Org. Chem.*, **25**, 1473 (1960).
183. T. Sato and H. Obase, *Tetrahedron Letters*, 1633 (1967).
184. R. K. Hill and R. T. Conley, *Chem. Ind. (London)*, 1314 (1956).
185. R. T. Conley and M. C. Annis, *J. Org. Chem.*, **27**, 1961 (1962).
186. R. Lukes and J. Hofman, *Collection Czech. Chem. Commun.*, **26**, 523 (1961).
187. R. T. Conley and B. E. Nowak, *J. Org. Chem.*, **27**, 3196 (1962).
188. R. T. Conley and B. E. Nowak, *J. Org. Chem.*, **27**, 1965 (1962).
189. R. T. Conley and L. J. Frainier, *J. Org. Chem.*, **27**, 3844 (1962).
190. R. T. Conley and R. J. Lange, *J. Org. Chem.*, **28**, 210 (1963).
191. R. K. Hill, R. T. Conley and O. T. Chortyk, *J. Am. Chem. Soc.*, **87**, 5646 (1965).
192. A. Campbell and J. Kenyon, *J. Chem. Soc.*, 25 (1946).
193. C. W. Shoppee, R. E. Lack and S. K. Roy, *J. Chem. Soc.*, 3767 (1963).
194. H. P. Fischer and C. A. Grob, *Helv. Chim. Acta*, **47**, 564 (1964).
195. R. T. Conley and T. M. Tencza, *Tetrahedron Letters*, 1781 (1963).
196. A. Hassner and E. G. Nash, *Tetrahedron Letters*, 525, (1965).
197. A. F. Ferris, *J. Org. Chem.*, **25**, 12 (1960).
198. A. F. Ferris, G. S. Johnson and F. E. Gould, *J. Org. Chem.*, **25**, 1813 (1960).

199. A. L. Green and B. Saville, *J. Chem. Soc.*, 3887 (1956).
200. J. P. Freeman, *J. Org. Chem.*, **26**, 3507 (1961).
201. A. Hassner and I. H. Pomerantz, *J. Org. Chem.*, **27**, 1760 (1962).
202. A. Hassner, W. A. Wentworth, and I. H. Pomerantz, *J. Org. Chem.*, **28**, 304 (1963).
203. A. Hassner and W. A. Wentworth, *Chem. Commun.*, 44 (1965).
204. H. P. Fischer, C. A. Grob and E. Renk, *Helv. Chim. Acta*, **45**, 2539, (1962).
205. H. P. Fischer and C. A. Grob, *Helv. Chim. Acta*, **46**, 936 (1963).
206. C. A. Grob, H. P. Fischer, H. Link and E. Renk, *Helv. Chim. Acta*, **46**, 1190 (1963).
207. C. A. Grob and A. Sieber, *Helv. Chim. Acta*, **50**, 2520 (1967).
208. C. A. Grob and A. Sieber, *Helv. Chim. Acta*, **50**, 2531 (1967).
209. C. W. Shoppee and S. K. Roy, *J. Chem. Soc.*, 3774 (1963).
210. T. E. Stevens, *Tetrahedron Letters*, 3017 (1967).
211. R. K. Hill, *J. Org. Chem.*, **27**, 29 (1962).
212. R. L. Autrey and P. W. Scullard, *J. Am. Chem. Soc.*, **87**, 3284 (1965).
213. M. Ohno and I. Terasawa, *J. Am. Chem. Soc.*, **88**, 5683 (1966).
214. E. C. Taylor, A. McKillop and R. E. Ross, *J. Am. Chem. Soc.*, **87**, 1990 (1965).
215. E. V. Crabtree and E. J. Poziomek, *J. Org. Chem.*, **32**, 1231 (1967).
216. A. Ahmad and I. D. Spenser, *Can. J. Chem.*, **38**, 1625 (1960).
217. M. Kuhara, K. Matsumiya and N. Matsunami, *Mem. Coll. Sci., Univ. Kyoto*, **1**, 105 (1914); *Chem. Abstr.*, **9**, 1613 (1915).
218. A. W. Chapman and F. A. Fidler, *J. Chem. Soc.*, 448 (1936).
219. C. A. Grob, H. P. Fischer, W. Raudenbusch and J. Zergenyi, *Helv. Chim. Acta*, **47**, 1003 (1964).
220. M. Kuhara and H. Watanabe, *Mem. Coll. Sci., Univ. Kyoto*, **9**, 349 (1916); *Chem. Abstr.*, **11**, 579 (1917).
221. H. P. Fischer, *Tetrahedron Letters*, 285 (1968).
222. W. Theilacker and H. Mohl, *Ann. Chem.*, **563**, 99 (1949).
223. K. Morita and Z. Suzuki, *J. Org. Chem.*, **31**, 233 (1966).
224. H. P. Fischer, *Helv. Chim. Acta*, **48**, 1279 (1965).
225. P. T. Lansbury and J. G. Colson, *J. Am. Chem. Soc.*, **84**, 4167 (1962).
226. P. T. Lansbury, J. G. Colson and N. R. Mancuso, *J. Am. Chem. Soc.*, **86** 5225 (1964).
227. P. T. Lansbury and N. R. Mancuso, *J. Am. Chem. Soc.*, **88**, 1205 (1966).
228. P. T. Lansbury and N. R. Mancuso, *Tetrahedron Letters*, 2445 (1965).
229. P. J. McNulty and D. E. Pearson, *J. Am. Chem. Soc.*, **81**, 612 (1959).
230. Y. Yukawa and M. Kawakami, *Chem. Ind. (London)*, 1401 (1961).
231. I. T. Glover and V. F. Raaen, *J. Org. Chem.*, **31**, 1987 (1966).
232. M. Kawakami and Y. Yukawa, *Bull. Chem. Soc. Japan*, **37**, 1050 (1964).
233. M. I. Vinnik and N. G. Zarakhani, *Russ. Chem. Rev.*, **36**, 51 (1967).
234. A. W. Chapman, *J. Chem. Soc.*, 1550 (1934).
235. W. Z. Heldt. *J. Org. Chem.*, **26**, 1695 (1961).
236. E. M. Kosower, *J. Am. Chem. Soc.*, **80**, 3253 (1958).
237. O. Mumm, H. Hesse and H. Volquartz, *Ber.*, **48**, 379 (1915).
238. A. W. Chapman, *J. Chem. Soc.*, 1992 (1925).

239. A. W. Chapman, *J. Chem. Soc.*, 2296 (1926).
240. A. W. Chapman, *J. Chem. Soc.*, 1743 (1927).
241. A. W. Chapman, *J. Chem. Soc.*, 569 (1929).
242. J. W. Schulenberg and S. Archer, *Org. Reactions*, **14**, 1 (1965).
243. K. B. Wiberg and B. I. Rowland, *J. Am. Chem. Soc.*, **77**, 2205 (1955).
244. E. R. H. Jones and F. G. Mann, *J. Chem. Soc.*, 786 (1956).
245. M. P. Lippner and M. L. Tomlinson, *J. Chem. Soc.*, 4667 (1956).
246. M. M. Jamison and E. E. Turner, *J. Chem. Soc.*, 1954 (1937).
247. R. Barclay, Jr., *Can. J. Chem.*, **43**, 2125 (1965).
248. G. Singh, S. Singh, A. Singh and M. Singh, *J. Indian Chem. Soc.*, **28**, 459 (1951).
249. G. Singh, A. Singh, S. Singh and M. Singh, *J. Indian Chem. Soc.*, **28**, 698 (1951).
250. G. Singh, S. Singh, A. Singh and M. Singh, *J. Indian Chem. Soc.*, **29**, 783 (1952).
251. J. Cymerman-Craig and J. W. Loder, *J. Chem. Soc.*, 4309 (1955).
252. W. Wislicenus and M. Goldschmidt, *Ber.*, **33**, 1467 (1900).
253. K. B. Wiberg, T. M. Shryne and R. R. Kinter, *J. Am. Chem. Soc.*, **79**, 3160 (1957).
254. K. B. Wiberg and B. I. Rowland, *J. Am. Chem. Soc.*, **77**, 1159 (1955).
255. T. Taguchi, Y. Kawazoe, K. Yoshira, H. Kanayama, M. Mori, K. Tabata and K. Harano, *Tetrahedron Letters*, 2717 (1965).
256. H. Hettler, *Tetrahedron Letters*, 1793 (1968).
257. O. Mumm and F. Moller, *Ber.*, **70**, 2214 (1937).
258. R. M. Roberts and F. A. Hussein, *J. Am. Chem. Soc.*, **82**, 1950 (1960).
259. F. A. Hussein and K. S. Al-Dulaimi, *J. Chem. U.A.R.*, **9**, 287 (1966).
260. F. A. Hussein and S. Y. Kazandji, *J. Indian Chem. Soc.*, **43**, 663 (1966).
261. G. D. Lander, *J. Chem. Soc.*, 406 (1903).
262. A. E. Arbuzov and V. E. Shishkin, *Dokl. Akad. Nauk. SSSR*, **141**, 349 (1961); *Chem. Abstr.*, **56**, 11491 (1962).
263. V. E. Shishkin, *Mater. Nauch. Konf. Sovnarkhoz. Nizhne-volzh. Ekon. Raiona Volgograd. Politekh. Inst.*, **2**, 69 (1965); *Chem. Abstr.*, **67**, 5020 (1967).
264. A. E. Arbuzov and V. E. Shishkin, *Dokl. Akad. Nauk. SSSR*, **141**, 611 (1961); *Chem. Abstr.*, **56**, 15424 (1962).
265. F. Cramer and K. Baer, *Chem. Ber.*, **93**, 1231 (1960).
266. C. L. Stevens and M. E. Munk, *J. Am Chem. Soc.*, **80**, 4065 (1958).
267. M. L. Ernst and G. L. Schmir, *J. Am. Chem. Soc.*, **88**, 5001 (1966).
268. E. Hedaya, R. L. Hinman and S. Theodoropulos, *J. Org. Chem.*, **31**, 1317 (1966).
269. R. F. Meyers, *J. Org. Chem.*, **28**, 2902 (1963).
270. E. Allenstein and R. Fuchs, *Chem. Ber.*, **100**, 2604 (1967).
271. K. Gulbins and K. Hamann, *Angew. Chem.*, **73**, 434 (1961).
272. T. Mukaiyama, T. Fujisawa, H. Nohira and T. Hyugaji, *J. Org. Chem.*, **27**, 3337 (1962).
273. P. W. Neber and A. Friedolsheim, *Ann. Chem.*, **449**, 109 (1926).
274. P. W. Neber and A. Uber, *Ann. Chem.*, **467**, 52 (1928).
275. P. W. Neber and A. Burgard, *Ann. Chem.*, **493**, 281 (1932).
276. P. W. Neber and G. Huh, *Ann. Chem.*, **515**, 283 (1935).

277. P. W. Neber, A. Burgard and W. Thier, *Ann. Chem.*, **526**, 277 (1936).
278. C. O'Brien, *Chem. Rev.*, **64**, 81 (1964).
279. D. J. Cram and M. J. Hatch, *J. Am. Chem. Soc.*, **75**, 33 (1953).
280. M. J. Hatch and D. J. Cram, *J. Am. Chem. Soc.*, **75**, 38 (1953).
281. D. H. R. Barton and L. R. Morgan, *J. Chem. Soc.*, 622 (1962).
282. H. O. House and W. F. Berkowitz, *J. Org. Chem.*, **28**, 2271 (1963).
283. R. F. Parcell, *Chem. Ind.* (*London*), 1396 (1963).
284. D. F. Morrow and M. E. Butler, *J. Heterocyclic Chem.*, **1**, 53 (1964).
285. D. F. Morrow, M. E. Butler and E. C. Y. Huang, *J. Org. Chem.*, **30**, 579 (1965).
286. R. Huisgen and J. Wulff, *Tetrahedron Letters*, 917 (1967).
287. D. W. Kurtz and H. Shechter, *Chem. Commun.*, 689 (1966).
288. A. Hassner and F. W. Fowler, *J. Am. Chem. Soc.*, **90**, 2869 (1968).
289. P. A. S. Smith, *The Chemistry of Open-Chain Nitrogen Compounds*, Vol. 2, W. A. Benjamin, New York, 1966, p. 55.
290. H. E. Baumgarten and F. A. Bower, *J. Am. Chem. Soc.*, **76**, 4561 (1954).
291. G. H. Alt and W. S. Knowles, *J. Org. Chem.*, **25**, 2047 (1960).
292. H. E. Baumgarten, J. M. Petersen and D. C. Wolf, *J. Org. Chem.*, **28**, 2369 (1963).
293. H. E. Baumgarten, J. E. Dirks, J. M. Petersen and R. L. Zey, *J. Org. Chem.*, **31**, 3708 (1966).
294. P. A. S. Smith and E. E. Most, Jr., *J. Org. Chem.*, **22**, 358 (1957).
295. K. R. Henery-Logan and T. L. Fridinger, *J. Am. Chem. Soc.*, **89**, 5724 (1967).
296. C. L. Stevens, R. D. Elliot, B. L. Winch and I. L. Klundt, *J. Am. Chem. Soc.*, **84**, 2273 (1962).
297. C. W. Shoppee and D. A. Prins, *Helv. Chim. Acta*, **26**, 185 (1943).
298. C. L. Stevens, A. Thuillier and F. A. Daniher, *J. Org. Chem.*, **30**, 2962 (1965).
299. C. L. Stevens, H. T. Hanson and K. G. Taylor, *J. Am. Chem Soc.*, **88**, 2769 (1966).
300. D. F. Morrow, M. E. Brokke, G. W. Moersch, M. E. Butler, C. F. Klein, W. A. Neuklis and E. C. Y. Huang, *J. Org. Chem.*, **30**, 212 (1965).
301. S. Yamada, H. Mizuno and S. Terashima, *Chem. Commun.*, 1058 (1967).
302. C. L. Stevens, A. B. Ash, A. Thuillier, J. H. Amin, A. Balys, W. E. Dennis, J. P. Dickerson, R. P. Glinski, H. T. Hanson, M. D. Pillai and J. W. Stoddard, *J. Org. Chem.*, **31**, 2593 (1966).
303. S. Robev, *Compt. Rend. Acad. Bulg. Sci.*, **7**, 37 (1954); *Chem. Abstr.*, **50**, 7073 (1956).
304. F. Cooper and M. Partridge, *J. Chem. Soc.*, 257 (1953).
305. I. I. Grandberg, Y. A. Naumov and A. N. Kost, *Zh. Obshch. Khim.*, **1**, 805 (1965).
306. S. Robev, *Compt. Rend. Acad. Bulg. Sci.*, **31**, 159 (1960); *Chem. Abstr.*, **55**, 18676 (1961).
307. S. Robev, *Dokl. Akad. Nauk SSSR*, **101**, 277 (1955); *Chem. Abstr.*, **50**, 3315 (1956).
308. S. Robev, *Chem. Ber.*, **91**, 244 (1958).
309. R. Huisgen, *Angew. Chem.*, **72**, 359 (1960).
310. P. A. S. Smith, *J. Am. Chem. Soc.*, **76**, 436 (1954).

311. P. A. S. Smith and E. Leon, *J. Am. Chem. Soc.*, **80**, 4647 (1958).
312. A. F. Hegarty, J. B. Aylward and F. L. Scott, *Tetrahedron Letters*, 1259 (1967).
313. A. Heesing and H. Schulze, *Angew. Chem. Intern. Ed. Engl.*, **6**, 704 (1967).
314. E. Haruki, T. Inaike and E. Imoto, *Bull. Soc. Chem. Japan*, **38**, 1806 (1965).

CHAPTER 10

Cleavage of the carbon–nitrogen double bond

ALBERT BRUYLANTS AND
MRS. E. FEYTMANTS-DE MEDICIS

Université de Louvain, Belgium

I. INTRODUCTION

Substituted derivatives of the $\rangle C{=}N{-}X$ structure can undergo different kinds of reaction:

(a) Addition

$$\rangle C{=}N{-}X \xrightarrow{H_2} \rangle CH{-}NHX$$

(b) Elimination

$$\rangle C{=}N{-}X \longrightarrow {-}C{\equiv}N$$

(c) Nucleophilic substitution and particularly hydrolysis, this reaction being the subject of the present work.

$$\rangle C{=}N{-}X \xrightarrow{H_2O} \rangle C{=}O + H_2NX$$

The nomenclature of the various compounds involved is set out below.

The problem of the hydrolysis scission of the $\rangle C{=}N{-}$ group is one of the most fruitful subjects that has been studied within the last few years. This reaction occurs in the most varied fields: scission of the peptide links in proteins and oligopeptides by way of the intermediate iminolactone[1–3]; the existence of the Schiff base rhodopsine in the visual pigment[4–6] the biological importance of Schiff bases in the enzymic aldolization[7–11], and decarboxylation[12–15] reactions, and in pyridoxal phosphate degradation[16–20]; and finally, in the field of colour photography[21].

Over the last few years, however, it has been the elucidation of the mechanism of this reaction which has particularly drawn the attention of specialists. Following on the chapter entitled 'Carbonyl addition reactions' by Hammett[22], we note Jencks' remarkable study 'Mechanism and Catalysis of Simple Carbonyl Group Reactions'[23]. The following reviews should also be consulted: Layer's 'The Chemistry of Imines'[24], Roger and Neilson's 'The Chemistry of Imidates'[25], 'Properties and Reactions of Azlactones' *Org. Reactions*, **3**, 213 (1946) and *Organic Reaction Mechanisms*[26].

The table on the right lists the relevant C=N nomenclature.

A. DERIVATIVES OF THE CARBONYL GROUP.

1. *Isonitriles* or carbylamines $\rangle C{=}N{-}R$

2. *Imines*

(a) Acyclic, derived from aliphatic or aromatic amines and which are called, depending on the nomenclature adopted, Schiff bases, azomethines or anils. Amongst these are to be found:

(i) the aldimines $R{-}CH{=}N{-}R$

(ii) the ketimines $RR'C{=}N{-}R$

(iii) imines proper, sometimes incorrectly called imides $RR'C{=}NH$

(iv) the quinonimides (mono- and di-)

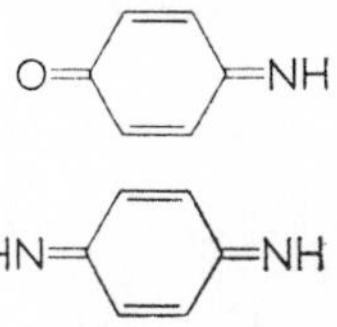

O=⟨ring⟩=NH

HN=⟨ring⟩=NH

(b) Cyclic, non-aromatic:
Five membered heterocycles:
oxazole, thiazole, thiazoline, isoxazole, imidazole, pyrazole, pyrazoline, furazanes, triazole, tetrazole, etc.

3. *Oximes*, derived from hydroxylamine

(i) aldoximes $RCH{=}NOH$

(ii) ketoximes $RR'C{=}NOH$

(iii) fulminic acid $\rangle C{=}N{-}OH$

4. *Hydrazones*, derived from hydrazine

$$\rangle C{=}N{-}N\langle$$

which include, in particular, the osazones

$$\begin{array}{l} CH{=}N{-}NH{-}Ph \\ | \\ C{=}N{-}NH{-}Ph \\ | \\ (CHOC)_n \\ | \\ CH_2OH \end{array}$$

5. *Semicarbazones*, derived from semicarbazide

$$\rangle C{=}N{-}NHCONH_2$$

B. Derivatives of the Carboxyl Group.

1. *Amidines and hydrazidines*

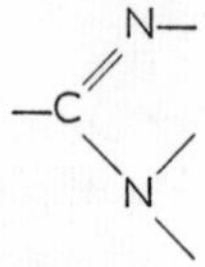

$$-C(=N-)-N<$$

2. *Imino ethers:* imidols and imidates

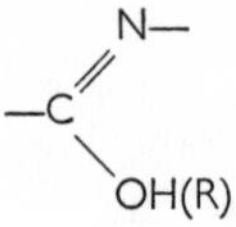

$$-C(=N-)-OH(R)$$

3. *Halimines*

$$-C(=N-)-X$$

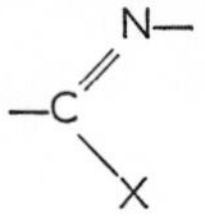

C. Derivatives of Carbonic acid.

1. *Guanidine* or iminourea

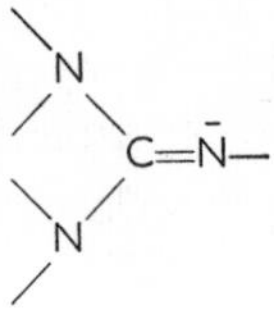

$$(>N)_2C=\bar{N}-$$

2. *Isoureas and isothioureas*

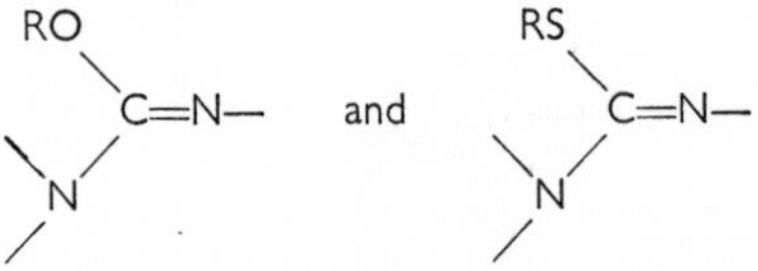

$$RO(>N)C=N- \quad \text{and} \quad RS(>N)C=N-$$

D. Structures with conjugated double bonds.

Azines

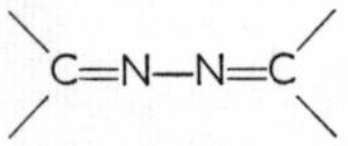

$$>C=N-N=C<$$

II. HYDROLYSIS OF SCHIFF BASES

The hydrolysis of the imine group takes place in two steps:

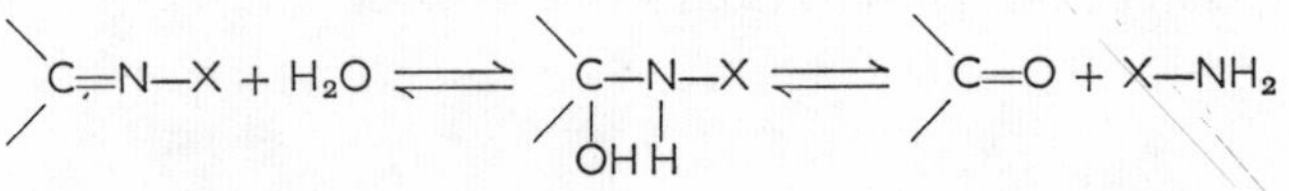

$$>C=N-X + H_2O \rightleftharpoons >C(OH)-NH-X \rightleftharpoons >C=O + X-NH_2 \qquad (1)$$

thus involving a tetrahedral carbinolamine intermediate. This intermediate product is unstable and can only be isolated in rare cases, as in the preparation of the hydrazone and semicarbazone from chloral[27–29], or the oxime and hydrazone from 2-formyl-1-methylpyridinium iodide[30]. The existence of the intermediate product has been polarographically demonstrated in the hydrolysis of benzylideneaniline[31], but it has not been spectrophotometrically confirmed, neither in the hydrolysis nor in the formation of the Schiff bases[32–34].

It would, however, appear reasonable to postulate the existence of a carbinolamine intermediate in all reactions of a given type in which the rate of reaction versus the pH curve assumes a bell-shaped form. In fact, a bell-shaped pH–rate curve which cannot be accounted for by the ionization of the reactants is diagnostic of a change in the rate-determining step and, for this to be possible, two consecutive steps and a reaction intermediate must necessarily exist. The rate–pH curve thus constitutes a very useful method for recognizing the existence of intermediates in reactions[23].

Since the earliest work of Lapworth, such curves, characteristic of reactions in which $\rangle C{=}N{-}$ groups participate, are well known: e.g. in the formation of acetoxime[35,36]; in the equilibrium, formation and hydrolysis of oximes[37,38], in the hydrolysis and polymerization of cyanimide[39–41] and in the formation and hydrolysis of semicarbazones[14,42].

Reaction schemes which can give rise to such a bell-shaped curve have been characterized[43], and include the following:

(a) Double ionization of a reagent, the intermediate ion leading to the reaction products; this case is not met with in the hydrolysis of the imine groups.

(b) Action of a base on an acid or of a conjugate base on a conjugate acid leading, in one rate-determining step, to the products. The bell-curve effect results from the opposed acidity effects on the nucleophilic and the electrophilic reactants. The two conjugate reaction schemes are kinetically indistinguishable and can only be resolved by complementary proof or by chemical intuition:

$$AH + B \longrightarrow \text{products}$$

$$A^- + BH^+ \longrightarrow \text{products}$$

(c) Pre-ionization equilibrium followed by nucleophilic attack on the protonated species to yield a reaction intermediate which itself

TABLE 1. Equilibrium constants for the formation and the hydrolysis of imine derivatives.

Derivatives	Experimental conditions	$K_B = \frac{[\rangle C{=}N{-}X]}{[\rangle C{=}O][H_2N{-}X]}$ (l/mole × 10^{-5})	$K_A = \frac{[\rangle C{=}\overset{+}{N}H{-}X]}{[\rangle C{=}O][NH_3 \pm X]}$ (l/mole)	Reference
p-$ClC_6H_4CH{=}NC_6H_5$	25°C H_2O + 13% EtOH $f_i = 0{\cdot}50$ − pH = 2·05 − 3·50		0·080	48
$(CH_3)_2C{=}N{-}OH$	20°C		71	37
$C_6H_5CH{=}N{-}NHCONH_2$	25°C H_2O + 25% EtOH; HCl 1N	6·9		50
	25°C H_2O	3·3	67	42
$(H_3C)(HOOC)C{=}NNHCONH_2$	25°C H_2O	1·96	1·000	42
(2-furyl)$CH{=}NNHCONH_2$	25°C H_2O	1·32	833	42
$(CH_3)_3CCH{=}NNHCONH_2$	25°C H_2O	0·54	–	42
$CH_3{-}CH{=}NNHCONH_2$	25°C H_2O	0·34	132	42
(cyclohexylidene)$C{=}NNHCONH_2$	25°C H_2O	0·0047	–	42
$(CH_3)_2C{=}NNHCONH_2$	25°C H_2O	0·0031	1·45	42

$CH_3(C(CH_3)_3)C{=}NNHCONH_2$	25°c H_2O	0·00079	–	42
$CH_3(HOOC)C{=}NCH_2COOH$			2·47	53
$CH_3(HOOC)C{=}NCH(CH_3)COOH$			0·93	53
$(CH_3)_2C{=}NH$			0·004	55
$(CH_3)_2C{=}NCH_2COOH$			0·068	55
o-$HOC_6H_4CH{=}NCH_3$	25°c H_2O	0·372		57
p-$ClC_6H_4CH{=}N{-}OH$	25°c H_2O	610		58
p-$ClC_6H_4C{=}\overset{+}{N}(CH_3){-}O^-$	25°c H_2O	9·4		58

gives rise to the reaction products; the bell-curve effect stems, in this case, from the change in the rate-determining step.

The first kinetic interpretation put forward by Hammett[22] was based on acid–base equilibria. More recent work, either corrected for buffer effects or carried out in the absence of general acid–base catalysis, has shown that it was the consequence of a change in the rate-determining step[23]. Proof for this is seen in the pH-dependence of the sensitivity of the reaction to general acid catalysis, in the regions of the pH situated on both sides of the maximum[44,45], in the effect of changing the substituents[45,46], and in the fact that, in the region of maximum rate, the rate-determining step can be modified by varying the concentration of catalyst[44,45].

The reaction is reversible and generally is in equilibrium[47–49], as follows:

$$\rangle C{=}O + NH_2{-}X \rightleftharpoons \rangle C{=}N{-}X + H_2O \tag{2}$$

Examination of the equilibrium constants shows that the formation of unsaturated derivatives, such as oximes[37], and semicarbazones[45,50], is always more favoured than the formation of Schiff bases derived from ammonia or from primary amines[51–55] (Table 1) and from *p*-toluidine[56].

It is difficult to determine the relative importance of the parts played in this phenomenon by the carbon affinity and the resonance stabilization

$$\rangle C{=}N{-}X \longleftrightarrow \rangle \bar{C}{-}N{=}\overset{+}{X} \tag{3}$$

Moreover, knowledge of the overall equilibrium constants:

$$\rangle C{=}O + RNH_3^+ \overset{K_1}{\rightleftharpoons} \rangle C{=}N^+HR + H_2O \tag{4}$$

$$RNH_3^+ \overset{K_2}{\rightleftharpoons} RNH_2 + H^+ \tag{5}$$

$$\rangle C{=}O + RNH_2 \overset{K_3}{\rightleftharpoons} \rangle C{=}N{-}R + H_2O \tag{6}$$

permits a satisfactory evaluation of the constant[48]:

$$\rangle C{=}N^+H{-}R \overset{K_4}{\rightleftharpoons} \rangle C{=}N{-}R + H^+ \tag{7}$$

$$K_4 = \frac{K_2K_3}{K_1}$$

A. Schiff Bases Derived from Aliphatic Amines

The kinetics of the hydrolysis of the Schiff bases derived from strongly basic amines such as aliphatic amines, have been studied recently, the mechanism of the reaction having been successfully elucidated.

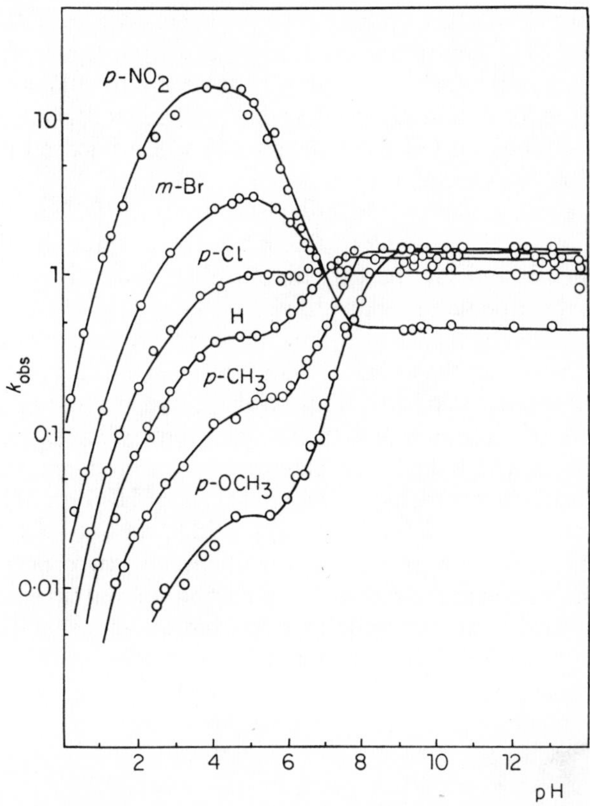

FIGURE 1. Logarithm of the rate constants of hydrolysis of substituted benzylidene-1,1-dimethyelthylamines against the pH (taken from reference 62).

In the case of the benzylidene amines carrying electron acceptor substituents, the rate initially increases as a function of the pH to a maximum value, after which it decreases, to become finally independent of it. The semilogarithmic rate–pH curve thus shows a bell-like form under neutral and acidic conditions. In the case of electron-donor substituents, the rate is lowered in this region (see Figure 1).

The portion of the curve over which the rate is independent of the pH can be explained either on the basis of an attack by the hydroxyl ion on the protonated Schiff base (SH^+) or by the action of water on the free base, both of which are kinetically equivalent. In fact, the rate of hydrolysis of a Schiff base which has its labile proton replaced by a non-labile group will increase with the pH: such is the case for the benzylidenedimethylammonium ion[60]. The actual mechanism is thus an attack of OH^- on SH^+ and the constant rate is due to the fact that the SH^+ concentration diminishes in the same proportion as the increase in the OH^- concentration, and as a function of the pH. Moreover, the formation of micelles which takes place in the presence of cationic, anionic or non-ionic detergents, will influence the rate of hydrolysis in the expected way[61].

While the calculated second order rate constant for the attack of hydroxide ion on the protonated Schiff base is substantially the same for the aqueous and for the cationic micellar phase, it is markedly reduced in the anionic micellar phase.

Further, in the region of neutral pH, the concentration of protonated Schiff base increases, the attack of water thus becoming preponderant over that of the hydroxyl ion. This results in an increase or in a decrease of the rate proportional to the percentage of ionization right up to the point where the Schiff base is totally protonated, the measured rate at this point being the rate of attack of H_2O on SH^+.

With conditions becoming even more acidic, the rate of hydrolysis decreases linearly with respect to the hydroxyl ion concentration. Such a modification in the rate–pH curve is interpreted as representing a transition in the rate-determining step from the nucleophilic attack on the protonated Schiff base to that of decomposition of the carbinolamine intermediate.

The intermediate possesses the dipolar structure

```
        H
        |
   —C—N—R
    |   |+
  -|O|  H
```

and the concentration of this zwitterionic species diminishes in proportion to the pH, under conditions in which the Schiff base is wholly protonated. This results in the rate of decomposition falling below the rate of attack of water on the protonated Schiff base, independent of pH, and there ensues a change of the rate-determining step together with a fall in the rate itself as the acidity increases.

The reaction scheme put forward to account for such behaviour is as follows: the substrate (S) undergoes protonation (equation 8), the protonated form (SH^+) can then undergo attack by a molecule of water (equation 9) and by the hydroxyl ion (equation 10), either simultaneously or separately, according to conditions, yielding an amino alcohol or carbinolamine (SHOH) which finally gives rise to products (equation 11).

$$S + H^+ \overset{K_{SH^+}}{\rightleftharpoons} SH^+ \tag{8}$$

$$SH^+ + H_2O \underset{k_{-1}}{\overset{k_1}{\rightleftharpoons}} SHOH + H^+ \tag{9}$$

$$SH^+ + OH^- \underset{k_{-2}}{\overset{k_2}{\rightleftharpoons}} SHOH \tag{10}$$

$$SHOH \xrightarrow{k_3} \text{products} \tag{11}$$

For an intermediate at low and constant concentration we have:

$$k_{obs} = \frac{k_1k_3[H^+] + 10^{-14}k_2k_3}{([H^+] + K_{SH^+})(k_{-1}[H^+] + k_{-2} + k_3)} \tag{12}$$

and assuming that for the hydroxyl ion attack on the protonated Schiff base under alkaline conditions,

$$k_{-2} + k_3 \gg k_{-1}[H^+]$$

we have

$$k_{obs} = \frac{k_1k_3[H^+]}{(K_{SH^+} + [H^+])(k_{-1}[H^+] + k_{-2} + k_3)} + \frac{10^{-14}k_2k_3}{K_{SH^+}(k_{-2} + k_3)} \tag{13}$$

The first term in equation (13) describes the kinetic behaviour at acidic and neutral pH, the second describing it at basic pH.

It is possible to calculate, from experiment, several of the parameters in these equations, together with their ratios, and the calculated rate–pH curve will coincide well with the experimental points (see Table 2).

In the case of Schiff bases derived from *o*-, *m*-, and *p*-hydroxybenzaldehydes and 2-aminopropane, an increase in the rate is obtained in alkaline conditions by comparison with the corresponding methoxy compounds. This increase is a consequence of the ionization of the hydroxyl groups and is in agreement with the predictions of the Hammett equation.

TABLE 2. Kinetic data for the hydrolysis of Schiff bases derived from aliphatic ami

Schiff base	Experimental conditions	pK_{SH^+}	$k_{(OH^- + SH^+)}$ mole^{-1}min^{-1}	$k_{(H_2O + SH^+)}$ min^{-1}	Refer
$C_6H_5CH{=}NC(CH_3)_3$	Water, 3% ethanol, 25°C, ionic strength 0·50	6·70	$3{\cdot}0 \times 10^7$	0·41	62
$(C_6H_5)_2C{=}NCH_3$	Water, 3% ethanol, 25°C, ionic strength 0·50	7·22	$4{\cdot}3 \times 10^4$	0·016	63
p-$CH_3OC_6H_4\cdot CH{=}NCH(CH_3)_2$	Water, 2% methanol, 30°C, ionic strength 0·10	7·1	$6{\cdot}2 \times 10^6$	0·2	64

Under neutral conditions, the *o*- and *p*-hydroxyl derivatives undergo a decrease in rate with respect to the corresponding methoxy derivatives. This phenomenon is attributed to a decrease in the substrate reactivity as a result of enolimine–keto amine tautomerization:

$$HO-C_6H_4-CH{=}N-CH(CH_3)_2 \rightleftharpoons O{=}C_6H_4{=}CH-NH-CH(CH_3)_2 \leftrightarrow {}^-O-C_6H_4-CH{=}\overset{+}{N}H-CH(CH_3)_2 \qquad (14)$$

Furthermore, the protonated anion is identical to the zwitterionic or keto amine tautomer:

$$^-O-C_6H_4-CH{=}NCH(CH_3)_2 + H^+ \overset{K_2}{\rightleftharpoons} {}^-O-C_6H_4-CH{=}\overset{+}{N}HCH(CH_3)_2 \qquad (15)$$

except in the case of the *m*-hydroxy derivative; it then becomes necessary to introduce a constant for the zwitterion formation, as follows:

$$m\text{-}HO-C_6H_4-CH{=}NCH(CH_3)_2 \overset{K_3}{\rightleftharpoons} m\text{-}{}^-O-C_6H_4-CH{=}\overset{+}{N}HCH(CH_3)_2 \qquad (16)$$

Assuming a single mechanism for the whole pH 7–14 range in which the hydroxide ion attacks the zwitterion as well as the protonated

Schiff base, the authors propose the following reaction scheme (see equations 8–11):

$$\mathrm{SH^+} \underset{}{\overset{K_1}{\rightleftharpoons}} \mathrm{S + H^+}$$

$$\mathrm{S} \underset{}{\overset{K_2}{\rightleftharpoons}} \mathrm{S^- + H^+}$$

$$\mathrm{SH^+ + H_2O} \underset{k_{-1}}{\overset{k_1}{\rightleftharpoons}} \mathrm{X + H^+}$$

$$\mathrm{SH^+ + OH^-} \underset{k_{-2}}{\overset{k_2}{\rightleftharpoons}} \mathrm{X}$$

$$\mathrm{S + OH^-} \underset{k_{-3}}{\overset{k_3}{\rightleftharpoons}} \mathrm{X}$$

$$\mathrm{X} \xrightarrow{k_4} \text{products}$$

The kinetic equation thus constructed perfectly describes the observed behaviour[64]:

$$k_{\text{obs}} = \frac{k_1k_4[\mathrm{H^+}] + k_2k_4 \times 10^{-14} + k_3k_4k_1[\mathrm{OH^-}]}{(K_{1+}[\mathrm{H^+}]+K_1K_2/[\mathrm{H^+}])(k_{-1}[\mathrm{H^+}]+k_{-2}+k_{-3}[\mathrm{OH^-}]+k_4)} \tag{17}$$

On the other hand, Reeves[65] has studied the hydrolysis of *o*- and *p*-hydroxybenzilideneanilines (bearing an hydroxyl or an *N*-trimethylammonium substituent on the aniline ring) and of *p*-*N*-trimethylammoniumbenzilidene-*o*- and *p*-hydroxyaniline. He finds an increase in rate in the case of the salicylidene derivatives

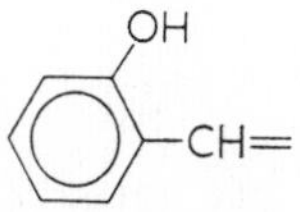

OH

—CH=

with respect to the *m*- and *p*-isomers. No increase is observed with OH on the aniline ring. This accelerating effect of an ionized *o*-OH group is interpreted in terms of an intramolecular general base catalysis.

The question may be raised now as to whether these are fundamental differences between Schiff bases derived from salicylideneaniline and salicylidene-2-aminopropane and whether the rate increase observed in the case of the salicylidene OH group, with respect to the *m*- and *p*- OH isomers, is not a rate decrease with respect to the corresponding methoxy derivatives.

B. Schiff Bases Derived from Aromatic Amines

In the case of Schiff bases derived from weakly basic aromatic amines, the expected displacement of the curve towards the lower regions of pH is observed.

In strongly basic conditions (pH 9 to 14), there appears, in the semilogarithmic rate–pH diagram, a straight line in which the rate of hydrolysis is directly proportional to the concentration of hydroxyl ions[31,48]. This suggests that the imine-carrying carbon atom of the neutral Schiff base is sufficiently positive to undergo attack by the hydroxyl ion, even without being protonated:

$$H\bar{O} \longrightarrow \;>\overset{\delta+}{C}=\overset{\delta-}{N}-R$$

Finally, several authors[32,48,65–67] have observed a point of inflexion in the bell-shaped part of the curve. They put forward a reaction scheme which includes a protonation equilibrium of the carbinolamine intermediate, with subsequent decomposition of these two forms into products:

$$S + H^+ \xrightleftharpoons{K_{SH^+}} SH^+$$

$$SH^+ + H_2O \underset{k_{-1}}{\overset{k_1}{\rightleftharpoons}} SHOH + H^+$$

$$SH^+ + OH^- \underset{k_{-2}}{\overset{k_2}{\rightleftharpoons}} SHOH$$

$$SHOH + H^+ \xrightleftharpoons{K_{SH_2OH^+}} SH_2OH^+$$

$$SHOH \underset{k_{-3}}{\overset{k_3}{\rightleftharpoons}} \text{products}$$

$$SH_2OH^+ \underset{k_{-4}}{\overset{k_4}{\rightleftharpoons}} \text{products} + H^+$$

The kinetic equation then becomes:

$$k_{obs} = \frac{[H^+]}{[H^+] + K_{SH^+}} \cdot \frac{k_2k_3K_{SH^+}/[H^+] + k_1k_3 + k_2k_4K_{SH^+}/K_{SH_2OH^+} + k_1k_4[H^+]/K_{SH_2OH^+}}{k_{-2} + k_3 + (k_{-1} + k_4/K_{SH_2OH^+})[H^+]} \tag{18}$$

This is of the general form:

$$k = \frac{A + B[H^+] + C[H^+]^2}{D + E[H^+] + F[H^+]^2} \tag{19}$$

Equation (19) describes the experimental curves for the hydrolysis of benzylideneaniline and its substituted derivatives, and only the general parameters A, B, . . . can be calculated from it. Depending on the relative values of these parameters, a point of inflexion will or will not appear in the curve, as a result of a change in the rate-determining step. In this case, the maximum in the bell-shaped curve does not any longer correspond to the rate of attack of the protonated Schiff base by water, and the reaction scheme adopted will account for the fact that the rate of hydrolysis does not exactly correspond to the fraction of the Schiff base that is ionized.

In the strongly acidic region ($H_0 < 0$), the rate decreases linearly with the acidity. The cationic intermediate decomposes by general base catalysis, as follows:

$$\text{B:} \rightarrow \text{H}-\text{O}-\text{C}-\overset{\text{H}}{\underset{\text{H}}{\overset{|+}{\underset{|}{\text{N}}}}}-\text{R} \rightarrow \text{BH}^+ + {>}\text{C}{=}\text{O} + \text{RNH}_2$$

The decrease in the rate of hydrolysis is no longer attributed to a decrease in the concentration of the zwitterionic carbinolamine, but rather to a decrease in the activity of the water which acts as a nucleophilic agent and as a catalyst for the proton extraction during the process of decomposition of the cationic intermediate[48]. Such a mechanism is present in the case of 2,3-dimethylanilino-pent-3-en-2-one[47] and of the acetylimidazolium ion[69]. Such behaviour is characteristic of a proton transfer by way of water; with increasing acidity the transfer becomes more difficult and the rate consequently decreases[70]. A rate decrease is also observed in a deuterated solvent[48].

At intermediate pH values, in the region in which the rate is independent of the pH, proof of the fact that it is the protonated form of the Schiff base that reacts is furnished by the fact that the addition of an amphiphylic compound (for instance detergents like sodium lauryl sulphate or cetyltrimethylammonium bromide) complexes this charged molecule and causes considerable decrease in the rate of hydrolysis[71].

C. Thermodynamic Behaviour

The activation energies usually obtained for the hydrolysis of Schiff bases are relatively low (see Table 3), which is understandable considering that they represent the balance between the ionization and the hydrolysis. Any derivation of activation entropies or enthalpies

TABLE 3. Thermodynamic parameters for the hydrolysis of Schiff bases.

Schiff base	Type of reaction	E_a(kcal/mole)	log A (s^{-1})	Reference
$C_6H_5CH{=}NC_6H_5$	$SH^+ + OH^-$	13·2		31
$Cl^- Me_3\overset{+}{N}C_6H_4CH{=}NC_6H_4OH$	$SH^+ + H_2O$	7·6		31
	$SH^+ + OH^-$	13·2	4·78	67
	$SH^+ + OH^-$	14·4		32
$C_6H_5CH{=}N{-}C_6H_4{-}\overset{+}{N}(CH_3)_3$-$p$$Cl^-$	$SH^+ + OH^-$	16·8	8·2	65
$C_6H_5CH{=}N{-}CH(CH_3)(CH_2)_5CH_3$	—	10·6	5·1	73

would be pointless without prior separation of the various steps. Moreover, a characteristic of the Petersen figure[65] is the absence of an isokinetic temperature.

A more detailed study of the thermodynamic behaviour of the benzylidene-1,1-dimethylethylamines (see Table 4)[72] has shown that the entropy factor, as also the enthalpy factor, varies with the substituent. It has also been noted that the electrostatic interactions between the protonated Schiff base and the hydroxyl ion constitute energy factors which promote the reaction.

TABLE 4. Thermodynamic parameters for the hydrolysis of benzylidene-1,1 dimethylethylamine[72].

Type of reaction	$\Delta F^\ddagger$ (kcal/mole)	$\Delta H^\ddagger$ (kcal/mole)	$\Delta S^\ddagger$(e.u.)
$SH^+ + OH^-$	10·7	0·4	−34.5
$SH^+ + H_2O$	23·3	13·4	−33·2
decomposition SHOH	24·4	18·2	−20·8

On the other hand, the decomposition of the carbinolamine intermediate is isoentropic, contrary to the isoenthalpic phenomenon which takes place in the hydrolysis of the ethylbenzimidates. The measured activation entropy (Table 4) corresponds to the ionization entropy of the initial product and that of its hydration to a carbinolamine. The actual entropy of decomposition is close to zero, as is to be expected for a monomolecular decomposition reaction.

D. Structure–Reactivity Relationships

With respect to the structure–reactivity relationships, we have already mentioned the fact that the overall equilibrium of formation and hydrolysis of imine compounds changes but little with the nature of the nucleophile. Unlike the equilibria, the rates of addition are highly dependent on the basic nature of the attacking nucleophile. One may justifiably predict that the rate of addition of the hydroxyl ion will be greater than that of the water molecule in the hydrolysis of Schiff bases (see Table 2).

Actually, in the hydrolysis of Schiff bases, the reactions of the >C=N— group are less sensitive to the polar effects of substituents than are those of the >C=O group in the formation of Schiff bases (nitrogen being less electronegative than oxygen), even though the former reaction includes a protonation step which renders the >C=N⁺< group more electrophilic[23].

A separate study of the protonation equilibria of the Schiff bases has led to an evaluation of the reaction constants for this step (see Table 5).

TABLE 5. Relation between structure and reactivity in the equilibrium of protonation of Schiff bases, from the Hammett equation.

Derivatives	$\rho(\log K_{S_XH^+}/K_{S_0H^+} = \rho\sigma)$	References
$X-C_6H_4CH=NC(CH_3)_3$	$(+2{\cdot}4)\ \rho^+ = 1{\cdot}6$[a]	62
$(X-C_6H_4)(C_6H_5)C=NCH_3$	$+2{\cdot}0$	63
$X-C_6H_4CH=NCH(CH_3)_2$	$+2{\cdot}4$	64
p-$(CH_3)_3N^+C_6H_4CH=NC_6H_4-X$	$+1{\cdot}5$	32
$(X-C_6H_4CO)(Y-C_6H_4NHCO)C=NC_6H_4N(CH_3)_2$	$\rho_X = 0{\cdot}288$; $\rho_Y = 0{\cdot}137$	33[a]

[a] Better correlation is obtained by the use of the Brown–Okamoto constants for polar effects of the mesomeric type (σ^+).

For the overall hydrolysis reaction, the free-energy rate correlations can be called upon for use as mechanistic criteria[68]. They confirm, for example, the fact that it is the hydroxyl ion which attacks the protonated Schiff base (not water reacting with the free base) in the pH-independent rate zone, thus resolving a kinetic ambiguity previously alluded to.

Considering the reaction schemes again as well as the kinetic equations which result from them, namely:

$$\left.\begin{array}{l} \mathrm{S} + \mathrm{H_2O} \underset{k_{-1}}{\overset{k_1'}{\rightleftharpoons}} \mathrm{SHOH} \\ \mathrm{SHOH} \xrightarrow{k_2} \text{products} \end{array}\right\} k'_{obs} = \frac{k_1' k_2}{(k'_{-1} + k_2)} \tag{20}$$

$$\left.\begin{array}{l} \mathrm{SH^+} + \mathrm{OH^-} \underset{k_{-1}}{\overset{k_1}{\rightleftharpoons}} \mathrm{SHOH} \\ \mathrm{SHOH} \xrightarrow{k_2} \text{products} \end{array}\right\} k_{obs} = \frac{k_1 k_2}{K_{SH^+}(k_{-1} + k_2)} \tag{21}$$

various possible cases become apparent: if $k_2 \gg k_{-1}$, $k'_{obs} = k_1'$ and $k_{obs} = k_1/K_{SH^+}$, the intermediate rapidly transforms into product and the nucleophilic attack determines the rate. In the reaction $S + H_2O$, k'_{obs} is identical to a rate constant characteristic of a simple step, and the Hammett diagram must show a continuous linear relationship, in which, nucleophilic character remaining constant, $\rho_{k'_{obs}}$ would be positive. In fact, this is not the case (see Table 6), and this type of mechanism must therefore be definitely abandoned. In the reaction $SH^+ + OH^-$, the overall rate constant breaks down into two factors, an equilibrium constant and a specific rate. The corresponding Hammett diagram also consists of a linear relationship, the value $\rho_{k_{obs}}$ which consists of the sum of the values ρ_{k_1} (or $\rho_{k_1} = \rho_{SH^+ + OH^-}$ in Table 6) and $\rho_{1/K_{SH^+}}$. As ρ_{k_1} is probably positive and $\rho_{1/K_{SH^+}}$ negative, the overall value $\rho_{k_{obs}}$ must either be positive, or less negative than $\rho_{1/K_{SH^+}}$. Observations indicate this to be the case.

If $k_{-1} \gg k_2$; $k_{obs} = k_2 \times (1/K_{SH^+}) \times (k_1/k_{-1})$, the rate-determining step is the decomposition of the carbinolamine. The overall rate constant breaks down into three factors, two equilibrium rate factors and a specific rate factor. A linear Hammett diagram is thus to be expected, but no value of ρ can be predicted, since it is a composite term in which the substituents will exert opposing effects on k_2, on the one hand, and on $1/K_{SH^+}$ and k_1/k_{-1} on the other. The only experimental value known is positive ($\rho_{k_{obs}}/[OH^-] = +2{\cdot}17$) for the hydrolysis of benzylidene-*t*-butylamines at a pH of 1 [62a].

TABLE 6. Hammett parameters for the hydrolysis of Schiff bases.

Derivatives	Range of pH-independent rates Mechanism $SH^+ + OH^-$ $\rho_{k_{obs}}$	$\rho SH^+ + OH^- = \rho K_{SH^+} + \rho_{k_{obs}}$	Mechanism $SH^+ + H_2O$ $\rho^+ SH^+ + H_2O$	References
$X—C_6H_4CH{=}NC(CH_3)_3$	$-0{\cdot}21$	$+1{\cdot}26$ (ρ^+)	$+1{\cdot}71$	62a
$X—C_6H_4$, C_6H_5 \> $C{=}NCH_3$	—	$+1{\cdot}8$ (″)	$+1{\cdot}1$	63
$X—C_6H_4CH{=}NCH(CH_3)_2$	$-0{\cdot}27$	$+2{\cdot}1$	—	64
$X—C_6H_4$, C_6H_5 \> $C{=}\overset{+}{N}H_2Cl^-$	—	—	$+2{\cdot}0$	62b
$C_6H_5CH{=}NC_6H_4—X$	$-1{\cdot}45$		no linearity	67
$C_6H_5CH{=}NC_6H_4—X$	$-0{\cdot}8$			75
p-$(CH_3)_3\overset{+}{N}C_6H_4CH{=}NC_6H_4X$	$-0{\cdot}71$	$+0{\cdot}8$	—	68

If $k_{-1} \simeq k_2$, experimental measurements are not satisfied by $(k_{-1} \simeq k_2) < k_1$, as the intermediate carbinolamine does not continue to accumulate in the course of the reaction. If $(k_1 \simeq k_2) > k_2$, one arrives at the overall equation (20), with the rate being determined by the hydroxyl ion attack.

As the terms cannot be factorized, the Hammett diagram is not linear. This is the case in the hydrolysis of the benzylideneanilines[67] and of the (*p*-*N*-dimethylaminophenyl)-iminobenzoylacetanilides[33a], and in the formation of semicarbazones from substituted benzaldehydes[44].

It must here be added that a non-linear Hammett diagram does not necessarily constitute proof of a change in the rate-determining step[76], and that it is sometimes risky to account for the non-linearity by a change in the rate-determining step rather than an unequal balance of the polar effects of the substituents in the nucleophilic attack as well as in the subsequent reaction of the carbinolamine intermediate. Several ambiguous examples of non-linear Hammett diagrams are known: condensation of n-butylamine with substituted piperonals[60], condensation of semicarbazide with substituted benzaldehydes[6b], formation of substituted benzophenone oximes[33b].

On the other hand, the variations of the reaction parameter, ρ, in the two different ranges of pH situated on either side of the maximum in the log rate–pH curve (see Figure 1), always indicate a change in the rate-determining step, such as shown in the case of the formation of the benzylideneanilines[48] and in the hydrolysis of the benzylidene-*t*-butylamines[62a].

A recent study was undertaken to relate in a general way the structure to the reactivity and the catalysis on theoretical grounds laid out by Leffler and Hammond.

If we have *i* substrates identified by σ_i, and *j* general acid catalysts identified by pK_{aj}, the Hammett and Brønsted relations for such a system are:

$$\log k_{ij} = \log k_{oj} + \sigma_i \rho_j$$

$$\log k_{ij} = \log G_i{}^A - \alpha_i pK_{aj}$$

Combining these, we obtain the constant expression:

$$\frac{pK_{a2} - pK_{a1}}{\rho_2 - \rho_1} = \frac{\sigma_i}{\alpha_0 - \alpha_i} = C_1 \qquad (21)$$

Thus, a variation in the substrate reactivity resulting from a change in the substituents will result in a linear variation of the Brønsted general acid catalysis parameter (α) as a function of σ. Moreover, if one modifies the nature of the acid catalyst, the sensitivity of the reaction rate to the substituent effect (ρ) will change in a linear manner with the acidity of the catalyst (pK_a). Experiments have shown that C_1 is large and positive ($C_1 \simeq 10$ to 100) in the reaction of the substituted benzaldehydes with semicarbazide[45]. In fact, ρ^+ changes from 0·71 to 0·94 when the catalyst is changed from proton to chloroacetic and acetic acids to water, whereas α is practically insensitive to variations in σ^+.

In the particular case in which the nucleophilic reactivity of a set of molecules lies in linear relationship to their basicity, and in which the Brønsted parameter β represents the sensitivity of the reaction to the basicity of the nucleophilic reactants, we obtain:

$$\frac{pK_{a2} - pK_{a1}}{\rho_2 - \rho_1} = \frac{\sigma_i}{\beta_0 - \beta_i} = C_2 \qquad (22)$$

The parameter of generalized basic catalysis, β, varies in linear fashion with the polar effect of the substrate substitutents (σ) and, if the nature of the catalyst is changed (pK_a), one observes a linear variation of the sensitivity of the reaction rates to the substituent effect (ρ). C_2 is large and negative. In the hydrolysis of the benzylidene-*t*-

butylamines[62], ρ^+ (H_2O) = 1·71 and ρ^+ (OH^-) = 1·26 (see Table 6). pK_a (OH^-) − pK_a (H_2O) = 14, which leads to $C_2 = -39$, together with a practically negligible variation of β as a function of the substitution.

By using the Brønsted relation for j general acids characterized by pK_{aj}, and the Swain–Scott relation for n nucleophilic reactants identified by their nucleophilic character n_K;

$$\log k_{jn} = \log G_n{}^A - \alpha_n pK_{aj}$$
$$\log k_{jn} = \log k_{j0} + s_j n_K$$

which leads to the constant expression:

$$\frac{pK_{a2} - pK_{a1}}{s_1 - s_2} = \frac{n_K}{\alpha_K - \alpha_0} = C_3 \qquad (23)$$

If the nature of the nucleophilic agent is modified, the consequent variation in α is linearly related to the reactivity of the nucleophilic agent (n_K). Moreover, if the nature of the acid catalyst is modified, the sensitivity of the reaction to nucleophilic attack (S) varies in linear relation to the pK_a of the acid catalyst. According to the Hammond hypothesis, the transition state will be reached more quickly with an increase in the reactivity of the nucleophilic agent, and α decreases with increase in the nucleophilic character. C_3 must, therefore, be negative. This is in fact observed in the attack by different nucleophiles (the pK_a of which is a measure of n_K) on the carbon atom of the $>$C=O group (see Table 7 reproduced from reference 45). The catalyst becomes less selective with respect to the nucleophilic agent when the reactivity of the latter increases.

TABLE 7. Variation of the catalytic parameter as a function of acidity of the nucleophilic agent[45].

Substrate	Nucleophilic agent	pK_a	α
Acetaldehyde	Water	−1·74	0·54
sym-Dichloroacetone	Water	−1·74	0·27
Benzaldehyde	Semicarbazide	3·65	0·25
p-chlorobenzaldehyde	Aniline	4·60	0·25
Benzaldehyde	Hydroxylamine	5·97	~0
Aliphatic aldehydes	Cyanide	9·40	~0
Formaldehyde	Trimethylamine	9·76	~0

E. General Catalysis

Workers who have studied the effect of general catalysis on the hydrolysis of imines as well as on the formation of $\rangle C{=}N-$ groups in the light of the principle of microscopic reversibility, distinguish between several catalytic mechanisms, according to the steps proposed and depending on the basicity of the amine.

I. Direct rate-determining attack of the hydroxyl ion on the unprotonated imine group

$$\overset{\delta-}{HO}\cdots \overset{|}{\underset{|}{C}} {\,\text{---}\,} \overset{\delta-}{NR}$$

The hydrolysis of Schiff bases derived from weak amines is subject to base catalysis. The reverse reaction is the dehydration of the negative carbinolamine ion. This is the case in the alkaline hydrolysis of benzilideneanilines [31,48], in the hydration of 2-hydroxypteridine [77], and in the dehydration step during the formation of oximes [35,44], of semicarbazones [46] and of hydrazones [41b].

It is not known whether these reactions proceed by specific base catalysis (a):

$$\text{(a)}\quad HO^- \rightarrow \rangle C{=}N{-}R \rightleftharpoons (\overset{\delta-}{HO}\cdots\overset{|}{\underset{|}{C}}{\,\text{---}\,}\overset{\delta-}{N}{-}R) \rightleftharpoons$$

$$HO{-}\overset{|}{\underset{|}{C}}{-}\bar{N}R \underset{\text{fast}}{\overset{\pm H^+}{\rightleftharpoons}} HO{-}\overset{|}{\underset{|}{C}}{-}NHR$$

or by general base catalysis (b and c):

$$\text{(b)}\quad A^-\ H{-}\underset{\underset{H}{|}}{O} \rightarrow \rangle C{=}NR \rightleftharpoons (\overset{\delta-}{A}\cdots H\cdots\underset{\underset{H}{|}}{O}\cdots\overset{|}{\underset{|}{C}}{\,\text{---}\,}NR) \rightleftharpoons$$

$$HO{-}\overset{|}{\underset{|}{C}}{-}\bar{N}R + AH \underset{\text{fast}}{\overset{\pm H^+}{\rightleftharpoons}} A^- + HO{-}\overset{|}{\underset{|}{C}}{-}NHR$$

$$\text{(c)}\quad A^- + H{-}\underset{\underset{H}{|}}{O} + \rangle C{=}NR \underset{\text{fast}}{\rightleftharpoons} HO^- \rightarrow \rangle C{=}\underset{\underset{R}{|}}{N}\ H{-}A \rightleftharpoons$$

$$(HO^-\cdots\overset{|}{\underset{|}{C}}{\,\text{---}\,}\underset{\underset{R}{|}}{N}\cdots H\cdots A^-) \rightleftharpoons HO{-}\overset{|}{\underset{|}{C}}{-}NHR + A^-$$

2. Rate-determining attack of the hydroxyl ion on the protonated imine group, without general basic catalysis

$$X^{-}\cdots C\overset{\cdots}{=}\overset{+}{N}(H)-R$$

This case applies to Schiff bases in the range of pH-independent rates (see Figure 1). In this type of mechanism, no specific or general basic catalysis is ever encountered:

$$H_2O + {>}C{=}NR \xrightleftharpoons{\text{fast}} HO^{-} \rightarrow {>}C{=}\overset{+}{N}(H)-R \rightleftharpoons$$

$$\left(HO^{-}\cdots\overset{|}{\underset{|}{C}}\overset{\cdots}{=}\overset{+}{N}(H)-R\right) \rightleftharpoons HO-\overset{|}{\underset{|}{C}}-NHR$$

This example is illustrated, for derivatives of strongly basic amines, by the hydrolysis of benzylidene-*t*-butylamines [62a], by the formation of Schiff bases derived from ammonia [78] and from n-butylamine [60], and for derivatives of weakly basic amines, by the hydrolysis and the formation of benzylideneanilines [31,32,48,79].

3. Rate-determining attack of the hydroxyl ion and of the water molecule on the protonated imine group, with general acid–base catalysis

$$A^{-}\cdots H\cdots X\cdots\overset{|}{\underset{|}{C}}\overset{\cdots}{=}\overset{+}{N}(H)-R$$

At intermediate pH values at which the mechanism of water addition increases in importance, a general acid–base catalysis is observed in the case of Schiff bases [32,48,62a,63], oximes [80], semicarbazones [42,46,81], and phenylhydrazones [80a].

In the case of the reverse reaction (the formation of an imine) in which the dehydration of the carbinolamine determines the rate, general acid catalysis is observed as in the formation of the benzylideneanilines in methanol [60,82], the formation of phenylhydrazones [38,80a], the hydrolysis of the alkoxymethylureas [6c] and in the addition of water to 2-hydroxypteridine [77].

In the hydrolysis reaction there exists some kinetic ambiguity as to the nature of the catalytic agent, due to the fact that general acid catalysis, in terms of the free Schiff base, is equivalent to general base catalysis in terms of the protonated Schiff base:

$$V = k[\mathrm{S}][\mathrm{AH}] = k'[\mathrm{SH}^+][\mathrm{A}^-] \text{ with } k' = k\frac{K_{\mathrm{SH}^+}}{K_{\mathrm{AH}}}$$

However, it is easily shown that the mechanism in fact consists of general base catalysis by operating it in conditions under which the Schiff base is totally protonated[62a].

The nature of the catalysis depends on the timing of the proton transfer:

$$\begin{array}{c} | \\ -\mathrm{C}{\cdots}\mathrm{N}-\mathrm{R} \\ \\ \mathrm{H_2O}_{\delta+}\mathrm{H}\cdots\mathrm{A}_{\delta-} \end{array}$$

If the two processes (of protonation and of nucleophilic attack) took place simultaneously, this would give rise to a general acid catalysis. If the proton is transferred prior to the nucleophilic attack, the origin of the proton would be of no importance[32].

Several mechanisms are possible:

(a) $\mathrm{A^-}\curvearrowright\mathrm{H{-}O(H)}\curvearrowright\mathrm{{>}C{=}\overset{+}{N}(R){-}H} \rightleftharpoons \left(\mathrm{A^{\delta-}}\cdots\mathrm{H}\cdots\mathrm{O(H)}\cdots\mathrm{C}{\cdots}\mathrm{N^{\delta+}(R){-}H}\right) \rightleftharpoons \mathrm{AH} + \mathrm{HO{-}\overset{|}{\underset{|}{C}}{-}NHR}$

(b) $\mathrm{H{-}O(H)}\curvearrowright\mathrm{{>}C{=}\overset{+}{N}(R){-}H}\curvearrowright\mathrm{A^-} \rightleftharpoons \left(\mathrm{H{-}O^{\delta+}(H)}\cdots\mathrm{C}{\cdots}\mathrm{N(R)}\cdots\mathrm{H}\cdots\mathrm{A^{\delta-}}\right) \rightleftharpoons \mathrm{-\overset{|}{C}(^+OH_2){-}\bar{N}R} + \mathrm{HA} \underset{\text{fast}}{\rightleftharpoons} \mathrm{-\overset{|}{C}(OH){-}NHR}$

(c) $\mathrm{H{-}O(H)}\curvearrowright\mathrm{{>}C{=}\overset{+}{N}(R){-}H}\curvearrowright\mathrm{A^-} \rightleftharpoons \left(\mathrm{H{-}O^{\delta+}(H)}\cdots\mathrm{C}{\cdots}\mathrm{N(R){-}H}\cdots\mathrm{A^{\delta-}}\right) \rightleftharpoons \mathrm{\overset{|}{C}(^+OH_2){-}NHR} + \mathrm{A^-} \rightleftharpoons \mathrm{-\overset{|}{C}(OH){-}NHR} + \mathrm{HA}$

In mechanism (c), the role of the catalyst is restricted to stabilizing the transition state by solvation with respect to the initial reagent. This hypothesis can be immediately eliminated, as the N—H group is less acidic in the transition state, and it is therefore less likely to give rise to the formation of hydrogen bonds. Moreover, it is observed that in the formation of oximes and nitrones[58], catalysis is as efficient with anionic acids (e.g. phosphate or succinate monoanion) as it is with cationic acids (e.g. morpholinium or imidazolinium ion).

The choice between the two remaining mechanisms (a) and (b), is the subject of much controversy.

Mechanism (a) appears, at first, to be more probable if one considers that in the case of (b), the H . . . A bond diminishes the carbonium ion nature of the electrophilic reagent. However, the proponents of mechanism (b)[80b] put forward the argument that 'chemists often draw perilous inferences by considering: what would I do in that particular situation if I were an unshared pair of electrons?'. These authors consider that it is the difference in free energy between the reagents and the transition state which governs the type of catalysis, and that the basic catalyst A^- should be placed in a position such that the energy of the transition state be reduced to a minimum, in other words close to the most acidic hydrogen, the immonium hydrogen, this being more acidic than the carbinolamine hydrogen.

They reason that in the mechanism (a), the transition state would resemble the products, the carbinolamine hydrogen would be closer to the catalyst than to the oxygen and, as the proton transfer would be more than half completed, according to Brønsted's catalysis law β would be $>0{\cdot}50$. On the other hand, in the case of mechanism (b), the transition state would resemble the reagents, the immonium hydrogen would be closer to the nitrogen than to the catalyst, as the Schiff base is a stronger base than the acetate ($pK_{a_{AcOH}} = 4{\cdot}7$ and, in the benzylidene-*t*-butylamines, $pK_{SH^+} = 6{\cdot}7$, see Table 2), the proton transfer would be less than half completed, and β would be $<0{\cdot}50$.

Moreover, for Schiff bases derived from weaker amines, the transition states would become more symmetrical; in the case of mechanism (a), the transition state would approach that of the reagents and β would decrease whereas, in the case of mechanism (b), the transition state would approach that of the products and β would increase.

In favour of mechanism (b), it is in fact observed that in the hydrolysis of the benzylidene-*t*-butylamines, $\beta = 0{\cdot}25$. The proponents of mechanism (a)[58,75] reply to this, in 'a defense of anthropomorphic electrons', that mechanism (b) is impossible in the hydrolysis of the benzhydrylidenedimethylammonium ion $(C_6H_5)_2C{=}\overset{+}{N}(CH_3)_2$ and of the nitrone $ClC_6H_4CH{=}\overset{+}{N}(CH_3)O^-$ for which β values of 0·27 and 0·23 are respectively obtained. They judge that Swain's reasoning is an oversimplified approach to the problem due to the presence of two concurrent mechanisms, and that the addition of a catalyst will disturb the structure of the transition state of which the overall stability and not only the stabilization by hydrogen bonding should be considered.

$$HO^- + \rangle C{=}\overset{+}{N}\langle \rightleftharpoons HO^-\cdots C{\cdots}\overset{+}{N}\langle \qquad (\mathbf{2c}) \quad (\mathbf{3c})$$

$$H_2O + \rangle C{=}\overset{+}{N}\langle \underset{}{\overset{\pm A^-}{\rightleftharpoons}} A^{\delta-}\cdots H\cdots \underset{H}{O}\cdots C{\cdots}\overset{\delta+}{N}\langle \overset{\pm HA}{\rightleftharpoons} HO{-}C{-}N\langle \qquad (\mathbf{1}) \quad (\mathbf{2a}) \quad (\mathbf{3a})$$

$$H_2\overset{\delta+}{O}\cdots C{\cdots}\overset{\delta+}{N}\langle \rightleftharpoons H_2\overset{+}{O}{-}C{-}N\langle \qquad (\mathbf{2b}) \quad (\mathbf{3b})$$

($\mathbf{1} \overset{\pm H^+}{\rightleftharpoons} \mathbf{2c}$; $\mathbf{3c} \rightleftharpoons \mathbf{3a}$; $\mathbf{1} \rightleftharpoons \mathbf{2b}$; $\mathbf{3b} \overset{\pm H^+}{\rightleftharpoons} \mathbf{3a}$)

The energy diagram (see Figure 2) and the mechanistic scheme include the three paths of reaction possible at a pH of 7: in the absence of catalyst (attack by either the hydroxyl ion or by water) and in the presence of general acid catalysis.

Under neutral conditions, the formation of the hydroxyl ion implies an initial energy consumption of 9·5 kcal/mole, and the subsequent attack by the OH^- on the protonated imine gives rise to a transition state **3c** with a structure closer to the initial than to the final state. In the case of an attack on the protonated imine by H_2O, the transition state **2b** is closer to the oxonium form, **3b**, than to the initial structure, and the formation of the oxonium ion from the intermediate carbinolamine (**3a**) takes up 12·6 kcal/mole.

Clearly, to the extent that the transition states resemble the unstable intermediates **2c** and **3b**, these reaction paths will be unfavourable in respect to the middle path, in which the formation of these intermediates is avoided by general acid-base catalysis. The free energies required for their formation constitutes a great deal of the free activation energy of the reaction.

One might then question the importance of the hydrogen bonds between the catalysts and the intermediates. The strength of a hydrogen bond is a function of the acidity of the acid HA and the basicity of the base B, the most favourable acid–base pair being one in which the basicities of A^- and B are equal (acetic acid–acetate).

However, in the course of the reaction there is a large variation in the acidity of the reacting groups; the difference in acidity between the carbinolamine and the hydroxyl ion, or between the oxonium ion

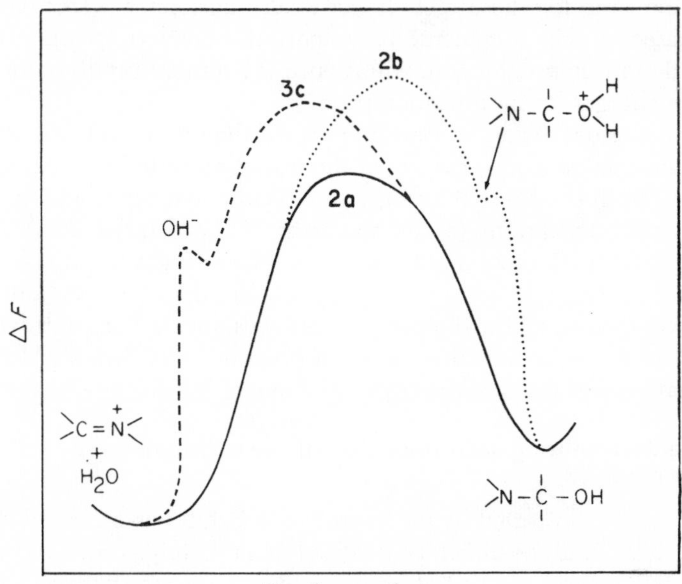

FIGURE 2. Energy diagram for the hydrolysis of Schiff bases (taken from reference 58).

and water is in the region of 18 p*K* units, whereas the difference between the acidity of the N—H groups in the immonium form and the carbinolamine can be estimated to be only 15 p*K* units. For this reason one may expect a greater stabilization by interaction between the catalyst and the carbinolamine group. Interaction of this type changes the structure of the transition state by changing the nature of the nucleophilic agents in such a manner that the transition state of the catalysed reaction (**2a**) will shift towards the centre of the reaction axis.

The role of the catalyst is not limited to lowering the free energy of activation by disposing the protons in the most favourable locations. The catalyst also serves to decrease that fraction of the activation energy which arises from the formation and the rupture of C—O and C—N bonds, which it does by decreasing the dipole strength of the transition state.

In the transition state for the expulsion of hydroxide ion from the carbinolamine, an acid catalyst, partially converting the leaving group to a water molecule, reduces charge separation. Similarly, in the transition state for the attack of the weak nucleophile water on the protonated imine, a general base, partially converting the water molecule into hydroxide ion, will reduce the mutual repulsion of the positive charges in the transition state.

In conclusion, Swain, in putting forward the mechanism (b), bases his arguments on a stabilization of the transition state by a hydrogen bond between the catalyst and the most acidic hydrogen, and Jencks favours the mechanism (a) on the basis of the catalyst combining preferentially with those protons whose acidity changes the most in the course of the reaction because, as he says, 'electrons, like most anthropoids will tend to follow the path of least resistance'. The controversy continues at the present time between proponents of Swain's 'static' explanation and those of the 'dynamic' one of Jencks.

4. Rate-determining decomposition of the carbinolamine intermediate

This case is illustrated by the hydrolysis of Schiff bases derived from weakly basic amines under generalized basic catalysis and the corresponding inverse reaction of formation of the intermediate carbinolamine, under acid catalysis[48]:

$$\text{B:}\frown\text{H}-\text{O}-\overset{|}{\underset{|}{\text{C}}}-\overset{\text{H}}{\underset{\text{H}}{\overset{|}{\underset{|}{\text{N}^+}}}}-\text{R} \rightleftharpoons \;>\!\text{C}{=}\text{O} + \text{RNH}_2 + \text{BH}^+$$

In the case of the Schiff bases derived from highly basic amines, there is no general catalysis:

$$\text{O}^- - \overset{|}{\underset{|}{\text{C}}}-\overset{\text{H}}{\underset{\text{H}}{\overset{|}{\underset{|}{\text{N}^+}}}}-\text{R} \rightleftharpoons \;>\!\text{C}{=}\text{O} + \text{RNH}_2$$

III. HYDROLYSIS OF THE IMINOLACTONES

A recent case of hydrolysis of the imine group has shown that iminolactones may yield different products according to the conditions of pH[68,83].

At acid pH, the iminolactone derived from tetrahydrofuran hydrolyses to aniline and the α-keto derivative of tetrahydrofuran:

$$\text{(iminolactone, } =NC_6H_5) + H_2O \longrightarrow \text{(lactone, } =O) + C_6H_5NH_2$$

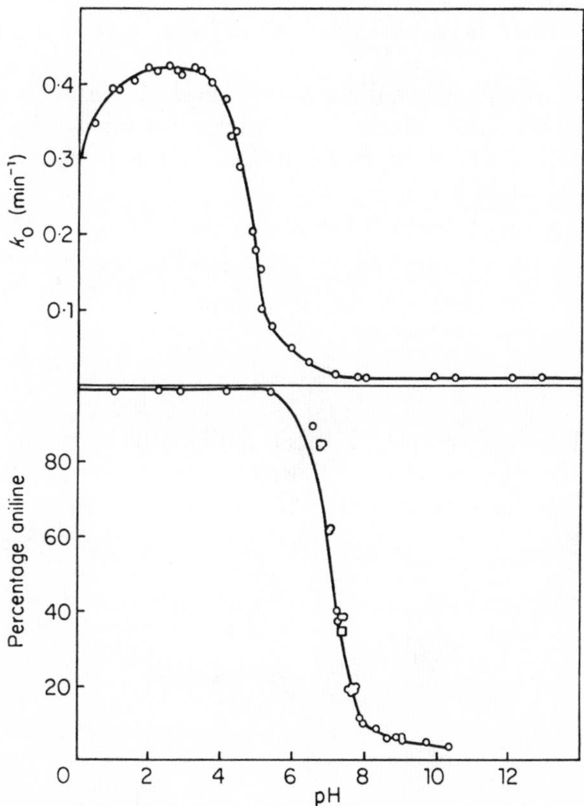

FIGURE 3. A: pH–rate profile for the hydrolysis of the iminolactone at 30° C; B: effect of pH on the nature of the products of iminolactone hydrolysis.

Under basic conditions the product of hydrolysis is γ-hydroxybutyranilide:

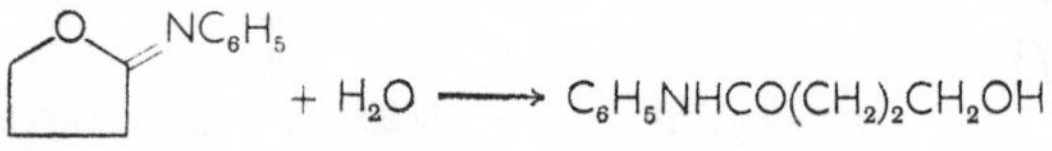

$$\text{(2-phenylimino-tetrahydrofuran)} + H_2O \longrightarrow C_6H_5NHCO(CH_2)_2CH_2OH$$

The rate–pH curve shows the classical bell-like shape under acidic conditions together with a pH-independent rate zone under basic conditions.

On the other hand, the percentage of aniline titrated by diazotization follows a normal titration curve with an inflexion situated at a pK = 7·07 (Figure 3).

Such behaviour may be explained by way of several mechanisms:

(a) The γ-hydroxybutyranilide could conceivably arise as a result of a nucleophilic displacement by the hydroxyl ion at the carbon atom $C_{(5)}$. Thus, following establishment of the protonation pre-equilibrium of the imine:

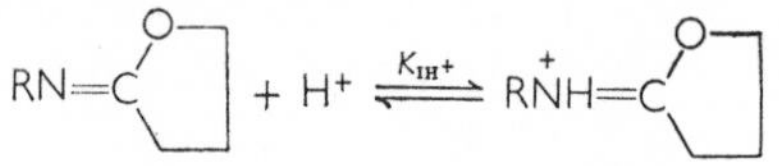

$$RN{=}C(\text{ring}) + H^+ \overset{K_{IH^+}}{\rightleftharpoons} R\overset{+}{N}H{=}C(\text{ring})$$

(I) (IH⁺)

there would arise a competitive nucleophilic attack on the carbon $C_{(2)}$ by a molecule of water, followed by breakdown, under acid conditions, of the carbinolamine intermediate in its zwitterionic form;

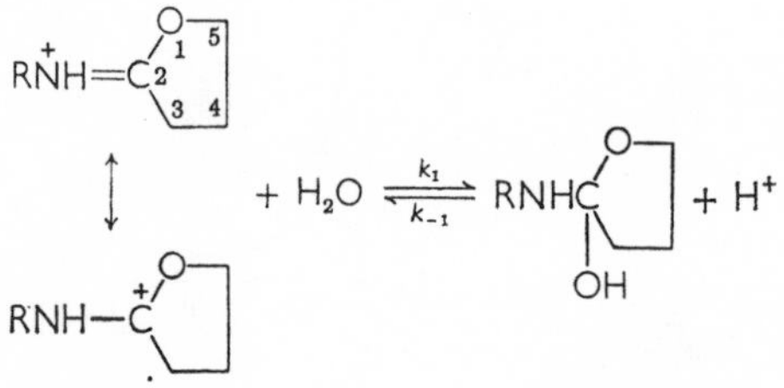

$$R\overset{+}{N}H{=}C_2(\text{ring } O_1, C_3, C_4, C_5) \leftrightarrow RNH{-}\overset{+}{C}(\text{ring}) \;(IH^+) + H_2O \underset{k_{-1}}{\overset{k_1}{\rightleftharpoons}} RNHC(OH)(\text{ring}) \;(IHOH) + H^+$$

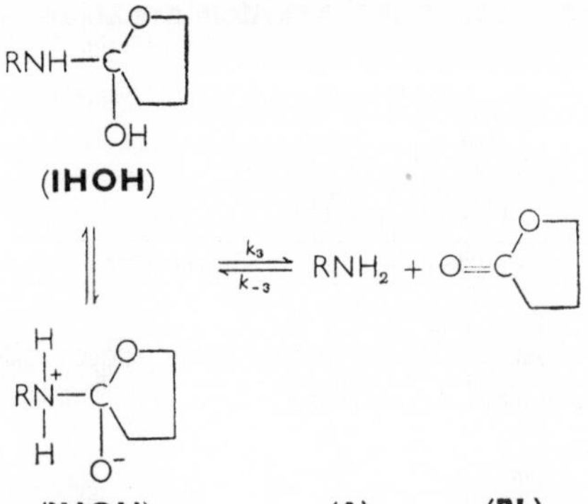

RNH—C(ring O) —OH (IHOH) ⇌ $\underset{k_{-3}}{\overset{k_3}{\rightleftharpoons}}$ RNH_2 + O=C(ring O)

$RN^{+}H_2$—C(ring O)—O^- (IHOH) (A) (BL)

Nucleophilic attack by the hydroxyl ion to the $C_{(5)}$ carbon, will take place under basic conditions:

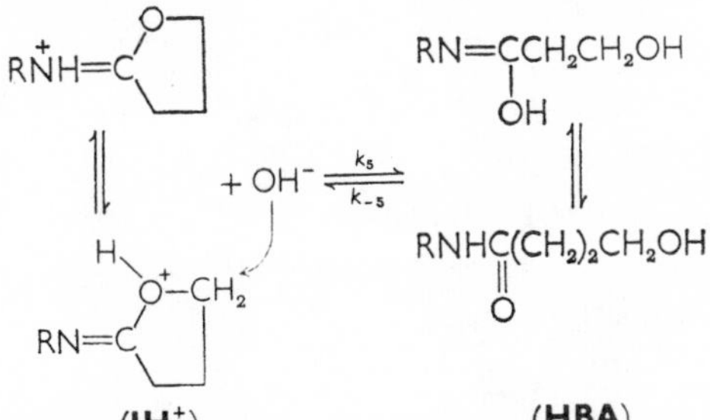

$RN^{+}H$=C(ring O) ⇌ H—O^{+}—CH_2 ring, RN=C (IH⁺) + OH^- $\underset{k_{-5}}{\overset{k_5}{\rightleftharpoons}}$ RN=C(OH)CH_2CH_2OH ⇌ RNHC(=O)$(CH_2)_2CH_2OH$ (HBA)

(b) The most reasonable explanation is based on the mechanism of the hydrolysis of Schiff bases derived from anilines[32]. The protonation equilibrium of the iminolactone is followed by a nucleophilic attack by water in acid medium, and by the hydroxyl ion in basic medium, on the $C_{(2)}$ carbon, to give rise to the carbinolamine intermediate:

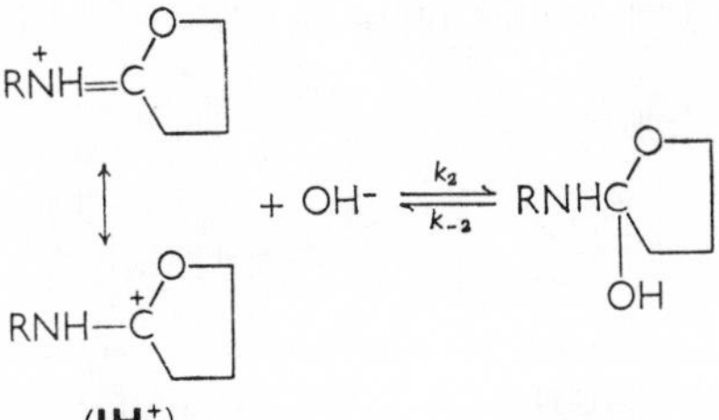

$RN^{+}H$=C(ring O) ↔ RNH—C^{+}(ring O) (IH⁺) + OH^- $\underset{k_{-2}}{\overset{k_2}{\rightleftharpoons}}$ RNHC(ring O)—OH

In this case one must assume the existence of an anionic intermediate:

$$\underset{\textbf{(IHOH)}}{\text{RNHC(OH)}\langle\text{O}\rangle} \underset{}{\overset{k_{IHOH}}{\rightleftharpoons}} \underset{\textbf{(IHO}^-\textbf{)}}{\text{RNHC(O}^-\text{)}\langle\text{O}\rangle}$$

which will decompose to different products according to the extent of protonation. Under acid conditions:

$$\text{RNHC(OH)}\langle\text{O}\rangle \rightleftharpoons \underset{\textbf{(IHOH)}}{\text{R—}\overset{+}{\text{N}}\text{H}_2\text{—C(O}^-\text{)}\langle\text{O}\rangle} \underset{k_{-3}}{\overset{k_3}{\rightleftharpoons}} \underset{\textbf{(A)}}{\text{RNH}_2} + \underset{\textbf{(BL)}}{\text{O=C}\langle\text{O}\rangle}$$

Under basic conditions:

$$\underset{\textbf{(IHO}^-\textbf{)}}{\text{RNH—C(O}^-\text{)}\langle\text{O}\rangle} \underset{k_{-4}}{\overset{k_4}{\rightleftharpoons}} \underset{\textbf{(HBA)}}{\text{RNHC(=O)(CH}_2)_2\text{CH}_2\text{O}^-}$$

Decomposition to aniline is inconceivable in the course of this step since this would involve the expulsion of an anilino anion leaving group, which is difficult to admit. However, another kinetically equivalent mechanism would include formation of a cationic intermediate:

$$\underset{\textbf{(IHOH)}}{\text{R—NH—C(OH)}\langle\text{O}\rangle} \overset{k_{IHOH_2^+}}{\rightleftharpoons} \underset{\textbf{(IHOH}_2^+\textbf{)}}{\text{R—NH—C(}^+\text{OH}_2\text{)}\langle\text{O}\rangle}$$

with decomposition, in acidic conditions:

$$\mathrm{RNH{-}C(OCH_2CH_2CH_2)({}^{+}OH_2)} \rightleftharpoons \mathrm{R{-}\overset{+}{N}H_2{-}C(OCH_2CH_2CH_2)(OH)} \underset{k''_{-3}}{\overset{k''_3}{\rightleftharpoons}} \underset{(\mathbf{A})}{\mathrm{RNH_2}} + \underset{(\mathbf{BL})}{\mathrm{O{=}C(OCH_2CH_2CH_2)}} + \mathrm{H^+}$$

$$\mathrm{R{-}\overset{+}{N}H_2{-}C(OCH_2CH_2CH_2)(OH)} \rightleftharpoons \underset{(\mathbf{IHOH_2})}{\mathrm{R{-}\overset{+}{N}H_2{-}C(O(H)CH_2CH_2CH_2)(O^-)}}$$

and in basic conditions:

$$\mathrm{RNH{-}C(OCH_2CH_2CH_2)(OH)} \rightleftharpoons \underset{(\mathbf{IHOH})}{\mathrm{R{-}\overset{+}{N}H_2{-}C(OCH_2CH_2CH_2)(O^-)}} \underset{k_{-4}}{\overset{k_4}{\rightleftharpoons}} \mathrm{RNHC(=O)(CH_2)_2CH_2OH}$$

The two mechanisms are kinetically equivalent, and one is thus unable to decide as to the type of ionization undergone by the intermediate.

The overall reaction scheme may now be written as:

$$\mathbf{I} + H^+ \underset{}{\overset{K_{HI^+}}{\rightleftharpoons}} \mathbf{IH}^+$$

$$\mathbf{IH}^+ + H_2O \underset{k_{-1}}{\overset{k_1}{\rightleftharpoons}} \mathbf{IHOH} + H^+$$

$$\mathbf{IH}^+ + OH^- \underset{k_{-2}}{\overset{k_2}{\rightleftharpoons}} \mathbf{IHOH}$$

$$\mathbf{IHOH} \underset{}{\overset{K_{IHOH}}{\rightleftharpoons}} \mathbf{IHO}^- + H^+$$

$$\mathbf{IHOH} \underset{k_{-3}}{\overset{k_3}{\rightleftharpoons}} \mathbf{A} + \mathbf{BL}$$

$$\mathbf{IHOH} \underset{k_{-4}}{\overset{k_4}{\rightleftharpoons}} \mathbf{HBA}$$

which, taking into account the following assumptions:

$$k_3 \gg k_{-2}$$

$$\text{at pH} < 3: \quad k_3 \gg \frac{K_{IHOH}k_4}{[H^+]}$$

$$\text{at pH} > 3: \quad k_3 + \frac{K_{IHOH}k_4/[H^+]}{k_{-1}} \gg [H^+]$$

$$\text{at pH} \gg 6: \quad K_{IH^+} \gg [H^+]$$

leads to the kinetic equation:

$$k_{obs} = \frac{k_1\, k_3/k_{-1}[H^+]}{(K_{IH} + [H^+])\{(k_3/k_{-1}) + [H^+]\}} + \frac{k_2\, K_W}{K_{IHOH}}$$

in which the first term applies to the bell-shaped curve and the second to the pH-independent rate zone.

The aniline yield follows from the breakdown rate constants:

$$\%\mathbf{A} = \frac{[H^+]}{[H^+] + K_{IHOH}\, k_4/k_3} \times 100$$

with a point of inflexion situated at pK', for

$$K' = \frac{K_{IHOH}\, k_4}{k_3}$$

These equations enable us to calculate the parameters which lead to a point by point establishment of the rate–pH diagram:

$$pK_{IH^+} = 5{\cdot}06$$

$$k_1 = 0{\cdot}415\ \text{min}^{-1}$$

$$\frac{k_3}{k_{-1}} = 2{\cdot}5\ \text{mole}$$

$$k_4 = 1{\cdot}4 \times 10^6\ \text{mole}^{-1}\ \text{min}^{-1}$$

$$pK' = 7{\cdot}07$$

These parameters are in agreement with those resulting from the hydrolysis of the Schiff bases (see Table 1) and of the Δ_2 thiazolines, as expected, considering that the mechanisms of the rate-determining steps are identical. The reaction is also sensitive to general base catalysis which modifies the reaction scheme at a pH < 7, under conditions in which the hydroxyl ion reaction is negligible, in the following manner:

$$\mathbf{IH}^+ + \mathbf{B} + H_2O \underset{k'_{-1}}{\overset{k'_1}{\rightleftharpoons}} \mathbf{IHOH} + \mathbf{BH}^+$$

$$\mathbf{IHOH} \underset{(B)}{\overset{k'_3}{\rightleftharpoons}} \text{products}$$

$$\mathbf{B} + H^+ \overset{K_{BH^+}}{\rightleftharpoons} \mathbf{BH}^+$$

$$k_{obs} = \frac{k_1[H^+]}{[H^+] + K_{SH^+}} + [\mathbf{B}]\frac{k'_1 K_{BH^+}[H^+]}{([H^+] + K_{SH^+})(K_{BH^+} + [H^+])}$$

The catalyst effect is maximum for:

$$pH = \frac{pK_{SH^+} + pK_{BH^+}}{2}$$

It will be seen that this catalytic scheme is well applicable to the hydrolysis of Schiff bases[62a].

However, the role of the catalyst is not limited to modifying the rate of hydrolysis. Without changing the kinetic picture, addition of catalyst strongly increases the aniline yield at values of pH > 7. The authors[68] have applied an empirical mathematical treatment to this astonishing phenomenon.

Assuming the aniline yield, at constant pH, to be proportional to the concentration of a complex formed between the iminolactone and the buffer:

$$\mathbf{S} + \text{buffer } (\mathbf{B}) \underset{}{\overset{K_{app}}{\rightleftharpoons}} \text{complex}$$

and, defining ΔA as the difference between the aniline yield at a given concentration of buffer (A_B), and the aniline yield in the absence of buffer (A_0), $I/\Delta A$ as a function of $I/[\mathbf{B}]$ should be a straight line with a slope of K_{app} and an intercept of $I/\Delta A_{max}$ (ΔA_{max} is the difference between the maximum yield of aniline obtained in the presence of a high buffer concentration and in its absence). The equation is confirmed and $\Delta A_{max} + A_0 = 94$ to 98%, which means that at all values of pH examined (7 to 9·5), and at sufficiently high concentrations of buffer the iminolactone can yield aniline exclusively. The values of K_{app} represent the buffer concentrations required to lead to an aniline yield of $\Delta A_{max}/2$ for a given value of pH. As A_0 varies with the pH, ΔA_{max} increases with increase in pH.

A particular study of phosphate, bicarbonate and acetate buffers has shown that the active forms of the bicarbonate and acetate buffers are the basic ones, whereas the phosphate acts either in its basic or in its acidic form (H_2PO^- and $HPO_4{}^-$).

As the rate of hydrolysis remains constant though the aniline yield increases, one must assume that the buffer influences the decomposition of the non-rate determining intermediate.

On the other hand, the activity of the buffers neither obeys the laws of a classical generalized acid–base catalysis (the phosphate buffer, $pK = 6{\cdot}77$, being 240 times more active than the imidazole, $pK = 7{\cdot}02$, with respect to aniline yield whereas, for the hydrolysis of Δ-thiovalerolactone and of ethyl dichloroacetate their activities are identical and, for the hydrolysis of iminolactone, the phosphate is twice as active as the imidazole), nor those of a nucleophilic catalysis (the phosphate dianion is 10^3 times less active than the imidazole in the nucleophilic attack on *p*-nitrophenylacetate or on acetylphenylphosphate)[68].

Finally, the common characteristic of these active buffers which lead to an increase in the yield of aniline, is that they possess acidic as well as basic groups attached to a central atom.

These observations have led the authors to suggest a mechanism of cyclic and concerted proton transfer, involving the neutral tetrahedral intermediate and catalysts such as mono- and di- phosphate anions, bicarbonate, acetic acid, arsenate, monophenylphosphate:

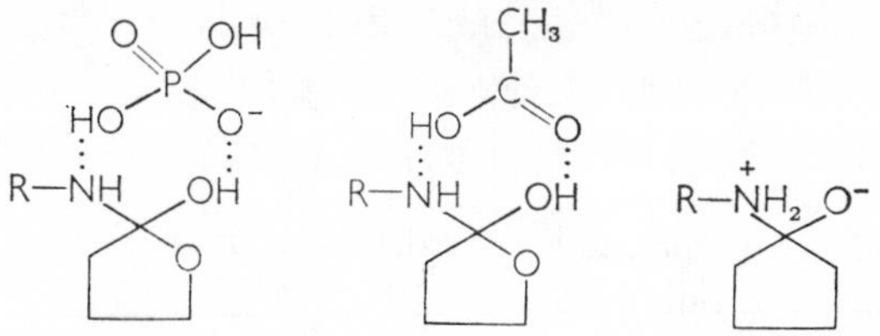

This will permit the formation of the zwitterion required to rupture the C—N bond, and yield aniline.

IV. HYDROLYSIS OF THE Δ_2 THIAZOLINES

An examination of the kinetic behaviour of a large number of substituted Δ_2 thiazolines

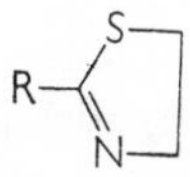

has led to an elaboration of the general laws[76] which agree with earlier investigations carried out on two alkylthiazolines[49,59,74] and which are similar to those governing the hydrolysis of Schiff bases.

The Δ_2 thiazolines give rise to two types of products according to whether it is the C—S or the C—N bond which breaks in the intermediate:

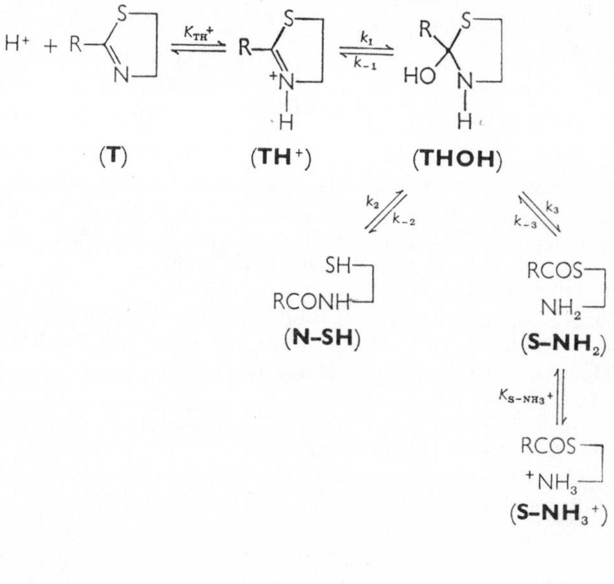

(T) (TH$^+$) (THOH)

(N-SH) (S-NH$_2$)

(S-NH$_3^+$)

The kinetic equation governing the bell-shaped curve is:

$$k_{obs} = \frac{k_1[H^+]\{(k_2 + k_3)/k_{-1}\}}{(K_{TH^+} + [H^+])\{[H^+] + (k_2 + k_3)/k_{-1}\}}$$

and the calculated values of K_{TH^+} calculated from kinetics correspond to the experimental ones.

If we plot the value of log k_1 against pK_1 (= pK_{TH^+}), we obtain several straight lines for various groups of the Δ_2 thiazoline families. One of these groups is composed of 2-aryl thiazolines, substituted in position 4, and fulfils the equation:

$$\log k_1 = -0{\cdot}91 \; pK_{TH^+} - 0{\cdot}48$$

The substituents in position 4 (*para* on the aryl group) will influence the electron density at the site of the nucleophilic attack, this effect being quantitatively accounted for by the dissociation constant of the thiazolinium ion.

The hydrolysis of the Δ_2 thiazolines is an equilibrium reaction. In the case of the 2-methylthiazoline, all equilibria have been experimentally determined[49]:

$$K_{ST} = \frac{[\mathbf{S{-}NH_3^+}]}{[\mathbf{TH^+}]} = \frac{k_3 k_1}{k_{-1}\, k_{-3}\, K_{S\text{-}NH_3^+}} = 12$$

$$K_{NS} = \frac{[H^+][\mathbf{N{-}SH}]}{[\mathbf{S{-}NH_3^+}]} = \frac{k_2\, k_{-3}\, K_{S\text{-}NH_3^+}}{k_{-2}\, k_3} = 0{\cdot}035\ \mu$$

$$K_{NT} = \frac{[\mathbf{N{-}SH}]}{[\mathbf{T}]} = \frac{k_1\, k_2}{K_{TH^+} k_{-1}\, k_{-2}} = 7 \times 10^4$$

V. REFERENCES

1. L. A. Cohen and B. Witkop, *Angew. Chem.*, **73**, 253 (1961).
2. B. Witkop, *Advan. Proteinchem.*, **16**, 221 (1961).
3. B. Witkop and L. K. Ramachandran, *Metabolism*, **13**, 1016 (1964).
4. R. Hubbard, *Proc. Natl. Phys. Lab., London* (*Symp. No. 8*), **1**, 151 (1958).
5. R. A. Morton and G. A. J. Pitt, *Biochem. J.*, **59**, 128 (1955).
6. (a) R. A. Morton and G. A. J. Pitt, *Progr. Chem. Org. Nat. Prod.*, **14**, 244 (1957).
   (b) D. Noyce, A. Bottini and S. Smith, *J. Org. Chem*, **23**, 452 (1958).
   (c) F. Nordhoy and J. Ugelstad, *Acta. Chem. Scand.*, **13**, 864 (1959).
7. E. Grazi, T. Cheng and B. L. Horecker, *Biochem. Biophys. Res. Commun.*, **7**, 250 (1962).
8. E. Grazi, P. T. Rowley, T. Cheng, O. Tchola and B. L. Horecker, *Biochem. Biophys. Res. Commun.*, **9**, 38 (1962).
9. B. L. Horecker, S. Pontremoli, C. Ricci and R. Cheng, *Proc. Natl. Acad. Sci. U.S.*, **47**, 1949 (1961).

10. J. C. Speck, Jr. and A. A. Forist, *J. Am. Chem. Soc.*, **79**, 4459 (1957).
11. J. C. Speck, Jr., P. T. Rowley and B. L. Horecker, *J. Am. Chem. Soc.*, **85**, 1012 (1963).
12. I. Fridovitch and F. H. Westheimer, *J. Am. Chem. Soc.*, **84**, 3208 (1962).
13. G. A. Hamilton and F. H. Westheimer, *J. Am. Chem. Soc.*, **81**, 6332 (1959).
14. F. H. Westheimer, *J. Am. Chem. Soc.*, **56**, 1962 (1934).
15. F. H. Westheimer, *Proc. Chem. Soc.*, 253 (1963).
16. A. E. Braunstein, P. D. Boyer, M. Lardy and K. Myrback, *The Enzymes*, Vol. 2, Academic Press, New York, 1960, p. 113.
17. E. H. Cordes and W. P. Jencks, *Biochemistry*, **1**, 773 (1962).
18. G. G. Hammes and P. Fasella, *J. Am. Chem. Soc.*, **84**, 4644 (1962).
19. W. P. Jencks and E. H. Cordes, *Symp. Pyridoxal Catalysis*, Pergamon Press, Rome, 1963, p. 57.
20. E. E. Snell, *The Mechanism of Action of Water-Soluble Vitamins*, Little, Brown, Boston, 1961, p. 18.
21. K. O. Ganguin, *J. Phot. Sci.*, **9**, 172 (1961).
22. L. P. Hammett; *Physical Organic Chemistry*, McGraw-Hill, New York, 1940, p. 329.
23. W. P. Jencks, *Progr. Phys. Org. Chem.*, **2**, 63 (1964).
24. R. W. Layer, *Chem. Rev.*, **63**, 489 (1963).
25. R. Roger and D. G. Neilson, *Chem. Rev.*, **61**, 179 (1962).
26. B. Capon, M. J. Perkins and C. W. Rees, *Organic Reaction Mechanisms*, John Wiley and Sons, London, 1967, p. 317–21.
27. A. Kling, *Compt. Rend.*, **148**, 569 (1909).
28. G. Knöpfer, *Monatsh. Chem.*, **32**, 768 (1911).
29. P. K. Chang and T. L. V. Ulbricht, *J. Am. Chem. Soc.*, **80**, 976 (1958).
30. E. J. Poziomek, D. N. Kramer, B. W. Fromm and W. A. Mosher, *J. Org. Chem.*, **26**, 423 (1961).
31. B. Kastening, L. Holleck and G. A. Melkonian, *Z. Elektrochem.*, **60**, 130 (1956).
32. R. L. Reeves, *J. Am. Chem. Soc.*, **84**, 3332 (1962) and **85**, 724 (1963).
33. (a) E. de Hoffmann and A. Bruylants, *Bull. Soc. Chim. Belg.*, **75**, 90 (1966). (b) J. Dickinson and C. Eaborn, *J. Chem. Soc.*, 3036 (1959).
34. E. Feytmans-de Medicis and A. Bruylants, *Bull. Soc. Chim. Belg.*, **75**, 691 (1966).
35. E. Barrett and A. Lapworth, *J. Chem. Soc.*, **93**, 85 (1908).
36. S. F. Acree and J. M. Johnson, *Am. Chem. J.*, **38**, 308 (1907).
37. A. Olander, *Z. Physik. Chem. (Leipzig)*, **291**, 1 (1927).
38. E. G. R. Ardagh and F. C. Rutherford, *J. Am. Chem. Soc.*, **57**, 1085 (1935).
39. G. H. Buchanan and G. Barsky, *J. Am. Chem. Soc.*, **52**, 195 (1930).
40. G. Barsky, *Chem. Ind.*, **28**, 1032 (1932).
41. (a) G. Barsky and G. H. Buchanan, *J. Am. Chem. Soc.*, **53**, 1270 (1931). (b) D. I. R. Barton, R. E. O'Brien and S. Sternhell, *J. Chem. Soc.*, 470 (1962).
42. J. B. Conant and P. D. Bartlett, *J. Am. Chem. Soc.*, **54**, 2881 (1932).
43. B. Zerner and M. L. Bender, *J. Am. Chem. Soc.*, **83**, 2267 (1961).
44. W. P. Jencks, *J. Am. Chem. Soc.*, **81**, 475 (1959); also **82**, 1773 (1960).
45. E. H. Cordes and W. P. Jencks, *J. Am. Chem. Soc.*, **84**, 4319 (1962).
46. B. M. Anderson and W. P. Jencks, *J. Am. Chem. Soc.*, **82**, 1773 (1960).
47. T. G. Bonner and M. Barnard, *J. Chem. Soc.*, 4176 (1958).

48. E. H. Cordes and W. P. Jencks, *J. Am. Chem. Soc.*, **84**, 832 (1962).
49. R. B. Martin, S. Lowey, E. L. Elson and J. T. Edsall, *J. Am. Chem. Soc.*, **81**, 5089 (1959).
50. R. Wolfenden and W. P. Jencks, *J. Am. Chem. Soc.*, **83**, 2267 (1961).
51. P. Zuman, *Collection Czech. Chem. Commun.*, **15**, 839 (1951).
52. P. Zuman, *Sbornik Mezina rod. Polarog. Sjezdu Praze, 1st Congr., 1951*, Pt. I. *Proc.* 704–11 (in Russian), 711–17 (in English). Pt. III *Proc.*, 520–6; discussion 527–9 (in Czech).
53. P. Zuman, *Chem. Listy*, **46**, 516 (1952).
54. P. Zuman, *Chem. Listy*, **46**, 521 (1952).
55. P. Zuman and M. Brezina, *Chem. Listy*, **46**, 599 (1952).
56. O. Bloch-Chaude, *Compt. Rend.*, **239**, 804 (1954).
57. R. W. Green and E. L. Lemesurier, *Australian J. Chem.*, **19**, 229 (1966).
58. J. E. Reinmann and W. P. Jencks, *J. Am. Chem. Soc.*, **88**, 3963 (1966).
59. R. B. Martin, R. I. Hedrick and A. Parcell, *J. Org. Chem.*, **29**, 3197 (1964).
60. G. M. Santerre, C. Hansrote and T. I. Cromwell, *J. Am. Chem. Soc.*, **80**, 1254 (1958).
61. M. T. A. Behme and E. H. Cordes, *J. Am. Chem. Soc.*, **87**, 260 (1965).
62. (a) E. H. Cordes and W. P. Jencks, *J. Am. Chem. Soc.*, **85**, 2843 (1963).
(b) J. B. Culbertson, *J. Am. Chem. Soc.*, **73**, 4818 (1951).
63. K. Koehler, N. Sandstrom and E. H. Cordes, *J. Am. Chem. Soc.*, **86**, 2413 (1964).
64. W. Bruyneel, J. J. Charette and E. de Hoffmann, *J. Am. Chem. Soc.*, **88**, 3808 (1966).
65. R. L. Reeves, *J. Org. Chem.*, **30**, 3129 (1965).
66. A. V. Willi and R. E. Robertson, *Can. J. Chem.*, **31**, 361 (1953).
67. A. V. Willi, *Helv. Chim. Acta*, **39**, 1193 (1956).
68. B. A. Cunningham and G. L. Schmir, *J. Am. Chem. Soc.*, **88**, 551 (1966).
69. S. Marburg and W. P. Jencks, *J. Am. Chem. Soc.*, **84**, 232 (1962).
70. J. F. Bunnett, *J. Am. Chem. Soc.*, **83**, 4956, 4968, 4973 (1961).
71. K. G. Van Senden and C. Koningsberger, *Tetrahedron*, **22**, 1301 (1966).
72. R. K. Chaturvedi and E. H. Cordes, *J. Am. Chem. Soc.*, **89**, 1230 (1967).
73. A. Geiseler, F. Asinger and G. Hennig, *Ber.*, **94**, 1008 (1961).
74. R. B. Martin and A. Parcell, *J. Am. Chem. Soc.*, **83**, 4830 (1961).
75. L. do Amaral, W. A. Sandstrom and E. H. Cordes, *J. Am. Chem. Soc.*, **88**, 2225 (1966).
76. G. L. Schmir, *J. Am. Chem. Soc.*, **87**, 2743 (1965).
77. Y. Inoue and D. D. Perrin, *J. Phys. Chem.*, **66**, 1689 (1962).
78. R. K. McLeod and T. I. Crowell, *J. Org. Chem.*, **26**, 1094 (1961).
79. G. Kresze and H. Goetz, *Z. Naturforsch.*, **10b**, 370 (1955).
78. R. K. McLeod and T. I. Cromwell, *J. Org. Chem.*, **26**, 1094 (1961).
80. (a) G. H. Stempel, Jr. and G. S. Schaffel, *J. Am. Chem. Soc.*, **66**, 1158 (1944).
(b) C. G. Swain, D. A. Kuhn and R. L. Schowen, *J. Chem. Soc.*, **87**, 1553 (1965).
81. J. A. Olson, *Arch. Biochem. Biophys.*, **85**, 225 (1959).
82. G. Kresze and H. Manthey, *Z. Elektrochem.*, **58**, 118 (1954).
83. G. L. Schmir and B. A. Cunningham, *J. Am. Chem. Soc.*, **87**, 5692 (1965).

CHAPTER **11**

Electrochemistry of the carbon–nitrogen double bond

HENNING LUND

University of Aarhus, Denmark

I. INTRODUCTION

The electrolytic method of reducing and oxidizing organic molecules has many inherent advantages and some disadvantages. Its merits ought to have secured for itself a status similar to that of, say, catalytic hydrogenation; however, the activation energy necessary to bring the average chemist to consider an electrolytic reaction as a possible solution to a chemical problem is still too high to ensure that the electrochemical process is used, even when it presents an advantage over other methods.

The electrolytic method presents a possibility to control over a wide range the activity of the reagent, the electron, by a proper choice of the electrode potential, i.e. the potential difference across the electrical double layer. The main part of this potential drop occurs within a distance of a few ångstroms from the electrode surface; the electrical gradient near the electrode is thus of the magnitude of 10^7–10^8 v/cm[1].

The transfer of electrons can occur at low temperatures and at a chosen pH, so that sensitive compounds, such as many biologically active molecules, can be reduced or oxidized under mild and well-defined conditions.

The electrolytic method is favoured by the absence of chemical reagents and their reaction products. This might facilitate the isolation of the product from the electrolytic reaction and make the development of a continuous process easier. The electrolytic process is also inherently easy to control automatically.

An obvious disadvantage of the method is that the reaction of 1 mole of a substance requires $n \times 96500$ coulomb, where n is the number of electrons in the electrode reaction. However, as high currents might be employed when proper design and well-chosen conditions are used, this is not a serious disadvantage. Furthermore, the electron is a very cheap reagent.

A more serious limitation may be caused by the necessity of employing a medium capable of conducting the electrical current. Water is a suitable solvent, but the reacting substances often require an organic solvent or a mixed solvent as medium. In some cases the use of 'hydrotropic'[2] solvents such as a strong aqueous solution of tetra-alkylammonium toluenesulphonate may be advantageous. In aprotic solutions the low concentration of protons must be taken into consideration by adding suitable proton donors, unless the scarcity of protons is important for the formation of the desired product which might not survive in a proton-rich medium.

Some older and newer monographs and review articles on electrochemical reactions are available[3–8]. The following chapter will treat electrolytic reactions which involve azomethine compounds either as starting material, intermediate, or product, and which have been performed at controlled electrode potential. The results are discussed in the light of the information obtained by polarographic[9–13] and voltammetric investigations.

II. EXPERIMENTAL CONDITIONS

A. Control of the Electrolytic Reaction

In the classical electrolytic reactions the current density, measured in A/dm^2, was controlled, possibly because it was the easiest factor to measure and keep constant. Nevertheless, Haber[14], in his famous papers on the stepwise reduction of nitro compounds, realized as early as 1898 that the potential of the working electrode was the proper quantity to control.

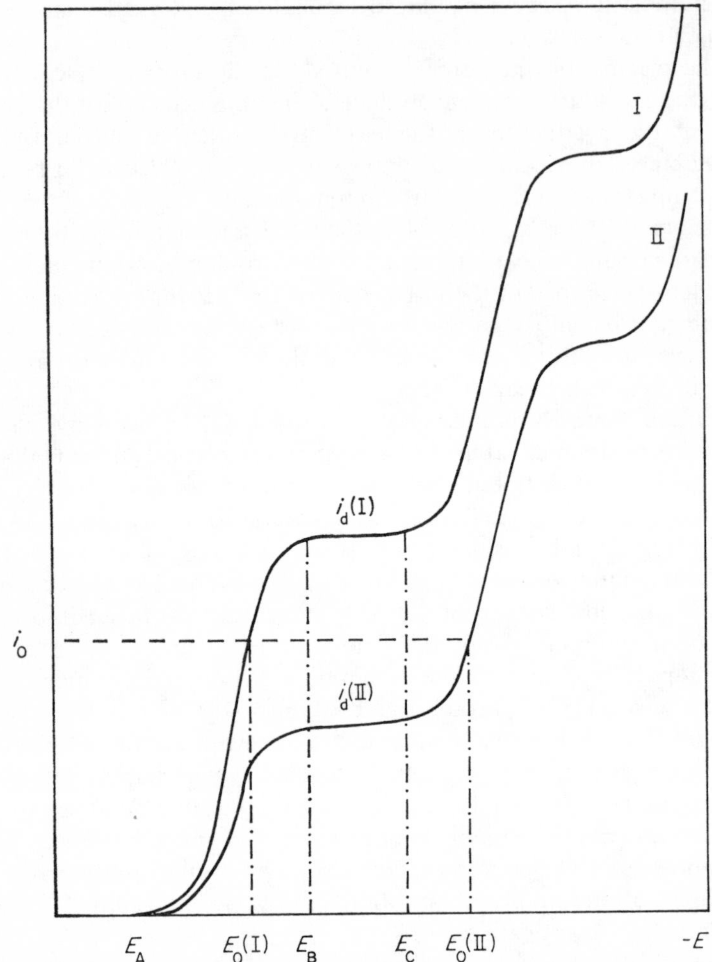

FIGURE 1. Schematic representation of the connection between the current and the potential of the working electrode in a solution containing a compound with two groups reducible at different potentials. (Curve I, before electrolysis; curve II, after the passage of some current; i_0 is an applied current, E_0(I) and E_0(II) the potentials corresponding to i_0; i_d is the limiting current, and E_A, E_B and E_C are applied potentials.)

The difference in the two ways of controlling the electrolytic reaction is illustrated in Figure 1. In this figure curve I depicts the connection between the current through the cell and the potential of the working electrode in the initial solution containing two reducible compounds or one compound with two groups reducible at different potentials. When the potential at the cathode is between 0 and E_A, no electron transfer across the electrical double layer can take place and thus no current flows through the cell. If the cathode potential is made more negative, electron transfer becomes possible, that is, the reduction of the most easily reducible compound or group starts. Between E_A and E_B the current rises in dependence on the potential, but when the value E_B has been reached, all the molecules that arrive to the electrode and which can undergo the first reduction are reduced as soon as they reach the electrode. In the potential interval E_A to E_C the current is limited by the transportation of the reducible compound to the cathode; this current is called the limiting current, i_d, and it is under fixed conditions proportional to the concentration of the electroactive compound.

A further diminishing of the electrode potential results in the occurrence of the second electrode reaction and the current rises; a similar S-shaped curve results from this reduction. At more negative potentials a third reaction or a reduction of the medium takes place.

If a suitable current i_0 [$i_0 < i_d(\mathrm{I})$] is sent through the cell, the cathode potential assumes the value $E_0(\mathrm{I})$, and when $i_0 < i_d(\mathrm{I})$ this is well below the potential (E_C) where the second electrode reaction starts; a selective reduction thus occurs at the beginning of the electrolysis. During the electrolysis the concentration of the reducible compound, and thus its limiting current, diminishes and after a while (curve II) the limiting current becomes smaller than the applied current [$i_0 > i_d(\mathrm{II})$]. The cathode potential has then, by necessity, reached the value $E_0(\mathrm{II})$ and at this potential the second electrode reaction takes place also; the electrolysis is no longer selective.

On the other hand, when the electrode potential is the controlled factor and it is kept at a suitable value, e.g. E_B, the second electrode process cannot take place, and the reduction remains selective to the end. The current through the cell is never higher than the limiting current corresponding to the first electrode reaction; this means that the current decreases during the reduction and becomes very small towards the end of the reaction, as the limiting current is proportional to the concentration of the electroactive material.

The reaction can thus be controlled by letting the reduction proceed at a suitable potential. The potential can be kept at the desired value in two ways: it can either be controlled directly or indirectly. The indirect control can be performed by a continuous addition of new, reducible material to and removal of product from the electrolysed solution, so the limiting current of the reducible material is always higher than the constant current which is sent through the solution. This has been done in the reductive dimerization of acrylonitrile to adiponitrile[15] in the cell shown in Figure 2.

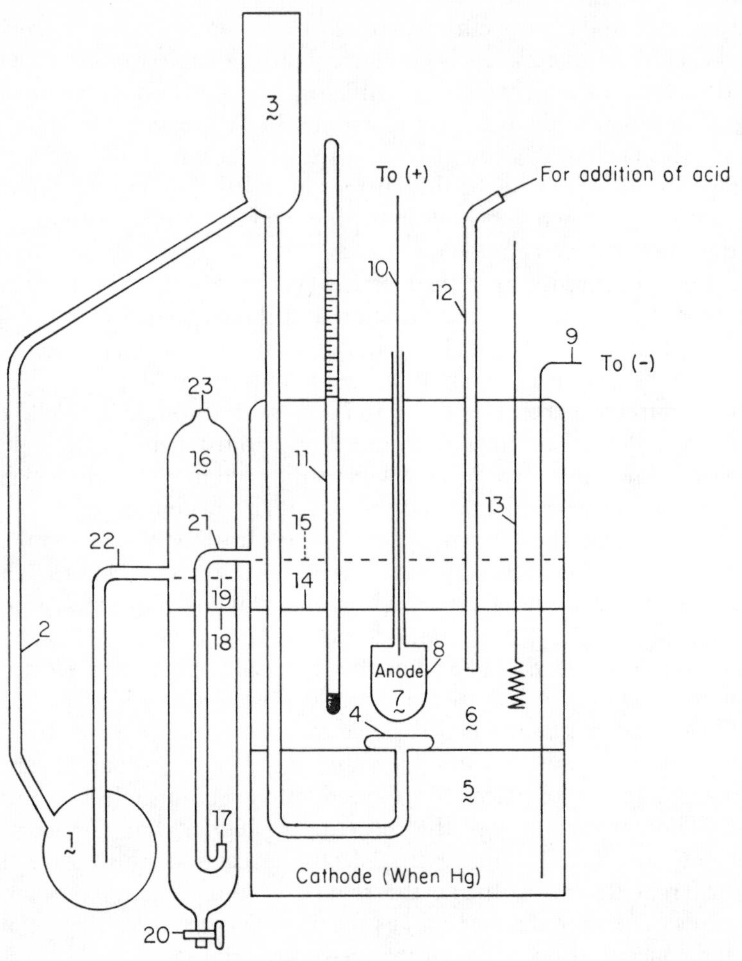

FIGURE 2. Schematic drawing of continuous laboratory cell: 1, boiler; 2, riser; 3, condenser; 4, disperser; 5, cathode when Hg; 6, catholyte solution; 7, anode chamber; 8, diaphragm; 9, lead to cathode; 10, lead to anode; 11, thermometer; 12, inlet tube; 13, stirrer; 14, catholyte level; 15, level of supernatant AN; 16, washer; 17, disperser; 18, water level; 19, supernatant level; 20, stopcock; 21, overflow tube; 22, overflow tube; 23, water inlet.

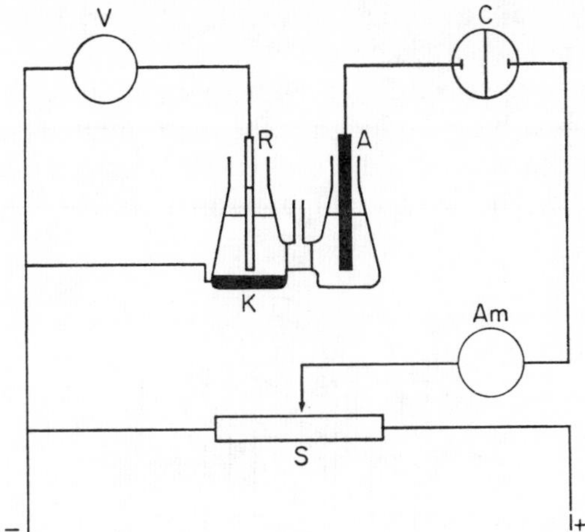

FIGURE 3. Circuit for constant potential reduction. (K cathode, A anode, R reference electrode, V potentiometer or pH meter, C coulometer, Am ammeter, S voltage adjuster. From reference 18.)

A direct manual control of the potential can be obtained using a circuit shown in Figure 3, which is made from components available in all laboratories. The manual control can be replaced by an automatic control by using a potentiostat. Such an apparatus is now

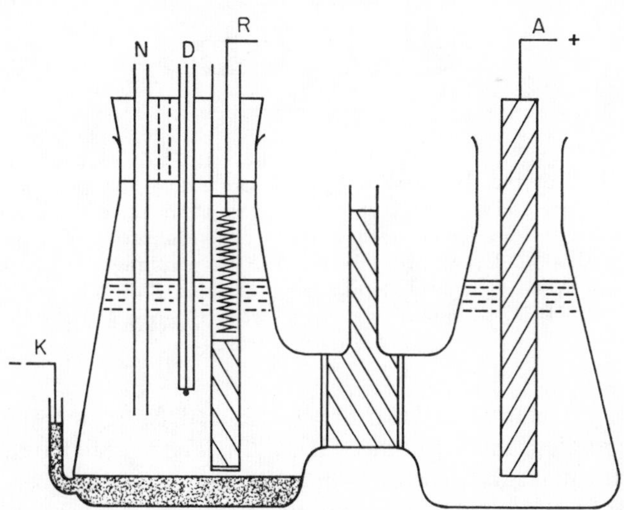

FIGURE 4. Cell for preparative reduction. (K mercury cathode, A carbon anode, R reference electrode, D dropping mercury electrode, N inlet for nitrogen. From reference 18.)

commercially available, often at a reasonable price, or can be built according to published diagrams[16,17].

The design of the electrolytic cell may vary widely, and every electrochemist thinks his own design is the most suitable one; the author has for many years employed[18] the two simple types of cells shown in Figures 4 and 5. The first one (Figure 4) is a divided, slightly modified

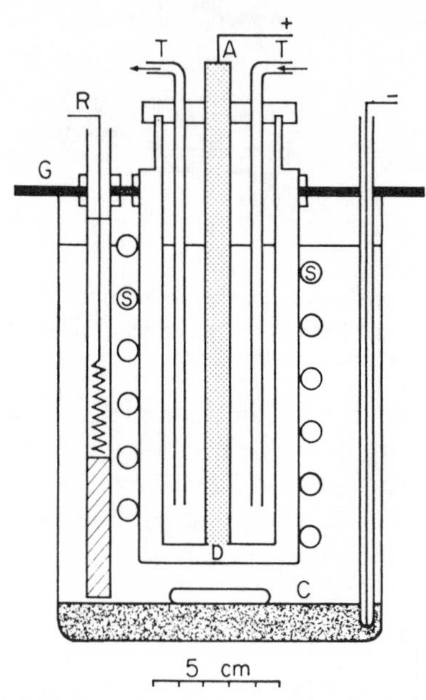

FIGURE 5. Cell for macro-scale electrolysis at controlled potential consisting of a 2 l beaker covered with a glass plate G, containing holes for a silver/silver chloride reference electrode R, the anode compartment, a cooling coil S, a thermometer, an inlet for nitrogen, and one for withdrawing of samples. (The mercury cathode C has an area of 125 cm^2. The diaphragm D consists of two porous clay cylinders separated by agar containing KCl. The anolyte (15% aqueous NaOH) is continuously renewed through T. Anode A of stainless steel. From reference 19.)

Lingane cell made from two 250 ml conical flasks, and the second one (Figure 5) is made from a beaker (2 l). The big cell has the anode compartment in the centre of the cell; the anode chamber is quite small and it is therefore necessary to circulate the anolyte continuously[19]. The cell has been used for large scale preparations (30 to 150 g) using currents up to 25 A. For the reduction of greater amounts of material it may be practical to circulate both the catholyte and the anolyte.

The higher the applied current is, the more critical becomes the design of the cell; the ohmic resistance must be kept low, and it is especially of importance that the tip of the reference electrode (the 'Luggins capillary')[14] ends close to the working electrode; otherwise the inevitable potential drop due to the ohmic resistance between the working electrode and the 'Luggins capillary' (the '*IR* drop') becomes intolerably great.

B. Factors Influencing the Electrolytic Reaction

Although the electrolytic reaction results in a reduction at a certain site in the molecule, it is the properties of the whole molecule which determine the energy, i.e. the reduction potential necessary for the transfer of the electrons to a suitable empty orbital. The presence of certain groups, however, makes the molecule reducible in most cases, and the rest of the molecule influences the reduction potential only to a minor degree. Such groups often contain double bonds, e.g. the nitro-, nitroso-, carbonyl- and azomethine groups, but a reductive cleavage of single bonds may also occur. The presence of electron-withdrawing groups facilitates the reduction, and the dependence of the reduction potential on the nature of the substituents in a series of compounds can often be represented by a Taft–Hammett type of equation[20].

An electrolytic oxidation is a transfer of electrons from the molecule to the anode, and in a given series of compounds, electron-donating groups, e.g. methoxy or amino groups, facilitate such a reaction.

As illustrated in Figure 1, the electrode potential determines under fixed conditions which electrode reaction may occur. It must, however, always be remembered that an electrolytic reaction can be controlled by means of the potential only until and including the potential determining step; besides the potential other factors influence the electrode reaction.

1. Effect of pH

Hydrogen ions are involved in most organic electrode reactions, and pH therefore affects the reduction potential. Besides that, a change in pH may change the course of the electrode reaction; the reduction may take place in different parts of the molecule in acid and alkaline solution, the number of electrons in the electrode reaction may depend on pH, or the stereochemistry of the product may be pH dependent. The protonated species is more readily reducible than the unprotonated one. Sufficient buffer capacity is necessary to ensure that the consumption of hydrogen ions in the electrode reaction does not change the pH in the immediate vicinity of the electrode.

In aprotic media, the use of too strong a proton donor may lead to a preferential reduction of protons. The scarcity of protons in aprotic solvents is valuable in the study of some electrode reactions, since intermediates, e.g. radical ions[21,22], which in aqueous solution would react rapidly with protons or water, may be sufficiently long-lived to be detected or trapped by reaction with a suitable reagent.

2. Effect of electrolyte

The accessible potential region at a certain electrode is dependent on the choice of supporting electrolyte. The alkali metal cations are reduced at about $-2{\cdot}0$ v (s.c.e.), whereas tetraalkylammonium ions can be used until about $-2{\cdot}5$ v (s.c.e.). These ions also interact with anion radicals to a smaller extent than the smaller alkali metal cations. This is of importance for the use of electrolytic reactions in the production of radicals[23] for e.s.r. measurements. The tetraalkylammonium ions are, however, more strongly adsorbed on the electrode than the metal ions, which may influence the kinetics of the reaction.

The choice of the anion is most important in anodic reactions. Perchlorates have been found very useful as they are difficult to oxidize and are often soluble both in water and non-aqueous solvents. High concentrations of tetraalkylammonium *p*-toluenesulphonates in water make the solubility of organic compounds higher than in pure water, and such solutions combine a low ohmic resistance with a good dissolving power.

3. Effect of electrode material

The electrode material is important in different ways. The magnitude of the hydrogen and oxygen overvoltage determines the accessible potential range; special surface properties, such as adsorptive and catalytic effects, may determine the course of the reduction. In the 'electrocatalytic' reactions (e.g. Section III.A.3) the electrochemical step consists in a reduction of hydrogen ions to adsorbed hydrogen which then reacts with the substrate as in a catalytic reaction. The study of the influence of the electrode material on the course of the reaction is an area in which further research is very much needed.

C. Determination of Optimal Conditions for an Electrosynthesis

In order to determine the optimum conditions for an electrosynthesis a series of current–voltage curves are produced using different electrode materials, solvents and pH; when mercury is used as electrode material, the ordinary polarographic technique[8,9] is applied. With some experience it is possible from such a series of experiments to choose conditions suitable for the reaction.

The results from a polarographic investigation may be found in the literature, where either a reproduction of the experimental curves (e.g. Figure 8, Section IV.C.2), a graphical plot of the half-wave potentials and the limiting currents as a function of pH (e.g. Figure 6, Section IV. B.5) or a table gives the required data. It must, however, be kept in mind that there are cases where differences between the results obtained in micro- and macroelectrolysis occur[24].

Sometimes the height of the polarographic wave points to an uptake of, say, two electrons, whereas a preparative reduction results in $n < 2$. This difference can be caused by two types of mechanism. One of these types operates when the reduction proceeds through a radical which either can be reduced further or can dimerize according to

$$\mathrm{R} \xrightarrow{e^- + H^+} {}^{\bullet}\mathrm{RH} \xrightarrow{e^- + H^+} \mathrm{RH_2}$$
$$\downarrow$$
$$\tfrac{1}{2}\,\mathrm{H{-}R{-}R{-}H}$$

The dimerization often, but not always, takes place at the surface of the electrode, where the radicals are stabilized by partly bonding to the electrode. With increasing concentration of the radicals, the rate of the dimerization (second-order reaction) increases faster than the further reduction, and the electron consumption decreases. This mode of reaction often operates when the radical formed is fairly stable.

Sometimes the radicals (perhaps in some cases in the form of organic mercury compounds) form a layer on the electrode which makes the surface less accessible for the unreduced molecules, so they require a slightly more negative potential for the reduction. This phenomenon is less noticeable at the low concentrations normally employed in polarography.

In such cases the dimerized compound can be prepared by employing a high concentration of the reducible compound and stirring the mercury electrode, so a fresh surface is produced, while the electrode potential is kept at a value corresponding to the foot of the polarographic wave. If the further reduced compound is the desired product, only the solution is stirred, and the potential is kept at a potential on the diffusion plateau of the wave.

If the radical formed is not stable or stabilized at the electrode, it is instantly reduced further. The carbanion thus produced may either react with hydrogen ions or with the reducible, unsaturated material according to

$$\begin{array}{ccccccccc} R & \xrightarrow{2e^- + H^+} & H\text{—}R^- & \xrightarrow{R} & H\text{—}R\text{—}R^- & \xrightarrow{R} & H\text{—}R\text{—}R\text{—}R^- & \xrightarrow{R} & \cdots \\ & & \downarrow{\scriptstyle H^+} & & \downarrow{\scriptstyle H^+} & & \downarrow{\scriptstyle H^+} & & \\ & & RH_2 & & HR\text{—}R\text{—}H & & HR\text{—}R\text{—}R\text{—}H & & \end{array}$$

Besides the simple reduction a di-, tri- or polymerization may thus result, and the overall electron consumption decreases.

The waves of irreversibly reduced compounds cover a greater potential range at higher concentrations than at lower concentrations. A reduction which in the microscale experiments gives two separate reduction waves may be difficult to carry out as a selective reaction. The best way to get a partial reduction in such a case is to use a potential at the foot of the composite wave.

Differences between the micro- and macroscale experiments may also be caused by differences in their duration. If a slow step occurs in the reaction after the uptake of some electrons, the reduced compound may diffuse away from the electrode before it is reduced further. At the microelectrode the concentration of the partly reduced species remains low and does not influence the polarographic curve visibly; in a macroscale electrolysis a higher concentration of the partly reduced species is built up and the compound may or may not be reduced further when it diffuses to the electrode, depending on its reduction potential. If the reduction potential of the partly reduced species is more negative than that of the starting material, the partly reduced species can be obtained in a macroscale reduction at a suitable potential, and the difference between the micro- and macroscale experiments is that the further reduction is not visible on the polarograms although the macro reduction shows that a reducible compound is formed.

Sometimes more waves are visible on the polarograms than can be realized by macroelectrolysis. Some may be catalytic waves, and sometimes it is found that the product from the first reduction does not give these waves. In these cases a tautomeric change similar to that described later in the reduction of phenylhydrazones and some cyclic azines may be operating, so it is the primarily formed species which is responsible for the observed waves whereas the more stable tautomeric form is reduced by another route.

When the partly reduced species is more easily reducible than the starting material, a macroelectrolysis will show a higher electron consumption than that corresponding to the height of the polargraphic wave. During a macroscale electrolysis the partly reduced species may be detectable polarographically in the reaction mixture; it will produce a small wave at a less negative potential than that of the starting material. The concentration of the intermediate will remain low as it is reduced in preference to the starting material. Only if it is possible to trap the intermediate as a non-reducible derivative can it be obtained as a product from the reduction.

Furthermore, as the polarographic curve can be influenced by certain compounds, such as inhibitors, adsorption phenomena can complicate the interpretation of the curves and 'catalytic' waves may suggest further reductions than found by macroelectrolysis, a certain caution must be exercised in evaluating the voltammetric data; however, in most cases no complications arise, and with a little experience the differences mentioned above are not serious draw-backs, but are of value as the combination of polarography and macroelectrolysis then throws light on one or more of the steps in the reaction.

III. ELECTROCHEMICAL PREPARATION OF AZOMETHINE COMPOUNDS

Azomethine compounds may be formed by electrolytic reduction as well as by oxidation of suitable nitrogen derivatives. When the nitrogen or the carbon atom of the potential azomethine group is in a high oxidation state, an electrolytic reduction may produce an azomethine compound. Likewise, oxidation of a hydrazine or a hydroxylamine may lead to such a compound.

A. Reduction of Nitro and Nitroso Compounds

1. Aliphatic nitro compounds

The electrolytic reduction of *aliphatic* nitro compounds is possible in acid and neutral solution, whereas the formation of the anion of the *aci* form in strongly alkaline solution prevents the reduction of primary and secondary nitroalkanes. The main product in acid solution is a hydroxylamine which can be reduced further only in a rather narrow pH interval around pH 4 at a more negative potential. Some amine is formed as a by-product during the electrolysis, and as the hydroxylamine is not reducible under these conditions to the amine, the following reduction scheme has been suggested[25,26], in which the intermediate nitroso compound tautomerizes to the oxime, and the reduction of the oxime is responsible for the formation of the amine.

$$\mathrm{RR'CHNO_2} \xrightarrow[2e\,+\,2H^+]{H^+} [\mathrm{RR'CHNO}]H^+ \xrightarrow{2e\,+\,2H^+} \mathrm{RR'CH\overset{+}{N}H_2OH}$$

$$\downarrow$$

$$[\mathrm{RR'C{=}NOH}]H^+ \xrightarrow{4e\,+\,4H^+} \mathrm{RR'CH\overset{+}{N}H_3}$$

$$\downarrow$$

$$\mathrm{RR'C{=}O} + \mathrm{H_3\overset{+}{N}OH}$$

2. α,β-Unsaturated nitro compounds

The oxime formed in the side reaction in the scheme above cannot be isolated, and its presence is only inferred from the isolation of the amine and the carbonyl compound as side products; when, however, the nitro compound is α,β-unsaturated or carries a suitable leaving group on the α- or β-carbon, an oxime is a likely product. The reduction of β-nitrostyrene (**1**) exemplifies this[27,28]:

$$\underset{(\mathbf{1})}{\mathrm{C_6H_5CH{=}CHNO_2}} \xrightarrow{4e\,+\,5H^+} \mathrm{C_6H_5CH{=}CH\overset{+}{N}H_2OH} \xrightarrow{-H^+} \mathrm{C_6H_5CH_2CH{=}NOH}$$

The unsaturated hydroxylamine rapidly forms the more stable tautomer, the oxime. The reduction might be formulated as a 1,4 reduction of the initially formed unsaturated nitroso compound; there is no conclusive evidence for either route, but the former has here been chosen as the protonation of the heteroatom would be expected to be faster than the protonation of the carbon atom.

3. α-Substituted nitroalkanes

The reduction of α-halogenated nitroalkanes may also lead to oximes[29]. For instance, 1-bromo-1-nitroethane (**2**) gives three waves in acid solution; the half-wave potential of the first wave is independent of pH, and the third wave corresponds to the reduction of acetaldoxime (**3**). Controlled potential reduction at pH 0·25 ($E = -0{\cdot}6$ v, $t = 10°$) of 2-chloro-2-nitropropane yielded 98% acetone, formed by acid hydrolysis of the *aci*-nitro compound, $(CH_3)_2C{=}NO_2H$. A reduction at a potential corresponding to the second polarographic wave of **2** gave 60% **3** and 30% acetaldehyde. The first two reductions of **2** can be formulated as

$$\text{1st wave}\quad \underset{(\mathbf{2})}{CH_3CHBrNO_2} \xrightarrow{2e} CH_3CH{=}NO_2^- + Br^- \xrightarrow[H^+]{\text{slow}} CH_3CH_2NO_2$$

$$CH_3CH{=}NO_2^- \xrightarrow{H^+,\ \text{fast}} \underset{(\mathbf{4})}{CH_3CH{=}NO_2H}$$

$$\text{2nd wave}\quad \underset{(\mathbf{4})}{CH_3CH{=}NO_2H} \xrightarrow{2e\,+\,2H^+} \underset{(\mathbf{3})}{CH_3CH{=}NOH} + H_2O$$

This reduction is another example of the faster protonation at an oxygen atom than at a carbon atom, which results in the primary formation of the less stable *aci*-nitro compound (**4**) rather than the stable nitroalkane.

A similar reduction route is found for dihalogenated nitroalkanes, e.g. 1,1-dichloro-1-nitroethane (**5**) [29]. This compound shows two polarographic two-electron waves, but the second wave does not correspond to the reduction of 1-chloro-1-nitroethane, which is the product from a controlled potential reduction at the potential of the first wave ($E = -0{\cdot}2$ v, pH $= 1$). A reduction at the plateau of the second wave ($E = -1{\cdot}0$ v, pH $= 0{\cdot}2$, $t = 5°$) produced chloroacetaldoxime, which was further characterized by its transformation into acetonitrile oxide at pH 6.

The reduction of trichloronitromethane (**6**) may give methylamine or methylhydroxylamine, but by working with a cooled tin cathode in 35% sulphuric acid at $t < 5°$ it was possible to obtain dichloroformoxime (**7**) as a product[30]; by employing a suitable extraction procedure the yield of this compound could be raised[31]. Brintzinger et al.[30] suggest essentially the following reduction path:

$$\underset{(\mathbf{6})}{CCl_3NO_2} \xrightarrow[-H_2O]{2e\,+\,2H^+} CCl_3NO \xrightarrow[-Cl^-]{2e\,+\,H^+} CHCl_2NO \rightarrow \underset{(\mathbf{7})}{CCl_2{=}NOH} \xrightarrow{2e\,+\,2H^+}$$

$$CHCl_2NHOH \xrightarrow[-Cl^-]{2e\,+\,H^+} CH_2ClNHOH \xrightarrow[-Cl^-]{2e\,+\,2H^+} CH_3\overset{+}{N}H_2OH \xrightarrow[-H_2O]{2e\,+\,2H^+} CH_3\overset{+}{N}H_3$$

A polarographic investigation of **6**[29] shows a stepwise removal of the chlorine under these conditions before the reduction of the nitro group occurs. Whether the difference in reduction path reflects the difference in the electrode material, electrode potential, or medium is not known; an alternative formulation of the formation of **7** could be

$$\underset{(\mathbf{6})}{CCl_3NO_2} \xrightarrow{2e} Cl^- + CCl_2{=}NO_2^- \xrightarrow{H^+} CCl_2{=}NO_2H \xrightarrow{2e+2H^+} \underset{(\mathbf{7})}{CCl_2{=}NOH}$$

The reduction of mono- and dihalogenated nitrosoalkanes[29] goes analogously to that of the nitroalkanes; thus 2-chloro-2-nitrosopropane (**8**) gives acetoxime (**9**) on reduction ($E = -0{\cdot}50$ v, pH 3) in 98% yield:

$$\underset{(\mathbf{8})}{(CH_3)_2CClNO} \xrightarrow{2e\,+\,H^+} \underset{(\mathbf{9})}{(CH_3)_2C{=}NOH} + Cl^-$$

and 1,1-dichloro-1-nitrosoethane produces chloroacetaldoxime in 76% yield on electrolytic reduction.

The first step in the polarographic reduction of *pseudo*-nitroles[29] in acid solution is a loss of the nitro group as a nitrite ion with the formation of an oxime, e.g.

$$(CH_3)_2C(NO)NO_2 \xrightarrow{2e} NO_2^- + (CH_3)_2C^-{-}NO \xrightarrow{H^+} \underset{(\mathbf{9})}{(CH_3)_2C{=}NOH}$$

Whereas 2,2-dinitropropane is reduced in a two-electron reduction to the *aci*-2-nitropropane[32,33], 1,1-dinitroethane (**10**) gives a 5–6 electron wave in acid solution. The following explanation has been suggested[29]:

$$\underset{(\mathbf{10})}{CH_3CH(NO_2)_2} \xrightarrow{2e} CH_3CH{=}\bar{N}O_2 + NO_2^- \xrightarrow{2H^+}$$

$$CH_3CH{=}NO_2H + HNO_2 \rightarrow \underset{(\mathbf{11})}{CH_3C(NO_2){=}NOH} + H_2O \xrightarrow{4e\,+\,4H^+}$$

$$\underset{(\mathbf{12})}{CH_3C(NHOH){=}NOH}$$

The ethylnitrolic acid **11** has been shown[34] to be electrolytically reducible to acethydroxyamidoxime (**12**), which can be oxidized anodically to ethylnitrosolic acid.

4. Aromatic nitro compounds

Aromatic nitro compounds are reducible in acid solution to the hydroxylamines which at a more negative potential are reduced to the amines. However, *o*- or *p*-nitrophenols or -anilines are in acid solution reducible in a six-electron reduction to the amine[35–37]. The reaction is believed to follow this path, exemplified by the reduction of *p*-nitrophenol.

$$HO{-}C_6H_4{-}NO_2 \xrightarrow{4e\,+\,5H^+} HO{-}C_6H_4{-}\overset{+}{N}H_2OH \xrightarrow{-H_2O}$$

$$O{=}C_6H_4{=}\overset{+}{N}H_2 \xrightarrow{2e\,+\,2H^+} HO{-}C_6H_4{-}\overset{+}{N}H_3$$

The loss of water is a slow step which is acid or base catalysed. About pH 5 the slow step can be established by classical polarography as the wave height in this region corresponds to a four-electron reduction. A preparative reduction at this pH would, however, show a six-electron reduction to the amine.

A reaction involving *o*-quinonediimine (**12**) as an intermediate has also been suggested[38] in the reduction of *o*-dinitrobenzene in acid solution, which resulted in the formation of 2,3-diamino-9,10-dihydrophenazine or 2-amino-3-hydroxy-9,10-dihydrophenazine besides *o*-phenylenediamine. The reaction can be formulated as:

NO_2 NO_2 $\xrightarrow{10e + 10H^+}$ NH NH **(12)**

(12) $\xrightarrow{12}$ **(13)**; **(12)** + NH/NOH $\rightarrow$ **(14)** $\xrightarrow{2e\ 2H^+}$ **(13)**

The quinonediimine could either condense with another molecule of **12** to form 2,3-diamino-9,10-dihydrophenazine (**13**) or with another intermediate of a higher oxidation state, e.g. *o*-dihydroxylaminobenzene or its dehydration product, to form the phenazine **14** followed by reduction to the dihydrophenazine.

B. Reduction of Acid Derivatives

The electrochemical preparation of azomethine compounds from acid derivatives involves the same difficulties as those found in the reduction of acid derivatives to aldehydes; the product is generally more easily reducible than the starting material. Two routes may lead to the desired result; one can choose a very easily reducible derivative or one can trap the product as a non-reducible derivative. For the reduction of acid derivatives the first possibility is found in the reduction of acid chlorides in non-aqueous medium[39] and the other one in the reduction of isonicotinic acid[40] or other heterocyclic acids[41,42] into the hydrated aldehydes.

Hydroxamic acid halides[43] such as the rather unstable iodide (**15**) in acid solution may be reduced to benzaldoxime (**16**).

$$\underset{\textbf{(15)}}{C_6H_5C(=NOH){-}I} \xrightarrow{2e\,+\,2H^+} \underset{\textbf{(16)}}{C_6H_5CH{=}NOH} + HI$$

Benzonitrile oxide[43] (**17**) in acid solution is reduced to benzaldoxime (**16**), although most reductions of benzonitrile oxides give benzonitriles. Nitrones are reduced in acid solution with an initial loss of the oxygen[44,45]. The same is found for most heteroaromatic *N*-oxides[46,47]; quinazoline-3-oxide[48] is, however, an exception. The reduction of **17** has been formulated[43] as:

$$\underset{\textbf{(17)}}{C_6H_5C{\equiv}\overset{+}{N}O^-} \longleftrightarrow C_6H_5\overset{+}{C}{=}N{-}O^- \xrightarrow[2e\,+\,H^+]{H^+} \underset{\textbf{(16)}}{C_6H_5CH{=}NOH}$$

The primary reduction of thioamides[49] results in a potential azomethine derivative which might either hydrolyse to an aldehyde or lose hydrogen sulphide to give the aldimine. The stability of the intermediate is high when the aldehyde is reactive towards nucleophilic reagents; thus isonicotinic thioamide[50] yields on reduction an intermediate which is quite stable in acid solution.

$$RCSNH_2 \xrightarrow[2e\,+\,2H^+]{H^+} RCH(SH)\overset{+}{N}H_3$$

$$RCH(SH)\overset{+}{N}H_3 \xrightarrow{H_2O} H_2S + NH_3 + RCHO \qquad RCH(SH)\overset{+}{N}H_3 \longrightarrow RCH{=}\overset{+}{N}H_2 + H_2S$$

$$H_2S + NH_3 + RCHO \xleftarrow{H_2O} RCH{=}\overset{+}{N}H_2 + H_2S$$

$$RCHO \xrightarrow{2e\,+\,2H^+} RCH_2OH \qquad RCH{=}\overset{+}{N}H_2 \xrightarrow{2e\,+\,2H^+} RCH_2NH_2$$

Benzaldimine has been shown to be an intermediate in oscillopolarographic reduction of benzamide[51] in acid solution and benzaldehyde hydrazone[52] has been detected under similar conditions in the reduction of benzhydrazide.

Cyanamide (**18**) does not give a polarographic wave, but it can be reduced in good yield electrolytically to formamidine on an electrolytically formed nickel-sponge cathode in an almost neutral phosphate buffer[53]. The formamidine **19** is not reduced further under these conditions. The reaction consists in this case not of a transfer of electrons directly to the substrate, but rather of an electrolytic generation of hydrogen which on the catalytically active nickel surface reacts with (**18**) as in a catalytic hydrogenation; such reactions are called electrocatalytic reductions and are mostly found at electrodes with low hydrogen overvoltage such as platinum and nickel cathodes.

The reaction is:

$$2H^+ + 2e \longrightarrow 2H(Ni)$$

$$\underset{(\mathbf{18})}{H_2NC{\equiv}N} \xrightarrow{2H(Ni)} \underset{(\mathbf{19})}{H_2NCH{=}NH}$$

At a spongy tin cathode **18** is reduced similarly in the first step, but under these conditions **19** may be reduced further[54,55].

C. Partial Reduction of Heterocyclic Compounds

The reduction of aromatic and other heterocyclic compounds will be discussed only if they are reduced to genuine azomethine compounds, cyclic or non-cyclic. Thus, compounds such as pyridine and pyrazine which are reduced to 1,4-dihydro derivatives without a carbon–nitrogen double bond are not treated here. Most of the compounds discussed below are derivatives of pyridazine or pyrimidine; the investigation of the stepwise reduction of these is often complicated by the lability of the partly reduced compounds. They are usually easy to reoxidize and to transform into other tautomeric forms.

1. Pyridazines

Aryl- and alkylsubstituted pyridazines can often under suitable conditions be reduced to dihydropyridazines; these are further reducible (see Section IV.B.2). 3,6-Diphenylpyridazine (**20**) gives in alkaline solution a two-electron polarographic wave; at pH 13 it is followed by a small second wave at −1·70 v (vs. s.c.e.). Reduction of a suspension of the slightly soluble (**20**) in aqueous alcoholic potassium hydroxide (0·2 m) at −1·75 v (vs. s.c.e.) gives 2,3,4,5,-tetrahydro-3,6-diphenylpyridazine (**21**) in a four-electron reaction at a mercury cathode[39,56,57]. The mechanism was in analogy to the reduction of other pyridazines and of benzalazine (Section IV.B.1) suggested to be:

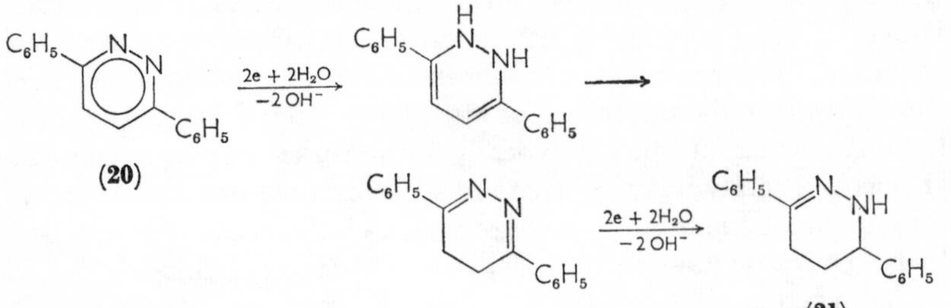

(20)

(21)

21 is a cyclic benzylhydrazone and is as such further reducible in acid solution (Section IV.B.3) but not in alkaline solution.

In acid solution a two-electron reduction of pyridazines takes place primarily, and from the reduction of 1-methyl-3,6-diphenylpyridazinium iodide (**22**) the dihydro derivative can be isolated. The n.m.r. spectrum in $CDCl_3$ of the product extracted from a slightly alkaline solution indicates that it is 1-methyl-1,4-dihydro-3,6-diphenylpyridazine (**23**), but in CF_3COOH the protonated 4,5-dihydro derivative (**24**) is found, so that the species present in the reduced acidic solution is probably the 4,5-dihydropyridazine (**24**).

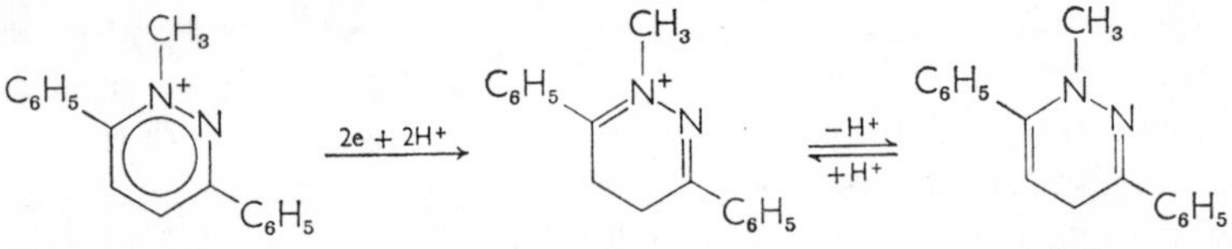

(22) (24) (23)

Such a reduction has been used to prove the site of the quaternization of 4-*t*-butyl-3,6-diphenylpyridazine (**25**). The quaternization with methyl iodide yielded only one compound, and from the n.m.r. spectrum of the two-electron reduction product of the quaternized pyridazine it was proved that the quaternization had yielded 1-methyl-5-*t*-butyl-3,6-diphenyl-pyridazine[58] (**26**).

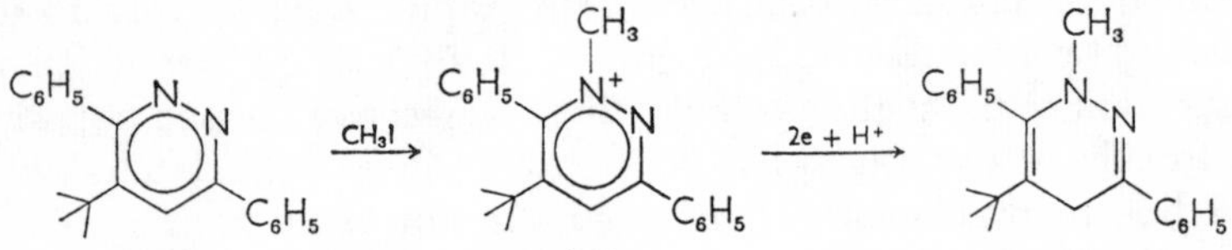

(25) (26)

Pyridazines substituted with halogens, amino-, hydroxyl- or methoxy groups are reduced in the ring to the 4,5-dihydro derivatives; exceptions are the 3-iodopyridazines which are first reduced to the pyridazines[39,56]. The reduction to the dihydro derivatives takes place both in acid and alkaline solution; the stability of the dihydropyridazines varies. 3-Phenyl-6-dimethylaminopyridazine (**27**) is reduced to the 4,5-dihydro derivative (**28**) which can be isolated as it is reasonably stable at room temperature; on heating or long storage at room temperature it loses dimethylamine.

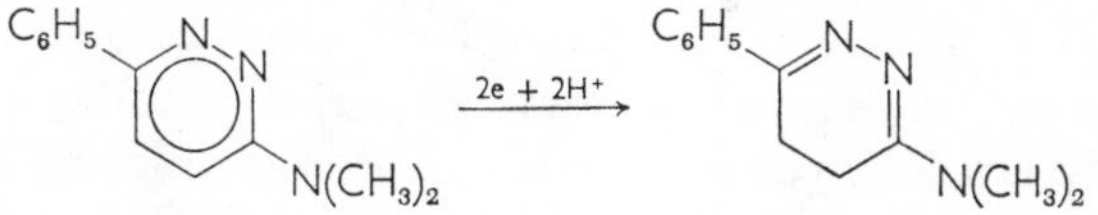

C_6H_5, N, N, $N(CH_3)_2$ — $\xrightarrow{2e + 2H^+}$ — C_6H_5, N, N, $N(CH_3)_2$

(**27**) (**28**)

The 4,5-dihydropyridazines obtained from the reduction of 3-methoxy or 3-chloropyridazines are easily hydrolysed and have not yet been isolated; the isolated product from the reduction is the pyridazinone (**29**). This compound is also obtained in the reduction of pyridazones[39,59] (**30**).

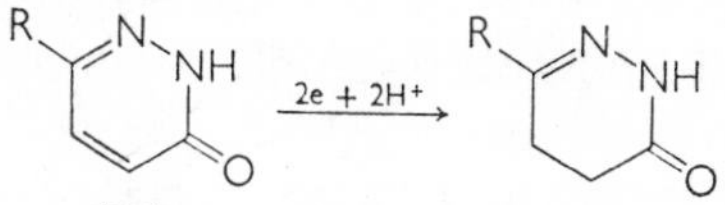

R, N, NH, O — $\xrightarrow{2e + 2H^+}$ — R, N, NH, O

(**30**) (**29**)

The product is formally an acylated hydrazone of a ketone and is, as most other compounds of this type, reducible in acid but not in alkaline solution.

2. Cinnolines

The reduction of cinnolines is similar to that of pyridazines; the benzene ring fused to the hetero ring acts as a non-reducible, non-displaceable unsaturated centre which determines the kind of dihydro derivative that is the product. Dihydrocinnolines corresponding to 4,5-dihydropyridazines can thus be formed.

Cinnoline and alkylcinnolines (**31**) yield a two-electron polarographic wave[60] which in acid solution is followed by further waves. Such cinnolines have been shown to be reduced to 1,4-dihydrocinnolines[60] (**32**) in acid solution; if the solution is very strongly acid, the 1,4-dihydrocinnolines rearrange to *N*-aminoindoles[61,62]. The dihydrocinnolines do not give an anodic polarographic wave as do dihydrobenzo[c]cinnolines, indicating that they are not as easily reoxidized as dihydrobenzo[c]cinnolines[60].

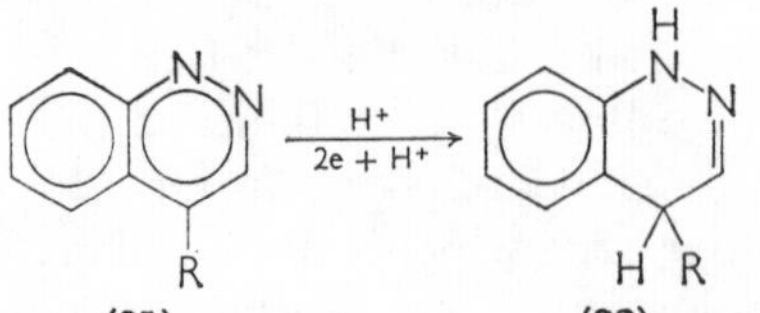

(31) (32)

3-Phenylcinnolines give in acid solution a two-electron polarographic wave followed by a four-electron wave[60]. In acid solution dimeric products may be formed besides the 1,4-dihydrocinnolines. The reduction to a 1,4-dihydrocinnoline has been used in the proof of the site of the quaternization of 3-phenylcinnoline. This compound gives on quaternization with methyl iodide a 72:28 mixture of two isomers. The major product was reduced to the dihydrocinnoline, and from its n.m.r. spectrum it was concluded that the major product from the quaternization was 1-methyl-3-phenylcinnolinium iodide.

1,4-Dihydrocinnoline is also formed in the reduction of some substituted cinnolines; thus 4-mercaptocinnoline (**33**), which gives a four-electron polarographic wave in acid and a two-electron wave in alkaline solution, is both at pH 0 and pH 9 reduced in a four-electron reaction to 1,4-dihydrocinnoline[60] (**34**).

The discrepancy between the results obtained by polarographic (two-electron wave) and preparative electrolysis (four-electron reduction) was explained by the following mechanism:

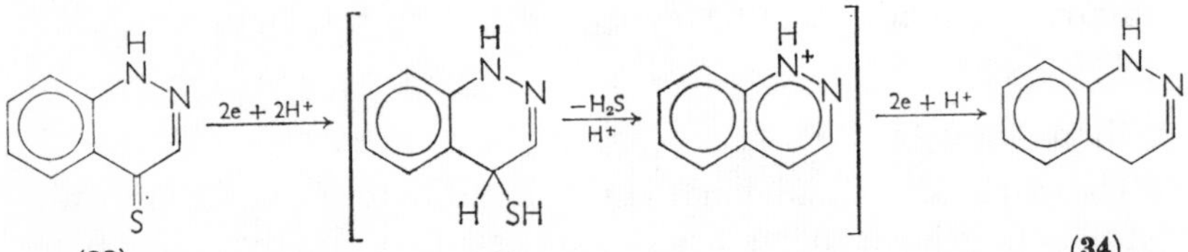

(33) (34)

In support of the reaction route suggested above the following arguments were used. The two-electron reduction occurring at the dropping mercury electrode in alkaline solution yielded an electroinactive species. This compound cannot be 1,2,3,4-tetrahydro-4-thiocinnolone, which can be regarded as a derivative of thioacetophenenone; this compound would be expected to be reducible at a rather positive potential. A reduction at a macro-mercury electrode yielded **34** even at pH 9; this means that a slow step occurs after the uptake of the first two electrons and an electroactive species is formed in this slow reaction. The formation of the electroactive species is too slow to influence the polarographic curve at pH > 9, but not the products from the preparative reduction. The slow reaction was suggested to be the loss of hydrogen sulphide, which seemed more reasonable than assuming that a slow step was involved in the reduction of the carbon–sulphur bond. The reduction of 4-mercapto-1,4-dihydrocinnoline would require a protonation which could be slow in alkaline solution, but the reduction of the protonated species would probably result in a cleavage of the nitrogen–nitrogen bond.

A further confirmation of the reduction route suggested in the scheme was found in the results from the preparative reduction of 4-mercaptocinnoline at pH 9. A small prewave, which was found in the non-reduced solution at the same potential as the reduction potential of cinnoline, grew during the earlier part of the reduction and became in the later part of the reaction nearly as high as the wave of the remaining 4-mercaptocinnoline.

The slope of the $E_{\frac{1}{2}}$–pH curve is the same from pH 0 to 11, suggesting that the same electrode reaction is the primary step both in acid and alkaline solution. The occurrence of a prewave in the whole pH region at the reduction potential of cinnoline points in the same direction. The explanation of the decrease of the limiting current around pH 7 would then be that the loss of hydrogen sulphide from the primarily formed product was acid catalysed and became a slow step in alkaline solution.

3. Phthalazines

In phthalazines the benzene ring fused to the hetero ring occupies the position of the 4,5 double bonds of the pyridazines, which means that compounds analogous to 4,5-dihydro- or 1,4-dihydropyridazines cannot be formed.

In acid solution phthalazine (**35**) gives a six-electron polarographic wave[63], and the reduction has as an intermediate probably either 1,2-dihydrophthalazine (**36**) or *o*-phthalaldehydediimine. In alkaline solution **35** can be reduced to **36** at −1·7 v (s.c.e.); it can be reduced[39,64] further to 1,2,3,4-tetrahydrophthalazine (**37**) at −1·85 v (s.c.e.); **37** can be oxidized anodically to **36** in the same medium. The electrolytical method thus presents an attractive way to prepare 1,2-dihydrophthalazines; if the reduction potential has not been kept at the optimal value and some **37** has been formed, the current is reversed and the tetrahydrophthalazine reoxidized to the desired product.

1,2-Dihydrophthalazine (m.p. 47–48°) has been claimed to be formed by reduction of (2H)-phthalazinone with lithium aluminium hydride[65]. The n.m.r. spectrum of the product (m.p. 85°) of the controlled potential reduction of phthalazine in alkaline solution proves, however, that it is 1,2-dihydrophthalazine.

$$\textbf{(35)} \xrightarrow[-OH^-]{e + H_2O} \text{radical} \xrightarrow[-OH^-]{e + H_2O} \textbf{(36)} \underset{-2e,\ -2H_2O}{\overset{2e + 2H_2O}{\rightleftharpoons}} \textbf{(37)} + 2\,OH^-$$

Dimer (**38**)

If the dimeric compound **38** is the desired product it is advisable to stir the mercury electrode rapidly and to keep the potential at the foot of the polarographic wave measured directly in the electrolysed solution. **36** is formed in a high yield when the solution, but not the electrode, is stirred and the potential is kept at a value where the limiting current is reached.

Substituted phthalazines may also be reduced to dihydro derivatives; thus 1-methylphthalazine is reduced to 1-methyl-3,4-dihydrophthalazine[57], whereas 1-methyl-4-methoxyphthalazine (**39**) and 1-methyl-4-dimethylaminophthalazine are reduced to 1,2-dihydro-1-methyl-4-methoxyphthalazine (**40**) and 1,2-dihydro-1-methyl-4-dimethylaminophthalazine, respectively[39]. These dihydrophthalazines, even **40**, are stable enough to be isolated.

$$\textbf{(39)} \xrightarrow{2e + 2H^+} \textbf{(40)}$$

4. Pyrimidines

The electrolytic reduction of pyrimidine (**41**) also occurs stepwise; in acid solution two one-electron reductions are found, whereas two two-electron reactions occur in neutral and one four-electron reduction in alkaline solution[66]. These reactions can be represented in the scheme below; the structure of the tetrahydropyrimidine was not proved owing to its hydrolysis in the alkaline solution.

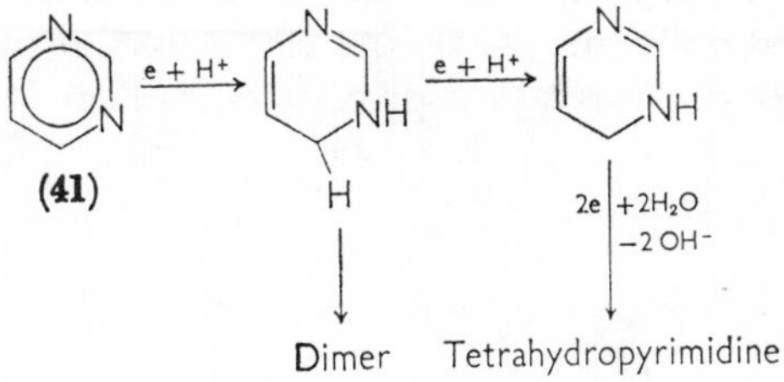

(41)

Dimer Tetrahydropyrimidine

2-Aminopyrimidine is reduced similarly, but the dihydro derivative is not further reducible[66,67]. The reduction of quinazoline[48] (**42**) and purine[68] follows the same pattern as that of pyrimidine. In alkaline solution the stepwise reduction of the former proceeds as follows:

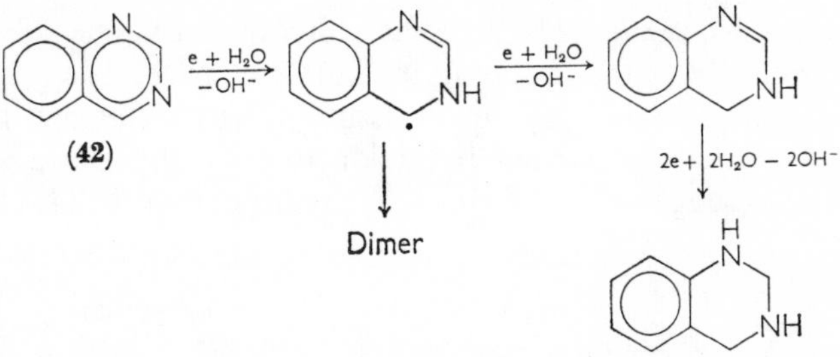

(42)

Dimer

On the basis of coulometric data the electrode reactions of adenine (**43**) shown below have been suggested[68], but it is not known whether the elimination of ammonia occurs before or after the second reduction. The coulometric investigation of large adenine concentrations is unreliable, since the reduction product exerts a strong catalytic effect on the reduction of hydrogen ions.

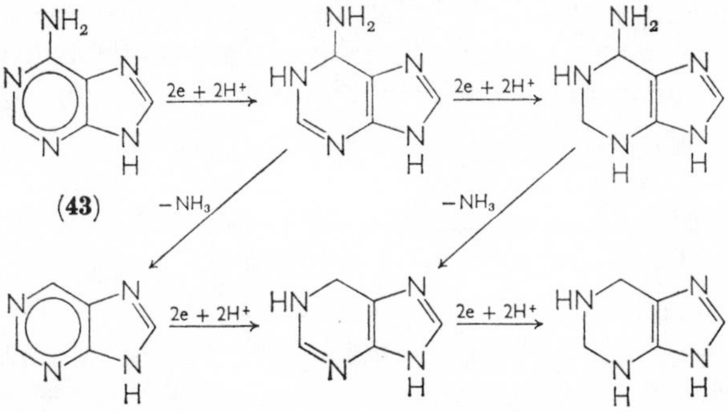

(43)

5. Quinoxaline

Quinoxaline (**44**) is reversibly reduced to 1,4-dihydroquinoxaline (**45**), whereas substituted quinoxalines mostly form 1,2-(or 3,4)-dihydroquinoxalines[69]. In N HCl **44** is reduced at the potential of the first one-electron wave to a protonated radical (**46**). Reduction of **44** in aqueous alkaline solution yields **45** and a dimer which on heating with acid produces **46**[56].

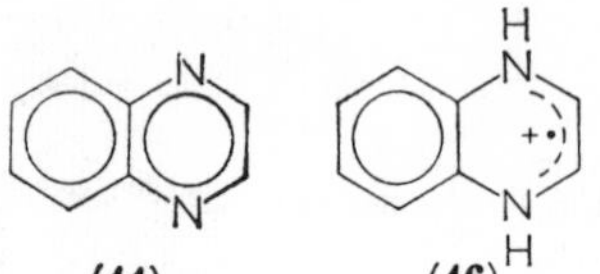

(**44**) (**46**)

6. Benzo-1,2,4-triazine

Benzo-1,2,4-triazine (**47**) and the dihydro derivative **48** behave nearly reversibly at the dropping mercury electrode. The position of the double bond in 3-phenyldihydrobenzo-1,2,4-triazine has not been proved, but from the further reduction to phenylbenzimidazole the compound is suggested to be the 1,4-dihydro derivative (**48**)[70].

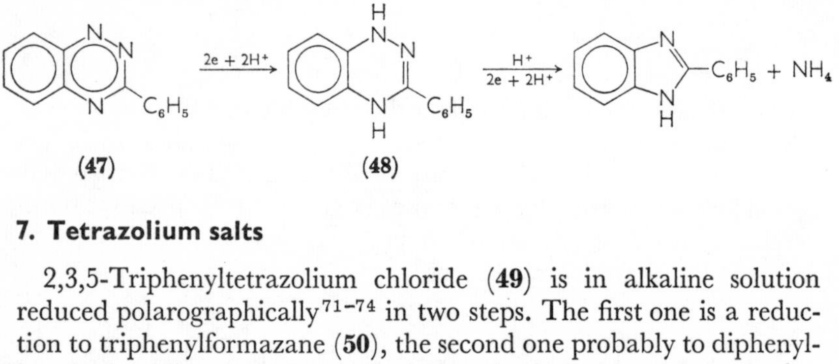

$\xrightarrow{2e\ +\ 2H^+}$ $\xrightarrow[2e\ +\ 2H^+]{H^+}$ $-C_6H_5 + NH_4$

(**47**) (**48**)

7. Tetrazolium salts

2,3,5-Triphenyltetrazolium chloride (**49**) is in alkaline solution reduced polarographically[71–74] in two steps. The first one is a reduction to triphenylformazane (**50**), the second one probably to diphenylbenzhydrazidine (**51**). In acid solution the isolated products from a

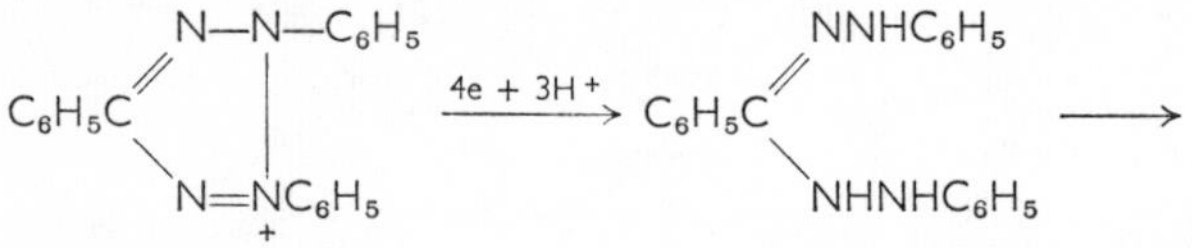

$$C_6H_5C(=N-N(C_6H_5)-)N=\overset{+}{N}C_6H_5 \ \text{(49)} \xrightarrow{4e\ +\ 3H^+} C_6H_5C(=NNHC_6H_5)NHNHC_6H_5 \ \text{(51)} \longrightarrow$$

$$\tfrac{1}{2}\,C_6H_5NH_2 \ \text{(53)} + \tfrac{1}{2}\,C_6H_5C(NH_2){=}NNHC_6H_5 \ \text{(52)} + \tfrac{1}{2}\,C_6H_5C(N{=}NC_6H_5){=}NNHC_6H_5 \ \text{(50)}$$

four-electron reduction are phenylbenzamidrazone (**52**), **50**, and aniline (**53**). It has been suggested[18] that **51** is formed initially and then disproportionates into the isolated products **50**, **52**, and **53**.

8. Oxaziridines

Oxaziridines (**54**) show in aqueous solution two waves[75]; the second one corresponds in acid solution to the reduction of the azomethine compound (**55**) from which the oxaziridine is derived. At higher pH **55** is hydrolysed rapidly and only the reduction of the carbonyl compound **56** is seen on the polarographic curve.

In aqueous solution it is generally not possible to isolate the Schiff base from a reduction owing to its rapid hydrolysis, but if the reduction is performed in a non-aqueous solution as acetonitrile, it might be possible to obtain the azomethine compound

$$\underset{(54)}{\text{RCH}\!-\!\text{NR}'\ (\text{O bridge})} \xrightarrow{2e\,+\,2H^+} \underset{\text{OH}}{\text{RCHNHR}'} \xrightarrow{-H_2O} \underset{(55)}{\text{RCH}\!=\!\text{NR}'}$$

$$\text{RCH(OH)NHR}' \xrightarrow{-H_2NR'} \underset{(56)}{\text{RCHO}} \xleftarrow[H_2O]{-H_2NR'} \text{RCH}\!=\!\text{NR}'$$

D. By Anodic Oxidation

I. Amines

An anodic oxidation of an amine to a quinoneimine is found in, for example, the oxidation of *o*-tolidine (**57**) in acid solution. The quinonediimine **58** is rather stable in acid solution, but loses ammonia at higher pH[76]. A single two-electron oxidation is found at $pH < 3$, but at $pH > 3$ and in a non-aqueous solution such as acetonitrile two one-electron waves are found[77]. Potentiometric data[78] suggest a free-radical semiquinone; an e.s.r. spectrum was obtained in acetonitrile by *in situ* electrolysis in the e.s.r. cavity[79], but as the spectrum lacked hyperfine structure, no definite structural identification of the radical was possible. In aqueous solution of pH 4 no e.s.r. spectrum could be obtained by using *in situ* electrolysis. By employing an optical transparent anode of 'doped' tin oxide[80], it was possible to obtain the spectrum of the one-electron oxidation product[77] from 300–800 mμ, and combined with other spectral data it led to the conclusion that the intermediate was present predominantly as a dimeric species (**59**). At pH 4 the oxidation of *o*-tolidine can thus be represented by:

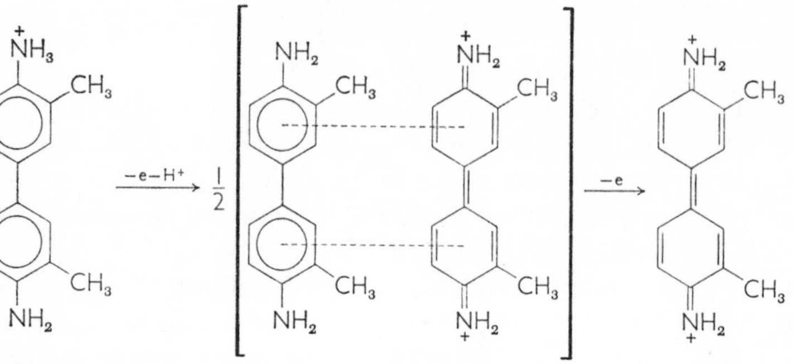

$$(57) \xrightarrow{-e-H^+} \frac{1}{2}[(59)] \xrightarrow{-e} (58)$$

(57) **(59)** **(58)**

2. Hydrazines

Anodic oxidations of aliphatic and aromatic *N,N'*-disubstituted hydrazines produce in the first step azo compounds; monoacylated hydrazines probably also form azo compounds initially, but the tautomeric equilibrium, azo derivative ⇌ hydrazone, is generally shifted towards the hydrazone. 1-Benzyl-2-benzoylhydrazine (**60**) thus yields benzaldehyde benzoylhydrazone (**61**) on anodic oxidation in alkaline solution[81].

$$\underset{(60)}{C_6H_5CH_2NHNHCOC_6H_5} \xrightarrow{-2e\ -\ 2H^+} C_6H_5CH_2N{=}NCOC_6H_5 \longrightarrow \underset{(61)}{C_6H_5CH{=}NNHCOC_6H_5}$$

Similarly 1-benzylsemicarbazide produces benzaldehyde semicarbazone[82] on anodic oxidation in alkaline aqueous alcoholic solution, and 4-methyl-3,4-dihydro-1(2H)-phthalazinone gives the phthalazinone[81].

3. Hydroxylamines

Phenylhydroxylamine produces on anodic oxidation in alkaline solution nitrosobenzene, which reacts with phenylhydroxylamine to give azoxybenzene. In the oxidation of aliphatic hydroxylamines the nitroso derivative may either react with excess of the hydroxylamine to give an azoxy compound or tautomerize to the oxime[83,84]. The relative amounts of the products will depend on the hydroxylamine and the experimental conditions. The anodic oxidation of benzylhydroxylamine can be formulated as:

$$C_6H_5CH_2NHOH \xrightarrow{-2e\,-\,2H^+} [C_6H_5CH_2NO] \xrightarrow{C_6H_5CH_2NHOH} C_6H_5CH_2\overset{O}{\overset{\uparrow}{N}}{=}NCH_2C_6H_5$$

$$[C_6H_5CH_2NO] \longrightarrow \underset{\textbf{(16)}}{C_6H_5CH{=}NOH}$$

in which **16** is the main product.

4. 2,4,6-Tri-t-butylaniline

An unusual reaction takes place when 2,4,6-tri-*t*-butylaniline (**62**) is oxidized anodically at a platinum electrode in acetonitrile containing pyridine[85]. This is a medium which previously has been found useful for the anodic oxidation of many types of organic compounds[86,87]. In the reaction two electrons are lost from the substituted aniline; pyridine makes a nucleophilic attack on the benzene ring *ortho* to the amino group, which results in the loss of a *t*-butyl cation; this in turn attacks acetonitrile in a Ritter-type reaction, and the cation thus formed reacts with another molecule of the substituted aniline to form the acetamidine shown (**63**) (or the other tautomeric form).

2 (**62**) [2,4,6-tri-t-butylaniline, NH_2] $\xrightarrow[C_5H_5N,\,CH_3CN]{-4e\,-\,4H^+}$ [pyrido-benzimidazole product] (60%) + [2,4,6-tri-t-butylphenyl]–$N{=}C(CH_3){-}NHC_4H_9$ (25%) (**63**)

IV. REDUCTION OF AZOMETHINE COMPOUNDS

Most compounds containing a carbon–nitrogen double bond are reducible electrolytically under suitable conditions; the azomethine group is generally more easily reducible than the corresponding carbonyl group. In several cases only the protonated azomethine compound is reducible in a convenient potential range.

When interpreting polarographic curves it must be kept in mind that the species reduced at the surface of the electrode is not necessarily the same as that present in the bulk of the solution. There might be differences with respect to, say, protonation and tautomeric forms.

At a potential where only the protonated compound is reducible, this form will be removed at the electrode from the equilibrium; protonation will then occur in order to re-establish the equilibrium, but the protonated form will be reduced as fast as it is formed at the electrode. The result is that the protonated form may be the reduced species even if it is present in a relatively small concentration in the bulk of the solution; the height of the polarographic wave due to the reduction of the protonated species will then in a certain pH interval be determined partly by the rate of the protonation; such polarographic waves are called 'kinetic waves'[9,88,89]. This type of consideration can be applied to other types of mobile equilibria where the reducibility of the compounds differs.

The reduction of azomethine compounds in acid solution differs depending on whether or not an electronegative atom is bonded to the nitrogen atom[44]. Thus, in a compound RR'C=NYR″ a two-electron reaction resulting in a saturation of the carbon–nitrogen double bond is generally found when Y is carbon; if, on the other hand, Y is oxygen or nitrogen, the protonated compound most probably will be reduced in a four-electron reaction and the following reduction scheme has been proposed[44]:

$$\mathrm{RR'C{=}NYR''} \xrightleftharpoons{\mathrm{H^+}} [\mathrm{RR'C{=}NYR''}]\mathrm{H^+} \xrightarrow{2e\,+\,2\mathrm{H^+}}$$

$$\mathrm{RR'C{=}\overset{+}{N}H_2} + \mathrm{HYR''} \xrightarrow{2e\,+\,2\mathrm{H^+}} \mathrm{RR'\overset{+}{C}HNH_3}$$

In Table 1 are compiled the half-wave potentials of some azomethine derivatives of benzaldehyde and of some related compounds. The medium was at pH 1; this was chosen to ensure that all the compounds were reduced in their protonated form.

From Table 1 it may be seen that all the azomethine derivatives of benzaldehyde are easier to reduce than the parent aldehyde itself. The half-wave potential of benzaldimine cannot be measured directly in 0·1N HCl, as the compound is rapidly hydrolysed in this medium, but if the reduction route suggested for benzaldehyde thiobenzoylhydrazone (Section IV.B.6) is correct, the half-wave potential of the second wave of this compound [−0·72 v (s.c.e.)] is that of benzaldimine; also the second wave of benzaldehyde benzylhydrazone would be caused by the reduction of this intermediate. Table 1 shows that

TABLE 1. Half-wave potentials (s.c.e.) at pH 1 of benzaldehyde and some of its azomethine derivatives. (Included also are a few azomethine derivatives of benzoic acid. The medium contains 40% alcohol.)

Compound		$-E_{\frac{1}{2}}$ v (s.c.e.)
Benzaldehyde		0·90
Benzaldehyde thiobenzoylhydrazone	1st wave	0·61
	2nd wave	0·72
Benzaldehyde hydrazone		0·72
Benzalazine		0·72
Benzaldehyde benzylhydrazone	1st wave	0·56
	2nd wave	0·73
Benzaldehyde phenylhydrazone		0·77
Benzaldehyde *N*-methylphenylhydrazone		0·75
Benzaldehyde semicarbazone		0·76
Benzaldehyde trimethylhydrazonium iodide		0·87[a]
Benzaldehyde benzenesulphonylhydrazone		0·72
Benzaldoxime		0·74
Benzaldoxime acetate		0·70
N-*t*-Butylaldoxime	1st wave	0·66
	2nd wave	0·78
N-*t*-Butylbenzaldimine		0·76
2-*t*-Butyl-3-phenyloxaziridine	2nd wave	0·76
Benzimino methyl ether		1·12
Isothiobenzanilide-*S*-methyl ether	1st wave	0·69

[a] Measured in water.

none of the azomethine derivatives of benzaldehyde, which are reduced in a single four-electron wave according to the general scheme presented above, are more easily reducible than benzaldimine, so from the point of view of reduction potentials there is nothing to exclude the possibility that the observed four-electron reaction follows the suggested route. The reaction scheme also implies that an azomethine compound which is more easily reducible than the corresponding imine gives two or more waves.

In alkaline solution a saturation of the carbon–nitrogen bond appears to be the most common reaction, but four-electron reactions are also found. In the middle of the pH range the reduction may occur partly as a four-electron and partly as a two-electron reaction, depending on many factors such as concentration and kind of buffer, medium, temperatures, concentration, and kind of azomethine compound.

In the following, the compounds will be treated in three main groups depending on whether the azomethine compound is derived from ammonia, hydrazine or hydroxylamine.

A. Derivatives of Ammonia

In this section the electrochemistry of compounds having the group RR′C=NY will be discussed, where the nitrogen is bonded to a carbon or hydrogen atom; R and R′ may stand for several types of groups, and thus Schiff bases, imino ethers, iminothio ethers, amidines and isothiocyanates are treated here.

I. Schiff bases

Imines of aldehydes and ketones are polarographically reducible in a wide pH range[44,90–93], but as they are usually easily hydrolysed both in acid and alkaline solution, they are difficult to investigate. Both classical[94,95] and controlled potential[44] reductions are reported to yield the expected secondary amine; many of the classical reductions were performed in 50% sulphuric acid at 0°c at a lead cathode.

The reduction potentials of ketimines are generally less negative than that of the parent ketone, and this difference can be exploited in the electrochemical preparation of amines from carbonyl compounds[44]. In a solution containing a mixture of carbonyl compound and amine (or ammonia) in equilibrium with the imine, the latter may be present in a rather low concentration. If, however, the electrolysis is performed at a potential where the imine but not the carbonyl compound is reducible, then the imine is removed from the equilibrium by reduction in the form of a secondary amine nearly as fast as it is formed from the carbonyl compound and the primary amine. Thus, *N*-methylcyclohexylamine has been prepared in a high yield by controlled potential reduction of a solution of cyclohexanone in aqueous methylamine[44].

$$RR'C{=}O + CH_3NH_2 \rightleftharpoons RR'C{=}NCH_3 + H_2O \xrightarrow{2e\,+\,2H^+} RR'CHNHCH_3$$

In some cases, especially where hydration of the carbonyl group competes successfully with the formation of the Schiff base, it is advantageous to work in a non-aqueous medium. Thus, thiazole-2-carbaldehyde in an aqueous solution containing aniline gives the Schiff base only to a small degree, but in acetonitrile containing acetic acid as a proton donor and sodium perchlorate as supporting electrolyte the addition of aniline shifts the polarographic wave of thiazole-2-carbaldehyde to a more positive value and no wave of the free aldehyde is visible. A preparative reduction in this medium produced the expected 2-anilinomethyl thiazole in good yield[96].

Sometimes the amine formed by reduction of a Schiff base may condense with another group in the molecule with the formation of a hetero ring. The reduction of phenolphthalein oxime[97] (**64**) and similar anils[98] may exemplify this; here the very stable Schiff base is reduced both in acid or alkaline solution to the amine which condenses with the carbonyl group to give the phthalimidine (**65**).

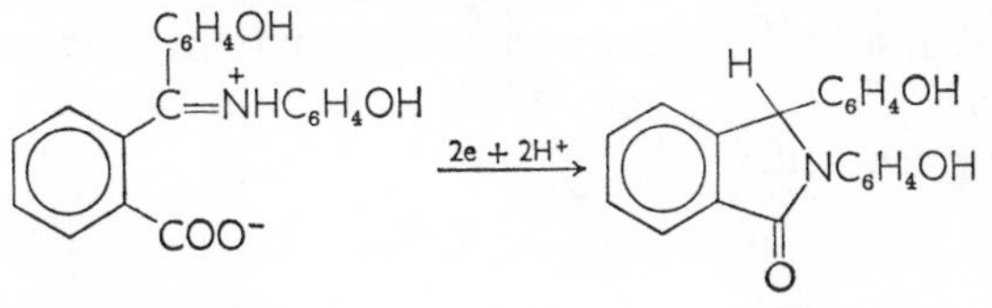

$$\text{C}_6\text{H}_4(\text{COO}^-)\text{—C(C}_6\text{H}_4\text{OH)=}\overset{+}{\text{N}}\text{HC}_6\text{H}_4\text{OH} \xrightarrow{2e\,+\,2H^+} \text{phthalimidine: CH(C}_6\text{H}_4\text{OH)—NC}_6\text{H}_4\text{OH—C=O}$$

(64) **(65)**

Some Schiff bases of aromatic aldehydes and ketones give two one-electron polarographic waves in suitable media, mostly in acid solutions[44,90]. The product of the first one-electron reduction is a radical which may either be reduced further to the amine or may dimerize. In a classical type of reduction benzalaniline (**66**) was found to dimerize to a small degree at a lead cathode in sulphuric acid[95].

$$2\,C_6H_5CH{=}NC_6H_5 \longrightarrow \begin{array}{c} C_6H_5CHNHC_6H_5 \\ | \\ C_6H_5CHNHC_6H_5 \end{array} (+C_6H_5CH_2NHC_6H_5)$$

(66)

In non-aqueous solution a high yield of the dimer can be obtained. Many Schiff bases give two polarographic waves in dimethylformamide and acetonitrile[99,100]; the primarily formed species have in some cases been shown by e.s.r. spectroscopy[99] to be radicals. This is also made plausible by the red colour produced on reduction of benzophenone anil in acetonitrile containing sodium perchlorate and some acetic acid as proton donor[101]; the colour faded on interruption of the current, but reappeared on continued electrolysis.

When **66** is reduced in the presence of excess of acrylonitrile, a mixture is formed in which 1,5-diphenyl-2-pyrrolidone (**67**) is also found. The following reaction sequence has been suggested[102]:

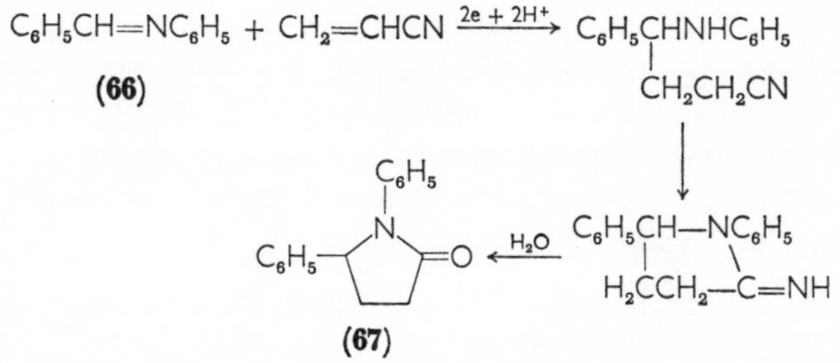

$$C_6H_5CH{=}NC_6H_5 + CH_2{=}CHCN \xrightarrow{2e\,+\,2H^+} \begin{array}{c} C_6H_5CHNHC_6H_5 \\ | \\ CH_2CH_2CN \end{array}$$

(66)

$$\downarrow$$

$$\text{1,5-diphenyl-2-pyrrolidone} \xleftarrow{H_2O} \begin{array}{c} C_6H_5CH{-}NC_6H_5 \\ | \qquad \backslash \\ H_2CCH_2{-}C{=}NH \end{array}$$

(67)

2. Imidic esters

These compounds are also apt to be hydrolysed in aqueous solution, but in a cold solution some are stable enough for a polarographic and preparative reduction. Classical reduction at a lead cathode in cold 2N sulphuric acid of benzimidic acid ethyl ester (**68**) gave benzylamine in 76% yield[103]. The reduction can probably be described by the following sequence.

$$\underset{(\mathbf{68})}{RC(OR'){=}NH} \xrightarrow[2e\,+\,2H^+]{H^+} [RCH(OR')\overset{+}{N}H_3] \xrightarrow{-HOR'} RCH{=}\overset{+}{N}H_2 \xrightarrow{2e\,+\,2H^+} RCH_2\overset{+}{N}H_3$$

In most cases it will probably be advantageous to use non-aqueous media such as acetonitrile, in which the hydrolysis of the imidic ester can be avoided.

3. Imidothio esters

The reduction of such compounds has only recently received attention, and only a few have been investigated polarographically[104]. *S*-Methylisothiobenzanilide (**69**) is polarographically reducible in acid solution in two steps; the product of the first two-electron reduction can be hydrolysed to benzaldehyde and could thus be either $C_6H_5CH{=}NC_6H_5$ or $C_6H_5CH(SCH_3)NHC_6H_5$. A small second wave is found at about the same potential as that of benzalaniline, so the result of the first two-electron reduction is probably a saturation of the carbon–nitrogen double bond; the possible reactions are shown below:

$$\underset{(\mathbf{69})}{C_6H_5C(SCH_3){=}NC_6H_5} \xrightarrow{2e\,+\,2H^+} C_6H_5CH(SCH_3)NHC_6H_5$$

$$C_6H_5C(SCH_3){=}NC_6H_5 \xrightarrow{2e\,+\,2H^+} CH_3SH + \underset{(\mathbf{66})}{C_6H_5CH{=}NC_6H_5}$$

$$C_6H_5CH(SCH_3)NHC_6H_5 \longrightarrow CH_3SH + C_6H_5CH{=}NC_6H_5$$

$$C_6H_5CH{=}NC_6H_5 \xrightarrow{H_2O} C_6H_5CHO + H_2NC_6H_5$$

$$C_6H_5CH(SCH_3)NHC_6H_5 \xrightarrow{H_2O} C_6H_5CHO + H_2NC_6H_5 + HSCH_3$$

4. Amidines

Amidines of aliphatic acids (**67**) are generally not reducible in buffered solutions. Those of aromatic acids (**68**) are reducible only at very negative potentials and in a rather narrow pH interval, i.e. in slightly acid to alkaline solution. The reduction is similar to that of imino ethers and yields the amine[105].

$$\underset{(\mathbf{68})}{C_6H_5C(NH_2){=}\overset{+}{N}H_2} \xrightarrow{4e\,+\,5H^+} C_6H_5CH_2\overset{+}{N}H_3 + \overset{+}{N}H_4$$

2-Phenoxyacetamidines (**69**) are polarographically reducible[105], but give only a cleavage of the carbon–oxygen bond and no reduction of the C=N bond.

$$C_6H_5OCH_2C(NH_2){=}\overset{+}{N}H_2 \xrightarrow{2e} C_6H_5O^- + CH_3C(NH_2){=}NH$$

(**69**) (**67**)

5. Cyclic amidines

The reduction of some cyclic amidines, such as dihydropyrimidines, has already been discussed (Section III.C.4); in addition the elucidation of the reduction path of 7-chloro-2-methylamino-5-phenyl-3H-1,4-benzodiazepine-4-oxide (**70**) has been the purpose of many investigations[106–110]. This compound can be regarded as a cyclic nitrone and a cyclic amidine; and by controlled potential reduction the following three steps were proved[108–110].

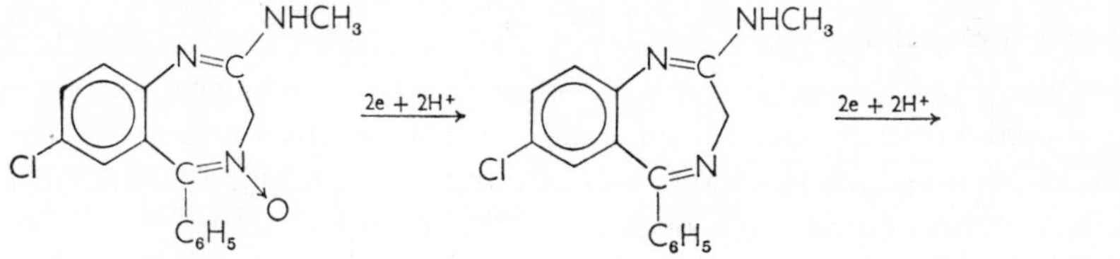

(**70**) $\xrightarrow{2e + 2H^+}$ $\xrightarrow{2e + 2H^+}$ (**71**) $\xrightarrow{2e + 3H^+}$ (**72**) + $CH_3\overset{+}{N}H_3$

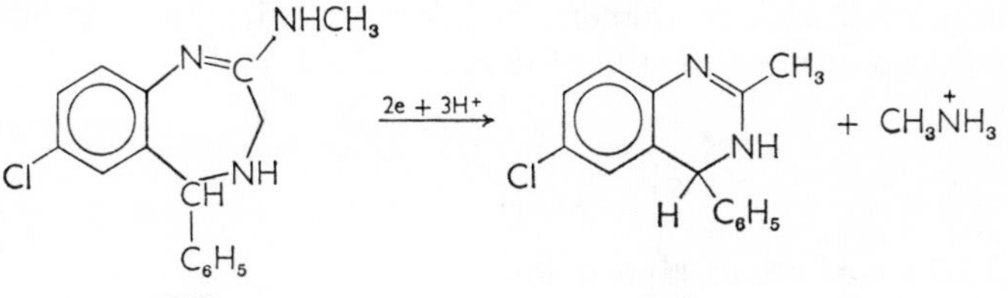

The first two steps are straightforward, but the reductive ring contraction of the 7-chloro-2-methylamino-5-phenyl-3H-4,5-dihydro-1,4-benzodiazepine (**71**) to 2-methyl-4-phenyl-6-chloro-3,4-dihydroquinazoline (**72**) is unusual and the following mechanism has been presented for discussion[110]:

71 $\xrightarrow{2e + 2H^+}$ $\xrightarrow{-CH_3NH_2}$ [ ] $\longrightarrow$ **72**

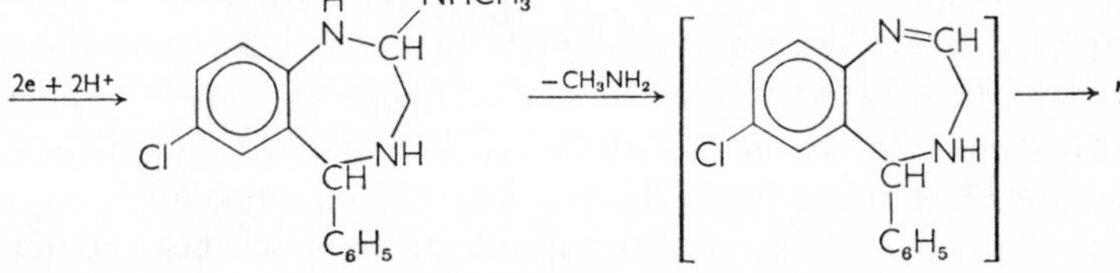

Another reaction path would be in analogy to the reduction of 2-phenoxyacetamidine (**69**) discussed above (Section IV.A.4) and with the first step in the reduction[111] of 2-aminoacetophenones. After the initial cleavage of the $C_{(3)}$—$N_{(4)}$— bond the primary amino group formed would attack the C=N bond of the substituted acetamidine (**73**) with ring closure and loss of methylamine. Future investigations may eventually show which reaction route is the more likely.

$$\mathbf{71} \xrightarrow{2e\,+\,2H^+} \text{4-Cl-C}_6\text{H}_3\text{(N=C(NHCH}_3\text{)CH}_3\text{)(CH(NH}_2\text{)C}_6\text{H}_5\text{)} \xrightarrow{-H_2NCH_3} \mathbf{72}$$

(73)

6. Isothiocyanates

Phenylisothiocyanate (**74**) and other aromatic isothiocyanates have been investigated polarographically; in acid solution a four-electron wave or two poorly separated two-electron waves are found, whereas a single two-electron wave occurs in alkaline solution[49,112,113]. Controlled potential reduction in alkaline solution showed that the reduction product was thioformanilide[18,49] (**75**); **75** is in acid solution reduced at the potential of the second wave of **74**.

$$\underset{(\mathbf{74})}{C_6H_5N{=}C{=}S} \xrightarrow{2e\,+\,2H_2O} \underset{(\mathbf{75})}{C_6H_5NHCH{=}S} + 2\,\bar{O}H$$

B. Derivatives of Hydrazine

The electrochemistry of hydrazones, azines, and other derivatives of hydrazine is complicated by the possibility of tautomerization. The following forms must be considered[114,115].

$$\underset{(\mathbf{76})}{RCH_2CH_2N{=}NR'} \rightleftharpoons \underset{(\mathbf{77})}{RCH_2CH{=}NNHR''} \rightleftharpoons \underset{(\mathbf{78})}{RCH{=}CHNHNHR'}$$

The hydrazone (**77**) is the predominant form, and in acid and neutral solution no indication of either the azo- (**76**) or the ene-hydrazine form (**78**) has been found. In alkaline solution, however, studies with tritium[116] have shown exchange of both the aldehydic hydrogen and of hydrogen atoms at the α-carbon, indicating the presence of both the forms **76** and **78**. The equilibrium constant in 0·1M ethanolic KOH for the equilibrium ω-benzeneazotoluene $\rightleftharpoons$

benzaldehyde phenylhydrazone is about 10^8, but for monoalkylhydrazones of aliphatic carbonyl compounds a higher proportion of the azo form is found. Acetaldehyde propylhydrazone in *t*-butanol containing 0·02M potassium *t*-butylate at 100° is in equilibrium with 3·7% of the azo form[117].

The different tautomeric forms may be reduced at different potentials, and the removal of one of them will lead to its continuous formation. The rate of the formation of the reducible tautomer is thus of importance for the electrode reaction.

I. Azines

Not many acyclic azines have been investigated electrochemically, which might be so partly because of their easy hydrolysis in acid solution. Benzalazine (**79**) and some other aromatic azines have been polarographed[44,118,119], but only benzalazine has been reduced by controlled potential[44]. In an aqueous acetate buffer containing 30% alcohol, benzalazine yielded on reduction at a mercury cathode mainly benzylamine (**80**) in a six-electron reduction. As both benzaldimine (**81**) and benzaldehyde benzylhydrazone (**82**) in acid and neutral solution are reduced at the same or at less negative potentials than **79** (Table 1), the reduction might either start with a hydrogenation of the nitrogen–nitrogen bond with the formation of two molecules of **81** or with a two-electron reduction to **82** followed by a four-electron reduction of this compound as illustrated below[44].

$$C_6H_5CH{=}\overset{+}{N}HN{=}CHC_6H_5 \xrightarrow{2e\,+\,3H^+} 2\,C_6H_5CH{=}\overset{+}{N}H_2$$

(**79**) (**81**)

$$\downarrow 2e + 2H^+ \qquad\qquad \downarrow 4e + 4H^+$$

$$C_6H_5CH{=}N\overset{+}{N}H_2CH_2C_6H_5 \xrightarrow{4e\,+\,5H^+} 2\,C_6H_5CH_2\overset{+}{N}H_3$$

(**82**) (**80**)

This reduction scheme has been criticized[119], as a microcoulometric determination of the number of electrons in the electrode reaction at pH 4 was found to be close to four for **79** and *p,p*'-dimethoxybenzalazine, and it was suggested that the reduction produced *N,N*'-dibenzylhydrazine (**83**).

The reason for this discrepancy is not clear. A reinvestigation of the reduction of **79** in an acetate buffer at 0° confirmed[82] the previously published results, i.e. **80** was the main reduction product; no anodic waves were detectable in the reduced solution after it was made alkaline, which showed that no **83** was present and that no detectable amounts of hydrazine was formed by hydrolysis of the azine.

In alkaline solution the reduction of **79** at a mercury cathode yields **82** in a two-electron reduction. Like most other hydrazones this compound is not further reducible in alkaline solutions containing metal cations, but might possibly be reduced to **83** at the very negative potentials obtainable in solutions only containing tetralkylammonium cations as supporting electrolyte.

2. Cyclic azines

One of the tautomeric forms of dihydropyridazine is a cyclic azine, and, for example, 4,5-dihydro-3,6-diphenylpyridazine (**24**) is reduced electrolytically in alkaline solution to 2,3,4,5-tetrahydropyridazine (**21**) quite analogously to the reduction of benzalazine. In acid solution the reduction of dihydropyridazines is more complicated; the scheme below may exemplify this[57].

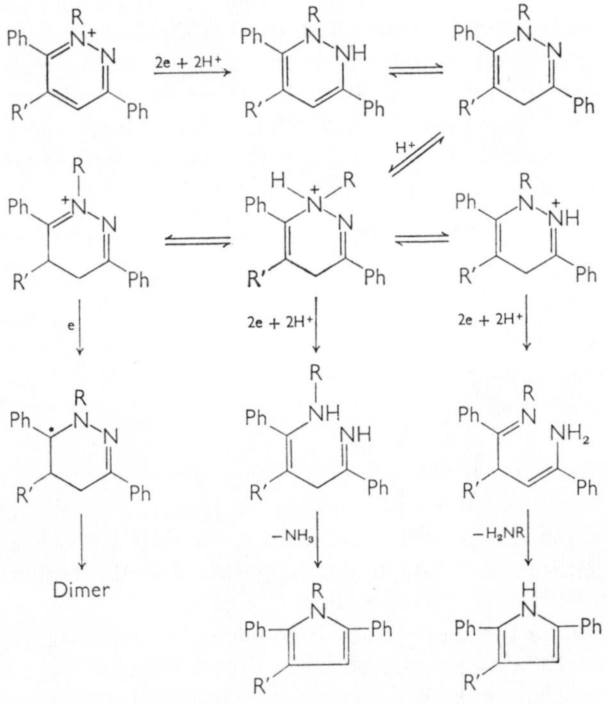

R = H or CH_3; R′ = H or *t*-butyl

When R = CH_3 and R′ = *t*-butyl, the reduction produces a mixture of **84** and **85**, whereas when R = R′ = H the reaction mixture consists mainly of dimeric products. Probably the bulky *t*-butyl groups force one of the phenyl groups out of the plane of the pyridazine ring, and thereby influence the possibility of conjugation and adsorption and thus the relative concentration at the electrode surface of the tautomeric dihydropyridazines.

The 1,4-dihydropyridazines are formally cyclic hydrazones and are as such reduced by an initial hydrogenation of the nitrogen–nitrogen bond; the amino group then attacks the imino group with ring closure to form the isolated pyrroles.

The 4,5-dihydropyridazines are cyclic azines; the formation of a dimeric product by saturation of one of the carbon–nitrogen double bonds may be taken as a slight indication that **79** is reduced through **82** to **80**; care must, however, be taken when comparing a cyclic and an acyclic compound.

3. Hydrazones

The typical polarographic behaviour of a phenylhydrazone is that it is reducible in acid but not in alkaline solution[44]; the main electrode reaction both at a lead electrode in 50% sulphuric acid[120] and at a mercury cathode in dilute hydrochloric acid[44] has been found to be a four-electron reduction to aniline and another amine. The following path has been suggested[44]:

$$RR'C{=}NNHC_6H_5 \xrightleftharpoons{H^+} (RR'C{=}NNHC_6H_5)H^+ \xrightarrow{2e\ +\ 3H^+}$$

$$RR'C{=}\overset{+}{N}H_2 + C_6H_5\overset{+}{N}H_3 \xrightarrow{2e\ +\ 2H^+} RR'CH\overset{+}{N}H_3$$

One of the arguments in favour of this reaction sequence is that the hydrazine, $RR'CHNHNHC_6H_5$, is not reducible under the conditions used for the reduction of the phenylhydrazone.

The reduction of benzaldehyde phenylhydrazone at a lead cathode gave besides benzylamine a small amount (12%) of benzylaniline[120]. This product might be formed by an attack of aniline on the intermediate benzaldimine with the formation of benzalaniline, which then is reduced to benzylaniline.

Benzaldehyde benzylhydrazine is reduced quite analogously to the phenylhydrazone[44], and it is likely that similar reductions will be found for other hydrazones. The reduction of some quaternized hydrazones has been shown to occur both in acid and alkaline solution in a four-electron reduction[57]; this supports the view that it is the protonated hydrazone which is normally the reduced species.

$$C_6H_5CH{=}N\overset{+}{N}(CH_3)_3 \xrightarrow{4e\,+\,5H^+} C_6H_5CH_2\overset{+}{N}H_3 + H\overset{+}{N}(CH_3)_3$$

Cinnamaldehyde phenylhydrazone is also polarographically reducible in alkaline solution. No preparative reduction has been made but it seems likely that the reduction is a saturation of the carbon–carbon double bond.

The polarographic curves of nitrophenylhydrazones have been interpreted as follows, assuming that the nitro group is reduced before the hydrazone[121]:

$$O_2NC_6H_4NHN{=}CR_2 \xrightarrow{4e\,+\,5H^+} HOH_2\overset{+}{N}C_6H_4NHN{=}CR_2 \xrightarrow{2e\,+\,3H^+}$$

$$H_3\overset{+}{N}C_6H_4NHN{=}CR_2 \xrightarrow{2e\,+\,4H^+} H_3\overset{+}{N}C_6H_4\overset{+}{N}H_3 + H_2\overset{+}{N}{=}CR_2 \xrightarrow{2e\,+\,2H^+} H_3\overset{+}{N}CHR_2$$

Cyclic benzylhydrazones, e.g. 3,6-diphenyl-2,3,4,5-tetrahydropyridazine (**21**), are reduced analogously to the acyclic phenylhydrazones[39,57].

$$\textbf{(21)} \xrightarrow[4e\,+\,5H^+]{H^+} C_6H_5CH(\overset{+}{N}H_3)CH_2CH_2CH(\overset{+}{N}H_3)C_6H_5$$

(21)

Another type of hydrazone is 1,2-dihydrophthalazine (**36**); in strongly acid solution it is reduced like other hydrazones to the amine, *o*-xylene-α-α′diamine, in a four-electron reduction, in which the 1,2,3,4-tetrahydrophthalazine (**37**) is not the intermediate as it is not reducible at a mercury electrode under these conditions. In alkaline solution **36** differs from most other non-acylated hydrazones in that it is reducible in a two-electron reduction to **37**[39,56].

4. Acylated hydrazones

In strongly acid solution acylated hydrazones are reduced in a four-electron reduction in a similar way to phenylhydrazones. In alkaline solution the acylated hydrazones are often reducible in a two-electron reaction, and from reduction of *N*-benzoyl-*N*-benzalhydrazine *N*-benzoyl-*N′*-benzylhydrazine was isolated, possibly formed by reduction of one of the other tautomeric forms of the hydrazone.

Several polarographic investigations of Girard hydrazones, semi- and thiosemicarbazones have been made[44,122–129], but relatively few controlled potential reductions have been reported[44,127,128]. From

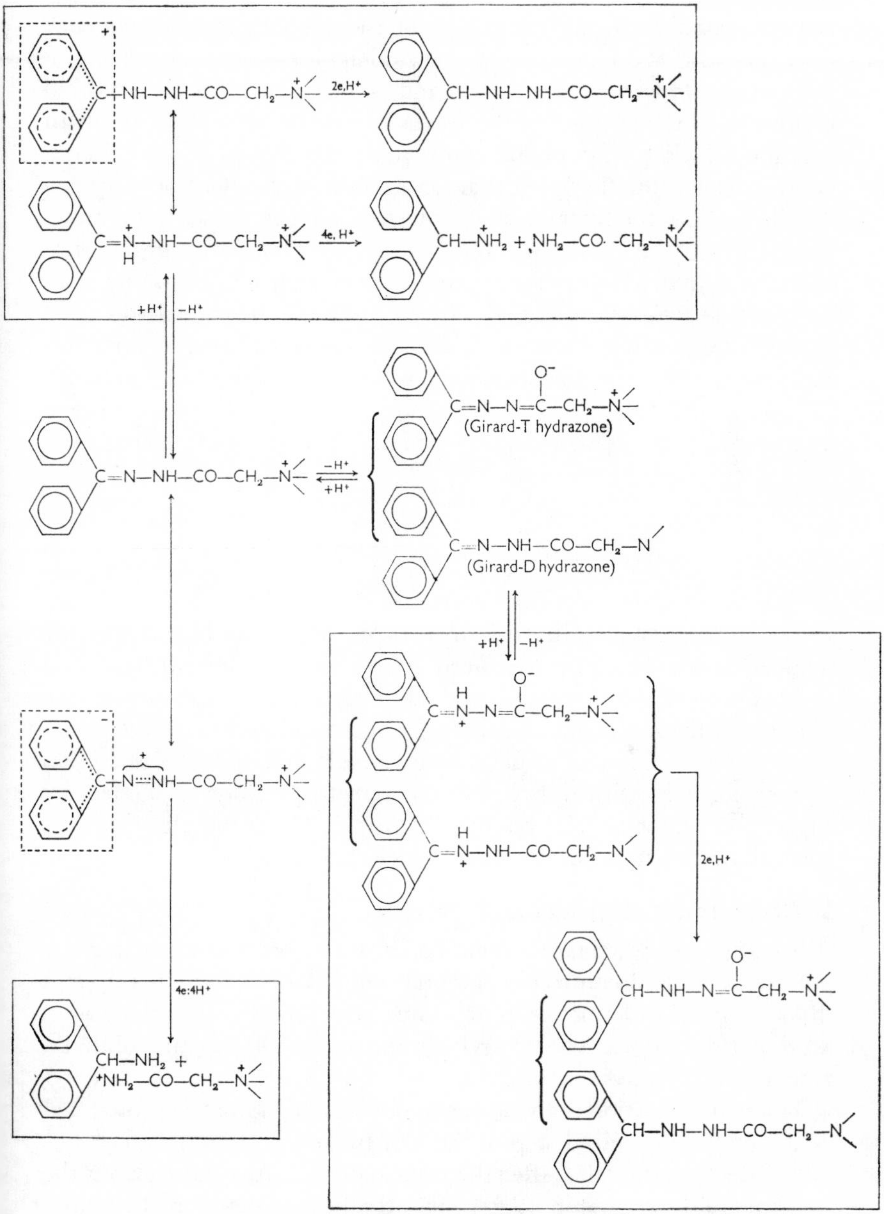

C–NH—NH—CO—CH₂—N⁺ → (2e, H⁺) CH—NH—NH—CO—CH₂—N⁺
C=N⁺H—NH—CO—CH₂—N⁺ → (4e, H⁺) CH—NH₂ + NH₂—CO—CH₂—N⁺
+H⁺ −H⁺
C=N—N=C(O⁻)—CH₂—N⁺
(Girard-T hydrazone)
C=N—NH—CO—CH₂—N⁺ ⇌ (−H⁺ / +H⁺)
C=N—NH—CO—CH₂—N
(Girard-D hydrazone)
+H⁺ −H⁺
C=N⁺H—N=C(O⁻)—CH₂—N⁺
C=N⁺H—NH—CO—CH₂—N
C–N⁺=NH—CO—CH₂—N⁺
2e, H⁺
CH—NH—N=C(O⁻)—CH₂—N⁺
CH—NH—NH—CO—CH₂—N
4e: 4H⁺
CH—NH₂ +
⁺NH₂—CO—CH₂—N⁺

these it appears that in acid solution a four-electron reduction takes place and that the first step is a hydrogenolysis of the nitrogen–nitrogen bond. In strongly acid solution there is often found a tendency to a stepwise reduction; thus, benzaldehyde thiosemicarbazone[127] gives at pH 0 two waves at −0·66 and −0·73 v (s.c.e.); when alcohol is present, the separation of the waves is poor or not visible at all. In alkaline solution some of the compounds are reducible, and then a two-electron reduction results in saturation of the carbon–nitrogen double bond to the hydrazine. At intermediate pH values both reduction paths may be followed simultaneously, and the relative importance of the two routes depends on many factors such as pH, buffer concentration, temperature and solvent. Thus, benzaldehyde and benzophenone semicarbazone[44], *p*-acetamidobenzaldehyde thiosemicarbazone (**86**)[127], *p*-acetamidobenzaldehyde *S*-methyl thiosemicarbazone (**87**)[127], and cyclopentanone and benzophenone Girard hydrazone[128] all yielded the expected amine at low pH; at high pH

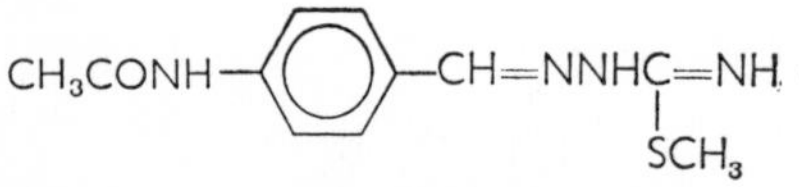

$CH_3CONH-C_6H_4-CH{=}NNHC({=}NH)SCH_3$

(**87**)

$CH_3CONH-C_6H_4-CH{=}NNHCSNH_2$

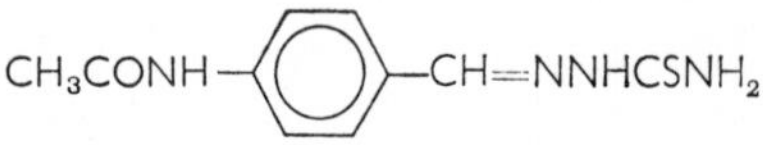

(**86**)

values benzophenone Girard-D hydrazone (**88**)[128] and benzaldehyde semicarbazone (**89**)[82] were reduced in high yield to the hydrazine, and at pH 6·6 **88**[128], **89**[44] at pH 4, and **86** at pH 8·3 gave mixtures of the amine and the hydrazine. Similarly, a microcoulometric investigation of some semicarbazones at pH 6 and 7 gave $n \approx 2$[125,126]. A reduction scheme for the Girard hydrazones of benzophenone suggested by Masui and Ohmori[128] is presented on p. 545. The rectangle represents electrolytic reactions occurring at the electrode surface.

5. Cyclic acylated hydrazones

Several cyclic acylated hydrazones have been investigated by polarography and controlled potential reductions in some detail, and different types of reduction routes have been found. The reduction of alkyl pyridazinones (**90**)[57], aryl pyridazinones (**91**)[57] and phthalazinones (**92**)[81] may illustrate this.

The hydrogenation of the nitrogen–nitrogen bond of the protonated compound is the initial step of the first two examples, and they thus follow the general rule. After the uptake of the first two electrons the structure of the compound determines the further reaction. 3-Methylpyridazin-6-one (**90**) is a derivative of a methyl alkyl ketone, and an

$$\textbf{(90)} \xrightarrow[2e\,+\,2H^+]{H^+} CH_3C(=\overset{+}{N}H_2)CH_2CH_2CONH_2 \xrightarrow{2H_2O} CH_3COCH_2CH_2COOH + 2NH_3$$

(90)

$$\textbf{(91)} \xrightarrow[4e\,+\,4H^+]{H^+} C_6H_5CH(\overset{+}{N}H_3)CH_2CH_2CONHR$$

(91)

$$\textbf{(92)} \underset{}{\overset{H^+,\,2e\,+\,2H^+}{\rightleftharpoons}} \textbf{(93)} \xrightarrow{2e\,+\,2H^+} \textbf{(94)} + \overset{+}{N}H_4$$

(92) (93) (94)

imine of this is not reducible under the conditions employed for the reduction; 3-phenylpyridazinone (**91**) is a derivative of a propiophenone, and the ketimine derived from this compound is reducible immediately.

The polarographic behaviour of 4-methyl-1(2H)-phthalazinone (**92**) and the 3,4-dihydro derivative (**93**) is depicted in Figures 6 and 7, respectively[81]. Figure 6 shows that **92** in alkaline solution is reduced in a pH-independent reaction, which means that the reducible species is the unprotonated molecule. The height of the wave suggests that it is a two-electron reduction. At pH 4 to 7 is found another wave with the same wave-height but with $E_{\frac{1}{2}}$ dependent on pH; this suggests that the protonated species is reduced in a two-electron reduction under these conditions. Between pH 7 and 9 there is a change in reaction mechanism from a reduction of the protonated to a reduction of the unprotonated form. Below pH 3 the primarily formed reduction product is further reducible in a two-electron reduction at a potential slightly more negative than that of the first reduction.

If the partly reduced compound **93** is the desired product one could either carry out the reduction in alkaline solution at −1·85 v (s.c.e.) or one could choose an acetate buffer (pH 5) and a cathode potential of −1·30 v (s.c.e.). A reduction at low pH would be more difficult, but probably still possible to stop at the dihydro stage, and in such a case it would be advisable to use a potential not more negative than the half-wave potential of the first wave in that medium. Figure 6 also shows

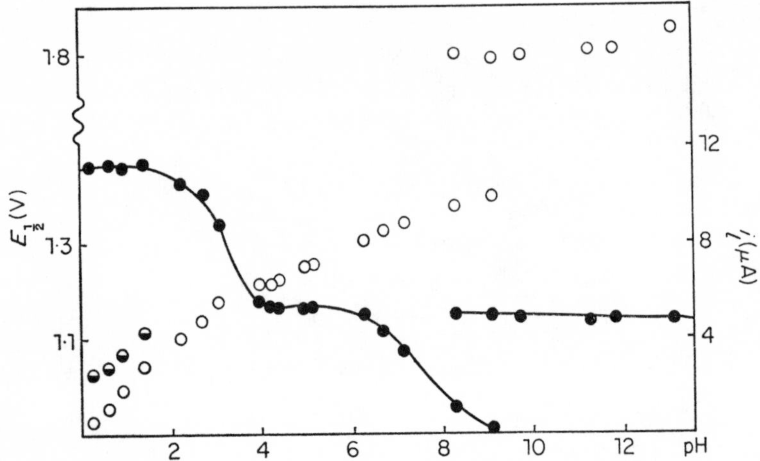

FIGURE 6. Dependence on pH of the Limiting Current (μA) ● and the half-wave potentials (s.c.e.) ◒, ○ of 4-methyl-1(2H)-phthalazinone (**92**). (Concentration 2·5 × 10^{-4}M. From reference 81.)

that a reduction to a phthalimidine (**94**) can only be performed at low pH.

From Figure 7 can be seen that **93** is only reducible at pH < 3 and that it can be oxidized anodically at pH > 6; the wave heights show

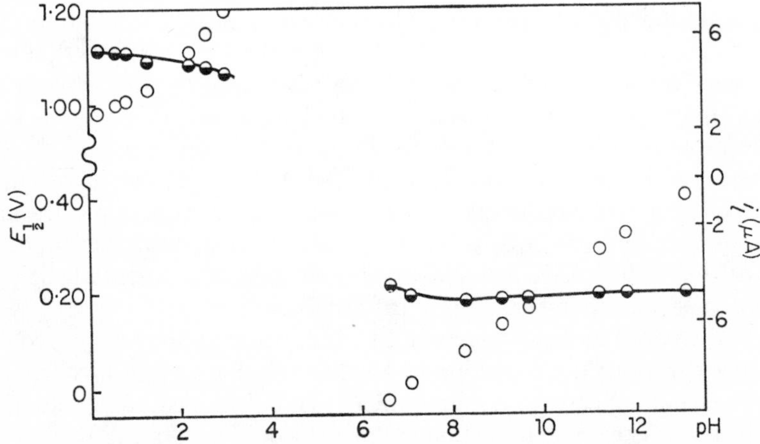

FIGURE 7. Dependence on pH of the limiting current (μA) ● and the half-wave potentials (s.c.e.) ○ of 3,4-dihydro-4-methyl-1(2H)-phthalazinone (**93**). (Concentration 2·5 × 10^{-4}M. From reference 81.)

that both reactions are two-electron reactions. The half-wave potentials of **93** are close to those of the second wave of **92**; the polarographic data thus suggest that **92** is reduced to **93** in the first step, and this has been confirmed by preparative reductions at controlled potential.

1(2H)-Phthalazinones[81] are so far the only azomethine compounds which have been shown by controlled potential reduction to be reduced differently from that predicted by the general rule[44]. The only other compound which could be considered in this connecton is 3-hydroxycinnoline[60], which is reduced as shown, but as the position of the tautomeric equilibrium is uncertain and its inclusion as an azomethine compound is dubious it will be disregarded here.

$$\text{3-hydroxycinnoline (OH)} \rightleftharpoons \text{3(2H)-cinnolinone (NH, =O)} \xrightleftharpoons{2e + 2H^+} \text{1,4-dihydro-3(2H)-cinnolinone (H-N–NH, =O)}$$

A satisfactory explanation for the fact that the reduction of the protonated phthalazinone goes differently from other azomethine compounds has not been found yet. Polarographic data show that it is the protonated form which is reduced in acid solution; in alkaline solution a pH-independent wave is found. It has been pointed out[57] that the difference in potential between that required for the reduction of the azomethine group and for the hydrogenolysis of the nitrogen–nitrogen bond is not great, and an apparently small structural change may alter the sequence of the reduction steps. Thus, the phenyl group fused to the pyridazinone ring acts as a non-reducible, non-displaceable, unsaturated centre in the hetero ring, which together with the steric requirements imposed on the azomethine compound by its ring structure may be reponsible for its departure from the main reduction route. The further reduction of dihydrophthalazinones is also unusual, as derivatives of hydrazines are generally not reducible; the isolated phthalimidines are formed by hydrogenation of the nitrogen–nitrogen bond followed by an attack of the amine on the amide group.

In alkaline solution phthalazinones can be reduced to 3,4-dihydrophthalazinones which can be reoxidized anodically to the phthalazinone. Aryl dihydrophthalazinones are generally not further reducible.

The reduction of substituted dihydrocinnolines[60] depends on whether the substituent is alkyl or aryl. The initial step is, however, in both cases a hydrogenolysis of the nitrogen–nitrogen bond. 4-Methyl-1,4-dihydrocinnoline (**95**) is in acid solution reduced to skatole (**96**) as follows:

$$\mathbf{(95)} \xrightarrow[2e + 2H^+]{H^+} \left[\mathbf{(97)}\ \text{(}o\text{-}\overset{+}{N}H_3\text{-}C_6H_4\text{-}CH(CH_3)CH{=}NH\text{)} \right] \xrightarrow{-\overset{+}{N}H_4} \mathbf{(96)}$$

(95) (97) (96)

The intermediate imine **97** is not isolated, but its reduction can be detected as a second wave on the polarographic curve of **95** in acid solution; the amino group attacks the aldimine with ring closure to give **96**. 4-Methylcinnoline yields on quaternization with methyl iodide a mixture consisting of approximately 8 parts of one isomer and one part of another one. Controlled potential reduction of the more abundant isomer yielded **96**, which proved that the methylation took place predominantly at $N_{(2)}$[60].

1,4-Dihydro-3-phenylcinnoline (**98**) is in acid solution reducible in a four-electron reduction to **99** according to

$$\mathbf{(98)} \xrightarrow[4e + 5H^+]{H^+} o\text{-}\overset{+}{N}H_3\text{-}C_6H_4\text{-}CH_2CH(\overset{+}{N}H_3)C_6H_5\ \mathbf{(99)}$$

(98) (99)

3-Phenylcinnoline (**100**) yields on quaternization with methyl iodide a mixture consisting of approximately 3 parts of one and 1 part of another isomer. Reduction of the dihydro compound of the more abundant isomer yielded a diamine which could not be diazotized thus proving that **100** quaternizes mainly at $N_{(1)}$.

6. Thiobenzoylhydrazones

Benzaldehyde thiobenzoylhydrazone (**101**) is reduced in acid solution, according to the general reduction scheme where the initial step is the hydrogenolysis of the nitrogen–nitrogen bond. At pH < 2 three polarographic waves are found[57], and on the basis of controlled potential reductions the following steps were suggested:

$$\text{1st wave}\quad C_6H_5CH{=}NNHCSC_6H_5 \xrightarrow[2e\,+\,2H^+]{H^+} C_6H_5CH{=}\overset{+}{N}H_2 + H_2NCSC_6H_5$$

$$\text{2nd wave}\quad C_6H_5CH{=}\overset{+}{N}H_2 \xrightarrow{2e\,+\,2H^+} C_6H_5CH_2\overset{+}{N}H_3$$

$$\text{3rd wave}\quad C_6H_5CSNH_2 \xrightarrow[2e\,+\,2H^+]{H^+} C_6H_5CH(SH)\overset{+}{N}H_3$$

followed[49] by hydrolysis and reduction of benzaldimine and benzaldehyde.

In alkaline solution a two-electron reduction is found which results primarily in a saturation of the carbon–nitrogen double bond; the reducible species is probably $C_6H_5CH{=}NN{=}C(S^-)C_6H_5$.

7. Diazoalkanes

Owing to their reactivity only few of these compounds have been investigated polarographically; diazoacetophenone (**102**) has been investigated by polarography and controlled potential reduction[130,131]. In neutral medium three waves are found, and the electrode reactions have been suggested to be

$$\underset{(\mathbf{102})}{C_6H_5COCH{=}N^+{=}N^-} \xrightarrow{6e\,+\,7H^+} C_6H_5COCH_2NH_2 + NH_4^+ \xrightarrow{2e\,+\,3H^+}$$

$$C_6H_5COCH_3 + NH_4^+ \xrightarrow{2e\,+\,2H^+} C_6H_5CHOHCH_3$$

C. Derivatives of Hydroxylamine

1. Oximes

Among the azomethine derivatives of hydroxylamine the oximes are the most important and many polarographic and electrolytic investigations of oximes have been made[44,132–134]. Most oximes are reducible in acid solution, much fewer in alkaline solution; polarographic data prove that the protonated oxime is the reducible species in acid and neutral solution. Classical reductions in sulphuric acid at lead cathodes or controlled potential reductions in acid solution at mercury cathodes yield the amine in a four-electron reduction. Since the hydroxylamine corresponding to the oxime is not reducible under the conditions employed, it has been suggested[44] that the reduction proceeds as follows:

$$RR'C{=}NOH \rightleftharpoons (RR'C{=}NOH)H^+ \xrightarrow{2e\,+\,2H^+}$$

$$(RR'C{=}\overset{+}{N}H_2) + H_2O \xrightarrow{2e\,+\,2H^+} RR'CH\overset{+}{N}H_3$$

As the site of the protonation on the species to which the transfer of electrons occurs is not known, protonation at both N and at O may be considered, although N is the most likely basic centre.

The postulated intermediate, the imine, can in most cases not be isolated or detected by classical polarography. The reduction of 2,4-dihydroxybenzophenone oxime occurs[135], however, in two steps, and reduction at the potential of the first wave produced the imine which in this case was sufficiently stable to permit its isolation.

If the reaction path shown above is so general as the available evidence suggests, attempts to reduce protonated oximes electrolytically to hydroxylamines are not likely to succeed. Some unprotonated oximes, e.g. benzaldoxime (**17**) and benzophenone oxime, are reducible in not too strongly alkaline solution; the oxime anion is not reducible. An investigation of the reduction of **17** in alkaline solution[82] showed that some benzylhydroxylamine (**103**) is formed under these conditions. In Table 2 the yield of **103** is given as a function of pH and temperature.

The mechanism of the hydroxylamine formation is not known, but protonation of the oxime anion could take place on carbon, nitrogen or oxygen forming

$$C_6H_5CH_2N{=}O, \quad C_6H_5CH{=}\overset{+}{N}HO^- \quad \text{or} \quad C_6H_5CH{=}NOH.$$

The reduction of these tautomeric forms may lead to different products. Another possibility is that the step after the uptake of the first electron or electron + proton may either be an uptake of a proton or another electron, which may lead to different products, and that the relative rates of these competing reactions are affected by pH and temperature.

TABLE 2. Yield of benzylhydroxylamine, determined by anodic polarography, in the reduction of *syn*-benzaldoxime at different pH and temperatures[82]. (The yield is corrected for a small amount of unreduced oxime.)

Buffer→	Borate	Borate	Phosphate	Phosphate	Phosphate
pH interval	9·2–9·5	10·1–10·3	12·25–12·45	12·25–12·45	12·6–12·8
E(vvs. s.c.e.)	–1·75	–1·75	–1·80	–1·80	–1·80
Temperature	25°	25°	25°	5°	25°
% benzyl-hydroxyl-amine	6	8	27	42	28

Polarographically there is a difference between the *syn*- and the *anti*-oxime in unbuffered solutions having tetraalkylammonium ions as supporting electrolyte[136–139]. *Syn*-benzaldoxime and other *syn*-oximes give two waves in this medium; the first wave at −1·84 v (s.c.e.) of *syn*-benzaldoxime is kinetically controlled, whereas the second one at −2·2 v (s.c.e.) is diffusion controlled. *Anti*-benzaldoxime gives only one wave [at −1·84 v (s.c.e.)] which is diffusion controlled. It has been suggested[137,138] that the two waves of the *syn*-form reflect the tautomeric equilibrium $\rangle C{=}NOH \rightleftharpoons \rangle C{=}\overset{+}{N}HO^-$.

In some cases a stereoselective electrochemical reduction of oximes has been found[140]. Reduction of camphor oxime (**104**) and norcamphor oxime (**105**) at a cathode potential of −2·0 v (Ag/AgCl) in 80% aqueous methanol containing lithium chloride gave yields of the corresponding amines in the range 50–70%, together with small amounts of unreacted oxime and ketone. The stereochemistry of these reactions is summarized[140] in Table 3 together with the stereochemistry of the products obtained by other reagents.

4e + 4H⁺ →

R = CH_3 (**104**) R = H (**105**) → R = CH_3 (**106**) R = H (**108**) + (**107**) (**109**)

α,β-Unsaturated oximes are in acid solution reduced preferentially to the amines, sometimes in two steps; a reduction at the potential of the first step has been shown to yield the α,β-unsaturated ketone formed

TABLE 3. Stereochemistry in reductions of bicyclic oximes. (From reference 140.)

Substrate	Reducing agent	Relative % products *exo*-Amine	*endo*-Amine
104	Mercury cathode	99 (**106**)	1 (**107**)
104	$LiAlH_4$	99 (**106**)	1 (**109**)
104	Na/EtOH	4 (**106**)	96 (**107**)
105	Mercury cathode	0	100 (**109**)
105	$LiAlH_4$	0	100 (**109**)
105	Na/EtOH	75 (**108**)	25 (**109**)

by hydrolysis of the intermediate ketimine[44]. In alkaline solution an oxime as benzalacetone oxime is reducible in a four-electron reaction and from a controlled potential reduction of this compound benzylacetone was the isolated product. As neither benzylacetone oxime nor the α,β-unsaturated hydroxylamine are reducible under these conditions, the following reaction has been suggested[44]:

$$RCH{=}CHC(CH_3){=}NOH \xrightarrow{2e\,+\,H_2O} RCH{=}CHC(CH_3){=}NH + 2\,OH^- \xrightarrow{2e\,+\,2H_2O}$$

$$RCH_2CH_2C(CH_3){=}NH + OH^- \xrightarrow{H_2O} RCH_2CH_2COCH_3 + NH_3$$

2. Alkylated oximes

Oximes may be alkylated at nitrogen or oxygen; both types are electrolytically reducible[44,45,141] in their protonated form; the nitrones are generally also reducible in alkaline solution, whereas the *O*-methyl ethers of oximes are not reducible in ordinary media containing alkali metal ions; in solutions containing quaternary ammonium *p*-toluenesulphonates some *O*-alkylated oximes give a polarographic wave.

N-Alkyl substituted oximes of both aliphatic and aromatic aldehydes are reduced in a single four-electron wave in slightly acid or alkaline solution[44,45,141]. *N*-Phenylbenzaldoxime (**110**) is reduced in two or more steps as shown in Figure 8; the first two-electron reduction results in the formation of benzalaniline (**66**), which in acid solution is reduced in two one-electron reactions[45].

$$C_6H_5CH{=}\overset{+}{N}(O^-)C_6H_5 \underset{}{\overset{H^+}{\rightleftharpoons}} C_6H_5CH{=}\overset{+}{N}(OH)C_6H_5 \xrightarrow{2e\,+\,2H^+}$$

$$\underset{(\mathbf{66})}{C_6H_5CH{=}\overset{+}{N}HC_6H_5} + H_2O \xrightarrow{e\,+\,H^+} C_6H_5\dot{C}H\overset{+}{N}H_2C_6H_5 \longrightarrow \text{dimer}$$

$$C_6H_5\dot{C}H\overset{+}{N}H_2C_6H_5 \xrightarrow{e\,+\,H^+} C_6H_5CH_2\overset{+}{N}H_2C_6H_5$$

An example of the structural effect on the reduction potentials of some *N*-substituted benzaldoximes is shown in Figure 9, where the half-wave potentials are plotted against Taft's polar substituent constants σ^* [142]. The half-wave potentials of both the protonated and unprotonated form show a linear dependence on σ^* which suggests that the substituents on nitrogen exert mainly an inductive effect and that the geometry of the transition state of the electrode process is not substantially affected.

The reduction of the *O*-methyl oximes[45] is probably analogous to the reduction of the oximes in that the nitrogen–oxygen bond is

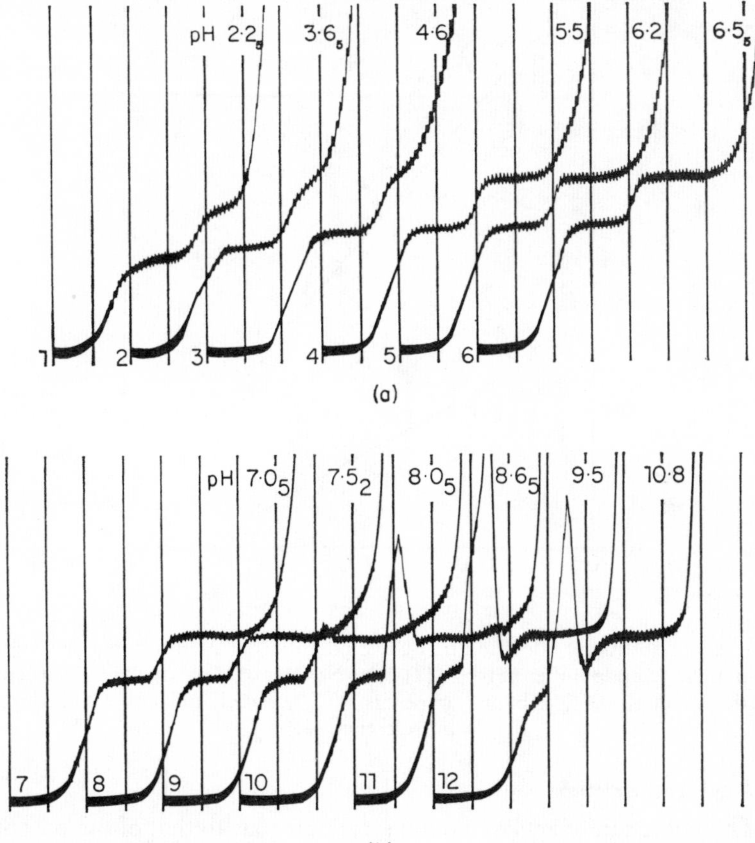

FIGURE 8. pH-dependence of reduction waves of $C_6H_5CH{=}N(O)C_6H_5$ (**110**). (2×10^{-4}M nitrone **110** Britton–Robinson buffers, pH given on the polarogram, 2% ethanol, Curves starting at: 1–3 0·2v; 4–10 0·4v; 11, 12 0·6v. s.c.e., 200mv/absc., h = 80 cm, full scale sensitivity 8·8 μA. From reference 45.)

cleaved before the saturation of the carbon–nitrogen double bond. The similarity in the reduction in acid solution of the oxime and its alkylated derivatives is understandable when the close resemblance of the electroactive forms is considered.

$$RCH{=}\overset{+}{N}H(OH) \qquad RCH{=}\overset{+}{N}R'(OH) \qquad RCH{=}\overset{+}{N}H(OR')$$

The reduction of *N*-oxides of certain 1,4-benzodiazepine derivatives was discussed in Section IV.A.5.

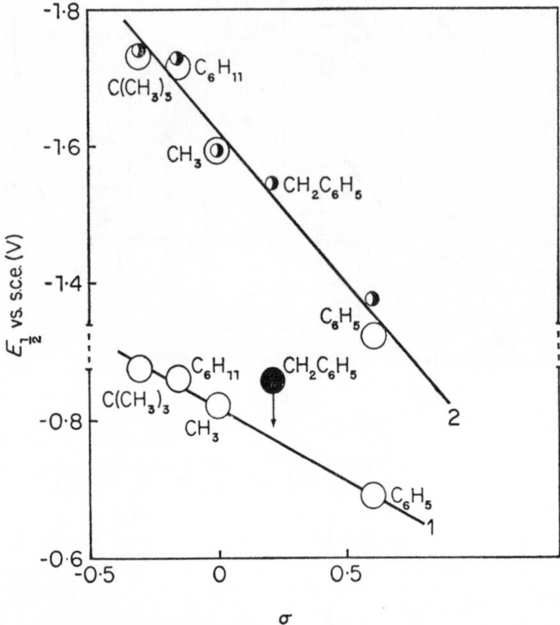

FIGURE 9. Substituent effects on half-wave potentials of *N*-substituted Benzald-oximes. (1) Acetate buffer pH 4·7; (2) 0·1M-NaOH; σ* Taft's polar substituent constants; circles 4% ethanol, crescents 40% ethanol, full circle deviating. (From reference 45.)

3. Acylated oximes

The reduction of acylated oximes is analogous to that of the parent oximes; in both acid and alkaline solution the amine is the product. Special cases are the benzoxazinones[135]. In acid solution these are mostly reduced in two two-electron steps, and from a controlled potential reduction of 4-(4′-methoxyphenyl)-2,3-benzoxazin-1-one (**111**) in 0·5N HCl at the potential of the first wave the ketimine of 2-(4′-methoxybenzoyl) benzoic acid (**112**) was isolated. The oxime of this ketone is reduced in a single four-electron reaction in acid solution.

$C_6H_4OCH_3$; $\xrightarrow[2e\,+\,2H^+]{H^+}$; $C_6H_4OCH_3$, $C{=}\overset{+}{N}H_2$, COOH ; $\xrightarrow{2e\,+\,2H^+}$; $C_6H_4OCH_3$, $CH\overset{+}{N}H_3$, COOH

(111) **(112)** **(113)**

4. Special reactions

The reduction of oximes and their derivatives described above follows a common route to the amine, but in a few cases special products have been found.

Methylglyoxal monoxime ('isonitrosoacetone') (**114**) has been found to yield dimethylpyrazine[143] on reduction in acid solution at a lead cathode by the classical procedure. The reaction may be interpreted as a 'normal' reduction to the amino ketone, followed by condensation to the dihydropyrazine (**115**), and oxidation to 2,5-dimethylpyrazine (**116**) during the isolation of the product.

$$CH_3COCH{=}NOH \xrightarrow{4e\,+\,4H^+} CH_3COCH_2NH_2 \xrightarrow{-H_2O} \tfrac{1}{2}\,(\mathbf{115}) \xrightarrow{\text{oxidation}} (\mathbf{116})$$

(115) **(116)**

Phenylglyoxal monoxime (**117**) is reduced differently, and in acid solution gives a six-electron polarographic wave followed by a smaller wave at the potential of acetophenone. A preparative reduction at pH 4 yielded acetophenone, and the reaction was formulated as [144]:

$$(\mathbf{117}) \xrightarrow{4e\,+\,5H^+} C_6H_5C(OH){=}CH\overset{+}{N}H_3 \rightleftharpoons C_6H_5COCH_2\overset{+}{N}H_3 \xrightarrow{2e\,+\,2H^+} C_6H_5COCH_3 + \overset{+}{N}H_4$$

(117) **(118)**

The reduction of 2-aminoacetophenone (**118**) to acetophenone has been investigated previously[111].

When glyoxal dioxime (**119**) is reduced in 60% sulphuric acid at a lead cathode a slightly soluble sulphate is obtained[120]. This decomposed on attempts to purify it, but it was suggested that it was the sulphate of 1,2-dihydroxylaminoethane (**120**), as it could be easily oxidized and only half of the amount of electricity expected for the formation of the diamine had been consumed. This would be the first example of a reduction of a protonated oxime to a hydroxylamine, and until further investigations substantiate this claim, it must be regarded with a certain caution; an alternative explanation could be that one of the oxime groups was reduced to the amine stage, and this amino group attacked an oxime group of a similar molecule with the formation of the oxidizable dihydropyrazine (**121**). At a mercury electrode dimethylglyoxime yields in acid solution the diamine in an eight-electron reduction[145].

A reaction where the interpretation is more difficult to give is the reduction of acetylacetone dioxime (**122**) in cold 30% sulphuric acid at a lead cathode to 3,5-dimethylpyrazolidine (**123**)[146]. The following reaction path in which the loss of water from the hydroxylamine might involve a sulphonation might be considered.

$$\underset{(\mathbf{122})}{\underset{\text{NOH}\ \ \text{NOH}}{CH_3\overset{\Vert}{C}CH_2\overset{\Vert}{C}CH_3}} \xrightleftharpoons{H^+} \underset{\text{HNOH}\ \ \text{HN}\overset{+}{\text{O}}\text{N}}{CH_3C{=}CHCCH_3} \xrightarrow{6e\,+\,6H^+}$$

$$\underset{\text{HNOH}\ \ \ NH_2}{CH_3CHCH_2CHCH_3} + H_2O \xrightarrow[-H_2O]{H^+} \underset{(\mathbf{123})}{\underset{\text{HN}\text{---}\text{NH}}{CH_3CHCH_2CHCH_3}}$$

5. Hydroxamic acids

N-acylated hydroxylamines may exist in several tautomeric forms; the neutral molecule is found predominantly as the 'keto' form (**125**). The protonation takes place mainly on oxygen and the loss of the proton occurs from nitrogen on forming the anion **126**[147]. The following species thus predominate in aqueous solution

$$\underset{(\mathbf{124})}{\underset{\text{OH}}{RC{=}\overset{+}{N}HOH}} \xrightleftharpoons{-H^+} \underset{(\mathbf{125})}{\underset{\overset{\Vert}{\text{O}}}{RCNHOH}} \xrightleftharpoons{-H^+} \underset{(\mathbf{126})}{\underset{\text{O}^-}{RC{=}NOH}}$$

As the protonated hydroxamic acid (**124**) which is the electrolytically reducible species is an azomethine derivative, its reduction will be discussed here.

Hydroxamic acids[148] are reducible in a convenient potential region only when they are substituted with an electron-attracting group such as a pyridine ring or a *p*-cyanophenyl group. Most hydroxamic acids are polarographically reducible at potentials more negative than −2·1 v (s.c.e.), which is experimentally accessible only in solutions containing tetraalkylammonium salts as supporting electrolyte[149–153].

The reduction of isonicotinic hydroxamic acid (**127**) in acid solution occurs in two steps, as follows:

$$\text{H}\overset{+}{\text{N}}\langle\text{C}_5\text{H}_4\rangle\text{—C(=O)NHOH} \underset{}{\overset{\text{H}^+}{\rightleftharpoons}} \text{H}\overset{+}{\text{N}}\langle\text{C}_5\text{H}_4\rangle\text{—C(OH)=}\overset{+}{\text{N}}\text{HOH}$$

(**127**)

$$\text{1st step} \xrightarrow{2e\,+\,2H^+} \text{H}\overset{+}{\text{N}}\langle\text{C}_5\text{H}_4\rangle\text{—C(OH)=}\overset{+}{\text{N}}\text{H}_2$$

(**128**)

2nd step:

$$\xrightarrow{2e\,+\,2H^+} \text{H}\overset{+}{\text{N}}\langle\text{C}_5\text{H}_4\rangle\text{—CH(OH)}\overset{+}{\text{N}}\text{H}_3$$

$$\xrightarrow{H_2O} \text{H}\overset{+}{\text{N}}\langle\text{C}_5\text{H}_4\rangle\text{—CH(OH}_2\text{)} + \overset{+}{\text{N}}\text{H}_4 \rightleftharpoons \text{H}\overset{+}{\text{N}}\langle\text{C}_5\text{H}_4\rangle\text{—CHO} + \text{H}_2\text{O}$$

$$\xrightarrow{2e\,+\,2H^+} \text{H}\overset{+}{\text{N}}\langle\text{C}_5\text{H}_4\rangle\text{—CH}_2\text{OH}$$

A controlled potential reduction in 0·2N HCl at the potential of the first wave of isonicotinic hydroxamic acid produced isonicotinic amide (**128**); the reduction of this compound has been investigated previously[50].

Also, in the reduction in acid solution of phenylglyoxal hydroxamic acid an amide, $C_6H_5CHOHCONH_2$, is formed together with dimeric products, but the primary attack seems to occur at the carbonyl group[153].

6. Amide oximes

Amide oximes are reduced in acid solution similarly to the other reducible derivatives of hydroxylamine[44,154–156]; the first step is the loss of the oxygen. From the reduction of benzamide oxime (**119**) benzamidine (**68**) was isolated[44]; the reduction of the latter has been discussed in Section IV.A.4.

$$\underset{(\mathbf{119})}{C_6H_5C(NH_2){=}NOH} \overset{H^+}{\rightleftharpoons} C_6H_5C(NH_2){=}\overset{+}{N}HOH \xrightarrow{2e\,+\,2H^+} \underset{(\mathbf{68})}{C_6H_5C(NH_2){=}\overset{+}{N}H_2} + H_2O$$

V. REFERENCES

1. G. J. Hoijtink, *Rec. Trav. Chim.*, **76**, 885 (1957).
2. R. M. McKee, *Ind. Eng. Chem.*, **38**, 382 (1946).
3. F. Fichter, *Organische Elektrochemie*, Verlag Theodor Steinkopff, Dresden, 1942.
4. M. J. Allen, *Organic Electrode Processes*, Chapman and Hall, London, 1958.
5. B. E. Conway, *Theory and Principles of Electrode Processes*, The Ronald Press Company, New York, 1965.
6. S. Swann, Jr., *Trans. Electrochem. Soc.*, **69**, 53 (1936); **77**, 40 (1940); **88**, 18 (1945).
7. F. D. Popp and H. P. Schultz, *Chem. Revs.* **62**, 19 (1962).
8. C. L. Perrin, in *Progress in Physical Organic Chemistry*, Vol. 3, (Eds. S. G. Cohen, A. Streitwieser, Jr., and R. W. Taft). Interscience, New York, 1965, pp. 165–316.
9. J. Heyrovský and J. Kůta, *Grundlagen der Polarographie*, Akademie-Verlag, Berlin, 1965.
10. I. M. Kolthoff and J. J. Lingane, *Polarography*, 2nd Ed., Interscience, New York, London, 1952.
11. S. Wawzonek, *Anal. Chem.* **28**, 638 (1965); **30**, 661 (1958); **32**, 144 R (1960); **34**, 182 R (1962).
12. S. Wawzonek and D. J. Pietrzyk, *Anal. Chem.*, **36**, 220 R (1964).
13. D. J. Pietrzyk, *Anal. Chem.*, **38**, 278 R (1966), **40**, 194 R (1968).
14. F. Haber, *Z. Elektrochem.*, **4**, 506 (1898); *Z. Physik. Chem.*, **32**, 193 (1900).
15. M. M. Baizer, *J. Elektrochem. Soc.*, **111**, 215 (1964).
16. J. J. Lingane, *Electroanalytical Chemistry*, 2nd Ed., Interscience, New York, London, 1958, p. 296.
17. G. L. Booman and W. B. Holbrook, *Anal. Chem.*, **35**, 1793 (1963).
18. H. Lund, *Studier over Elektrodereaktioner i Organisk Polarografi og Voltammetri*, Aarhuus Stiftsbogtrykkerie A/S, Aarhus, 1961.
19. P. E. Iversen and H. Lund, *Acta Chem. Scand.*, **19**, 2303 (1965).
20. P. Zuman, *Substituent Effects in Organic Polarography*, Plenum Press, New York, 1967.
21. D. H. Geske and A. H. Maki, *J. Am. Chem. Soc.*, **82**, 2671 (1960).
22. P. H. Rieger and G. K. Fraenkel, *J. Chem. Phys.*, **37**, 2975 (1962).
23. T. Kitagawa, T. Layoff and R. N. Adams, *Anal. Chem.*, **36**, 925 (1964).
24. H. Lund, *Abstr. 19th Intern. Congr. Pure Appl. Chem., London* (1963), p. 466.
25. M. Masui, H. Sayo and K. Kishi, *Tetrahedron*, **21**, 2831 (1965).
26. P. E. Iversen and H. Lund, *Tetrahedron Letters* 4027 (1967).
27. M. Masui and H. Sayo, *Pharm. Bull.* (*Tokyo*), **4**, 332 (1956).
28. M. Masui, H. Sayo and Y. Nomura, *Pharm. Bull.* (*Tokyo*), **4**, 337 (1956).
29. J. Armand, *Bull. Soc. Chim. France*, 543 (1966).
30. H. Brintzinger, H. W. Ziegler and E. Schneider, *Z. Elektrochem.*, **53**, 109 (1949).
31. J. H. Madaus and H. B. Urbach, *U.S. Pat.* 2,918,418 (1959); *Chem. Abstr.*, **54**, 11774 (1960).
32. J. T. Stock, *J. Chem. Soc.*, 4532 (1957).
33. M. Masui and H. Sayo, *J. Chem. Soc.*, 4773 (1961).
34. J. Armand, *Bull. Soc. Chim. France*, 1658 (1966).

35. D. Stocesova, *Collection Czech. Chem. Commun.*, **14**, 615 (1949).
36. H. Lund, *Acta Chem. Scand.*, **12**, 1444 (1958).
37. M. LeGuyader, *Bull. Soc. Chim. France.*, 1858 (1966).
38. M. Breant and J.-C. Merlin, *Bull. Soc. Chim. France*, 53 (1964).
39. H. Lund, *Österr. Chem. Z.*, **68**, 43 (1967).
40. H. Lund, *Acta Chem. Scand.*, **17**, 972 (1963).
41. P. E. Iversen and H. Lund, *Acta Chem. Scand.*, **21**, 279 (1967).
42. P. E. Iversen and H. Lund, *Acta Chem. Scand.*, **21**, 389 (1967).
43. J. Armand, *Bull. Soc. Chim. France*, 882 (1966).
44. H. Lund, *Acta Chem. Scand.*, **13**, 249 (1959).
45. P. Zuman and O. Exner, *Collection Czech. Chem. Commun.*, **30**, 1832 (1965).
46. T. Kubota and H. Miyazaki, *Bull. Chem. Soc. Japan*, **35**, 1549 (1962).
47. T. Kubota and H. Miyazaki, *Bull. Chem. Soc. Japan*, **39**, 2057 (1966).
48. H. Lund, *Acta Chem. Scand.*, **18**, 1984 (1964).
49. H. Lund, *Collection Czech. Chem. Commun.*, **25**, 3313 (1960).
50. H. Lund, *Acta Chem. Scand.*, **17**, 2325 (1963).
51. R. Kalvoda and G. Budnikov, *Collection Czech. Chem. Commun.*, **28**, 838 (1963).
52. Yu. P. Kitaev and G. Budnikov, *Dokl. Akad. Nauk SSSR*, **154**, 1379 (1964).
53. G. Trümpler and H. E. Klauser, *Helv. Chim. Acta*, **42**, 407 (1959).
54. K. Odo and K. Sugino, *J. Elektrochem. Soc.*, **104**, 160 (1957).
55. K. Odo, E. Ichikawa, K. Shimogai and K. Sugino, *J. Electrochem. Soc.*, **105**, 598 (1958).
56. H. Lund, Lecture, *IV. Intern. Congr. Prague Polarography*, 1966.
57. H. Lund, *Disc. Faraday Soc.*, **45**, 193 (1968).
58. H. Lund and P. Lunde, *Acta Chem. Scand.*, **21**, 1067 (1967).
59. P. Pflegel, G. Wagner and O. Manousek, *Z. Chem.*, **6**, 263 (1966).
60. H. Lund, *Acta Chem. Scand.*, **21**, 2525 (1967).
61. L. S. Besford, G. Allen and J. M. Bruce, *J. Chem. Soc.*, 2867 (1963).
62. L. S. Besford and J. M. Bruce, *J. Chem. Soc.*, 4037 (1964).
63. C. Furlani, S. Bertola and G. Morpurgo, *Ann. Chim. (Rome)*, **50**, 858 (1960).
64. H. Lund, *12. Nordiske Kemikermøde, June 1965. Abstr. of Papers*, p. 33.
65. Yu. S. Shabarov. N. I. Vasil'ev and R. Ya. Levina, *Zh. Obshch. Khim.*, **31**, 2478 (1961).
66. D. L. Smith and P. J. Elving, *J. Am. Chem. Soc.*, **84**, 2741 (1962).
67. K. Sugino, K. Shirai, T. Sekine and K. Odo, *J. Electrochem. Soc.*, **104**, 667 (1957).
68. D. L. Smith and P. J. Elving, *J. Am. Chem. Soc.*, **86**, 1412 (1962).
69. J. Pinson and J. Armand, *Compt. Rend. Ser. C*, **268**, 629 (1969).
70. S. Kwee and H. Lund, *Acta Chem. Scand.*, **23**, (1969).
71. B. Jambor, *Acta Chim. Acad. Sci. Hung.*, **4**, 55 (1954).
72. C. Campbell and P. O. Kane, *J. Chem. Soc.*, 3110 (1956).
73. P. Kivalo and K. K. Mustakallio, *Suomen Kemistilehti*, **B29**, 154 (1956).
74. P. Kivalo and K. K. Mustakallio, *Suomen Kemistilehti*, **B30**, 214 (1957).
75. H. Lund, *Acta Chem. Scand.* **23**, 563 (1969).
76. L. K. J. Tong, M. C. Glesmann and R. L. Bent, *J. Am. Chem. Soc.*, **82**, 1968 (1960).
77. T. Kuwana and J. W. Strojek, *Disc. Faraday Soc.*, **45**, 134 (1968).

78. W. M. Clark, *Oxidation Reduction Potentials of Organic Systems*, Williams and Wilking, Baltimore, 1960, p. 397.
79. L. H. Piette, P. Ludwig and R. N. Adams, *J. Am. Chem. Soc.*, **83**, 3909 (1961).
80. T. Kuwana, R. K. Darlington and D. W. Leedy, *Anal. Chem.*, **36**, 2023 (1964).
81. H. Lund, *Collection Czech. Chem. Commun.*, **30**, 4237 (1965).
82. H. Lund, *Tetrahedron Letters*, 3651 (1968).
83. A. Hill and R. G. A. New, *Brit. Pat.* 869,773 (1961); *Chem. Abstr.* **55**, 23129 (1961).
84. P. E. Iversen and H. Lund, to be published.
85. G. Cauquis, private communication (1968).
86. H. Lund, *Acta Chem. Scand.*, **11**, 491 (1957).
87. H. Lund, *Acta Chem. Scand.*, **11**, 1323 (1957).
88. K. Wiesner, *Z. Elektrochem.*, **49**, 164 (1943).
89. J. Koutecky, *Collection Czech. Chem. Commun.*, **18**, 311, 597 (1953).
90. L. Holleck and B. Kastening, *Z. Elektrochem.*, **60**, 127 (1956).
91. B. Kastening, L. Holleck and G. A. Melkonian, *Z. Elektrochem.*, **60**, 130 (1956).
92. V. N. Dmitrieva, V. B. Smelyakova, B. M. Krasovitskii and V. D. Bezuglyi, *Zh. Obshch. Khim.*, **36**, 405 (1966).
93. V. N. Dmitrieva, N. I. Mal'tseva, V. D. Bezuglyi and B. M. Krasovitskii, *Zh. Obshch. Khim.*, **37**, 372 (1967).
94. K. Brand, *Ber.*, **42**, 3461 (1909).
95. H. D. Law, *J. Chem. Soc.*, 154 (1912).
96. P. E. Iversen and H. Lund, unpublished observation.
97. H. Lund, *Acta Chem. Scand.*, **14**, 359 (1960).
98. H. Lund, P. Lunde and F. Kaufmann, *Acta Chem. Scand.*, **20**, 1631 (1966).
99. J. M. W. Scott and W. H. Jura, *Can. J. Chem.*, **45**, 2375 (1967).
100. N. F. Levchenko, L.Sh. Afanasiadi, and V. D. Bezuglyi, *Zh. Obshch. Khim.*, **37**, 666 (1967).
101. H. Lund, unpublished observation.
102. M. M. Baizer, J. D. Anderson, J. H. Wagenknecht, M. R. Ort and J. P. Petrovich, *Electrochem. Acta.*, **12**, 1377 (1967).
103. H. Wenker, *J. Am. Chem. Soc.*, **57**, 772 (1935).
104. H. Lund, to be published.
105. P. O. Kane, *Z. Anal. Chem.*, **173**, 50 (1960).
106. H. Oelschläger, *Arch. Pharm.*, **296**, 396 (1963).
107. B. Z. Senkowski, M. S. Levin, J. R. Urbigkit and E. G. Wollish, *Anal. Chem.* **36**, 1991 (1964).
108. H. Oelschläger, J. Volke and H. Hoffmann, *Collection Czech. Chem. Commun.*, **31**, 1264 (1966).
109. H. Oelschläger, J. Volke, H. Hoffmann and E. Kurek, *Arch. Pharm.*, **300**, 250 (1967).
110. H. Oelschläger and H. Hoffmann, *Arch. Pharm.*, **300**, 817 (1967).
111. P. Zuman, and V. Horak, *Collection Czech. Chem. Commun.*, **26**, 176 (1961).
112. R. Zahradnik, *Chem. Listy* **49**, 764 (1955).
113. A. M. Kardos, J. Volke and P. Kristian, *Collection Czech. Chem. Commun.*, **30**, 931 (1965).

114. Yu. P. Kitaev and A. E. Arbuzov, *Izv. Akad. Nauk SSSR, Otdel Khim. Nauk*, 1037 (1957).
115. Yu. P. Kitaev and T. V. Troepol'skaya, *Izv. Akad. Nauk SSSR, Otdel Khim. Nauk*, 454, 465 (1963).
116. H. Simon and W. Moldenhauer, *Chem. Ber.*, **100**, 1949 (1967).
117. B. V. Ioffe and V. S. Stopskij, *Tetrahedron Letters*, 1333 (1968).
118. G. C. Whitnack, J. E. Young, H. H. Sisler, and E. St. C. Gantz, *Anal. Chem.*, **28**, 833 (1956).
119. V. D. Bezuglyi and N. P. Shimanskaya, *Zh. Obshch. Khim.*, **35**, 17 (1965).
120. J. Tafel and E. Pffermann, *Ber.*, **35**, 1510 (1902).
121. Yu. P. Kitaev and I. M. Skrebkova, *Zh. Obshch. Khim.*, **37**, 1204 (1967).
122. V. Prelog and O. Häfliger, *Helv. Chim. Acta*, **32**, 2088 (1949).
123. M. Brezina, V. Volkova and J. Volke, *Collection Czech. Chem. Commun.*, **19**, (1954).
124. P. Souchay and M. Graizon, *Chim. Anal.*, **36**, 85 (1954).
125. Yu. P. Kitaev, G. K. Budnikov, and A. E. Arbuzov, *Izv. Akad. Nauk SSSR, Otdel. Khim. Nauk*, 824 (1961).
126. Yu. P. Kitaev and G. K. Budnikov, *Zh. Obshch. Khim.*, **33**, 1396 (1963).
127. Y. Asahi, *Chem. Pharm. Bull.* (*Tokyo*) **11**, 930 (1963).
128. M. Masui and H. Ohmori, *Chem. Pharm. Bull.*, **12**, 877 (1964). M. Masui, private communication (1968).
129. B. Fleet and P. Zuman, *Collection Czech. Chem. Commun.*, **32**, 2066 (1967).
130. A. Foffani, L. Salvagnini and C. Pecile, *Ann. Chim.* (*Rome*), **49**, 1677 (1959).
131. D. M. Coombs and L. L. Levison, *Anal. Chim. Acta*, **30**, 209 (1964).
132. P. Souchay and S. Ser, *J. Chim. Phys.*, **49**, C 172 (1952).
133. R. M. Elofsen and J. G. Atkinson, *Can. J. Chem.*, **34**, 4 (1956).
134. H. J. Gardner and W. P. Georgans, *J. Chem. Soc.*, **4180** (1956).
135. H. Lund, *Acta Chem. Scand.*, **18**, 563 (1964).
136. N. Tütülkoff and St. Buduroff, *Compt. Rend. Acad. Bulgare Sci.*, **6**(3), 5 (1953).
137. N. Tütülkoff and I. Bakardziev, *Compt. Rend. Acad. Bulgare Sci.*, **12**, 133 (1959).
138. N. Tütülkoff and E. Paspaleev, *Compt. Rend. Acad. Bulgare Sci.*, **14**, 159 (1961).
139. E. Paspaleev, *Monatsh. Chem.*, **97**, 230 (1966).
140. A. J. Fry and J. H. Newberg, *J. Am. Chem. Soc.*, **89**, 6374 (1967).
141. T. Kubota, H. Miyazaki and Y. Mori, *Bull. Chem. Soc. Japan*, **40**, 245 (1967).
142. R. W. Taft, Jr., *Separation of Polar, Steric, and Resonance Effects in Reactivity in Steric Effects in Organic Chemistry* (Ed. M. S. Newman), John Wiley and Sons, New York, 1956.
143. F. B. Ahrens and G. Meissner, *Ber.*, **30**, 532 (1897).
144. Ya. P. Stradins, I. Ya. Kravis, and N. O. Saldabol, *Zh. Obshch. Khim.*, **37**, 977 (1967).
145. M. Spritzer and L. Meites, *Anal. Chim. Acta*, **26**, 53 (1962).
146. J. Tafel and E. Pffermann, *Ber.*, **36**, 219 (1903).
147. O. Exner and B. Kakac, *Collection Czech. Chem. Commun.*, **28**, 1656 (1963).
148. H. Lund, *Talanta*,**12**, 1065 (1965).
149. M. Prytz and T. Østerrud, *Acta Chem. Scand.*, **11**, 1530 (1957).

150. M. Prytz and T. Østerrud, *Acta Chem. Scand.*, **15**, 1285 (1961).
151. T. Østerrud and M. Prytz, *Acta Chem. Scand.*, **15**, 1923 (1961).
152. B. V. Mateev and G. G. Tsybaeva, *Zh. Obshch. Khim.*, **34**, 2491 (1964).
153. J. Armand, P. Souchay and F. Valentini, *Compt. Rend.*, **265**, 1267 (1967).
154. M. Kuras and J. Mollin, *Collection Czech. Chem. Commun.*, **24**, 290 (1959).
155. J. Mollin and F. Kasparek, *Collection Czech. Chem. Commun.*, **25**, 451 (1960).
156. J. Mollin and F. Kasparek, *Collection Czech. Chem. Commun.*, **26**, 2438 (1961).

CHAPTER **12**

Photochemistry of the carbon–nitrogen double bond

GUNNAR WETTERMARK

Institute of Physical Chemistry, University of Uppsala, Sweden

I. SPECTRAL PROPERTIES

An isolated azomethine group gives rise to two absorption peaks in the ultraviolet region of the spectrum, at about 2400 and 1800 Å (see Figure 1). The band at 2400 Å has been assigned to a $n \rightarrow \pi^*$ transition because its molar absorptivity is about 2×10^2 and it shows a bathochromic shift on decreasing the polarity of the solvent[1]. The band at 1800 Å has a higher molar absorptivity of about 10^4, and shows a hypsochromic shift. It is therefore designated as a $\pi \rightarrow \pi^*$ transition[2]. Protonation of the non-bonded electrons on the nitrogen atom causes the weak band to be replaced by a strong absorption, presumably due to a $\pi \rightarrow \pi^*$ transition[1]. When the azomethine bond is part of a conjugated system of double bonds the spectrum may become very complex[3–5] (see Figure 1). Jaffé, Yeh and Gardner[3] isolated four bands from the spectrum of *N*-benzylideneaniline in the

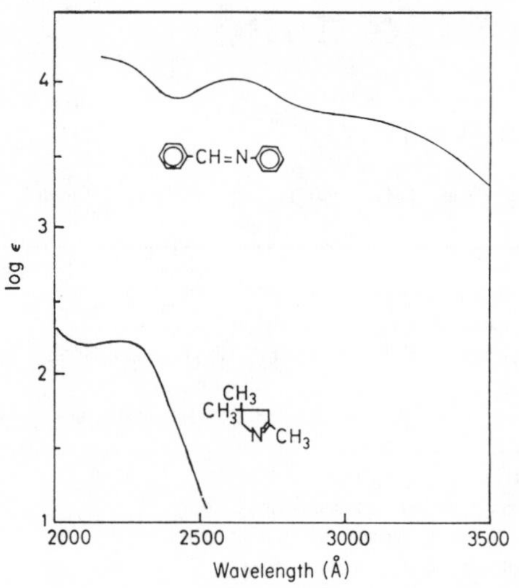

FIGURE 1. Absorption spectrum in ethanol solution of an isolated azomethine group and of a Schiff base where the azomethine group is part of a conjugated system[1,3].

region 2300–4000 Å. Difficulties in assigning bands largely arise from the fact that the lone pair electrons are appreciably delocalized in these compounds. An account of the unusual properties resulting from this delocalization can be found in a recent review article[6].

II. SOLID-STATE PHOTOCHROMISM

A great number of C=N containing compounds exhibit photochromism in the solid state. A list of such compounds includes among others anils, hydrazones, osazones and semicarbazones. In some cases the reactions responsible for the colour change are known, but for several reactions the mechanism is totally unknown.

Detailed studies have been undertaken with Schiff bases which can be thought of as derivatives of salicylaldehyde, i.e. Schiff bases having a hydroxyl group *ortho* to the C=N bond on the benzaldehyde ring. It has long been known that several of these compounds give rise to crystals which change their colour from yellow to red upon exposure to light[7–10]. The red form has an absorption peak at about 4800 Å (Figure 2). The process is reversible; heating in the dark or irradiation within the new absorption band reverts the system to the initial state.

$$\text{yellow crystals} \underset{h\nu_2 \text{ or heat}}{\overset{h\nu_1}{\rightleftharpoons}} \text{red crystals} \qquad (1)$$

The crystals have been investigated by x-ray and infrared methods during the course of the reaction, but no systematic changes were noted[11–13]. Some anils give rise to dimorphism but both crystal modifications are not necessarily photochromic. The anil of salicylaldehyde and *o*-methylaniline exists for instance in two crystal forms, of which only one is photochromic[14]. Several theories have been proposed to explain the phenomenon of photochromism with anils. Corresponding colour reactions take place in solution and it is largely from such studies that the reaction mechanism has been confirmed. The enol form of the hydroxy anil changes into a quinonoid species.

(2)

Detailed studies have been undertaken at the Weizmann Institute to find the conditions necessary for this reaction[13,15–17]. It was confirmed that no correlation exists between photochromic activity and the chemical nature of the substituents R^1 and R^2. For instance, exchanging the bromine atom for chlorine in the compound 4-bromo-*N*-salicylideneaniline changes a highly photochromic compound into a substance showing no photochromism at all[19]. The molecular packing arrangement in the crystalline lattice was shown to be the operative factor. The crystals have been divided into two groups, denoted α and β. The α-crystals are pale yellow at all temperatures but change into a red state upon exposure to light. The β-type crystals are thermochromic and have always a more or less reddish colour at room temperature due to an absorption band at about 4800 Å. This band disappears when the crystals are cooled. It is assumed that a tautomerism of the type shown in equation (2) is responsible for the thermochromic behaviour. It has been stated that photochromism and thermochromism are mutually exclusive properties, i.e. crystalline modifications of anils can be either photochromic or thermochromic but not both. It is also likely that the thermochromic crystals photo-isomerize enol → quinonoid, but the reformation of enol is so rapid that a colour change is not observable with the currently available techniques. The emission spectrum

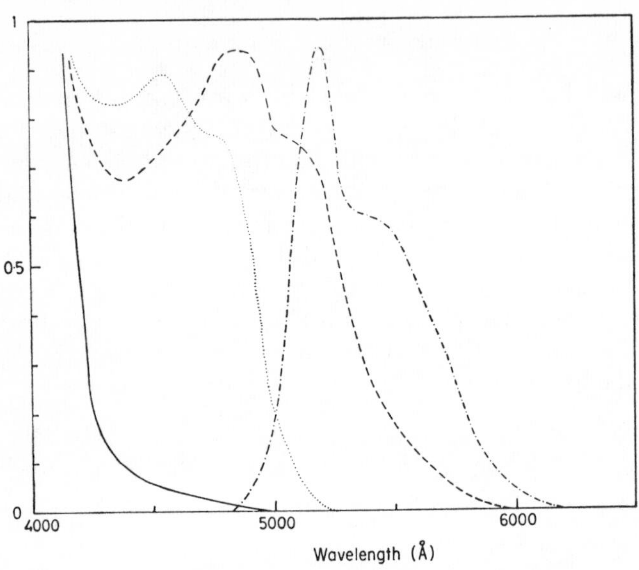

FIGURE 2. Spectra of anil crystals[4,5,13,15,18]. (Optical density or relative intensity in arbitrary units.)

———————	absorption of the uncoloured state of *N*-benzylideneaniline
– – – – – – –	absorption of the photo-coloured state of *N*-benzylideneaniline
. . . .	absorption of the thermo-coloured state of *N*-(5-chlorosalicylidene)aniline
–.–.–.–.–.–.	fluorescence emission from *N*-(5-chlorosalicylidene)aniline

of fluorescence from thermochromic anils indicates that a quinonoid state emits the light. Ideally the 0–0 transition observed in fluorescence and absorption should coincide. Furthermore the vibrational levels are often similarly spaced in the ground state and the excited state giving rise to a more or less pronounced mirror-image relationship between absorption and fluorescence spectra. In the case of the thermochromic anils, the absorption of the thermo-coloured state bears a much closer relation to the fluorescence spectrum than the spectrum of the uncoloured state. This is illustrated in Figure 2 by the spectra of *N*-(5-chlorosalicylidene)aniline. The uncoloured state of this compound absorbs essentially as *N*-benzylideneaniline shown by the solid line in Figure 2.

The kinetics for the fading of photo-coloured anil crystals has been investigated by numerous researchers. Some compounds fade by a first-order process and the fading can be described by the rate constant, k, given for 30°c in Table 1.

$$\text{red crystals } (\lambda_{max} \sim 4800\ \text{Å}) \xrightarrow{k} \text{yellow crystals} \tag{3}$$

From measurements at various temperatures Arrhenius activation energies have been calculated and are included in the table. Other compounds do not fade by first-order kinetics and it has been claimed that the process is second order in the case of *N*-salicylidene-β-naphthylamine[22]. Both first and second order kinetics failed to describe the fading of 2-chloro-*N*-salicylideneaniline and 2-bromo-*N*-salicylideneaniline[13].

TABLE 1. Kinetics of fading of some photochromic crystals.

Compound	Rate constant at 30°c[a] (s^{-1})	Activation energy (kcal/mole)	Method of analysis
N-salicylidene-aniline	4×10^{-4}	20·05	Transmission[13]
	$\left\{\begin{array}{l} 5 \times 10^{-3\,b} \\ 4 \times 10^{-4\,b} \end{array}\right.$	38·0 22·8	Transmission[18]
N-salicylidene-*m*-toluidine	7×10^{-5}	25	Reflectance[20]
	9×10^{-5}	29·84	Transmission[13]
	9×10^{-5}	29·5	Transmission[18]
N-2-cyanopropyl--salicylideneamine	3×10^{-4}	21·23	Transmission[13]
Benzaldehyde phenylhydrazone	$1{\cdot}1 \times 10^{-5}$	15·7	Reflectance[21]
Cinnamaldehyde semicarbazone	7×10^{-6}	18·7	Reflectance[21]

[a] Estimated value from measurements at other temperatures.

[b] The experimental fading curves were resolved into two first-order decay curves assuming concurrent competing decay reactions.

With C=N containing compounds other than anils, the knowledge of the photochromism is often rudimentary; in many cases all that is known is that a particular compound undergoes a certain colour change. The photochromic reactions have been reviewed by Brown and Shaw[10]. In a few cases the kinetics of the fading reaction has been studied quantitatively and Table 1 contains results from an investigation on benzaldehyde phenylhydrazone and cinnamaldehyde semicarbazone. In the case of the hydrazone, the rate constant refers to the

C_6H_5–CH=NNH–C_6H_5 C_6H_5–CH=CHCH=NNHCONH$_2$

benzaldehyde phenylhydrazone cinnamaldehyde semicarbazone

fading of the photo-coloured red form, λ_{max} = 4900 Å, to the original whitish state. For the semicarbazone the rate of formation of a yellow form, C, from a photo-activated state, B, was measured (equation 4)

$$A \xrightarrow{h\nu} B(\lambda_{max} = 4000\ \text{Å}) \xrightarrow{k} C(\lambda_{max} = 4300\ \text{Å}) \tag{4}$$

Reflectance spectrum measurements showed a narrow absorption peak at 4000 Å for species B and a broad band with a maximum at 4300 Å for species C. The data refer to the difference spectrum with respect to the colourless substance A.

III. TAUTOMERISM OF HYDROXYL-SUBSTITUTED SCHIFF BASES IN SOLUTION

The photochromic reactions observed in certain anil crystals may also take place in solution. In this state, however, the reaction is of wider occurrence and it has been stated that all anils of salicylaldehydes undergo photo-induced colour changes in solid solution at low temperature[23]. Also the anils from 1-hydroxy-2-naphthaldehyde and 2-hydroxy-1-naphthaldehyde are photo-colourable in this environment[24]. The light-produced absorption band usually extends from about 4000 Å to about 5000 Å with the maximum near 4800 Å. The effect is reversed when the solid solution is allowed to melt. The reaction also occurs in ethanol solution at room temperature, but the reverse reaction is so fast that the colour changes cannot be observed with the 'naked eye'[25,26]. Reaction rates and activation energies for the reverse reaction are shown in Table 2. It is assumed that the reaction involves an intramolecular hydrogen transfer in a six-membered ring to form coloured quinonoid-type compounds.

BLE 2. Kinetic data for the fading of the photo-produced absorption band near 4800 Å[26]. e data refer to acetate-buffered ethanol solutions, [HOAc]/[Ac$^-$] = 1·4. The fading s accelerated at increasing buffer strength and the kinetics therefore separated into a ffer-catalysed and a non-catalysed portion according to the equation $k_{\text{exptl}} = k_0 + k_2$ OAc]. k_{exptl} is the experimentally determined rate constant and [HOAc] the molarity of acetic acid.

Compound	Non-catalysed portion		Buffer-catalysed portion	
	Rate constant k_0, at 30°C (s^{-1})	Activation energy (kcal/mole)	Rate constant k_2, at 30°C ($mole^{-1}\ s^{-1}$)	Activation energy (kcal/mole)
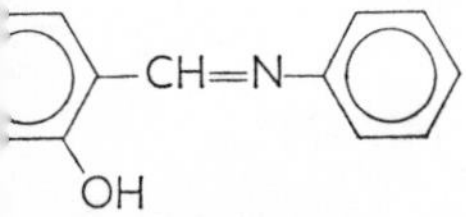 *N*-salicylideneaniline	$5{\cdot}0 \times 10^3$	6·7	$3{\cdot}7 \times 10^7$	3·4
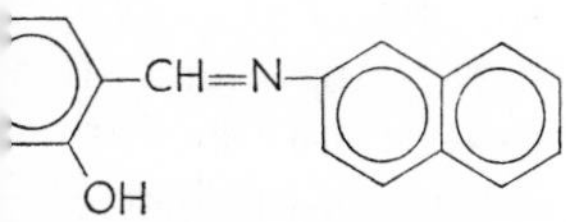 *N*-salicylidenenaphthylamine	$4{\cdot}0 \times 10^3$	7·7	$4{\cdot}4 \times 10^7$	2·6

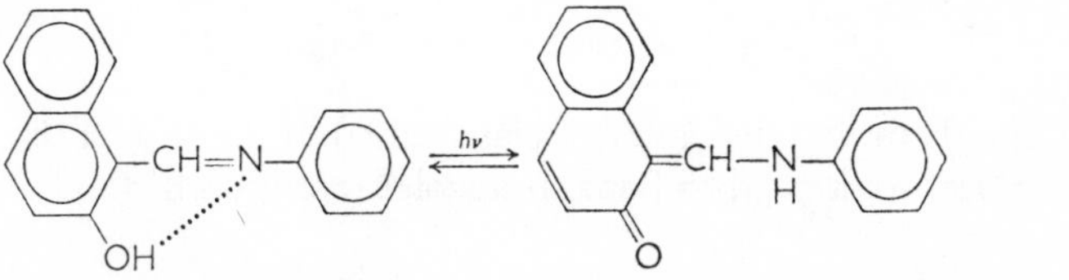

(5)

Evidence has been brought forward showing that *cis–trans* isomerism in the quinonoid state plays an important role for the photochemical reactions[27]. Equation (6) describes the findings at 77°K in 3-methylpentane. The equilibrium **1** ⇌ **2** becomes displaced towards **2** upon addition of acid and in polar solvents, and also when substituents on the salicylaldehyde ring make the phenolic proton more acidic. This gives rise to exceptions to the rule that all salicylidene anils are photochromic in solution. Certain nitro derivatives fall in this class.

(1) enol ⇌ (2) *cis*-quinonoid; (1) —$h\nu$→ (3) *trans*-quinonoid —thermal reaction→ (2) (6)

IV. TRANSPOSITION OF RING ATOMS IN FIVE-MEMBERED RINGS

Five-membered ring heterocyclic compounds can undergo photo-isomerization to form a different five-membered ring. The overall reaction can be described as a transposition of two ring atoms. It has, for instance, been shown that pyrazoles can be subjected to such a reaction in several solvents whereby the photo-product is imidazole[28,29].

pyrazole —$h\nu$→ imidazole (7)

The substituent dependence was studied in the case of the indazoles which are isomerized to benzimidazoles (reaction 8[29]).

R-indazole —$h\nu$→ R-benzimidazole (8)

Electron donors in the 5, 6 or 7 position enhance the reaction, while acceptors have an inhibitory action. A substituent on nitrogen atom 1 prevents the reaction but a substituent on nitrogen atom 2 makes the reaction proceed with greater ease.

Indazoles substituted in the 1-*N* position undergo instead another reaction. For instance, 1-methylindazole forms the isomer 2-cyano-*N*-methylaniline.

$$\text{1-methylindazole} \xrightarrow{h\nu} \text{2-cyano-}N\text{-methylaniline} \quad (9)$$

Corresponding reactions take place in an isoxazole ring, i.e. in a heterocyclic compound with a five-membered ring containing neighbouring oxygen and nitrogen atoms. Thus benzisoxazole is photoisomerized to benzoxazole and salicylonitrile[30].

$$\text{benzisoxazole} \xrightarrow{h\nu} \text{benzoxazole} + \text{salicylonitrile} \quad (10)$$

The mechanism of transposition of ring atoms has been somewhat clarified from experiments with a derivative of isoxazole, namely 3,5-diphenylisoxazole[31]. A compound containing the three-membered heterocyclic azirine ring was demonstrated to be an intermediate in the photo-reaction. In fact it was observed that the azirine was photosensitive and the reactions of the azirine dramatically dependent on the wavelength of the exciting light. The oxazole is formed with 2537 Å light, but the isoxazole is reformed with light of longer wavelengths (> 3000 Å).

$$\text{3,5-diphenylisoxazole} \underset{>3000\ \text{Å}}{\overset{2537\ \text{Å}}{\rightleftharpoons}} \text{2-benzoyl-3-phenyl azirine} \xrightarrow{2537\ \text{Å}} \text{2,5-diphenyl-oxazole} \quad (11)$$

It has been proposed that the wavelength dependence is caused by selective excitation of the chromophores (the benzoyl and the azomethine group), longer wavelengths leading only to excitation of the benzoyl chromophore.

V. *cis–trans* ISOMERIZATION

The research on the *cis–trans* interconversion about the carbon–nitrogen double bond has recently been reviewed[6]. It was pointed out that *cis–trans* isomerization has long been known for oximes but that recent research has shown that azomethine compounds are more generally subject to such isomerism. The reason why isomerism about the C=N link was largely unknown for a long period of time is that compared with the azo or the ethylenic equivalent, the thermal relaxation around the C=N link is, as a rule, considerably faster. At room temperature it is often necessary to employ special experimental techniques, such as flash photolysis, in order to observe the isomerism.

Even before the turn of the century, there was evidence that ultraviolet light may cause rearrangement in oximes from one geometric isomer to the other[32,33]. Sporadic studies of this reaction have since been undertaken[34–41]. Its use for the chemical synthesis of the pharmacologically active isomer has even been described for the preparation of *trans*-isonicotinaldehyde oxime[41].

cis isomer $\xrightarrow{h\nu}$ *trans* isomer (12)

Some cases have also been described in which the *O*-ethers of oximes undergo this isomerism[35,37,39,42].

$$R^1CH{=}NOR^2 \underset{}{\overset{h\nu}{\rightleftharpoons}} R^1CH{=}NOR^2 \tag{13}$$

The all-*trans* form of cinnamaldehyde azine (cinnamalazine) is photo-converted to two different mono-*cis* isomers[43]. One of these is believed to have a *cis* configuration with respect to the C=N linkage.

Photochemical *cis–trans* isomerization of hydrazones has been reported[44–46]. The quinonoid system 1,2-naphthoquinone-2-diphenylhydrazone gives rise to remarkably large spectral changes[45,46]: a wavelength shift of the main absorption peak of about 1000 Å has been reported. The structure of the compound is shown in Table 3, together with data for the kinetics of the thermal *cis–trans* relaxation.

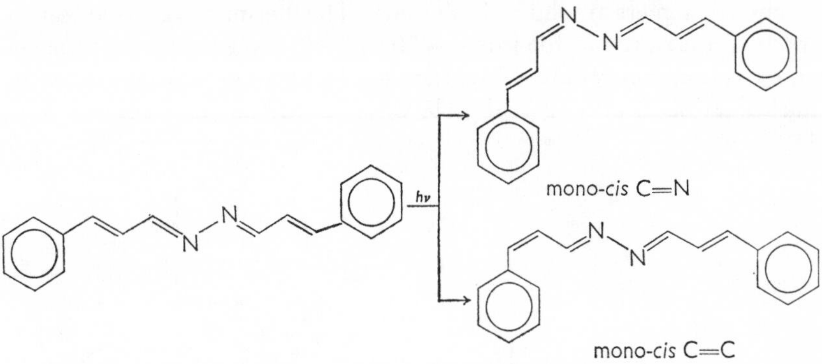

hν

mono-*cis* C=N

mono-*cis* C=C

(14)

Triphenylformazone undergoes a rearrangement which is thought to be a *cis–trans* isomerization about the C=N bond. A benzene or toluene solution of this compound changes colour from red to yellow upon irradiation with visible light[44,47–49]. A quantum yield of about 0·02 has been measured for this process[44].

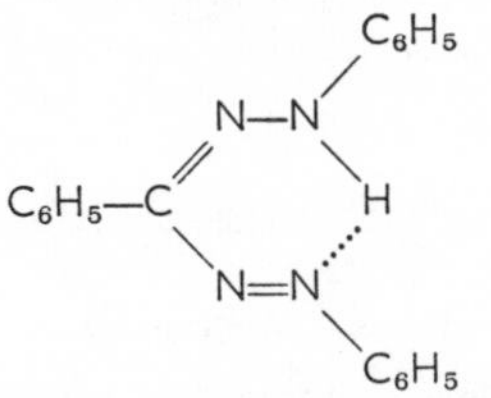

C_6H_5, N—N, C_6H_5—C, H, N=N, C_6H_5

The knowledge about the *cis–trans* isomerization of Schiff bases is all of recent origin[6]. From time to time there have been reports in the literature claiming that geometric isomers were isolated. Soon afterwards, however, other authors have stated that they were unable to repeat the experiment or that the isomerism in question was a case of dimorphism. This unfruitful search for *cis–trans* isomerism gradually caused the development of theories according to which *cis–trans* isomerism should not exist with compounds of the benzylideneaniline type.

In 1957, in work at very low temperatures with solutions of aromatic Schiff bases, changes were observed which were characteristic of *cis–trans* isomerization[50]. The compounds in question were *N*-benzylideneaniline, *N*-(α-naphthylidene)-α-naphthylamine and related compounds. When irradiated at −100°c and below, these gave changes in the absorption spectrum which could be reversed by increasing the temperature of the solution. Absorption spectra estimated for two geometric isomers are shown in Figure 3. The thermal relaxation which could be measured in the range -70 to -40°C was estimated to have an activation energy of 16–17 kcal/mole, which should be compared with the values of 23 and 42 kcal/mole which apply to the π-isoelectronic

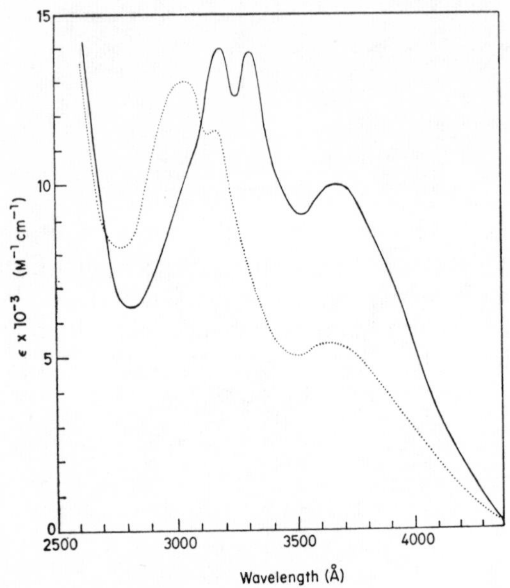

FIGURE 3. Absorption spectrum of *N*-benzylideneaniline in methylcyclohexane/isopentane (1:1) at -140°C[50].
———————— *trans* isomer
· · · · · · *cis* isomer

molecules azobenzene and stilbene, respectively. *N*-benzohydrylideneaniline showed no spectral changes under the same circumstances, and this was taken as additional proof for the interpretation of *cis–trans* isomerism.

C=N

N-benzohydrylideneaniline

Later it became possible to demonstrate *cis–trans* isomerism with *N*-benzylideneaniline and similar compounds even at room temperature[25,26]. At this temperature the thermal relaxation proceeds with a half-life of about one second. Table 3 shows the range of activation energies obtained for this reaction for different derivatives of *N*-benzylideneaniline, all *para* substituted. It is assumed that the *trans* form of *N*-benzylideneaniline as a rule is thermally favoured and that irradiation leads to an increased population of the *cis* species. The recorded rate constant, k, would then refer to a *cis* → *trans* interconversion.

$$R^1C_6H_4CH{=}NC_6H_4R^2 \ (trans) \underset{k}{\overset{h\nu}{\rightleftharpoons}} R^1C_6H_4CH{=}NC_6H_4R^2 \ (cis) \qquad (15)$$

The aniline ring in equilibrium (15) is probably rotated with respect to the plane of the rest of the molecule[6,58–64]. The explanation is based on the tendency of the lone pair electrons of the nitrogen atom to enter into conjugation with the phenyl ring. This causes marked differences in the effect of a substitution on the aldehyde ring and the aniline ring. Thus the thermal relaxation of photo-isomerized solutions of *para*- or *meta*-substituted *N*-benzylideneaniline follows the Hammett equation, $\log k = \rho\sigma + \log k_0$, but the ρ value is considerably higher for substitution on the nitrogen side than on the carbon side[55,65] (Figure 4). The two values are 1·85 and 0·41 (0·35), respectively. An even larger difference in the effect of substituents has been found for Schiff bases derived from benzophenone[57]. A ρ value of about +0·1 for substitution on a phenyl ring attached to the carbon atom of the azomethine bond should be compared with a value of about +1·7 for substitution on the nitrogen side. The positive value of ρ indicates that the low electron density at the C=N bond facilitates the reaction. This is in line with recent findings that the rate constant for the thermal relaxation of photo-isomerized aromatic Schiff bases follows the bond order of the azomethine bond[66,67]. The Schiff bases in question have the following structure:

$$Ar^1CH{=}NAr^2$$

TABLE 3. Kinetic data for the thermal *cis–trans* relaxation at the C=N bond.

Compound	Solvent	Method	Half life at 30°c (s)	Activation energy (kcal/mole)
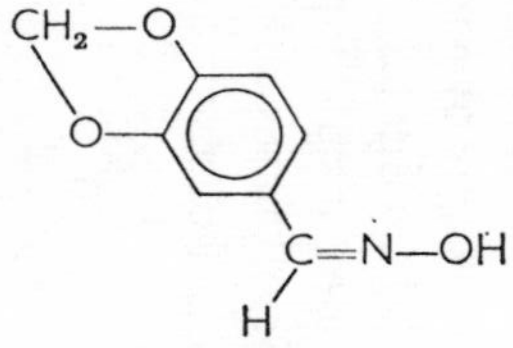 CH_2—O, O, C=N—OH, H	benzene	dilatometry[51]	$\sim 10^4$	22·6
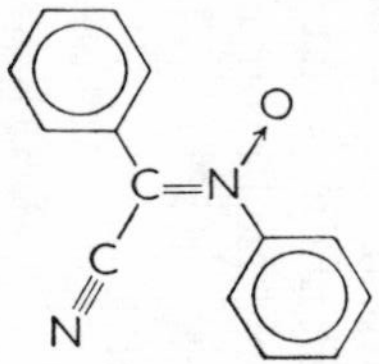 O, C=N, C≡N	n-butanol	spectro-photometry[52]	$\sim 10^6$	24·6
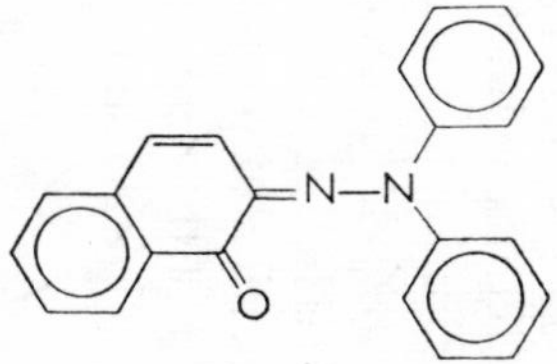 =N—N, O	methyl cyclohexane	spectro-photometry[45]	$\sim 10^3$	10
	ethanol	spectro-photometry[45]	$\sim 10^3$	16–17

Compound	Solvent	Method		
$(F_3C)_2C{=}N{-}CF(CF_3)_2$	none	n.m.r.[53]	$\sim 10^{-3}$	13 ± 3
$(H_3C)_2C{=}N{-}C_6H_5$	quinoline	n.m.r.[54]	1	21
$o\text{-}HOC_6H_4CH{=}N{-}C_6H_5$ (OH···N)	ethanol (acetate buffered)	flash photolysis[26]	10^{-1}	14·2
$o\text{-}HOC_6H_4CH{=}N{-}C_{10}H_7$ (2-naphthyl; OH···N)	ethanol (acetate buffered)	flash photolysis[26]	10^{-1}	15·4

(continued)

Compound	Solvent	Method	Half life at 30°C (s)	Activation energy (kcal/mole)
R^1-C_6H_4-CH=N-C_6H_5; R^1 = O^-, $N(CH_3)_2$, OH, OCH_3, CH_3, H, Cl, Br or NO_2 C_6H_5-CH=N-C_6H_4-R″; R″ = $N(CH_3)_2$, OCH_3, CH_3, H, Cl or Br	ethanol (acetate buffered)	flash photolysis[55]	10^{-1}–10	14–18
R-C_6H_4-C(C_6H_5)=N—CH_3; R = Cl or NO_2	cyclohexane	ultraviolet spectro-photometry[56]	10^5	$25 \pm 0{\cdot}3$ (R = Cl) $27{\cdot}1 \pm 0{\cdot}5$ (R = NO_2)

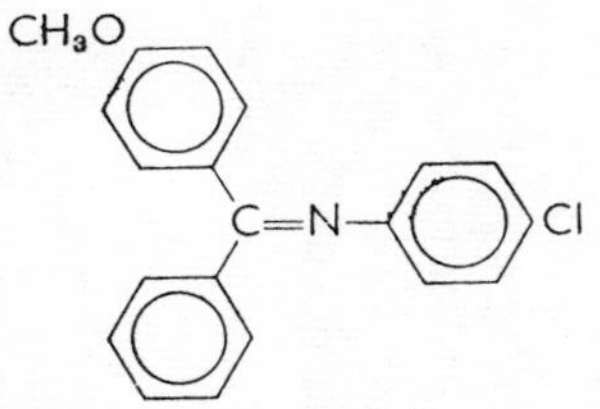 CH_3O C=N Cl	carbon tetrachloride	ultraviolet spectrophotometry[57]	10	19·7 ± 0·4
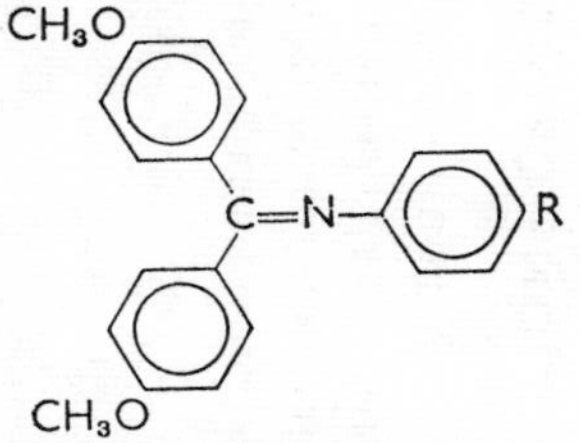 CH_3O C=N R CH_3O $R = N(CH_3)_2$, CH_3, H or $COOC_2H_5$	carbon tetrachloride	n.m.r.[57]	10^{-1}–10	18·1 ± 0·1(R = H)

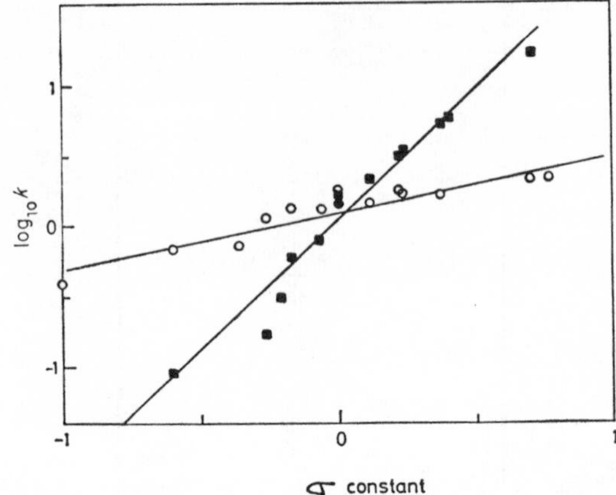

FIGURE 4. *Cis–trans* isomerization. Hammett plot for the thermal relaxation of photo-isomerized anils at 30°C[55,65].

(1)—CH=N—(2)

○ substitution on ring 1

■ substitution on ring 2

where Ar^1 and Ar^2 are phenyl, naphthyl or anthranyl groups. The bond orders were obtained from MO–LCAO calculations using the Hückel approximation.

Little is known about which electronic states are involved in the photochemical *cis–trans* isomerization. The *cis* → *trans*, as well as the *trans* → *cis*, isomerization of α-cyano-α-phenyl-*N*-phenylnitrone takes place by eosine and uranine photosensitization, showing that the conversion proceeds via a triplet state in this case[52]. The reaction is discussed in more detail in Section VI.A. Data for the thermal isomerization *cis* → *trans* in n-butanol are given in Table 3.

VI. REACTIONS OF *N*-OXIDES

Among the *N*-oxides it is common to separate out a group of compounds, the so-called nitrones. The definition of a nitrone may be somewhat obscure but usually means an azomethine *N*-oxide in which canonical forms **4a** and **4b** contribute most to the structure[68].

$R^1R^2C{=}\overset{+}{N}(R^3)O^-$ **(4a)** $R^1R^2\overset{+}{C}{-}\ddot{N}(R^3)O^-$ **(4b)**

Compounds in which the positive charge is markedly delocalized are not included, for example, pyridine-*N*-oxide is not a nitrone.

pyridine-N-oxide

A. Rearrangement of Nitrones

Irradiation of nitrones may produce isomers having the three-membered oxaziridine ring[69,70].

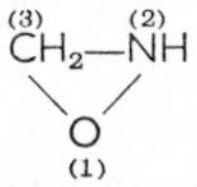

(3) CH_2—NH (2), O (1)

One of the first syntheses of this kind concerned 5,5-dimethyl-1-pyrroline-1-oxide, which gives a bicyclic oxaziridine when irradiated in a cyclohexane–ethanol solution[71].

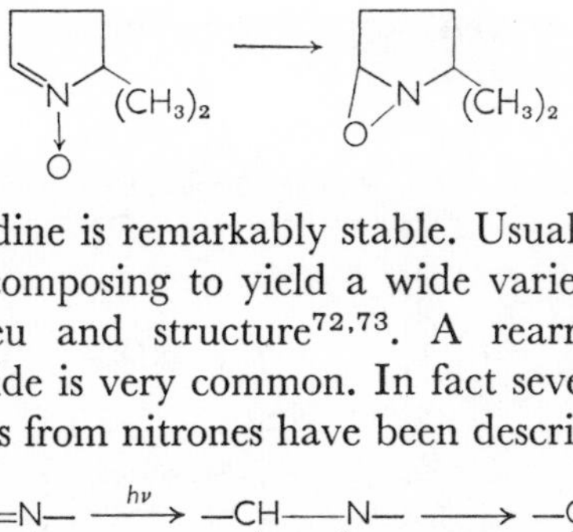

(16)

This fused oxaziridine is remarkably stable. Usually oxaziridines are very unstable, decomposing to yield a wide variety of products depending on milieu and structure[72,73]. A rearrangement to the corresponding amide is very common. In fact several photochemical syntheses of amides from nitrones have been described[74–76].

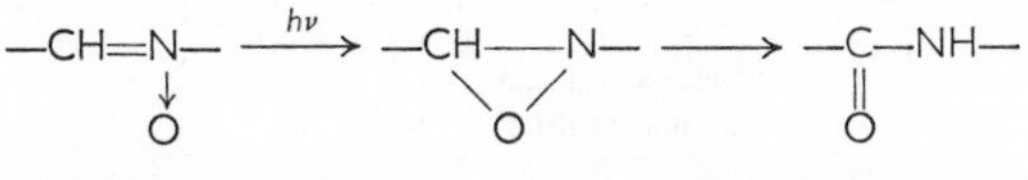

$$—CH{=}N(\rightarrow O)— \xrightarrow{h\nu} —CH—N—(\text{via }O) \longrightarrow —C(=O)—NH— \qquad (17)$$

The reactions of certain *N*-phenylnitrones appear to be strongly solvent dependent[77,78]. Thus it has been described that irradiation in benzene yields oxaziridine which reacts back to nitrone in the dark. In ethanol, on the other hand, the amide is obtained as a photo-product.

Investigations on *N*,α-diphenylnitrone indicate that the oxygen transfer to form oxaziridine takes place in the lowest excited singlet state before any *cis–trans* isomerization has time to occur (equation 18)[79]. A quantum yield of 0·28 in cyclohexane and 0·18 in ethanol was recorded for the reaction with 3130 Å light in the range 30–35°c.

$$\xrightarrow{h\nu} \qquad (18)$$

The oxaziridine isomerized rapidly in its turn to form benzanilide.

Further understanding of the mechanism of the nitrone rearrangement comes from experiments with the two geometric isomers of α-cyano-α-phenyl-*N*-phenylnitrone[52]. The *trans* isomer has the lowest energy and the thermal *cis* → *trans* isomerization was recorded, see Table 3. Both geometric isomers photo-isomerize to oxaziridine without giving rise to *cis–trans* isomerization. On the other hand *cis* → *trans*, as well as *trans* → *cis* isomerization can be brought about photochemically if a suitable photosensitizer is present such as uranine or eosine, both having a long-lived triplet state. All this has led to the conclusion that the triplet state is the intermediate in the *cis-trans* isomerization of a nitrone. In the excited singlet state, however, the nitrone rearranges to oxaziridine by oxygen atom transfer before any triplet state is formed (Figure 5).

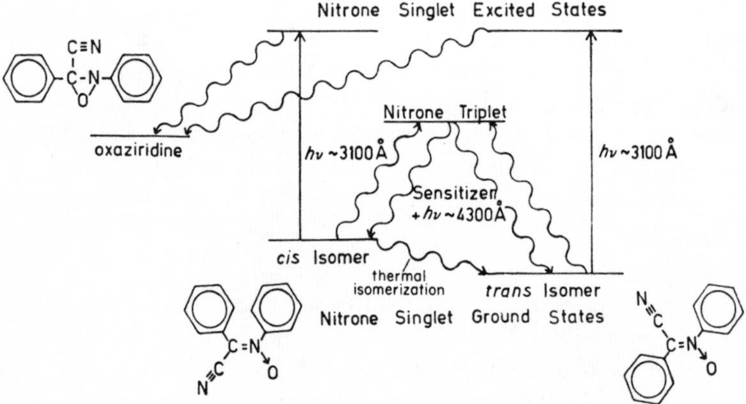

FIGURE 5. Scheme for the photochemistry of a nitrone[52].

Irradiation of nitrones has been reported to yield azo-compounds in the following reaction[80].

$$2\,C_6H_5\text{–}CH{=}N(\rightarrow O)\text{–}C_6H_5 \xrightarrow{h\nu} C_6H_5\text{–}N{=}N\text{–}C_6H_5 + 2\,C_6H_5\text{–}CHO \quad (19)$$

The reaction has a similarity to the rearrangement of an aromatic Schiff base to azobenzene and stilbene structures, where a four-membered ring compound has been suggested as an intermediate (see Section VIII.C).

B. Reactions of Aromatic N-Oxides

Aromatic *N*-oxides can also undergo photochemical conversion to the corresponding amide type compounds[81,82]. It has been postulated that the oxaziridine ring is an intermediate in this reaction too, as in the photolysis of nitrones[83–85]. This helps to explain the shift of a substituent which one obtains for instance with 2-methylquinoline-*N*-oxide, where 3-methylcarbostyril is the principal product[83].

$$\text{2-methylquinoline-}N\text{-oxide} \xrightarrow{h\nu} \text{3-methylcarbostyril} \quad (20)$$

Aromatic *N*-oxides are liable to undergo a series of other light-induced reactions. The path of the reactions appears to be strongly dependent on environmental conditions. For instance 2-methylquinoline-*N*-oxide yields as reaction products, besides 3-methylcarbostyril, the following compounds: *N*-methylcarbostyril, *N*-acetylindole, 2-methylquinoline and *N*-acetyl-2-hydroxy-2,3-dihydroindole[84,86]. An important photo-reaction of the *N*-oxides is the loss of oxygen. In the gas phase this appears to be the most important reaction pathway[87–89]. Thus, pyridine-*N*-oxide vapour can be photolysed to pyridine, but in ethanol solution the compound is reported to be stable towards ultraviolet light[84].

$$\text{pyridine-}N\text{-oxide} \xrightarrow{h\nu} \text{pyridine} \quad (21)$$

A quite remarkable case of wavelength dependence has been reported in the vapour phase photolysis of 2-picoline-*N*-oxide[87]. In this compound there is a methyl group adjacent to the $N \rightarrow O$

group and thus a possibility of intramolecular hydrogen bonding. With 2537 Å light 2-picoline is obtained, but when the molecule is excited with 3261 Å light the product is 2-pyridylmethanol. The different pathways are thought to originate from different electronic transitions, at 2537 Å a $\pi \rightarrow \pi^*$ and at 3261 Å a $n \rightarrow \pi^*$ transition.

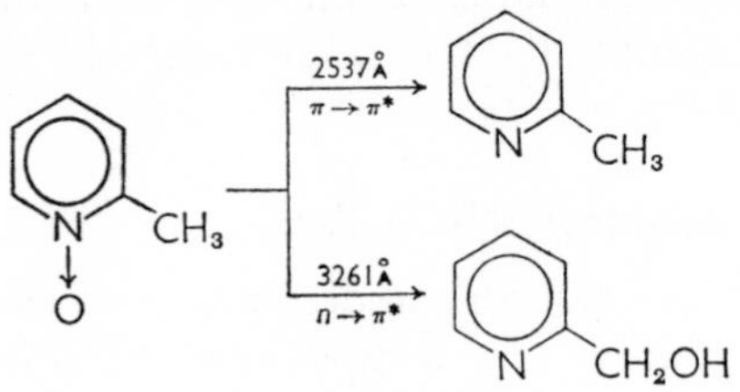

CH_3 CH_3 2537 Å $\pi \rightarrow \pi^*$ 3261 Å $n \rightarrow \pi^*$ CH_2OH (22)

Purine bases are as a rule quite insensitive to light but their *N*-oxides react according to a number of pathways[90–92]. For adenosine-1-*N*-oxide the following reaction scheme has been drawn[92]:

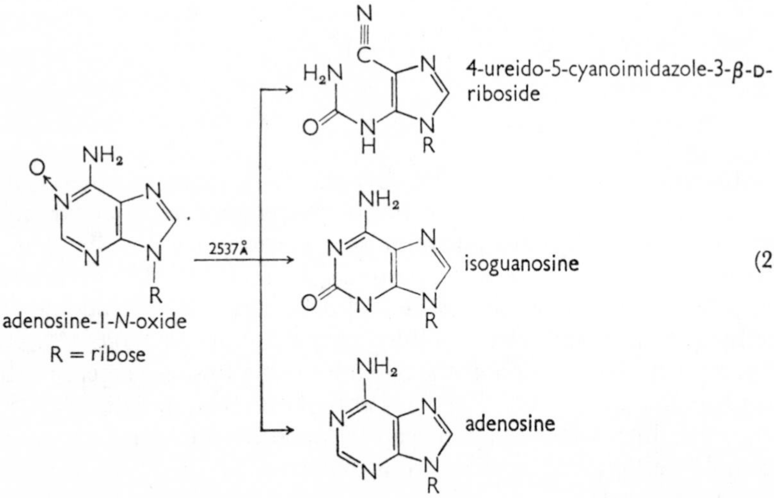

adenosine-1-*N*-oxide
R = ribose
2537 Å
4-ureido-5-cyanoimidazole-3-β-D-riboside
isoguanosine (23)
adenosine

In studies of the decomposition of adenine-1-*N*-oxide it was observed that the absorption peak at 2300 Å of the purine oxide showed a simple dependence on the amount of absorbed light; the peak disappeared proportionally to the amount absorbed. Furthermore, a fairly constant quantum yield of 0·10 was obtained for the wavelength region 2300–2800 Å. This has led to the suggestion that adenine-1-*N*-oxide could serve as a simple actinometre for irradiation with the biologically important wavelengths near 2600 Å[91].

VII. OXIME–AMIDE REARRANGEMENT

Irradiation of aryl aldoximes leads to the formation of amides. The reaction involves a transfer of oxygen from nitrogen to carbon and is analogous to a reaction described above for *N*-oxides. The similarity to the photochemical rearrangement of azoxybenzene to hydroxyazobenzene is also striking. Benzaldoxime substituted at the phenyl ring may show a very marked substituent dependence with respect to the oxime–amide rearrangement[93]. Some compounds react violently while others do not react at all. The reason for this is not understood.

$$R\text{-}C_6H_4\text{-}CH{=}NOH \xrightarrow{h\nu} R\text{-}C_6H_4\text{-}C(=O)\text{—}NH_2 \qquad (24)$$

VIII. CYCLIZATION OF SCHIFF BASES

A. Oxidative Ring Closure

The synthesis of dibenzophenanthridine from *N*-(1-naphthylidene)-1-naphthylamine represents probably the first observation of a photo-oxidative ring closure of an aromatic Schiff base to form a phenanthridine[94]. In this case the compound was irradiated in an ethanol solution in the presence of air; a hypothetical intermediate, dihydrophenanthridine, was suggested.

$$C_{10}H_7\text{-}CH{=}N\text{-}C_{10}H_7 \xrightarrow{h\nu} \text{dihydrophenanthridine (CH–CH, CH–N)} \xrightarrow[\text{oxidation}]{-H_2} \text{dibenzophenanthridine} \qquad (25)$$

The reaction is analogous to the well-known oxidative ring closure of stilbene and azobenzene and derivatives of these compounds. In the case of stilbene one obtains phenanthrene, and in the case of azobenzene the product is phenazone.

Also the simple aromatic Schiff base, *N*-benzylideneaniline, can undergo a photo-reaction to form phenanthridine[95]. An oxidizing agent must be present, for instance oxygen or iodine. Furthermore it was proved necessary to decrease the temperature to about 10°c in order to obtain this reaction in cyclohexane, benzene or ethanol solution. At higher temperature the anil decomposes. The explanation for this has been that the reaction proceeds via the *cis* isomer in an analogous way to what has been found for the photochemical reaction stilbene → phenanthrene. In this case it is necessary to excite the *cis* isomer with light. It is known that the photo-stationary concentration of *cis*-anil is very low at room temperature. Half-lives of about one second have been recorded for the thermal isomerization process, compare Section V. Thus a photon is unlikely to find a *cis* molecule and cause cyclization at room temperature.

$$(H)(C_6H_5)C{=}N{-}C_6H_5 \underset{}{\overset{h\nu_1}{\rightleftharpoons}} (C_6H_5)(H)C{=}N{-}C_6H_5 \xrightarrow{h\nu_2} \text{phenanthridine} \tag{26}$$

If this mechanism is right the photo-cyclization of *N*-benzohydrylideneaniline ought to be more efficient since this molecule is always in a favourable conformation.

$$(C_6H_5)_2C{=}N{-}C_6H_5 \xrightarrow[\text{oxidation}]{h\nu} \text{6-phenylphenanthridine} \qquad \Phi \sim 3 \times 10^{-5} \tag{27}$$

This was proved to be the case; reaction (27) proceeds more efficiently and with higher quantum yields than reaction (26). Nevertheless the quantum yield of reaction (27) is low, 3×10^{-5} at 25°c on excitation with 3130 Å light. The corresponding reaction at the C=C bond, reaction (28), has a quantum yield of 5×10^{-2}, keeping the environmental conditions the same. This difference in quantum yield has received the following explanation. The photo-cyclization takes place via a π,π^* excited state both with Schiff base and stilbene. This state is the lowest excited state in the stilbene case but a Schiff base has a lower n,π^* state. In this latter case a $\pi \rightarrow \pi^*$ excited molecule may therefore be rapidly converted into the n,π^* state and the time allowed for cyclization would be 'almost prohibitively short'. If the nitrogen atom

$$(C_6H_5)_2C{=}CH(C_6H_5) \xrightarrow[\text{oxidation}]{h\nu} \text{9-phenylphenanthrene} \qquad \Phi = 5 \times 10^{-2} \tag{28}$$

is protonated a π,π^* state will become the lowest state and the photocyclization proceeds with a higher quantum yield[96].

B. *Incorporation of a Solvent Fragment*

Irradiation of a Schiff base in ethanol solution may lead to cyclization in which a C_2 fragment from the solvent is incorporated[97]. The compound 3-phenyl-5,6-benzoquinoline was obtained when *N*-benzylidene-2-naphthylamine was exposed to ultraviolet light in the presence of air. The C_2 fragment probably stems from acetaldehyde formed in the photo-oxidation of ethanol.

$$C_6H_5CH{=}N\text{-}C_{10}H_7 \xrightarrow[\text{ethanol}]{h\nu} \text{3-phenyl-5,6-benzoquinoline} \tag{29}$$

Substituted benzoquinolines with a side chain in the 2-position are obtained when the Schiff base is photolysed in the appropriate higher alcohol[98]. The solvent may also replace the aldehyde part of the original Schiff base and thus participate in the reaction twice over. (Schiff bases are commonly thought of as consisting of an aldehyde or ketone part together with an amine part, as they may be produced in the reaction: $R^1R^2CO + R^3NH_2 \rightarrow R^1R^2C{=}NR^3 + H_2O$.)

C. *Dimerization*

It has been reported that Schiff bases may photochemically dimerize across the C=N bond to form a four-membered ring[99]. The evidence for the formation of such a ring, 1,2-diazetidine, rests on product analysis. Photolysis of *N*-(*p*-dimethylaminobenzylidene)aniline gave essentially *trans*-azobenzene and *cis*-4,4′-bis-(dimethylamino)stilbene, plus some 9-dimethylaminophenanthridine. A similar reaction has been observed in the photolysis of nitrones (see Section VI.A).

$$2\,R^1CH{=}NR^2 \xrightarrow{h\nu} \begin{array}{c} R^1\qquad R^2 \\ CH{-}N \\ |\quad\ | \\ CH{-}N \\ R^1\qquad R^2 \end{array} \longrightarrow R^1CH{=}CHR^1 + R^2N{=}NR^2 \qquad (30)$$

IX. ADDITION OF AN ALDEHYDE OR METHYL GROUP TO PHENANTHRENE QUINONIMINE

It is known that aldehydes add to phenanthraquinone monoimine photochemically[100]. The product is believed to be a hydroxyl-, amido-substituted phenanthrene or the corresponding cyclic compound, a derivative of dihydrooxazole.

C=NH, C=O $\xrightarrow[\text{RCHO}]{h\nu}$ H N—C(=O)—R, OH *or* H N, R, C, O, OH (31)

A similar addition reaction is also described for methyl substituted aromatic compounds, in which 2-arylphenanthroxazoles are formed[101]. The reaction was obtained both with ultraviolet light and ^{60}Co-γ-radiation.

NH, O $\xrightarrow[\text{CH}_3\text{Ar}]{h\nu}$ N, —Ar, O (32)

X. HYDROGEN ABSTRACTION

The photo-excited carbon–nitrogen double bond is known to abstract hydrogen when a suitable hydrogen donor is present; in many cases the solvent itself provides a suitable donor. Irradiation of benzophenone methylimine in propanol yields an almost quantitative conversion to benzhydryl methylamine and acetone[102]. The reaction is completely inhibited by adding small amounts of compounds which

$$(C_6H_5)_2C{=}NCH_3 \xrightarrow[\text{isopropanol}]{h\nu} (C_6H_5)_2CH{-}NHCH_3 \qquad (33)$$

have a triplet energy less than 61·8 kcal/mole. This indicates that the chemical activity stems from a triplet state at about this energy level.

The C=N bond can also be photochemically reduced when it is a part of a quinonoid system. *N,N′*-diphenyl-*p*-phenylenediimine is reduced to a semiquinone imine when a suitable hydrogen donor is present[103]. The reduction of the diimine to diamine with 4630 Å light and allylthiourea as donor has been investigated in various solvents of different hydrogen-bonding strength[104]. A correlation was noticed between the extent of the blue shift of the fundamental absorption band and the quantum yield of the process. The quantum yield decreased to $\frac{1}{5}$ when going from benzene to methanol as λ_{max} changed from 4520 to 4390 Å. The existence of both a n,π^* and a π,π^* state is assumed, where only the n,π^* state leads to reaction (34). Interaction

$$C_6H_5{-}N{=}C_6H_4{=}N{-}C_6H_5 \xrightarrow[\text{allylthiourea}]{h\nu} C_6H_5{-}NH{-}C_6H_4{-}NH{-}C_6H_5 \qquad (34)$$

with the non-bonding electron pair on the nitrogen atom increases the n,π^* level relative to the π,π^* state in hydrogen bonding solvents and may reverse their order. A fast internal conversion $n,\pi^* \rightarrow \pi,\pi^*$ would then prevent the chemical reaction (Figure 6).

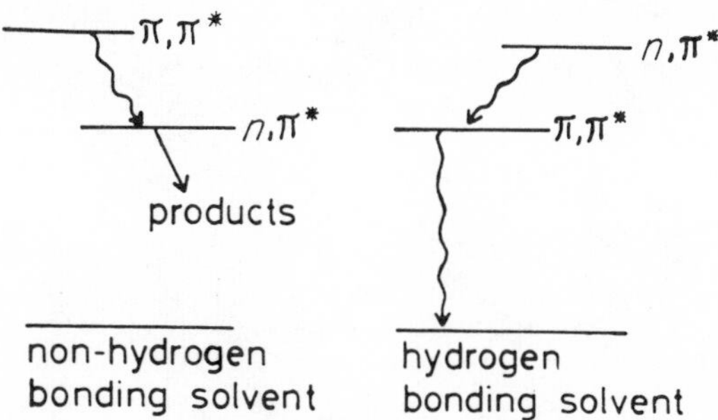

FIGURE 6. Solvent dependence of reaction (34)[104].

Several aromatic *N*-heterocycles have a tendency to undergo photo-reduction. A reaction of this kind frequently forms the base for photogalvanic cells. A good example is the group of phenothiazine dyes, such as thionine and methylene blue, which together with ferrous ions provide the environment for the electrode[105,106]. The detailed reaction scheme appears to be quite complex and involves a photo-induced reduction to a semiquinone radical which subsequently abstracts an electron to yield a two-electron reduction scheme[107]. The reaction is highly reversible in the absence of oxygen and represents a case of photochromism where exposure to light results in bleaching instead of colour formation.

$$2Fe^{2+} + H^{+} + (CH_3)_2N\text{-[phenothiazinium]}=\overset{+}{N}(CH_3)_2 \underset{}{\overset{h\nu}{\rightleftharpoons}} 2Fe^{3+} + (CH_3)_2N\text{-[10H-phenothiazine]}\text{-}N(CH_3)_2 \quad (35)$$

Abstraction of hydrogen from the solvent has been studied for several aromatic *N*-heterocycles. Characteristic examples are phenazine and acridine which yield more or less stable products with different degrees of hydrogenation[108–122]. The reduction is frequently accompanied by a dimerization. Diacridanes arise from acridines, and during the photolysis of an acridine solution the dimerization is noticed as a white precipitation. The hydrogen abstraction by acridine seems to take place in the excited singlet state[122].

$$2\,\text{acridine} \xrightarrow{h\nu} \text{(acridan-9-yl)CH–CH(acridan-9-yl)} \quad (36)$$

XI. REFERENCES

1. R. Bonnett, *J. Chem. Soc.*, 2313 (1965).
2. D. A. Nelson and J. J. Worman, *Tetrahedron Letters*, 507 (1966).
3. H. H. Jaffé, S.-J. Yeh and R. W. Gardner, *J. Mol. Spectr.*, **2**, 120 (1958).
4. J. Schulze, F. Gerson, J. N. Murrell and E. Heilbronner, *Helv. Chim. Acta*, **44**, 428 (1961).

5. J. D. Margerum and J. A. Sousa, *Appl. Spectr.*, **19**, 91 (1965).
6. G. Wettermark, *Svensk Kem. Tidskr.*, **79**, 249 (1967).
7. H. Stobbe, *Ber. Verhandl. Sächs. Akad. Wiss. Leipzig*, **74**, 161 (1922).
8. L. Chalkley, *Chem. Rev.*, **6**, 217 (1929).
9. S. S. Bhatnagar, P. L. Kapur and M. S. Hashuni, *J. Indian Chem. Soc.*, **15**, 573 (1938).
10. G. H. Brown and W. G. Shaw, *Rev. Pure Appl. Chem.*, **11**, 2 (1961).
11. V. de Gaouck and R. J. W. Le Févre, *J. Chem. Soc.*, 1457 (1939).
12. J. M. Tien and I. M. Hunsberger, *Chem. Ind.* (*London*), 119 (1955).
13. M. D. Cohen, G. M. J. Schmidt and S. Flavian, *J. Chem. Soc.*, 2041 (1964).
14. M. D. Cohen and G. M. J. Schmidt, *J. Chem. Soc.*, 1996 (1964).
15. M. D. Cohen and G. M. J. Schmidt, *J. Phys. Chem.*, **66**, 2442 (1962).
16. M. D. Cohen, Y. Hirshberg and G. M. J. Schmidt in *Hydrogen Bonding* (Ed. D. Hadzi), Pergamon Press, London, 1959, p. 293.
17. M. D. Cohen and G. M. J. Schmidt in *Reactivity of Solids* (Ed. J. H. de Boer), Elsevier, Amsterdam, 1961, p. 556.
18. A. A. Burr, E. J. Llewellyn and G. F. Lothian, *Trans. Faraday Soc.*, **60**, 2177 (1964).
19. A. Senior and F. G. Shepheard, *J. Chem. Soc.*, **95**, 1943 (1909).
20. G. Lindemann, *Z. Wiss. Phot. Photophysik Photochem.*, **50**, II, 347 (1955).
21. G. Wettermark and A. King, *Photochem. Photobiol.*, **4**, 417 (1965).
22. M. Padoa and T. Minganti, *Atti Accad. Nazl. Lincei*, **II 22**, 500 (1913); *Chem. Abstr.*, **8**, 1368 (1914).
23. M. D. Cohen, Y. Hirshberg and G. M. J. Schmidt, *J. Chem. Soc.*, 2051 (1964).
24. M. D. Cohen, Y. Hirshberg and G. M. J. Schmidt, *J. Chem. Soc.*, 2060 (1964).
25. G. Wettermark and L. Dogliotti, *J. Chem. Phys.*, **40**, 1486 (1964).
26. D. G. Anderson and G. Wettermark, *J. Am. Chem. Soc.*, **87**, 1433 (1965).
27. R. S. Becker and W. F. Richey, *J. Am. Chem. Soc.*, **89**, 1298 (1967).
28. H. Tiefenthaler, W. Dörscheln, H. Göth and H. Schmid, *Tetrahedron Letters*, 2999 (1964).
29. H. Göth, H. Tiefenthaler and W. Dörscheln, *Chimia* (*Aarau*), **19**, 596 (1965).
30. H. Göth and H. Schmid, *Chimia* (*Aarau*), **20**, 148 (1966).
31. E. F. Ullman and B. Singh, *J. Am. Chem. Soc.*, **88**, 1844 (1966).
32. A. Hantzsch, *Ber.*, **23**, 2325 (1890).
33. A. Hantzsch, *Ber.*, **24**, 51 (1891).
34. G. Ciamician and P. Silber, *Ber.*, **36**, 4266 (1904).
35. R. Ciusa, *Atti Accad. Nazl. Lincei*, [5] **15**, **II**, 721 (1907).
36. R. Stoermer, *Ber.*, **44**, 637 (1911).
37. O. L. Brady and F. P. Dunn, *J. Chem. Soc.*, **103**, 1619 (1913).
38. O. L. Brady and G. P. McHugh, *J. Chem. Soc.*, **125**, 547 (1924).
39. J. H. Amin and P. de Mayo, *Tetrahedron Letters*, 1585 (1963).
40. R. Calas, R. Lolande, F. Moulines and J.-G. Faugere, *Bull. Soc. Chim. France*, 121 (1965).
41. E. J. Poziomek, *J. Pharm. Sci.*, **54**, 333 (1965).
42. O. L. Brady and L. Klein, *J. Chem. Soc.*, **129**, 874 (1927).

43. J. Dale and L. Zechmeister, *J. Am. Chem. Soc.*, **75**, 2379 (1953).
44. D. Schulte-Frohlinde, *Ann. Chem.*, **622**, 47 (1959).
45. E. Fischer and M. Kaganowich, *Bull. Res. Council Israel, Sect. A*, **10**, 138 (1961).
46. E. Fischer in *Fortschritte der chemischen Forschung*, Band 7, Springer Verlag, Berlin, Heidelberg, New York, 1967, p. 605.
47. I. D. Hausser, *Naturwissenschaften*, **36**, 313 (1949).
48. I. Hausser, D. Jerchel and R. Kuhn, *Chem. Ber.*, **82**, 515 (1949).
49. R. Kuhn and H. M. Weitz, *Chem. Ber.*, **86**, 1199 (1953).
50. E. Fischer and Y. Frei, *J. Chem. Phys.*, **27**, 808 (1957).
51. R. J. W. Le Févre and J. Northcott, *J. Chem. Soc.*, 2235 (1949).
52. K. Koyano and I. Tanaka, *J. Phys. Chem.*, **69**, 2545 (1965).
53. S. Andreades, *J. Org. Chem.*, **27**, 4163 (1962).
54. H. A. Staab, F. Vögtle and A. Mannschreck, *Tetrahedron Letters*, 697 (1965).
55. G. Wettermark, J. Weinstein, J. Sousa and L. Dogliotti, *J. Phys. Chem.*, **69**, 1584 (1965).
56. D. Y. Curtin and J. W. Hausser, *J. Am. Chem. Soc.*, **83**, 3474 (1961).
57. D. Y. Curtin, E. J. Grubbs and C. G. McCarty, *J. Am. Chem. Soc.*, **88**, 2775 (1966).
58. V. A. Izmailskii and E. A. Smirnov, *Zh. Obshch. Khim.*, **26**, 3042 (1956).
59. N. Ebara, *Bull. Chem. Soc. Japan*, **33**, 534 (1960).
60. N. Ebara, *Bull. Chem. Soc. Japan*, **34**, 1151 (1960).
61. P. Brocklehurst, *Tetrahedron*, **18**, 299 (1962).
62. V. I. Minkin, E. A. Medyantseva and A. M. Simonov, *Dokl. Akad. Nauk SSSR*, **149**, 1347 (1963).
63. W. F. Smith, *Tetrahedron*, **19**, 445 (1963).
64. V. A. Izmailskii and Yu. A. Federov, *Zh. Fiz. Khim.*, **39**, 768 (1965).
65. G. Wettermark, *Arkiv. Kemi.* **27**, 159 (1967).
66. G. Wettermark and E. Wallström, *Acta Chem. Scand.*, **22**, 675 (1968).
67. G. Wettermark in *Nobel Symposium Series*, Vol. 5 (Ed. S. Claesson), Almquist & Wiksell, Stockholm, 1967.
68. G. R. Delpierre and M. Lamchen, *Quart. Rev. (London)*, **19**, 329 (1965).
69. F. Kröhnke, *Ann. Chem.*, **604**, 203 (1957).
70. M. Kamlet and L. Kaplan, *J. Org. Chem.*, **22**, 576 (1957).
71. R. Bonnett, V. M. Clark and A. Todd, *J. Chem. Soc.*, 2102 (1959).
72. W. E. Emmons, *J. Am. Chem. Soc.*, **79**, 5739 (1957).
73. H. Shindo and B. Umezawa, *Chem. Pharm. Bull.* (*Tokyo*), **10**, 492 (1962).
74. L. Chardonners and P. Heinrich, *Helv. Chim. Acta*, **32**, 656 (1949).
75. B. M. Mikhailov and G. S. Ter-Sarkisyan, *Bull. Acad. Sci. USSR, Div. Chem. Sci. Engl. Transl.*, 559 (1954); *Chem. Abstr.*, **49**, 10953 (1955).
76. A. Schönberg, *Preparative Organische Photochemie*, Springer Verlag, Berlin, 1958.
77. J. S. Splitter and M. Calvin, *J. Org. Chem.*, **20**, 1086 (1955).
78. J. S. Splitter and M. Calvin, *J. Org. Chem.*, **23**, 651 (1958).
79. K. Shinzawa and I. Tanaka, *J. Phys. Chem.*, **68**, 1205 (1964).
80. J. Kosar, *Light-Sensitive Systems*, John Wiley and Sons, New York, 1965, p. 360.

81. J. K. Landquist, *J. Chem. Soc.*, 2830 (1953).
82. O. Buchardt, *Acta. Chem. Scand.*, **17**, 1461 (1963).
83. O. Buchardt, J. Becher and C. Lohse, *Acta Chem. Scand.*, **19**, 1120 (1965).
84. M. Ishikawa, S. Yamada and C. Koneko, *Chem. Pharm. Bull* (*Tokyo*), **13**, 747 (1965).
85. M. Ishikawa, S. Yamada, H. Hotta and C. Kaneko, *Chem. Pharm. Bull.* (*Tokyo*), **14**, 1102 (1966).
86. O. Buchardt, J. Becher, C. Lohse and J. Møller, *Acta Chem. Scand.*, **20**, 262 (1966).
87. N. Hata, *Bull. Chem. Soc. Japan*, **34**, 1440 (1961).
88. N. Hata, *Bull. Chem. Soc. Japan*, **34**, 1444 (1961).
89. N. Hata and I. Tanaka, *J. Chem. Phys.*, **36**, 2072 (1962).
90. G. B. Brown, G. Levin and S. Murphy, *Biochemistry*, **3**, 880 (1964).
91. G. Levin, R. B. Setlow and G. B. Brown, *Biochemistry*, **3**, 883 (1964).
92. F. Cramer and G. Schlingloff, *Tetrahedron Letters*, 3201 (1964).
93. J. H. Amin and P. de Mayo, *Tetrahedron Letters*, 1585 (1963).
94. M. P. Cava and R. H. Schlessinger, *Tetrahedron Letters*, 2109 (1964).
95. F. B. Mallory and C. S. Wood, *Tetrahedron Letters*, 2643 (1965).
96. G. M. Badger, C. P. Joshua and G. E. Lewis, *Tetrahedron Letters*, 3711 (1964).
97. J. S. Shannon, H. Silberman and S. Sternhell, *Tetrahedron Letters*, 659 (1964).
98. P. J. Collin, H. Silberman, S. Sternhell and G. Sugowdz, *Tetrahedron Letters*, 2063 (1965).
99. S. Searles, Jr. and R. A. Clasen, *Tetrahedron Letters*, 1627 (1965).
100. A. Schönberg and W. I. Awad, *J. Chem. Soc.*, 197 (1945).
101. G. Pfundt and W. M. Hardham, *Tetrahedron Letters*, 2411 (1965).
102. M. Fischer, *Tetrahedron Letters*, 5273 (1966).
103. H. Linschitz, J. Rennert and T. M. Korn, *J. Am. Chem. Soc.*, **76**, 5839 (1954).
104. J. Rennert and J. Wiesenfeld, *Photochem. Photobiol.*, **5**, 337 (1966).
105. E. Rabinowitch, *J. Chem. Phys.*, **8**, 551 (1940).
106. E. Rabinowitch, *J. Chem. Phys.*, **8**, 560 (1940).
107. C. A. Parker, *J. Phys. Chem.*, **63**, 26 (1959).
108. C. Dufraisse, A. Etienne and E. Toromanoff, *Compt. Rend.*, **235**, 759 (1952).
109. A. Kellman, *J. Chim. Phys.*, **54**, 468 (1957).
110. V. Zanker and P. Schmid, *Z. Physik. Chem.* (*Frankfurt*), **17**, 11 (1958).
111. A. Kellman, *J. Chim. Phys.*, **56**, 574 (1955).
112. V. Zanker and H. Schnith, *Chem. Ber.*, **92**, 2210 (1959).
113. A. Kellmann, *J. Chim. Phys.*, **57**, 1 (1960).
114. S. Kato, S. Minagawa and M. Koizumi, *Bull. Chem. Soc. Japan*, **34**, 1026 (1961).
115. M. Giurgea, V. Topa and S. Haragea, *J. Chim. Phys.*, **58**, 705 (1961).
116. A. Kellmann, *Bull. Soc. Chim. Belges*, **71**, 811 (1962).
117. S. J. Ladner and R. S. Becker, *J. Phys. Chem.*, **67**, 2481 (1963).
118. M. Giurgea, G. Mihai, V. Totar and M. Musa, *J. Chem. Phys.*, **61**, 619 (1964).

119. F. Mader and V. Zanker, *Chem. Ber.*, **97**, 2418 (1964).
120. A. Kellmann and J. T. Dubois, *J. Chem. Phys.*, **42**, 2518 (1965).
121. V. Zanker, E. Erhardt, F. Mader and J. Thies, *Z. Physik. Chem.* (*Frankfurt*), **48**, 179 (1966).
122. E. Vander Donckt and G. Porter, *J. Chem. Phys.*, **46**, 1173 (1967).

CHAPTER **13**

Imidoyl halides

R. Bonnett

Queen Mary College, London, England

I. INTRODUCTION AND NOMENCLATURE

The imidoyl halides are characterized by the function —C(Hal)=N— and thus they are structurally related to the aldoazomethines in the same way as the acyl halides are related to the aldehydes. It is probably true to say that, in general, the chemistry of the imidoyl halides is unfamiliar. A comprehensive review does not appear to have been written, although a summary[1] by von Braun appeared in 1934. The topic is somewhat obscured by a variety of nomenclature. One practice is to append the term 'iminochloride' (or imidochloride, imide chloride or imidyl chloride) to the name of the parent amide, so that PhCCl=NPh becomes benzanilide iminochloride. While this nomenclature is convenient on occasion, it is preferred here to adopt a more systematic procedure and to name the compounds as derivatives of the *imidic acids* RC(=NH)OH. Thus PhCCl=NPh becomes *N*-phenylbenzimidoyl chloride.

Cyclic compounds are exceptions to this treatment since they are most usefully named after the parent ring system (e.g. 1-pyrroline,

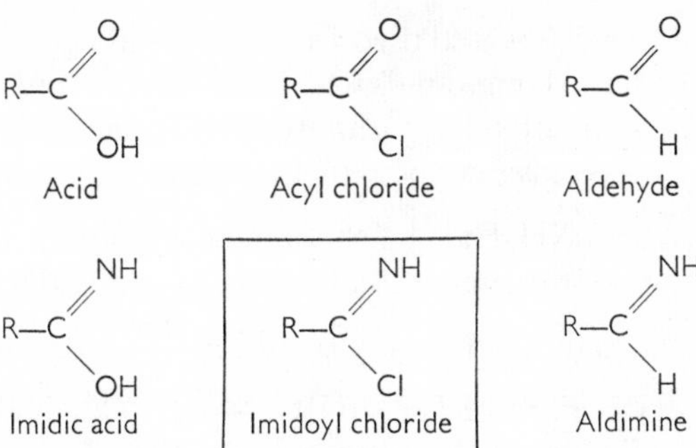

1-piperideine). Some representative structures and names are given in Chart 1.

PhCCl=NPh
N-Phenylbenzimidoyl chloride [2]
large colourless plates
m.p. 39–40°, b.p. 310°

PhCCl=NMe
N-Methylbenzimidoyl chloride [3]
colourless mobile liquid
b.p. 112°/30 mm

$Me_2CHCCl{=}NC_6H_4Me$-*p*
N-(*p*-Tolyl)isobutyrimidoyl chloride [4]
b.p. 80–85°/0·8 mm, n_D^{25} 1·5299

n-BuEtCHCCl=NBu-*n*
N-(*n*-Butyl)-2-ethylhexanimidoyl chloride [4]
b.p. 72–76°/0·7 mm, n_D^{25} 1·4485
d^{25} 0·893

$Cl_3CCCl{=}NPh$
N-Phenyltrichloroacetimidoyl chloride [5]
m.p. 37°, b.p. 136–138°/14 mm

Ph(F)C=N(Bu-*t*)
syn-*N*-(*t*-Butyl)benzimidoyl fluoride [6]
Decomposes at 25°

2,3,3-Trichloro-1-piperideine [7]
m.p. 27°

1,3,3,4,5,6,7-Heptachloroisoindolenine [8]
colourless rhombs, m.p. 167–168°,
b.p. 208°/6 mm

$PhCCl{=}NSO_2Ph$
N-Benzenesulphonylbenzimidoyl chloride [9]
large transparent plates, m.p. 79–80°

MeCOCCl=NOH
Chloroisonitrosoacetone [10]
(*N*-hydroxypyruvimidoyl chloride)
colourless crystals, m.p. 104·5–105·5°

CHART 1. Imidoyl halides—nomenclature.

The term *amidohalide* is also used frequently and needs definition. It is used here to refer to compounds which may be regarded as amides in which the oxygen function has formally been replaced by two halogen atoms: such compounds may be either covalent (e.g. $RCHal_2NH_2$) or ionic (e.g. $RCHal{=}\overset{+}{N}H_2Hal^-$). The ionic compounds themselves fall into two classes: (a) the imidoyl halide hydrohalides derived by reaction of hydrogen halide with nitriles ($RCN \xrightarrow{2HCl} RCCl{=}\overset{+}{N}H_2\ Cl^-$) or with imidoyl halides ($RCCl{=}NR \xrightarrow{HCl} RCCl{=}\overset{+}{N}HR'\ Cl^-$) and (b) the *N,N*-dialkylamidochlorides, i.e. the ternary immonium salts derived from tertiary amides ($RCONR_2' \xrightarrow{PCl_5} RCCl{=}\overset{+}{N}R_2'\ Cl^-$).

Two limitations are being placed on the scope of this review. First, most of the work reported in the literature deals with the imidoyl *chlorides*: this is because, on the one hand, the bromides (and, presumably, the iodides, which have received very little attention) are less stable or are less readily isolated in a pure condition[11], while, on the other hand, the study of the imidoyl fluorides has begun only quite recently. There are few definitive data on comparative reactivity as the halogen is varied, and the bulk of the reactions to be discussed here will refer to the chlorides. Secondly, it is considered appropriate to devote most attention to the typical imidoyl halides, i.e. those in which the substituents at the characteristic function are aryl or alkyl groups. However, compounds with other substituents (e.g. $RCCl{=}NSO_2R'$, $RCCl{=}NOH$, $RCCl{=}NNHR'$, $Hal_2C{=}NR$, $RCOCCl{=}NR'$) are not without interest and will also be mentioned where appropriate.

Whether or not the imidoyl halides are well known today, they are certainly well established in the annals of organic chemistry and have been studied for well over a century, although some of the early work was confused by the sensitivity to moisture and by the instability of many of the compounds. In 1847 Cahours[12] observed the effect of phosphorus pentachloride on amides: he obtained the corresponding nitriles, but did not observe intermediate compounds. Other investigators attempted similar reactions[13], but the first reasonably pure imidoyl chloride does not appear to have been obtained until 1858, when Gerhardt[14], who had the fortune to chose an amide which gave a tractable product, isolated *N*-phenylbenzimidoyl chloride. Wolkoff[15] confirmed part of Gerhardt's work and prepared other imidoyl halides. At about this time interest was developing in the reaction of the nitriles with hydrogen halides. Crystalline adducts were obtained by Gautier[16] and the structures of these compounds have subsequently been the subject of much investigation and controversy. Only recently has a satisfactory conclusion (based, in part, on a neutron diffraction analysis) been possible (see Section II.E.1).

Towards the close of the last century Wallach placed the chemistry of the imidoyl halides on a somewhat firmer basis: he prepared new compounds and explored new reactions. This study was taken up by Julius von Braun and his colleagues, who, for a period of about 30 years from 1904, devoted a considerable effort to extending imidoyl halide chemistry. Much of our knowledge today is based on these investigations. While it is always necessary to step with caution through the older literature (notably because spectroscopic methods were not then available to check proposed chemical structures) it appears to the author that the bulk of the observations made earlier in this area, and those presented here, are sound. It is pertinent to note that there has lately been a marked reawakening of interest in these compounds, for example in 1,3-dipolar addition reactions (Huisgen), and in the imidoyl fluorides which had escaped study in the earlier years.

II. PREPARATION

Undoubtedly the most important route to the imidoyl chlorides starts with the corresponding amide and proceeds in a manner analogous to the familiar preparation of acyl halides from acids, thus:

$$\text{RCOOH} \xrightarrow{\text{PCl}_5\text{ etc.}} \text{RCOCl}$$

$$\text{RCONHR}' \xrightarrow{\text{PCl}_5\text{ etc.}} \text{RCCl}{=}\text{NR}'$$

Several subsidiary routes are available, however, and for compounds with substituents other than alkyl and aryl special approaches may be required.

A. Imidoyl Halides from Amides

This method has been used successfully for the chlorides and bromides, although the latter tend to decompose more readily and are difficult to isolate[11]. In a fairly general procedure a secondary amide (1 mole) is heated with phosphorus pentachloride (1 mole) either alone[14], or in an anhydrous inert solvent such as benzene[18], toluene[19], xylene[7] or nitrobenzene[21]. The solvent and phosphorus oxychloride are removed by distillation, and the residue, which in a multistage synthesis is often sufficiently pure to be used directly[22], may be purified by crystallization or by fractional distillation under reduced pressure. Although complications arise in certain cases (see p. 619) the yields are often good.

Reagents equivalent to the phosphorus pentahalides have been employed (PCl_3/Cl_2[8], PBr_3/Br_2[23], $PhPCl_4$[24]). Thionyl chloride is also a useful reagent; although reported to be ineffective in early work[27,28] it, in fact, gives excellent results, especially with amides derived from aromatic acids[29]. Moreover the by-products are gaseous (contrast $POCl_3$, b.p. 105°) and the imidoyl halide is the more readily purified. 2·5–3 equivalents of thionyl chloride are recommended[30], and the reagent may be used either neat or in a diluent such as nitromethane. Phosgene has also been employed[31], and is the preferred reagent for imidoyl chlorides derived from aliphatic acids.

Comparative studies on the reaction of the three most useful reagents (PCl_5, $SOCl_2$, $COCl_2$) with benzanilide[32] and dimethylformamide[33] have been published. Phosgene did not appear to react with the former substrate, and other failures with this reagent and *N*-aryl amides have been noted[34].

Chart 2 summarizes some preparations of this general type.

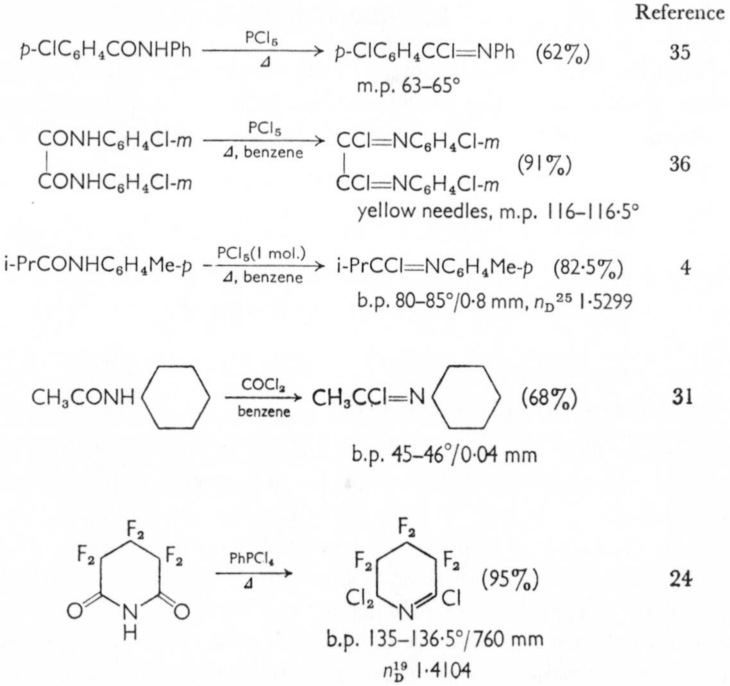

Reaction	Reference
p-$ClC_6H_4CONHPh$ $\xrightarrow[\Delta]{PCl_5}$ *p*-$ClC_6H_4CCl{=}NPh$ (62%), m.p. 63–65°	35
$CONHC_6H_4Cl$-*m* / $CONHC_6H_4Cl$-*m* (oxamide) $\xrightarrow[\Delta,\ \text{benzene}]{PCl_5}$ $CCl{=}NC_6H_4Cl$-*m* / $CCl{=}NC_6H_4Cl$-*m* (91%), yellow needles, m.p. 116–116·5°	36
i-$PrCONHC_6H_4Me$-*p* $\xrightarrow[\Delta,\ \text{benzene}]{PCl_5(1\ \text{mol.})}$ i-$PrCCl{=}NC_6H_4Me$-*p* (82·5%), b.p. 80–85°/0·8 mm, n_D^{25} 1·5299	4
CH_3CONH-cyclohexyl $\xrightarrow[\text{benzene}]{COCl_2}$ $CH_3CCl{=}N$-cyclohexyl (68%), b.p. 45–46°/0·04 mm	31
3,3,4,4,5,5-hexafluoroglutarimide (F_2, F_2, F_2; O, NH, O) $\xrightarrow[\Delta]{PhPCl_4}$ ring with F_2, F_2, F_2, Cl_2, Cl, N (95%), b.p. 135–136·5°/760 mm, n_D^{19} 1·4104	24

$$PhCONHPh \xrightarrow[\Delta]{SOCl_2} PhCCl{=}NPh \quad (100\%)$$ 29

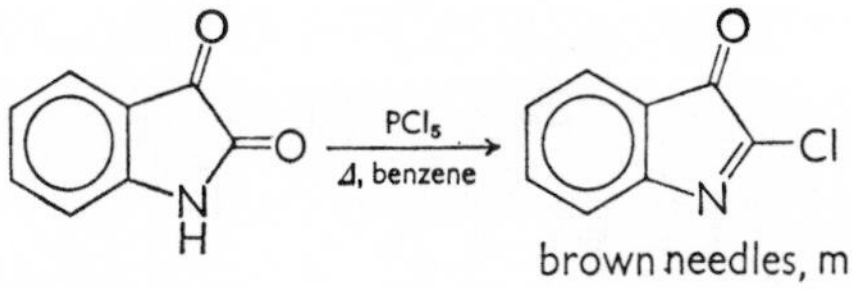

[Isatin] $\xrightarrow[\Delta,\ benzene]{PCl_5}$ [2-chloro-3H-indol-3-one] —Cl (73%) 37

brown needles, m.p. 180° (decomp.)

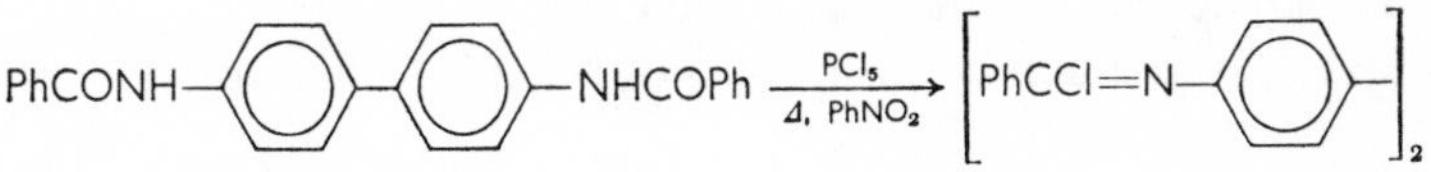

PhCONH–C₆H₄–C₆H₄–NHCOPh $\xrightarrow[\Delta,\ PhNO_2]{PCl_5}$ $[PhCCl{=}N\text{–}C_6H_4\text{–}]_2$ 21

yellow crystals, m.p. 212°

$$PhCONHNHCOPh \xrightarrow{PCl_5} PhCCl{=}NN{=}CClPh$$ 38

prisms, m.p. 123°

$$Ph_2CClCONHPOCl_2 \xrightarrow{PCl_5} Ph_2CClCCl{=}NPOCl_2 \quad (88\%)$$ 39

prisms, m.p. 81–83°

CHART 2. Some examples of the preparation of imidoyl chlorides from amides.

Although the examples in Chart 2 look most satisfactory, certain limitations and difficulties should be mentioned. Primary amides ($RCONH_2$) and tertiary amides ($RCONR_2$) give other products. Thus benzamide[40,41] when heated in dry benzene with phosphorus pentachloride gives *N*-benzoylphosphorimidic trichloride (**1**) which, in a process reminiscent of the Wittig reaction, is pyrolysed to benzonitrile and phosphorus oxychloride:

$$PhCONH_2 \xrightarrow[\Delta,\ benzene]{PCl_5} \underset{(\mathbf{1})}{PhCON{=}PCl_3} \xrightarrow{\Delta} PhCN + POCl_3$$

The isomeric structure, $PhCCl{=}NPOCl_2$, which was for long entertained[42–44] as the intermediate in this type of reaction, has been discounted by experiments with ^{18}O-labelled materials[40]. Tertiary amides on the other hand give amidochlorides (see p. 608).

$$ArCONR_2 \xrightarrow{PCl_5} ArCCl{=}\overset{+}{N}R_2\ Cl^-$$

$$HCONMe_2 \xrightarrow{COCl_2} HCCl{=}\overset{+}{N}Me_2Cl^-$$

Imidoyl chlorides derived from alkanoic acids offer a special difficulty since they are much less stable than aromatic derivatives, and they may undergo self-condensation under certain conditions to give oily products, which contain amidines and heterocyclic bases (see

Section IV.A). Moreover with phosphorus pentachloride the process is frequently (though not always—see item 3 in Chart 2) complicated by α-halogenation, e.g.[5,45]

$$CH_3CH_2CONHPh \xrightarrow{PCl_5,\ \Delta} CH_3CCl_2CCl{=}NPh$$

$$Cl_2CHCONHEt \xrightarrow[\Delta,\ CCl_4]{PCl_5} CCl_3CCl{=}NEt$$

As generally used, phosgene does not have this effect, and is the preferred reagent with alkanoic acid amides[31].

A conjugated double bond is not attacked in the phosphorus pentachloride reaction, e.g.[46]

$$PhCH{=}CHCONHPh \xrightarrow[100°,\ PhMe]{PCl_5} PhCH{=}CHCCl{=}NPh$$

but a similarly situated acetylenic group may suffer conjugate addition of hydrogen chloride[47]:

$$PhC{\equiv}CCONHPh \xrightarrow[\text{room temp. } C_6H_6]{PCl_5} PhCCl{=}CHCCl{=}NPh$$

Diacylamines usually give products of the expected type. Thus dibenzamide[48] gives the imidoyl chloride (**2**) which has also been obtained by the action of phosphorus pentachloride on α-benzil monoxime at 0°[49].

$$PhCONHCOPh \xrightarrow[\text{room temp. } CHCl_3]{PCl_5} \underset{(\mathbf{2})}{PhCCl{=}NCOPh} \xleftarrow[Et_2O]{PCl_5} PhC({=}N{-}OH){-}COPh$$

Formally related compounds from urethane and sulphonamide derivatives have also been reported[43]:

$$ClCH_2CONHCOOEt \xrightarrow{PCl_5} Cl_2CHCCl{=}NCOOEt$$

$$ClCH_2CONHSO_2Ph \xrightarrow{PCl_5} ClCH_2CCl{=}NSO_2Ph$$

On the other hand glutarimides have been observed to aromatize to give 2,6-dichloropyridines, e.g.[50]

$$\text{3-methylglutarimide (Me; O, N–H, O)} \xrightarrow[\Delta]{PCl_5} \text{pyridine with Cl, Me, Cl, Cl substituents}$$

Wallach[42] believed that ethyl oxamate gave an amidochloride and then an imidoyl chloride with phosphorus pentachloride, thus:

$$COOEtCONH_2 \xrightarrow{PCl_5} COOEtCCl_2NH_2 \longrightarrow COOEtCCl{=}NH$$

However, Kirsanov[217] has re-examined this type of reaction and has reformulated the products as alkoxydichloroacetamides.

$$COOEtCONH_2 \xrightarrow{PCl_5} EtOCCl_2CONH_2$$

$$COOEtCONMe_2 \xrightarrow{PCl_5} EtOCCl_2CONMe_2$$

An interesting complication of the phosphorus pentachloride reaction has been reported fairly recently[51]. From *N*-methyltrifluoroacetamide one obtains, in addition to the expected imidoyl halide (**3**), a compound which is formulated with a four-membered ring system (**4**).

$$CF_3CONHMe \xrightarrow[\Delta]{PCl_5} CF_3CCl{=}NMe \quad 44\% \quad (\mathbf{3})$$

$$CF_3CONHMe \xrightarrow[\Delta]{PCl_5} CF_3COCl + Me{-}N\langle P(Cl_3)\rangle_2 N{-}Me + HCl \quad (\mathbf{4})\ 52\%$$

The mechanism of imidoyl halide formation deserves further investigation. It seems most likely that attack occurs at oxygen to give an intermediate such as **5**: there are occasional reports of the isolation of such species[24]. Rearrangement of **5** to generate the ion-pair, loss of a neutral molecule ($POCl_3$) from the anion, and collapse of the residual ion-pair would then furnish the products. A similar mechanism would apply to reactions with thionyl chloride and phosgene (loss of sulphur dioxide and carbon dioxide, respectively).

$$R{-}C({=}O){-}NHR' + PCl_4{-}Cl \xrightarrow{-HCl} R{-}C(O{-}PCl_4){=}NR'\ (\mathbf{5}) \longrightarrow \left[\bar{O}{-}P{-}Cl_3\ \ Cl\ \ R{-}\overset{+}{C}{=}NR'\right] \longrightarrow POCl_3 + RCCl{=}NR'$$

The early workers, following Wallach[42], tended to regard the amidochloride as a precursor of the imidoyl chloride; the former, however, is not properly regarded as an intermediate, although it could arise by the interaction of the products (i.e. hydrogen halide and imidoyl halide)[52]. Thus the imidoyl chloride **3** forms an adduct with hydrogen chloride[51] which is stable up to $-20°$. Amidochlorides are also generated from tertiary amides, and presumably arise in a way analogous to that described above, i.e. via the intermediate cation $RC(=\overset{+}{N}R_2)OPCl_4$. The chemistry of the amidochlorides has been reviewed[53].

It has been noted above that primary amides give intermediates with N—P linkages, and it is pertinent to enquire if secondary amides may do the same. There is little evidence for this, but the formation of product **4** has been interpreted[51] via such a system.

$$CF_3CON(Me)(PCl_3Cl) \longrightarrow CF_3COCl + MeN{=}PCl_3 \xrightarrow{\text{dimerization}} \mathbf{4}$$

B. Halogenation of Aldimines

The direct halogenation of azomethines is rather difficult to control, and has been little used as a preparative method. Thus the low temperature fluorination of benzylidene-*t*-butylamine in trichlorofluoromethane in the presence of sodium fluoride (HF scavenger) gives a mixture of products including *N*-(*t*-butyl)benzimidoyl fluoride[6].

$$PhCH{=}NBu\text{-}t \xrightarrow[-78°]{1{\cdot}5\ F_2} PhCF_2NFBu\text{-}t + PhCF{=}NBu\text{-}t \quad (CCl_3F,\ NaF)$$

This reaction is thought to proceed by an addition–elimination mechanism: only the *syn*-isomer (**7**) is observed, and this has been rationalized in terms of a *trans*-elimination from the favoured conformation of the initial adduct (**6**).

$$PhCH{=}NR \xrightarrow{F_2} \underset{(\mathbf{6})}{PhCHF{-}NFR} \xrightarrow{-HF} \underset{(\mathbf{7})}{Ph(F)C{=}NR} \xrightarrow{F_2} PhCF_2NFR$$

Chlorination of aldimines may be effected, however, with *t*-butyl hypochlorite[54]: the initial product is regarded as an immonium salt which eliminates *t*-butanol to give the imidoyl chloride, thus:

$$t\text{-}BuCH{=}NBu\text{-}n \xrightarrow{t\text{-}BuOCl} [t\text{-}BuCH{=}\overset{+}{N}(Cl)Bu\text{-}n]\ t\text{-}BuO^-$$

$$\xrightarrow{-t\text{-}BuOH} [t\text{-}BuC{\equiv}\overset{+}{N}Bu\text{-}n]\ Cl^- \longrightarrow t\text{-}BuCCl{=}NBu\text{-}n$$

The imidoyl halides prepared by this route were not isolated but were characterized using amidine and imidate derivatives.

In the special case of the aldoximes and closely related compounds direct chlorination is often the method of choice.

$$\text{PhCH=NOH} \xrightarrow[\text{CHCl}_3 \text{ or Et}_2\text{O}]{\text{Cl}_2} \text{PhCCl=NOH} \qquad \text{(refs. 55, 56)}$$

$$\text{PhCOCH=NOH} \xrightarrow{\text{Cl}_2} \text{PhCOCCl=NOH} \qquad \text{(ref. 57)}$$

There is n.m.r. evidence that this reaction proceeds through the dimer of the corresponding nitroso compound[59].

$$\text{MeCH=NOH} \xrightarrow[\text{Et}_2\text{O},\ -60^\circ]{\text{Cl}_2} \text{MeCHClN(}\rightarrow\text{O)=N(}\rightarrow\text{O)CHClMe} \rightleftharpoons \text{MeCHClNO} \rightarrow \text{MeCCl=NOH}$$

With fulminic acid and its salts, addition occurs to give dichloroformoxime in good yield[60]. Bromine behaves similarly[61].

$$\text{HCNO} \xrightarrow{\text{Cl}_2} \text{CCl}_2\text{=NOH}$$

Nitrosyl chloride has also been employed as a chlorinating agent. Acetone suffers both nitrosation and chlorination[62].

$$\text{PhCH=NOH} \xrightarrow[\text{H}_2\text{O}]{\text{NOCl}} \text{PhCCl=NOH}$$

$$\text{Me}_2\text{CO} \xrightarrow[\text{CCl}_4]{\text{NOCl}} \text{MeCOCCl=NOH (and other products)}$$

Halogenation of phenylhydrazones and of azines occurs in an analogous fashion; ring halogenation may also occur unless the phenyl of the phenylhydrazone moiety is deactivated.

$$\text{PhCH=NNH-C}_6\text{H}_4\text{-NO}_2 \xrightarrow[\text{HOAc}]{\text{Br}_2} \text{PhCBr=NNH-C}_6\text{H}_4\text{-NO}_2 \qquad \text{(ref. 63)}$$

$$\text{PhCH=NNHPh} \xrightarrow[\text{HOAc}]{\text{Br}_2} \text{PhCBr=NNH-C}_6\text{H}_3\text{Br}_2 \qquad \text{(ref. 64)}$$

$$\text{PhCH=NN=CHPh} \xrightarrow[\text{CCl}_4]{\text{Cl}_2} \text{PhCCl=NN=CHPh} \qquad \text{(ref. 58)}$$

C. Imidoyl Halides from Amidohalides

Occasionally amidochlorides from secondary amides appear to have been isolated and converted thermally into the imidoyl chlorides, e.g.[65] (but see ref. 217)

$$\text{COOEtCCl}_2\text{NHPh (or COOEtCCl=}\overset{+}{\text{N}}\text{HPh Cl}^-\text{)} \xrightarrow[-\text{HCl}]{110^\circ} \text{COOEtCCl=NPh}$$

The dehalogenation of a *N*-fluoroamidofluoride over glass has been observed[6].

$$\mathrm{PhCF_2NFBu\text{-}}t \xrightarrow{\text{glass}} \mathrm{Ph(F)C{=}N\text{-}Bu\text{-}}t$$

Dealkylation of *N*,*N*-dialkylamidochlorides from tertiary amides has been examined, although not principally with imidoyl halide synthesis in view[66]. The reactions (von Braun degradation) are thought to be of the following type:

$$\mathrm{PhCONRR'} \xrightarrow{\mathrm{PCl_5}} \mathrm{PhCCl{=}\overset{+}{N}R'R\ Cl^-} \xrightarrow{140°} \mathrm{PhCCl{=}NR'} + \mathrm{RCl}$$

von Braun showed that the ease of elimination of the alkyl residue increases in the series Bu < Pr < Et < Me < $PhCH_2$. The same order was observed for the cyanogen bromide dealkylation:

$$\mathrm{R_3N} + \mathrm{BrCN} \xrightarrow{0°} \mathrm{RBr} + \mathrm{R_2NCN}$$

Both reactions are regarded as S_N2 processes, the rates of which would be expected to be sensitive to the bulk of the α_N-substituent[30,67] (see Section VI.F).

D. Imidoyl Halides from Isonitriles

In his extensive work on isonitriles, Nef[68] studied their interaction with various halides, and obtained products which were formulated as imidoyl halides. Many of the reactions proceed readily in the cold and give products of α-addition as shown in Chart 3. Alkyl halides which might, by such a process, give imidoyl halides of a more familiar sort, do not appear to give identifiable products.

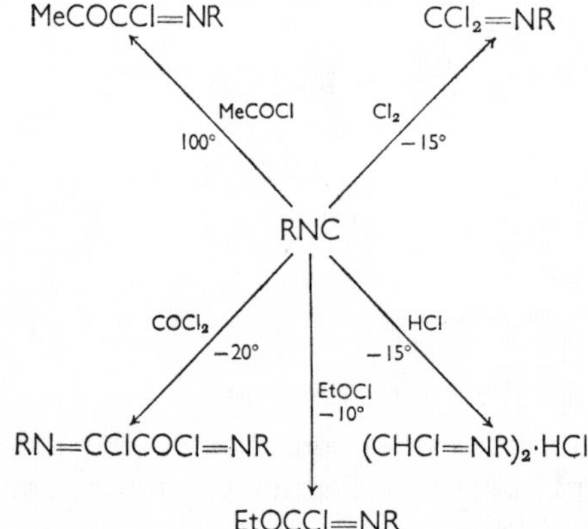

CHART 3. α-Additions to phenyl isonitrile: products formulated with the —CHal=N— function (Nef, 1892–1895).

Certain of the reactions have been confirmed by recent work[69]. The reaction with acyl chlorides is an excellent route to imidoyl chlorides derived from α-keto acids.

$$\text{n-PrCOCl} + C_6H_{11}NC \longrightarrow \underset{77\%}{\text{n-PrCOCCl}{=}NC_6H_{11}}$$

The synthesis of iminophosgenes (isonitrile dihalides, $CHal_2{=}NR$) has been reviewed recently[70].

E. Nitriles as Precursors

I. Addition of hydrogen halides to nitriles

The adducts formed between nitriles and hydrogen halides were first investigated by Gautier[16], who obtained crystalline compounds which appeared to have constitutions such as 2MeCN·3HBr and EtCN·HCl. Michael and Wing[71] suggested that the products could be divided into three classes:

$$\text{RCN·HHal} \quad \text{or} \quad \text{RCHal}{=}\text{NH}$$

$$\text{RCN·2HHal} \quad \text{or} \quad \text{RCHal}_2\text{NH}_2$$

$$\text{2 RCN·3HHal} \quad \text{or} \quad \text{RCHal}\begin{cases}\text{N}{=}\text{CHalR}\\ \text{NH}_2\text{·HHal}\end{cases}$$

and reinvestigated the adduct EtCN·HCl. They put forward the imidoyl chloride structure for this since it gave the expected amidine with aniline; however, it was reported to crystallize unchanged from cold water which does not accord with the imidoyl halide formulation (see Section IV.E.2). Nevertheless, the imidoyl halide or covalent amidohalide structures appear to have been generally assumed, and, indeed, accounted in a straightforward manner for the formation of imidates when the reaction was carried out in the presence of an anhydrous alcohol (Pinner imidate synthesis). A change of opinion developed when, on the basis of cryoscopic and optical measurements (carried out largely on sulphuric acid–nitrile systems), Hantzsch[73] proposed that the adducts were nitrilium salts:

$$[\text{RC}{\equiv}\overset{+}{\text{N}}\text{H}]\text{X}^- \qquad [\text{RC}{\equiv}\overset{+}{\text{N}}\text{H}]\text{XHX}^-$$

Usually the dihalide precipitated, and no monochloride was observed; monobromo compounds were reported, however. The adducts were very sensitive to moisture and readily lost hydrogen halide on standing. With water the nitrile was regenerated (but see Section

IV.E.2). However, in 1945 Hinkel and Treharne[74] repeated Gautier's experiment on the MeCN–HCl system, but at a lower temperature (−17°). A dichloride was obtained which was formulated as the covalent amidochloride $MeCCl_2NH_2$; an analogous covalent structure had been suggested much earlier by Lander and Laws[75] for the adduct of *N*-phenylbenzimidoyl chloride and hydrogen iodide.

The past few years have seen much clarification of this problem (review to 1962)[20]. The present position can be summarized as follows. Evidence has been obtained for various complexes in solution; freezing point diagrams[76] have indicated MeCN·HCl, 2MeCN·3HCl, MeCN·5HCl and MeCN·7HCl—but surprisingly not MeCN·2HCl —in the acetonitrile–hydrogen halide system, while conductivity data have been interpreted in terms of the initial formation of unionized molecular compounds ('outer complexes') followed (and sometimes quite slowly) by their conversion to ionic species ('inner complexes')[82]. However, it appears that the crystalline adducts obtained are generally of the composition RCN·2HHal. Although 1:1 adducts have been reported even in fairly recent times[77], on balance the evidence suggests that such compounds are unstable and that they readily disproportionate[78,79]:

$$2RCN\cdot HHal \rightarrow RCN + [RCHal{=}\overset{+}{N}H_2]\ Hal^-$$

a process which reflects the greater basicity of the imino function *vis-à-vis* the cyano group. Thus a reinvestigation[78] of three examples where Hantzsch reported monohydrobromides ($PhCH_2CN$, Cl_3CCN and PhCH=CHCN) has led to the isolation of the dihydrobromides*. The formulation of these amidohalides (see also Section III.B) as imidoyl halide hydrohalides

$$[RCHal{=}\overset{+}{N}H_2]\ Hal^-$$

is reasonable, since the typical imidoyl halides are known to be basic enough to form adducts with hydrogen halides which have salt-like properties (see Section IV.C). Amongst the many adducts which may now be formulated in this way the following may be mentioned: Hal = iodine; R = Me[82], $PhCH_2$[25], (*o*, *m*-, *p*-) MeC_6H_4[25], (*o*-, *m*-, *p*-) $NO_2C_6H_4$[25]: Hal = bromine; R = Me[82], $PhCH_2$[78], Cl_3C[78], PhCH=CH[78], Ph[73], *p*-MeC_6H_4[73]: and Hal = chlorine; R = Me[74,82,84], $BrCH_2$[26], FCH_2[26], PhCH=CH[73], *p*-MeC_6H_4[73]. The ionic formula-

* However simple nitrilium salts $RC{\equiv}\overset{+}{N}H\ \overset{-}{X}$ with complex anions can be isolated, especially if suitably low temperatures are employed[80] [e.g. $(PhC{\equiv}\overset{+}{N}H)_2SnCl_6^{2-}$].

tion accounts for the observations of electrical conductivity[82] and for the appearance of a C=N stretching mode in the infrared spectrum (e.g.[72] 1626 cm^{-1} in $MeCBr{=}\overset{+}{N}D_2\ Br^-$). Amidochlorides derived from tertiary amides have analogous structures and show similar absorption (e.g.[83] $CH_3CCl{=}\overset{+}{N}Me_2\ Cl^-$, $\nu_{C=N}$ 1653 cm^{-1}). Although such a band was not observed[82] in the 1:2 acetonitrile–hydrogen chloride adduct, a neutron diffraction analysis[84] at $-5°$ of crystals of this adduct grown from anhydrous MeCN–HCl at $-16°$ under nitrogen provides definitive evidence for the imidoyl chloride hydrochloride structure in the solid state.

Dimer hydrohalides of the general composition 2RCN·*x*HCl are also formed under certain conditions, and are considered in Section IV.A.

2. Addition of acyl halides to nitriles

It has been shown by Meerwein and his colleagues[81] that nitriles react with acyl halides in the presence of Lewis acids to give nitrilium salts. Certain cyanamides, however, have been reported to give adducts with acyl halides in the absence of Lewis acids; these are formulated as imidoyl halides and in some cases are isolable compounds, e.g.[86]

$$Me_2NCN + p\text{-}NO_2C_6H_4COCl \xrightarrow{70°} Me_2NCCl{=}NCOC_6H_4NO_2\text{-}p$$

$$Me_2NCN + COCl_2 \xrightarrow{CH_2Cl_2} Me_2NCCl{=}NCCl{=}NCCl{=}\overset{+}{N}Me_2Cl^-$$

These compounds hold some interest as intermediates in heterocyclic synthesis[86]. There is evidence[88] that the reaction of malonyl chloride with nitriles, which eventually gives pyridines, proceeds via imidoyl halide intermediates, e.g.

$$EtCN + CH_2(COCl)_2 \longrightarrow [EtCCl{=}NCOCH_2COCl] \longrightarrow$$

$$\left[MeCH{=}CClN{=}\overset{OH}{\overset{|}{C}}CH_2COCl\right] \longrightarrow \text{(pyridine: 2-HO, 6-Cl, 5-Me, 4-OH)}$$

F. High-temperature Chlorination of Amines and Amides

Chlorination of tertiary amines and of the acyl derivatives of primary and secondary amines at high temperature (~200°) gives imidoyl chlorides (review)[89]. Free-radical chlorination is followed by an elimination analogous to the von Braun degradation to afford the products. If reaction conditions are carefully arranged, remarkably good yields of a single product can often be obtained. The following examples are illustrative.

$$Me_2NPh \xrightarrow[\text{temperature increased } <60° \rightarrow 190°]{Cl_2} Cl_2C{=}N{-}C_6H_2Cl_3 \quad 82\text{–}87\%$$

$$Me_2NPh \rightarrow \left[Cl_3C{-}N(CHCl_2){-}C_6H_2Cl_3 \right] \rightarrow \left[Cl^-\ Cl_2C{=}\overset{+}{N}(CHCl_2){-}C_6H_2Cl_3 \right] \xrightarrow[-CHCl_3]{\Delta} Cl_2C{=}N{-}C_6H_2Cl_3$$

$$Et_2NPh \xrightarrow[200°]{Cl_2} CCl_3CCl{=}N{-}C_6H_2Cl_3 + \text{1-(2,4,6-trichlorophenyl)tetrachloropyrrole}$$

$$Me_2NCH_2CH_2NMe_2 \xrightarrow[200°]{Cl_2} Cl_2C{=}NCCl_2CCl_2N{=}CCl_2 \quad 30\text{–}71\%$$

$$\text{N-chlorocarbonylmorpholine} \xrightarrow[200°]{Cl_2} \text{hexachloro-dihydro-oxazine (}Cl_2, Cl_2, Cl_2, Cl\text{)} \quad 64\%$$

$$(CH_2Cl)_2NCOCl \xrightarrow[>180°]{Cl_2} CCl_3{-}N{=}CCl_2 \quad 82\%$$

G. Miscellaneous Routes

1. From imidates[90]

$$PhC(OEt){=}NPh \xrightarrow[\Delta]{PCl_5} PhCCl{=}NPh + EtCl + POCl_3$$

2. From isothiocyanates[91]

$$PhNCS \xrightarrow[CHCl_3]{Cl_2} CCl_2{=}NPh + SCl_2 \quad \text{(and other products)}$$

3. From thioamides[45,92]

$$PhCSNHMe \xrightarrow{Cl_2 \text{ or } NOCl} PhCCl{=}NMe$$

4. From diazotates[93]

$$CHCl(COOK)_2 + PhN{=}NOH \xrightarrow[0^\circ]{NaOAc} PhN{=}NCCl{=}NNHPh$$

5. From trichloronitrosomethane[94]

$$Cl_3CSO_2Na \xrightarrow[KNO_3,\ NaNO_2]{H_2SO_4} Cl_3CNO$$

$$Cl_3CNO \xrightarrow{SnCl_2} Cl_2C{=}NOH$$

$$Cl_3CNO \xrightarrow{AgNHCN} NCNHCCl{=}NOH$$

$$Cl_3CNO \xrightarrow{\Delta} CCl_2{=}NCCl_3 + CCl_3NO_2 + NOCl$$

$$Cl_3CNO \xrightarrow{PhNH_2} PhN{=}NCCl{=}NPh$$

6. From oximes by Beckmann rearrangement[95,96,227]

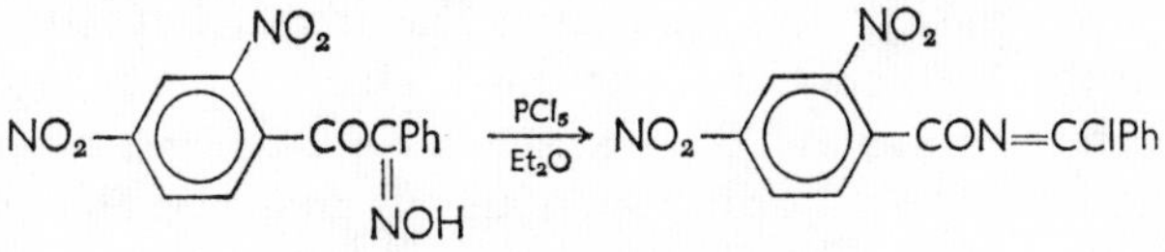

$$2,4\text{-}(NO_2)_2C_6H_3C({=}NOH)Ph \xrightarrow[Et_2O]{PCl_5} 2,4\text{-}(NO_2)_2C_6H_3CON{=}CClPh$$

$$Ph_2C{=}NOH \xrightarrow[MeCN,\ 30^\circ-60^\circ]{COCl_2} PhCCl{=}NPh$$

7. From trichloroacetamides[97]

$$CCl_3CONHPh \xrightarrow[C_6H_6,\ 40^\circ]{Bu_3P} Cl_2CHCCl{=}NPh + Bu_3PO$$

8. From difluoroamine[98]

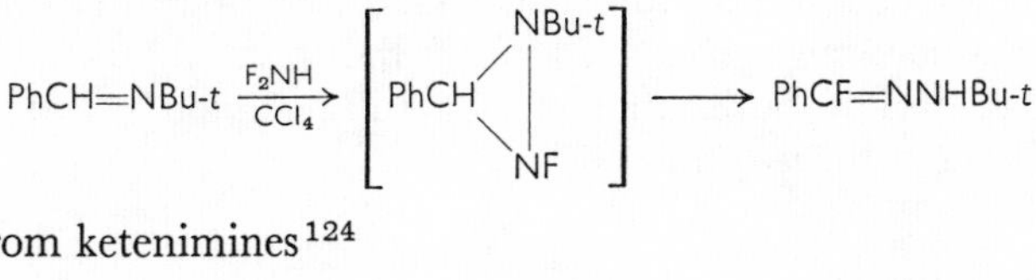

$$PhCH{=}NBu\text{-}t \xrightarrow[CCl_4]{F_2NH} \left[PhCH\langle N(Bu\text{-}t)\text{–}NF \rangle \right] \longrightarrow PhCF{=}NNHBu\text{-}t$$

9. From ketenimines[124]

$$Ph_2C{=}C{=}NC_6H_4Me\text{-}p \xrightarrow[\text{petroleum ether}]{Cl_2} Ph_2CClCCl{=}NC_6H_4Me\text{-}p$$

III. PHYSICAL PROPERTIES AND RELATED MATTERS

A. General

The imidoyl halides are usually colourless liquids or solids, soluble in inert solvents such as chloroform, benzene and light petroleum. The last two solvents are often useful for crystallization purposes. Some specific physical properties have been noted in Charts 1 and 2. Many of the imidoyl halides, especially the liquids, have unpleasant irritating odours; certain of them have been used as war gases (for example $CCl_2{=}NPh$ in 1917)[99].

The typical imidoyl halides are highly reactive substances and must often be prepared as required: they are readily hydrolysed and must be protected from moisture.

B. The Bonding of the Halogen Atom

This topic has been touched on earlier in reference to the nitrile-hydrogen halide adducts (see Section II.E.1) but has at times been rather confused in the literature. A summarizing statement of the present interpretation is therefore not out of place.

The *imidoyl halides* give every indication of being covalently bonded substances. They are low-melting solids or volatile liquids, soluble in solvents of low polarity (e.g. benzene). The infrared spectra show absorption in the 1680 cm^{-1} region: this is rather higher than $\nu_{C=N}$ for simple azomethines as would be expected. Nevertheless the carbon–halogen bond is highly polarized: spontaneous and reversible ionization to give an ion-pair has been postulated, and the relationship between structure and reactivity is rather analogous to that found in the *t*-alkyl halides[31].

$$\mathrm{RCCl{=}NR'} \rightleftarrows \left[\begin{array}{c}\mathrm{RC{\equiv}NR'}\\ +\\ \mathrm{Cl^-}\end{array}\right]$$

The tendency for ionization to occur is increased by certain factors: thus electron-donating groups stabilize the resulting cation (see Section IV.E.2) while stabilization of the resulting anion by complexation can also be achieved, thus[80,221]:

$$\mathrm{PhCCl{=}NPh} \xrightarrow[\mathrm{PhNO_2}]{\mathrm{FeCl_3}} [\mathrm{PhC{\equiv}\overset{+}{N}Ph}]\ \mathrm{FeCl_4}^-$$

$$\mathrm{PhCCl{=}NPh} \xrightarrow[-\mathrm{AgCl}]{\mathrm{AgBF_4}} [\mathrm{PhC{\equiv}\overset{+}{N}Ph}]\ \mathrm{BF_4}^-$$

The *amidochlorides* are of three types (derived from RCN, RCONHR′, and $RCONR'_2$) but all three appear to be predominantly ionic substances, i.e. immonium salts.

$$[RCCl{=}\overset{+}{N}R'R'']Cl^- \longleftrightarrow RCCl_2{-}NR'R'' \quad R'\ R'' = \text{aryl, alkyl, H}$$

They are solids which generally do not have sharp melting points; they are much less soluble in solvents of low polarity than are the corresponding imidoyl halides. Thus *N*-methylbenzimidoyl chloride is soluble in ether, but treatment with anhydrous hydrogen chloride causes the amidochloride to precipitate[78].

In the infrared (Table 1) the amidochlorides show bands at ~1640 cm^{-1} attributed to the C=N stretching mode; in the case of crystalline acetamidochloride neutron diffraction[84] establishes the immonium salt structure beyond dispute, and there is little reason to doubt that other simple amidochlorides have analogous structures. The amidobromides and amidoiodides are probably similarly constituted, the C—Br and C—I bond strengths being lower than that of the C—Cl bond. The C—F bond strength is higher, however, and the available evidence suggests that although the nitrile–hydrogen fluoride adducts are ionic (the anion being stabilized by complexation, e.g.[100] $MeCF{=}\overset{+}{N}H_2\ (HF)_nF^-$) the other amidofluorides behave as covalent compounds. Thus the amidochloride $HCCl{=}\overset{+}{N}Me_2\ Cl^-$ is a solid, m.p. 140–145°, which shows a pronounced band in the infrared attributed to C=N stretching (Table 1), whereas the corresponding amidofluoride[101], which is prepared in an analogous manner, i.e.

$$HCONMe_2 \xrightarrow[-CO_2]{COF_2} CF_2HNMe_2$$

is a volatile liquid, b.p. 49–51°. The infrared spectrum shows no pronounced absorption in the double bond region above ~1500 cm^{-1}, while n.m.r. spectroscopy shows that the fluorine atoms are equivalent[101]. Hence the covalent structure is indicated for this compound and is likely for other amidofluorides (e.g. $PhCF_2NMe_2$[101] b.p. 55–59°/12 mm; $(CF_2)_5NH$[102] b.p. 73·4°).

C. Spectroscopy

1. Infrared spectra

The C=N stretching frequencies of some imidoyl halides and related compounds are summarized in Table 1. Just as acyl chlorides

Table 1. Infrared spectra of imidoyl halides and related compounds.

Structure	Phase (where stated)	C—N Stretching Vibrations (cm^{-1}) $\nu_{C=N}$	$\nu_{C\equiv N}$	Reference
Imidoyl Halides				
$HCCl{=}NMe$	CH_2Cl_2	1689		32
$CH_3CCl{=}N{-}C_6H_{11}$ (cyclohexyl)		1705		31
$PhCCl{=}NPh$	CH_2Cl_2	1672		32
$CF_3CCl{=}NMe$		1695		51
$Ph(F)C{=}N{-}Bu\text{-}t$		1718		6
$F(Ph)C{=}N{-}Bu\text{-}t$		1686		6
ring: F_2, F_2, F_2, F_2, N=C–F		1754		104
$CF_3CF{=}NCF_3$		1786		104
$F_2C{=}NR$		~1790		70
$Cl_2C{=}NR$		1645–1660		70
$Br_2C{=}NR$		1650–1680		70
Amidohalides				
$MeCBr{=}\overset{+}{N}H_2\ Br^-$	Solid	1664, 1531		72
$MeCBr{=}\overset{+}{N}D_2\ Br^-$	,,	1626		72
$MeCI{=}\overset{+}{N}H_2\ I^-$	,,	1637, 1503		72
2-furyl–$CCl{=}\overset{+}{N}HMe\ \bar{C}l$	,,	1623		34
$PhCCl{=}\overset{+}{N}HMe\ Br^-$	,,	1618		34
$(CF_2)_5NH$	Vapour	No band reported in double bond region		102
$HCCl{=}\overset{+}{N}Me_2\ Cl^-$	CH_2Cl_2	1680		83
$MeCCl{=}\overset{+}{N}Me_2\ Cl^-$	,,	1653		83
$PhCCl{=}\overset{+}{N}Me_2\ Cl^-$	Solid	1642		34
	CH_2Cl_2	1634		32
HCF_2NMe_2	Liquid film	No major absorption double bond region $> 1500\ cm^{-1}$		101
Nitrilium Salts				
$PhC{\equiv}\overset{+}{N}Ph\ SbCl_6^-$	Solid		2315	105
$PhC{\equiv}\overset{+}{N}Pr\text{-}i\ SbCl_6^-$	,,		2326	105
	CH_2Cl_2		2326	105
$PhC{\equiv}\overset{+}{N}Ph\ AlCl_4^-$	CH_2Cl_2		2309	32

($\nu_{C=O} \sim$ 1770–1815 cm^{-1})[103] show carbonyl absorption at higher frequencies than do the aldehydes ($\nu_{C=O} \sim$ 1720–1740 cm^{-1}) so the bands for the imidoyl chlorides ($\nu_{C=O} \sim$ 1670–1710 cm^{-1}) appear somewhat higher than do those for aldimines ($\nu_{C=N} \sim$ 1640–1690 cm^{-1}). As expected, the values for imidoyl fluorides are higher than those for the chlorides. The infrared spectra of $MeCBr{=}\overset{+}{N}H_2\,Br^-$, $MeCI{=}\overset{+}{N}H_2\,I^-$ and $MeCCl{=}\overset{+}{N}H_2\,SbCl_6^-$ and their deuterated analogues have been presented and discussed in detail[72].

2. Nuclear magnetic resonance spectra

Both proton and fluorine n.m.r. spectroscopy have been applied to the solution of configurational problems (see below) but no comprehensive survey in this area has been made.

D. Geometrical Isomerism

In principle the imidoyl halides can exist in two possible configurations, which will here be denoted *syn* and *anti* with respect to *N*-substituent and halogen atoms, following the model of the aldoximes.

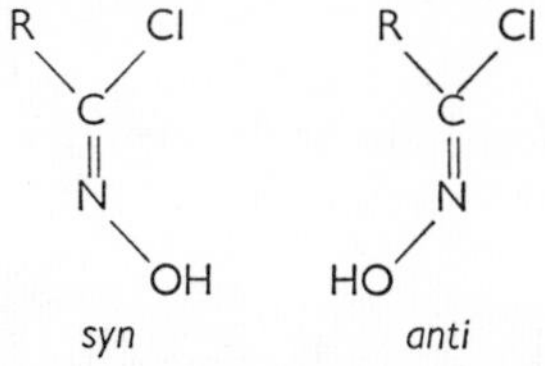

R(Cl)C=N–OH (*syn*)   R(Cl)C=N–OH (*anti*)

syn   *anti*

The earliest example appears to be the chloroxime system, stereoisomers of which have been interconverted[106].

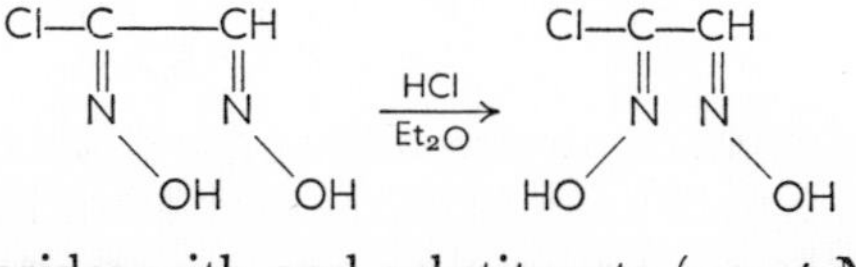

Cl–C(=N–OH)–CH=N–OH $\xrightarrow[Et_2O]{HCl}$ Cl–C(=N–OH)–CH=N–OH

For imidoyl chlorides with aryl substituents (e.g. p-$NO_2C_6H_4CCl{=}NC_6H_4NO_2$-$p$) dipole moment measurements suggest that the *syn*-configuration predominates[107]. Some n.m.r. studies on imidoyl fluorides have proved valuable as the following three examples show.

(a) The n.m.r. spectrum of

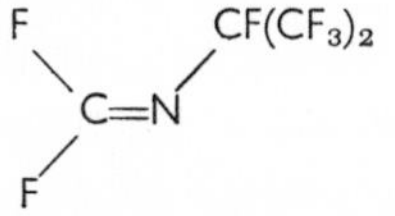

$F_2C{=}N{-}CF(CF_3)_2$

is temperature dependent[108]. At room temperature the *N*-substituent oscillates from one position to the other so quickly that the fluorine atoms of the perfluoromethylene group appear to be equivalent: at $-63°$, however, this process is slowed sufficiently for non-equivalence to be observed, and an AB quartet is recorded with $J_{F\ gem} = 84{\cdot}5$ Hz

(b) The *anti* isomer (**8**) of *N*-(*t*-butyl)-benzimidoyl fluoride is stable at room temperature whereas the *syn* compound (**9**) readily decomposes, possibly by a cyclic rearrangement (arrows) to give benzonitrile, isobutene and isobutyl fluoride.

Ph(F)C=N–Bu-t: τ8·5 (doublet J_{HF}=1 Hz); 1686 cm⁻¹

(**8**) *anti*

Ph(F)C=N–C(Me)₂CH₂–H: τ8·6 (singlet); 1718 cm⁻¹

(**9**) *syn*

The *syn*-imidoyl fluoride is produced by fluorination of the corresponding azomethine (see Section II.B): the *anti* isomer formed slowly in pentane solutions of $PhCF_2NFBu$-*t* in the presence of silica gel at room temperature[6].

(c) The addition of perfluorohydrazine to vinyl fluoride, followed by elimination of hydrogen fluoride over caesium fluoride, gives a mixture of geometrical isomers, thus[109]:

$$CH_2{=}CHF \xrightarrow[o\text{-}C_6H_4Cl_2]{N_2F_4} \left[\begin{matrix} NF_2 & NF_2 \\ | & | \\ CH_2 & -CHF \end{matrix}\right] \xrightarrow[-3HF]{CsF} \underset{\substack{anti \\ (1\ \text{part}) \\ J_{FF} = 265\ \text{Hz}}}{CN{-}C(F){=}N{-}F} + \underset{\substack{syn \\ (3\ \text{parts}) \\ J_{FF} = 52\ \text{Hz}}}{NC{-}C(F){=}N{-}F}$$

These have been separated by preparative gas-liquid chromatography. The configurations were assigned on the basis of the fluorine couplings, the system with the greater coupling constant having the *trans* (in this case, *anti*) arrangement.

IV. CHEMICAL REACTIONS

A. *Self-condensation*

Those imidoyl halides having one or more hydrogens at the α_C-carbon atom tend to be unstable in the special sense that they undergo self-condensation. The products may be polymeric, but are often identifiable amidines and heterocyclic compounds. The self-condensation is susceptible to steric hindrance on either side of the azomethine function: thus $Me_2CHCCl{=}NC_6H_4Me\text{-}p$[4], n-BuCHEtCCl=NBu-n[4], $CH_3CCl{=}NC_6H_4Br\text{-}o$[110] and $ClCH_2CCl{=}NC_6H_4Br\text{-}o$[110] have been isolated by distillation and analysed.

Two types of self-condensation might be expected:

Claisen-type condensation:

$$RCH_2CCl{=}NR' \xrightarrow{-HCl} RCH_2C({=}NR')\text{—}CH(R)\text{—}CCl{=}NR' \longrightarrow \text{products}$$

Amidine formation:

$$RCCl{=}NR' \longrightarrow \left\{ [R\text{—}C({=}NR')\text{—}\overset{+}{N}(R'){=}CClR]\,Cl^- \rightleftharpoons R\text{—}C({=}NR')\text{—}N(R')\text{—}CCl_2R \right\} \longrightarrow \text{products}$$

and processes of both types are known, although the state of the literature in this area is not altogether satisfactory.

von Braun[1,111] regarded the self-condensation as involving a rearrangement to the enamine (**10**) followed by condensation of the two species to give an *N*-chlorovinylamidine (**11**):

$$R_2CHCCl{=}NR' \longrightarrow \underset{(\mathbf{10})}{R_2C{=}CClNHR'} \longrightarrow \underset{(\mathbf{11})}{R_2CHC({=}NR')\text{—}R'NCCl{=}CR_2}$$

Other schemes can be written[52] and, in particular at higher temperatures, ketenimines may be formed[4].

Wallach and his colleagues[112,113] made the first determined foray (1874) into this difficult area, observing bases of high molecular weight where imidoyl chlorides were expected. The following bases may result directly from processes which formally lead to the imidoyl halide, and are presumed to result from its self-condensation, although condensation with the original amide may offer an alternative pathway[218]. In certain cases the self-condensation has been demonstrated[74].

I. Amidines

The reaction of acetanilide with phosphorus pentachloride gives a crystalline base $C_{16}H_{15}ClN_2$ which Wallach[114] formulated ($CH_3C({=}NPh)CH_2CCl{=}NPh$) as a condensation product of the Claisen type. The base resisted the action of water, however, and was reformulated[111] as an *N*-chlorovinylamidine (**12**) on the basis of the chemical evidence shown below.

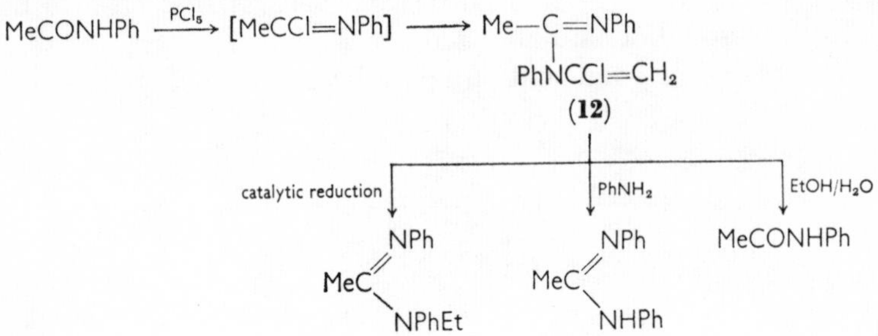

$$\text{MeCONHPh} \xrightarrow{\text{PCl}_5} [\text{MeCCl=NPh}] \longrightarrow \text{Me—C(=NPh)—PhNCCl=CH}_2 \quad \textbf{(12)}$$

(**12**): catalytic reduction → MeC(=NPh)NPhEt; $PhNH_2$ → MeC(=NPh)NHPh; $EtOH/H_2O$ → MeCONHPh

However, the amidochlorides derived from tertiary amides do react by a Claisen-type condensation[121].

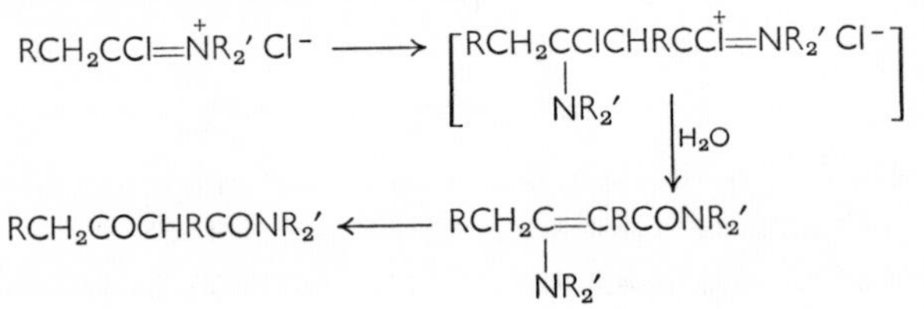

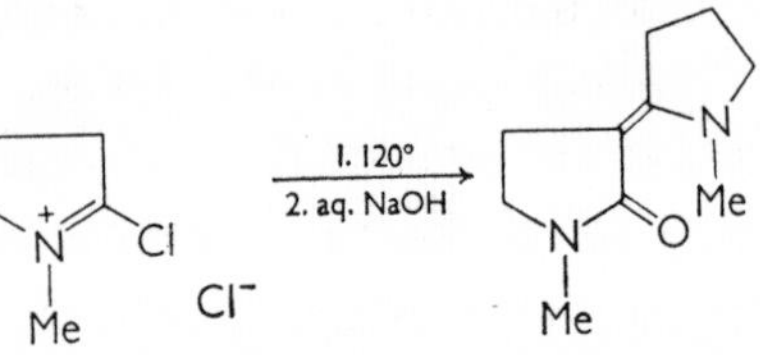

$$\text{RCH}_2\text{CCl=}\overset{+}{\text{N}}\text{R}_2'\ \text{Cl}^- \longrightarrow \left[\text{RCH}_2\text{CCl(NR}_2')\text{CHRC}\overset{+}{\text{Cl}}\text{=NR}_2'\ \text{Cl}^-\right] \xrightarrow{\text{H}_2\text{O}} \text{RCH}_2\text{C(NR}_2')\text{=CRCONR}_2' \longrightarrow \text{RCH}_2\text{COCHRCONR}_2'$$

1-Methyl-2-chloro-1-pyrrolinium chloride $\xrightarrow{\text{1. 120°; 2. aq. NaOH}}$ 1-methyl-3-(1-methylpyrrolidin-2-ylidene)pyrrolidin-2-one

In the studies on the interaction of nitriles with hydrogen halides a second type of product besides the imidoyl halide hydrohalide (Section II.E.1) has been encountered. This has the general composition 2RCN·*x*HHal and appears to be an *N*-(α-haloalkylidene)amidine derivative[74,87] which may be regarded as arising by the self-condensation of the imidoyl halide hydrohalide.

$$2\,\text{RCCl=}\overset{+}{\text{N}}\text{H}_2\ \text{Cl}^- \xrightarrow{-\text{HCl}} \left[\text{RCCl=}\overset{+}{\text{N}}\text{H—C(R)=}\overset{+}{\text{N}}\text{H}_2\right] 2\,\text{Cl}^- \xrightarrow{-2\text{HCl}} \text{RCCl=NC(R)=NH}$$

$$\rightleftharpoons \left[\text{RCCl}_2\text{NH—C(R)=}\overset{+}{\text{N}}\text{H}_2\right] \text{Cl}^-$$

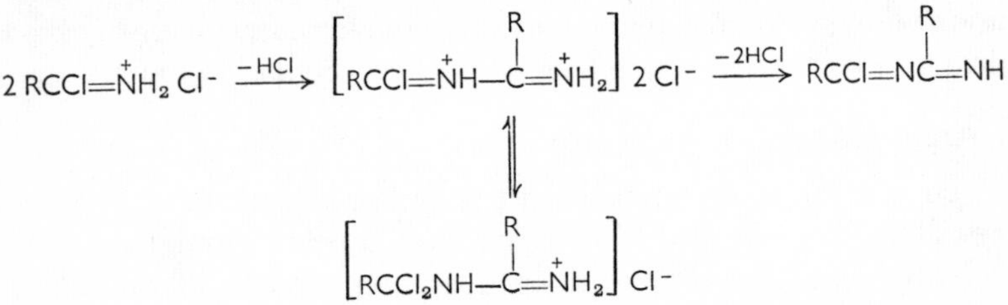

Indeed such a reaction using acetimidoyl chloride hydrochloride as a starting material has been reported[74]. Both normal (i.e. imidoyl halide hydrohalide) and dimer adducts may be obtained from the same system under different reaction conditions. Thus, bromo- and fluoroacetonitrile give imidoyl halide hydrohalides at low temperatures ($-50°$ to $-5°$) but at higher temperatures ($\sim 20°$) the dimer adduct is formed[26].

$$FCH_2CN$$

$$\xrightarrow{HCl/Et_2O,\ -20°} [FCH_2CCl{=}\overset{+}{N}H_2]\ Cl^- \xrightarrow{H_2O/Et_2O} FCH_2CONH_2$$

$$\xrightarrow{HCl/Et_2O,\ 15°} \left[FCH_2CCl{=}N\underset{|}{\overset{CH_2F}{C}}{=}\overset{+}{N}H_2\right] Cl^-\cdot HCl \xrightarrow[-30°]{EtOH} FCH_2C(={\overset{+}{N}H_2})NH_2\ Cl^-$$

The acetonitrile–HCl system also gives products of both types[74].

2. Iminazoles

In certain cases if the reaction is carried out at higher temperatures cyclization of the *N*-chlorovinylamidine is reported to lead to the iminazole system[115].

$$C_6H_{13}CONHEt \xrightarrow{PCl_5} C_6H_{13}C(=NEt)\text{—}N(Et)\text{—}CCl{=}CHC_5H_{11} \xrightarrow[-EtCl]{\Delta} \text{1-Et-2-}C_6H_{13}\text{-4-}C_5H_{11}\text{-iminazole}$$

Iminazoles are also formed from certain unstable bis-imidoyl halides based on oxalic acid: here the reaction proceeds under mild conditions[42,116].

$$\begin{matrix}CONHEt\\|\\CONHEt\end{matrix} \xrightarrow{PCl_5} \text{1-Et-5-Cl-2-Me-iminazole}$$

This reaction can be rationalized in terms of ylid formation from the bis-imidoyl halide, thus:

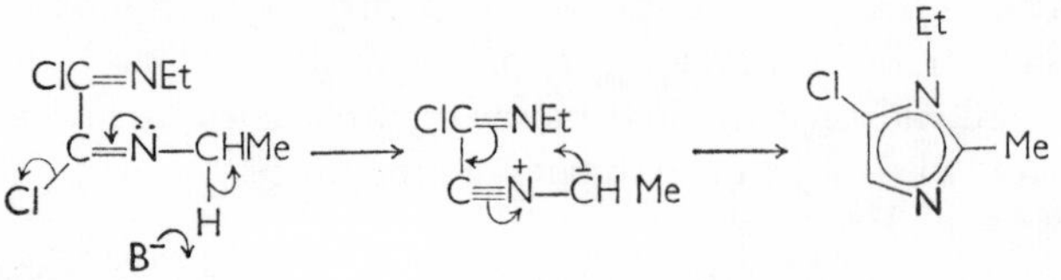

ClC=NEt C=N–CHMe Cl H B⁻ ⟶ ClC=NEt C≡N⁺–C̄H Me ⟶ Cl N Et Me N

3. Quinolines

Quinolines are produced by self-condensation of certain *N*-aryl imidoyl halides. Thus Bischoff and Walden[118] reported that phosphorus pentachloride and glycollic anilide gave a yellow crystalline compound, $C_{16}H_{13}Cl_3N_2$, the hydrochloride of a colourless quinoline. The reaction may be rationalized in the following way[119]:

$$HOCH_2CONHPh \xrightarrow{PCl_5} [ClCH_2CCl{=}NPh] \rightleftharpoons [ClCH{=}CClNHPh]$$

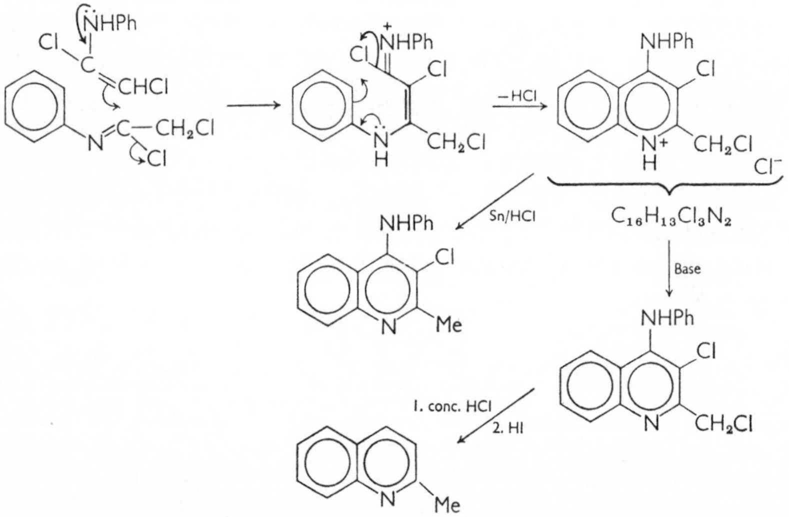

NHPh Cl C CHCl N C–CH₂Cl Cl ⟶ NHPh Cl Cl N H CH₂Cl $\xrightarrow{-HCl}$ NHPh Cl N⁺ H CH₂Cl Cl⁻

$C_{16}H_{13}Cl_3N_2$

Sn/HCl ⟶ NHPh Cl N Me

Base ⟶ NHPh Cl N CH₂Cl

1. conc. HCl 2. HI ⟶ N Me

the structural formulation resting on the chemical reactions shown.

Several other examples of this reaction have been reported[120], e.g.

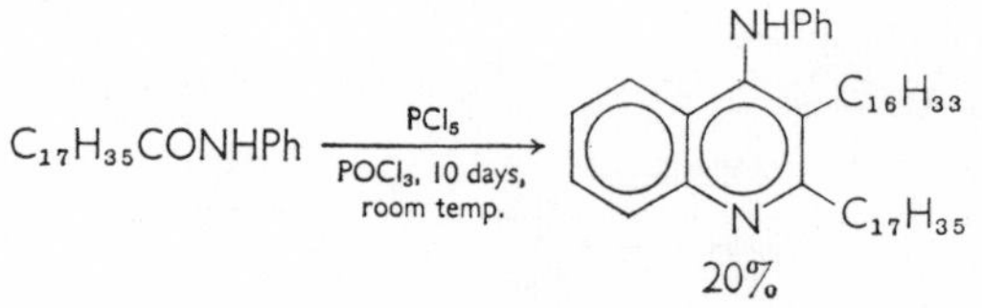

$$C_{17}H_{35}CONHPh \xrightarrow[\text{POCl}_3,\ 10\ \text{days, room temp.}]{PCl_5}$$ NHPh $C_{16}H_{33}$ N $C_{17}H_{35}$

20%

B. Elimination Reactions

1. Pyrolysis

Reactions of the type

$$\text{PhCCl}{=}\text{NR}' \xrightarrow{\Delta} \text{R}'\text{Cl} + \text{PhCN}$$

and

$$\text{PhCCl}{=}\overset{+}{\text{N}}\text{R}'\text{R}''\ \text{Cl}^- \xrightarrow{\Delta} \text{R}'\text{Cl} + \text{PhCCl}{=}\text{NR}'' \xrightarrow{\Delta} \text{PhCN} + \text{R}''\text{Cl}$$

constitute the von Braun degradation. When R′ is hydrogen (R″ being alkyl) hydrogen halide is eliminated very readily, indeed so readily that compounds of the type RCCl=NH have not been isolated, while imidoyl halide hydrohalides of this type have been observed to lose hydrogen halide in refluxing benzene[34] or on melting[52]. When both R′ and R″ are present and are hydrogen (i.e. the imidoyl halide hydrohalides from nitriles and hydrogen halides) the compounds are stable enough to be isolated, but again undergo ready thermal dissociation: an intermediate imidoyl halide stage has not been demonstrated. When both R′ and R″ are alkyl groups the pyrolytic degradation is somewhat less readily achieved, and requires temperatures $< 100°$, and generally $\sim 150°$. *N-t*-Butyl imidoyl halides appear to be an exception to this[6,31], possibly because reaction via a cyclic transition state can occur in the *syn*-species (cf. **9**, Section III.D but see Section VI.F).

This reaction is discussed further elsewhere (Section II.C, Section VI.F). Some examples are given below.

$$\text{PhCCl}{=}\text{NBu-n} \xrightarrow{>100°} \text{PhCN} + \text{n-BuCl} \qquad \text{(ref. 30)}$$

$$\text{PhCCl}{=}\text{NCH}_2\text{Ph} \xrightarrow{\sim 175°} \text{PhCN} + \text{PhCH}_2\text{Cl} \qquad \text{(ref. 122)}$$

$$\text{PhCCl}{=}\text{NSO}_2\text{Ph} \longrightarrow \text{PhCN} + \text{PhSO}_2\text{Cl} \qquad \text{(ref. 112)}$$

$$\text{Ph}_2\text{CHCCl}{=}\text{NMe} \xrightarrow{\sim 190°} \text{Ph}_2\text{CHCN} + (\text{MeCl}) + (\text{Ph}_2\text{C}{=}\text{C}{=}\text{NMe})_2 \qquad \text{(ref. 4)}$$

$$\text{CCl}_2{=}\text{NCOMe} \xrightarrow[\text{room temp.}]{\text{stand,}} \text{ClCN} + \text{MeCOCl} \qquad \text{(ref. 70)}$$

2. α_C-Elimination*: ketenimine and ynamine formation

Treatment of imidoyl halides containing an α_C-H atom with a tertiary base causes elimination of hydrogen halide to generate the corresponding ketenimine, e.g.[4]

$$\text{Me}_2\text{CHCCl}{=}\text{NC}_6\text{H}_4\text{Me-}p \xrightarrow{\text{Et}_3\text{N}} \text{Me}_2\text{C}{=}\text{C}{=}\text{NC}_6\text{H}_4\text{Me-}p$$

$$\text{Ph}_2\text{CHCCl}{=}\text{NBu-n} \xrightarrow{\text{Et}_3\text{N}} \text{Ph}_2\text{C}{=}\text{C}{=}\text{NBu-n}$$

* The α-position with respect to an azomethine group may refer to substituents on either carbon or nitrogen: to distinguish these possibilities the appropriate atom is indicated as a subscript.

The reaction has also been observed as a thermal process[4]. *N,N*-Dialkylamidochlorides cannot generate ketenimines for structural reasons: under strongly basic conditions elimination occurs to give the ynamines[226]. The addition of hydrogen halide to the latter regenerates the amidochlorides.

$$RCH_2CCl{=}\overset{+}{N}R'_2Cl^- \underset{HCl}{\overset{LiNEt_2}{\rightleftharpoons}} RC{\equiv}C{-}NR'_2$$

An alternative route to ketenimines starts with the α-chloroimidoyl chlorides; Staudinger attempted to dehalogenate such compounds with zinc, but had little success[123]. More recently it has been shown that dehalogenation occurs readily in certain instances with sodium iodide in acetone[124]:

$$Ph_2CClCCl{=}NC_6H_4Me\text{-}p \xrightarrow[Me_2CO]{NaI} Ph_2C{=}C{=}NC_6H_4Me\text{-}p$$

3. α_N-Elimination to give 1,3-dipolar systems

In suitable circumstances elimination of hydrogen halide may also involve a proton from the α_N-atom:

$$\underset{Cl\quad H}{RC{=}NA} \xrightarrow[-HCl]{Base} R\overset{+}{C}{=}N{-}\overset{-}{A} \longleftrightarrow RC{\equiv}\overset{+}{N}{-}\overset{-}{A}$$

Examples where A is carbon, nitrogen and oxygen are familiar, e.g.

$$RCCl{=}NCH_2R \xrightarrow{-HCl} R{-}\overset{+}{C}{=}N{-}\overset{-}{C}H{-}R \quad \text{Nitrile ylid}$$

$$RCCl{=}NNHR \xrightarrow{-HCl} R{-}\overset{+}{C}{=}N{-}\overset{-}{N}{-}R \quad \text{Nitrile imine}$$

$$RCCl{=}NOH \xrightarrow{-HCl} R{-}\overset{+}{C}{=}N{-}\overset{-}{O} \quad \text{Nitrile oxide}$$

The products undergo cycloaddition reactions ('1,3-dipolar addition') which have been reviewed elsewhere[125]. Chart 4 summarizes an example which demonstrates the value of this process in heterocyclic synthesis (see also section V.C.).

Huisgen and Raab[127] observed that the two isomeric imidoyl chlorides $PhCCl{=}NCH_2C_6H_4NO_2\text{-}p$ and $p\text{-}NO_2C_6H_4CCl{=}NCH_2Ph$ gave the same mixture of the *cis*- and *trans*-1-pyrroline isomers (**13**) when treated with triethylamine and methyl acrylate. This led to the remarkable discovery that the two imidoyl chlorides equilibrate with

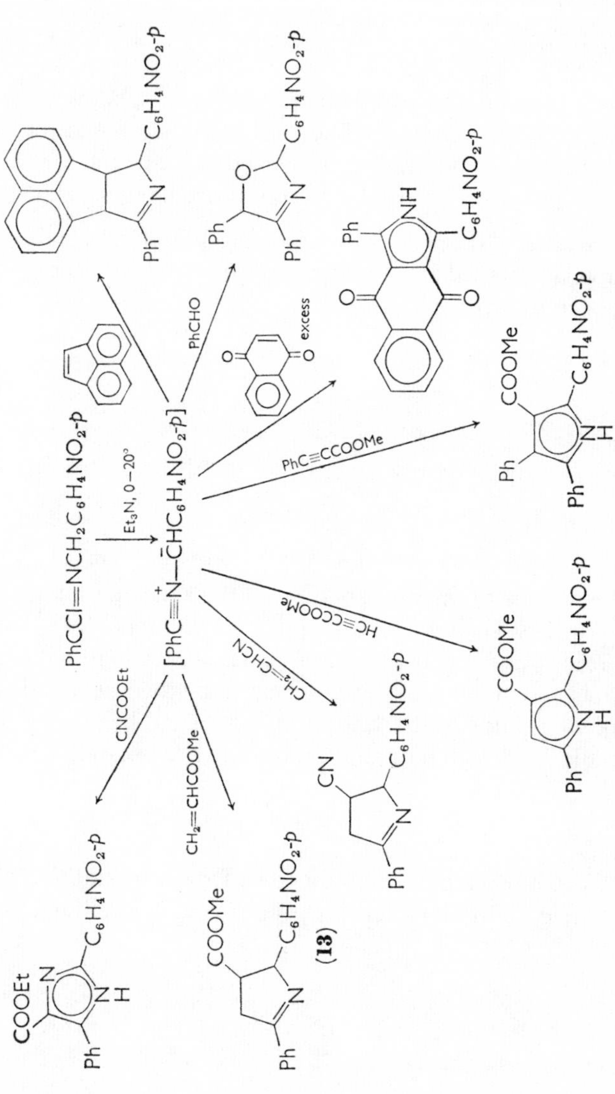

Chart 4. Some 1,3-dipolar cycloadditions of a nitrile ylid[126].

one another in benzene solution in the presence of triethylamine (40 mol. %) at 20°.

$$\underset{\substack{|\\ \mathrm{Cl}}}{\mathrm{PhC}}{=}\mathrm{NCH_2C_6H_4NO_2}\text{-}p \overset{\mathrm{Et_3N}}{\rightleftarrows} \mathrm{PhCH_2N{=}CClC_6H_4NO_2}\text{-}p$$

92% 8%

A hydrogen shift in the intermediate nitrilium ylid is presumably involved: the equilibration does not occur in the absence of base.

In a rather analogous way it has been shown that the *N*-phenyl hydrazidoyl chloride PhCCl=NNHPh exchanges chlorine isotopically (with $Et_3\overset{+}{N}H^{38}Cl^-$ in benzene or chloroform) via the anion: no exchange is observed at room temperature in the absence of bases[128]. This has been interpreted as follows:

$$\mathrm{PhCCl{=}NNHPh} \overset{\mathrm{Et_3N}}{\rightleftharpoons} \mathrm{PhCCl{=}N\bar{N}Ph} + \mathrm{Et_3\overset{+}{N}H}$$

$$\Updownarrow$$

$$\mathrm{Adduct} \xleftarrow{\text{Dipolarophile}} \mathrm{PhC{\equiv}\overset{+}{N}{-}\bar{N}Ph} + \mathrm{Et_3\overset{+}{N}H} + \mathrm{Cl^-}$$

C. Salt Formation

Treatment of a typical imidoyl halide with hydrogen halide in an inert solvent leads to the precipitation of the expected immonium salt (amidochloride). Grdinic and Hahn[34] have prepared a variety of hydrochlorides and hydrobromides (partial exchange with covalent halogen may occur[78]) in this way. These derivatives generally analyse as monohydrohalide salts, and lose hydrogen halide on heating. Occasionally imidoyl halide dihydrohalides are isolated, e.g.[32,52]

$$\mathrm{PhCCl{=}NPh} \underset{75^\circ}{\overset{\mathrm{HCl,\ Et_2O}}{\rightleftharpoons}} [\mathrm{PhCCl{=}\overset{+}{N}HPH}]\mathrm{Cl^-{\cdot}HCl}$$

Hydroiodides have also been obtained[75]. However, addition of hydrogen fluoride to an imidoyl fluoride gives a covalent adduct, not a salt, e.g.[102]

$$\text{(perfluoro-2-fluoro-3,4,5,6-tetrahydropyridine: } F_2, F_2, F_2, F_2, \text{ N, F)} \underset{\text{pyrex}}{\overset{\text{HF, 40°}}{\rightleftharpoons}} \text{(perfluoropiperidine: } F_2, F_2, F_2, F_2, F_2, \text{ NH)}$$

Two other types of salt can be generated directly. *N*-Alkylation of an imidoyl halide is reported to occur with Meerwein's reagent to give an immonium salt[131],

$$\text{3,3-dimethyl-2-chloro-3H-indole} \xrightarrow[\text{ClCH}_2\text{CH}_2\text{Cl}]{\text{Et}_3\text{OBF}_4} \text{1-ethyl-3,3-dimethyl-2-chloro-3H-indolium (N}^+\text{–Et) } BF_4^-$$

while Lewis acids capable of complexing halide ion give nitrilium salts[32,81,80]:

$$PhCCl{=}NPh \xrightarrow[CH_2Cl_2]{AlCl_3} PhC{\equiv}\overset{+}{N}Ph \quad AlCl_4^-$$

$$PhCCl{=}NPr\text{-}i \xrightarrow[ClCH_2CH_2Cl]{SbCl_5} PhC{\equiv}\overset{+}{N}Pr\text{-}i \quad SbCl_6^-$$

$$PhCCl{=}NPh \xrightarrow[PhNO_2]{FeCl_3} PhC{\equiv}\overset{+}{N}Ph \quad FeCl_4^-$$

However, if the basicity of the system is reduced by electron-withdrawing substituents, salt formation may no longer occur. Thus although *N*-phenylbenzimidoyl chloride forms a nitrilium salt with boron trichloride, *N*-benzoylbenzimidoyl chloride gives a covalent adduct[228].

D. Reduction

The reduction of imidoyl halide to aldimine salt, followed by hydrolysis, is a classical route from acid to aldehyde, due to Sonn and Müller[46] (for review, see reference 132).

$$RCOOH \rightarrow RCONHPh \rightarrow RCCl{=}NPh \xrightarrow{\text{reduction}} RCH{=}NPh{\cdot}HX \xrightarrow{H_2O} RCHO$$

Both stannous chloride and chromous chloride have been used in this reduction, the latter being the more effective reagent[133]. The method is not suitable for preparing aliphatic aldehydes of the type $R'CH_2CHO$ or $R_2'CHCHO$ (see Section IV.A) but has found some success with $\alpha\beta$-unsaturated aldehydes. Thus it was employed by Kuhn and Morris[134] in their synthesis of vitamin A.

$$\text{(trimethylcyclohexenyl)-CH=CH-C(Me)=CH-CONHC}_6\text{H}_4\text{Me-}o \xrightarrow[0°,\ C_6H_6]{PCl_5} \text{(trimethylcyclohexenyl)-CH=CH-C(Me)=CH-CCl=NC}_6\text{H}_4\text{Me-}o \xrightarrow[N_2]{CrCl_2}$$

$$\text{(trimethylcyclohexenyl)-CH=CH-C(Me)=CH-CHO} \xrightarrow[\text{2. Al(OPr-i)}_3]{\text{1. Me}_2\text{C=CHCHO, base}} \text{(trimethylcyclohexenyl)-CH=CH-C(Me)=CH-CH=CH-C(Me)=CH-CH}_2\text{OH}$$

Vitamin A_1

Other examples of the Sonn and Müller reduction are shown below: as a preparative method, however, it has largely been eclipsed by controlled reduction with metal hydride reagents[132].

$$PhCH{=}CHCCl{=}NPh \xrightarrow{SnCl_2} \underset{92\%}{PhCH{=}CHCHO} \qquad \text{(ref. 46)}$$

$$n\text{-}C_3H_7CH{=}CHCCl{=}NC_6H_4Me\text{-}o \xrightarrow{CrCl_2} \underset{50\%}{n\text{-}C_3H_7CH{=}CHCHO} \qquad \text{(ref. 133)}$$

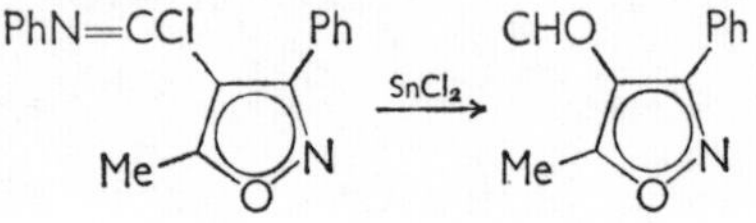

$$\text{4-(PhN=CCl)-3-Ph-5-Me-isoxazole} \xrightarrow{SnCl_2} \text{4-CHO-3-Ph-5-Me-isoxazole} \qquad \text{(ref. 135)}$$

Reduction of *N,N*-dialkylamidochlorides with lithium aluminium hydride in ether gives the corresponding tertiary amines as the major products[223].

$$[PhCCl{=}\overset{+}{N}Me_2]\ Cl^- \xrightarrow[Et_2O]{LiAlH_4} PhCH_2NMe_2$$

E. Nucleophilic Substitution at C=N

The most characteristic reaction of the imidoyl halides, and one in which they show many similarities to the acyl halides, is substitution of halogen by a nucleophile Y, which probably occurs in general in a manner approximating to an S_N1 process, as follows

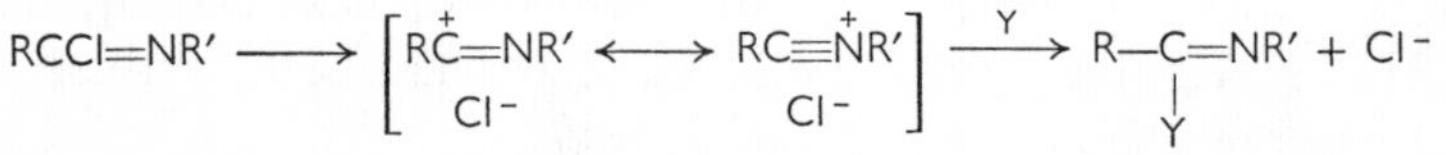

$$RCCl{=}NR' \longrightarrow \left[\underset{Cl^-}{R\overset{+}{C}{=}NR'} \longleftrightarrow \underset{Cl^-}{RC{\equiv}\overset{+}{N}R'}\right] \xrightarrow{Y} R{-}\underset{\underset{Y}{|}}{C}{=}NR' + Cl^-$$

An analogous reaction occurs with the immonium salts which, due to the charged system, are even more reactive: since an S_N1 step in this case requires a doubly charged intermediate, an S_N2 process seems more likely:

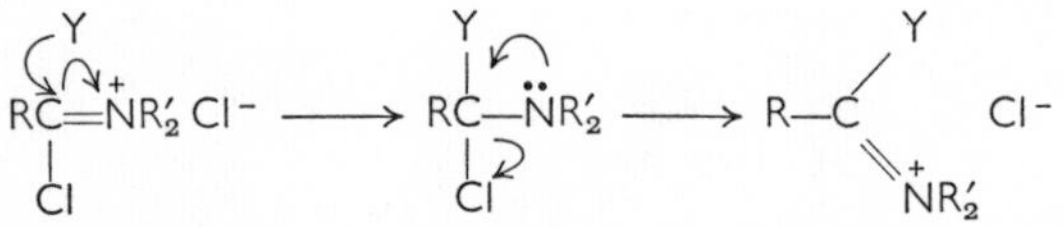

$$\underset{\underset{Cl}{|}}{R\overset{Y}{C}}{=}\overset{+}{N}R'_2\ Cl^- \longrightarrow \underset{\underset{Cl}{|}}{\overset{\overset{Y}{|}}{RC}}{-}\ddot{N}R'_2 \longrightarrow R{-}C(Y){=}\overset{+}{N}R'_2\quad Cl^-$$

The reactions with water (→ amides) and amines (→ amidines) are commonly used to prepare derivatives for characterization purposes. In the following summary the reactions with various nucleophilic reagents are considered in turn.

1. Halide exchange

Little work has been done on this, but an exchange process has occasionally been used to prepare imidoyl fluorides, e.g.[24,136]

$$\text{(perhalogenated tetrahydropyridine: } F_2, F_2, F_2, Cl_2 \text{ ring with } N{=}C\text{–Cl)} \xrightarrow{AgF} \text{(ring with } F_2, F_2, F_2, F_2 \text{, } N{=}C\text{–F)}$$

$$CBr_2{=}N{-}N{=}CBr_2 \xrightarrow[75°]{AgF} CF_2{=}N{-}N{=}CBrF \xrightarrow[125°]{AgF} CF_2{=}N{-}N{=}CF_2$$

2. Water

Hydrolysis is undoubtedly the most carefully studied substitution: the typical imidoyl halides react rapidly and the characteristic product is the amide:

$$RCCl{=}NR' \xrightarrow{H_2O} RCONHR'$$

$$RCCl{=}\overset{+}{N}R'_2\ Cl^- \xrightarrow{H_2O} RCONR'_2$$

However, in the special case of amidochlorides derived from nitriles, an alternative pathway is open. The water may function merely as a base to remove hydrogen halide, and the nitrile may be regenerated[73,74]. It is often accompanied by some amide, however, especially if the water is not in excess[82,137]. Amide hydrochlorides are

$$MeCCl{=}\overset{+}{N}H_2\ Cl^- \xrightarrow{H_2O} \begin{cases} \xrightarrow{-2HCl} MeCN \\ \xrightarrow{\text{hydrolysis}} MeCONH_2 \end{cases}$$

frequently isolable as intermediates and their formation and properties have been reviewed[20]. The iminophosgenes are also rather a special case: thus *N*-phenyliminophosgene does not react noticeably with cold water, but on heating gives the corresponding amine and *sym*-urea, thus[138]:

$$PhN{=}CCl_2 \xrightarrow{H_2O} [PhNHCOCl] \longrightarrow [PhNHCOOH] \xrightarrow{-CO_2} PhNH_2$$

$$[PhNHCOCl] + PhNH_2 \longrightarrow PhNHCONHPh$$

The kinetics of hydrolysis of a series of imidoyl chlorides have been measured in aqueous acetone by Ugi, Beck and Fetzer[31], and some of their results are given in Table 2 (initial pseudo-unimolecular rate

TABLE 2. Rate constants for hydrolysis of imidoyl halides.

(10^{-3}–10^{-4} mole/l) in aqueous acetone ($\gamma H_2O = 0{\cdot}333$) Imidoyl chloride	Temp. (°C)	Initial rate const. $k_{H_2O,0} \times 10^4\ s^{-1}$
1. $CH_3CCl{=}N$–cyclohexyl	−50	8800
$CCl_3CCl{=}N$–cyclohexyl	−20	0·000003
2. $EtCCl{=}N$–cyclohexyl	−50	8000
i-$PrCCl{=}N$–cyclohexyl	−50	6300
t-$BuCCl{=}N$–cyclohexyl	−50	3200
3. $MeCOCCl{=}N$—Bu-n	−20	0·15
$MeCOCCl{=}N$—Bu-*t*	−20	115
4. $PhCCl{=}N$—Bu-n	−20	400
$PhCCl{=}N$—Bu-*t*	−20	550
$PhCCl{=}N$–cyclohexyl	−20	420
p-$NO_2C_6H_4CCl{=}N$–cyclohexyl	−20	28
5. $PhCCl{=}NPh$	−20	3·8
$PhCCl{=}NC_6H_4NO_2$-*p*	−20	0·18
$PhCCl{=}NC_6H_4OMe$-*p*	−20	25
p-$NO_2C_6H_4CCl{=}NC_6H_4NO_2$-*p*	−20	0·00237

constants). The rate is not much affected by steric hindrance at the α_C position (e.g. item 2) and a possible steric acceleration is observed on increasing the bulk of the *N*-substituent (items 3, 4). Electronic effects are, however, very important (items 1, 4, 5): electron-withdrawing substituents at carbon or nitrogen (CCl_3, MeCO, *p*-$NO_2C_6H_4$—) generally cause a pronounced decrease in reactivity, while electron donors (Me, p-$MeOC_6H_4$—) increase the rate of hydrolysis. These observations accord with an S_N1 mechanism: and in some cases first-order kinetics are reported (e.g. hydrolysis of $CCl_3CCl{=}N$—cyclohexyl; aminolysis of PhCCl=NPh). In most cases, however, the hydrolysis deviates from first-order as the reaction progresses, due to inhibition by chloride ion. Ugi, Beck and Fetzer have suggested a two-stage mechanism involving a nitrilium ion pair intermediate to account for these observations:

$$RCCl{=}NR' \underset{k_{-1}}{\overset{k_1}{\rightleftharpoons}} \left[\begin{matrix} RC{\equiv}\overset{+}{N}R' \\ Cl^- \end{matrix}\right] \xrightarrow{k_2,\ H_2O} RCONHR' + HCl$$

The reactivity order measured here and observed qualitatively in other such nucleophilic substitutions is thus as follows:

$$AlkCCl{=}NAlk \gg ArCCl{=}NAlk > ArCCl{=}NAr$$

3. Alcohols and phenols

The imidoyl halides react with alcohols and phenols (preferably as the alkoxides or phenoxides) to give the corresponding imidates, e.g.

$$\begin{matrix} CCl{=}NPh \\ | \\ CCl{=}NPh \end{matrix} \xrightarrow[PhMe]{NaOEt} \begin{matrix} EtOC{=}NPh \\ | \\ EtOC{=}NPh \end{matrix} \qquad \text{(ref. 139)}$$

$$\begin{matrix} CCl{=}NPh \\ | \\ CCl{=}NPh \end{matrix} \xrightarrow{NaOPh} \begin{matrix} PhOC{=}NPh \\ | \\ PhOC{=}NPh \end{matrix} \qquad \text{(ref. 139)}$$

$$PhCCl{=}NPh \xrightarrow[EtOH]{NaOEt} PhC(OEt){=}NPh \qquad \text{(ref. 140)}$$

$$CCl_2{=}NPh \xrightarrow{NaOEt} EtOCCl{=}NPh \qquad \text{(ref. 141)}$$

$$PhCCl{=}N(\beta\text{-Naphthyl}) \xrightarrow[NaOEt,\ EtOH]{PhOH} PhC(OPh){=}N(\beta\text{-naphthyl}) \qquad \text{(ref. 142)}$$

Phenols may, under different conditions, undergo ring substitution (see Section IV.E.15).

N,N-Dialkylamidochlorides behave differently: the product of the initial substitution is a *N,N*-dialkylimmonium salt (**14**) which undergoes addition to give the amino acetal, e.g.[143]

$$HCCl{=}\overset{+}{N}Me_2\ Cl^- \xrightarrow{NaOMe} \underset{\mathbf{(14)}}{HC(OMe){=}\overset{+}{N}Me_2\ Cl^-} \xrightarrow{NaOMe} (MeO)_2CHNMe_2$$

Alcohols readily substitute amidochlorides of this type to give the same type of immonium salt as **14**: this reacts with water to give the corresponding ester, and with hydrogen sulphide to give the corresponding thioester[146].

$$EtCCl{=}\overset{+}{N}C_5H_{10}\ Cl^- \xrightarrow[-HCl]{EtOH} EtC(OEt){=}\overset{+}{N}C_5H_{10}\ Cl^-$$

$$\xrightarrow{H_2O} EtCOOEt\ (88\%) \qquad \xrightarrow{H_2S} EtCSOEt\ (72\%)$$

4. Amines

Ammonia and primary and secondary amines react with imidoyl halides to give amidines. Pechmann used this route in his studies on the tautomerism of amidines[3,147]. Even the labile imidoyl halides, if they are reacted with an amine as soon as they are formed, give the expected compound, e.g.[17]

$$CH_3CCl{=}NC_6H_4Me \xrightarrow{C_{10}H_7NH_2} CH_3C({=}NC_6H_4Me)NHC_{10}H_7$$

Many other examples are available, e.g.

$$PhCCl{=}NPh \xrightarrow{NH_3} PhC(NH_2){=}NPh \qquad \text{(ref. 90)}$$

$$ArCCl{=}NAr \xrightarrow[Et_3N]{\text{aziridine (NH)}} ArC({=}NAr)\text{-N(aziridinyl)} \xrightarrow{I^-} Ar\text{-}C{=}N\text{-}CH_2CH_2\text{-}N(Ar) \text{ (1-aryl-2-imidazoline)} \qquad \text{(ref. 144)}$$

$$MeCCl{=}NC_6H_4OEt\text{-}p + p\text{-}NH_2C_6H_4COOEt \xrightarrow{C_6H_6} MeC({=}NC_6H_4OEt\text{-}p)NHC_6H_4COOEt\text{-}p \qquad \text{(ref. 18)}$$

$$MeCOCCl{=}NOH \xrightarrow[Et_2O]{PhNH_2} MeCOC(NHPh){=}NOH \qquad \text{(ref. 145)}$$

5. Urethanes

The condensation of *N*-(*p*-tolyl)benzimidoyl chloride with sodio-*N*-phenylurethane gives the expected substitution product[149]:

$$PhCCl{=}NC_6H_4Me\text{-}p \xrightarrow[Et_2O]{Ph\bar{N}COOEt\ Na^+} PhC(=NC_6H_4Me\text{-}p)N(Ph)COOEt$$

Certain products of this type undergo thermal cyclization to give 4-quinazolones (4-hydroxyquinazolines), thus:

$$PhCCl{=}NPh + NaNHCOOEt \xrightarrow{Et_2O} C_6H_5N{=}C(Ph)NHCOOEt \xrightarrow{180^\circ} \text{4-hydroxy-2-phenylquinazoline}$$

(or tautomer)

6. Amidines

$$PhCCl{=}NPh + PhC(NH_2){=}NH \xrightarrow{Et_2O} PhC(=NH){-}NH{-}C(=NPh){-}Ph \quad \text{(ref. 150)}$$

$$PhCCl{=}NNHPh + PhC(NH_2){=}NH \xrightarrow{Et_2O} \text{1,3,5-triphenyl-1,2,4-triazole} \quad \text{(ref. 151)}$$

7. Hydroxylamines

Hydroxylamine itself gives the amidoxime system.

$$PhCCl{=}NPh \xrightarrow{NH_2OH} PhC({=}NOH)NHPh \quad \text{(ref. 152)}$$

$$PhCCl{=}NOH \xrightarrow{NH_2OH} PhC({=}NOH)NHOH \quad \text{(ref. 153)}$$

$$\text{1-methyl-2-chloro-1-pyrrolinium } Cl^- \xrightarrow{NH_2OH} \text{1-methyl-2-pyrrolidinone oxime (=NOH)} \quad \text{(ref. 146)}$$

N-Substituted hydroxylamines give the expected substitution products which cannot, of course, tautomerize to oximes. Thus the products from the two reactions

$$PhCCl{=}NPh + p\text{-}MeC_6H_4NHOH \xrightarrow{Et_2O} PhC({=}NPh)N(OH){-}C_6H_4Me\text{-}p \quad \text{m.p.} \sim 175^\circ$$

$$PhCCl{=}NC_6H_4Me\text{-}p + PhNHOH \xrightarrow{Et_2O} PhC({=}NC_6H_4Me\text{-}p)N(OH){-}Ph \quad \text{m.p. } 191^\circ$$

are not identical[122].

8. Hydrazines

Hydrazine itself reacts with imidoyl halides to give the hydrazidine–amidrazone system, e.g.[154]

$$\text{PhCCl=N–N(5-phenyltetrazol-1-yl)} \xrightarrow[\text{Et}_2\text{O}]{\text{N}_2\text{H}_4,\ \text{H}_2\text{O}} \text{PhC(NHNH}_2\text{)=N–N(tetrazolyl)} \rightleftharpoons \text{Ph–C(=NNH}_2\text{)–NH–N(tetrazolyl)}$$

or to give azines, e.g.[146]

$$\text{1-methyl-2-chloro-pyrrolinium chloride} \xrightarrow{\text{N}_2\text{H}_4\cdot\text{H}_2\text{O}} \text{(1-methylpyrrolidin-2-ylidene)=N–N=(1-methylpyrrolidin-2-ylidene)}$$

Both nitrogen atoms of phenylhydrazine serve as nucleophilic centres, but attack at the exposed electron-rich β-nitrogen predominates[155].

$$\text{PhCCl=NPh} + \text{PhNHNH}_2 \xrightarrow{0°,\ \text{EtOH}} \underset{\text{major product}}{\text{PhC(=NPh)NHNHPh}} + \text{PhC(=NPh)N(NH}_2\text{)Ph}$$

9. Hydrazoic acid

Schroeter[156] observed that treatment of *N*-phenylbenzimidoyl chloride with sodium azide in amyl ether gave 1,5-diphenyltetrazole, thus:

$$\text{PhCCl=NPh} \xrightarrow{\text{NaN}_3} \left[\text{PhC(=NPh)N}_3\right] \longrightarrow \text{1,5-diphenyltetrazole} \qquad \text{(ref. 156)}$$

$$\text{PhCCl=NOH} \xrightarrow[\text{Et}_2\text{O}]{\text{NaN}_3} \text{1-hydroxy-5-phenyltetrazole} \qquad \text{(ref. 148)}$$

Hydrazoic acid has also been used as the reagent, and this constitutes a valuable synthesis of 1,5-disubstituted tetrazoles[154,157]. Such tetrazoles are convenient precursors for certain carbodiimides, e.g.[85]

$$\text{1,5-diphenyltetrazole} \xrightarrow{210\text{–}220°} \underset{70\%}{\text{PhN=C=NPh}}$$

10. Hydrogen cyanide

Substitution by cyanide occurs satisfactorily even in aqueous or alcoholic solution, e.g.

$$PhCCl{=}NPh \xrightarrow[Et_2O/H_2O]{KCN} PhC(CN){=}NPh \qquad \text{(refs. 158, 159)}$$

$$PhCCl{=}NC_6H_4Br\text{-}p \xrightarrow[20\ h,\ Et_2O/H_2O]{KCN} PhC(CN){=}NC_6H_4Br\text{-}p \qquad \text{(ref. 160)}$$

N,N-Dialkylamidochlorides give dicyano derivatives[161],

$$CHCl{=}\overset{+}{N}Me_2\ Cl^- \xrightarrow{HCN} [HC(CN){=}\overset{+}{N}Me_2]\ Cl^- \xrightarrow{HCN} (CN)_2CHNMe_2$$

a reaction which parallels the formation of aminoacetals (see Section IV.E.3).

In the presence of a suitable heterocyclic base (pyridine, quinoline, isoquinoline) *N*-phenylbenzimidoyl chloride and hydrogen cyanide give an immonium salt (e.g. **15** from quinoline)[159,162]. The reaction is analogous to the formation of Reissert compounds using acyl halides.

Quinoline $\xrightarrow[HCN]{PhCCl{=}NPh}$ **(15)** [1-(C(Ph)=$\overset{+}{N}$HPh)-2-cyano-1,2-dihydroquinoline] Cl⁻

(15) $\xrightarrow{H_3O^+}$ PhCHO + quinoline-2-COOH

(15) $\xrightarrow[PhNO_2]{\Delta}$ quinoline-2-CN

The product **15** shows no band in the 2200 cm^{-1} region, and probably exists in the isomeric cyclized form **16**.

[Structure: NH, =N⁺, Ph, Ph, Cl⁻]

(16)

11. Carboxylic acids

Salts of aliphatic and aromatic acids give initially the isoimide which readily rearrange to give a diacylamine, thus[130,160]:

$$PhCCl{=}NPh + PhCH{=}CHCOONa \longrightarrow [PhC({=}NPh)OCOCH{=}CHPh] \longrightarrow PhCONPhCOCH{=}CHPh$$

$$PhCCl{=}NC_6H_4Br\text{-}p \xrightarrow[Et_2O/H_2O]{NaOOCC_6H_4Br\text{-}p} PhCON(C_6H_4Br\text{-}p)COC_6H_4Br\text{-}p$$

In a recent example (with 2,4-dinitrophenyl as the *N*-substituent) it has been possible to isolate the intermediate isoimide and to study its rearrangement to the diacylamine[164]. The rearrangement may be rationalized in terms of the following steps:

$$PhCCl{=}NC_6H_3(NO_2)_2 \xrightarrow{ArCOOAg} Ph(OCOAr)C{=}NC_6H_3(NO_2)_2 \rightleftharpoons$$

$$Ph(OCOAr)C{=}NC_6H_3(NO_2)_2 \longrightarrow [\text{cyclic intermediate: } Ph{-}C{=}\overset{+}{N}C_6H_3(NO_2)_2,\ O{-}C(Ar){-}O^-] \longrightarrow PhCON(C_6H_3(NO_2)_2)COAr$$

Salts of dialkylphosphate esters behave rather similarly, and O → N migrations have been observed[117]. The imidoyl dialkylphosphates can be isolated in certain cases, however, and function as phosphorylating agents, e.g.

$$PhCCl{=}NPh + (PhCH_2O)_2PO_2Ag \longrightarrow Ph{-}C(OP(O)(OCH_2Ph)_2){=}NPh \xrightarrow{(PhO)_2PO_2H} (PhO)_2P(O){-}O{-}P(O)(OCH_2Ph)_2$$

$$PhCCl{=}NMe + Et_3\overset{+}{N}H(PhCH_2O)_2PO_2^- \xrightarrow[\text{room temp.}]{} (PhCH_2O)_2P(O){-}O{-}P(O)(OCH_2Ph)_2$$

When the free carboxylic acid is used the reaction takes a different course and the corresponding acyl chloride may be isolated[165]. Presumably halide ion attacks the intermediate protonated isoimide thus:

$$PhCCl{=}NPh \xrightarrow{PhCOOH} PhC(OC(O)Ph){=}\overset{+}{N}HPh\ \ \bar{C}l \longrightarrow PhCONHPh + PhCOCl \quad (90\%\ \text{yield isolated})$$

This process has been evaluated[165] as a mild method for generating the acyl halide function for subsequent peptide synthesis, but it appears to have found little application.

$$t\text{-BuCCl=NPh} \xrightarrow[\text{10 h, room temp.}]{\text{TosNHCHMeCOOH}} \left[\text{TosNHCH(Me)COCl}\right] \xrightarrow[\text{2. } H_2/Pd]{\text{1. } NH_2CH(Me)COOCH_2Ph} \text{Tosylalanyl alanine}$$

The catalytic effect exerted by dimethylformamide on the formation of acid chlorides from acids and thionyl chloride[33] may be attributed to the intermediacy of the amidochloride $CHCl{=}\overset{+}{N}Me_2\ Cl^-$.

$$Me_2NCHO \xrightarrow{SOCl_2} Me_2\overset{+}{N}{=}CHOSOCl\ Cl^- \xrightarrow{-SO_2} Me_2\overset{+}{N}{=}CHCl\ \bar{Cl}$$

$$Me_2\overset{+}{N}{=}CHOSOCl\ Cl^- \xrightarrow[-SO_2]{RCOOH} RCOCl + Me_2NCHO \qquad Me_2\overset{+}{N}{=}CHCl\ \bar{Cl} \xrightarrow{RCOOH} RCOCl + Me_2NCHO$$

12. Hydrogen sulphide, thiols, thioacids

The reactions are broadly similar to those of the oxygen analogues. Thus hydrogen sulphide gives thioamides:

$$PhCCl{=}NPh \xrightarrow{H_2S} PhCSNHPh \qquad \text{(ref. 166)}$$

$$\text{2-chloro-1-methyl-azepinium chloride (ring } \overset{+}{N}(Me){=}C(Cl)\text{, } Cl^-) \xrightarrow[CHCl_3]{H_2S} \text{N-methyl-azepane-2-thione (ring } N(Me)C{=}S) \qquad \text{(ref. 146)}$$

In spite of an early report to the contrary[166], thiols appear to react in the expected manner. The amidochlorides $RCCl{=}\overset{+}{N}R_2\ Cl^-$ give intermediates which may be cleaved with water or with hydrogen sulphide[146].

$$PhCCl{=}\overset{+}{N}Me_2\ Cl^- \xrightarrow[Et_2O]{PhSH} PhC(SPh){=}\overset{+}{N}Me_2\ Cl^- \xrightarrow{H_2O} PhCOSPh \quad (61\%)$$

$$EtCCl{=}\overset{+}{N}C_5H_{10}\ Cl^- \xrightarrow{EtSH} EtC(SEt){=}\overset{+}{N}C_5H_{10}\ Cl^- \xrightarrow{H_2S} EtCSSEt \quad (86\%)$$

Taken with the analogous reactions of the ternary imidate salts (see Section IV.E.3) this provides an elegant route to each of the three sulphur analogues of carboxylic acid esters (i.e. RCOSR′, RCSOR′, RCSSR′) from a common intermediate.

Salts of the thioacids react in an analogous way to those of carboxylic acids: the initial substitution product rearranges to give the diacylamine analogue, thus:

$$\text{PhCCl=NPh} \xrightarrow[\substack{C_6H_6\\ \text{room temp.}}]{\text{PhCOSK}} \left[\begin{array}{c}\text{PhC=NPh}\\ |\\ \text{SCOPh}\end{array}\right] \longrightarrow \begin{array}{c}\text{PhCSNPh}\\ |\\ \text{COPh}\end{array} \qquad \text{(ref. 167)}$$

$$\text{EtOCCl=NPh} \xrightarrow[C_6H_6/CHCl_3]{(PhCSS)_2Pb} \begin{array}{c}\text{EtOC=NPh}\\ |\\ \text{SCSPh}\end{array} \longrightarrow \begin{array}{c}\text{EtOCSNPh}\\ |\\ \text{CSPh}\end{array} \qquad \text{(ref. 168)}$$

13. Phosphites[169] (Michaelis–Arbuzov reaction)

N-Phenyl and *N*-methylbenzimidoyl chlorides react with trialkyl phosphites to give the corresponding dialkyl phosphonates, thus:

$$p\text{-MeC}_6\text{H}_4\text{CCl=NPh} + (\text{EtO})_3\text{P} \xrightarrow{\sim 160^\circ} \left[\begin{array}{c}\text{Cl}^- \\ (\text{EtO})_2\overset{+}{\text{P}}\text{—O—CH}_2\text{Me}\\ |\\ p\text{-MeC}_6\text{H}_4\text{—C=NPh}\end{array}\right] \longrightarrow \begin{array}{c}(\text{EtO})_2\text{PO}\\ |\\ p\text{-MeC}_6\text{H}_4\text{C=NPh}\end{array}$$

The products are readily hydrolysed to the benzoyl phosphonates which may be isolated as the 2,4-dinitrophenylhydrazone derivatives.

14. Carbanion reagents

Compounds containing activated methylene groups readily condense with imidoyl halides in the presence of base. The products from *β*-dicarbonyl compounds have generally been formulated with *C*-substitution (e.g. **17**) but there is evidence that *O*-substituted compounds (e.g. **18**) are also formed in appreciable amounts in certain cases[175], and this may not always have been recognized in the past. Either type of substitution product appears to suffer cyclization to the quinoline system on heating.

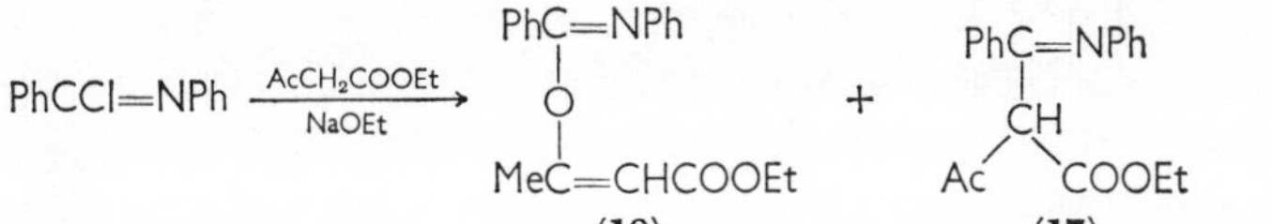

PhCCl=NPh $\xrightarrow[\text{NaOEt}]{\text{AcCH}_2\text{COOEt}}$ PhC=NPh–O–MeC=CHCOOEt (**18**) + PhC=NPh–CH(Ac)COOEt (**17**)

(**18**) $\xrightarrow{\Delta}$ 4-OH, 3-COPh, 2-Me quinoline; (**17**) $\xrightarrow{\Delta}$ 4-OH, 3-COMe, 2-Ph quinoline

This route to 4-hydroxyquinolines has been of some importance[174]; the following examples are illustrative:

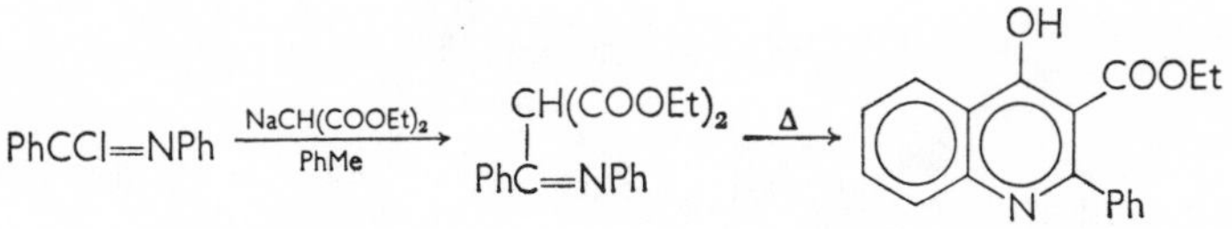

PhCCl=NPh $\xrightarrow[\text{PhMe}]{\text{NaCH(COOEt)}_2}$ CH(COOEt)$_2$–PhC=NPh $\xrightarrow{\Delta}$ OH, COOEt, Ph, N (ref. 174)

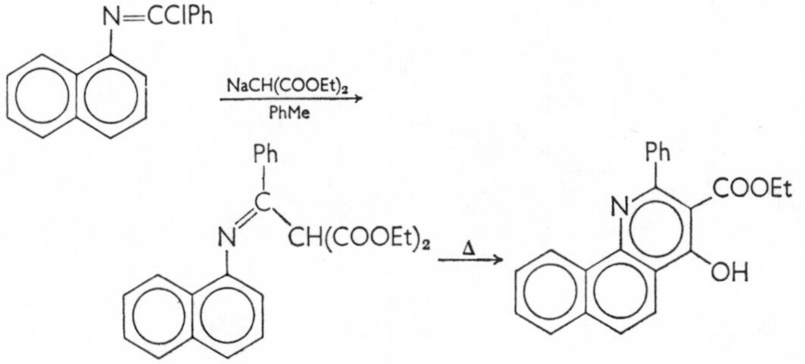

N=CClPh $\xrightarrow[\text{PhMe}]{\text{NaCH(COOEt)}_2}$ Ph–C(=N)–CH(COOEt)$_2$ $\xrightarrow{\Delta}$ Ph, COOEt, N, OH (ref. 171)

PhCCl=N–C$_6$H$_4$–N=CClPh $\xrightarrow[\text{2. }\Delta]{\text{1. NaCH(COOEt)}_2\text{, PhMe}}$ Ph, EtOOC, OH, N, OH, COOEt, N, Ph

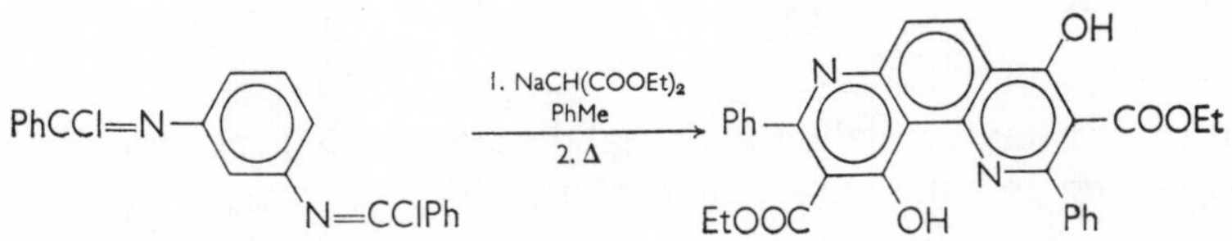

(ref. 172)

N-Phenylhydrazidoyl chlorides give pyrazoles under mild conditions in a reaction which can be rationalized as a 1,3-dipolar cycloaddition.

$$PhCCl{=}NNHPh \xrightarrow[NaOEt]{AcCH_2COOEt} \text{(pyrazole: 4-EtOOC, 3-Ph, 5-Me, 1-Ph)}$$ (ref. 170)

$$PhCCl{=}NNHPh \xrightarrow[Base]{CH_2(CN)_2} \text{(pyrazole: 4-CN, 3-Ph, 5-}H_2N\text{, 1-Ph)}$$ (ref. 173)

15. Friedel–Crafts reactions

Activated aromatic systems (phenols, phenolic ethers, amines) readily undergo Friedel–Crafts reactions with imidoyl halides, usually in the presence of a Lewis acid catalyst such as aluminium chloride, e.g.

$$\text{2-naphthol} \xrightarrow[AlCl_3,\ Et_2O]{PhCCl{=}NPh} \text{1-(PhC=NPh)-2-naphthol}$$ (ref. 176)

$$\text{2-methoxynaphthalene} \xrightarrow[2.\ 10\%\ HCl]{1.\ PhN{=}CClCCl{=}NPh,\ C_6H_6/AlCl_3} \text{methoxyacenaphthenequinone}$$ (ref. 177)

$$PhNMe_2 \xrightarrow[AlCl_3,\ CS_2]{PhCCl{=}NPh} Me_2N{-}C_6H_4{-}CPh{=}NPh \xrightarrow{H_2O} Me_2N{-}C_6H_4{-}COPh$$ (ref. 178)

Neither the imidoyl halide hydrohalides nor the ternary immonium salts require an aluminium halide catalyst, e.g.

$$PhCCl{=}\overset{+}{N}HPh\ Cl^- \cdot HCl \xrightarrow[2.\ hydrolysis]{1.\ PhNMe_2} PhCOC_6H_4NMe_2\text{-}p$$ (ref. 32)

5%

$$Me_2CHCCl{=}\overset{+}{N}\text{(pyrrolidinium)}\ Cl^- \xrightarrow[2.\ hydrolysis]{1.\ PhN(Bu\text{-}n)_2} Me_2CHCO{-}C_6H_4{-}N(Bu\text{-}n)_2$$ (ref. 146)

This, together with the related Vilsmeier–Haack synthesis (see Section VI.D), constitutes an important synthetic approach to aromatic aldehydes and ketones.

F. Substitution at the α_C-Carbon Atom

Imidoyl halides with hydrogen at the α_C-carbon are readily substituted by chlorine and bromine: indeed α_C-halogenated products often result directly in preparations from amides with phosphorus pentachloride (see Section II.A).

$$[Me_2CHCCl{=}NEt] \xrightarrow{Br_2} Me_2CBrCCl{=}NEt + Me_2CBrCBr{=}NEt \quad \text{(ref. 111)}$$

$$MeCH_2CCl{=}\overset{+}{N}Me_2\ \bar{Cl} \xrightarrow[CHCl_3]{Cl_2} MeCCl_2CCl{=}\overset{+}{N}Me_2\ \bar{Cl} \xrightarrow{H_2O} MeCCl_2CONMe_2 \quad \text{(ref. 121)}$$

Fluorine has been observed to exchange halogen and to add to the azomethine group[6].

$$PhCCl{=}NBu\text{-}s \xrightarrow[-78^\circ]{F_2} PhCF_2NFBu\text{-}s$$

Photochemical chlorination results in extensive or complete halogenation of alkyl groups, often with a shift of the azomethine bond, e.g.[45]

$$PhCH_2N{=}CCl_2 \xrightarrow[h\nu,\ CCl_4]{Cl_2} PhCCl_2N{=}CCl_2 \xleftarrow[h\nu,\ CCl_4]{Cl_2} PhCCl{=}NMe$$

$$PhCCl{=}NEt \xrightarrow[h\nu,\ CCl_4]{Cl_2} PhCCl_2N{=}CClCCl_3$$

Substitution at α_C-methylene or α_C-methine groups has also been observed in self-condensation processes (see Section IV.A); and in reactions with diazonium salts[121].

$$EtCCl{=}\overset{+}{N}C_5H_{10}\ Cl^- \xrightarrow[DMF]{p\text{-}NO_2C_6H_4\overset{+}{N}_2BF_4^-} \left[\begin{array}{l} MeCHCCl{=}\overset{+}{N}C_5H_{10} \\ \quad | \\ \quad N{=}NC_6H_4NO_2\text{-}p \end{array}\right] Cl^- \xrightarrow{H_2O} \begin{array}{l} MeC{-}CON C_5H_{10} \\ \quad \| \\ \quad N \\ \quad NHC_6H_4NO_2\text{-}p \end{array}$$

G. Miscellaneous

Imidoyl halides appear to give Grignard derivatives, but little study has been made of these. Thus *N*-phenylbenzimidoyl chloride gives a magnesium derivative which reacts with water to give the corresponding azomethine[179].

$$PhCCl{=}NPh \xrightarrow{Mg} PhCMgCl{=}NPh \xrightarrow{H_2O} PhCH{=}NPh$$

However, complications may be expected to arise since imidoyl halides are known to react with Grignard reagents[163].

$$PhCCl{=}NPh \xrightarrow{PhMgBr} PhC({=}NPh)Ph$$

$$PhCCl{=}NPh \xrightarrow{PhCH_2MgCl} PhC({=}NPh)CH_2Ph$$

N,*N*-Dialkylamidochlorides[223] react with Grignard reagents to give tertiary amines accompanied by the aldehyde (or ketone) corresponding to mono-substitution, e.g.

$$[CHCl{=}\overset{+}{N}Me_2]\,Cl^- \xrightarrow[2.\ H_2O]{1.\ PhMgBr,\ Et_2O} \underset{38\%}{Ph_2CHNMe_2} + \underset{18\%}{PhCHO}$$

V. HETEROCYCLIC SYNTHESIS

It will perhaps be evident from the chemistry discussed in Section IV that imidoyl halides may frequently serve as precursors for heterocyclic systems, and it is pertinent to outline some of the more useful approaches.

A. *Intramolecular Condensation*

Several of the syntheses involve intramolecular condensation, and examples have already been met (see Section IV.A) leading to iminazoles and quinolines, e.g.[120]

$$PhNHCO(CH_2)_5CONHPh \xrightarrow{PCl_5}$$ NHPh (tetrahydroacridine, N) 50%

Imidoyl halide hydrohalides derived as intermediates by treating appropriate dinitriles with hydrogen halide (usually HBr or HI) may also give cyclic products, the cyclization being visualized as follows:

CN CN $\xrightarrow{HBr}$ Br, C, H_2N^+, Br^-, C, N, Br $\longrightarrow$ $H_2\overset{+}{N}$, N, Br, Br^-

(or tautomeric arrangement)

It has been used to prepare pyridines, isoquinolines and benzazepines.

$(NCCH_2)_2CHOH \xrightarrow[HOAc]{HBr}$ 2-amino-6-bromopyridine (ref. 181)

o-(cyanomethyl)benzonitrile $\xrightarrow{HBr}$ 3-amino-1-bromoisoquinoline (ref. 182)

o-phenylenediacetonitrile $\xrightarrow{HBr}$ 2-amino-4-bromo-3H-3-benzazepinium bromide (^+N Br^-) (ref. 183)

Examples of intramolecular Friedel–Crafts reactions are also available[215].

4,4'-dinitro-2-(N=CClPh)biphenyl $\xrightarrow[PhNO_2]{\Delta,\ POCl_3}$ dinitro-6-phenylphenanthridine (NO_2, NO_2, N, Ph)

B. Nucleophilic Substitution followed by Cyclization

The most general route involves nucleophilic substitution of halogen, followed by a cyclization at either the nitrogen atom (**19**) or an *N*-substituent (**20**).

R(Cl)C=NR' + A–B ⟶ R(A–B)C=NR' ⟶ ring R–C(–A)–N(R')–B (**19**)

R(A–B)C=NR' ⟶ ring R–C(–A)=N–R'–B (**20**)

Examples of type **20** leading to quinazolines (from urethanes, Section IV.E.5) and quinolines (from β-ketoesters and dialkyl malonates, Section IV.E.14) have already been encountered. Thus

Elderfield and his colleagues[184] used this route—without isolation of the intermediate—in work on synthetic antimalarials, e.g.

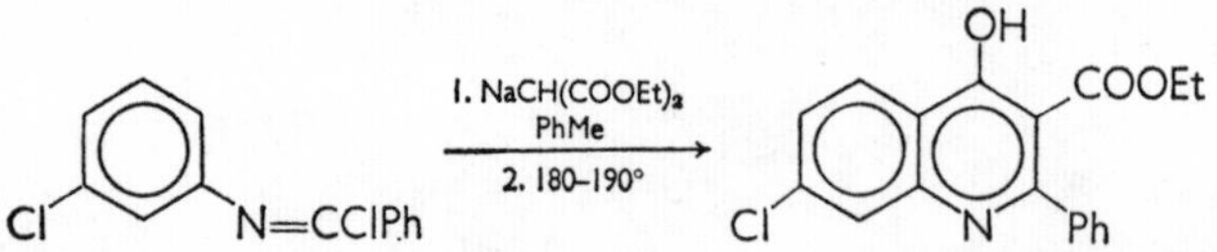

while Meerwein and his colleagues have shown that the reaction of *N*-aryl imidoyl chlorides with nitrile–Lewis acid complexes is an excellent route to the quinazolines[224]. *N*-Substituents other than aryl may be involved in the cyclization, e.g.[185]

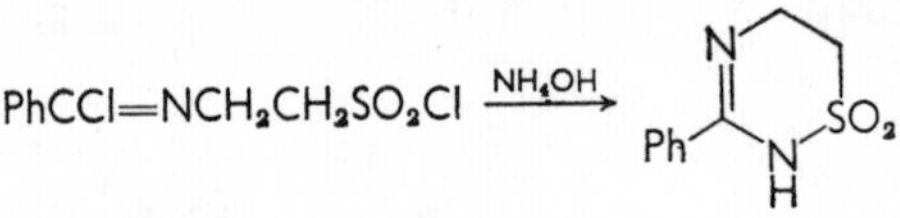

$$PhCCl{=}NCH_2CH_2SO_2Cl \xrightarrow{NH_4OH}$$

Products of type **19** have been met in the 1,5-disubstituted tetrazoles (from azides, Section IV.E.9), whilst 1,2,4-triazoles are formed in a similar way from hydrazides, e.g.[180]

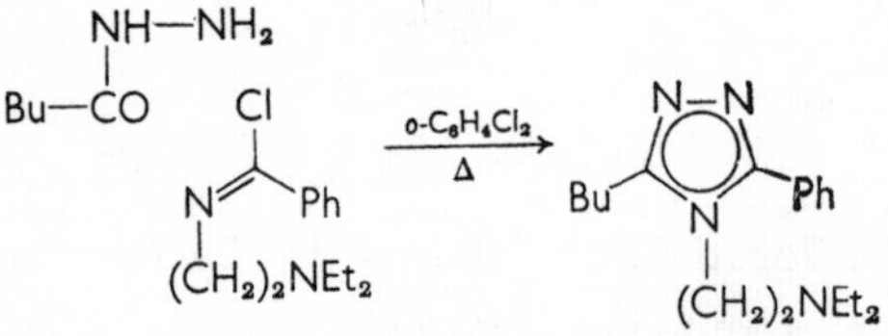

1,2,4-Triazoles are available in another and intriguing way from 5-phenyltetrazoles. The latter react with imidoyl halides, presumably to give the substitution product **21**. This, however, is unstable and loses nitrogen to give the 1,2,4-triazole in good yield (~80%), possibly in the manner shown. Many examples of this reaction have been given[216].

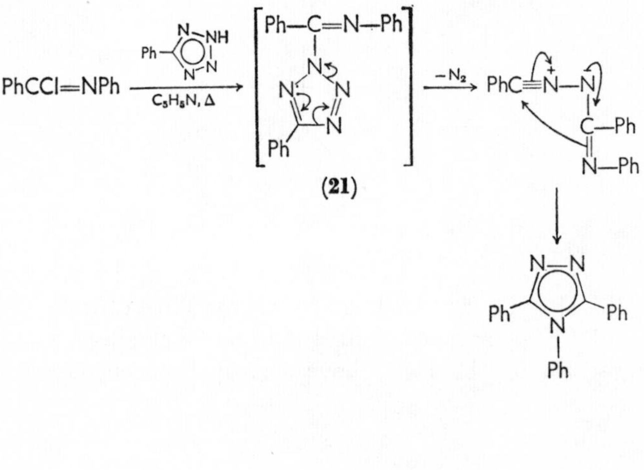

(**21**)

The *N,N*-dialkylamidochlorides salts do not react in the generalized manner outlined above, since the nitrogen atom may be eliminated as secondary amine, i.e.

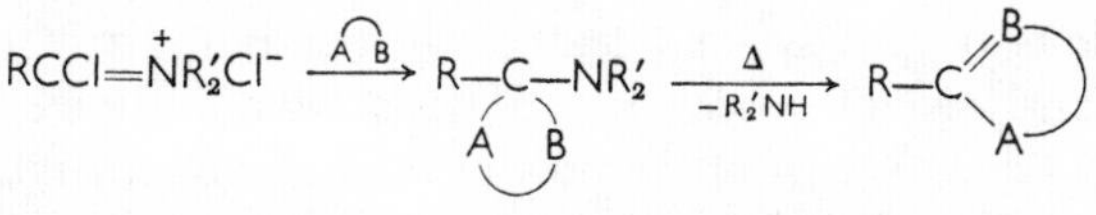

$$RCCl{=}\overset{+}{N}R'_2Cl^- \xrightarrow{A\frown B} R{-}C{-}NR'_2 \xrightarrow[-R'_2NH]{\Delta} R{-}C\langle{}^{B}_{A}\rangle$$

(contrast, for example, the formation of the 1,3,4-oxadiazole system in Chart 5 with that of the 1,2,4-triazole system (above)—both from acylhydrazines). Chart 5 illustrates some reactions of this type which proceed by way of the imidate salt[146].

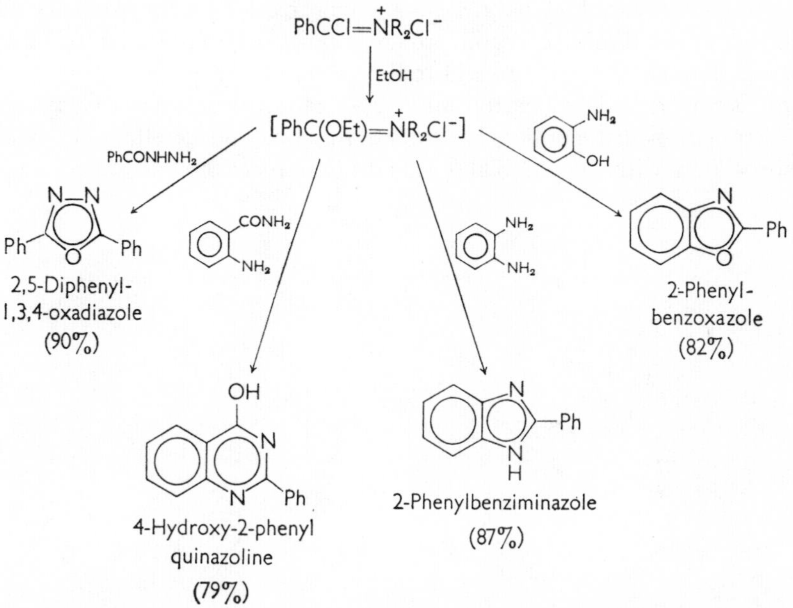

$PhCCl{=}\overset{+}{N}R_2Cl^-$ →(EtOH) $[PhC(OEt){=}\overset{+}{N}R_2Cl^-]$

- PhCONHNH$_2$ → 2,5-Diphenyl-1,3,4-oxadiazole (90%)
- o-aminobenzamide (CONH$_2$, NH$_2$) → 4-Hydroxy-2-phenyl quinazoline (79%)
- o-phenylenediamine (NH$_2$, NH$_2$) → 2-Phenylbenziminazole (87%)
- o-aminophenol (NH$_2$, OH) → 2-Phenylbenzoxazole (82%)

CHART 5. Elaboration of heterocyclic systems from *N,N*-dialkylimidoyl halide salts[146].

The addition of 1,3-dienes to imidoyl chlorides (postulated as intermediates but not characterized) is reported to give dihydropyridines[188].

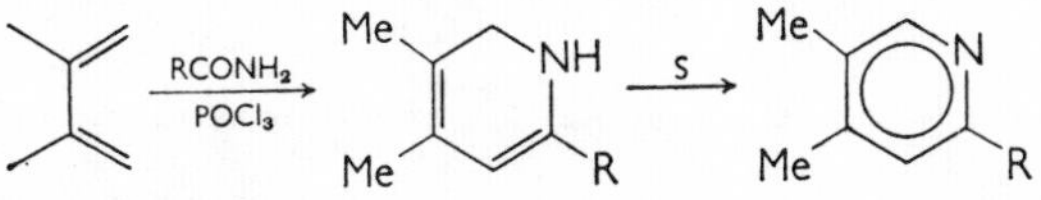

2,3-dimethylbutadiene $\xrightarrow[POCl_3]{RCONH_2}$ 4,5-dimethyl-2-R-dihydropyridine (Me, Me, NH, R) $\xrightarrow{S}$ 4,5-dimethyl-2-R-pyridine (Me, Me, N, R)

C. 1,3-Dipolar Cycloaddition[125]

The elimination of hydrogen halide from *N*-benzylbenzimidoyl halides gives the corresponding nitrile ylides; some cycloaddition reactions of this system, leading to compounds of the pyrrole, iminazole and oxazole series have already been outlined (see Section IV.B.3 and Chart 4). In an analogous way the *N*-phenylhydrazidoyl chlorides ArCCl=NNHAr eliminate hydrogen halide to give the nitrile imine: this may dimerize to the dihydro-1,2,4,5-tetrazine, but in the presence of dipolarophiles it gives heterocyclic systems (pyrazole, 1,2,4-triazole and 1,3,4-oxadiazole nuclei) as shown in Chart 6.

N-Hydroxyimidoyl halides give nitrile oxides in the presence of base. These dimerize readily to give furoxans, but under suitable conditions can be made to add to dipolarophiles to give compounds of the isoxazole, 1,3,4-dioxazole and 1,2,4-oxadiazole series (see Chart 7). There can be little doubt that the broad scope and excellence of yield associated with the dipolar cycloaddition reaction has added con-

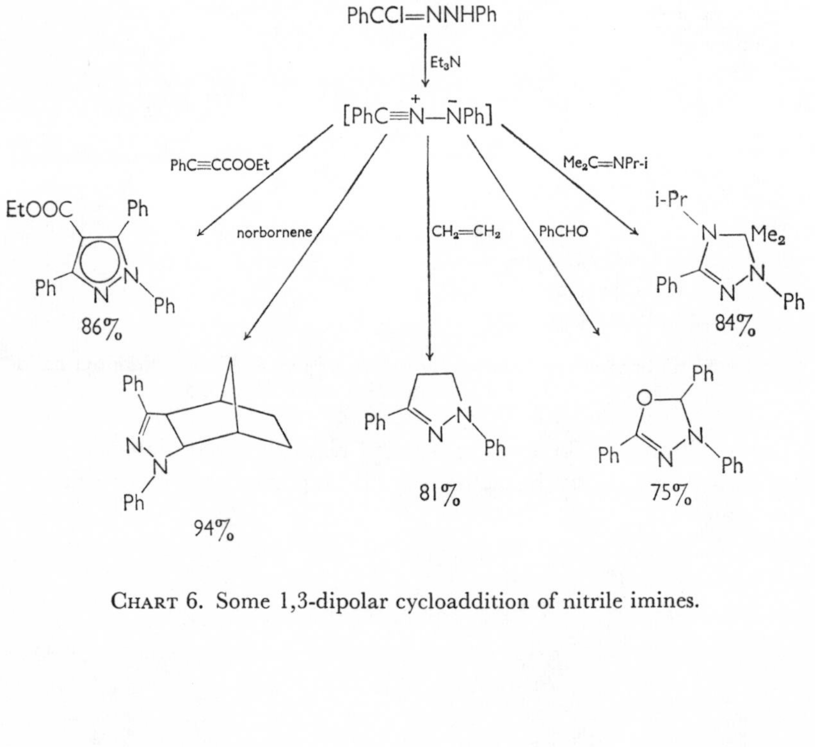

PhCCl=NNHPh
Et_3N
$[PhC\equiv\overset{+}{N}-\overset{-}{N}Ph]$
$PhC\equiv CCOOEt$ | norbornene | $CH_2=CH_2$ | PhCHO | $Me_2C=NPr$-i
EtOOC, Ph, Ph, N, N, Ph 86%
i-Pr, N, Me_2, Ph, N, N, Ph 84%
Ph, N, N, Ph 94%
Ph, N, N, Ph 81%
Ph, O, Ph, N, N, Ph 75%

CHART 6. Some 1,3-dipolar cycloaddition of nitrile imines.

siderably to the value of imidoyl halides as precursors in heterocyclic synthesis.

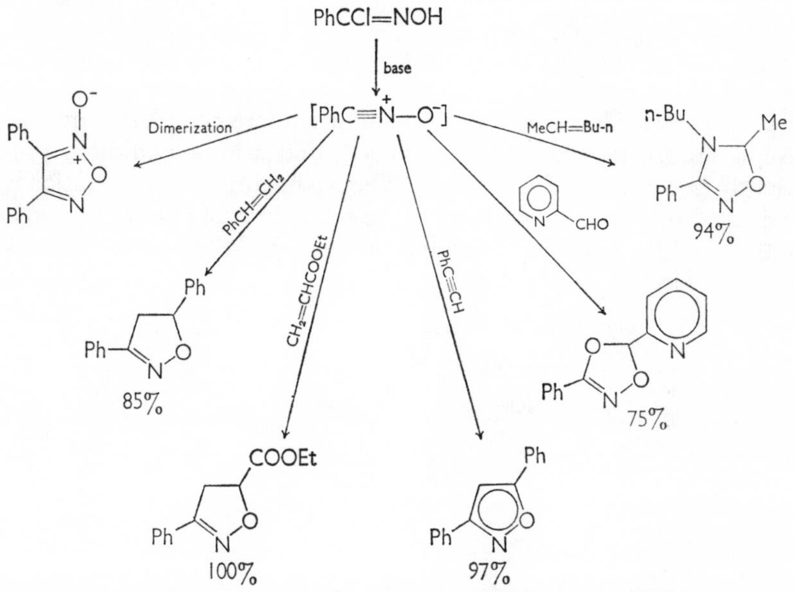

CHART 7. Some 1,3-dipolar cycloadditions of nitrile oxides.

VI. IMIDOYL HALIDES AS REACTIVE INTERMEDIATES

It is as unstable, sometimes hypothetical, intermediates that the imidoyl halides are most commonly met, and it is perhaps this role which has led to a general feeling that they are best avoided as well-defined precursors. In the following reactions the intermediacy of an imidoyl halide has been proved, or is at least quite reasonable; since with this series of reactions we are once more on familiar ground only an outline treatment will be given.

A. The Beckmann Rearrangement

When the Beckmann rearrangement is carried out with phosphorus pentachloride, phosgene or thionyl chloride the imidoyl halide may be isolated as an intermediate, so satisfactorily, indeed, that this constitutes a minor preparative method (see Section II.G.6). The reaction might have been expected to proceed as follows[186]:

$$\underset{\text{HON}}{\overset{}{R{-}\underset{\|}{C}{-}R'}} \xrightarrow[-HCl]{PCl_5} \underset{PCl_4O}{R{-}\underset{\underset{N}{\|}}{C}{-}R'} \xrightarrow[-POCl_3]{\text{Beckmann}} RCCl{=}NR' \xrightarrow{H_2O} RCONHR'$$

but there is evidence[225] the mechanism is more complex than this. Amidines are formed as by-products[52], and arise as indicated previously (see sections II.A, IV.A.1). The *trans* nature of the migration is indicated by the rearrangement of the two oximes of phenyl 2-pyridyl ketone and the isolation of the appropriate imidoyl chlorides (as their hydrohalides)[97].

$$\text{2-pyridyl}{-}\underset{\underset{NOH}{\|}}{C}{-}Ph \xrightarrow[CHCl_3]{SOCl_2} \text{2-pyridyl}{-}N{=}CClPh\cdot HCl$$

$$\text{2-pyridyl}{-}\underset{\underset{HON}{\|}}{C}{-}Ph \xrightarrow[CHCl_3]{SOCl_2} \text{2-pyridyl}{-}CCl{=}NPh\cdot HCl$$

The intermediacy of imidoyl halides is also suggested by trapping experiments, although these are generally less definitive. Thus 2-chloropiperideine from cyclopentanoxime has been trapped with *o*-aminobenzyl chloride hydrochloride[187]:

$$\text{cyclopentanone oxime (NOH)} \xrightarrow{PCl_5} [\text{2-chloropiperideine (N, Cl)}] \xrightarrow{o\text{-}CH_2Cl\text{-}C_6H_4\text{-}NH_3^+\ Cl^-} \text{tricyclic quinazoline product (N, N)}$$

B. The Stephen Reduction

This aldehyde synthesis (for a review, see reference 132) was interpreted by its discoverer[189] as proceeding via the imidoyl chloride; while this must still be regarded as an acceptable explanation (although it seems preferable to write the imidoyl chloride as its hydrohalide: see Section II.E.1)

$$RCN + 2\,HCl \longrightarrow RCCl{=}\overset{+}{N}H_2\,\overset{-}{Cl} \xrightarrow[SnCl_2,\ Et_2O]{HCl} (RCH{=}\overset{+}{N}H_2)_2\,SnCl_6^{2-} \xrightarrow{H_2O} RCHO$$

it must be noted that, since a Lewis acid is present, a nitrilium ion intermediate $(RC{\equiv}\overset{+}{N}H)_2\ Sn^{II}Cl_4^{2-}$ is also a reasonable possibility. In this case the overall reaction would be:

$$(RC{\equiv}\overset{+}{N}H)_2\ Sn^{II}Cl_4^{2-} + 4\,HCl \xrightarrow{SnCl_2} (RCH{=}\overset{+}{N}H_2)_2\ Sn^{IV}Cl_6^{2-} + SnCl_4$$

C. Interaction of Amines with Trihalomethyl Functions

Primary and secondary amines appear to react with compounds containing a trichloromethyl group with the intermediate formation of imidoyl chlorides. Thus benzotrichloride reacts with aniline to give *N,N'*-diphenylbenzamidine, presumably via *N*-phenylbenzimidoyl chloride[190].

$$PhCCl_3 \xrightarrow[-HCl]{PhNH_2} [PhCCl{=}NPh] \xrightarrow{PhNH_2} PhC(NHPh){=}NPh$$

With secondary amines the reaction may be used to go directly from methyl to *N,N*-dialkyl amide[191].

$$\text{PhCH}_2\text{OOC-pyrrole(3-Me, 4-Et, 5-Me, NH)} \xrightarrow[Et_2O]{SO_2Cl_2} \left[\text{PhCH}_2\text{OOC-pyrrole(3-Me, 4-Et, 5-CCl}_3\text{, NH)}\right] \xrightarrow[C_6H_6]{Me_2NH}$$

$$\left[\text{PhCH}_2\text{OOC-pyrrole(3-Me, 4-Et, 5-CCl=}\overset{+}{N}\text{Me}_2\text{, NH)}\ Cl^-\right] \xrightarrow{H_2O} \text{PhCH}_2\text{OOC-pyrrole(3-Me, 4-Et, 5-CONMe}_2\text{, NH)}$$

D. Reactions with Nitrile–Hydrogen Halide Systems

It has been shown (Section II.E.1) that, under certain conditions, imidoyl halide hydrohalides may be isolated from nitrile–hydrogen halide systems. It is often assumed that when such systems are used directly as a reagent the imidoyl halide is formed and functions as the reactive intermediate. The Stephen reduction (see Section VI.B) is such a case. The formation of imidate hydrohalides from nitriles, hydrogen halides and alcohols under anhydrous conditions is another example (although it is alternatively shown as a one-step process[192]); as is the hydrogen halide-catalysed polymerization of nitriles to *s*-triazines[87]. Again the reaction of carboxylic acids:

$$RCN + HCl + 2\,R'COOH \longrightarrow RCONH_2{\cdot}HCl + (R'CO)_2O$$

$$RCN + 2\,HCl + R'COOH \longrightarrow RCONH_2{\cdot}HCl + R'COCl$$

rather resembles their reaction with imidoyl chlorides (see Section IV.E.11), although in the present case mixtures of anhydrides and acyl halides are often observed as products[193].

Imidoyl halides may react in the Friedel–Crafts manner with aromatic nuclei (see Section IV.E.15) and related substitution processes—usually requiring an activated aromatic system—are known where a nitrile–hydrogen halide mixture is the reagent. Both the Hoesch ketone synthesis[194] and the Gattermann aldehyde synthesis[195] may be envisaged as proceeding via an imidoyl halide intermediate. This mechanism appears to be reasonable (although, again, in those cases where Lewis acids are used the nitrilium cation may be the attacking species). Thus it has been shown that interaction of acetimidoyl chloride hydrochloride and resorcinol gives the substitution product directly[27,196].

$$\mathrm{MeCN} \xrightarrow{\mathrm{HCl}} \mathrm{MeCCl{=}\overset{+}{N}H_2\ Cl^-} \xrightarrow{\text{resorcinol (HO, OH)}}$$

$$\mathrm{MeC(=\overset{+}{N}H_2\ Cl^-)}\text{–}C_6H_3(OH)_2 \xrightarrow{\mathrm{H_2O}} \mathrm{MeCO}\text{–}C_6H_3(OH)_2$$

Low temperature conditions and the formation of the imidoyl halide hydrohalide *in situ* as a preliminary step appear to improve the reaction, as would be expected on the basis of this mechanism[196]. The use of a zinc chloride catalyst also increases the yield; an examination[197] of the complexes formed in the system $PhCN–ZnCl_2–HCl$ appears to suggest that (at $-5°$) the nitrilium trichlorozincate is not an important intermediate.

The Gattermann reaction is more complex, and the present position is still somewhat confused[195]. Hinkel and his colleagues put forward various proposals[198] and concluded[199] that the intermediate was a derivative of the sesquichloride $2HCN \cdot 3HCl$ which was itself written as $CHCl_2NHCHClNH_2$. This substance has been formulated[200] as the hexahydro-*s*-triazine derivative (**22**)[200], but recent infrared work[219] accords with the earlier[195] amidinium salt structure $Cl_2CHNHCH{=}\overset{+}{N}H_2\ Cl^-$.

$$HCN + HCl \longrightarrow [HCCl{=}NH] \longrightarrow \left[\text{(Cl-substituted hexahydro-1,3,5-triazine)} \right] 3HCl \xrightarrow{\text{Quinoline}} \text{1,3,5-triazine}$$

(**22**)

Moreover, aluminium chloride is known to form complexes with the reagents[198]. Nevertheless, the formation of formimidoyl chloride as an unstable intermediate appears reasonable: whether it is the only, or even the major, electrophilic species in the Gattermann reaction remains open to doubt.

An alternative route to imidoyl halides required as intermediates in aromatic substitutions starts with amides. Thus the interaction of formamide and phosphorus oxychloride has been envisaged to generate formimidoyl chloride which then attacks β-naphthylamine as follows[201]:

$$\text{2-naphthylamine} \xrightarrow[POCl_3]{HCONH_2} [\text{intermediate}] \longrightarrow \text{benzo[f]quinazoline}$$

The Vilsmeier–Haack synthesis[202] also falls into this category: typically (but not exclusively) an activated aromatic system is formylated with dimethylformamide and phosphorus oxychloride, although DMF/$COCl_2$ and DMF/$SOCl_2$ may also be used[32]. In the last two cases the amidochloride ($CHCl{=}\overset{+}{N}Me_2Cl^-$)is a likely intermediate: indeed recent n.m.r. studies[220] have indicated that this is formed rapidly in the reaction with phosgene, but that with thionyl chloride an intermediate salt, formulated as $CHCl{=}\overset{+}{N}Me_2OSOCl^-$, may be obtained. This loses sulphur dioxide, giving the amidochloride, at 40–50° *in vacuo*. The first intermediate in the amide/$POCl_3$ reaction

$$HCONMe_2 \xrightarrow[-CO_2]{COCl_2} CHCl{=}\overset{+}{N}Me_2Cl^-$$

$$HCONMe_2 \xrightarrow{SOCl_2} HCCl{=}\overset{+}{N}Me_2\ OSOCl^- \xrightarrow[-SO_2]{40\text{-}50^\circ} CHCl{=}\overset{+}{N}Me_2Cl^-$$

is more stable, but there has been some controversy about the detail of its structure. Of the two reasonable alternatives, **23a** and **23b**, the former has been generally favoured[32,203] although the evidence for it is not compelling: n.m.r. evidence has been presented[220] which has been interpreted in favour of **23b** ($R = H, R^1 = Me$) for the product

$$\left[\begin{array}{c} RC{=}\overset{+}{N}R^1_2 \\ | \\ OPOCl_2 \end{array}\right] Cl^- \qquad \left[\begin{array}{c} RC{=}\overset{+}{N}R^1_2 \\ | \\ Cl \end{array}\right] OPOCl_2^-$$

(23a) **(23b)**

from dimethylformamide and phosphorous oxychloride. However, it would not be surprising if both **23a** and **23b** were present in proportions depending on the nucleophilicities of the anions concerned, on the substituents, and on the conditions. It is of interest that the corresponding intermediate from formamide (**23a** and/or **23b**, $R = R^1 = H$) leads to adenine in remarkably good yield (43·5%) when formamide and phosphorus oxychloride are heated together in a sealed tube[222].

E. Reaction of Triphenylphosphine with N-Haloamides

Trippett and Walker[204] observed that *N*-bromo derivatives of primary amides, when treated with triphenylphosphine (or an equivalent reagent), gave the corresponding nitrile together with triphenylphosphine oxide. They interpreted the reaction as a nucleophilic attack at enolate *oxygen* with concomitant loss of bromide, thus:

$$\begin{array}{c} \curvearrowleft \dot{P}Ph_3 \\ O^- \\ | \\ Ph{-}C{=}N{-}Br \end{array} \xrightarrow[\text{room temp. 15 min}]{C_6H_6} PhCN + Ph_3PO + Br^-$$

More recently, Speziale and Smith[79] have found that when *N*-halo derivatives of secondary amides are treated with triphenylphosphine the imidoyl halide is formed, e.g.

$$PhCONClEt \xrightarrow[\text{benzene, } 50°]{Ph_3P} PhCCl{=}NEt + Ph_3PO$$

and, since in this example enolization is not possible, they have put forward a reaction mechanism in which attack is initiated at halogen, thus:

$$Ph{-}\overset{O}{\overset{\|}{C}}{-}N(Et){-}Cl \;\; \dot{P}Ph_3 \longrightarrow Ph{-}C(O{-}PPh_3{-}Cl){=}NEt \xrightarrow{S_Ni} PhCCl{=}NEt$$

F. The von Braun Degradation (cf. Sections II.C and IV.B.I)

When secondary or tertiary amides are heated with phosphorus pentachloride or phosphorus pentabromide dealkylation occurs, thus[30]:

$$ArCONR'_2 \xrightarrow[\Delta]{PX_5} R'X + ArCX{=}NR' \xrightarrow{\Delta} ArCN + R'X$$

a reaction discovered by Pechmann[129], but generally given von Braun's name since he made the first extensive study of it. The reaction can be employed (a) to dealkylate secondary amines, e.g.[66,205]

$$PhCONMePh \xrightarrow[120°]{PCl_5} MeCl + PhCCl{=}NPh \xrightarrow[hydrolysis]{} PhNH_2$$

(b) to prepare alkyl halides, especially α,ω-dihalides, e.g.

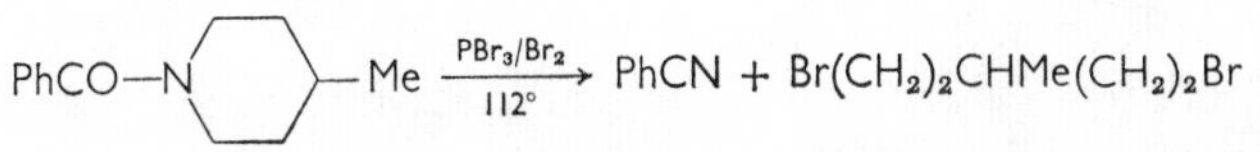

$$PhCO{-}N\langle\text{piperidine}\rangle{-}Me \xrightarrow[112°]{PBr_3/Br_2} PhCN + Br(CH_2)_2CHMe(CH_2)_2Br \quad \text{(ref. 206)}$$

65%

$$PhCONHCH_2\overset{Me}{\underset{*}{C}}HCH_2CH_2NHCOPh \xrightarrow[\Delta]{PCl_5} ClCH_2\overset{Me}{\underset{*}{C}}HCH_2CH_2Cl \quad \text{(ref. 207)}$$

28%

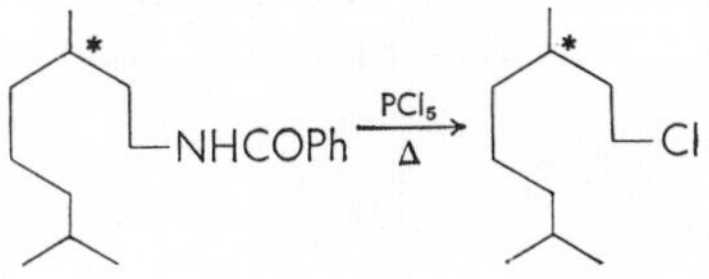

NHCOPh $\xrightarrow[\Delta]{PCl_5}$ Cl (ref. 208)

40%

and (c) to effect ring cleavage in cyclic secondary amines, e.g.

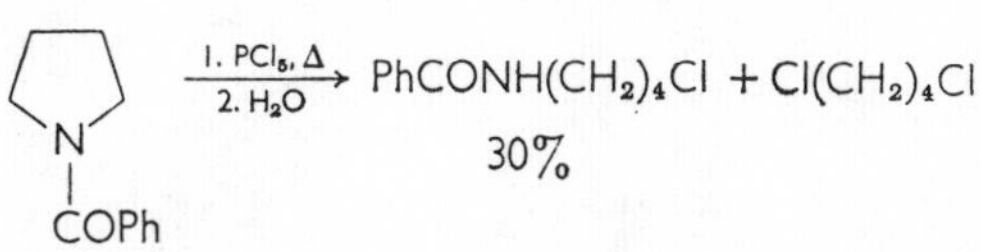

N-COPh (pyrrolidine) $\xrightarrow[2.\ H_2O]{1.\ PCl_5,\ \Delta}$ $PhCONH(CH_2)_4Cl + Cl(CH_2)_4Cl$ (ref. 209)

30%

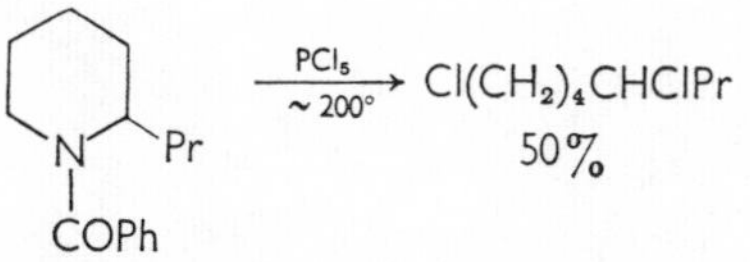

N-COPh, 2-Pr (piperidine) $\xrightarrow[\sim 200°]{PCl_5}$ $Cl(CH_2)_4CHClPr$ (ref. 210)

Benzoyl coniine

50%

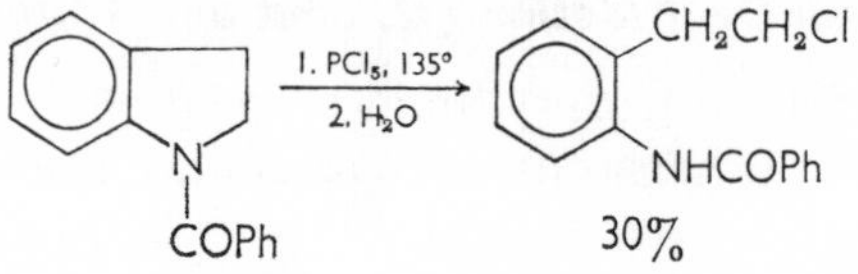

(ref. 211)

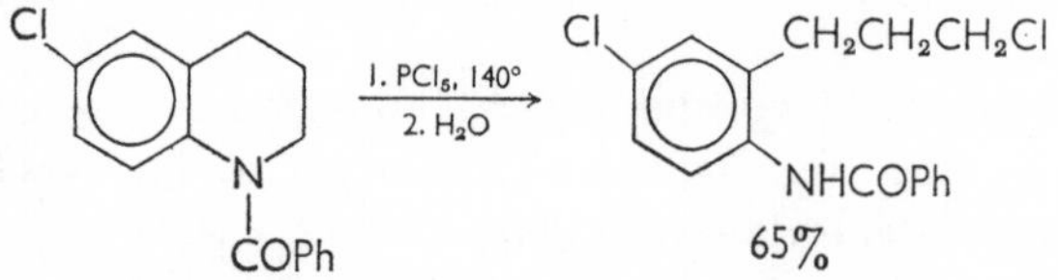

(ref. 212)

The aryl nitrile which is formed concomitantly may be removed by hydrolysis or by fractional distillation[213], or by converting it into the imidate hydrochloride from which the required compound can be extracted with ether[214].

The reaction is evidently a nucleophilic substitution at a saturated carbon atom and has attracted some attention from a mechanistic viewpoint. Leonard and Nommensen[67] studied (mainly) the phosphorus pentabromide reaction, and, probably because the imidoyl bromides are rather unstable, did not observe or even postulate such intermediates. The following observations were made.

(a) The yields of halide decreased as steric hindrance about the α_N-carbon atom increased. von Braun had earlier observed that attack occurred at the smaller of the *N*-alkyl groups (see Section II.C).

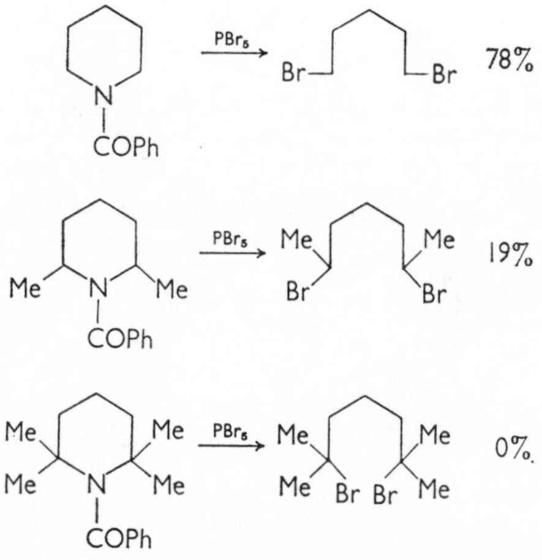

(b) The substitution proceeded with inversion.

$$\text{PhCONH—CH(Me)(Et)} \xrightarrow{PBr_5} \text{H—C(Me)(Et)—Br}$$

(However, with *N*-benzyl substitution racemization is observed[30].)

(c) Steric hindrance at the α_C-carbon atom did not much affect the reaction, e.g.

$$\text{2,4,6-}Me_3C_6H_2\text{—CONHBu-n} \xrightarrow[\Delta]{PCl_5} \text{n-BuCl} \quad 54\%$$

These data accord with an S_N2 type displacement of nitrogen and Leonard and Nommensen put forward the following type of mechanism:

Br, N, C, Ph, O, $\overset{+}{P}Br_4$ ⟶ Br, N, Br, Br_3P, O, C, Ph ⟶ Br—$(CH_2)_5$—Br, $POBr_3$, PhCN

Vaughan and Carlson[30] have studied the dealkylation of secondary amides with thionyl chloride; here imidoyl chlorides have been isolated under the conditions of the von Braun degradation. Thus when *N*-(n-butyl)benzamide was treated with thionyl chloride, the initial reaction afforded sulphur dioxide and hydrogen chloride. Excess thionyl chloride could be distilled off leaving the imidoyl chloride, which did not decompose to give benzonitrile and n-butyl chloride until it was heated above 100°. It is reasonable to conclude that imidoyl halides are likely intermediates in the von Braun degradation, thus:

$$\text{ArCONMeBu} \xrightarrow[\text{etc.}]{PCl_5} \text{ArC(Cl)=}\overset{+}{\text{N}}\text{(Me)—Bu} \quad Cl^- \longrightarrow \text{MeCl + ArCCl=NBu}$$

$$\text{ArCCl=NBu} \rightleftharpoons \left[\text{ArC}\equiv\overset{+}{\text{N}}\text{—CH}_2\text{Pr} \quad Cl^-\right] \longrightarrow \text{ArCN + BuCl}$$

Such a mechanism is, however, not obligatory since direct elimination from intermediate stages [such as $ArC(OPCl_4){=}\overset{+}{N}MeBu\ Cl^-$, $ArC(OPCl_4){=}NBu$, $ArC(OSOCl){=}NBu$] in imidoyl halide formation (see Section II.A) must be regarded as a reasonable alternative. Such a process is essentially that shown in the second stage of the mechanism proposed by Leonard and Nommensen. When the *N*-substituent is readily stabilized as a carbonium ion (e.g. *t*-butyl, benzyl), it is possible that it may be expelled as such from the ion pair:

$$\left[\begin{matrix} Cl^- \\ ArC{\equiv}\overset{+}{N}{-}\overset{*}{C}H{-}Ph \\ \quad\quad | \\ \quad\quad Me \end{matrix}\right] \longrightarrow ArCN + Ph\overset{+}{C}HMe \begin{matrix} \nearrow PhCHMeCl\ \text{inactive} \\ \searrow PhCH{=}CH_2 \end{matrix}$$

Such a process would account for the racemization observed when (−)-*N*-α-methylbenzylacetamide is submitted to the von Braun degradation[30], and for the occasional reports[6] of olefin formation during the thermal elimination of *N*-*t*-alkyl imidoyl halides.

VII. REFERENCES

1. J. von Braun, *Angew. Chem.*, **47**, 611 (1934).
2. O. Wallach and M. Hoffmann, *Ber.*, **8**, 313 (1875).
3. H. von Pechmann, *Ber.*, **28**, 2362 (1895).
4. C. L. Stevens and J. C. French, *J. Am. Chem. Soc.*, **76**, 4398 (1954).
5. J. von Braun, F. Jostes and W. Münch, *Ann. Chem.*, **453**, 113 (1927).
6. R. F. Merritt and F. A. Johnson, *J. Org. Chem.*, **32**, 416 (1967).
7. J. von Braun and A. Heymons, *Ber.*, **63**, 502 (1930).
8. F. Baumann, B. Bienert, G. Rösch, H. Vollmann and W. Wolf, *Angew. Chem.*, **68**, 133 (1956).
9. O. Wallach and A. Gossmann, *Ber.*, **11**, 753 (1878).
10. R. Behrend and J. Schmitz, *Ber.*, **26**, 626 (1893).
11. J. von Braun and C. Müller, *Ber.*, **39**, 2018 (1906).
12. A. Cahours, *Compt. Rend.*, **25**, 724 (1847).
13. H. Limpricht and L. von Usler, *Ann. Chem.*, **106**, 32 (1858); W. Henke, *Ann. Chem.*, **106**, 272 (1858); R. Fittig, *Ann. Chem.*, **106**, 277 (1858).
14. C. Gerhardt, *Ann. Chem. Phys.*, [3], **53**, 302 (1858); *Ann. Chem.*, **108**, 214 (1858).
15. A. Wolkoff, *Ber.*, **5**, 139 (1872).
16. A. Gautier, *Compt. Rend.*, **63**, 920 (1866); A. Gautier, *Ann. Chim. Phys.* [4], **17**, 103 (1869).
17. O. Wallach, *Ber.*, **9**, 1214 (1876).
18. A. J. Hill and M. V. Cox, *J. Am. Chem. Soc.*, **48**, 3214 (1926).
19. R. Bauer, *Ber.*, **40**, 2650 (1907).
20. E. N. Zilberman, *Russ. Chem. Rev.*, **31**, 615 (1962).

21. H. K. S. Rao and T. S. Wheeler, *J. Chem. Soc.*, 1643 (1937).
22. J. W. Williams, C. H. Witten and J. A. Krynitsky, *Org. Syn.*, Coll. Vol. 3, 818 (1955).
23. J. A. Arvin and R. Adams, *J. Am. Chem. Soc.*, **50**, 1983 (1928).
24. H. Ulrich, E. Kober, H. Schroeder, R. Rätz and C. Grundmann, *J. Org. Chem.*, **27**, 2585 (1962).
25. H. Blitz, *Ber.*, **25**, 2533 (1892).
26. A. Y. Lazaris, E. N. Zilberman and O. D. Strizhakov, *J. Gen. Chem. USSR*, **32**, 890 (1962).
27. H. Stephen, *J. Chem. Soc.*, 1529 (1920).
28. R. C. Shah, *J. Indian Inst. Sci.*, **7**, 205 (1924).
29. J. von Braun and W. Pinkernelle, *Ber.*, **67**, 1218 (1934).
30. W. R. Vaughan and R. D. Carlson, *J. Am. Chem. Soc.*, **84**, 769 (1962).
31. I. Ugi, F. Beck and U. Fetzer, *Chem. Ber.*, **95**, 126 (1962); but see R. Buyle and H. G. Viehe, *Tetrahedron*, **24**, 4217 (1968).
32. H. H. Bosshard and H. Zollinger, *Helv. Chim. Acta*, **42**, 1659 (1959); cf. Z. Arnold, *Chem. Listy*, **52**, 2013 (1958), Z. Arnold and F. Sörm, *Chem. Listy*, **51**, 1082 (1957).
33. H. H. Bosshard, R. Mory, M. Schmid and H. Zollinger, *Helv. Chim. Acta*, **42**, 1653 (1959).
34. M. Grdinic and V. Hahn, *J. Org. Chem.*, **30**, 2381 (1965).
35. S. A. Kulkarni and R. C. Shah, *J. Indian Chem. Soc.*, **27**, 111 (1950).
36. C. C. Price and B. H. Velzen, *J. Org. Chem.*, **12**, 386 (1947).
37. A. von Baeyer, *Ber.*, **12**, 456 (1879).
38. R. Stollé, *J. Prakt. Chem.*, (2), **73**, 277 (1906); *J. Prakt. Chem.*, [2], **75**, 416 (1907).
39. A. V. Kirsanov and G. I. Derkach, *J. Gen. Chem. USSR*, **27**, 3284 (1957).
40. A. Lapidot and D. Samuel, *J. Chem. Soc.*, 2110 (1962).
41. A. V. Kirsanov, *J. Gen. Chem. USSR*, **22**, 329 (1952); A. V. Kirsanov and R. G. Makirta, *J. Gen. Chem. USSR*, **26**, 1033 (1956); A. A. Kropacheva, G. I. Derkach and A. V. Kirsanov, *J. Gen. Chem. USSR*, **31**, 1489 (1961).
42. O. Wallach, *Ann. Chem.* **184**, 1 (1877).
43. J. von Braun and W. Rudolph, *Ber.*, **67**, 1762 (1934).
44. A. W. Titherley and E. Worral, *J. Chem. Soc.*, 1143 (1909).
45. H. Holtschmidt, E. Degener and H. G. Schmelzer, *Ann. Chem.*, **701**, 107 (1967).
46. A. Sonn and E. Müller, *Ber.*, **52**, 1927 (1919).
47. J. von Braun and H. Ostermayer, *Ber.*, **70**, 1002 (1937).
48. A. W. Titherley and E. Worral, *J. Chem. Soc.*, 839 (1910).
49. E. Beckmann and K. Sandel, *Ann. Chem.*, **296**, 279 (1897).
50. W. W. Crouch and H. L. Lochte, *J. Am. Chem. Soc.*, **65**, 270 (1943); cf. O. Bernheimer, *Gazz. Chim. Ital.*, **12**, 281 (1882).
51. W. P. Norris and H. B. Jonassen, *J. Org. Chem.*, **27**, 1449 (1962).
52. H. Stephen and W. Bleloch, *J. Chem. Soc.*, 886 (1931).
53. H. Eilingsfeld, M. Seefelder and H. Weidinger, *Angew. Chem.*, **72**, 836 (1960).
54. H. Paul, A. Weise and R. Dettmar, *Chem. Ber.*, **98**, 1450 (1965).
55. A. Werner and H. Buss, *Ber.*, **27**, 2193 (1894).
56. H. Ley and M. Ulrich, *Ber.*, **47**, 2941 (1914).

57. L. Claisen and O. Manasse, *Ann. Chem.*, **274**, 95 (1893).
58. R. Stollé and F. Helwerth, *Ber.*, **47**, 1132 (1914).
59. G. Casnati and A. Ricca, *Tetrahedron Letters*, 327 (1967).
60. L. Birkenbach and K. Sennewald, *Ann. Chem.*, **489**, 7 (1930); G. Endres, *Ber.*, **65**, 65 (1932).
61. I. de Paolini, *Gazz. Chim. Ital.*, **60**, 700 (1930).
62. H. Rheinboldt, *Ann. Chem.*, **451**, 161 (1927); H. Rheinboldt and O. Schmitz-Dumont, *Ann. Chem.*, **444**, 113 (1925).
63. F. L. Scott and J. B. Aylward, *Tetrahedron Letters*, 841 (1965).
64. F. D. Chattaway and A. J. Walker, *J. Chem. Soc.*, 975 (1925).
65. H. Klinger, *Ber.*, **8**, 310 (1875).
66. J. von Braun and J. Weismantel, *Ber.*, **55**, 3165 (1922).
67. N. J. Leonard and E. W. Nommensen, *J. Am. Chem. Soc.*, **71**, 2808 (1949).
68. J. U. Nef, *Ann. Chem.*, **270**, 267 (1892); *Ann. Chem.*, **280**, 291 (1894); *Ann. Chem.*, **287**, 265 (1895).
69. I. Ugi and U. Fetzer, *Chem. Ber.*, **94**, 1116 (1961).
70. E. Kühle, B. Anders and G. Zumach, *Angew. Chem. Intern. Ed. Engl.*, **6**, 649 (1967).
71. A. Michael and J. F. Wing, *Am. Chem. J.*, **7**, 71 (1885).
72. E. Allenstein and A. Schmidt, *Spectrochim. Acta*, **20**, 1451 (1964).
73. A. Hantzsch, *Ber.*, **64**, 667 (1931); A. Hantzsch, *Ber.*, 1219 (1931).
74. L. E. Hinkel and G. J. Treharne, *J. Chem. Soc.*, 866 (1945).
75. G. D. Lander and H. E. Laws, *J. Chem. Soc.*, 1695 (1904).
76. F. E. Murray and W. G. Schneider, *Can. J. Chem.*, **33**, 797 (1955).
77. *U.S. Pat.* 2,411,064 (1946); *Chem. Abstr.*, **41**, 1236 (1947); W. E. Elstrow and B. C. Platt, *Chem. Ind. (London)*, 449 (1952); E. Ronwin, *Can. J. Chem.*, **35**, 1031 (1957).
78. F. Klages and W. Grill, *Ann. Chem.*, **594**, 21 (1955).
79. A. J. Speziale and L. R. Smith, *J. Am. Chem. Soc.*, **84**, 1868 (1962).
80. F. Klages, R. Ruhnau and W. Hauser, *Ann. Chem.*, **626**, 60 (1959).
81. H. Meerwein, P. Laasch, R. Mersch and J. Spille, *Chem. Ber.*, **89**, 209 (1956); cf. R. R. Schmidt, *Chem. Ber.*, **98**, 334 (1965), *Tetrahedron Letters*, 3448 (1968).
82. G. J. Janz and S. S. Danyluk, *Chem. Rev.*, **60**, 209 (1960); G. J. Janz and S. S. Danyluk, *J. Am. Chem. Soc.*, **81**, 3846, 3850, 3854 (1959).
83. H. H. Bosshard, E. Jenny and H. Zollinger, *Helv. Chim. Acta*, **44**, 1203 (1961).
84. S. W. Peterson and J. M. Williams, *J. Am. Chem. Soc.*, **88**, 2866 (1966).
85. P. A. S. Smith, *J. Am. Chem. Soc.*, **76**, 436 (1954); P. A. S. Smith and E. Leon, *J. Am. Chem. Soc.*, **80**, 4647 (1958).
86. K. Bredereck and R. Richter, *Chem. Ber.*, **99**, 2454, 2461 (1966).
87. C. Grundmann, G. Weisse and S. Seide, *Ann. Chem.*, **577**, 77 (1952).
88. S. J. Davis, J. A. Elvidge and A. B. Foster, *J. Chem. Soc.*, 3638 (1962).
89. H. Holtschmidt, *Angew. Chem. Intern. Ed. Engl.*, **1**, 632 (1962).
90. W. Lossen, *Ann. Chem.*, **265**, 129 (1891).
91. E. Sell and G. Zierold, *Ber.*, **7**, 1228 (1874); O. Helmers, *Ber.*, **20**, 786 (1887); G. M. Dyson and T. Harrington, *J. Chem. Soc.*, 191 (1940); 150 (1942).
92. K. Heyns and W. von Bebenburg, *Chem. Ber.*, **89**, 1303 (1956).

93. R. Fusco and R. Romani, *Gazz. Chim. Ital.*, **76**, 419 (1946); **78**, 332 (1948).
94. W. Prandtl and K. Sennewald, *Ber.*, **62**, 1754 (1929); W. Prandtl and W. Dollfus, *Ber.*, **65**, 754 (1932).
95. G. Bishop and O. L. Brady, *J. Chem. Soc.*, 2364 (1922); 810 (1926).
96. *Brit. Pat.* 1,007,413(1965); *Chem. Abstr.*, **64**, 8158 (1966).
97. E. H. Huntress and H. C. Walter, *J. Am. Chem. Soc.*, **70**, 3702 (1948).
98. W. H. Graham, *J. Am. Chem. Soc.*, **88**, 4677 (1966).
99. R. S. Bly, G. A. Perkin and W. L. Lewis, *J. Am. Chem. Soc.*, **44**, 2896 (1922).
100. K. Wiechert, H. H. Heilmann and P. Mohr, *Z. Chem.*, **3**, 308 (1963).
101. F. S. Fawcett, C. W. Tullock and D. D. Coffmann, *J. Am. Chem. Soc.*, **84**, 4275 (1962).
102. R. E. Banks, W. M. Cheng and R. N. Haszeldine, *J. Chem. Soc.*, 2485 (1964).
103. L. J. Bellamy, *The Infrared Spectra of Complex Molecules*, Methuen, London, 1958.
104. R. E. Banks, W. M. Cheng and R. N. Haszeldine, *J. Chem. Soc.*, 3407 (1962).
105. C. A. Grob, H. P. Fischer, W. Raudenbusch and J. Zergenyi, *Helv. Chim. Acta*, **47**, 1003 (1964).
106. A. Hantzsch, *Ber.*, **25**, 705 (1892); W. Steinkopf and B. Jurgens, *J. Prakt. Chem.*, **83**, 453 (1910); G. Ponzio and F. Baldracco, *Gazz. Chim. Ital.*, **60**, 415 (1930); G. Ponzio, *Gazz. Chim. Ital.*, **60**, 886 (1930).
107. B. Greenberg and J. G. Aston, *J. Org. Chem.*, **25**, 1894 (1960).
108. P. H. Ogden and G. V. D. Tiers, *Chem. Commun.*, 527 (1967).
109. A. L. Logothetis and G. N. Sausen, *J. Org. Chem.*, **31**, 3689 (1966).
110. J. von Braun and H. Silberman, *Ber.*, **63**, 498 (1930).
111. J. von Braun, F. Jostes and A. Heymons, *Ber.*, **60**, 92 (1927).
112. O. Wallach, *Ber.*, **7**, 326 (1874); G. Kresze and W. Wucherpfennig, *Chem. Ber.*, **101**, 365 (1968).
113. O. Wallach and I. Kamenski, *Ber.*, **13**, 516 (1880); O. Wallach, *Ann. Chem.*, **214**, 193 (1882).
114. O. Wallach and M. Hoffmann, *Ber.*, **8**, 1567 (1875).
115. A. Heymons, *Ber.*, **65**, 320 (1932).
116. O. Wallach, *Am. Chem.*, **214**, 257 (1882); E. F. Godefroi, C. A. M. Eycken and P. A. J. Janssen, *J. Org. Chem.*, **32**, 1259 (1967).
117. F. R. Atherton, A. L. Morrison, R. J. W. Cremlyn, G. W. Kenner, Sir Alexander Todd, and R. F. Webb, *Chem. Ind.* (*London*), 1183 (1955).
118. C. A. Bischoff and P. Walden, *Ann. Chem.* **279**, 45 (1894).
119. J. von Braun and A. Heymons, *Ber.*, **63**, 3191 (1930).
120. J. von Braun, A. Heymons, and G. Manz, *Ber.*, **64**, 227 (1931).
121. H. Eilingsfeld, M. Seefelder and H. Weidinger, *Chem. Ber.*, **96**, 2899 (1963).
122. H. Ley and E. Holzweissig, *Ber.*, **36**, 18 (1903).
123. H. Staudinger, *Ann. Chem.*, **356**, 51 (1907).
124. C. L. Stevens and J. C. French, *J. Am. Chem. Soc.*, **75**, 657 (1953).
125. R. Huisgen, *Angew. Chem. Intern. Ed. Engl.*, **2**, 565 (1963).
126. R. Huisgen, H. Stangl, H. J. Sturm and H. Wagenhofer, *Angew. Chem. Intern. Ed. Engl.*, **1**, 50 (1962).
127. R. Huisgen and R. Raab, *Tetrahedron Letters*, 649 (1966).
128. J. S. Clovis, A. Eckell, R. Huisgen and R. Sustmann, *Chem. Ber.*, **100**, 60 (1967).
129. H. von Pechmann, *Ber.*, **33**, 611 (1900).

130. H. L. Wheeler and T. B. Johnson, *Am. Chem. J.*, **30**, 24 (1903).
131. S. Hünig and H. Balli, *Ann. Chem.*, **609**, 160 (1957).
132. E. Mosettig, *Org. Reactions*, **8**, 218 (1954).
133. J. von Braun and W. Rudolph, *Ber.*, **67**, 269, 1735 (1934).
134. R. Kuhn and C. J. O. R. Morris, *Ber.*, **70**, 853 (1937).
135. A. Quilico and L. Panizzi, *Gazz. Chim. Ital.*, **68**, 411 (1938).
136. R. A. Mitsch and P. H. Ogden, *J. Org. Chem.*, **31**, 3833 (1966).
137. A. E. Kulikova, E. N. Zilberman and N. A. Sazanova, *J. Gen. Chem. USSR*, **30**, 2159 (1960).
138. E. Sell and G. Zierold, *Ber.*, **7**, 1228 (1874).
139. V. R. Heeramaneck and R. C. Shah, *J. Univ. Bombay*, **6**, II, 80 (1937); [*Chem. Abstr.*, **32**, 3761 (1938)].
140. G. D. Lander, *J. Chem. Soc.*, 591 (1902).
141. W. R. Smith, *Am. Chem. J.*, **16**, 372 (1894).
142. A. E. Arbuzov and V. E. Shishkin, *J. Gen. Chem. USSR*, **34**, 3628 (1964).
143. H. Bredereck and K. Bredereck, *Chem. Ber.*, **94**, 2278 (1961).
144. H. W. Heine and H. S. Bender, *J. Org. Chem.*, **25**, 461 (1960).
145. G. Ponzio and G. Charrier, *Gazz. Chim. Ital.*, **37**, (2), 65 (1907).
146. H. Eilingsfeld, M. Seefelder and H. Weidinger, *Chem. Ber.*, **96**, 2671 (1963).
147. H. von Pechmann and B. Heinze, *Ber.*, **30**, 1783 (1897).
148. M. O. Forster, *J. Chem. Soc.*, 184 (1909).
149. R. C. Shah and M. B. Ichaporia, *J. Chem. Soc.*, 431 (1936); H. P. Ghadiali and R. C. Shah, *J. Indian Chem. Soc.*, **26**, 117 (1949).
150. H. Ley and F. Müller, *Ber.*, **40**, 2957 (1907); F. C. Cooper, M. W. Partridge and W. F. Short, *J. Chem. Soc.*, 391 (1951).
151. R. Fusco and C. Musante, *Gazz. Chim. Ital.*, **68**, 147 (1938).
152. H. Ley, *Ber.*, **31**, 240 (1898).
153. H. Ley, *Ber.*, **31**, 2126 (1898).
154. H. Behringer and H. J. Fischer, *Chem. Ber.*, **95**, 2546 (1962).
155. M. Busch and R. Ruppenthal, *Ber.*, **43**, 3001 (1910).
156. G. Schroeter, *Ber.*, **42**, 3356 (1909).
157. J. von Braun and W. Rudolph, *Ber.*, **74**, 264 (1941).
158. O. Mumm, *Ber.*, **43**, 886 (1910).
159. O. Mumm, H. Volquartz and H. Hesse, *Ber.*, **47**, 751 (1914).
160. A. H. Lamberton and A. E. Standage, *J. Chem. Soc.*, 2957 (1960); O. Mumm, H. Hesse and H. Volquartz, *Ber.*, **48**, 379 (1915).
161. Z. Arnold, *Chem. Ind.* (*London*), 1478 (1960).
162. P. Davis and W. E. McEwen, *J. Org. Chem.*, **26**, 815 (1961).
163. M. Busch and M. Fleischmann, *Ber.*, **43**, 2553 (1910); M. Busch and F. Falco, *Ber.*, **43**, 2557 (1910).
164. D. Y. Curtin and L. L. Miller, *Tetrahedron Letters*, 1869 (1965).
165. F. Cramer and K. Baer, *Chem. Ber.*, **93**, 1231 (1960).
166. H. Leo, *Ber.*, **10**, 2133 (1877).
167. G. S. Jamieson, *J. Am. Chem. Soc.*, **26**, 177 (1904).
168. H. Rivier and J. Schalch, *Helv. Chim. Acta*, **6**, 605 (1923).
169. K. Zieloff, H. Paul and G. Hilgetag, *Chem. Ber.*, **99**, 357 (1966); N. Kreutzkamp and G. Cordes, *Ann. Chem.*, **623**, 103 (1959).
170. R. Fusco, *Gazz. Chim. Ital.* **69**, 344 (1939).

171. V. R. Heeramaneck and R. C. Shah, *J. Chem. Soc.*, 867 (1937); K. D. Kulkarni and R. C. Shah, *J. Ind. Chem. Soc.*, **26**, 171 (1949).
172. H. K. S. Rao and T. S. Wheeler, *J. Chem. Soc.*, 476 (1938).
173. R. Justoni and R. Fusco, *Gazz. Chim. Ital.*, **68**, 59 (1938).
174. F. Just, *Ber.*, **18**, 2632 (1885); *Ber.*, **19**, 1462, 1541 (1886); R. Seka and W. Feuchs, *Monatsh. Chem.*, **57**, 52 (1931); R. C. Shah and V. R. Heeramaneck, *J. Chem. Soc.*, 428 (1936).
175. G. Singh and G. V. Nair, *J. Am. Chem. Soc.*, **78**, 6105 (1956).
176. R. Phadke and R. C. Shah, *J. Indian Chem. Soc.*, **27**, 349 (1950).
177. H. Staudinger, H. Goldstein and E. Schlenker, *Helv. Chim. Acta*, **4**, 342 (1921).
178. R. C. Shah and J. S. Chaubal, *J. Chem. Soc.*, 650 (1932).
179. H. Staudinger, *Ber.*, **41**, 2217 (1908).
180. *Brit. Pat.* 970,480(1964); *Chem. Abstr.*, **62**, 567 (1965).
181. F. Johnson, J. P. Panella, A. A. Carlson and D. H. Hunneman, *J. Org. Chem.* **27**, 2473 (1962).
182. F. Johnson and W. A. Nasutavicus, *J. Org. Chem.*, **27**, 3953 (1962).
183. J. H. Osborne, *Dissertation Abstr.*, **19**, 2475 (1959); J. Gardent, *Compt. Rend.*, **259**, 4724 (1964); W. A. Nasutavicus and F. Johnson, *J. Org. Chem.*, **32**, 2367 (1967); L. G. Duquette and F. Johnson, *Tetrahedron*, **23**, 4517; 4539 (1967); for a review of heterocyclic syntheses with dinitriles and hydrogen halides see F. Johnson and R. Madronero, *Advan. Heterocyclic Chem.*, **6**, 128 (1966).
184. R. C. Elderfield, W. J. Gensler, T. H. Bembry, C. B. Kremer, J. D. Head, F. Brody and R. Frohardt, *J. Am. Chem. Soc.*, **68**, 1272 (1946).
185. R. Winterbottom, J. W. Clapp, W. H. Miller, J. P. English and R. O. Robin, *J. Am. Chem. Soc.*, **69**, 1393 (1947).
186. G. H. Coleman and R. E. Pyle, *J. Am. Chem. Soc.*, **68**, 2007 (1946).
187. G. G. Munoz and R. Madronero, *Chem. Ber.*, **95**, 2182 (1962).
188. M. Lora-Tamayo, G. G. Munoz and R. Madronero, *Bull. Soc. Chim. France*, 1331 (1958).
189. H. Stephen, *J. Chem. Soc.*, 1874 (1925).
190. H. Limpricht, *Ann. Chem.*, **135**, 80 (1865); S. P. Joshi, A. P. Khanolkar and T. S. Wheeler, *J. Chem. Soc.*, 793 (1936).
191. A. H. Jackson, G. W. Kenner and D. Warburton, *J. Chem. Soc.*, 1328 (1965).
192. R. Roger and D. G. Neilson, *Chem. Rev.*, **61**, 179 (1961).
193. A. Colson, *Compt. Rend.*, **121**, 1155 (1895); E. N. Zilberman, *J. Gen. Chem. USSR*, **30**, 1302 (1960); A. Y. Lazaris and E. N. Zilberman, *Tr. Khim. i Khim. Technol.* (*Gorkii*), **3**, 434 (1960), [*Chem. Abstr.*, **56**, 4669 (1962)].
194. K. Hoesch, *Ber.*, **48**, 1122 (1915). For a recent review see W. Rushe, *Friedel–Crafts and Related Reactions*, Vol. III. 1 (Ed. G. A. Olah), Interscience, New York, 1964, p. 383.
195. L. Gattermann, *Ber.*, **31**, 1149 (1898). *Ann. Chem.*, **357**, 313 (1907). For recent review see G. A. Olah and S. J. Kuhn, *Friedel–Crafts and Related Reactions*, Vol. III. 2 (Ed. G. A. Olah), Interscience, New York, 1964, p. 1153.
196. E. N. Zilberman and N. A. Rybakova, *J. Gen. Chem. USSR*, **30**, 1972, (1960).
197. E. N. Zilberman and N. A. Rybakova, *J. Gen. Chem. USSR*, **32**, 581 (1962); E. N. Zilberman and N. A. Rybakova, *Kinetics Catalysis* (*USSR*), **5**, 468 (1964).

198. L. E. Hinkel, E. E. Ayling and W. H. Morgan, *J. Chem. Soc.*, 2793 (1932); L. E. Hinkel, E. E. Ayling and J. H. Beynon, *J. Chem. Soc.*, 184 (1936); L. E. Hinkel and T. I. Watkins, *J. Chem. Soc.*, 407 (1940).
199. L. E. Hinkel and R. P. Hullin, *J. Chem. Soc.*, 1593 (1949).
200. C. Grundmann and A. Kreutzberger, *J. Am. Chem. Soc.*, **76**, 5646 (1954).
201. C. E. Loader and C. J. Timmons, *J. Chem. Soc.(C)*, 1343 (1967).
202. A. Vilsmeier and A. Haack, *Ber.*, **60**, 119 (1927). For recent review see G. A. Olah and S. J. Kuhn, *Friedel–Crafts and Related Reactions*, Vol. III. 2 (Ed. G. A. Olah), Interscience, New York, 1964, p. 1153.
203. H. Lorenz and R. Wizinger, *Helv. Chim. Acta*, **28**, 600 (1945).
204. S. Trippett and D. M. Walker, *J. Chem. Soc.*, 2976 (1960); cf. F. Hampson and S. Trippett, *J. Chem. Soc.*, 5129 (1965).
205. J. von Braun, *Ber.*, **37**, 2812 (1904).
206. N. J. Leonard and Z. W. Wicks, *J. Am. Chem. Soc.*, **68**, 2402 (1946).
207. J. von Braun and F. Jostes, *Ber.*, **59**, 1091, 1444 (1926).
208. J. von Braun, H. Kröper and W. Reinhardt, *Ber.*, **62**, 1301 (1929).
209. J. von Braun and E. Beschke, *Ber.*, **39**, 4119 (1906).
210. J. von Braun and E. Schmitz, *Ber.*, **39**, 4365 (1906); J. von Braun and J. Pohl, *Ber.*, **57**, 480 (1924).
211. J. von Braun and W. Sobecki, *Ber.*, **44**, 2158 (1911); J. von Braun and G. Kirschbaum, *Ber.*, **45**, 1263 (1912).
212. J. von Braun, A. Grabowski and M. Rawicz, *Ber.*, **46**, 3169 (1913).
213. J. von Braun and W. Sobecki, *Ber.*, **43**, 3596 (1910).
214. J. von Braun and W. Sobecki, *Ber.*, **44**, 1464 (1911).
215. H. J. Barber, L. Bretherick, E. M. Eldridge, S. J. Holt, and W. R. Wragg, *J. Soc. Chem. Ind.*, **69**, 82 (1950).
216. R. Huisgen, J. Sauer and M. Seidel, *Chem. Ber.*, **93**, 2885 (1960).
217. A. V. Kirsanov, *Bull. Acad. Sci. USSR Div. Chem. Sci.*, 551 (1954); A. V. Kirsanov and V. P. Molosnova, *J. Gen. Chem. USSR*, **28**, 31 (1958); **29**, 981 (1959).
218. W. Jentzsch, *Chem. Ber.*, **97**, 1361; 2755 (1964).
219. E. Allenstein, A. Schmidt and V. Beyl, *Chem. Ber.*, **99**, 431 (1966).
220. G. Martin and M. Martin, *Bull. Soc. Chim. France*, 1637 (1963); M. L. Filleux-Blanchard, M. T. Quemeneur and G. J. Martin, *Chem. Commun.*, 836 (1968).
221. R. N. Loeppky and M. Rotman, *J. Org. Chem.*, **32**, 4010 (1967).
222. M. Ochiai, R. Marumoto, S. Kobayashi, H. Shimazu and K. Morita, *Tetrahedron*, **24**, 5731 (1968).
223. R. Oda, S. Katsuragawa, Y. Ito and M. Okano, *J. Chem. Soc. Japan*, **87**, 1236 (1966).
224. H. Meerwein, P. Laasch, R. Mersch and J. Nentwig, *Chem. Ber.*, **89**, 224 (1956).
225. H. Stephen and B. Staskun, *J. Chem. Soc.*, 980 (1956).
226. R. Buijle, A. Halleux and H. G. Viehe, *Angew. Chem. Intern. Ed. Engl.*, **5**, 584 (1966).
227. M. Masaki, K. Fukui and M. Ohta, *J. Org. Chem.*, **32**, 3564 (1967).
228. D. Hall, P. K. Ummat and K. Wade, *J. Chem. Soc. (A)*, 1612 (1967).

CHAPTER **14**

Quinonediimines and related compounds

K. Thomas Finley and L. K. J. Tong

Eastman Kodak Company, Rochester, New York, U.S.A.

I. INTRODUCTION

Early in the twentieth century, more than eighty years after the analogous *p*-benzoquinone had been prepared, Willstätter began extensive and productive studies of *p*-benzoquinonemono- and -diimine (**1** and **2**) and related compounds[1,2]. This long delay, in spite of a great deal of effort, can be understood from his observation that these compounds are unstable to light, water, and acids. The difficulty of

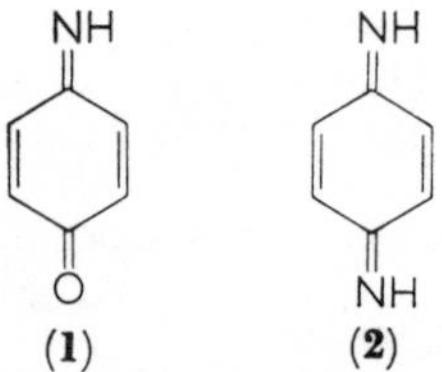

preparing and handling the quinonimines and diimines also accounts for the fact that no further substantial studies of their chemistry were made until midway through the present century.

It is our purpose in this chapter to examine the mechanisms of quinonediimine reactions. We are not presenting a comprehensive review, but a selection of the literature which in our opinion sheds light on the chemistry of this class of compounds. The quinonediimine chemistry studied prior to 1950 has been adequately reviewed[3,4] with the exception of the following brief reports.

At approximately the same time as Willstätter's work on the quinonediimines an interesting approach was taken by Jackson at Harvard[5]. He reasoned that heavy substitution of the ring might make the quinonoid form more stable. An attempted bromination of 2,6-dibromo-*p*-phenylenediamine produced a dark green precipitate having the characteristics of an equimolar mixture of the hydrobromides of the quinonediimine and the starting diamine (equation 1).

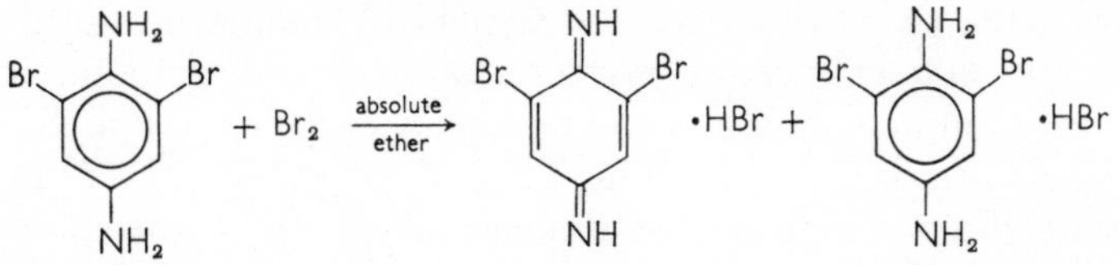

(1)

Similar results were obtained with *p*-phenylenediamine. Not one of the compounds was obtained analytically pure and the work was not continued.

The rearrangement of *N*-nitroanilines to the ring-substituted isomer, often with the displacement of halogen (equation 2), is

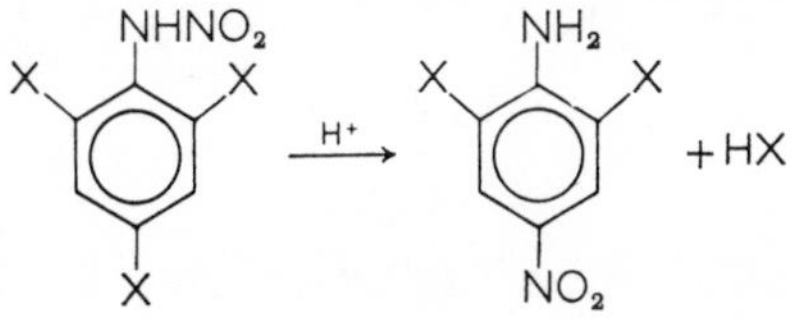

(2)

accompanied by side reactions having magenta coloured-products. A study of this reaction has been reported[6]. In the case of chloro groups, which resist displacement, the anil **3** was obtained. A later study of the

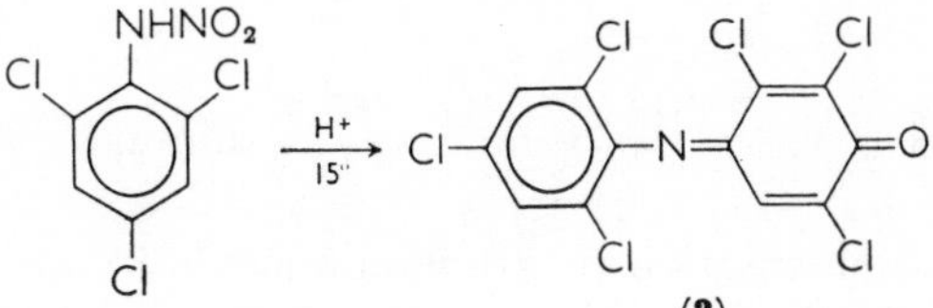

(**3**) (3)

corresponding tribromo compound produced a mixture of **4** and **5**[7]. The suggestion was made that these quinonoid structures may be

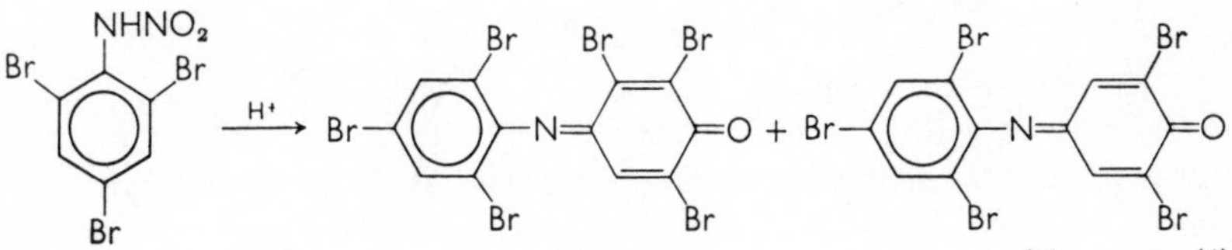

(**4**) (**5**) (4)

related to intermediates in the rearrangement to nitroanilines.

During the past twenty years a number of studies have appeared which represent significant contributions to our understanding of the chemistry of quinonediimines and related compounds. We shall begin by presenting a brief outline of the types of molecules and reactions to be discussed along with the leading references.

The first modern investigation of quinonediimines emphasizes the kinetic approach and deals largely with the reactions of the *N,N*-dialkylquinonediimines (**6**)[8]. The second examines a tremendous number of reactions of the *N,N'*-quinonediimides **7** and **8**[9].

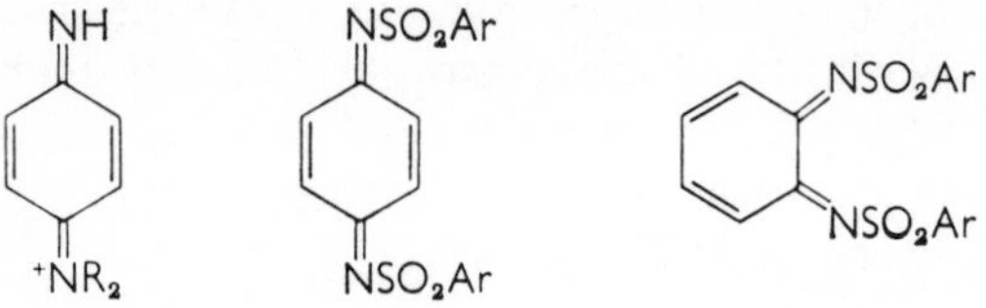

NH / $^{+}NR_2$ (**6**) NSO_2Ar / NSO_2Ar (**7**) NSO_2Ar / NSO_2Ar (**8**)

In addition to these areas, important work has been reported on quinonemonoximes (**9**) especially in regard to their tautomerism with nitrosophenols[10].

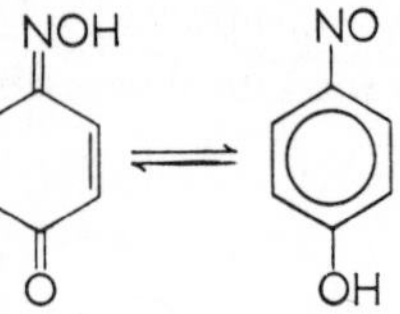

NOH / O (**9**) ⇌ NO / OH (5)

A study of the condensation of *o*-quinonimines (**10**) with aldehydes and similar compounds to produce various heterocycles, e.g. oxazoles (**11**), has been reported[11].

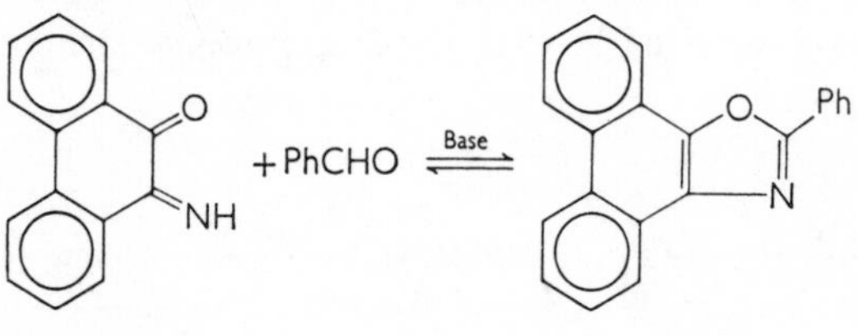

(**10**) + PhCHO $\xrightleftharpoons{\text{Base}}$ (**11**) (6)

Finally a number of interesting papers will be cited, but their limited scope precludes discussion.

II. THE CHEMISTRY OF QUINONEMONO- AND -DIIMINES

A. Background and Preparation

Long before the structure of the intermediates were determined, dye chemists had oxidized *p*-phenylenediamines with potassium dichromate, ferricyanide and permanganate[12–14]. In fact a wide variety of oxidizing agents can be used for the preparation of quinonediimines. Willstätter's use of silver ion[1,2] is interesting in view of its subsequent use by photographic chemists.

Present-day colour photography is based on the discovery by Fischer that exposed silver halide enhances the rate of the oxidative condensation of *p*-phenylenediamines with suitable molecules to form dyes[15,16]. A wide variety of substituted *p*-phenylenediamines and *p*-aminophenols show these characteristics, but for practical photographic reasons the *N,N*-dialkyl-substituted compounds have been most extensively investigated[17]. The results reported in this section will be derived mostly from these compounds with reference to other related types in certain special cases.

Regardless of the oxidizing agent, an *N,N*-dialkylquinonediimine bearing a formal positive charge is produced. Equation (7) shows that the oxidation takes place in two steps with the involvement of an intermediate ion radical corresponding to 'semiquinone'[18,19]. In this reaction, first studied by Michaelis, **R**, **S**, and **T** represent reduced, 'semiquinone', and totally oxidized (i.e. quinonediimine) respectively[20].

$$\underset{\textbf{(R)}}{H_2\ddot{N}-C_6H_4-\ddot{N}R_2} \underset{}{\overset{-e}{\rightleftharpoons}} \underset{\textbf{(S)}}{{}^{+}H_2\dot{N}-C_6H_4-\ddot{N}R_2} \overset{-e-H^{+}}{\rightleftharpoons} \underset{\textbf{(T)}}{HN{=}C_6H_4{=}\overset{+}{N}R_2} \qquad (7)$$

The instability of quinonediimines in acidic and neutral media was mentioned earlier[1] and it is also known that they will not survive in alkaline solutions[21]. For this reason they are seldom isolated, but generated *in situ* for the study of their reactions. The study of these subsequent rapid reactions imposes certain restrictions on the choice of oxidant. First, the oxidation must be quantitative; second, it must be very rapid; third, the oxidant must be mild enough so that it does not decompose the quinonediimine. Studies in two separate laboratories indicate that potassium ferricyanide meets these criteria and provides meaningful kinetic data[8,22].

B. Reactions with Nucleophiles

1. Deamination

The deamination of quinonemono- and -diimines in acidic methanol is well known[23]. Tong has shown that *N,N*-dialkylquinonediimines undergo basic deamination as well (equation 8)[8]. Evidence for this

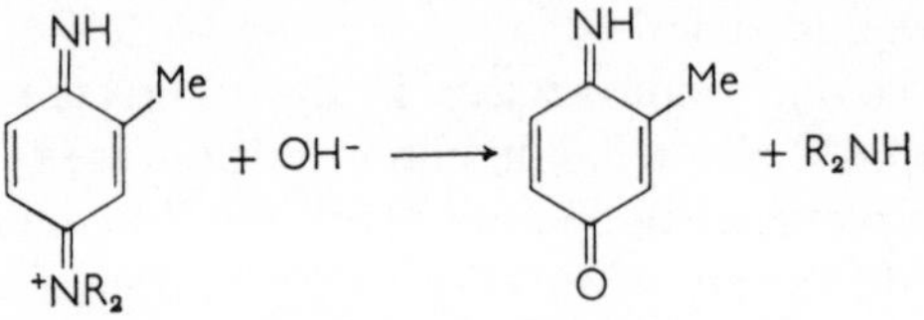

NH, Me, $^+NR_2$ + OH^- ⟶ NH, Me, O + R_2NH (8)

(**12**)

was obtained by allowing the quinonemonoimine **12** to couple with 1-naphthol to form the indonaphthol dye **13**. Elemental analysis of **13** is consistent with the structure shown and its visible spectrum is

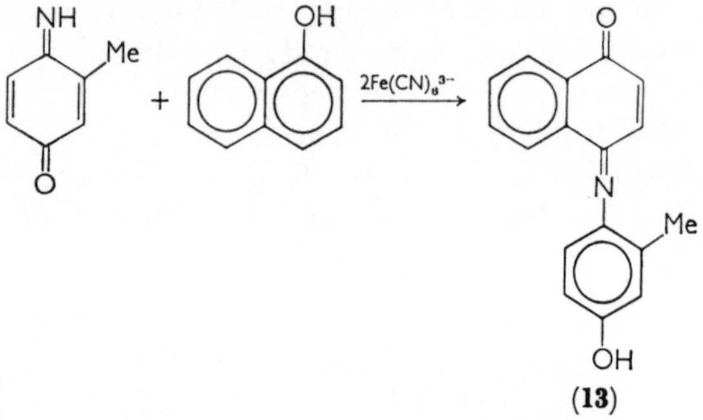

NH, Me, O + OH $\xrightarrow{2Fe(CN)_6^{3-}}$ O, N, Me, OH (9)

(**13**)

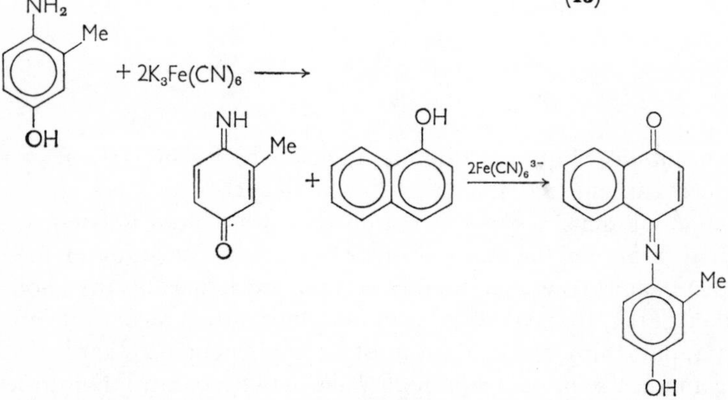

NH_2, Me, OH + $2K_3Fe(CN)_6$ ⟶

NH, Me, O + OH $\xrightarrow{2Fe(CN)_6^{3-}}$ O, N, Me, OH (10)

identical with that of the dye formed by the oxidative coupling of 2-methyl-4-hydroxyaniline with 1-naphthol.

2. Coupling*

Those reactions which lead to dyes are fundamental to colour photography and have been studied extensively. A number of excellent reviews are available[17,24,25]. While a large number of couplers have been studied and a much larger number suggested in the patent literature, they can be represented by a limited number of chemically distinct classes.

Coupling reactions of quinonemono- and -diimines with anilines (equations 11 and 12) lead to the formation of indamine (**14**) and indoaniline (**15**) dyes in a manner analogous to indonaphthol dye (**13**) formation.

$$HN{=}C_6H_4{=}NR + C_6H_5NR_2 \xrightarrow{-2e} R_2N{-}C_6H_4{-}N{=}C_6H_4{=}NR \qquad (11)$$

(**14**)

$$HN{=}C_6H_4{=}O + C_6H_5NR_2 \xrightarrow{-2e} R_2N{-}C_6H_4{-}N{=}C_6H_4{=}O \qquad (12)$$

(**15**)

The latter have the practical disadvantage of a slow rate of formation which make them unsuitable for photographic systems[17] and consequently they have not been studied extensively.

One general class of couplers which usually give yellow dyes are active methylene compounds, e.g. acylacetamides, $RCOCH_2CONHR'$. Equation (13) shows the coupling reaction where X and Y represent activating (electron-withdrawing) groups. Most of the studies of these compounds have been directed towards varying the colour or hue of the dye formed and not an elucidation of the fundamental chemistry involved.

* In this chapter coupling (and coupler) will be understood as reactions (and molecules) involving condensation with quinonemono- or -diimines usually leading to dye formation.

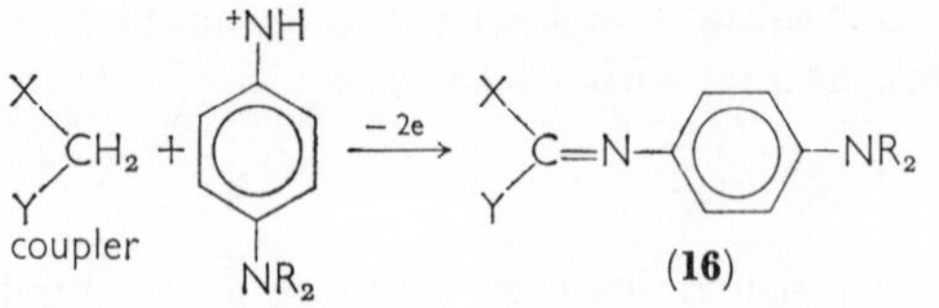

coupler (16) (13)*

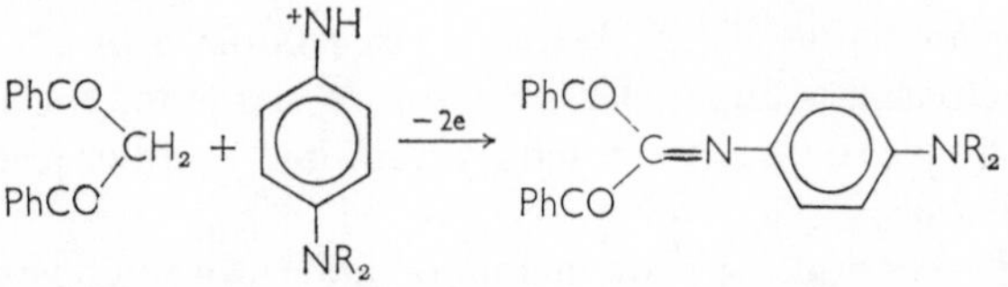

Yellow dyes can also be prepared using a class of couplers which do not possess an active methylene group. These are the benzisoxazolones (**17**) which react by ring opening[26]. This reaction differs in that the product formed is an azo dye (**18**).

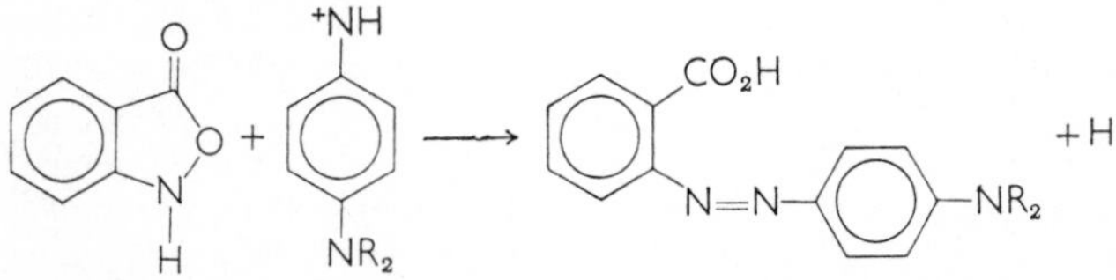

(17) (18) (14)

The benzisoxazolones are closely related to another class—the indazolones (**19**) which lead to magenta dyes[27,28]. In this case there is no ring opening and the product is termed mesoionic.

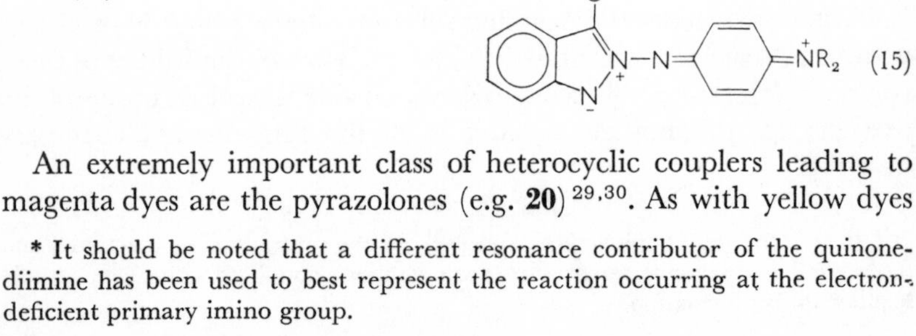

(19) (15)

An extremely important class of heterocyclic couplers leading to magenta dyes are the pyrazolones (e.g. **20**)[29,30]. As with yellow dyes

* It should be noted that a different resonance contributor of the quinonediimine has been used to best represent the reaction occurring at the electron-deficient primary imino group.

many derivatives have been examined without introducing new chemistry.

$$\text{PhN}\langle\text{N=CMe},\ \text{C(=O)}-\text{CH}_2\rangle\ (\mathbf{20}) + {}^{+}\text{NH}=\text{C}_6\text{H}_4=\text{NR}_2 \xrightarrow{-2e} \text{PhN}\langle\text{N=CMe},\ \text{C(=O)}-\text{C}=\text{N}-\text{C}_6\text{H}_4-\text{NR}_2\rangle \quad (16)$$

The final class of couplers to be considered is the methine or phenolic type which produces cyan dyes[31,32]. Here the dye formed is the same as that which results from the condensation of quinonemono-

$$\text{C}_6\text{H}_5\text{OH} + {}^{+}\text{NH}=\text{C}_6\text{H}_4=\text{NR}_2 \xrightarrow{-2e} \text{O}=\text{C}_6\text{H}_4=\text{N}-\text{C}_6\text{H}_4-\text{NR}_2 \quad (17)$$

imines with anilines (equation 12). The simple phenols are of limited practical importance; however, they have been useful in mechanistic studies to be discussed later in this chapter.

The phenolic couplers also form dye by displacement of certain *para* substituents[32,33]. Examples of *para* substituents studied are given in Table 1.

$$\text{HO}-\text{C}_6\text{H}_4-\text{X} + {}^{+}\text{NH}=\text{C}_6\text{H}_4=\text{NMe}_2 \xrightarrow{-2e} \text{O}=\text{C}_6\text{H}_4=\text{N}-\text{C}_6\text{H}_4-\text{NMe}_2 + \text{HX} \quad (18)$$

The pyrazolones also undergo dye-forming displacement reactions involving the conversion of an azo dye to an azomethine dye[34].

$$\text{PhN}\langle\text{N=CX},\ \text{C(=O)}-\text{CHN=NAr}\rangle + {}^{+}\text{NH}=\text{C}_6\text{H}_4=\text{NR}_2 \xrightarrow{-2e} \text{PhN}\langle\text{N=CX},\ \text{C(=O)}-\text{C}=\text{N}-\text{C}_6\text{H}_4-\text{NR}_2\rangle + \text{ArN}_2^{+} \quad (19)$$

$X = \text{R}, \text{NH}_2$, etc.

TABLE 1. The displacement of *para* substituents in the coupling of quinonediimines with phenols.

Substituent (X)	Formed dye	Did not form dye
alkyl		×
$-CHO$		×
$-CO_2R$		×
$-NR_2$		×
$-OH$		×
$-Cl$, $-Br$	×	
$-CO_2H$	×	
$-SO_3H$	×	
$-OR$	×	

An analogous reaction has been observed with aldehyde condensation products of the pyrazolones[35].

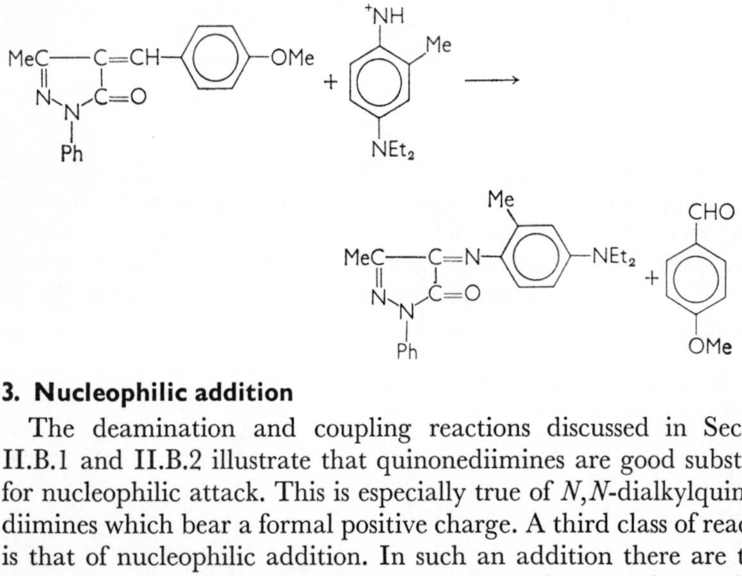

(20)

3. Nucleophilic addition

The deamination and coupling reactions discussed in Sections II.B.1 and II.B.2 illustrate that quinonediimines are good substrates for nucleophilic attack. This is especially true of N,N-dialkylquinonediimines which bear a formal positive charge. A third class of reaction is that of nucleophilic addition. In such an addition there are three sites of interest—the primary imino nitrogen and two nuclear positions (**21**).

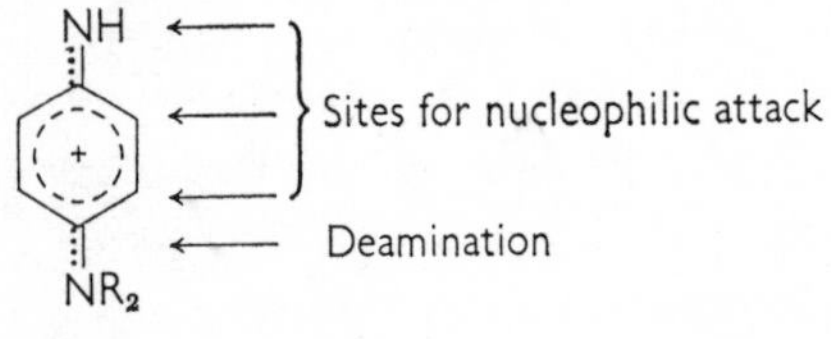

Sites for nucleophilic attack

Deamination

(**21**)

The actual site of attack depends upon several factors including the nucleophile and the structure of the quinonediimine.

The reaction of sulphite anion with *N,N*-dialkylquinonediimines (equation 21) provides one of the most striking illustrations of how

NH ... $^{+}NR_2$ + SO_3^{2-} ⟶ NH_2 ... NR_2 —SO_3^- (21)

the structure of the substrate controls the site of attack: e.g. where R = Me the sulphite ion attacks the position *meta* to the primary imino group (**22**); where R = Et attack is *ortho* (**23**)[36,37]. In the case of R = Me available evidence indicates that at least some of the

NH_2, SO_3^-, NMe_2 (**22**) NH_2, SO_3^-, NEt_2 (**23**) NH_2, SO_3^-, NMe_2 (**24**)

isomeric sulphonate (**24**) is formed[38] (see Section II.D.4).

The addition of aryl sulphinic acids (**25**) to quinonediimines has been studied and the products determined[39]. The relative yield of

NH, $^{+}NR_2$ + SO_2^-, X (**25**) ⟶ $NHSO_2$—X, NR_2 (**26**) + NH_2, SO_2—X, NR_2 (**27**) (22)

sulphonamide (**26**) and sulphone (**27**) is sensitive to a variety of reaction conditions. In this instance the position of ring attack is not changed by the structure of the quinonediimine. The preparation of sulphones by the addition of sulphinic acids to quinonediimines has been described although no proof of structure was reported[40a].

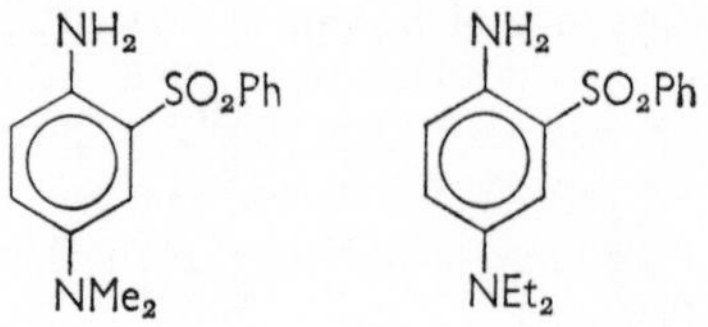

The prediction of orientation in nucleophilic addition reactions of quinonediimines by Hückel molecular orbital (HMO) theory is not unambiguous. One must remember that HMO theory is concerned only with changes in the π-electronic energy between the reacting molecules and the transition state, and not with changes in entropy, solvation forces, zero-point vibrational energy, or σ-electronic energy. In addition, one is confronted in the case of quinonediimines with a proper choice of Coulomb and resonance integrals for a variety of nitrogen valence states. Although some reactivity indices are rather insensitive to a choice of these integrals, others are found to be quite sensitive. Inevitably, some assumptions must be made about the values of these integrals. If one assumes that all resonance integrals in a quinonediimine and its intermediates in nucleophilic addition are equal to the benzene resonance integral, β_0, then a systematic variation of Coulomb integrals results in the following conclusions:

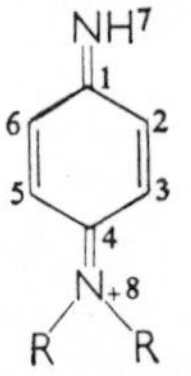

(a) Position 4 has the lowest electron density.

(b) The imino nitrogen (=N—) has the highest free valence towards nucleophilic attack.

(c) Frontier electron densities for nucleophilic attack do not give a clear prediction of reactivity. Any conclusion based upon this reactivity index is very dependent upon the choice of Coulomb and resonance integrals.

(d) Position 2 has the lowest atom stabilization energy for nucleophilic attack, ϵ^-, and therefore according to this index should be the most reactive position in such reactions.

(e) The localization energy for nucleophilic attack is lowered at the imino nitrogen.

HMO theory does not give a clear prediction of orientation in these reactions, since one is confronted with a crossing of energy curves[40b]. The results, especially the localization energy, tend to favour the imino nitrogen as the preferred point of nucleophilic attack[40c].

C. Experimental Techniques for the Study of Quinonediimine Reactions

The reactions of quinonediimines, especially the *N,N*-dialkyl derivatives, with nucleophiles are generally too rapid to measure without special instrumentation. A flow apparatus which can initiate and follow the reaction in the millisecond to second time range greatly extends the scope of investigations. The design and use of flow apparatus has been reviewed[41].

The type of flow apparatus most suitable for the study of quinonediimine reactions is one that measures absorptions under stationary state conditions. The flow velocity and distance between the region of mixing and that of measurement are both used to control the time. While requiring reasonable quantities of materials, the stationary state approach has several advantages: simplicity and the possibility of using quenches to arrest the reactions for subsequent analyses are important examples. Often the several components required in these studies must be mixed less than a few seconds prior to the initiation of the main reaction. This can be accomplished conveniently with multiple mixing chambers connected in sequence. Finally the perturbation of a reaction by the introduction of a competing reactant at a precise time is frequently useful. These features have been incorporated in the apparatus designed by Ruby[42] and used with modifications by these authors[43]. More recently an electrochemical method which is suitable for opaque systems has been devised to measure the rate of deamination and coupling[44].

D. Nucleophilic Addition to Quinonediimines—Kinetics and Mechanisms

1. General considerations

The reactions described qualitatively in Section II.B are those of the electrophilic sites of quinonediimines with reagents classified as nucleophiles. All of the studies we shall describe in this section have been conducted in buffered aqueous solution. The mechanisms in organic media may be different. It is a well known that the order of reactivity of nucleophiles depends on the electrophilic sites they are attacking. Bunnett's comprehensive review on the subject of nucleophilic reactivity lists the order of nucleophilicity of various reagents towards different classes of reaction sites[45]. In general these orders are different. In the reactions of quinonediimines, there are several reaction sites and in principle each would show a different order of reactivity. Because of the competitive nature of the reactions, the result with a given nucleophile will be reaction(s) at the site(s) having the highest rate(s). The situation here is similar to the reactions of nucleophiles with 2-chloro-4-nitrodiphenylsulphone, where the groups: phenylsulphonyl, nitro and chloro are replaceable. Loudon and Shulman[46] found that the groups displaced depended on the nucleophile as indicated in the following example:

$SO_2C_6H_4Me$-p, Cl, NH_2 ← (NH_3) — $SO_2C_6H_4Me$-p, Cl, NO_2 — (piperidine, N–H) → $SO_2C_6H_4Me$-p, $C_5H_{10}N$, NO_2

↓ Na^+ ^-OMe (23)

OMe, Cl, NO_2 + $SO_2C_6H_4Me$-p, Cl, OMe + $SO_2C_6H_4Me$-p, MeO, NO_2

The quinonediimines also possess more than one possible reaction site and reaction at the different sites may require different leaving groups, i.e. either another nucleophile or a proton. For example, some of the overall reactions are:

$$HN{=}C_6H_4{=}\overset{+}{N}R_2 + OH^- \longrightarrow HN{=}C_6H_4{=}O + NHR_2 \qquad (24)$$

$$HN{=}C_6H_4{=}\overset{+}{N}R_2 + SO_3^{2-} + H^+ \longrightarrow NH_2\text{–}C_6H_3(SO_3^-)\text{–}NR_2 \qquad (25)$$

$$\text{(p-}^{+}\text{HN=C}_6\text{H}_4\text{-NR}_2) + (\text{p-O}^{-}\text{-C}_6\text{H}_4\text{-Cl}) \longrightarrow \text{R}_2\text{N-C}_6\text{H}_4\text{-N=C}_6\text{H}_4\text{=O} + \text{HCl} \tag{26}$$

For a given nucleophile the position of attack is determined by the stability of the transition state leading to final products and by the electrophilic character of each reaction site.

Before discussing the individual reactions in detail, it may be appropriate to point out some general features of these and related reactions:

(a) By the proper choice of reagents and conditions each of the above reactions can be demonstrated to take place in two steps with the first step being reversible. The general scheme of the reaction is:

$$\text{X}^- + \text{HN=C}_6\text{H}_4\text{=N}^+\text{R}_2 \rightleftharpoons \underset{\textbf{(28)}}{\text{(Intermediate)}} \longrightarrow \text{Product} \tag{27}$$

There will be one intermediate and one product for each reaction site, although there may be additional 'intermediates' which are completely reverted and produce no products.

The product distribution may be treated quantitatively as follows. If n reactions are possible between a nucleophile X^- and a quinonediimine T^+, one for each electrophilic site, we have the following equations with I and P representing the intermediate and product respectively:

$$\begin{array}{ccccc}
X^- + T^+ & \underset{k_{1,2}}{\overset{k_{1,1}}{\rightleftharpoons}} & I_1 & \xrightarrow{k_{1,3}} & P_1 \\
\vdots & & \vdots & & \vdots \\
X^- + T^+ & \underset{k_{i,2}}{\overset{k_{i,1}}{\rightleftharpoons}} & I_i & \xrightarrow{k_{i,3}} & P_i \\
\vdots & & \vdots & & \vdots \\
X^- + T^+ & \underset{k_{n,2}}{\overset{k_{n,1}}{\rightleftharpoons}} & I_n & \xrightarrow{k_{n,3}} & P_n
\end{array}$$

If the products are examined after a sufficiently long time, so that we can assume that T^+ has been completely converted to the products with concentration $[P_i]_\infty$, i.e.

$$[T^+]_{\text{initial}} = \sum_{i=1}^{n} [P_i]_\infty, \quad \sum_{i=1}^{n} [I_i]_\infty = 0$$

it can be shown that the distribution of products is represented by the following equation, in which the index i designates the product for which a yield is desired and the running index j designates all products from 1 to n.

$$\frac{[\mathrm{P}_i]_\infty}{\sum_{j=1}^{n}[\mathrm{P}_j]_\infty} = \frac{k_{i,1}\,k_{i,3}/(k_{i,2}+k_{i,3})}{\sum_{j=1}^{n}(k_{j,1}k_{j,3}/(k_{j,2}+k_{j,3})}$$

This equation was derived without any special assumption of rapid equilibrium or steady state conditions.

(b) If the quinonediimine is neutral, the kinetics require an intermediate containing an additional proton, presumably on one of the nitrogens:

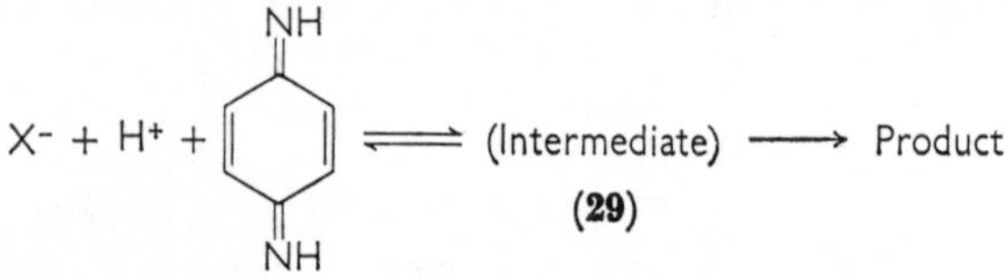

$$X^- + H^+ + \text{(quinonediimine, HN=}C_6H_4\text{=NH)} \rightleftharpoons \text{(Intermediate)} \longrightarrow \text{Product} \qquad (28)$$

(29)

This intermediate **29** can therefore be assumed to have the same electronic structure as the previous one (**28**).

(c) The species in the 'semiquinone state', coexisting with the quinonediimine, can be shown to make no contribution to the reaction rates.

(d) The predominant species of quinonediimines above pH 8 (where most reactions have been studied) was postulated by Michaelis to be:

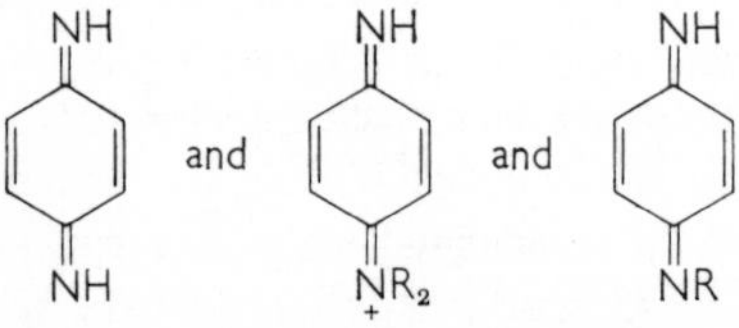

HN=C₆H₄=NH and HN=C₆H₄=$\overset{+}{N}R_2$ and HN=C₆H₄=NR

This has been further substantiated by the salt effect on the rate of deamination[47a] and by studies of 'semiquinone' dismutation equilibria[19].

(e) Some nucleophiles (specifically SCN^- and I^-) which have high reactivities in other reactions have no measurable rates with quinonediimines.

The above discussion was restricted to one reagent–quinonediimine reaction at a time. In kinetic experiments with these reactants, complications often arise which produce misleading results unless precautions are taken to avoid them. We suspect that the paucity of kinetic information on these systems results from the awareness of these complications by most kineticists and lack of special interest in these systems. On the other hand, photographic chemists have in the past used gelatin and silver halide which introduce complications from the standpoint of kinetic interpretation. Even with buffered aqueous solutions and ferricyanide as an oxidant, the studies may be complicated by side reactions. Typical of these side reactions are condensation of oxidized *p*-phenylenediamines to azo dyes[47] and polymers when the substrate concentration is high and deamination when the pH is high. At low pH and in the presence of diamine, autocoupling and 'semiquinone' formation are observed. The latter is especially interesting in terms of the early studies of the kinetics of these reactions. Discussions on the question of whether the quinonediimine or the 'semiquinone' is the reactive species in these reactions has been reviewed by Weissberger and Vittum for work up to 1953[17]. Later experimental investigations made use of the 'semiquinone' formation equilibrium and the predicted mass action due to the addition of either oxidant or reductant. The indirect effect produced on the reaction rates determined the species whose concentration should appear in the kinetics expression. Early workers carried out coupling reactions at relatively low pH in order to obtain measurable rates and in the presence of diamine to simulate photographic developers. They were impressed by the appearance of 'semiquinone' and assumed that this free radical should be very reactive. Their conclusion that the 'semiquinone' was the reactive species has proven to be incorrect[48]. The use of the multiple stage rapid flow machine, described in Section II.C, along with dilute solutions and at least an equivalent quantity of oxidant produced clear-cut reactions. Further, the reactions to be discussed in this section (quinonediimines with hydroxide, sulphite, and coupler ions) give products which are almost unique. This indicates that for each of these reactions there is one predominant rate $\{k_{i,1}k_{i,3}/(k_{i,2} + k_{i,3})\}$ corresponding to the most favourable reaction site. There is evidence, however, that there can be more than one significant product in reactions with some nucleophiles, for example, the aryl sulphinates cited earlier[39].

The kinetic data obtained under conditions having negligible concentrations of intermediates are expected to yield second-order rate constants. Unless simplification is justified on theoretical or experimental grounds, these constants should be regarded as being composite in nature, e.g.

$$k_{\text{obs}} = \frac{k_1 k_3}{k_2 + k_3}$$

The elementary constants involved are defined by the scheme:

$$\text{quinonediimine} \underset{k_2}{\overset{k_1[\text{X}^-]}{\rightleftharpoons}} \text{(intermediate)} \xrightarrow{k_3} \text{product} \qquad (29)$$

where $[\text{X}^-]$ represents the concentration of the nucleophile. In special cases the expression can be simplified:

1. when $k_3 \gg k_2$, $k_{\text{obs}} \approx k_1$
2. when $k_2 \gg k_3$, $k_{\text{obs}} \approx Kk_3$ where K is the equilibrium constant for the formation of the intermediate.

It is obvious that each of these elementary constants may have a different dependence on such variables as the structure of the quinonediimine, pH, etc. Therefore, in evaluating the observed rate constants, one must keep in mind their composite nature and interpret them in terms of their relationship to the elementary constants.

The subject of intermediates in reactions has been adequately described in standard texts to which one should refer for details[49]. The existence of intermediates has been assumed in the following discussion. The question here is mainly their stability and whether or not their concentrations are sufficiently low for the application of the steady state treatment. As will be pointed out in the appropriate sections to follow, intermediates of sufficiently high concentration have been detected kinetically. This has been accomplished in some cases by following the rate of disappearance of the quinonediimine and in others the rate of formation of products. In extreme cases the reactions have been shown to proceed in two distinct steps.

2. Deamination

Some of the statements offered without evidence in the preceding section can be substantiated by studies of the kinetics of deamination conducted in these laboratories[8,19,50]. Specifically an attempt will be made in this section to establish three points:

(a) The oxidation state of the reactive species in deamination is quinonediimine (not 'semiquinone').

(b) The deamination reactions proceed in two steps, with the first being reversible.

(c) Although different quinonediimines may be charged or uncharged depending on the substituents, the hydroxide ion always attacks a positively charged intermediate. The charge may result directly from oxidation or through the preliminary involvement of a proton.

It has been found that the deamination rate constant does not change in the presence of excess ferricyanide or excess diamine when the

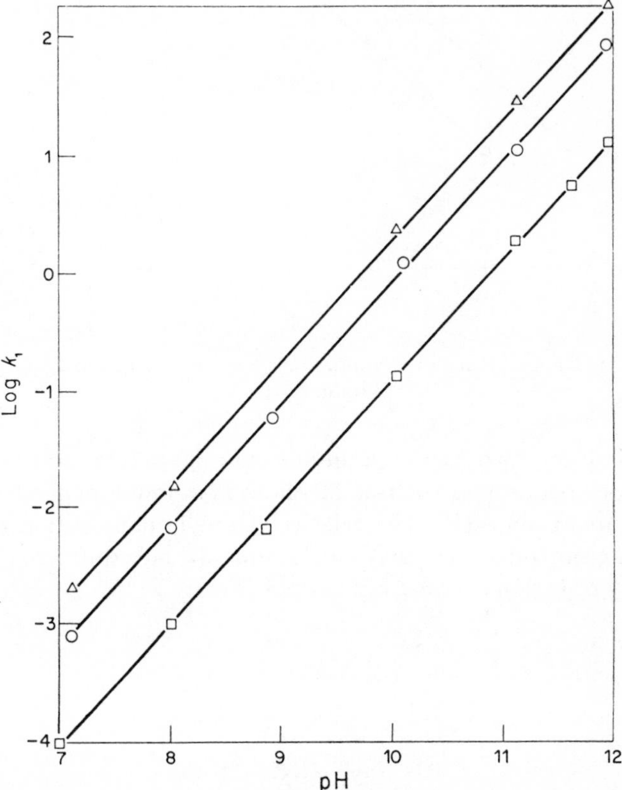

FIGURE 1. Pseudo first-order rate constants for deamination of alkyl substituted quinonediimines (**30**).

calculations are made on the basis of the diimine[8]. Because the equilibrium is rapid, and at high pH favours dismutation (see section II.D.1), the rate should be very sensitive to the concentration of oxidant and substrate if it is dependent on 'semiquinone'. The above observation requires that the rate be dependent on quinonediimine concentration and implies that the latter is the reactive species.

The structure of the quinonediimine can have a marked effect on the rate of deamination. Figures 1 and 2 show the dependence of this rate k

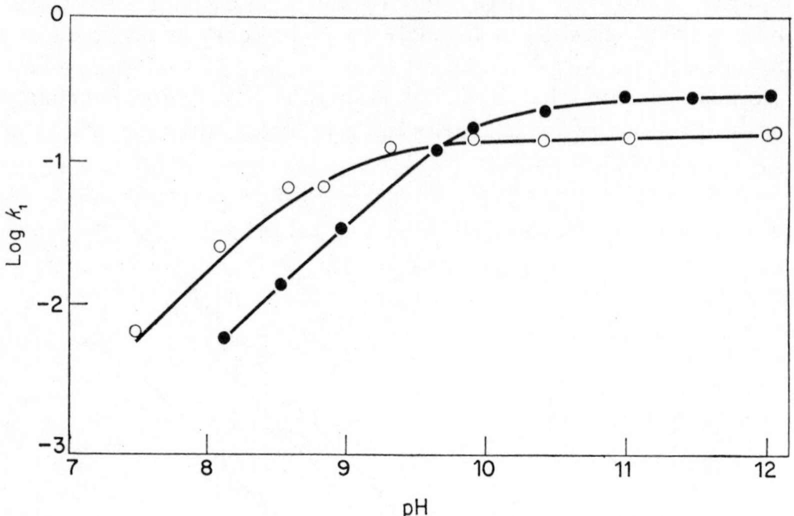

FIGURE 2. pH dependence of deamination rates for β-hydroxy-alkyl quinonediimines (**31**).

on [OH^-] for two types of quinonediimines. The simple *N*,*N*-dialkylquinonediimines such as **30** show first-power dependence on [OH^-] up to pH 12[19]. The relatively minor modification of a β-hydroxyl group on one of the *N*-alkyl chains (**31**) introduces an [OH^-]-independent region above a certain pH (Figure 2)[50].

NH ... $^+NEt_2$ (**30**) NH ... $EtNCH_2CH_2OH$ (+) (**31**)

In terms of the general scheme, the deamination reactions for these two types can be represented as:

NH=C₆H₄=N⁺RR′ + OH^- $\underset{k_2}{\overset{k_1}{\rightleftharpoons}}$ NH=C₆H₄(OH)(NRR′) (**32**) $\longrightarrow$ NH=C₆H₄=O + NHRR′ (30)

When R and/or R′ are —C_2H_4OH, the intermediate is unusually stable and the first step can be assumed to be an established equilibrium with constant $K = k_1/k_2$. The presence of an intermediate has been confirmed and its concentrations determined by absorption measurements in the flow machine under steady state conditions. The constants K for various diimine intermediates like **32** have been calculated from both kinetic and absorption data. The stability of this type of compound has been attributed to intramolecular hydrogen bonding (**33**). It is also possible that an intermediate having the spiro structure (**34**) may be involved. Such a compound would not deaminate directly, but would revert to **33** or **32** first.

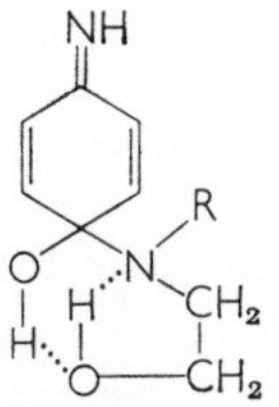

(**33**) 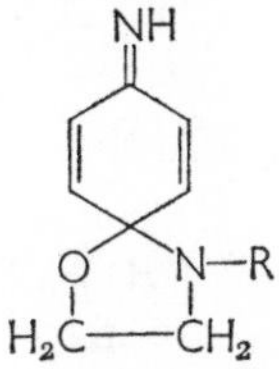 (**34**)

Although intermediates have not been directly observed for diimines of type **30** where both R's are alkyl, their presence must be implied from these observations since the two types would be expected to differ in degree, but not in kind.

The quinonediimine of unsubstituted *p*-phenylenediamine is uncharged between pH 8–12 where the deamination rate constants shown in Table 2 were obtained. This independence of pH is in sharp contrast to the *N,N*-dialkyl derivatives discussed above. The difference can be accounted for if we assume that both diimines form intermediates having analogous structures. Writing equilibria in two parts (equation 31) makes it apparent that after protonation the reaction of the unsubstituted quinonediimine is parallel to that of the

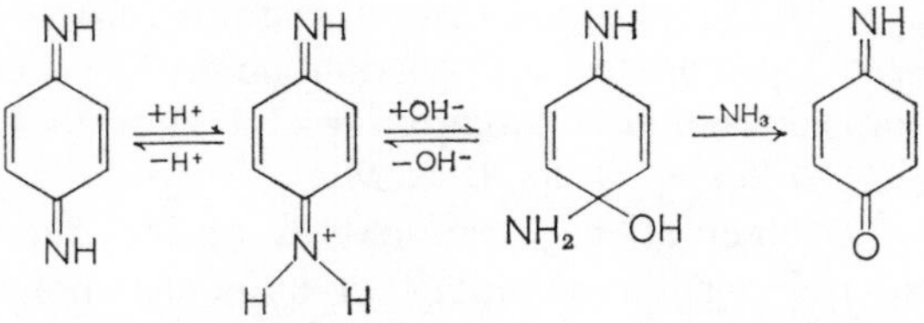

(31)

(**Q**) (**QH**$^+$)

disubstituted derivative. If the concentration of the intermediate is low, the rate would be independent of pH.

$$-\frac{d[\mathbf{QR}^+]}{dt} = k_1[\mathbf{QR}^+][\mathrm{OH}^-]$$

$$-\frac{d[\mathbf{Q}]}{dt} = k_2[\mathbf{QH}^+][\mathrm{OH}^-]$$

QR$^+$ indicates the *N,N*-dialkylquinonediimine.

TABLE 2. Deamination rate constants[8].

	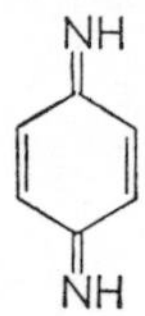		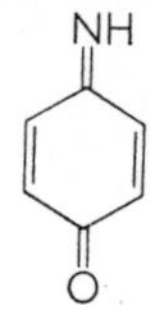
pH	$k(\mathrm{s}^{-1} \times 10^3)$	pH	$k(\mathrm{s}^{-1} \times 10^3)$
6·06	6·34	6·11	1·54
7·03	1·04	7·09	0·55
8·04	0·43	7·95	0·192
8·88	0·35	8·69	0·131
10·25	0·33	9·78	0·115
11·30	0·33	10·95	0·154
12·15	0·33	11·89	0·384

The above hypothesis concerning the structures of the intermediates **QH**$^+$ and **QR**$^+$ can be subjected to a more quantitative test. If we set up the same kinetic expression for the two types of quinonediimines, and substitute the appropriate constants, we obtain the equation:

$$\frac{-\,d\ln[\mathbf{Q}]}{dt} = \frac{k_2 k_w}{K} = 0{\cdot}3 \times 10^{-3}$$

in which $k_w = 10^{-14}$ and a dissociation constant $K = 10^{-6}$ has been obtained for the protonated quinonediimine[19]. The value of $k_2 = 3 \times 10^4$ obtained from this equation is very close to $k_1 = 10^4$ for R = Et and $k_1 = 2{\cdot}5 \times 10^4$ for R = Me.

Most of the observations and analysis of the deamination of unsubstituted quinonediimine apply to quinonemonoimine, which also deaminates with rates nearly independent of [OH$^-$] above pH 8 (see Table 2). Using similar arguments we write the following equation:

$$\text{HN=}C_6H_4\text{=O} \underset{-H^+}{\overset{+H^+}{\rightleftharpoons}} {}^+H_2N\text{=}C_6H_4\text{=O} \underset{-OH^-}{\overset{+OH^-}{\rightleftharpoons}} (H_2N)(HO)C_6H_4\text{=O} \xrightarrow[\text{determining step}]{\text{rate}} O\text{=}C_6H_4\text{=O} \qquad (32)$$

which accounts for the observations.

The so-called 'water rate' in the deamination of *N,N*-dialkyl-quinonediimine, which is observed only at relatively low pH, can also be written as a reaction preceded by protonation:

$$\text{HN=}C_6H_4\text{=}{}^+NR_2 \underset{-H^+}{\overset{+H^+}{\rightleftharpoons}} {}^+H_2N\text{=}C_6H_4\text{=}{}^+NR_2 \underset{-OH^-}{\overset{+OH^-}{\rightleftharpoons}} \begin{cases} (H_2N)(HO)C_6H_4\text{=}{}^+NR_2 \longrightarrow O\text{=}C_6H_4\text{=}{}^+NR_2 \\ {}^+H_2N\text{=}C_6H_4(R_2N)(OH) \longrightarrow {}^+H_2N\text{=}C_6H_4\text{=O} \end{cases} \qquad (33)$$

In this reaction, the deamination of the unsubstituted imino group makes a significant contribution and may in fact be the sole reaction.

In strong acid solutions Fieser measured the deamination rates of some quinonediimines and found that the reactions were catalysed by acid with the removal of the unsubstituted imino group[23]. Assuming the rates to be proportional to $[H_3O^+]$, a reaction scheme for the dialkyl derivative consistent with the above arguments would be:

$$\text{HN=}C_6H_4\text{=}{}^+NR_2 + H_3O^+ \rightleftharpoons (H_3N^+)(HO)C_6H_4\text{=}{}^+NR_2 \longrightarrow O\text{=}C_6H_4\text{=}{}^+NR_2 \qquad (34)$$

and for quinonediimine:

$$\text{HN=}C_6H_4\text{=NH} + H_3O^+ \rightleftharpoons (H_2N)(HO)C_6H_4\text{=}{}^+NH_2 \longrightarrow O\text{=}C_6H_4\text{=}{}^+NH_2 \qquad (35)$$

It would now be interesting to summarize the deamination reactions by comparing the structures of the intermediates and the acidic or basic substance which is kinetically significant. In Table 3 the dialkyl derivatives are regarded as permanently protonated quinonediimine and the reactions are classified according to the reaction intermediates. Note, however, the same intermediate may require different pH dependence according to the predominant species in that acidity region. Such a scheme describes a pH profile with several plateaux.

TABLE 3. Summary of deamination reaction schemes. (Intermediates within braces are similarly charged species.)

Reactants (Quinonediimine aqueous)		Intermediate		Product
$^{+}NR_2$ / NH + OH^-	⇌	R_2N OH / NH	⟶	O / NH
NH / NH + HOH	⇌	H_2N OH / NH	⟶	O / NH
NH / $^{+}NR_2$ + HOH	⇌	H_2N OH / $^{+}NR_2$	⟶	O / $^{+}NR_2$
NH / NH + H_3O^+	⇌	H_2N OH / $^{+}NH_2$	⟶	O / $^{+}NH_2$
NH / $^{+}NH_2$ + HOH	⇌ (to intermediate above)			

(*continued*)

TABLE 3 (*continued*)

Reactants (Quinonediimine aqueous)	Intermediate	Product
NH, $^{+}NR_2$ + H_3O^+ ⇌	H_3N^+ OH, $^{+}NR_2$ →	O, $^{+}NR_2$
$^{+}NH_2$, $^{+}NR_2$ + HOH ⇌		
NH, $^{+}NH_2$ + H_3O^+ ⇌	H_3N^+ OH, $^{+}NH_2$ →	O, $^{+}NH_2$
$^{+}NH_2$, $^{+}NH_2$ + HOH ⇌		

3. Coupling

The study of coupling is especially important to the elucidation of the general mechanism of nucleophilic reactions of quinonediimines [48]. This is because data for a wide variety of couplers and quinonediimines are available. The couplers' reaction characteristics can be varied in large jumps by making changes among the classes, e.g. phenols, naphthols, pyrazolones, acetoacetanilides, etc., or finer variation by changes of substituents within a particular class.

As in the case of deamination, the reactive species were deduced from kinetic data. Unlike the earlier studies, coupling introduces the additional need to identify the ionic species of the coupler as well as the oxidation states and ionic species of the oxidized diamine. Early results on the effect of pH on rates, restricted to low pH, led to the conclusion that coupler anion was the reactive species [17]. This was substantiated by results obtained using the flow method, which permitted extension of the experimental pH range beyond the pK of the coupler [48].

The oxidation state of the reactive species of the diamine was easily narrowed down to either the 'semiquinone' or the quinonediimine from the fact that oxidation was necessary, but which of these two was reactive was a matter of controversy for many years. The first conclusive evidence that quinonediimine was reactive came from the work of Hünig and Daum [22]. These authors interpreted their data as supporting the 'semiquinone' mechanism, at least in acid solutions where they were obtained, but stated that the quinonediimine mechanism might be followed in alkaline solution. Soon after the publication of this article, the data were reinterpreted by both Egger and Frieser [51] and Tong and Glesmann [52,53], to show that they actually supported the quinonediimine mechanism even in acid solutions where the 'semiquinone' is the principal species. The quinonediimine mechanism in alkaline solution was substantiated by results obtained using the flow method [48].

The equilibrium relationship for 'semiquinone' formation ($K = [\mathrm{S}]^2/[\mathbf{R}][\mathbf{T}]$) was used to distinguish between the two possible mechanisms. In acid solution the fact that K has a large value demands that the rate be suppressed by added diamine (**R**). In alkaline solution K has a small value and neither excess diamine or oxidant should have any significant effect on the rate. The fact that both of these predictions were borne out by experiment was taken as conclusive evidence that the coupling rate is proportional to quinonediimine.

The reactive ionic species of the quinonediimine was deduced in the following manner. Studies of pH rate profiles in the alkaline region revealed that the *N,N*-dialkyl compounds show rates proportional to the coupler ion, but the rates of unsubstituted quinonediimine are proportional to undissociated coupler. In the pH range studied it would be expected that the principal species would have structures **35** and **36** respectively*.

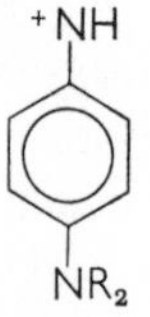

$^{+}$NH / NR$_2$ **(35)**   NH / NH **(36)**

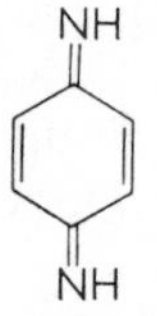

* Since the coupling reaction occurs at the monosubstituted imino nitrogen the resonance contributor shown in **35** will be used in this section.

The equilibrium actually studied (equation 36) and the greater nucleophilic reactivity of the coupler anion allows **35** (with R being

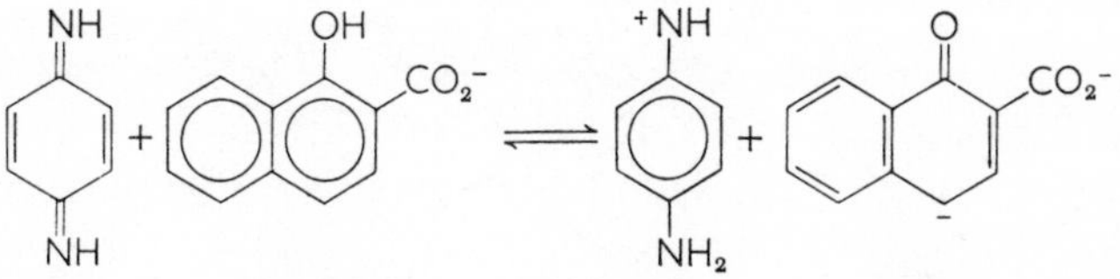

(T) (HC) (HT+) (C-) (36)*

either alkyl or hydrogen) to be considered as the reactive ionic species. Employing K_1 and K_2 as the acid dissociation constants of quinonediimine and coupler respectively, we can write the following equation describing these observations.

$$\frac{d[\text{dye}]}{dt} = k[\mathbf{HT}^+][\mathbf{C}^-] = \frac{kK_2}{K_1}[\mathbf{T}][\mathbf{CH}]$$

Some of the properties of the coupling reaction can be deduced by comparing the kinetics of couplers substituted at the coupling position and having the general structure **37**[54].

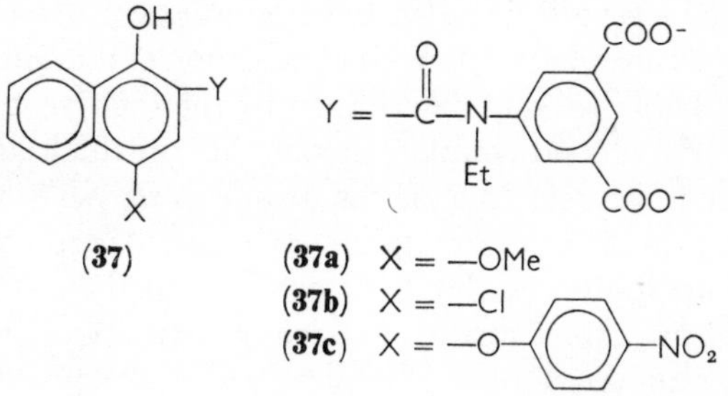

(**37**)

(**37a**) X = —OMe
(**37b**) X = —Cl
(**37c**) X = —O—C_6H_4—NO_2

* The contributors shown are those which most clearly indicate the course of the reaction; it should be remembered that the more important contributors are:

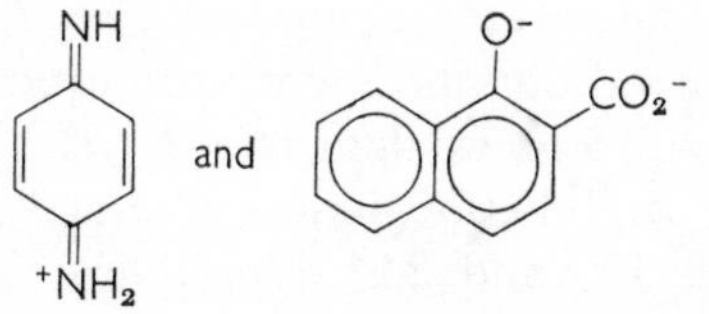

and

All three couplers were expected to react by the same scheme, shown in equation (37):

$$\mathbf{T}^+ + \mathbf{C}^- \underset{k_2}{\overset{k_1}{\rightleftharpoons}} \text{Intermediate } (\mathbf{I}) \xrightarrow{k_3} \text{Dye } (\mathbf{P}) + H^+ + X^- \text{ or } (HX) \quad (37)$$

However, with N,N-diethylquinonediimine the intermediate of coupler **37a** (X=OMe) was the only one present in sufficient concentration to be detected. In fact, with this pair of reactants, pseudo first-order rates were observed for dye formation. The constants were found to be independent of coupler, when it was used in excess. This indicates that $k_1[\mathbf{C}^-]$ is large in comparison with $(k_2 + k_3)$, so that shortly after mixing, $\mathbf{T}^+$ was almost completely converted to **I** which slowly decomposes to **P** with the first order rate constant k_3. An alternative interpretation that the quinonediimine and coupler anion had undergone a reversible redox reaction was disproved by the observation that the rate was unchanged by addition of diamine. The values of k_1 and k_2 (Table 4) were determined by the relative dye yield upon the addition of a second coupler, the kinetics of which had previously been examined (see reference 54 for detailed analysis).

In contrast to **37a**, the couplers **37b** and **37c** react with N,N-diethylquinonediimine without detectable accumulation of intermediates. Using the steady state approximation, the observed second-order rate constant is:

$$k_{obs} = \frac{k_1 k_3}{k_2 + k_3}$$

A lower limit for k_3 can be calculated by introducing the concentration of coupler anion and rearranging to $k_3 = k_{obs}[\mathbf{C}^-]\ \{(k_2 + k_3)/k_1[\mathbf{C}^-]\}$. Since the condition for low concentration of intermediate is $(k_2 + k_3 \gg k_1[\mathbf{C}^-]$, it follows that $k_3 > k_{obs}[\mathbf{C}^-]$. For reactions of **37b** and **37c** with N,N-diethylquinonediimine, $k_{obs}[\mathbf{C}^-] = 100$ and 200, therefore $k_3 > 100$ and 200 respectively. The conclusion that

TABLE 4. Elementary rate constants for coupling reactions of **37a**.

Diimine	k_1(l mole^{-1} s^{-1})	k_2(s^{-1})	k_3s($^{-1}$)
NH=C$_6$H$_4$=$^+$NEt$_2$	$3 \cdot 3 \times 10^7$	3·2	3·2
NH=C$_6$H$_4$=$^+$NMe$_2$	$6 \cdot 4 \times 10^7$	1·3	2·4
NH=C$_6$H$_3$(OMe)=$^+$NEt$_2$	$3 \cdot 1 \times 10^4$	4·6	1·5
NH=C$_6$H$_3$(Cl)=$^+$NEt$_2$	$5 \cdot 3 \times 10^7$	1·2	0·46

elimination of both —Cl and $-O-C_6H_4-NO_2$ are faster than that of —OMe is reasonable on the basis of the stability of the ions being eliminated. This is in agreement with the proposal of Bunnett and Zahler[55] for the elimination step of nucleophilic displacement reactions, although the site from which the ions are eliminated is not

$$X^- + Y\text{-}C_6H_4\text{-}NO_2 \rightleftharpoons (X)(Y)C_6H_4{=}N^+(O^-)_2 \longrightarrow X\text{-}C_6H_4\text{-}NO_2 + Y^- \quad (38)$$

equivalent in the two reactions.

When an N,N-diethylquinonediimine was coupled with a series of 2,6-dimethylphenols, which form dyes by elimination of another

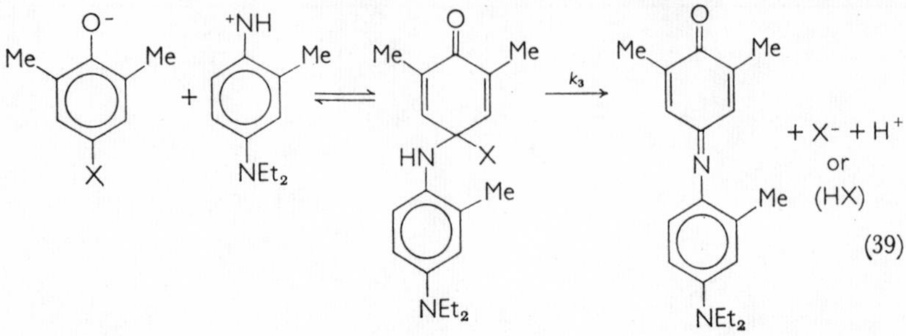

$$+ X^- + H^+ \text{ or } (HX) \qquad (39)$$

nucleophile, the rates were found to vary with the pK_a of the couplers as shown in Figure 3. In this plot, k_{C^-} is the observed second-order rate constant based on the concentration of the coupler ion. The relationship $k_{C^-} = \rho K_a{}^{\beta}$ obtained from the straight line (with $\beta < 0$) shows that the reactivity increases with the basicity of the coupler anion. This kind of correlation is quite common for reactions with nucleophiles (see Ibne-Rasa [56] and Bunnett [45] for summaries).

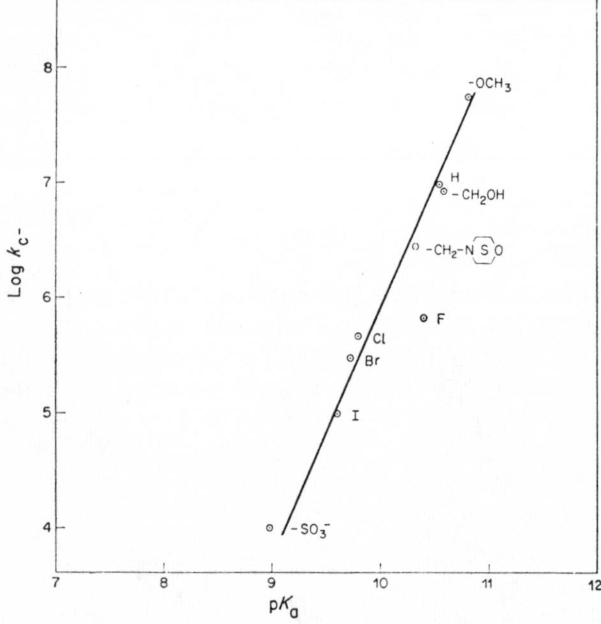

FIGURE 3. Effect of coupler pK_a on rate of coupling (equation 39).

It is significant that the result for unsubstituted coupler (X = H), in which the second step is an oxidation reaction instead of an elimination, also fits on the same line.

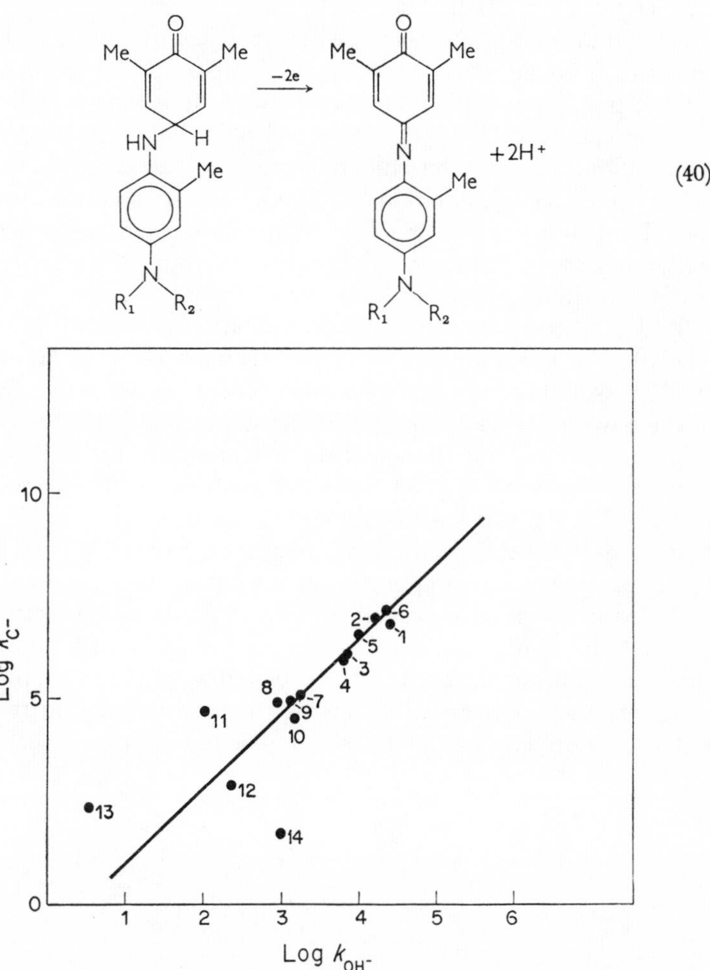

FIGURE 4. Variation of coupling and deamination rates with quinonediimine structure. The reactions of coupler **37** (X = H) with the following quinonediimines: 1. 4-amino-*N*, *N*-dimethylaniline, 2. 4-amino-*N*-ethyl-*N*-β-hydroxyethylaniline, 3. 4-amino-3-methyl-*N*-ethyl-*N*-β-sulphoethylaniline, 4. 4-amino-3-methyl-*N*-ethyl-(*N'*-methyl-β-methyl-sulphonamidoethyl)-aniline. All others are 4-amino-*N*, *N*-diethylanilines with the indicated substituents in the 3 and 5 positions: 5. H,H; 6. Cl, H; 7. Me, H; 8. Et, H; 9. n-Pr, H; 10. i-Pr, H; 11. MeO, H; 12. Me, Me; 13. MeO, MeO; 14. *t*-Bu, H.

Furthermore the overall reaction rate was found to be independent of the concentration of oxidant, which suggests that k_3 is very large, and the observed bimolecular rate constant is approximately equal to k_1. The combination of the two observations suggest that all reactions described in this section proceed with a rate-determining addition, and that this is solely determined by the basicity of the coupler anion. This hypothesis is further supported by the following arguments. As described earlier in the coupling of naphthol derivatives, where elimination was rate-determining, the elimination of Cl is much faster than the elimination of MeO. The observation that the overall reaction of the MeO derivative is faster than the Cl derivative when they are on 2,6-dimethylphenol would indicate that elimination is not rate limiting in these cases.

When the structure of the quinonediimine is altered, both the coupling rate (with a given coupler) and the deamination rate (at constant $[OH^-]$) would be expected to change. A comparison of the variations in these two rates with quinonediimine substituent would give a comparison of the relative change in the free energy of the two transition state complexes **38** and **39** since the changes in the reactants are common for the two reactions. Figure 4 shows such a correlation on both rates[54]. The slope, which is roughly a ratio of ρ's for the two reactions, shows that coupling is more sensitive to substituent effects than deamination. It was assumed that in deamination reactions the intermediate is in rapid equilibrium with the reactants and the decomposition of the former is rate-determining; while in coupling, the addition reaction is rate-determining. The corresponding transition states therefore have structures **38** and **39**.

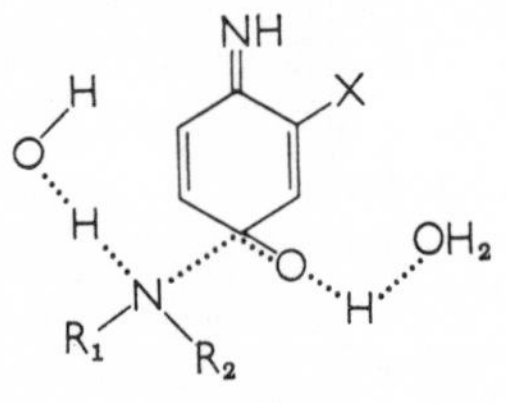

(38)

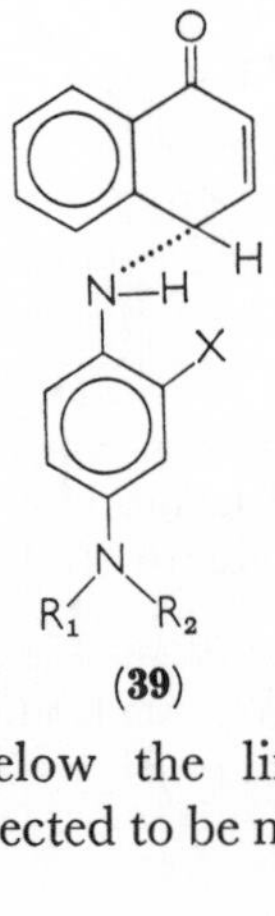

(39)

Those quinonediimines with points below the line have large groups for X and steric effects might be expected to be more important in coupling than in deamination. It is interesting to note that the effect is not great until X reaches the size and bulkiness of an isopropyl group or until both positions *ortho* to the reacting nitrogen atom are substituted. The two points lying above the line were obtained from quinonediimines with X = OMe. This indicates extra stability produced in transition state **39** over **38** by the OMe group; probably by a +R effect.

4. Sulphonation[38]

Reactions between sulphite and quinonediimines are easily followed by measuring the absorption of the latter in the flow apparatus. The difficulty in this problem is usually the identification of products. Such structure proof must accompany the kinetic measurements for each quinonediimine, because sulphonation does not take place at the same site for all substrates. The problem is especially difficult due to the fact that until recently there has been no simple analytical method available[57]. Consequently, although the rates for many quinonediimines have been measured, only three were done in sufficient detail for mechanistic interpretation. The danger of assuming the same reaction for all substrates was illustrated in Section II.B.3. Results with *N*,*N*-dimethylquinonediimine show evidence for small amounts of other products being formed in the relatively low pH region[38]. The rate of disappearance of quinonediimine, with low sulphite concentration, was found to be bimolecular involving [$\mathbf{T}^+$] and [SO_3^{2-}] each to the first power. When a large excess of sulphite was present, kinetic complications arose which suggested the accumulation of reaction intermediates. An acidic intermediate seems most reasonable since the effect appeared at lower sulphite concentrations when the pH was lower. The diimine appears to be the reactive species since in alkaline solution, where it is most stable, introduction of excess diamine or ferricyanide did not produce any change in the rate.

The kinetic evidence for the formation of intermediates is the observation that with a large excess of sulphite and at relatively low pH the first-order plot for the decomposition of the quinonediimine showed a break, i.e. a rapid initial rate followed by a much slower rate. However, when low and equivalent concentrations of sulphite and quinonediimine were used, good second-order rates were obtained. Evidently the accumulation of intermediates is a borderline situation depending on the concentration of sulphite. Therefore, except for those experiments designed to detect intermediates, the kinetic data were obtained at low sulphite concentrations to suppress the concentration

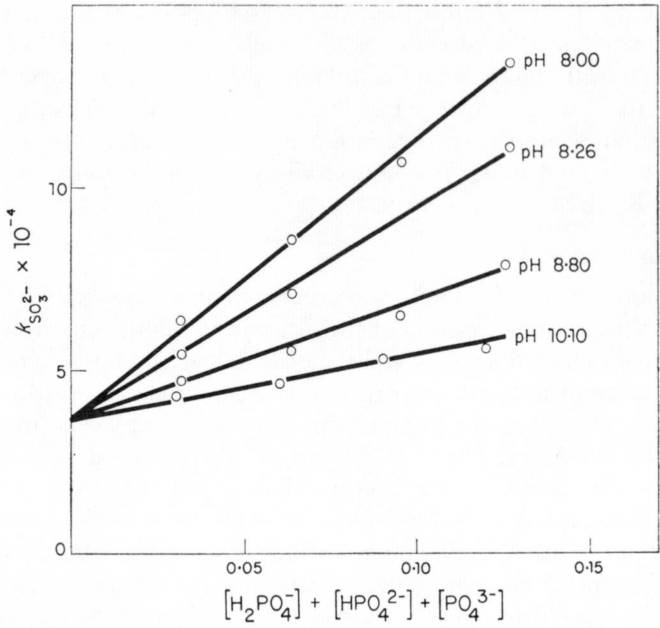

FIGURE 5. Phosphate buffer catalysis of the sulphonation of *N,N*-diethylquinonediimine.

of the intermediates. We should note that there may be more than one intermediate present and the predominant intermediate may not be that which leads to the predominant product.

Figure 5 is a plot of the second-order rate constant $k_{SO_3^{2-}}$ versus total phosphate concentration for four sets of experiments. Each of these was carried out at a constant pH and thus has a constant ratio of the buffering species; the ionic strength was kept constant with KCl. It is apparent that the dependence on the concentration of phosphates is greater than the dependence on pH; in fact, when extrapolated to zero phosphate concentration, all lines meet at a common point,

AH, NH, SO_3^-, H, NEt_2 **(40)**

H^+, NH, SO_3^-, H, A^-, NEt_2 **(41)**

showing independence of the rate on pH. This could be interpreted as the general acid-catalysed decomposition of intermediate **40** or the general base-catalysed decomposition of the conjugate acid **41**.

The second interpretation is preferred, since it is more reasonable to assume the removal of a proton from a carbon atom than the addition of a proton to a nitrogen atom as the rate-determining step. The mechanism suggested is a reversible equilibrium with **41** followed by general base-catalysed deprotonation to the product:

$$\text{p-HN}{=}\text{C}_6\text{H}_4{=}\text{N}^+\text{Et}_2 + \text{H}^+ + \text{SO}_3^{2-} \underset{\text{fast}}{\overset{K}{\rightleftharpoons}} \textbf{(41)} \; [\text{H}_2\text{N}^+{=}\text{C}_6\text{H}_4(\text{H}\cdots\text{A}^-)(\text{SO}_3^-)\text{–NEt}_2] \xrightarrow{k_{\text{A}^-}} \text{NH}_2\text{–C}_6\text{H}_3(\text{SO}_3^-)\text{–NEt}_2 \qquad (41)$$

The sulphonation of *N*,*N*-dimethylquinonediimine is more complex than that of the diethyl derivative. The graphs in Figure 6 show *k* versus pH profiles at constant phosphate concentrations and ionic strength. It is apparent that the rates at constant pH increase with phosphate as in the above example, i.e. more rapidly at lower pH. However, the difference is that the extrapolated curve for zero phosphate concentration is not independent of pH. The phosphate concentration dependence can be interpreted as general base catalysis; however, the terms for the solvent obtained by extrapolation to zero phosphate require a more complex interpretation. The spectrophotometric curves showed that at lower pH the reaction products were mixtures, which may be responsible for the complexity of the solvent term.

By assuming that the phosphate terms, at least, are due to general base catalysis leading to the principal product, we may write the equation:

$$\text{p-HN}{=}\text{C}_6\text{H}_4{=}\text{N}^+\text{Me}_2 + \text{H}^+ + \text{SO}_3^{2-} \rightleftharpoons [\text{H}_2\text{N–C}_6\text{H}_4(\text{SO}_3^-)(\text{H}\cdots\text{Base}){=}\text{N}^+(\text{C}{\equiv}(\text{H})_3)_2] \longrightarrow \text{H}_2\text{N–C}_6\text{H}_3(\text{SO}_3^-)\text{–NMe}_2 \qquad (42)$$

The structure of the intermediate is drawn as that required by the work of Bauer, Meyer, and Ulbricht[36,37], but as pointed out earlier (see Section II.B.3) there is evidence that this may not be the only product. The assumption of the protonated intermediate facilitates the explanation of different orientation of the sulpho group for the diethyl and dimethyl derivatives, because the difference between the two positions decreases after protonation and minor variations could alter the relative stability of the two possible intermediates (i.e. that

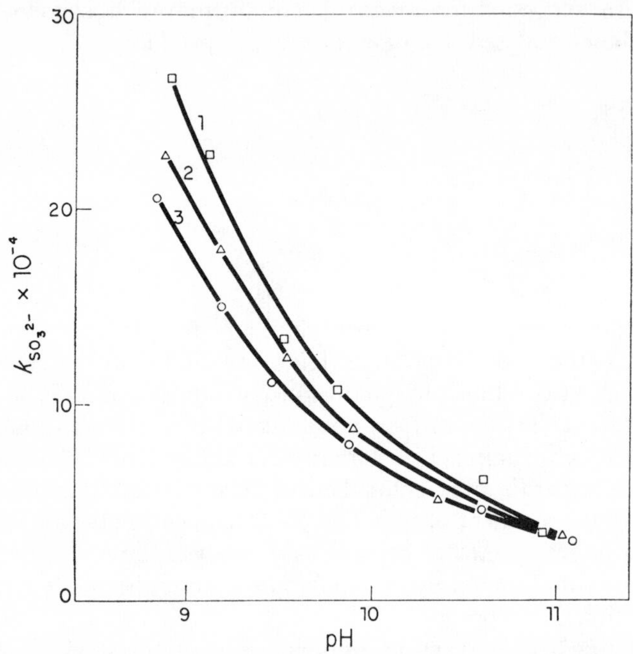

FIGURE 6. Sulphonation rate–pH profiles for *N,N*-dimethylquinonediimine at various phosphate buffer concentrations. 1. μ(phosphate) = 0·375; 2. μ(phosphate) = 0·1875, μ(KCl) = 0·1875; 3. μ(phosphate) = 0·0938, μ(KCl) = 0·2812.

shown in equation (42) and its isomer analogous to **41**). The difference in orientation can be rationalized by saying that steric hindrance of the ethyl groups prevents sulphonation at the position *ortho* to the tertiary nitrogen but additional stabilization by hyperconjugation and reduction of steric hindrance in the case of the dimethyl derivative favour sulphonation at this position. Without assuming the protonated intermediate, it is difficult to account for this selectivity in orientation because the quinonediimine molecule is electronically unsymmetrical with respect to the imino groups.

The unsubstituted quinonediimine, in contrast to the *N,N*-dialkyl derivatives, reacts with sulphite at rates having first and second-power

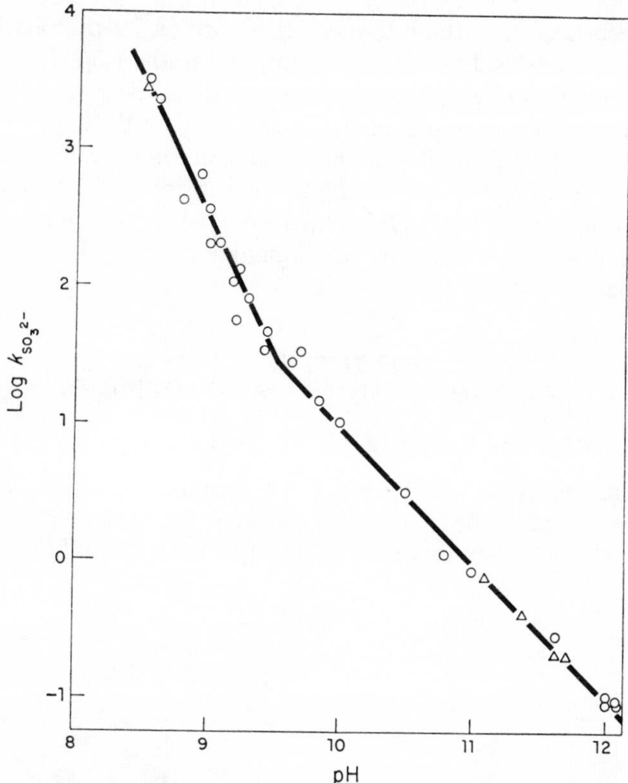

FIGURE 7. Sulphonation of quinonediimine at various pH's.

dependence on $[H^+]$ as shown in Figure 7. The kinetic equation contains no phosphate term:

$$\frac{-\mathrm{d}[\mathbf{T}]}{\mathrm{d}t} = (k_{OH^-}[OH^-] + k_{H_2O})K_I[H^+]^2[SO_3^{2-}][\mathbf{T}^+]$$

K_I is the equilibrium constant for the formation of the intermediate **42**, k_{OH^-} and k_{H_2O} are the rate constants for its catalytic decomposition by OH^- and H_2O respectively. Equation (43) is consistent with the scheme:

NH=C₆H₄=NH $+ 2H^+ + SO_3^{2-} \rightleftharpoons$ (**42**) $\xrightarrow[\text{pH} > 9{\cdot}5]{\text{catalysis by } OH^-}$ 2,5-diaminobenzenesulphonate (NH_2, SO_3^-, NH_2) (43)

(**42**): H_2N^+=, SO_3^-, H, NH_2 — catalysis by H_2O, pH < 9·5 → product which postulates that the intermediate is formed by protonation on both nitrogens as well as addition of sulphite to the ring. In this case there is no question as to the orientation of the sulpho group because of symmetry. The absence of phosphate catalysis on the deprotonation step requires a large β in the Brønsted equation for general base catalysis $k_\beta = G_B K_B^\beta$, so that at high concentration of phosphate the catalysis was dominated by OH^-; at low pH, where the OH^- catalysis was suppressed, the reaction was dominated by H_2O because the weaker bases were too unreactive.

III. THE CHEMISTRY OF *N,N'*-QUINONEMONO- AND -DIIMIDES

A. Preparation and Reactions

The instability of quinonediimines prompted Adams and his students to examine the chemistry of the analogous quinoneimides (**43**) and (**44**)[58,59]. All of this work, reported in some 50 publications, is

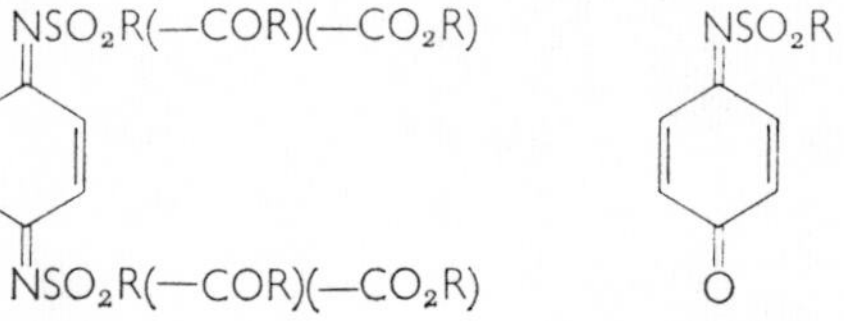

$NSO_2R(—COR)(—CO_2R)$ / $NSO_2R(—COR)(—CO_2R)$ (**43**) NSO_2R / O (**44**)

reviewed definitively[9] and only a brief selection of major points will be recorded here.

In contrast to quinonediimines the quinonediimides were isolated, recrystallized, analysed, and generally handled as ordinary stable organic molecules. For simplicity only the benzoquinonedisulphonimide will be shown with the understanding that a large variety of other

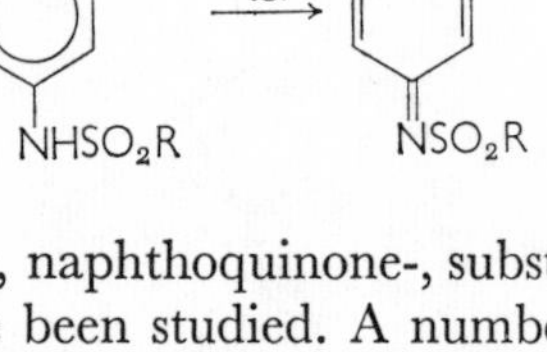

$$p\text{-}(NHSO_2R)_2C_6H_4 \xrightarrow{[O]} RSO_2N{=}C_6H_4{=}NSO_2R \qquad (44)$$

imides (e.g. benzimides, naphthoquinone-, substituted benzoquinonemonoimides, etc.) have been studied. A number of oxidizing agents were tried and found to be more or less useful, but the vast majority of the quinonediimides were prepared using lead tetraacetate in acetic acid. More sensitive compounds, e.g. acylquinonediimides, and those which reacted with acetic acid were prepared by oxidation in benzene or chloroform. Investigators in another laboratory have also prepared some compounds of these types[60,61].

The reactions of the quinonediimides may be divided into two broad classes—reduction and addition. Those reagents commonly recognized to be reducing agents, e.g. hydrogen and catalysts, zinc and acetic acid, hydriodic acid, etc. reduced the quinonediimides rapidly and quantitatively to the corresponding amides, i.e. the reverse of equation (44). In a number of cases reagents which are not usually thought of as reducing agents did reduce the quinonediimides; presumably because the latter are strong oxidizing agents. Examples of such reducing agents are aqueous sodium hydroxide and dilute sulphuric acid.

A wide variety of compounds were added to the quinonediimines as indicated by equations (45) to (51).

NSO_2R / NSO_2R (quinonediimide)

+ HCl ⟶ $NHSO_2R$, Cl, $NHSO_2R$ (45)

+ RCO_2H ⟶ $NHSO_2R$, O_2CR, $NHSO_2R$ (46)

+ ROH ⟶ $NHSO_2R$, OR, $NHSO_2R$ (47)

In most cases it was possible to reoxidize the substituted *p*-diamide product and make a second addition of either the same or a new reagent.

Certain special cases are of sufficient interest to point out individually. The mercaptans which are good reducing agents showed this

NSO$_2$R / NSO$_2$R

+ PhSO$_2$H ⟶ NHSO$_2$R, SO$_2$Ph, NHSO$_2$R (48)

+ RSH ⟶ NHSO$_2$R, SR, NHSO$_2$R (49)

+ CH$_2$(X)(Y) (active methylene compounds) ⟶ NHSO$_2$R, CH(X)(Y), NHSO$_2$R (50)

+ CH$_2$=CH–CH=CH$_2$ ⟶ NHSO$_2$R, NHSO$_2$R (51)

property, especially with imides which were sterically crowded. The addition could be promoted by the use of acidic and basic catalysts (to be discussed later).

RSSR + NHSO$_2$R, NHSO$_2$R $\xleftarrow{RSH}$ NSO$_2$R, NSO$_2$R $\xrightarrow{RSH}$ NHSO$_2$R, SR, NHSO$_2$R (52)

One group of reagents showed the interesting property of adding smoothly to the naphthoquinonediimides, but of either reducing or giving complicated mixtures with benzoquinonediimides. Grignard reagents, nitroalkanes, and hydrogen cyanide are examples of such compounds. Investigators in another laboratory have also studied the addition of Grignard reagents to quinonediimides and made similar observations[62–64].

$$\text{1,4-naphthoquinone bis(}N\text{-phenylsulfonyl)imine} + RMgX \longrightarrow \text{2-R-1,4-bis(}NHSO_2Ph\text{)naphthalene} \quad (53)$$

$$\text{1,4-benzoquinone bis(}N\text{-phenylsulfonyl)imine} + RMgI \longrightarrow \text{1,4-bis(}NHSO_2Ph\text{)benzene (after hydrolysis)} \quad (54)$$

The reactions of amines with the quinonediimides is complicated and very sensitive to reaction conditions and the class of amine being added. A few examples will serve to illustrate this.

$$\text{1,4-benzoquinone bis(}N\text{-phenylsulfonyl)imine} + \text{morpholine} \xrightarrow{Et_2O} \text{2-}N(CH_2CH_2)_2O\text{-1,4-bis(}NHSO_2Ph\text{)benzene } \mathbf{(45)} + \text{some } \mathbf{46}$$

$$\xrightarrow{C_6H_6} \text{2,5-bis[}N(CH_2CH_2)_2O\text{]-1,4-bis(}NHSO_2Ph\text{)benzene } \mathbf{(46)} + \text{some } \mathbf{45} \quad (55)$$

If chloroform was used as the solvent, no **45** was obtained. The complex product mixture contained **46** along with at least the following:

$$\text{1,4-bis(}NHSO_2Ph\text{)benzene} + \text{4-}N(CH_2CH_2)_2O\text{-}C_6H_4\text{-}NHSO_2Ph + PhSO_2NH_2 + \text{2,5-bis[}N(CH_2CH_2)_2O\text{]-1,4-benzoquinone bis(}N\text{-phenylsulfonyl)imine}$$

Aliphatic primary amines all seemed to show 1,2 addition followed by elimination of benzenesulphonamide.

$$\text{(1,4-}NSO_2Ph\text{ quinonediimide)} + 2\,n\text{-}BuNH_2 \longrightarrow \left[\text{(1,4-}NBu\text{-}n\text{ quinonediimide)}\right] + 2PhSO_2NH_2 \xrightarrow[H_2]{PtO_2} \text{(1,4-}NHBu\text{-}n\text{ benzene)} \qquad (56)$$

The aromatic amines followed yet another reaction path, as shown in equation (57):

$$\text{(1,4-}NSO_2Ph\text{ quinonediimide)} + PhNH_2 \longrightarrow \text{(2,5-}NHPh\text{-1,4-}NPh\text{ quinonediimide, major product)} + PhSO_2NH_2 + \text{(1,4-}NHPh\text{ benzene)} \qquad (57)$$

The 1,2-benzo- and naphthoquinonediimides were also investigated and, except for differences caused by their special structural arrangement, showed chemistry very similar to that described for the 1,4 cases.

B. The Mechanism of Ring Addition to Quinonediimides

The extensive synthetic studies cited in the preceding section do not lend themselves to rigorous mechanistic interpretation. In spite of this we feel that some qualitative remarks should be made. These remarks are purely speculative and are intended solely as tenuous hypotheses for anyone who might undertake additional studies of these reactions.

Adams and his students studied the quinonediimides with the expectation that the compounds would possess greater stability while still showing much of the potentially interesting chemistry of the quinonediimines[58]. The observed reactivity towards a wide variety of nucleophiles suggests that these reactions are related to the large body of 1,4 addition–elimination reactions typical of the α,β-unsaturated carbonyl group. This point of view is further substantiated by the observed acid and base catalysis. A considerable amount of work has been published concerning the mechanism of such reactions and has been reviewed recently[65,66]. In general, we should recall that at least two steps are involved, both of which are reversible and either of which may be rate determining. These facts clearly illustrate the impossibility of drawing mechanistic conclusions from yield data. Until the equilibria have been studied and the rate-determining step identified, we must content ourselves with pointing out those observations which pose questions which the eventual mechanism must satisfactorily answer.

Before looking at the usual 1,4 ring addition reactions we should examine the cases of 1,2 addition followed by loss of sulphonamide. These reactions might be considered analogous to the deamination reaction of the quinonediimines. The greater stability of the imides may, in part, account for the limited number of such reactions[67–69]. As shown in equations (58) and (59), hydrolysis was observed only with polycyclic aromatic or highly substituted systems.

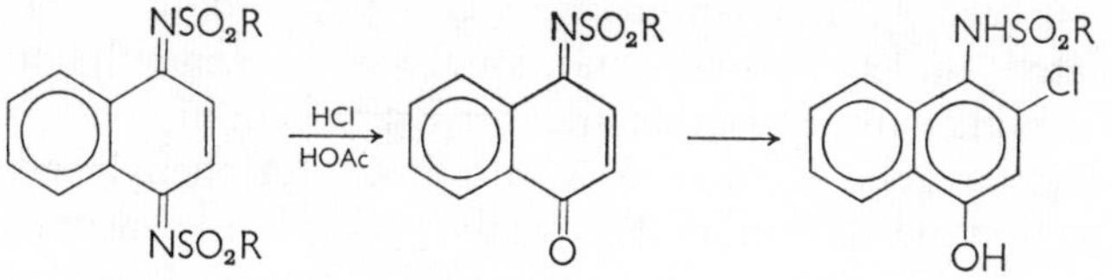

NSO₂R; HCl, HOAc; NSO₂R; O; NHSO₂R; Cl; OH (58)

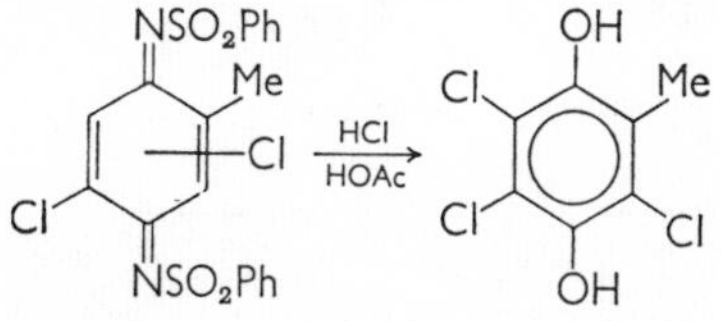

NSO₂Ph; Me; Cl; Cl; NSO₂Ph; HCl, HOAc; OH; Cl; Me; Cl; Cl; OH (59)

The latter reaction (equation 59) was also shown to result in only monodeamination under slightly different conditions[69].

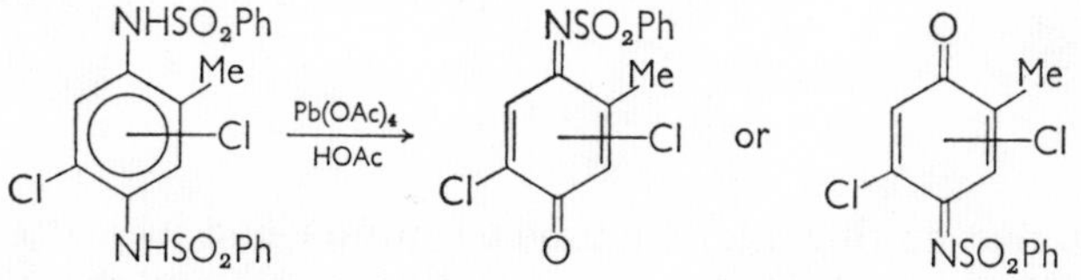

NHSO₂Ph; Me; Cl; Cl; NHSO₂Ph; Pb(OAc)₄, HOAc; NSO₂Ph; Me; Cl; Cl; O; or; O; Me; Cl; Cl; NSO₂Ph (60)

The acidic conditions required for these reactions indicate that the mechanism might be very similar to that which we have found with quinonediimines except that the uncatalysed rate with water is very slow.

NSO_2Ph (ring) NSO_2Ph + HCl ⇌ $^{+}NHSO_2Ph$ (ring) NSO_2Ph + Cl^- $\xrightleftharpoons{HOH}$ $PhSO_2HN$, OH (ring) $PhSO_2HN^+$ + Cl^- ⟶

$PhSO_2NH_2$ + O (ring) $PhSO_2HN^+$ + H^+ $\xrightleftharpoons{HOH}$ ^{+}OH (ring) $PhSO_2HN$, OH ⟶ O (ring) O (61)

In reaction sequence (61) the other ring substituents have been omitted for simplicity.

Like the quinonediimines, the quinonediimides show a basic deamination reaction. This was observed in the reactions of amines and is therefore more properly called a transamination. Perhaps the clearest example is in the case of n-butylamine, where the only product which could be isolated under normal conditions was benzenesulphonamide[70]. By carrying out the reaction under reducing conditions it was possible to isolate a good yield of *N,N′*-di-n-butyl-1,4-phenylenediamine, showing that the alkylimine (**47**) was the probable intermediate, and suggesting a 1,2 addition across the azomethine bonds followed by elimination of the weaker base.

NSO_2Ph (ring) NSO_2Ph + 2 n-$BuNH_2$ ⇌ $PhSO_2NH$, NHBu-n (ring) $PhSO_2NH$, NHBu-n ⟶

NBu-n (ring) NBu-n (**47**) + $2PhSO_2NH_2$ $\xrightarrow{[H]}$ NHBu-n (benzene ring) NHBu-n (62)

The fact that no ring addition products were found indicates that the transamination step is fast with respect to addition and that the *N,N′*-dialkylquinonediimine decomposes rapidly in accordance with Willstätter's observations[1,2,59].

The variety of products which are found when aniline reacts with quinonediimides in a non-polar solvent (see Section III.A[67]) can also be understood in terms of the above discussion. The scheme is clearly presented by Adams as a series of 1,2 and 1,4 additions[9]. The greater stability of the *N,N'*-diarylquinonediimines, reported by Willstätter, also makes this an attractive explanation of the experimental results.

Only a few amine additions were carried out in acidic media, but they are of special interest in view of the differences in product composition brought about by small changes in structure and/or reaction media and the possibilities for fruitful study they suggest. The case of *o*-toluidine is probably close to the border line in that the absence of acid produced only transamination[71].

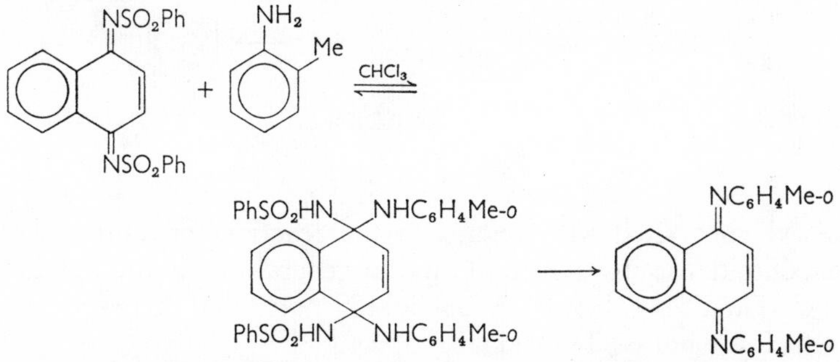

(63)

When the same reaction was carried out in acetic acid, ring addition became competitive with transamination and the product shown in equation (64), or the other ring isomer, was obtained.

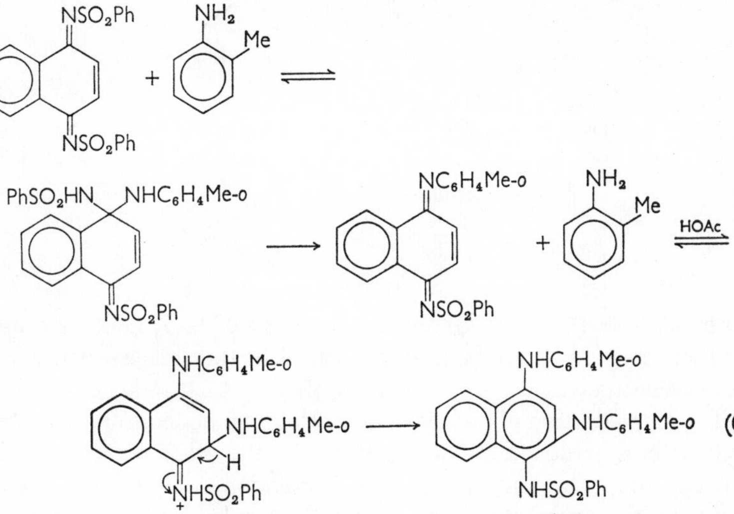

(64)

(48)

The structure which Adams thought most probable sterically (**48**) would also be the most likely on mechanistic grounds, assuming that the 1,2 addition and loss of benzenesulphonamide occurred first.

The addition of aniline[67], morpholine[67], methylaniline[71] and *p*-toluidine[71] to 1,4-naphthoquinonedibenzenesulphonimide in acetic acid also appear to be consistent with an acid-catalysed 1,4 addition. In all of these cases good yields of the 2-adduct were obtained, e.g. equation (65).

NSO$_2$Ph / NSO$_2$Ph + PhNH$_2$ $\xrightleftharpoons{\text{HOAc}}$ $\overset{+}{\text{N}}$HSO$_2$Ph, H, NHPh / NHSO$_2$Ph ⟶ NHSO$_2$Ph, NHPh / NHSO$_2$Ph (65)

The evidence cited with regard to the reactions of amines with quinonediimides seems to fit quite well a general mechanistic picture of carbonyl additions. However, it should be noted that there are other data which are not easily rationalized, e.g.[71,72]

NSO$_2$Ph / NSO$_2$Ph + piperidine (H, N) $\xrightarrow{\text{CHCl}_3}$ NHSO$_2$Ph, NC$_5$H$_{10}$ / NHSO$_2$Ph (66)

NSO$_2$Ph / NSO$_2$Ph + PhNHMe $\xrightarrow{\text{C}_6\text{H}_6}$ NHSO$_2$Ph, C$_6$H$_4$NHMe-*p* / NHSO$_2$Ph (67)

Both of these reactions give yields in excess of 90% and the reasons for their course will not be answered until further mechanistic studies have been made.

The ring addition of hydrogen chloride to quinonediimides has been studied in a greater number of cases than the other reagents. This is due to the ease and scope of the reaction[9]. Hydrogen chloride serves as its own catalyst even in such non-polar solvents as benzene.

In fact, some of the more reactive quinonimides reacted with hydrochloric acid in a heterogeneous mixture[59].

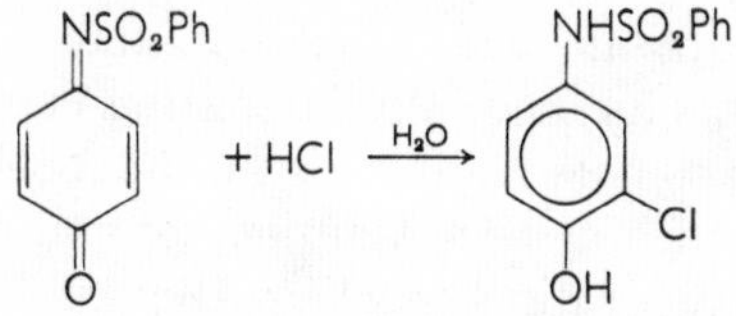

(68)

It seems reasonable to expect the following pathway to obtain:

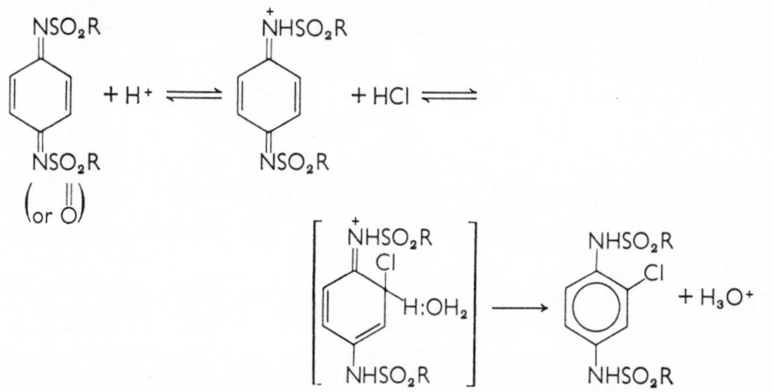

(69)

The organic acids and alcohols also add to quinonediimides, but are much less reactive as is illustrated by the fact that they are useful reaction and recrystallization solvents. The presence of strong Lewis acids such as boron trifluoride, hydrogen fluoride and concentrated sulphuric acid (especially the first of these) leads to smooth addition in good yield[73–75].

The action of these catalysts is typical of the effect of acids on reactions involving two reversible steps[65]. Whereas the reaction in strong acid leads to a high concentration of protonated intermediate, the Lewis acid may facilitate the removal of a nuclear proton in the product forming step:

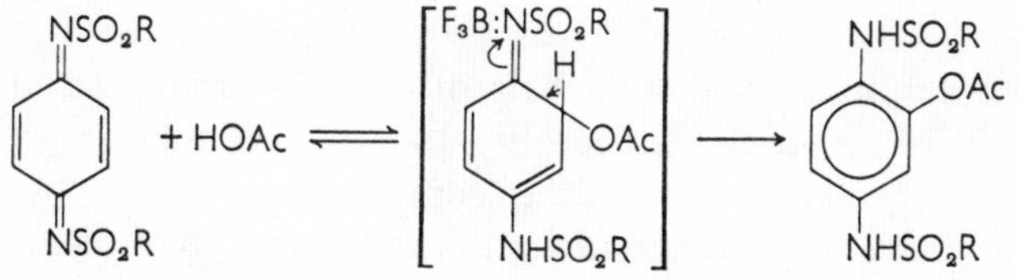

(70)

This would also be consistent with the pre-protonation which we have observed in the sulphonation of quinonediimines (see Section II.D.4).

The addition of organic acids to 1,4-naphthoquinonedibenzenesulphonimide represents an especially interesting case. The Lewis acids which effectively promoted addition in the benzoquinonediimides failed, while bases such as triethylamine, sodium acetate, and potassium cyanide were very satisfactory[76,77]. General base catalysis is also frequently observed in carbonyl additions[65]. In this case it may also involve such factors as steric hindrance by the naphthalene *peri*-hydrogens or a smaller tendency to reform the aromatic system, resulting in a stronger carbon–hydrogen bond which must be broken in the product-forming step:

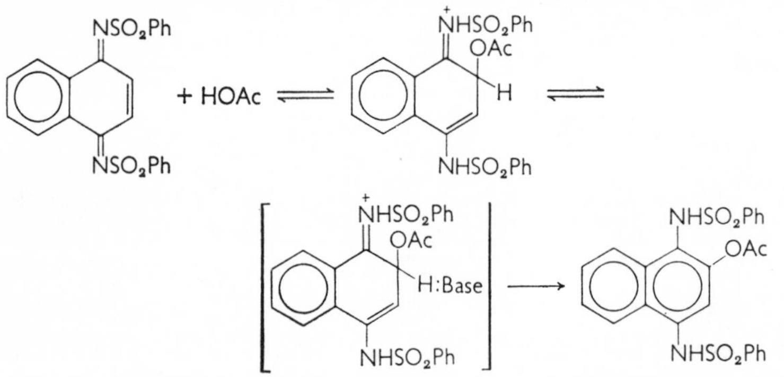

NSO₂Ph + HOAc ⇌ [N⁺HSO₂Ph, OAc, H, NHSO₂Ph] ⇌ [N⁺HSO₂Ph, OAc, H:Base, NHSO₂Ph] ⟶ NHSO₂Ph, OAc, NHSO₂Ph (71)

Studies of the addition of mercaptans provide additional evidence for the reasoning used in the above examples of general acid and base catalysis. Using thiophenol as a typical example, it is found that both acidic and basic catalysts produce good yields of ring addition[78]. Thus the strong acid catalyst leads to a high concentration of intermediate **49** in path (72a), while the weak acid thiophenol produces a low concentration of the same intermediate in path (72b). This latter circumstance is offset by the presence of a basic catalyst which facilitates proton removal in the activated complex **50** and both systems have overall rates which lead to practical synthetic routes.

Benzenesulphinic acid, which is of course a much stronger acid ($pK_a = 1{\cdot}21$[79]), adds smoothly to quinonimides without a catalyst[76,80,81]. Thus the cycle is completed and we return to the situation which was postulated for the addition of hydrogen chloride.

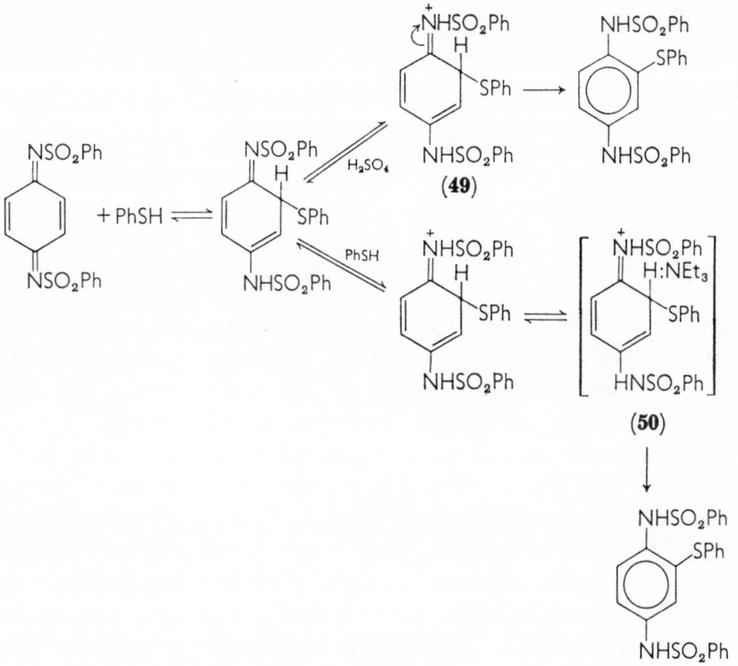

(72a)

(72b)

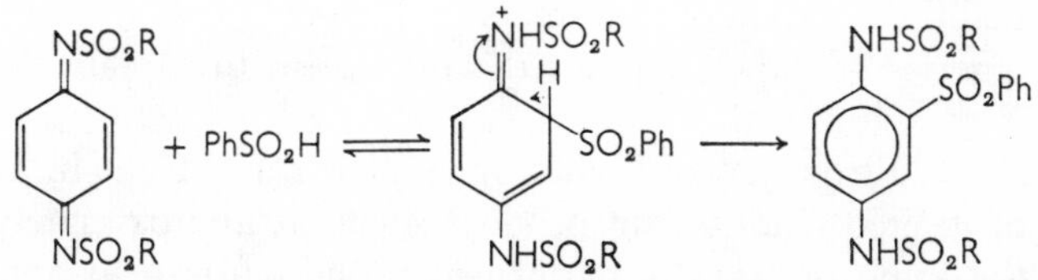

(73)

Two other types of data useful for mechanistic considerations are provided by these product studies of the reactions of quinonediimides—the effect of different classes of *N*-substituents and the orientation influence of the first ring substituent on the second entering group. In the former there is just sufficient data to suggest that further study would be profitable and the latter has been well reviewed and discussed[9].

IV. QUINONEMONOXIMES

A. Preparation and Tautomerism

One of the earliest observations of tautomeric equilibrium was that of benzoquinonemonoxime and *p*-nitrosophenol[82]. The two common methods of preparing these compounds both produce the same product mixture (equation 74).

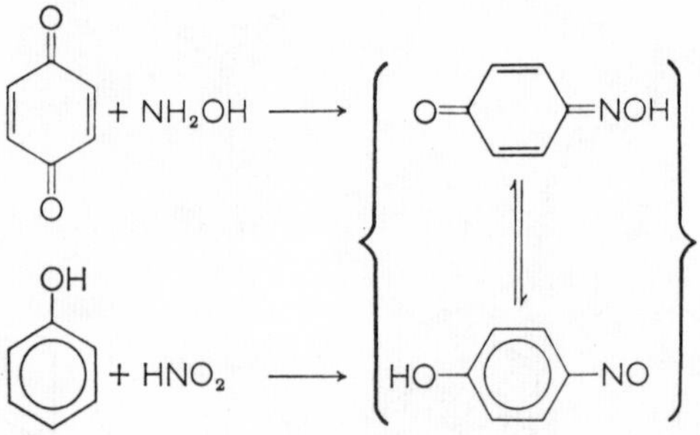

$$\text{benzoquinone} + NH_2OH \longrightarrow \left\{ \begin{array}{c} O{=}C_6H_4{=}NOH \\ \updownarrow \\ HO{-}C_6H_4{-}NO \end{array} \right\} \longleftarrow \text{phenol} + HNO_2 \qquad (74)$$

This system has been the subject of a great amount of study since the early part of the twentieth century. Extensive early studies of Hodgson[83,84] are well summarized in the literature[85]. They also provided data which led to some interesting recent research.

In 1923 Hodgson and Moore reported studies which gave chemical evidence that they had succeeded in isolating 3-chlorobenzoquinone-4-oxime and 3-chloro-4-nitrosophenol as individual compounds[83]. A number of reinvestigations have been made of this problem and the intriguing possibilities presented by it[86–92]. These latter studies have generally confirmed Hodgson's conclusions concerning the tautomerism of most quinonemonoximes with *p*-nitrosophenols, but strongly questioned the isolation of tautomers in the 3-chloro case. In the early 1950's a series of papers appeared which seem to explain this difficulty[10,93–96].

When the nitrosation of 3-chlorophenol is carried out in aqueous media of limited acidity, the product melts at approximately 140°. When the same reaction is performed in sulphuric acid a product melting at 184° is obtained. These are the materials assumed to have benzenoid (**51**) and quinonoid (**52**) structures respectively[80]:

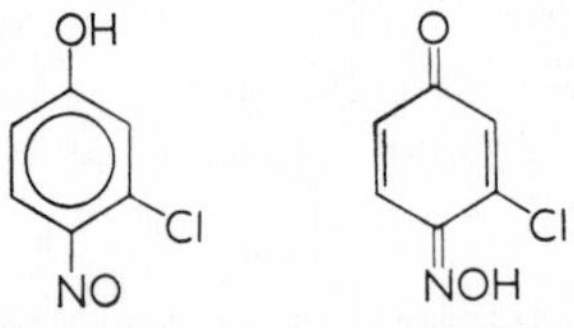

OH, Cl, NO — **(51)** 140°

O, Cl, NOH — **(52)** 184°

The low-melting product is unstable to light and can be transformed into the higher by treatment with acid or alkali[93]. All the evidence indicates that the high-melting material is a homogeneous substance which in solution is an equilibrium mixture of benzenoid and quinonoid tautomers[10,83–94]. The spectra of the low-melting material at various pH's indicate that it is a mixture. The application of counter-current distribution showed the presence of at least three components[94]. The major fraction was identical with the high-melting substance or tautomeric mixture.

Using column chromatography it was possible to obtain larger amounts of the components of the low-melting mixture[93]. Consideration of the possibility of nitrosation at the 6-position (and the product's sensitivity to light) led to the suggestion that one of the additional components might be 3-chloro-6-diazocyclohexadienone (**53**). This compound was prepared by an independent route and proved to be

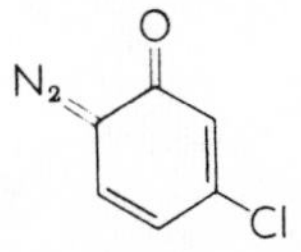

(53)

identical with that isolated from the nitrosation reaction mixture. Its marked sensitivity to light suggests that the one or more additional compounds found in counter-current separations are decomposition products. The implied diazotization of a nitroso group is well supported by the literature[97] and by the group reporting this finding[95].

B. The Beckmann Rearrangement

Ernst Beckmann himself attempted the rearrangement of benzoquinonemonoxime under rather unusual basic conditions[98].

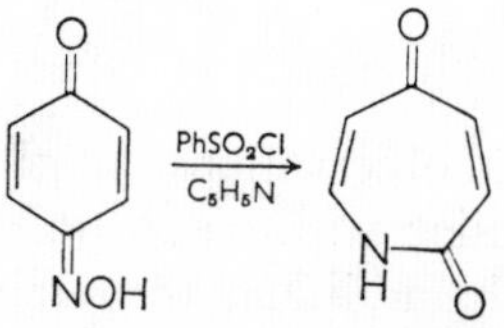

$$\xrightarrow[C_5H_5N]{PhSO_2Cl} \qquad (75)$$

(54)

The product obtained gave an acceptable analysis and some chemical evidence that it was the expected product (**54**). Because some of the tautomeric forms of this compound can be considered aza-γ-tropolone (equation 76), Leonard reinvestigated its structure[99].

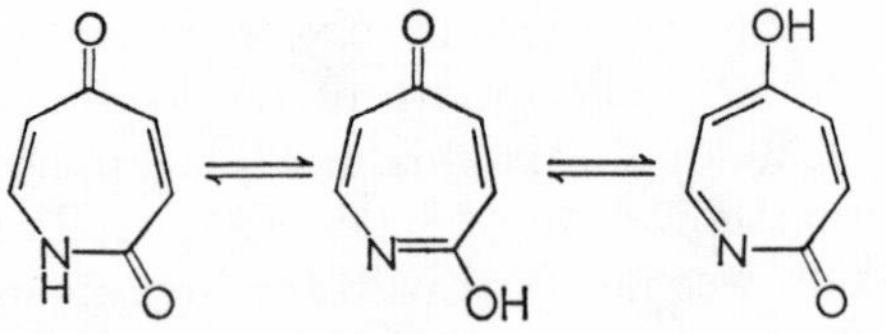

(76)

Chemical and spectral evidence obtained by these workers together with comparison with an authentic sample established the structure of the product as 4-azoxyphenol (**55**). The analytical and molecular weight data of Beckmann, which cannot fit the true product **55**, are regarded as being either fortuitous or in error. All the chemical observations are equally well understood for either **54** or **55**. In efforts to optimize the yield of product a number of significant observations

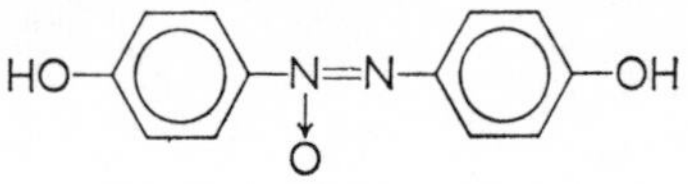

(**55**)

were made which led to the following proposed reaction pathway:

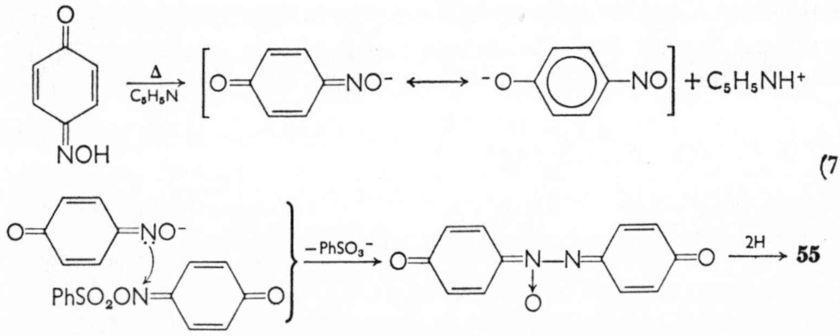

(77)

Concurrent with these observations, another laboratory published exactly the same conclusion with regard to the product from benzoquinonemonoxime[100]. In addition, they reinvestigated Beckmann's work with anthroquinonemonoxime and concluded that this did in fact undergo the usual rearrangement. As is indicated (equation 78), the product **56** underwent hydrolysis to a compound which is easily understood in terms of the expected Beckmann rearrangement precursor. Compound **56** acted as a normal lactam and showed no evidence of tropolone character.

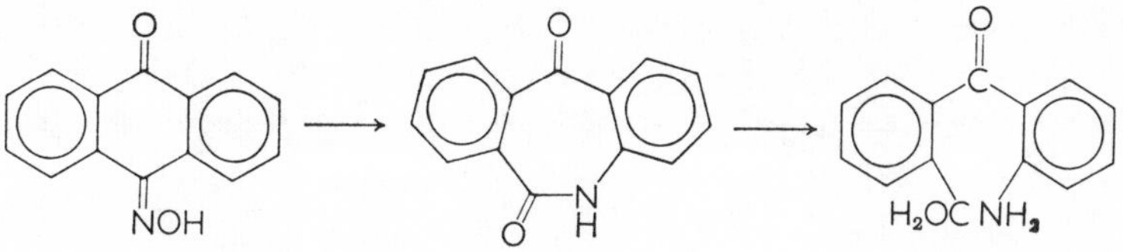

(78)

(56)

C. *Brief Reports Involving Quinone Oximes*

There are in the literature several interesting, but very limited studies, which will be cited with only brief discussion.

1. Nitrosation of primary aromatic amines

Diazotization with a solution of $NaNO_2$ in concentrated sulphuric acid, while not a general reaction, has been useful in certain cases. This reagent with a number of primary aromatic amines, e.g. *m*-toluidine, *m*-anisidine, 1-naphthylamine-2-sulphonic acid, etc., results in nitrosation[101]. These products, e.g. **57**, are in tautomeric equilibrium with the corresponding quinonimineoximes and this in many

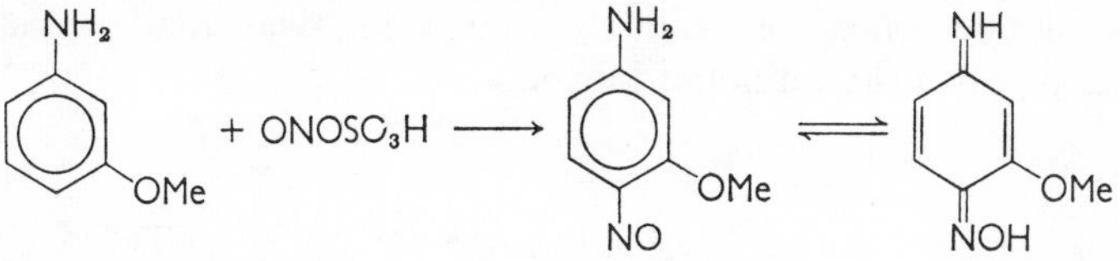

$$\text{+ ONOSO}_3\text{H} \longrightarrow \rightleftharpoons \qquad (79)$$

(57)

cases is the preferred route for their synthesis.

2. Inner complexes

Phenanthrenequinonemonoxime and related compounds were found to give extremely stable complexes with a variety of metallic ions, e.g. Ni, Cd, Cu, etc.[102] Spectral characteristics for these complexes are reported[103].

3. Hydrolysis

The preparation of substituted quinones by hydrolysis of the appropriate nitrosophenol is an attractive synthetic route since the latter can be prepared in good yield. Three methods of hydrolysis were examined, with that shown in equation (80) appearing most useful[104]. Some observations of mechanistic interest were made: (a) leaving out the acetone reduced the yield slightly while omission of the cuprous oxide resulted in a very low yield; (b) steric hindrance of both the

$$\text{R-C}_6\text{H}_3(\text{NO})(\text{OH}) \rightleftharpoons \text{R-quinonemonoxime} + \text{Me}_2\text{CO} + \text{Cu}_2\text{O} \xrightarrow[\text{cellosolve}]{\text{methyl-}} \text{R-quinone} \quad (80)$$

nitroso and hydroxyl groups made the hydrolysis difficult while partial hindrance of the nitroso group alone seemed to slightly favour reaction.

4. Reactions of diazomethane with 4-nitrophenols

With the parent compound, 4-nitrophenol, diazomethane gave only the expected ether[105].

$$\text{HO-C}_6\text{H}_4\text{-NO}_2 + \text{CH}_2\text{N}_2 \longrightarrow \text{MeO-C}_6\text{H}_4\text{-NO}_2 \quad (81)$$

However, reaction with 10-nitro-9-anthrone gave a 60% yield of anthraquinonemonoxime, and the suggestion was made that the reaction involves the quinonoid form.

$$\text{10-nitro-9-anthrone } (\text{O}_2\text{N}, \text{H}) + \text{CH}_2\text{N}_2 \xrightarrow[\text{C}_6\text{H}_6]{\text{dry Et}_2\text{O}} \text{anthraquinone monoxime (NOH)} \quad (82)$$

5. Imines of anthrone

In the course of a brief study of the stability of imines of anthracene a novel reaction involving benzoquinonemonoxime was observed[106].

$$\text{anthrone} + \text{benzoquinonemonoxime (NOH)} \xrightarrow{\text{C}_5\text{H}_5\text{N}} \text{anthraquinone imine (NC}_6\text{H}_4\text{OH-}p) \quad (83)$$

6. Molecular orbital discussion of the nitrosophenol quinonemonoxime tautomeric equilibrium

Using a perturbation method the difference in π-electron energy between the two tautomers was calculated and their relative stabilities estimated[107]. In the *para* case the quinonoid form is predicted to be favoured by 4–6 kcal/mole, in general agreement with experimental observations.

7. Polarography of quinone oximes

A study of the polarographic reduction of quinonemono- and -dioximes over a broad range of acidities has been reported[108]. The four-electron reduction of the monoxime takes place in one step (equation 84). Therefore the direct path through the quinonimine is

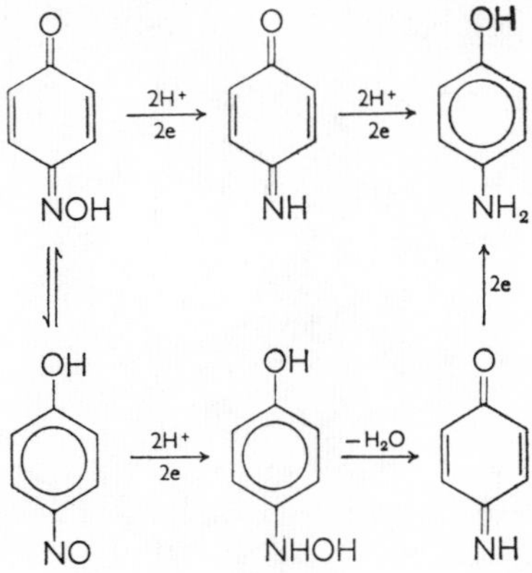

(84)

(58)

favoured over the alternative involving 4-hydroxylphenylhydroxylamine (**58**). This is by analogy with the reduction of 4-nitrophenol which proceeds in two steps through intermediate **58**.

The same general scheme accounts for the reduction of the quinonedioximes, with the additional complication that in neutral media a kinetic wave is observed which requires an isomerization at the electrode prior to reduction. It is suggested that nitrone tautomer **59** may be involved. In both acid and alkaline solution the isomerization is thought to be so rapid that a one-step process is observed. The work of Ramart–Lucas[90] is cited as offering some evidence for the existence of the nitrone. However, requirement of a third form (β)[88] no longer appears necessary (see Section IV.A).

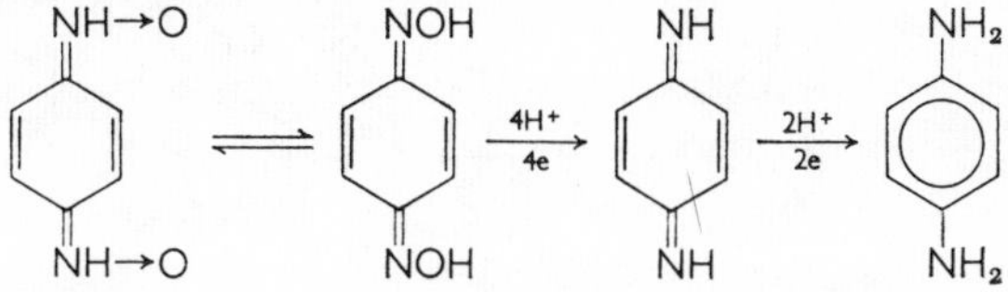

(85)

(59)

V. STUDIES OF MORE LIMITED SCOPE

A. The Diels–Alder Reaction of Quinonimine and Diimine

Unlike the quinones, which have an extensive literature of Diels–Alder reactions[109], only a single paper has appeared describing this reaction of the corresponding imines[110]. The reaction with cyclopentadiene produced only polymer except when catalysed by concentrated hydrochloric acid. This produced the dicyclopentadienequinonimine and -diimine (**60** and **61**) as hydrochlorides. The hydrochloride **60** was very stable and the free imine also showed reasonable

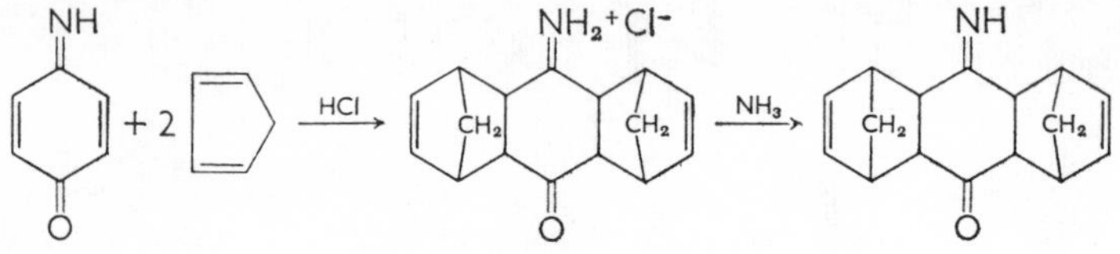

(86)

(**60**)

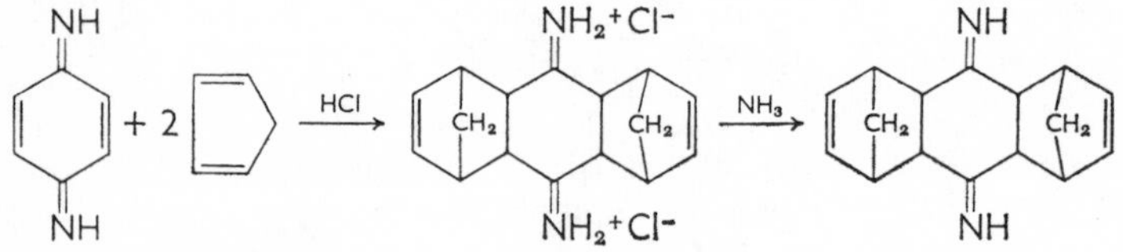

(87)

(**61**)

stability although it did decompose after several months. No comment was made regarding the stability of the diimine adducts. The structures of **60** and **61** were substantiated by analysis and hydrolysis to the known dicyclopentadienequinone.

An effort was made to obtain cyclopentadienequinonimine by reactions with less than two equivalents of cyclopentadiene. There were experimental difficulties associated with the very low yields of product which was only poorly purified, but the weight of evidence indicated that the desired compound **62** was obtained. All experiments directed towards the synthesis of cyclopentadienequinonediimine were unsuccessful.

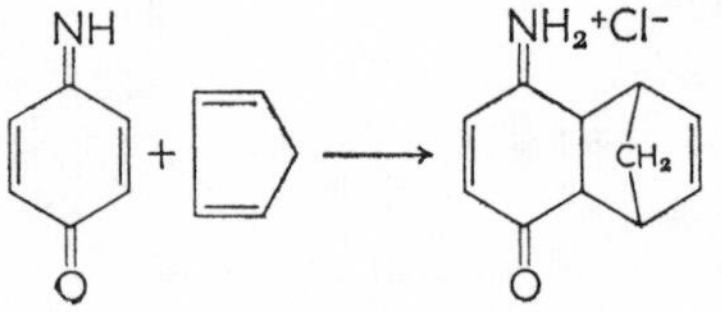

(88)

(**62**)

The catalytic effect of hydrochloric acid was discussed and rationalized in terms of a prior protonation of the quinonimine followed by a two-step cyclization reaction.

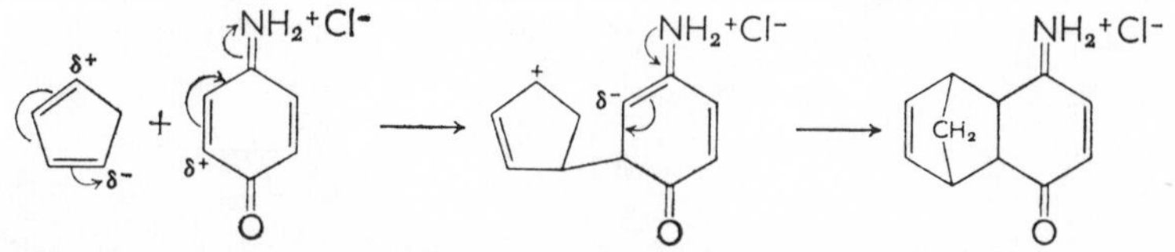

$NH_2^+Cl^-$ $NH_2^+Cl^-$ $NH_2^+Cl^-$ δ+ δ− δ+ δ− CH_2 O O O (89)

More recent studies of the mechanism of the Diels–Alder reaction[111] would tend to support this as a plausible mechanism, although it might be described as unequal bond formation in the transition state rather than two distinct steps.

B. Quinonimines as Intermediates in a Nitrile Cyclization

It was found that in concentrated sulphuric acid the intramolecular cyclization shown in equation (90) took place in good yield[112]. The reaction also took place with simple alkyl groups in place of an aroyl

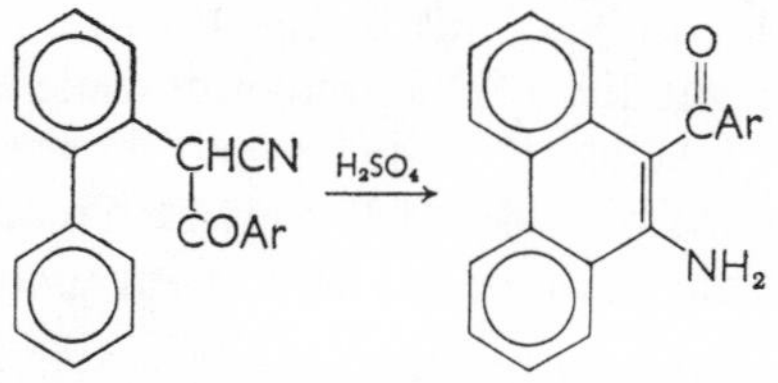

CHCN COAr $\xrightarrow{H_2SO_4}$ CAr O NH_2 (90)

group. The most reasonable mechanism for these reactions appears to involve the conjugate acid and the imine, which rapidly tautomerizes. When tautomerism was prevented by alkyl substitution, the correspond-

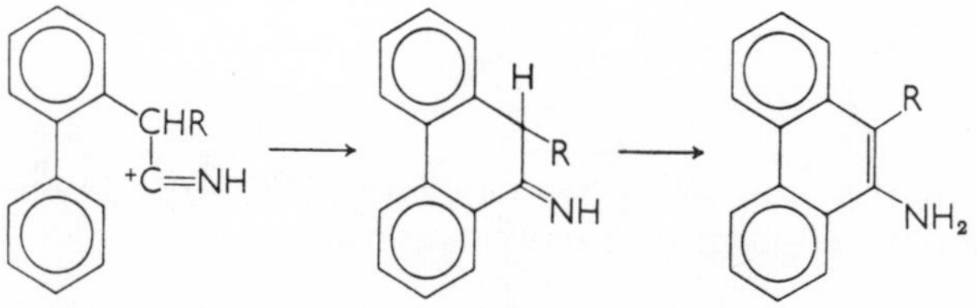

CHR +C=NH ⟶ H R NH ⟶ R NH_2 (91)

ing carbonyl compound **64** was isolated. This can easily be understood as the hydrolysis of the imine **63**; a reaction which is not at all surprising under the reaction conditions.

These studies were later extended to the synthesis of 9-aminoanthracenes by reaction of 2-benzylbenzonitriles[113]. When the reaction of the parent compound (**65**, R = H) was attempted, a large amount

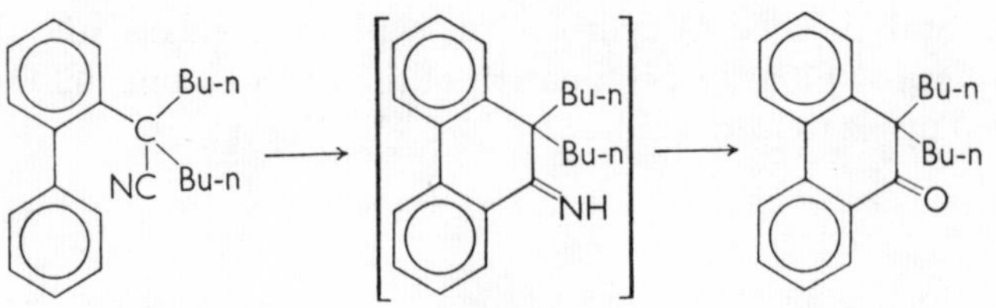

(92)

(63) **(64)**

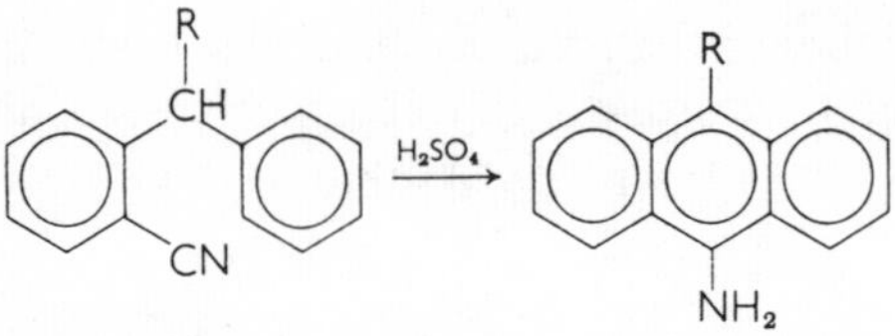

(93)

(65)

of sulphonation took place, but the product mixture gave some evidence of containing the desired amine. After numerous recrystallizations pure anthraquinonimine was obtained, giving clear evidence of the presence of the amine. This is true because it has been demonstrated that the instability of 9-aminoanthracene is due to its ease of autoxidation[114,115]. When R in compound **65** was phenyl or methyl the 9-imino-10-hydroperoxides (**66**) were stable and isolated in good yield. By preventing tautomerization with two substituents, the

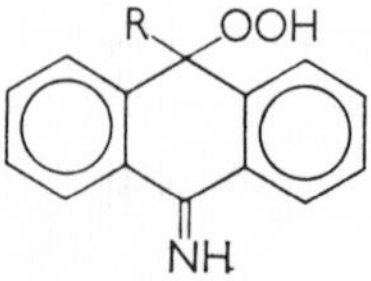

(66)

carbonyl compound **67** (presumably the hydrolysis product of the imino compound formed initially) was isolated.

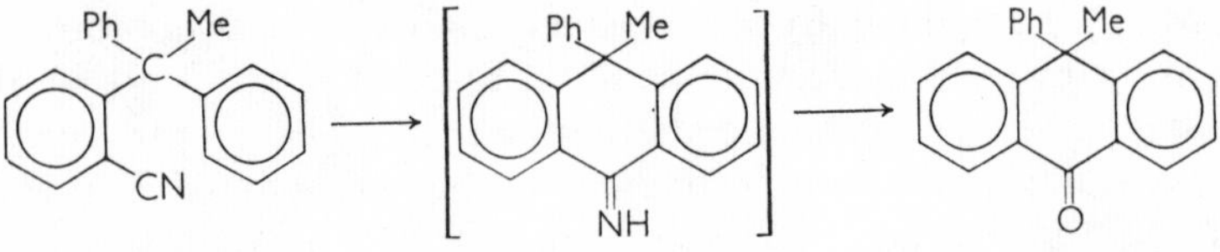

(94)

(67)

C. An Aldol Condensation of Aldehydes and Related Compounds with o-Quinonimines

The formation of 2-phenylphenanthroxazole (**68**) from phenanthraquinone, benzaldehyde and aqueous ammonia was observed nearly 90 years ago[116]. It has stimulated a good deal of experimental work and speculation, but no systematic mechanistic study appeared until 1941. The first of a series of papers provides a comprehensive review of the earlier work in which a quinonimine intermediate was postulated[11]. On the basis of a detailed study of the effect of reactants and

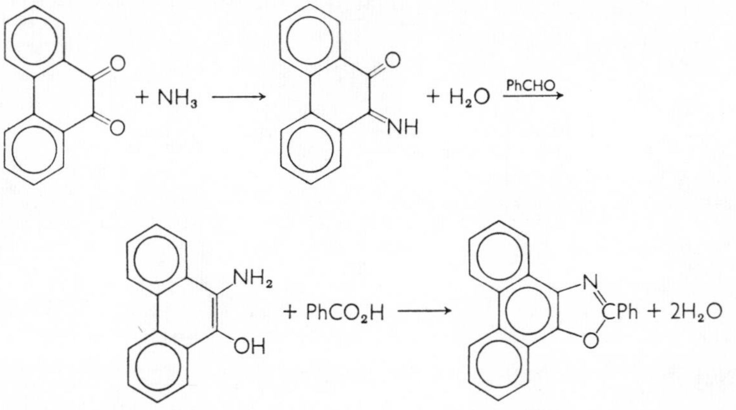

+ NH_3 ⟶ + H_2O $\xrightarrow{PhCHO}$

NH_2 OH + $PhCO_2H$ ⟶ N CPh O + $2H_2O$ (95)

(**68**)

reaction conditions on product yield it was shown that both the quinonimine and hydrobenzamide (**69**) are necessary intermediates. The structure of the retenoxazole (**70**) was not proven, but was considered most probable on the basis of steric influence. A mechanism

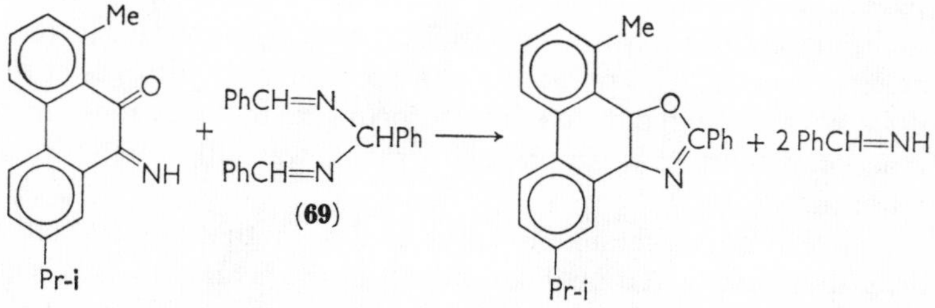

Me O NH Pr-i + PhCH=N, PhCH=N CHPh (**69**) ⟶ Me O CPh N Pr-i + 2 PhCH=NH (96)

(**70**)

was proposed, but was later revised and will be discussed in its more recent form below.

The next extension of this study involved the use of amines as catalysts for cyclization[117]. When retenequinonimine and benzaldehyde react in the presence of n-butylamine, good yields of 2-phenylretenoxazole (**70**) are obtained. It was recognized that two possible mechanisms might obtain: (a) an aldol condensation catalysed by the amine or (b) a preliminary reaction of the aldehyde and amine to form a Schiff base followed by condensation with the quinonimine.

In order to test the first mechanism, amines which cannot form Schiff bases were employed. Triethylamine and piperidine gave excellent yields of 2-phenylretenoxazole while the weaker base, pyridine, gave no oxazole. A mechanism closely related to the usual aldol condensation was proposed (B = base).

$$\text{C(=O)–C(=NH)} + \text{B} \rightleftharpoons [\text{C(=O)–C(=N)}]^- + \text{BH}^+ \xrightleftharpoons{\text{PhCHO}} \text{C(=O)–C(=NCHPhO}^-) \xrightleftharpoons{\text{BH}^+} \text{C(=O)–C(=NCHPhOH)} \rightleftharpoons \text{C(OH)=C(N=CPhOH)} \longrightarrow \text{C–O–CPh=N–C (oxazole)} + H_2O \qquad (97)$$

The irreversibility of the final step is well supported by earlier demonstration of great difficulty in hydrolysis of the oxazole[11] and accounts for the high yields of products. Reactions of phenanthraquinonimine showed similar results.

The second possible reaction pathway for primary amines, i.e. preliminary reaction with the aldehyde to form a Schiff base, was also studied[118]. Benzylidene-n-butylamine reacted rapidly with retenequinonimine under anhydrous conditions, to give a good yield of 2-phenylretenoxazole. The possibility of a preliminary hydrolysis of the Schiff base was excluded by its failure to form a phenylhydrazone under the reaction conditions when phenylhydrazine was substituted for the quinonimine.

The mechanism suggested is very similar to that for the aldol condensation path (equation 97), except that the Schiff base acts both as the basic catalyst and as the aldehyde component of the reaction. The cyclization step could conceivably occur by two different routes. The one-step path is favoured on the basis of the close analogy between Schiff bases and aldehydes in addition reactions, but the two-step

(98)

path could not be ruled out. An investigation of the reaction and physical characteristics of a series of Schiff bases indicated that their reactivity was influenced by both the basicity and state of aggregation. The Schiff bases which are most basic and predominantly monomeric gave the best yields of oxazole. As before, essentially the same observations were made with phenanthraquinonimine.

A brief investigation of the reaction between quinonimines and primary amines has been made[119]. The fact that oxazoles rather than imidazoles were formed indicated that the initial reaction took place at the imino group. It was also noted that only primary amines with two α-hydrogens reacted and the following mechanism was proposed.

(71) (99)

The final step, involving the oxidation of intermediate **71**, is interesting in that it requires a hydrogen acceptor. In fact, hydroquinone was isolated when the reaction was carried out in the presence of *p*-benzoquinone.

Still another class of compounds, the alkylidenebisamines, possess a structural similarity to the aldehydes and Schiff bases and their reaction with quinonimines has been studied[120]. When benzalbispiperidine and methylenebismorpholine were allowed to react with retenequinonimine, nearly quantitative yields of 2-phenylretenoxazole and retenoxazole were obtained. The mechanisms proposed above also account nicely for these reactions. One interesting abnormality was the reaction between phenanthraquinonimine and methylenebismorpholine which produced only 2-morpholinophenanthroxazole. The detailed reason for this was not investigated, but it is understandable in terms of the mechanism proposed for the reactions of primary amines (equation 99).

The first compound studied by this group, hydrobenzamide, may be considered either a Schiff base or an alkylidenebisamine. Its reaction with retenequinonimine was re-examined and the mechanism proposed earlier[11] recast in the light of later experimental facts. This new mechanism is essentially the same as those proposed above. This series of papers represents a model of the non-kinetic investigation of reaction mechanisms and can be profitably read by any physical-organic chemist. The studies are neatly summarized and analogies to other reactions cited in one publication[121].

D. Brief Reports

1. Quinonediimine-*N,N'*-dioxides

It was found that this class of compounds (**72**) could be prepared by the oxidation of either substituted quinonediimines or *p*-phenylenediamines[122]. The latter method also produced a by-product, a *p*-nitroaniline (**73**), when R and/or R' = unsubstituted alkyl. Some of the physical properties of these compounds were reported and the observation that they are light-sensitive was made.

$$\text{(NR)}=\text{C}_6\text{H}_4=\text{(NR')} + 2\,\text{PhCO}_3\text{H} \longrightarrow \text{(RN}\rightarrow\text{O)}=\text{C}_6\text{H}_4=\text{(R'N}\rightarrow\text{O)} \quad \textbf{(72)} \qquad (100)$$

$$\text{RHN}-\text{C}_6\text{H}_4-\text{NHR'} + 3\,\text{PhCO}_3\text{H} \longrightarrow \text{(RN}\rightarrow\text{O)}=\text{C}_6\text{H}_4=\text{(R'N}\rightarrow\text{O)} + \text{RHN}-\text{C}_6\text{H}_4-\text{NO}_2 \quad \textbf{(73)} \qquad (101)$$

A more complete study of the products of the photochemical reaction of the *N,N'*-dioxides revealed that the course of the reaction was strongly influenced by the substituent on nitrogen[123]. The first reaction is a rapid and quantitative decomposition into *p*-quinoneimine-*N*-oxides (e.g. **74**) and azo compounds. This is followed by the conversion of the *N*-oxides into *p*-quinone and azo compounds. The further reaction of the intermediate **74** can be separated from the initial

$$\text{PhN}(\rightarrow\text{O})=\text{C}_6\text{H}_4=\text{NPh}(\rightarrow\text{O}) \xrightarrow[>390\text{ m}\mu]{h\nu} \underset{(\mathbf{74})}{\text{PhN}(\rightarrow\text{O})-\text{C}_6\text{H}_4=\text{O}} + \text{PhN}=\text{NPh} + \text{O}=\text{C}_6\text{H}_4=\text{O} \quad (102)$$

reaction because it requires a shorter wavelength for activation. If R and R′ are different aryl groups, all three possible azo compounds are formed, while if one is an alkyl group only the symmetrical aryl azo compound was found. The aliphatic imine substituents were assumed to produce aliphatic azo compounds although none were isolated.

2. The Gibbs reaction

The formation of indophenols from 2,6-dihalobenzoquinone chloroimide has long been known as a sensitive test for phenols[124]. It was

$$\text{HO}-\text{C}_6\text{H}_4-\text{Y} + \text{ClN}=\text{C}_6\text{H}_2\text{X}_2=\text{O} \longrightarrow \text{O}=\text{C}_6\text{H}_2\text{X}_2=\text{N}-\text{C}_6\text{H}_4-\text{OH} + \text{YCl} \quad (103)$$

originally felt that very few phenols in which the *para* position was blocked (Y ≠ H) would give the characteristic blue colour. However, two more recent studies have shown that a wide variety of *para* substituents can be displaced[125,126]. The substituents included halo, carboxyl, alkoxyl, and amino groups. The spectral characteristics and melting points (of the leuco dyes) are given for a large number of the compounds studied.

3. A preparation of quinone sulphenimines

An attempted preparation of the mercaptoindophenol **5** resulted in a quantitative yield of the isomeric quinone sulphenimine (**76**)[127]. A number of these compounds were prepared and it was found that the desired compound could be obtained by an intramolecular rearrangement in refluxing acetic acid (equation 104).

SH
NCl
HS— —N= =O (104)
(**75**)
Δ
HOAc
—S—N= =O
(**76**)

4. Hydrolysis of 1,2-naphthoquinone-1-imine

This appears to be the first report of quantitative data for imine hydrolysis[128]. It was found that 1,2-naphthoquinone-1-imine was stable for several hours in 95% ethanol, but with more than 25% water rapid hydrolysis took place. It was observed that the corresponding benzimide was much more stable towards hydrolysis although it reacted with 95% ethanol to form an adduct.

VI. REFERENCES

1. R. Willstätter and E. Mayer, *Ber.*, **37**, 1494 (1904).
2. R. Willstätter and A. Pfannenstiehl, *Ber.*, **37**, 4605 (1904).
3. Z. E. Jolles in *Chemistry of Carbon Compounds*, Vol. IIIB (Ed. E. H. Rodd), Elsevier, Amsterdam, 1956, pp. 714–729.
4. N. V. Sidgwick, T. W. J. Taylor and W. Baker, *The Organic Chemistry of Nitrogen*, 2nd Ed., revised, Oxford University Press, London, 1942, pp. 97–104.
5. C. L. Jackson and O. F. Calhane, *Am. Chem. J.*, **31**, 209 (1904).
6. K. J. P. Orton and A. E. Smith, *J. Chem. Soc.*, **87**, 389 (1905).
7. A. E. Smith and K. J. P. Orton, *Proc. Chem. Soc.*, **23**, 14 (1907).
8. L. K. J. Tong, *J. Phys. Chem.*, **58**, 1090 (1954).
9. R. Adams and W. Reifschneider, *Bull. Soc. Chim. France*, **5**, 23 (1958).
10. E. Havinga and A. Schors, *Rec. Trav. Chim.*, **69**, 547 (1950).
11. S. I. Kreps and A. R. Day, *J. Org. Chem.*, **6**, 140 (1941).
12. H. Koechlin and O. N. Witt, *Ger. Pat.* 15,915 (1881).
13. P. Ehrlich and F. Sachs, *Ber.*, **32**, 2341 (1899).
14. F. Sachs, *Ber.*, **33**, 959 (1900).
15. R. Fischer, *Ger. Pat.* 253,335 (1912).
16. R. Fischer and H. Siegrist, *Phot. Korr*, **51**, 18 (1914).
17. P. W. Vittum and A. Weissberger, *J. Phot. Sci.*, **2**, 81 (1954).
18. L. Michaelis, *Chem. Rev.*, **16**, 243 (1945).
19. L. K. J. Tong and M. C. Glesmann, *Phot. Sci. Eng.*, **8**, 319 (1964).
20. L. Michaelis, *Ann. N.Y. Acad. Sci.*, **40**, 37 (1940).

21. L. Michaelis, M. P. Schubert and S. Granick, *J. Am. Chem. Soc.*, **61**, 1981 (1939).
22. S. Hünig and W. Daum, *Ann. Chem.*, **595**, 131 (1955).
23. L. F. Fieser, *J. Am. Chem. Soc.*, **52**, 4915 (1930).
24. P. W. Vittum and A. Weissberger, *J. Phot. Sci.*, **6**, 157 (1958).
25. J. R. Thirtle and D. M. Zwick, *Encyclopedia of Chemical Technology*, Vol. 5, 2nd Ed. (Eds. H. F. Mark, J. J. McKetta, Jr. and D. F. Othmer), Interscience, New York, 1964, pp. 822–826.
26. J. M. Woolley, *Brit. Pat.* 778,089 (1957).
27. J. Jennen, *Ind. Chim. Belge*, **16**, 472 (1951).
28. J. Jennen, *Chim. Ind. (Paris)*, **67**, 356 (1952).
29. M. W. Seymour, *U.S. Pat.* 1,969,479 (1934).
30. A. Weissberger and H. D. Porter, *J. Am. Chem. Soc.*, **64**, 2133 (1942).
31. P. W. Vittum and G. H. Brown, *J. Am. Chem. Soc.*, **68**, 2235 (1946).
32. P. W. Vittum and G. H. Brown, *J. Am. Chem. Soc.*, **69**, 152 (1947).
33. P. W. Vittum and G. H. Brown, *J. Am. Chem. Soc.*, **71**, 2287 (1949).
34. P. W. Vittum, G. W. Sawdey, R. A. Herdle and M. K. Scholl, *J. Am. Chem. Soc.*, **72**, 1533 (1950).
35. G. W. Sawdey, M. K. Ruoff and P. W. Vittum, *J. Am. Chem. Soc.*, **72**, 4947 (1950).
36. K. Meyer and H. Ulbricht, *Z. Wiss. Phot. Photophysik Photochem.*, **45**, 222 (1950).
37. K. H. Bauer, *J. Prakt. Chem.*, **4**, 65 (1958).
38. L. K. J. Tong, M. C. Glesmann and R. Andrus, unpublished results.
39. K. T. Finley, R. S. Kaiser, R. L. Reeves and G. Werimont, *J. Org. Chem.*, **34**, 2083 (1969).
40. (a) S. Pickholz, *J. Chem. Soc.*, 685 (1946).
(b) R. D. Brown, *Quart. Rev. (London)*, **6**, 63 (1952).
(c) W. F. Smith, Jr., Eastman Kodak Co., private communication.
41. F. J. W. Roughton and B. Chance, *Technique of Organic Chemistry*, Vol. 3, part 2, 2nd Ed., revised (Ed. A. Weissberger), Interscience, New York, 1963, pp. 703–757.
42. W. R. Ruby, *Rev. Sci. Instr.*, **26**, 460 (1955).
43. C. A. Bishop, R. F. Porter and L. K. J. Tong, *J. Am. Chem. Soc.*, **85**, 3991 (1963).
44. L. K. J. Tong, K. Liang and W. R. Ruby, *J. Electroanal. Chem.*, **13**, 245 (1967).
45. J. F. Bunnett, *Ann. Rev. Phys. Chem.*, 271 (1963).
46. J. D. Loudon and M. Shulman, *J. Chem. Soc.*, 722 (1941).
47. (a) L. K. J. Tong and M. C. Glesmann, *J. Am. Chem. Soc.*, **78**, 5827 (1956).
(b) C. A. Bishop and L. K. J. Tong, *Phot. Sci. Eng.*, **11**, 30 (1967).
48. L. K. J. Tong and M. C. Glesmann, *J. Am. Chem. Soc.*, **79**, 583 (1957).
49. M. L. Bender, *Technique of Organic Chemistry*, Vol. 8, part 2, 2nd Ed. revised (Ed. A. Weissberger), Interscience, New York, 1963, pp. 1427–1517.
50. L. K. J. Tong, M. C. Glesmann, and R. L. Bent, *J. Am. Chem. Soc.*, **82**, 1988 (1960).
51. J. Egger and H. Frieser, *Z. Electrochem.*, **60**, 372, 376 (1956).
52. P. W. Vittum and A. Weissberger, *Intern. Conf. Photo., Cologne, September, 1956* (see reference 24).

53. L. K. J. Tong and M. C. Glesmann, *J. Am. Chem. Soc.*, **79**, 592 (1957).
54. L. K. J. Tong and M. C. Glesmann, *J. Am. Chem. Soc.*, **90**, 5164 (1968).
55. J. F. Bunnett and R. E. Zahler, *Chem. Rev.*, **49**, 273 (1951).
56. K. M. Ibne-Rasa, *J. Chem. Educ.*, **44**, 89 (1967).
57. J. S. Parsons, *J. Gas Chromatog.*, **5**, 254 (1967).
58. R. Adams and A. S. Nagarkatti, *J. Am. Chem. Soc.*, **72**, 4601 (1950).
59. R. Adams and J. H. Looker, *J. Am. Chem. Soc.*, **73**, 1145 (1951).
60. S. I. Burmistrov and E. A. Titov. *J. Gen. Chem. USSR* (*Eng. Transl.*), **22**, 1053 (1952).
61. E. A. Titov and S. I. Burmistrov, *J. Gen. Chem. USSR* (*Eng. Transl.*), **30**, 643 (1960).
62. A. Mustafa and M. Kamel, *J. Am. Chem. Soc.*, **75**, 2939 (1953).
63. A. Mustafa and M. Kamel, *J. Am. Chem. Soc.*, **77**, 5630 (1955).
64. A. Mustafa and M. Kamel, *J. Org. Chem.*, **22**, 157 (1957).
65. W. P. Jencks in *Progress in Physical Organic Chemistry*, Vol. 2 (Eds. S. G. Cohen, A. Streitwieser, Jr. and R. W. Taft), Interscience, New York, 1964, pp. 63–128.
66. R. L. Reeves in *The Chemistry of the Carbonyl Group* (Ed. S. Patai), Interscience, New York, 1966, pp. 567–619.
67. R. Adams and R. A. Wankel, *J. Am. Chem. Soc.*, **73**, 131 (1951).
68. R. Adams and W. Moje, *J. Am. Chem. Soc.*, **74**, 2593 (1952).
69. R. Adams, E. F. Elslager and K. F. Heumann, *J. Am. Chem. Soc.*, **74**, 2608 (1952).
70. R. Adams and K. A. Schowalter, *J. Am. Chem. Soc.*, **74**, 2597 (1952).
71. R. Adams, B. H. Braun and S. H. Pomerantz, *J. Am. Chem. Soc.*, **75**, 4642 (1953).
72. R. Adams and S. H. Pomerantz, *J. Am. Chem. Soc.*, **76**, 702 (1954).
73. R. Adams and D. S. Acker, *J. Am. Chem. Soc.*, **74**, 3657 (1952).
74. R. Adams and D. S. Acker, *J. Am. Chem. Soc.*, **74**, 5872 (1952).
75. R. Adams and J. W. Way, *J. Am. Chem. Soc.*, **76**, 2763 (1954).
76. R. Adams and W. Moje, *J. Am. Chem. Soc.*, **74**, 5560 (1952).
77. R. Adams and W. Moje, *J. Am. Chem. Soc.*, **74**, 5562 (1952).
78. R. Adams, E. F. Elslager and T. E. Young, *J. Am. Chem. Soc.*, **75**, 663 (1953).
79. C. D. Ritchie, J. D. Saltier and E. S. Lewis, *J. Am. Chem. Soc.*, **83**, 4601 (1961).
80. R. Adams and M. D. Nair, *J. Am. Chem. Soc.*, **78**, 5927 (1956).
81. R. Adams, T. E. Young and R. W. P. Short, *J. Am. Chem. Soc.*, **76**, 1114 (1954).
82. C. Laar, *Ber.*, **18**, 648 (1885).
83. H. H. Hodgson and F. H. Moore, *J. Chem. Soc.*, **123**, 2499 (1923).
84. H. H. Hodgson, *J. Chem. Soc.*, 89 (1943).
85. H. H. Hodgson, *J. Chem. Soc.*, 520 (1937).
86. L. C. Anderson and M. B. Geiger, *J. Am. Chem. Soc.*, **54**, 3064 (1932).
87. L. C. Anderson and R. L. Yanke, *J. Am. Chem. Soc.*, **56**, 732 (1934).
88. P. Ramart-Lucas, M. Martynoff, M. Grumez and M. Chauvin, *Bull. Soc. Chim. France*, **15**, 571 (1948).
89. P. Ramart-Lucas and M. Martynoff, *Compt. Rend.*, **227**, 906 (1948).

90. P. Ramart-Lucas, M. Martynoff, M. Grumez and M. Chauvin, *Bull. Soc. Chim. France*, **16**, 53 (1949).
91. P. Ramart-Lucas, M. Martynoff, M. Grumez and M. Chauvin, *Bull. Soc. Chim. France*, **16**, 901 (1949).
92. P. Ramart-Lucas, M. Martynoff, M. Grumez and M. Chauvin, *Bull. Soc. Chim. France*, **16**, 905 (1949).
93. A. Kraaijeveld and E. Havinga, *Rec. Trav. Chim.*, **73**, 537 (1954).
94. E. Havinga and A. Schors, *Rec. Trav. Chim.*, **70**, 59 (1951).
95. A. Kraaijeveld and E. Havinga, *Rec. Trav. Chim.*, **73**, 549 (1954).
96. A. Schors, A. Kraaijeveld and E. Havinga, *Rec. Trav. Chim.*, **74**, 1243 (1955).
97. F. H. Westheimer, E. Segel and R. Schramm, *J. Am. Chem. Soc.*, **69**, 773 (1947).
98. E. Beckmann and O. Liesche, *Ber.*, **56B**, 1 (1923).
99. N. J. Leonard and J. H. Curry, *J. Org. Chem.*, **17**, 1071 (1952).
100. R. A. Raphael and E. Vogel, *J. Chem. Soc.*, 1958 (1952).
101. L. Blangey, *Helv. Chim. Acta*, **21**, 1579 (1938).
102. H. M. Haendler and G. McP. Smith, *J. Am. Chem. Soc.*, **61**, 2624 (1939).
103. H. M. Haendler and G. McP. Smith, *J. Am. Chem. Soc.*, **62**, 1669 (1940).
104. W. T. Sumerford and D. N. Dalton, *J. Am. Chem. Soc.*, **66**, 1330 (1944).
105. W. G. H. Edwards, *Chem. Ind. (London)*, 112 (1951).
106. M. L. Stein and H. v. Euler, *Gazz. Chim. Ital.*, **84**, 290, (1954).
107. H. H. Jaffé, *J. Am. Chem. Soc.*, **77**, 4448 (1955).
108. R. M. Elofson and J. G. Atkinson, *Can. J. Chem.*, **34**, 4 (1956).
109. L. W. Butz in *Org. Reactions*, **5**, 136 (1949).
110. C. J. Sunde, J. G. Erickson, and E. K. Raunio, *J. Org. Chem.*, **13**, 742 (1948).
111. R. B. Woodward and T. J. Katz, *Tetrahedron*, **5**, 70 (1959).
112. C. K. Bradsher, E. D. Little, and D. J. Beavers, *J. Am. Chem. Soc.*, **78**, 2153 (1956).
113. C. K. Bradsher and D. J. Beavers, *J. Org. Chem.*, **21**, 1067 (1956).
114. J. Rigaudy and G. Izoret, *Compt. Rend.*, **238**, 824 (1954).
115. J. Rigaudy, G. Cauquis, G. Izoret, and J. Baranne-Lafont, *Bull. Soc. Chim. France*, 1842 (1961).
116. F. R. Japp and E. Wilcock, *J. Chem. Soc.*, **37**, 661 (1880).
117. C. W. C. Stein and A. R. Day, *J. Am. Chem. Soc.*, **64**, 2567 (1942).
118. C. W. C. Stein and A. R. Day, *J. Am. Chem. Soc.*, **64**, 2569 (1942).
119. G. McCoy and A. R. Day, *J. Am. Chem. Soc.*, **65**, 1956 (1943).
120. G. McCoy and A. R. Day, *J. Am. Chem. Soc.*, **65**, 2157 (1943).
121. G. McCoy and A. R. Day, *J. Am. Chem. Soc.*, **65**, 2159 (1943).
122. C. J. Pedersen, *J. Am. Chem. Soc.*, **79**, 2295 (1957).
123. C. J. Pedersen, *J. Am. Chem. Soc.*, **79**, 5014 (1957).
124. H. D. Gibbs, *J. Biol. Chem.*, **72**, 649 (1927).
125. F. E. King, T. J. King and L. C. Manning, *J. Chem. Soc.*, 563 (1957).
126. H. Inove, Y. Kanaya and Y. Murata, *Chem. Pharm. Bull.*, **7**, 573 (1959); through *Chem. Abstr.*, **54**, 19559 (1960).
127. D. N. Kramer and R. M. Gamson, *J. Org. Chem.*, **24**, 1154 (1959).
128. C. C. Irving, *J. Org. Chem.*, **25**, 464 (1960).

Author Index

This author index is designed to enable the reader to locate an author's name and work with the aid of the reference numbers appearing in the text. The page numbers are printed in normal type in ascending numerical order, followed by the reference numbers in parentheses. The numbers in *italics* refer to the pages on which the references are actually listed.

Subject Index